Probability and Mathematical Physics

A Volume in Honor of Stanislav Molchanov

Volume 42

CRM
PROCEEDINGS & LECTURE NOTES

Centre de Recherches Mathématiques
Université de Montréal

Probability and Mathematical Physics

A Volume in Honor of Stanislav Molchanov

Donald A. Dawson
Vojkan Jakšić
Boris Vainberg
Editors

The Centre de Recherches Mathématiques (CRM) of the Université de Montréal was created in 1968 to promote research in pure and applied mathematics and related disciplines. Among its activities are special theme years, summer schools, workshops, postdoctoral programs, and publishing. The CRM is supported by the Université de Montréal, the Province of Québec (FCAR), and the Natural Sciences and Engineering Research Council of Canada. It is affiliated with the Institut des Sciences Mathématiques (ISM) of Montréal, whose constituent members are Concordia University, McGill University, the Université de Montréal, the Université du Québec à Montréal, and the École Polytechnique. The CRM may be reached on the Web at www.crm.umontreal.ca.

American Mathematical Society
Providence, Rhode Island USA

The production of this volume was supported in part by the Fonds pour la Formation de Chercheurs et l'Aide à la Recherche (Fonds FCAR) and the Natural Sciences and Engineering Research Council of Canada (NSERC).

2000 *Mathematics Subject Classification.* Primary 60–06, 35–06.

Library of Congress Cataloging-in-Publication Data

Probability and mathematical physics : a volume in honor of Stanislav Molchanov / Donald A. Dawson, Vojkan Jakšić, Boris Vainberg, editors.
p. cm. — (CRM proceedings & lecture notes, ISSN 1065-8580 ; v. 42)
Includes bibliographical references.
ISBN 978-0-8218-4089-4 (alk. paper)
1. Probabilities—Congresses. 2. Mathematical physics—Congresses. I. Molchanov, S. A. (Stanislav A.) II. Dawson, Donald Andrew, 1937– III. Jakšić, Vojkan, 1964– IV. Vainberg, B. R. (Boris Rufimovich)

QC20.7.P7P77 2007
530.15—dc22 2007060768

∞ The paper used in this book is acid-free and falls within the guidelines
established to ensure permanence and durability.
This volume was submitted to the American Mathematical Society
in camera ready form by the Centre de Recherches Mathématiques.
Visit the AMS home page at `http://www.ams.org/`

10 9 8 7 6 5 4 3 2 1 12 11 10 09 08 07

Contents

Preface

A conference celebrating Stanislav Molchanov's 65th birthday took place from June 27 to July 1, 2005 at the CRM in Montreal. The theme of the conference, "Probability and Mathematical Physics," reflected the wide range of Molchanov's scientific interests and in a similar spirit the present proceedings cover an exceptionally broad range of fields. As conference organizers and editors of the volume, we wish to thank all authors for their contributions. We also wish to thank Jacques Hurtubise and François Lalonde for making the conference and this volume possible, and to the CRM staff for the superb conference organization.

Don Dawson, Vojkan Jaksic, Boris Vainberg
February 2007

Stanislav Molchanov

Stanislav Molchanov (Stas) was born on December 21, 1940, in Snetinovo, a small village in the Ivanovo region of Russia. His parents were high school teachers. After finishing high school in 1958, he became a student of Moscow State University. Starting from his sophomore year, Stas was a regular participant (together with his classmates N. Krylov, M. Malutov and some older students, such as M. Freidlin and A. Wentzel) in a student research seminar chaired by Professor E. Dynkin. Later Dynkin became his Ph.D. adviser. Stas defended his Ph.D. thesis, entitled *Martin boundary for Markov processes*, in 1966. He then became an Assistant Professor in the probability theory division at the Department of Mechanics and Mathematics of Moscow State University. Stas received the advanced degree of Doctor of Physical and Mathematical sciences in 1983. His advanced thesis was entitled *Spectral theory of one-dimensional Schrödinger operators*. He taught at Moscow State University for 25 years, first as an Assistant, then as an Associate and eventually Full Professor. Stas arrived in the U.S. in 1991. He worked for three semesters at the University of California (Irvine) and a year at the University of Southern California (Los Angeles). Since 1994 Stas has been Professor at the University of North Carolina (Charlotte).

Moscow State University provided a unique opportunity for undergraduate and graduate students as well as young professors to learn not only from senior professors, but also from each other. Stas has very warm memories of scientific discussions with M. Freidlin after which his interest turned toward mathematical physics and (stochastic) differential equations. Later, under the influence V. Arnol′d he started working in the field of spectral theory. Stas published a very important paper [1] in 1975 on exact asymptotic expansions (for small t) for fundamental solutions of parabolic problems on Riemannian manifolds. His work in localization theory was triggered by his collaboration with I. Lifshitz and L. Pastur. In 1980, he published (together with I. Goldsheid and L. Pastur) a well-known paper [2] containing the first results on localization for one-dimensional Schrödinger operators. Later he also proved the exponential decay of eigenfunctions and the Poisson statistics for the eigenvalues for these operators [3, 4]. In the 90s, together with M. Aizenmann, Stas extended these results to the multidimensional case [7]. Now their approach to prove the localization is called the Aizenman–Molchanov method, and it is one of the most commonly used methods in the field.

At the end of the 80s, S. Molchanov, in collaboration with J. Zeldovich, A. Ruzmaikin, and D. Sokolov, developed the intermittency theory for non-stationary random media [5, 6]. Their works started at the physical level of rigor and had applications to magneto-hydrodynamic of stars and intermittency of the temperature field of the ocean. The mathematical theory of intermittency was developed later by S. Molchanov in collaboration with Jü. Gärtner, R. Carmona, and others [8, 9].

We mentioned only some contributions made by S. Molchanov in different areas of mathematics. In fact, he published more than 250 papers on stochastic differential equations, spectral theory for deterministic and random operators, localization and intermittency, mathematical physics and optics, and other topics. He was the Ph.D. adviser of more than 40 students, but a much wider circle of people has learned from him and appreciated the chance to work with him.

Vojkan Jaksic, Boris Vainberg

References

1. *Diffusion processes and Riemannian geometry*, Soviet Uspekhi Math, Sci. **30** (1975), no. 1, 2–59.
2. *A pure point spectrum for 1D Schrödinger operator*, Funct. Anal. and Appl **11** (1977), 1–10 (with I Goldsheid and L. Pastur).
3. *The structure of eigenfunctions for 1D disordered systems*, Math. USSR Izvestia **12** (1978), 69–101.
4. *Local structure of the spectrum of 1D random Schrödinger operator*, Comm. Math. Phys. **78** (1981), 429–446.
5. *Intermittency, diffusion and generation in nonstationary random media*, Sov. Sci. Reviews, Section C Math. Phys. **7** (1988), 1–110 (with Ja. Zeldovich, A. Ruzmaikin, D. Sokolov).
6. *Ideas in the theory of random media*, Acta Appl. Math. 12 (1991), 139–182.
7. *Localization at large disorder and extreme energies: an elementary derivation*, Comm. Math. Phys. **157** (1993), 245–278 (with M. Aizenman).
8. *Lectures in random media*, Lecture notes, vol. 1581, 1994, Springer (with D. Bakry, R. Grill).
9. *Parabolic Anderson model and intermittency*, Memoir AMS 158 (1995), (with R. Carmona).

Centre de Recherches Mathématiques
CRM Proceedings and Lecture Notes
Volume **42**, 2007

Transition Asymptotics for Reaction – Diffusion in Random Media

Gérard Ben Arous, Stanislav Molchanov, and Alejandro F. Ramírez

ABSTRACT. We describe a universal transition mechanism between annealed and quenched regimes in the context of reaction – diffusion in random media. We study the total population size for random walks which branch and annihilate on $\mathbb{Z}^d$, with time-independent random rates. The random walks are independent, continuous time, rate $2d\kappa$, simple, symmetric, with $\kappa \geq 0$. A random walk at $x \in \mathbb{Z}^d$, binary branches at rate $v_+(x)$, and annihilates at rate $v_-(x)$. The random environment w has coordinates $w(x) = (v_-(x), v_+(x))$ which are i.i.d. We identify a natural way to describe the annealed-Gaussian transition mechanism under mild conditions on the rates. Indeed, we introduce the exponents $F_\theta(t) := \big(H_1\big((1+\theta)t\big) - (1+\theta)H_1(t)\big)/\theta$, and assume that $\big(F_{2\theta}(t) - F_\theta(t)\big)/\big(\theta\log(\kappa t + e)\big) \to \infty$ for $|\theta| > 0$ small enough, where $H_1(t) := \log\langle m(0,t)\rangle$ and $\langle m(0,t)\rangle$ denotes the average of the expected value of the number of particles $m(0,t,w)$ at time t and an environment of rates w, given that initially there was only one particle at 0. Then the empirical average of $m(x,t,w)$ over a box of side $L(t)$ has different behaviors: if $L(t) \geq e^{F_\epsilon(t)/d}$ for some $\epsilon > 0$ and large enough t, a law of large numbers is satisfied; if $L(t) \geq e^{F_\epsilon(2t)/d}$ for some $\epsilon > 0$ and large enough t, a CLT is satisfied. These statements are violated if the reversed inequalities are satisfied for some negative ϵ. As corollaries, we obtain more explicit statements under regularity conditions on the tails of the random rates, including examples in the four universality classes defined in [14]: potentials which are unbounded of Weibull type, of double exponential type, almost bounded, and bounded of Fréchet type. For them we also derive sharper results in the nonannealed regime.

2000 *Mathematics Subject Classification.* Primary 82B41, 82B44; Secondary 60J45, 60J65, 82C22.

Key words and phrases. parabolic Anderson model, random walk, branching processes, Feynman – Kac formula, principal eigenvalue.

This work was partially supported by Fondo Nacional de Desarrollo Científico y Tecnológico grant 7020686.

The work of the last author is partially supported by Fondo Nacional de Desarrollo Científico y Tecnológico grant 1020686.

This is the final form of the paper.

1. Introduction

When studying the long time behavior of Markovian dynamics in random media, one is faced with an important distinction: quenched vs. averaged estimates. Should one work in the so-called quenched regime, where the randomness of the medium is frozen, i.e., where the dynamics are studied in one fixed random realization of the medium/environment? Or should one work in the averaged regime where both the randomness of the dynamics and of the medium are considered, i.e., when one studies the dynamics in a given realization but then also averages over the randomness of the medium?[1]

The true significance of this distinction "quenched vs. averaged" is important when these two regimes give different answers, which is the case in many situations where the extreme values of the random environment might play an important role. A good generic class of examples where this distinction is significant is given by models of reaction – diffusion in random media ([12, 13, 17]).

There are two opposing views about this distinction. The first approach is to think that the relevant and important asymptotic long-time estimates are the quenched ones. But, recognizing the obvious fact that the quenched estimates are the most intractable ones mathematically, the averaged/annealed results are seen as a welcome first approximation. The second and opposing view is that, in many applications, the averaged asymptotic estimates are the only relevant ones. The quenched results, though mathematically more challenging, are not seen as useful or relevant. This second view is naturally based on the idea that some mechanism must be at work, which allows for averaging in the medium randomness.

Some years ago, the authors of this paper held the two opposing opinions expressed above, based on their former collaborations with different domains of applications (specifically physics of pollution by underground waste storage on the one hand and chemical kinetics on the other). In the recent years we have built a common answer, which we believe provides not only a natural resolution of the scientific debate, but also introduces the idea that there exists in fact a very rich universal transition between the averaged and quenched results, which goes far beyond the reaction – diffusion context. This transition also explains, in our view,the true relevance of these regimes in various applications.

The key idea is the following: one should work with a fixed realization of the medium but introduce a new parameter, say L, which will be the scale of the spatial extent of the initial distribution of the dynamics. Then, depending on the respective sizes of the time scale t and the space scale L (when both t and L diverge), one should see the transition we mentioned between the quenched and averaged asymptotic results. More precisely one should expect the following transition: for time scales $t(L)$ short enough the averaged asymptotic results should be valid, whereas for very long time scales $t(L)$ the quenched results should hold, and our new intermediate asymptotics should emerge in between. The mechanism of this transition is the following: for short time scales $t(L)$, or equivalently for large space scales $L(t)$ for the initial distribution, the spatial ergodic theorem should ensure the needed averaging mechanism in order to enforce the validity of averaged/annealed results. For very large time scales no averaging is possible and the extreme values

[1]This regime is often called the annealed regime in the mathematical literature. This is an entrenched but misleading vocabulary since it is not the usual convention in physics

of the environment play the prominent role. The transition regimes are then easily understood, they consist in regions of parameters t and L where one sees a gradual emergence of the extreme values versus the average.

This scheme has been first established in the simplest possible context, i.e., i.i.d. samples [4]. More precisely sums of exponential of i.i.d. random variables are shown to exhibit this transition. In the context of reaction diffusion this could simply be seen as reaction with no diffusion! In this simple context a full transition is given from the Gaussian asymptotics to extreme value theory: one sees in [4] the gradual emergence of the importance of extreme values of the i.i.d. sample which gradually destroys the validity of the Central Limit Theorem and then of the Law of Large Numbers by enforcing α-stable fluctuations where the exponent α decreases through the whole possible range, i.e., from 2 to 0. This mechanism is analogous to the phase transition description of mean-field spin-glass equilibrium models such as the Random Energy Model ([8, 9]).

We then proceeded ([5]) to one important case of reaction – diffusion, i.e., annihilation (or absorption) for random walks in a random environment, more precisely random walks killed on random obstacles, building on the work of [17]. There, we studied the same transition mechanism for the natural quantity, which is the probability of survival. Our picture was less precise than in the i.i.d. context, in the sense that even though we get the proper scales for the intermediate regimes we cannot establish the stable nature of the fluctuations, due to a lack of precise enough understanding of the edge of the spectrum (for the generator of the dynamics which is the discrete Dirichlet Laplacian on a random domain of $\mathbb{Z}^d$).

In this paper we address a rather general case of reaction – diffusion in random environments, i.e., of a system of noninteracting continuous-time Random Walks on the lattice $\mathbb{Z}^d$ branching and annihilating with stationary random rates ([12, 13]).

Let us first describe the random environment $\{w(x) : x \in \mathbb{Z}^d\}$, with $w(x) := (v_-(x), v_+(x))$ where $v_+ := \{v_+(x) : x \in \mathbb{Z}^d\}$ represents the branching rates with $v_+(x) \in [0, \infty)$, while $v_- := \{v_-(x) : x \in \mathbb{Z}^d\}$ represents the annihilation rates, with $v_-(x) \in [0, \infty]$ so that we admit the possibility of hard core obstacles. We assume that the random variables $\{w(x) : x \in \mathbb{Z}^d\}$ are i.i.d. and call their distribution μ.

We consider the following dynamics in a fixed random realization of this environment. We start with one particle at each site of the box Λ_L of side L in $\mathbb{Z}^d$. Each random walk moves independently of the others according to a continuous time simple symmetric rate $2d\kappa$ dynamics for some $\kappa \geq 0$ (we admit the possibility that $\kappa = 0$, i.e., no diffusion at all, as in [4]). A random walk at a site $x \in \mathbb{Z}^d$, branches at rate $v_+(x)$, disappearing and producing two new independent offsprings, and annihilates at rate $v_-(x)$ (note that $v_-(x) = \infty$ means the annihilation is instantaneous and certain as in [5]).

We study the asymptotic behavior, when both t and L go to infinity, of the following observable of our system of random walkers in random environment:

$$m_L(0, t, w) := \frac{1}{|\Lambda_L|} \sum_{x \in \Lambda_L} m(x, t, w).$$

Note that $m_L(0, t, w)$ is simply the (normalized) size of the total population.

Our main result, Theorem 2.3, describes sharp conditions on the scales of t and L for the validity of annealed asymptotics of $m_L(0, t, w)$ and for these annealed asymptotics to cease to be true. It also describes the scales for t and L where the fluctuations about these annealed asymptotics are Gaussian and when they cease to be so. We intend to give a full and complete picture of the expected stable fluctuations, all the way to quenched asymptotics, for a wide class of branching rates in a forthcoming work.

The results we obtain here are general, in the sense that they are valid for a large class of product distributions for the random environment, under mild conditions on the branching and annihilation rates. They include examples in the four universality classes recently introduced by van den Hofstad et al. [14] describing all the cases of the random environment. For example, potentials which are unbounded of Weibull type, of double exponential type, almost bounded and bounded of Fréchet type. Furthermore, our results include and generalize both[4, Theorems 2.1 and 2.2] and [5, Theorem 1(ii), (iii) and Theorem 2(i)].

Let us now be more precise about our assumptions about the distribution of the environment. In the recent paper [14], it is shown that under regularity assumptions on the tail of the law of the *effective potential* $v(0) := v_+(0) - v_-(0)$, exactly four universality classes of environments can occur. Their assumptions are formulated in terms of the *cumulant generating function*,

$$H(t) = \log\langle \mathrm{e}^{v(0)t}\rangle, \quad t \geq 0, \tag{1.1}$$

of the law of the effective potential $v(0)$, where for any function of the environment f, we define $\langle f\rangle := \int f\,\mathrm{d}\mu$. Their basic assumption is that this function is defined and finite for every $t \geq 0$. Then, under two regularity assumptions on H they show that four universality classes can occur: (1) a first class where v is unbounded and has "heavy tails" at infinity, and which includes Weibull-type tails; (2) a second class of unbounded potentials with "lighter" tails which includes the double exponential law; (3) a class containing bounded and unbounded potentials; (4) a class of bounded potentials including those which have Fréchet-type tails near their essential supremum, and the degenerate case of random walks on hard core random obstacles with $\mu[v(0) = -\infty] = p < 1$. In this paper we generalize [5, Theorem 1(ii) and (iii)] describing the passage to an annealed and Gaussian regime, to the previously described system of random walks on the random environment w, under mild conditions which include cases in these four universality classes.

Throughout this article the following will be assumed.

Assumption E. The law of the effective potential is such that $\mu[v(0) = -\infty] < 1$ and

$$\langle \mathrm{e}^{v(0)t}\rangle < \infty, \quad t \geq 0.$$

Assumption E ensures that μ-a.s. the stochastic process of random walks on the random environment w can be constructed on infinite volume ([12]), as a limit of processes defined on finite boxes corresponding to continuous time Markov branching processes, as defined in Athreya and Ney in [3]. Furthermore, if $\zeta(t)$ denotes the total number of random walks at time t on a random environment w, given that initially there was only a single one at site x, and E_x^w the expectation defined by its law, Condition E ensures the existence of the first moment $m(x,t,w) = \mathrm{E}_x^w[\zeta(t)]$. This is the content of Proposition 2.1 of this paper. This first moment will be the central object of our study. We will see that Assumption E ensures the finiteness for $t \geq 0$ of the annealed first moments $\langle m(0,t)\rangle = \int m(0,t,w)\,\mathrm{d}\mu$, with which we will state our main assumptions. We will need to define the *growth functions*

$\{H_1(t) : t \geq 0\}$ by

$$H_1(t) := \log\langle m(0,t)\rangle. \tag{1.2}$$

Our main assumption will be formulated with the help of a family of functions $\{F_\theta : \theta \in \mathbb{R}\}$ which we call the *intermittency exponents*, defined for every $\theta \neq 0$ and $t \geq 0$ as,

$$F_\theta(t) := \frac{H_1\big((1+\theta)t\big) - (1+\theta)H_1(t)}{\theta}. \tag{1.3}$$

Assumption MI. For all $|\theta| > 0$ small enough,

$$\lim_{t\to\infty} \frac{F_{2\theta}(t) - F_\theta(t)}{\theta \log t} = \infty.$$

As it will be shown in Corollary 4.6, this assumption implies the occurrence of the so called *intermittent* behavior of the random field w [12]. It encompasses examples falling in the four universality classes of [14].

We will show that it is possible to formulate an assumption directly in terms of the cumulant generating function H, which is sufficient for Assumption MI to be satisfied, and includes the first class of [14]. For this we need to define the *cumulant exponents*, parametrized by $\theta \neq 0$, for $t \geq 0$ as,

$$G_\theta(t) := \frac{H\big((1+\theta)t\big) - (1+\theta)H(t)}{\theta}. \tag{1.4}$$

By Jensen's inequality it can be seen that $G_\theta(t) \geq 0$ whenever $0 < |\theta| \leq 1$. We will see that Condition MI is satisfied whenever the following happens.

Assumption SI. For all $|\theta| > 0$ small enough,

$$\lim_{t\to\infty} \frac{G_{2\theta}(t) - G_\theta(t)}{\theta t} = \infty.$$

Condition SI includes the first universality class of [14] and can be viewed as a strong intermittency requirement. It implies $H(t)/t \gg 1$. Hence, using the bounds $\mathrm{e}^{H(t)-2d\kappa t} \leq \langle m(0,t)\rangle \leq \mathrm{e}^{H(t)}$ (see, e.g., Theorem 3.1 of Gärtner and Molchanov [12]), we see that if Assumption SI is satisfied, we have the asymptotics

$$\log\langle m(0,t)\rangle \sim H(t), \tag{1.5}$$

which is much faster than the a.s. one (see [13]).

As already mentioned the interest of the results of the present paper are their generality. Namely, their are valid only under the Assumptions E and MI. Part (i) of Theorem 2.3, states that if for some $\epsilon > 0$ we have $L(t) \geq \mathrm{e}^{F_\epsilon(t)/d}$ eventually in t, the law of large numbers $m^L/\langle m\rangle \sim 1$ is satisfied in probability: hence we have the annealed behavior $\log m^L(0,t) \sim \log\langle m(0,t)\rangle$. On the other hand, if for some $\epsilon > 0$ we have $L(t) \leq \mathrm{e}^{F_{-\epsilon}(t)/d}$ eventually in t, in probability $m^L/\langle m\rangle \ll 1$. Part (ii) says that if for some $\epsilon > 0$ we have $L(t) \geq \mathrm{e}^{F_\epsilon(2t)/d}$ eventually in t, then $(m^L - (2L+1)^d\langle m\rangle)/\operatorname{Var}_\mu m^L$ converges in distribution to a centered normal law of unit variance $\mathcal{N}(0,1)$, where Var_μ denotes the variance. Also, if for some $\epsilon > 0$ we have $L(t) \leq \mathrm{e}^{F_{-\epsilon}(2t)/d}$ eventually in t, in probability $(m^L - (2L+1)^d\langle m\rangle)/\operatorname{Var}_\mu m^L \ll 1$. This discussion is summarized in Table 1.

Under an additional regularity assumption on the intermittency exponents (Assumption RI of Section 2.4) it will be shown in Corollary 2.6 that there exist two constants γ_1 and γ_2, called *transition exponents* and a function $J(t)\colon [0,\infty) \to [0,\infty)$

TABLE 1. Large time asymptotic behavior of the averaged first moments

Annealed behavior	$d\log L(t) \geq F_\epsilon(t)$	$m^L/\langle m\rangle \sim 1$
Non-annealed behavior	$d\log L(t) \leq F_{-\epsilon}(t)$	$m^L/\langle m\rangle \ll 1$
Gaussian behavior	$d\log L(t) \geq F_\epsilon(2t)$	$\frac{m^L - (2L+1)^d\langle m\rangle}{\mathrm{Var}_\mu m^L} \to \mathcal{N}(0,1)$
Non-Gaussian behavior	$d\log L(t) \leq F_\epsilon(2t)$	$\frac{m^L - (2L+1)^d\langle m\rangle}{\mathrm{Var}_\mu m^L} \ll 1$

with $J(t) \gg 1$, called the *growth exponent*, describing more explicitly the transition of Theorem 2.3. Indeed, in this case, a law of large numbers is satisfied when $d\log L(t) \geq \gamma J(t)$ for some $\gamma > \gamma_1$; the CLT when $d\log L(t) \geq \gamma J(t)$ for some $\gamma > \gamma_2$. Furthermore, if $d\log L(t) \leq \gamma J(t)$ for some $\gamma < \gamma_1$, the law of large numbers is not satisfied, while if $d\log L(t) \leq \gamma J(t)$ for some $\gamma < \gamma_2$, the CLT is not satisfied. Propositions 2.7, 2.8, 2.9 and 2.10 give the explicit value of γ_1, γ_2 and J in the case of unbounded potentials with Weibull-type tails, unbounded potentials with double exponential type tails, potentials in the third universality class of [14] and bounded potentials with Fréchet-type tails including the pure hard core case. Table 2 below summarizes those results.

Also, in Theorem 2.11, we obtain sharper upper bounds for the order of magnitude of the averaged first moments in the nonannealed regime for the examples treated in Propositions 2.7, 2.8, 2.9 and 2.10. This theorem generalizes Case 3 of [5, Theorem 2].

The special case in which Assumption SI is satisfied expressed as Corollary 2.4. This includes the case $\kappa = 0$, corresponding to sums of i.i.d. random exponentials where Condition MI reduces to,

$$\lim_{t\to\infty} \frac{G_{2\theta}(t) - G_\theta(t)}{\theta} = \infty, \tag{1.6}$$

for $|\theta| > 0$ large enough, and $H_1(t) = H(t)$. Corollary 2.4 is a result complementary to theorem of [4], generalizing [4, Theorems 2.1 and 2.2] where Weibull and Fréchet-type tails are assumed on $v(0)$.

One of the main ingredients of the proof of Theorem 2.3 is a coarse graining technique, necessary to reduce the asymptotics of the averaged first moments, to a sum of independent random exponentials. This technique, was introduced in [5],

TABLE 2. Transition and growth exponents in the four universality classes. In the pure hard core case $\mu[v(0) = -\infty] = p$ and c_2 is a constant depending on p and d

POTENTIAL	$-\log \mu[v(0) > x]$	$J(t)$	γ_1	γ_2
Weibull	x^ρ, $\rho > 1$	$H(t)$	$\frac{1}{1-\rho}$	$2^{1-\gamma_1}\gamma_1$
Double exponential	$e^{x/\rho}$	t	ρ	2ρ
Third class example	e^{x^2}	$\frac{t}{2\sqrt{\log t}}$	1	2
Pure hard core	—	$c_2 t^{d/(d+2)}$	$\frac{2}{d+2}$	$2^{1-\gamma_1}\gamma_1$

but here it faces the extra difficulty that the terms of the sum defining the averaged probabilities are not uniformly bounded with respect to the time variable t or the scale L (whereas in [5], such a bound existed having the value 1). This requires more careful estimates on these quantities, which are performed, via the Feynman – Kac formula and spectral estimates. Once the reduction to a sum of i.i.d. exponentials is achieved, an analysis based on von Bahr – Esseen inequality finishes the proof (see also [4, 5]).

Besides this introduction, this paper has four other sections. The main results are stated in Section 2. We first introduce in Section 2.1 the main definitions. In Section 2.2 we formulate Proposition 2.1, stating that a growth of the form $\limsup_{|x|\to\infty} v_+(x)/(|x|\log|x|) = 0$ is enough to ensure well-defined first moments for the total number of particles. When this proposition is combined with [12, Proposition 2], one concludes that under the condition E, the reaction – diffusion process on the lattice $\mathbb{Z}^d$ is such that the total number of particles at any given time has a finite first moment, for initial conditions with a finite total number of particles. In particular, there is no explosion, and the process is well defined. We then state Theorem 2.3 in Section 2.3. Corollary 2.4, under the Assumption SI is stated and proved next. The applications of Theorem 2.3 are given in Section 2.4. First, the regularity condition RI is introduced. This is applied to the case of unbounded effective potentials with Weibull-type tails, through Proposition 2.7, using the Kasahara exponential Tauberian theorem [6]. Next, Corollary 2.6 is applied to the case of unbounded potentials with double exponentially decaying type tails, through Proposition 2.8. Then, we treat the case of potentials falling in the third universality class (almost bounded) of [14] through Proposition 2.9. We end Section 2.4 considering the case of bounded potentials with tails near their upper-bound which are of the Fréchet type (Proposition 2.10). In Section 2.5, we state Theorem 2.11, which improves the upper bounds describing the order of magnitude of the empirical average for the examples considered in Section 2.4. The proof of Proposition 2.1 is the content of the third section. In Section 4, the truncated first moments are introduced. These are the first moments of a reaction – diffusion process defined on a finite box, with Dirichlet boundary conditions. They are then used to approximate some important quantities related to the averaged first moments. Then, several important estimates for the moments and correlations of the first moments are derived. The proof of Theorems 2.3 and 2.11 are given in Section 5. In Section 5.1, the partition analysis method of [5] is recalled. This and together with the estimates of Section 4, and the von Bahr – Esseen inequality, is subsequently applied to prove Theorem 2.3. The paper finishes with Section 5.7, where Theorem 2.11 is proved.

2. Notation and results

Here we will state the results of this paper. After introducing most of the notation and giving the main definitions in the first subsection, in Section 2.2 we state Proposition 2.2, which ensures that a.s. there is no explosion for the reaction – diffusion process under Assumption E. Then, the principal result of this paper, Theorem 2.3 is stated in Section 2.3, together with Corollary 2.4. In Section 2.4, we state Corollary 2.6, giving the form of Theorem 2.3, under certain regularity assumptions. Here we will consider applications of this results to several specific examples of distributions of the effective potential. We end the presentation of our results with Section 2.5, where we state Theorem 2.11.

2.1. Definition of the reaction – diffusion process. We begin defining a reaction – diffusion model corresponding to a set of random walks on the lattice $\mathbb{Z}^d$ branching and annihilating at rates depending on the their position. Consider the set of natural numbers $\mathbb{N}$ endowed with the discrete topology. Define the set $\Omega := \mathbb{N}^{\mathbb{Z}^d}$ representing the possible configuration of particles on the lattice. In this paper we will be interested only on the subset of configurations $\Omega_0 \subset \Omega$ characterized by the property that $\{x : \eta(x) > 0\}$ has finite cardinality whenever $\eta \in \Omega_0$. Let $v_+ := \{v_+(x) : x \in \mathbb{Z}^d\}$ and $v_- := \{v_-(x) : x \in \mathbb{Z}^d\}$, where $v_+(x) \in [0,\infty)$ and $v_-(x) \in [0,\infty]$. Here $v_+(x)$ and $v_-(x)$ represent the rate at which particles branch and annihilate at site x, respectively. Note that we admit the value ∞ for the annihilation rate: this represents a hard core obstacle, where particles are instantly annihilated. Call an ordered pair $w := (v_-, v_+) \in W$, with coordinates $w(x) = \big(v_-(x), v_+(x)\big)$, a *field configuration*, where $W := ([0,\infty) \times [0,\infty])^{\mathbb{Z}^d}$. We will denote the set of *hard core obstacle sites* by $\mathcal{G}(w) := \{x \in \mathbb{Z}^d : v_-(x) = \infty\}$. Given $r \in [0,\infty)$ and $x \in \mathbb{Z}^d$ we will call $\Lambda(x,r) := \{y \in \mathbb{Z} : \|x-y\| \leq r\}$ the ball of radius r centered at x under the norm $\|x\| := \sup_{1 \leq i \leq d} |x_i|$, where x_i are the coordinates of x. We will furthermore use the notation Λ_r in place of $\Lambda(0,r)$. In this subsection we will construct a stochastic process as the limit as $L \to \infty$ of processes defined on the boxes Λ_L. Throughout the sequel, given a subset $U \subset \mathbb{Z}^d$, we will denote by U^c the complement of U and $\delta U := \{x \in U^c : \inf_{y \in U} |x-y| = 1\}$, where $|\cdot|$ denotes the Euclidean distance. So, for each finite subset $U \subset \mathbb{Z}$, we want to consider a process with state space $\Omega_0^w := \{\eta \in \Omega_0 : \eta(x) = 0, x \in \mathcal{G}(w)\}$ defined formally by the infinitesimal generator,

$$
\begin{aligned}
L_U f(\eta) := & \sum_{x \in U \cap \mathcal{G}(w)^c} \sum_{\substack{y \in \mathcal{G}(w)^c \\ \|x-y\|=1}} \kappa \eta(x) \big(f(\eta^{x,y}) - f(\eta)\big) \\
& + \sum_{x \in U \cap \mathcal{G}(w)^c} \sum_{\substack{y \in \mathcal{G}(w) \\ \|x-y\|=1}} \kappa \eta(x) \big(f(\eta^{x,-}) - f(\eta)\big) \\
& + \sum_{x \in U \cap \mathcal{G}(w)^c} v_+(x) \eta(x) \big(f(\eta^{x,+}) - f(\eta)\big) \\
& + \sum_{x \in \mathcal{G}(w)^c} v_-(x) \eta(x) \big(f(\eta^{x,-}) - f(\eta)\big),
\end{aligned} \tag{2.1}
$$

acting on an appropriate dense subset $\mathcal{D}_U$ of the space of real bounded functions $B(\Omega_0^w)$, defined on Ω_0^w, endowed with the uniform norm. In the above expression, $\kappa \geq 0$, $\eta^{x,y}$ is the configuration where a particle from site x has jumped to site y so that $\eta^{x,y}(z) = \eta(z)$ if $z \neq x, y$, $\eta^{x,y}(x) = \eta(x) - 1$, and $\eta^{x,y}(y) = \eta(y) + 1$; $\eta^{x,+}$ is the configuration where there is an extra particle at site x and $\eta^{x,-}$ the configuration where one particle at site x has disappeared. It is a well known fact that it is possible to construct a strong Markov process, denoted by $\eta^U := \{\eta^U(t) : t \geq 0\}$, corresponding to an infinitesimal generator of the form (2.1), and taking values on the Skorokhod space $\mathcal{S} := D([0,\infty); \Omega_0^w)$. In fact, such a process falls in the category called $|U|$*-dimensional continuous time Markov branching process* by Athreya and Ney (see Chapter V, Sections 7.1 – 7.2 of Athreya – Ney [3]). Furthermore, it can be shown that a.s. the expected value of each coordinate of such a process $\{\eta^U(t) : t \geq 0\}$, is finite, ensuring that there cannot be infinitely many particles produced in a finite time (see [3, Section 7.1]). Let us now call $\mathcal{P}(\Omega_0^w)$ the set of probability measures defined on Ω_0^w endowed with the Borel σ-algebra associated to the subspace topology of Ω_0^w as a subset of Ω with the product topology. Then, for each field configuration $w \in W$ and probability measure $\nu \in \mathcal{P}(\Omega_0^w)$ denote by $P_\nu^{U,w}$ the law of the process $\{\eta^U(t) : t \geq 0\}$ defined on $\mathcal{S}$ endowed with its Borel σ-field $\mathcal{B}(\mathcal{S})$. We will call this process the *reaction – diffusion process on U with field w and initial condition ν*. In the particular case in which $U = \Lambda_n$ we will use the obvious notations L_n and $P_\nu^{n,w}$. Furthermore, we will call the process on Λ_n, the *reaction – diffusion process at scale n*. Now, note that using the natural coupling [15] and Kolmogorov's extension theorem, it is possible to define for each field configuration $w \in W$ and initial condition $\nu \in \mathcal{P}(\Omega_0^w)$ a probability measure Q_ν^w on the product space $\mathcal{S}^{\mathbb{N}}$, endowed with its Borel σ-algebra induced by the product topology, in such a way that if $\eta^n \in \mathcal{S}$ denotes the nth coordinate of an element $\eta \in \mathcal{S}^{\mathbb{N}}$, $\eta^n(t) \in \Omega_0^w$ its value at time $t \geq 0$ and $\eta^n(t,x) \in \mathbb{N}$ the value at time t of the x-coordinate of $\eta^n(t)$, then,

(i) for every $A \in \mathcal{B}(\mathcal{S})$ and $n \geq 1$,
$$Q_\nu^w[\eta^n \in A] = P_\nu^{n,w}[A].$$
In particular, for every $B \in \mathcal{B}(\Omega_0)$ we have that, $Q_\nu^w[\eta^n(0) \in B] = \nu[B]$.

(ii) for every $n \geq 1$,
$$Q_\nu^w[\eta^{n+1}(t) \geq \eta^n(t)] = 1.$$

(iii) for each $n \geq 1$ define the first exit time from the box Λ_n as
$$T_n := \inf\{t \geq 0 : \sup_{x \in \Lambda_n^c} \eta^n(x,t) > 0\}.$$
Then, for every $n \geq 1$,
$$Q_\nu^w[\eta^n(t) = \eta^{n+1}(t), T_n > t] = Q_\nu^w[T_n > t].$$

Let us now remark that due to property (ii), for every $t \geq 0$ and $x \in \mathbb{Z}^d$ the limit,
$$\eta(t,x) := \lim_{n \to \infty} \eta^n(t,x),$$
exists, possibly taking the value ∞. Define $\eta(t) := \{\eta(t,x) : x \in \mathbb{Z}^d\}$. We denote the stochastic process $\{\eta(t) : t \geq 0\}$, taking values on the space $\overline{\mathbb{N}}^{\mathbb{Z}^d}$, where $\overline{\mathbb{N}}$ is the Alexandrov compactification of the natural numbers, and distributed according to the measure Q_ν^w, the *reaction – diffusion process with field w and initial law ν*. We will denote by E_ν^w the corresponding expectation. In addition, for each $t \geq 0$, we define the *total number of particles at time t* by
$$\zeta(t) := \sum_{x \in \mathbb{Z}^d} \eta(t,x).$$
Also, whenever it is true that,
$$Q_\nu^w[\eta(x,t) < \infty, \forall\, x \in \mathbb{Z}^d, t \geq 0] = 1,$$
we will say that with probability one there is *no explosion*. In the sequel we define for each $x \in \mathbb{Z}^d$ the probability measure δ_x on $(\Omega_0^w, \mathcal{B})$ which assigns probability 1 to configurations with one particle at site x and none elsewhere. We will be interested in initial configurations where $\nu = \delta_x$. In such a case we will use the

notation P_x^w instead of $P_{\delta_x}^w$ and E_x^w for the corresponding expectation. In the case where $x \in \mathcal{G}(w)$, we adopt the convention that P_x^w is the probability measure which has a unique atom at the configuration $\eta \equiv 0$ ($\eta(x) = 0$ for every $x \in \mathbb{Z}^d$).

Let us denote by $\mathcal{P}(W)$ the set of probability measures defined on the space $W = ([0,\infty) \times [0,\infty])^{\mathbb{Z}^d}$ endowed with its natural σ-algebra. In the sequel we will take fields v_+ and v_- which are random, assigning a probability measure $\mu \in \mathcal{P}(W)$ in such a way that the field configuration $\{w(x) : x \in \mathbb{Z}^d\}$ has independent coordinates with respect to μ. Furthermore, we will use the notation $\langle\cdot\rangle$ to denote expectation with respect to this law and $Var_\mu(\cdot)$ variance. Now, let us define the *quenched first moment* on $\mathbb{Z}^d$ of the total number of particles at time t starting from site x as, $m(x,t,w) := \mathrm{E}_x^w[\zeta(t)]$, and the *annealed first moment* on $\mathbb{Z}^d$ of the total number of particles at time t starting from site x as, $\langle m(x,t)\rangle := \int m(x,t,w)\,\mathrm{d}\mu$. Furthermore, we call the sets $\{m(x,t,w) : x \in \mathbb{Z}^d\}$ and $\{\langle m(x,t)\rangle : x \in \mathbb{Z}^d\}$, the *fields of quenched first moments* and the *field of annealed first moments*, respectively. Depending on the context, we might write m or $m(x,t)$ in place of $m(x,t,w)$, dropping the dependence on the field configuration w, and $\langle m\rangle$ instead of $\langle m(x,t)\rangle$.

The quantity which will give us a transition mechanism between of the quenched first moments is the *averaged first moment* at scale L and time t defined for a reaction – diffusion process starting from site x as,

$$m^L(x,t,w) := \frac{1}{|\Lambda(x,L)|} \sum_{y\in\Lambda(x,L)} m(y,t,w).$$

2.2. Results giving conditions for no explosion. Here we will give a criteria on the field configuration w, stated as Proposition 2.1, which ensures that there is no explosion in the reaction – diffusion process with field w.

Proposition 2.1. *Consider the reaction diffusion process with field w and initial law ν. Assume that,*

$$\limsup_{|x|\to\infty} \frac{v_+(x)}{|x|\log|x|} = 0. \tag{2.2}$$

Then for every $t \geq 0$ we have that

$$\mathrm{E}_\nu^w[\zeta(t)] < \infty.$$

Hence, there is no explosion.

We state below with the name of Proposition 2.2, a result of Gärtner – Molchanov [12, Lemma 2.5] giving a sufficient condition on the law μ in order that the first moment of the total number of particles $\zeta(t)$ at time t, exists μ-a.s., and hence that there is no explosion. Given a field configuration $w = (v_+, v_-)$, we now need to introduce the *effective potential* $\{v(x) : x \in \mathbb{Z}^d\}$, defined by $v(x) := v_+(x) - v_-(x)$. Furthermore, set $\log_+ x = \log x$ if $x > e$ and $\log_+ x = 1$ otherwise, while define the positive part $x^+ := \max(0,x)$.

Proposition 2.2. *Consider the reaction diffusion process with field w and initial law ν. Assume that the field configuration w is distributed according to a product probability measure $\mu \in \mathcal{P}(W)$. Suppose that*

$$\left\langle \left(\frac{v^+(0)}{\log_+ v(0)}\right)^d \right\rangle < \infty. \tag{2.3}$$

Then μ-a.s. it is true that,

$$\limsup_{|x|\to\infty} \frac{v_+(x)}{|x|\log|x|} = 0,$$

and therefore, μ-a.s. there is no explosion for the reaction – diffusion process with field w and arbitrary initial law in $\mathcal{P}(\Omega_0)$.

Note that the last statement of Proposition 2.2 follows from Proposition 2.1. Furthermore, Assumption E is enough for (2.3) to be satisfied, and hence to ensure no explosion.

2.3. The Gaussian-annealed transition results. Here we will state the main result of this paper, which shows how under different growth of scales, the averaged first moment has an asymptotic behavior where a law of large numbers is satisfied, and a central limit theorem can describe the fluctuations around this law of large numbers. We will assume Condition E, ensuring the existence of the annealed first moments. We will also need to consider the *growth functions* $\{H_1(t) = \log\langle m(0,t)\rangle : t \geq 0\}$ already defined in equation (1.2) of the introduction and the *intermittency exponents* $\{F_\theta : \theta \in \mathbb{R}\}$, defined in equation (1.3). Let us now state the main result of this paper.

Theorem 2.3. *Consider a reaction – diffusion process with initial law δ_0 and field $w = (v_+, v_-)$ distributed according to a product measure $\mu \in \mathcal{P}(W)$. Consider the intermittency exponents $\{F_\theta : \theta \in \mathbb{R}\}$ defined in equation (1.3) and the growth functions $\{H_1(t) : t \geq 0\}$ defined in equation (1.2). Assume that Conditions* E *and* MI *are satisfied. Then the following statements are true,*

(i) **Law of large numbers.** *Assume that there is an $\epsilon > 0$ such that eventually in t, $L(t) \geq \exp\{F_\epsilon(t)/d\}$. Then in μ-probability we have*

$$\frac{m^L(0,t,w)}{\langle m(0,t)\rangle} \sim 1, \tag{2.4}$$

as $t \to \infty$. Furthermore, assume that there is an $\epsilon > 0$ such that eventually in t, $L(t) \leq \exp\{F_{-\epsilon}(t)/d\}$. Then in μ-probability we have

$$\frac{m^L(0,t,w)}{\langle m(0,t)\rangle} \ll 1, \tag{2.5}$$

as $t \to \infty$.

(ii) **Central limit theorem.** *Assume that there is an $\epsilon > 0$ such that eventually in t, $L(t) \geq \exp\{F_\epsilon(2t)/d\}$. Then,*

$$\lim_{t\to\infty} \frac{m^L(0,t,w) - \langle m(0,t)\rangle}{\sqrt{\mathrm{Var}_\mu(m(0,t))}} = \mathcal{N}(0,1), \tag{2.6}$$

where $\mathcal{N}(0,1)$ is a centered normalized normal law and the convergence is in the sense of distributions. Furthermore, assume that there is an $\epsilon > 0$ such that eventually in t, $L(t) \leq \exp\{F_{-\epsilon}(2t)/d\}$. Then in μ-probability we have that,

$$\frac{m^L(0,t,w) - \langle m(0,t)\rangle}{\sqrt{\mathrm{Var}_\mu(m(0,t))}} \ll 1. \tag{2.7}$$

To provide a better insight on the meaning of Theorem 2.3, we will see the form that it takes under the stronger Assumption SI, which includes the first universality class of [14]. This will be formulated as a corollary, which in the case $\kappa = 0$

generalizes [4, Theorems 2.1 and 2.2] to include distributions μ of the field, which not necessarily have regularly varying log-tails.

Corollary 2.4. *Consider a reaction – diffusion process with initial law δ_0 and field $w = (v_+, v_-)$ distributed according to a product measure $\mu \in \mathcal{P}(W)$. Consider the cumulant intermittency exponents $\{G_\theta : \theta \in \mathbb{R}\}$ defined in equation* (1.4) *and the cumulant generating function $\{H(t) : t \geq 0\}$ defined in equation* (1.1). *Assume that Conditions* E *and* SI *are satisfied. Then the following statements are true,*

(i) **Law of large numbers.** *Assume that there is an $\epsilon > 0$ such that eventually in t, $L(t) \geq \exp\{G_\epsilon(t)/d\}$. Then in μ-probability we have*

$$\frac{m^L(0,t,w)}{\langle m(0,t)\rangle} \sim 1,$$

as $t \to \infty$. In particular, $\log m^L(0,t,w)/H(t) \sim 1$. Furthermore, assume that there is an $\epsilon > 0$ such that eventually in t, $L(t) \leq \exp\{G_{-\epsilon}(t)/d\}$. Then in μ-probability we have

$$\frac{m^L(0,t,w)}{\langle m(0,t)\rangle} \ll 1,$$

as $t \to \infty$.

(ii) **Central limit theorem.** *Assume that there is an $\epsilon > 0$ such that eventually in t, $L(t) \geq \exp\{G_\epsilon(2t)/d\}$. Then,*

$$\lim_{t\to\infty} \frac{m^L(0,t,w) - \langle m(0,t)\rangle}{\sqrt{\mathrm{Var}_\mu(m(0,t))}} = \mathcal{N}(0,1),$$

where the convergence is in the sense of distributions. Furthermore, assume that there is an $\epsilon > 0$ such that eventually in t, $L(t) \leq \exp\{G_{-\epsilon}(2t)/d\}$. Then in μ-probability we have that,

$$\frac{m^L(0,t,w) - \langle m(0,t)\rangle}{\sqrt{\mathrm{Var}_\mu(m(0,t))}} \ll 1.$$

The proof of Corollary 2.4 in the case $\kappa = 0$ follows from the observation that $G_\theta(t) = F_\theta(t)$ in this case. The case $\kappa > 0$ is a direct consequence of the fact that $e^{H(t)-2d\kappa t} \leq \langle m(0,t)\rangle \leq e^{H(t)}$, stated in [12, Theorem 3.1], and the observation that $G_\theta(t) \geq 0$ for $\theta > 0$ and $G_\theta(t) \leq 0$ for $\theta < 0$, which follows from Jensen's inequality. Now, the following proposition, which will be proved in Section 4, shows that the condition,

$$\lim_{t\to\infty} tH''(t) = \infty, \tag{2.8}$$

is sufficient for Assumption SI to be true. This condition implies a kind of domination of the branching over the annihilation.

Proposition 2.5. *Consider the cumulant exponents $\{G_\theta : \theta \in \mathbb{R}\}$. Assume that condition* (2.8) *is satisfied. Then, Condition* SI *is satisfied. Furthermore:* (i) *for every $\theta \neq 0$, there is a $t_0 \geq 0$, such that the function $G_\theta(t)$ is monotone in t, for $t \geq t_0$;* (ii) *there is a $t_1 \geq 0$, such that the function $G_\theta(t)$ is monotone in θ for $t \geq t_1$.*

Condition (2.8) implies that the branching dominates the annihilation, in the sense that $H(t)/t \to \infty$ as $t \to \infty$, which implies that the essential supremum of the random variable $v(0)$ is infinite (see [12]).

2.4. Regularity assumptions on the intermittency exponents. For the purpose of applications, it will be important to identify cases where the assumptions in Theorem 2.3 can be formulated in a more explicit way. As it will be shown, the following assumption on the intermittency exponents, turns out to fall in one of these situations.

Assumption RI. The intermittency exponents $\{F_\theta(t) : \theta \neq \mathbb{R}\}$ satisfy the mild intermittency condition MI. In addition, there exist two increasing functions $f_1, f_2 \colon \mathbb{R} \setminus \{0\} \to \mathbb{R}$ and a function $J(t) \colon [0, \infty) \to [0, \infty)$, such that for $\theta \neq 0$ small enough,

(i) $F_\theta(t) \sim f_1(\theta) J(t)$, and $F_\theta(2t) \sim f_2(\theta) J(t)$.
(ii) There exists two constants γ_1 and γ_2, such that $\lim_{\theta\to 0} f_1(\theta) = \gamma_1$ and $\lim_{\theta\to 0} f_2(\theta) = \gamma_2$.

Throughout the sequel, the constants γ_1 and γ_2 will be called *transition exponents* and the function J *growth exponent.* It will be shown that there exist several important cases of random fields which fall in this category. Furthermore, the following corollary of Theorem 2.3 shows the convenience of Assumption RI.

Corollary 2.6. *Consider a reaction – diffusion process with initial law δ_0 and field $w = (v_+, v_-)$ distributed according to a product measure $\mu \in \mathcal{P}(W)$. Suppose that Assumptions* E *and* RI *are satisfied with transition exponents γ_1 and γ_2 and growth function J. Then the following statements are true,*

(i) **Law of large numbers.** *Assume that there is a $\gamma > \gamma_1$ such that eventually in t, $\log L(t) \geq \gamma J(t)/d$. Then in μ-probability we have*

$$\frac{m^L(0,t,w)}{\langle m(0,t)\rangle} \sim 1,$$

as $t \to \infty$. Furthermore, assume that there is a $0 < \gamma < \gamma_1$ such that eventually in t, $\log L(t) \leq \gamma J(t)/d$. Then in μ-probability we have

$$\frac{m^L(0,t,w)}{\langle m(0,t)\rangle} \ll 1,$$

as $t \to \infty$.

(ii) **Central limit theorem.** *Assume that there is a $\gamma > \gamma_2$ such that eventually in t, $\log L(t) \geq \gamma J(t)/d$. Then,*

$$\lim_{t\to\infty} \frac{m^L(0,t,w) - \langle m(0,t)\rangle}{\sqrt{\mathrm{Var}_\mu\big(m(0,t)\big)}} = \mathcal{N}(0,1),$$

where the convergence is in the sense of distributions. Furthermore, assume that there is a $0 < \gamma < \gamma_2$ such that eventually in t, $\log L(t) \leq \gamma J(t)/d$. Then in μ-probability we have that,

$$\frac{m^L(0,t,w) - \langle m(0,t)\rangle}{\sqrt{\mathrm{Var}_\mu\big(m(0,t)\big)}} \ll 1.$$

In the next subsection we will apply Corollary 2.6 to four situations each one falling in one of the universality classes described by van den Hofstad et al.in [14]. These classes encompass all possible situations under three conditions. The first is Condition E, ensuring the existence of the positive moments defining the cumulant generating functions (1.1). The second and third condition avoids different qualitative behaviors of the potential at different scales. Let us formulate next the second condition of [14].

Assumption H. The function $H(t)/t$ is in the de Haan class.

A function f is said to be in the de Haan class if for some regularly varying function $g\colon (0,\infty) \to \mathbb{R}$, we have that $\big(f(\lambda t) - f(t)\big)/g(t)$ converges to a limit different from 0 as $t \to \infty$, for $\lambda > 0$. Let us recall that a function h is regularly varying at infinity with index ρ, if for any $a > 0$ we have $\lim_{x\to\infty} h(ax)/h(x) = a^\rho$. This property will be stated as $h \in R_\rho$. Whenever H is satisfied, then $H(t)$ is regularly varying with index $\gamma \geq 0$. Furthermore, in [14, Proposition 1.1], it is proved that under Assumption H there exist a function $\widehat{H}\colon (0,\infty) \to \mathbb{R}$ and a continuous function $k(t)\colon (0,\infty) \to (0,\infty)$ such that,

$$\lim_{t\to\infty} \frac{H(ty) - yH(t)}{k(t)} = \widehat{H}(y) \neq 0, \tag{2.9}$$

for $y \in (0,1) \cup (1,\infty)$. It is also shown that $k(t)$ is regularly varying of index γ. We can now recall the third assumption of [14].

Assumption K. The limit $k^* = \lim_{t\to\infty} k(t)/t$ exists in $[0,\infty]$.

Under Assumptions E, H, and K, the four universality classes defined in [14] are:

(1) $\gamma > 1$, or $\gamma = 1$ and $k^* = \infty$.
(2) $\gamma = 1$ and $k^* \in (0,\infty)$.
(3) $\gamma = 1$ and $k^* = 0$.
(4) $\gamma < 1$.

In what follows we exhibit examples in each one of these classes satisfying Assumption RI so that Corollary 2.6 can be applied, and the transition exponents can be explicitly written. In the sequel, following [14], we will call the third class the class of *almost bounded potentials.*

2.4.1. *Unbounded potentials with Weibull-type tails.* Our first application of Corollary 2.4 will be to an example falling in the first universality class of [14]. We will assume that the essential supremum of $v(0)$ is ∞, and the tails at ∞ of $v(0)$ follow a Weibull-type law, $\mu[v(0) > x] = \exp\{-h(x)\}$ for $x > 0$, with $h \in R_\rho$ for some $1 < \rho < \infty$. In the terminology of [4] in the context of i.i.d. random exponentials ($\kappa = 0$ in our situation), this is called *Case* B. The following proposition shows that Assumption RI is satisfied in this situation, and hence Corollary 2.6, which generalizes [4, Theorems 2.1 and 2.2] in Case B form $\kappa = 0$ to $\kappa > 0$.

Proposition 2.7. *Suppose that the essential supremum of the effective potential $v(0)$ is ∞ with Weibull-type tails $\mu[v(0) > x] = \exp\{-h(x)\}$ for $x > 0$, and $h \in R_\rho$ for some $1 < \rho < \infty$. Then Assumption* RI *is satisfied with transition exponents*

$$\gamma_1 = \frac{1}{\rho - 1}, \qquad \gamma_2 = 2^{\frac{\rho}{\rho-1}} \frac{1}{\rho - 1},$$

and growth exponent,

$$J(t) = H(t).$$

Let us now prove Proposition 2.7. Note that under the conditions on the tail of the law of $v(0)$, the cumulant generating function $H(t)$ of $v(0)$ is well defined, smooth, nondecreasing and tends to infinity as $t \to \infty$. Furthermore, by the Kasahara exponential Tauberian theorem (see Bingham et al. [6, Theorem 4.12.7]), we know that $H \in R_{\rho'}$, where the index ρ' is defined by the equation,

$$\frac{1}{\rho} + \frac{1}{\rho'} = 1.$$

From this observation it is easy to check that Assumption SI of Corollary 2.4 is satisfied, and that for every $\epsilon \neq 0$, the cumulant growth exponents $G_\epsilon(t)$ and $G_\epsilon(2t)$ satisfy

$$\frac{G_\epsilon(t)}{H(t)} \sim \frac{(1+\epsilon)^{\rho'} - (1+\epsilon)}{\epsilon}$$

and

$$\frac{G_\epsilon(2t)}{H(t)} \sim 2^{\rho'} \frac{(1+\epsilon)^{\rho'} - (1+\epsilon)}{\epsilon}.$$

Now note that $\big((1+\epsilon)^{\rho'} - (1+\epsilon)\big)/\epsilon$ is increasing in ϵ and converges to $\gamma_1 = 1/(\rho-1)$ as $\epsilon \to 0$, and similarly $2^{\rho'}\big((1+\epsilon)^{\rho'} - (1+\epsilon)\big)/\epsilon$ is increasing in ϵ and converges to $\gamma_2 = 2^{\rho/(\rho-1)}\gamma_1$ as $\epsilon \to 0$.

2.4.2. *Unbounded potentials with double exponential type tails.* Here we consider the second universality class so that H is regularly varying with index $\gamma = 1$ and $k^* \in (0,\infty)$. As shown in [14, Proposition 1.1], this is equivalent to the existence of a constant $\rho \in (0,\infty)$ such that

$$\lim_{t\to\infty} \frac{H(yt) - yH(t)}{t} = \rho y \log y, \tag{2.10}$$

for all $y \in (0,1) \cup (1,\infty)$ (this and $0 < \rho < \infty$ is called Assumption H in [13]). Furthermore, under assumption (2.10) it is true that,

$$\lim_{t\to\infty} \frac{H(t)}{t} = \infty.$$

This second universality class includes the case of unbounded potentials which are double exponentially distributed with parameter ρ, $0 < \rho < \infty$,

$$\mu[v(0) > x] = \exp\{-\mathrm{e}^{x/\rho}\},$$

for $x \in \mathbb{R}$. We then have the following interesting proposition.

Proposition 2.8. *Suppose that assumption* (2.10) *is satisfied for some* $\rho \in (0,\infty)$. *Then Assumption* RI *is satisfied with transition exponents*

$$\gamma_1 = \rho, \quad \gamma_2 = 2\rho,$$

and growth exponent,

$$J(t) = t.$$

To prove Proposition 2.8, we quote [13, Theorem 1.2], which shows that under assumption (2.10) we have that,

$$\langle m(0,t)\rangle = \exp\Big\{H(t) - 2d\kappa\chi\Big(\frac{\rho}{\kappa}\Big)t + o(t)\Big\},$$

for $\kappa > 0$, where $\chi(x) := \frac{1}{2}\inf_{p\in\mathcal{P}(\mathbb{Z})}[S(p) + \rho I(p)]$ for $x \geq 0$, $\mathcal{P}(\mathbb{Z})$ is the space of probability measure on $\mathbb{Z}$, $S\colon \mathcal{P}(\mathbb{Z}) \to [0,\infty)$ is the Donsker–Varadhan functional defined by $S(p) := \sum_{x\in\mathbb{Z}}\big(\sqrt{p(x+1)} - \sqrt{p(x)}\big)^2$, while $I\colon \mathcal{P}(\mathbb{Z}) \to [0,\infty)$ is the entropy functional defined by $I(p) := -\sum_{x\in\mathbb{Z}} p(x)\log p(x)$. On the other hand, from the previous discussion we can conclude that, for every $\epsilon \neq 0$,

$$\frac{F_\epsilon(t)}{t} \sim \rho(1+\epsilon)\frac{\log(1+\epsilon)}{\epsilon},$$

and

$$\frac{F_\epsilon(2t)}{t} \sim 2\rho(1+\epsilon)\frac{\log(1+\epsilon)}{\epsilon}.$$

From these limiting behaviors, we see that Assumption MI is satisfied. Furthermore, form the fact that $(1+\epsilon)\log(1+\epsilon)/\epsilon$ is increasing for ϵ small enough and converges to 1 as $\epsilon \to 0$, we obtain the transition exponents at ρ and 2ρ of Proposition 2.8.

2.4.3. *Almost bounded potentials.* We now focus on the third universality class, where $\gamma = 1$ and $k^* = 0$. As shown in [14, Theorem 1.4], in this case it is true that,

$$\log\langle m(0,t)\rangle \sim \frac{H(t\alpha_t^{-d})}{\alpha_t^{-d}}, \tag{2.11}$$

where $\alpha_t\colon [0,\infty) \to [0,\infty)$ is the so called *scaling function*, which is implicitly defined for all $t > 0$ sufficiently large by the equation,

$$\frac{k(t\alpha_t^{-d})}{t\alpha_t^{-d}} = \frac{1}{\alpha_t^2}. \tag{2.12}$$

As shown in [14, Proposition 1.2], this function is unique up to asymptotic equivalence. Furthermore, for the third universality class, it is a slowly varying function (regularly varying of index 0).

Proposition 2.9. *Suppose that H is regularly varying of index 1 and that (2.9) is satisfied for $k(t)$ such that $k(t) \ll t$. Then Assumption RI is satisfied with transition exponents*

$$\gamma_1 = \rho, \quad \gamma_2 = 2\rho,$$

and growth exponent,

$$J(t) = \frac{t}{\alpha_t^2}. \tag{2.13}$$

The proof of Proposition 2.9 follows now applying the definition of α_t through (2.12), the asymptotics (2.11) and (2.9).

A specific example of a distribution falling in the third universality class is given by the *squared double exponential* law. In other words, an effective potential $v(0)$ which is unbounded, with law,

$$\mu[v(0) > x] = \exp\{-e^{x^2}\},$$

for $x \geq 0$. As it can be deduced from the discussion of [14, Example 1.4.3], in this case the scale function is given by $\alpha_t \sim 2^{1/2}(\log t)^{1/4}$ and $k(t) \sim t/(2\sqrt{\log t})$. Furthermore, $\rho = 1$ and the growth exponent can be chosen as $J(t) \sim t/(2(\log t)^{1/2})$ (see Table 2 of the introduction).

2.4.4. *Bounded potentials with Fréchet-type tails.* We now continue with an example falling in the universality class (4) of [14]. We will consider the case in which the essential supremum of $v(0)$ is 0 and the tails are of Fréchet type: $\mu[v(0) > -x] = \exp\{-h(x^{-1}\}$ for $x > 0$, with $h \in R_\rho$ for some $0 < \rho < \infty$. Using the terminology of [4] in the context of i.i.d. random exponentials, this is *Case* A. To state appropriately this result, we will need to recall the work of Biskup and König [7], who studied the asymptotics of the annealed and quenched first moments in the case of product environments μ such that the cumulant generating function $H(t)$ is in the so called γ-class for some $\gamma \in [0, 1)$. We say that H *is in the* γ-*class* if the essential supremum of $v(0)$ is 0 and if there is a nondecreasing function $\alpha_t \in (0, \infty)$ and a function $\widetilde{H}: [0, \infty) \to (-\infty, 0]$, $\widetilde{H} \neq 0$, such that

$$\lim_{t\to\infty} \frac{\alpha_t^{d+2}}{t} H\left(\frac{t}{\alpha_t^d} y\right) = \widetilde{H}(t), \tag{2.14}$$

for $y \geq 0$, uniformly on compact sets in $(0, \infty)$. We will denote the function α_t the *scale function.* It is not difficult to show, using the de Bruijn exponential Tauberian theorem [6], that if $v(0)$ is in Case A for some $\rho > 0$, then H is in the ρ'-class for $\rho' = \rho/(\rho + 1)$, or $1/\rho' - 1/\rho = 1$. We can now sate the following proposition.

Proposition 2.10. *Suppose that the essential supremum of the effective potential $v(0)$ is 0 with Fréchet-type tails $\mu[v(0) > -x] = \exp\{-h(x^{-1})\}$ for $x > 0$, and $h \in R_\rho$ for some $0 < \rho < \infty$. Then Assumption* RI *is satisfied with transition exponents*

$$\gamma_1 = \left(\frac{1}{d+2+2\rho}\right)^2, \quad \gamma_2 = 2^{1-\gamma_1}\gamma_1,$$

and growth exponent,

$$J(t) = \chi \frac{t}{\alpha_t^2},$$

for some constant $\chi \in (0, \infty)$.

As a consequence of Proposition 2.10 we can now apply Corollary 2.6, generalizing Case A of [4, Theorems 2.1 and 2.2] from $\kappa = 0$ to $\kappa \geq 0$. In contrast to Proposition 2.7, where there is no change in the value of the transition exponents γ_1 and γ_2 from $\kappa = 0$ to $\kappa > 0$, here there is.

Let us now prove Proposition 2.10. In [7, Proposition 2.1], it is shown that whenever μ is in the ρ'-class, then the scaling function $\alpha_t \in R_\nu$, for,

$$\nu := \frac{1-\rho'}{d+2-d\rho'}.$$

Furthermore, [7, Theorem 1.2] states that when μ is in the ρ'-class, there exists a $\chi \in (0, \infty)$ such that,

$$\log\langle m(0,t)^\beta\rangle \sim -\chi \frac{\beta t}{\alpha_{\beta t}^2},$$

for every $\beta \in (0, \infty)$. Then Condition MI of Theorem 2.3 is satisfied, and for every $\epsilon \neq 0$, the growth exponents $F_\epsilon(t)$ and $F_\epsilon(2t)$ satisfy

$$\frac{F_\epsilon(t)}{t/\alpha_t^2} \sim \frac{(1+\epsilon) - (1+\epsilon)^{1-\nu^2}}{\epsilon}$$

and

$$\frac{F_\epsilon(2t)}{t/\alpha_t^2} \sim 2^{1-\nu^2} \frac{(1+\epsilon)-(1+\epsilon)^{1-\nu^2}}{\epsilon}.$$

As in the proof of Proposition 2.7, we can show that $\big((1+\epsilon)-(1+\epsilon)^{1-\nu^2}\big)/\epsilon$ is increasing in ϵ and converges to $\gamma_1 = \nu^2$ as $\epsilon \to 0$, and similarly the expression $2^{1-\nu^2}\big((1+\epsilon)-(1+\epsilon)^{1-\nu^2}\big)/\epsilon$ is increasing in ϵ and converges to $\gamma_2 = 2^{1-\nu^2}\gamma_1$ as $\epsilon \to 0$.

2.5. The critical regime. For the examples discussed in the four previous sections, it is possible to obtain the following improvement of Theorem 2.3.

Theorem 2.11. *Consider a reaction – diffusion process with initial law δ_0 and field $w = (v_+, v_-)$ distributed according to a product measure $\mu \in \mathcal{P}(W)$. Then the following statements are satisfied.*

(i) **Weibull type.** *Assume that μ has Weibull-type tails so that $\mu[v(0) > x] = \exp\{-h(x)\}$ for $x > 0$, $h \in R_\rho$ for $1 < \rho < \infty$. Then if $d \log L(t) \leq \gamma H(t)$ eventually in t, and $0 < \gamma < \gamma_1 = 1/(\rho - 1)$, we have that for every $\delta > 0$ in μ-probability,*

$$\frac{m_L(0,t)}{\mathrm{e}^{(a_{\mathrm{W}}(\gamma)+\delta)H(t)}} \ll 1,$$

where

$$a_{\mathrm{W}}(\gamma) := \frac{\rho}{\rho-1}[(\rho-1)\gamma]^{1/\rho} - \gamma.$$

(ii) **Double exponential type.** *Assume that μ satisfies assumption* (2.10) *for some constant $\rho \in (0,\infty)$. Then if, $d \log L(t) \leq \gamma t$ eventually in t, and $\gamma < \gamma_1 = \rho$, we have that for every $\delta > 0$ in μ-probability,*

$$\frac{m_L(0,t)}{\exp\big\{H\big((a_{\mathrm{D}}(\gamma)+\delta)t\big)/(a_{\mathrm{D}}(\gamma)+\delta)\big\}} \ll 1,$$

where

$$a_{\mathrm{D}}(\gamma) := \gamma \mathrm{e}^{(\gamma-\rho)/\rho}.$$

(iii) **Almost bounded potentials.** *Assume that μ is such that H is regularly varying of index* 1 *and that* (2.9) *is satisfied for $k(t)$ such that $k(t) \ll t$. Consider the growth exponent $J(t)$ as defined in equation* (2.13). *Then if, $d \log L(t) \leq \gamma J(t)$ eventually in t, and $\gamma < \gamma_1 = 1$, we have that for every $\delta > 0$ in μ-probability,*

$$\frac{m_L(0,t)}{\exp\big\{H\big((a_{\mathrm{A}}(\gamma)+\delta)t\big)/(a_{\mathrm{D}}(\gamma)+\delta)\big\}} \ll 1,$$

where

$$a_{\mathrm{A}}(\gamma) := \gamma \mathrm{e}^{(\gamma-\rho)/\rho}.$$

(iv) **Fréchet type.** *Assume that μ is such that essupv$(0) = 0$ and is of Fréchet type so that $\mu[v(0) > -x] = \exp\{-h(x^{-1})\}$ for $x > 0$, and $h \in R_\rho$ for some $\rho \in (0,\infty)$. Then if $d \log L(t) \leq \gamma J(t)$ eventually in t, and $\gamma < \gamma_1 = \nu^2$, we have that for every $\delta > 0$ in μ-probability,*

$$\frac{m_L(0,t)}{\mathrm{e}^{-(a_{\mathrm{F}}(\gamma)-\delta)J(t)}} \ll 1,$$

where

$$a_{\mathrm{F}}(\gamma) := (1-\nu^2)\Big(\frac{\gamma}{\nu^2}\Big)^{-\nu^2/(1-\nu^2)} + \gamma.$$

Let us remark that Theorem 2.11(iv) includes as a particular case, Case 3 of [5, Theorem 2(i)]. We believe that the four functions a_{W}, a_{D}, a_{A}, a_{F} are sharp, in the sense that the quantities of the four parts of Theorem 2.11 diverge if the sign of δ is changed. Also, these four functions have as maximum value 1, which is reached at γ_1.

In the special case in which μ is a double exponential law so that $\log \mu[v(0) > x] = -e^{x^2}$ in Theorem 2.11(ii), the function $\exp\{H((a_{\mathrm{D}}(\gamma)+\delta)t)/(a_{\mathrm{D}}(\gamma)+\delta)\}$ takes the form $\exp\{t\log((a_{\mathrm{D}}(\gamma)+\delta)t/e)\}$, showing how the transition mechanism takes place at a logarithmic order in the exponent in contrast to the polynomial one of parts (i) and (iv). The whole picture suggested by Theorem 2.11 seems to indicate the presence of a phase transition type behavior, as it is found in some mean field statistical mechanics models like the Random Energy Model [8, 9]. Indeed, when combined with Theorem 2.3(i), we conclude that $\log m_L(0,t,w) \sim \bar{a}(\gamma)J(t)$, where $\bar{a}(\gamma)$ equals a_{W}, a_{D}, a_{A} or a_{F} depending on the potential for $\gamma < \gamma_1$, while $\bar{a}(\gamma) = 1$ for $\gamma > 1$. Thus, there is nonanalyticity at $\gamma = 1$ of a quantity playing the role of a "free energy."

3. The conditions for no explosion

In this section we will prove Proposition 2.1. Since the initial conditions $\nu \in \mathcal{P}(V)$ are concentrated on configurations with a finite number of particles, and by translation invariance of the dynamics of the reaction – diffusion process on $\mathbb{Z}^d$, note that it is enough to consider the case where $\nu = \delta_0$.

3.1. Preliminary lemmas. Let us consider the reaction – diffusion process at scale n with field w satisfying condition (2.2), and with the initial condition δ_0. Define the quantities,

$$\zeta^n(t) := \sum_{x\in\Lambda_n} \eta^n(t,x),$$

representing the total number of particles produced at time t and,

$$\bar{\zeta}^n(t) := \sum_{x\in\Lambda_n^c} (\eta^n(t,x) - \eta^n(0,x)),$$

representing the total number of particles which have touched Λ_n^c in the time interval $[0,t]$. From this definition, we can conclude that for $m > n \geq 1$ it is true that,

$$\zeta^m(t) = \zeta^n(t) + \sum_{k=1}^{\bar{\zeta}^n(t)} \zeta^m_{x_k}(t-\tau_k), \tag{3.1}$$

where for $1 \leq k \leq \bar{\zeta}^n(t)$, $x_k \in \delta\Lambda_n$ is the set of exit sites from Λ_n of the random walks which have touched the set Λ_n^c in the time interval $[0,t]$, $0 \leq \tau_k \leq t$ the exit times of each one of these random walks and $\{\zeta^m_{x_k}(s) : s \geq 0\}$ is a set of independent processes such that $\zeta^m_{x_k}(s) = 0$ for $s < 0$, $\zeta^m_{x_k}(s) = \delta_{x_k}$ for $s = 0$ while $\{\zeta^m_{x_k}(s) : s \geq 0\}$ has the law $P^{m,w}_{x_k}$. Let us now define for each $n \in \mathbb{N}$ the maximum value of the field v_+ on the box Λ_n by $v_n := \max_{x\in\Lambda_n} v_+(x)$.

Lemma 3.1. *Consider the reaction – diffusion process at scale n with field w and initial condition δ_0. Then,*

(i) *For every* $t \geq 0$ *and* $n \geq 1$,

$$E_0^w[\zeta^n(t)] \leq \exp\{v_n t\}.$$

(ii) *For every* $t \geq 0$ *and* $n \geq 2\kappa t$,

$$E_0^w[\bar{\zeta}^n(t)] \leq 4d \exp\left\{(v_n - 2\kappa)t - n \log\left(\frac{n}{2e\kappa t}\right)\right\}.$$

(iii) *For every* $t \geq 0$ *and* $m > n \geq 1$,

$$E_0^w[\zeta^m(t)] \leq E_0^w[\zeta^n(t)] + E_0^w[\bar{\zeta}^n(t)]\, E_0^w[\zeta^{2n+m}(t)].$$

The following elementary lemma will be used to prove Lemma 3.1. We will need to define,

$$I(y) := \sup_{\lambda \in \mathbb{R}}\{\lambda y - (\cosh \lambda - 1)\} = y \sinh^{-1} y - \sqrt{1 + y^2} + 1, \tag{3.2}$$

Note that $I\colon [0, \infty) \to [0, \infty)$ is one to one and that $I(x) > 0$ for $x > 0$.

Lemma 3.2. *Let* $\{X_t : t \geq 0\}$ *be a simple symmetric continuous time random walk on* $\mathbb{Z}$ *of total jump rate* $2\kappa > 0$ *starting from* 0. *For a nonnegative real* x *define* $\tau_x := \inf\{t \geq 0 : |X_t| \geq x\}$ *as the first exit time of this random walk from the interval* Λ_x. *Then, if* P *is its law, we have*

$$P[\tau_x < t] \leq 4 \exp\left\{ - 2\kappa t I\left(\frac{x}{2\kappa t}\right)\right\} \leq 4 \exp\left\{-2\kappa t - x \log\left(\frac{x}{2e\kappa t}\right)\right\}, \tag{3.3}$$

the second inequality being satisfied only for $x \geq 2\kappa t$.

The proof of Lemma 3.2 is a simple large deviation estimate and will be omitted.

Proof of Lemma 3.1. (i) Let us remark that for every bounded nondecreasing function $f\colon \mathbb{N} \to \mathbb{R}$ which is eventually constant we have,

$$L_n f(\zeta^n) \leq v_n \zeta^n \big(f(\zeta^n + 1) - f(\zeta^n)\big).$$

For a natural $N \geq 1$ fixed, choose $f(m) = m \wedge N$. Using the fact that $\zeta^n(0) = 1$, we then conclude that,

$$\zeta^n(t) \wedge N - 1 - v_n \int_0^t \zeta^n(s)\theta_{[0,N-1]}\big(\zeta^n(s)\big)\, ds, \tag{3.4}$$

where for $A \subset \mathbb{R}$, θ_A is the indicator function of the set A, is a super-martingale. Hence, since the integrand of the integral in (3.4) is a positive function, by Fubini's theorem,

$$E_0^w\big[\zeta^n(t)\theta_{[0,N-1]}\big(\zeta^n(t)\big)\big] \leq 1 + v_n \int_0^t E_0^w\big[\zeta^n(s)\theta_{[0,N-1]}\big(\zeta^n(s)\big)\big]\, ds.$$

Therefore, by Gronwall's lemma,

$$E_0^w[\zeta^n(t)\theta_{[0,N-1]}(\zeta^n(t)] \leq \exp\{v_n t\}.$$

Taking the limit when $N \to \infty$ and using the monotone convergence theorem we conclude the proof of part (i) of the lemma.

(ii) Let us note the following identity,

$$E_0^w[\bar{\zeta}^n(t)] \le d e^{v_n t} P[\tau_n < t], \tag{3.5}$$

where τ_n is the first exit time of a simple symmetric continuous time random walk of total jump rate $2d\kappa$, starting from the origin 0, from the box Λ_n and P is its law. Now, from the second inequality of equation (3.3) applied to each of the d coordinates of such a random walk, we conclude that,

$$P[\tau_n < t] \le 4d \exp\left\{-2\kappa t - n \log\left(\frac{n}{2e\kappa t}\right)\right\},$$

substituting the corresponding expression back in (3.5) and using part (i) of the proposition we conclude the proof of the lemma.

(iii) From (3.1) we have that,

$$E_0^w[\zeta^m(t)] = E_0^w[\zeta^n(t)] + E_0^w\left[\sum_{k=1}^{\bar{\zeta}^n(t)} E_0^w[\zeta_{x_k}^m(t-\tau_k) \mid \bar{\zeta}^n(t)]\right].$$

But, $E_0^w[\zeta_{x_k}^m(t-\tau_k) \mid \bar{\zeta}^n(t)] \le E_0^w[\zeta^{2n+m}(t)]$, which concludes the proof. □

3.2. Proof of Proposition 2.1. We will now prove Proposition 2.1 with the help of Lemma 3.1. Let $\delta = \frac{1}{5}$ and choose N so that whenever $n \ge N$ it is true that $v_n \le \delta\big(n\log(n/(2e\kappa t)) - 4n\log 4\big)/t$ whenever $n \ge N$. By part (i) of the lemma we have that,

$$E_x^w[\zeta^n(t)] \le \exp\left\{\delta\left(n\log\left(\frac{n}{2e\kappa t}\right) - 4n\log 4\right)\right\},$$

while by part (ii) we have,

$$E_x^w[\bar{\zeta}^n(t)] \le 4d\exp\left\{-(1-\delta)n\log\left(\frac{n}{2e\kappa t}\right) - 4\delta n\log 4\right\},$$

whenever $n \ge N$. Choosing $m = 2n > N$ in part (iii) of the same lemma, it follows that,

$$E_x^w[\zeta^{2n}(t)] \le A_n + B_n E_x^w[\zeta^{4n}(t)],$$

where $A_n := e^{\delta n\log(n/(2e\kappa t))}$ and $B_n := 4de^{-(1-\delta)n\log(n(2e\kappa t)) - 4\delta n\log 4}$. Repeating the bound for $2n$ and $m = 4n$ and substituting back we get,

$$E_x^w[\zeta^{2n}(t)] \le A_n + B_n A_{2n} + B_n B_{2n} E_x^w[\zeta^{8n}(t)].$$

Now, by induction on m, we get that,

$$E_x^w[\zeta^{2n}(t)] \le \sum_{k=0}^{m-1} c_k + c_m \frac{E_x^w[\zeta^{n2^{m+1}}(t)]}{A_{n2^m}}, \tag{3.6}$$

where $c_0 := A_n$ and for $k \ge 0$, $c_{k+1} := c_k B_{n2^k} A_{n2^{k+1}}/A_{n2^k}$. Now,

$$\frac{B_{n2^k}A_{n2^{k+1}}}{A_{n2^k}} \le 2de^{-(1-2\delta)n2^k\log(n2^k/(2e\kappa t))}.$$

Hence, by d'Alembert test and the fact that $1-2\delta > 0$ we know that the series $\sum_{k=0}^{\infty} c_k$ is convergent. On the other hand we have,

$$c_m \frac{E_x^w[\zeta^{n2^{m+1}}(t)]}{A_{n2^m}} \le c_{m-1} e^{\delta n2^m\log(n2^m/(2e\kappa t)) - (1-2\delta)n2^{m-1}\log(n2^{m-1}/(2e\kappa t))},$$

which tends to 0 since $2\delta < 1 - 2\delta$ and $c_m < 1$ for m large enough. Taking the limit when $m \to \infty$, then when $n \to \infty$ and using the monotone convergence theorem in inequality (3.6), we deduce that

$$\mathrm{E}_x^w[\zeta(t)] \le \sum_{k=0}^{\infty} c_k < \infty.$$

4. Moment and correlation estimates

Here we will obtain some important bounds for the large time asymptotic behavior of the field of quenched first moments $\{m(x,t,w) : x \in \mathbb{Z}^d\}$, the annealed first moment field $\{\langle m(x,t)\rangle : x \in \mathbb{Z}^d\}$, and their correlations. In the first subsection, we will prove Proposition 2.5.

4.1. Proof of Proposition 2.5. Our first lemma states a useful super-additivity and convexity property of the cumulant generating function of the random variable $v(0)$.

Lemma 4.1. *Consider the cumulant generating function $H(t)\colon [0,\infty) \to \mathbb{R}$ of the random variable $v(0)$, defined in equation (1.1). Then, the following statements are true.*

(i) *H is super-additive. In other words, if $t_1, \dots, t_n$ are nonnegative reals then,*

$$H(t_1 + \dots + t_n) \ge H(t_1) + \dots + H(t_n).$$

(ii) *Assume that $\lim_{t\to\infty} tH''(t) = \infty$. Then, for every $\alpha > 1$,*

$$\lim_{t\to\infty} \frac{H(\alpha t) - \alpha H(t)}{t} = \infty. \tag{4.1}$$

PROOF. (i) Let $t_1, t_2 \ge 0$. From Hölder's inequality, we have that $\phi(t_1)\phi(t_2) \le \phi(t_1 + t_2)$. Hence, $H(t_1 + t_2) \ge H(t_1) + H(t_2)$. By induction on n we conclude the proof.

(ii) From the assumption, note that we can write $H''(t) = f(t)/t$, where $\lim_{t\to\infty} f(t) = \infty$. Integrating the function H'' from t to αt, it follows that,

$$H'(\alpha t) - H'(t) \ge \inf_{s \ge t} f(s) \cdot \log \alpha. \tag{4.2}$$

Integrating again we obtain for $u > t$ that,

$$H(\alpha t) - \alpha H(t) \ge (t - u) \inf_{s \ge u} f(s) \cdot \alpha \log \alpha + c(u),$$

where $c(u) = H(\alpha u) - \alpha H(u)$. Dividing by t, taking the limit when $t \to \infty$ and then the limit when $u \to \infty$, we obtain (4.1). □

Let us now prove Proposition 2.5(i) and (ii). Note that,

$$\frac{\partial G}{\partial t} = \frac{(1+\theta)H'((1+\theta)t) - (1+\theta)H'(t)}{\theta}.$$

Then, inequality (4.2) implies part (i) of the lemma. On the other hand we have that,

$$\frac{\partial G}{\partial \theta} = \frac{\theta t H'((1+\theta)t) - H((1+\theta)t) + H(t)}{\theta^2},$$

By the mean value theorem there exists a $\bar{\theta}$ such that $0 < \bar{\theta} < \theta$ and $H\big((1+\theta)t\big) - H(t) = \theta t H'\big((1+\bar{\theta})t\big)$ and hence

$$\theta t H'\big((1+\theta)t\big) - H\big((1+\theta)t\big) + H(t) = \theta t\big[H'\big((1+\theta)t\big) - H'\big((1+\bar{\theta})t\big)\big],$$

which by inequality (4.2) is positive if $t \geq t_1$, where t_1 is independent of θ and $\bar{\theta}$. This proves that $G_\theta(t)$ is monotone in θ for $t \geq t_1$.

We now show that condition (2.8) ensures SI. We will without loss of generality assume that $\kappa > 0$. Let us define real valued functions f, g by $f(x) := H(\theta t + xt) - H(xt)$ and $g(x) := 2x$ for real x. By the generalized mean value theorem applied in the interval $[1, 1+\theta]$, there exists a $\theta_1 \in (0, \theta)$, such that $\big(f(1+\theta) - f(1)\big)/\big(g(1+\theta) - g(1)\big) = f'(1+\theta_1)/g'(1+\theta_1)$. In other words, the expression $G_{2\theta}(t) - G_\theta(t) = \big(H\big((1+2\theta)t\big) - 2H\big((1+\theta)t\big) + H(t)\big)/(2\theta)$, equals,

$$\frac{t}{2}\left(H'\big((1+\theta+\theta_1)t\big) - H'\big((1+\theta_1)t\big)\right) = \frac{\theta t^2}{2} H''\big((1+\theta_1+\theta_2)t\big),$$

where in the last equality we have applied the mean value theorem and $\theta_2 \in (0, \theta)$. It therefore follows that there is a function $\bar{\theta} \colon [0, \infty) \to (0, 2\theta)$ such that,

$$\frac{G_{2\theta}(t) - G_\theta(t)}{t} = \frac{\theta t}{2} H''((1+\bar{\theta}(t))t).$$

Our hypothesis $\lim_{t\to\infty} tH''(t) = \infty$, shows that the expression above tends to ∞ as $t \to \infty$. The proof of the case in which $\theta < 0$ is similar and the details will be omitted.

We now continue in Section 4.2, defining the truncated quenched first moments, and then describing the parabolic Anderson equation satisfied by the quenched first moments and the corresponding Feynman–Kac representations.

4.2. Truncated quenched first moments. In the sequel, given a real function $f(x)$ defined on $\mathbb{Z}^d$, we will define the discrete Laplacian by,

$$\Delta f(x) := \sum_{e \in \mathbb{Z}^d : |e|=1} \big(f(x+e) - f(x)\big). \tag{4.3}$$

Let us now for each finite set $U \subset \mathbb{Z}^d$ and environment $w \in W$, define the field $w^U := (v_-^U, v_+)$ with $v_-^U(x) = v_-(x)$ for $x \in U$, while $v_-^U(x) = \infty$ for $x \notin U$. We now define for $x \in \mathbb{Z}^d$ and $t \geq 0$,

$$\widetilde{m}_U(x, t, w) := m(x, t, w^U).$$

As it will be seen later, this expression satisfies the parabolic Anderson equation with Dirichlet boundary conditions. We will denote this quantity the *truncated quenched first moment* on U at time t for a reaction–diffusion process starting from x. Also, we will call the set $\{\widetilde{m}_U(x, t, w) : x \in \mathbb{Z}^d\}$, the *field of truncated first moments* on U at time t. Now, in the particular case in which $U = \Lambda(x, r)$ for some $r > 0$, we will use the notation $\widetilde{m}_r(x, t, w)$ instead of $\widetilde{m}_U(x, t, w)$. We will refer to this quantity as the *truncated quenched first moment* at scale r at time t for a reaction–diffusion process starting from site x. Furthermore, we will call the sets $\{\widetilde{m}_r(x, t, w) : x \in \mathbb{Z}^d\}$, the *field of truncated quenched first moments* at scale r at time t.

4.3. The parabolic Anderson equations. Here we will recall the moment equations satisfied by the field of quenched first moments $\{m(x,t,w)\}$ and by the corresponding truncated fields. Following [12], we have the proposition.

Proposition 4.2. *Let $U \subset \mathbb{Z}^d$ be a finite set and $w \in W$ an environment. Consider the field of quenched first moments $\{m(x,t,w) : x \in \mathbb{Z}^d\}$ on $\mathbb{Z}^d$ at time t and the field of truncated quenched first moments $\{\widetilde{m}_U(x,t,w) : x \in \mathbb{Z}^d\}$ on U at time t. Then the following statements are true.*

(i) *The field of truncated quenched first moments $\{\widetilde{m}_U(x,t,w) : x \in \mathbb{Z}^d\}$ of the total number of particles at time t on U, satisfies the equation,*

$$\begin{aligned} \frac{\partial \widetilde{m}_U}{\partial t} &= \kappa\Delta\widetilde{m}_U + v(x)\widetilde{m}_U, && \text{for } x \in U \cap \mathcal{G}(w)^c \\ \widetilde{m}_U(x,0,w) &= 1, && \text{for } x \in \mathbb{Z}^d \\ \widetilde{m}_U(x,t,w) &= 0, && \text{for } x \notin U \cap \mathcal{G}(w)^c,\ t > 0. \end{aligned}$$

(ii) *The field of quenched first moments $\{m(x,t,w) : x \in \mathbb{Z}^d\}$ of the total number of particles at time t on $\mathbb{Z}^d$, satisfies the equation,*

$$\begin{aligned} \frac{\partial m}{\partial t} &= \kappa\Delta m + v(x)m, && \text{for } x \in \mathcal{G}(w)^c \\ m(x,0,w) &= 1, && \text{for } x \in \mathbb{Z}^d \\ m(x,t,w) &= 0, && \text{for } x \notin \mathcal{G}(w)^c,\ t > 0. \end{aligned}$$

PROOF. Consider the family of functions $\{u_z(x,t) := \mathrm{E}_x^w[z^{\zeta(t)}]\}$, parametrized by complex z such that $0 < |z| \leq 1$. It is easy to see that,

$$\begin{gathered} \frac{\partial u_z}{\partial t} = \kappa\Delta u_z + v_+(x)u_z^2 - (v_+(x) + v_-(x))u_z + v_-(x), \\ u_z(x,0) = z, \end{gathered}$$

for $x \in \mathcal{G}(w)^c$, while $u_z(x,t) = 1$ for $t \geq 0$ and $x \in \mathcal{G}(w)$. Differentiating the above equation with respect to z we obtain part (i). A similar proof can be carried out for part (ii). □

4.4. Bounds on the quenched first moments. We will now obtain upper and lower bounds for the annealed moments of the quenched first moments. Let us first recall two elementary inequalities. For n natural, let $a_1, \ldots, a_n$ be arbitrary real numbers. Then, for $r \geq 1$, we have Jensen's inequality,

$$\left|\sum_{i=1}^n a_i\right|^r \leq n^{r-1}\sum_{i=1}^n |a_i|^r, \tag{4.4}$$

while for $0 \leq r \leq 1$ we have,

$$\left|\sum_{i=1}^n a_i\right|^r \leq \sum_{i=1}^n |a_i|^r. \tag{4.5}$$

We will also need to introduce for $L \geq 0$ the notation,

$$M_L := \max_{x \in \Lambda_L} |v(x)|. \tag{4.6}$$

Let us recall the following lemma, contained in the statement of [12, Theorem 2.1].

Lemma 4.3. *Consider a finite subset $U \subset \mathbb{Z}^d$ and $\mu \in \mathcal{P}(W)$. Assume that μ satisfies Condition* E. *Then, μ-a.s. for every $x \in \mathbb{Z}^d$ and $t \geq 0$, the quenched first moment $m(x,t,w)$ on $\mathbb{Z}^d$ at time t starting from site x admits the Feynman–Kac representation*

$$m(x,t,w) = \mathrm{E}_x\Big[\mathrm{e}^{\int_0^t v(X_s)\,\mathrm{d}s} 1\big(\tau_{\mathcal{G}(w)} > t\big)\Big], \tag{4.7}$$

and the truncated quenched first moment $\widetilde{m}_U(x,t,w)$ on U at time t starting from x also,

$$\widetilde{m}_U(x,t,w) = \mathrm{E}_x\Big[\mathrm{e}^{\int_0^t v(X_s)\,\mathrm{d}s} 1\big(\tau_{U^c \cup \mathcal{G}(w)} > t\big)\Big], \tag{4.8}$$

where in both (4.7) *and* (4.8), *$\{X_t : t \geq 0\}$ is a simple symmetric random walk of total jump rate $2d\kappa$ starting from x, of law P_x, E_x is the expectation related to this law, and for $A \subset \mathbb{Z}^d$ we define $\tau_A := \inf\{t \geq 0 : X_t \notin A\}$.*

We can now apply Lemma 4.3 to obtain the first estimates on the quenched first moments.

Proposition 4.4. *Consider a finite subset $U \subset \mathbb{Z}^d$ and $\mu \in \mathcal{P}(W)$. Assume that μ satisfies Condition* E. *Then,*

(i) *For each $x \in \mathbb{Z}^d$, $t \geq 0$ and $\beta > 0$ there exists a constant C such that,*

$$\mathrm{e}^{H(\beta t) - 2d\kappa t} \leq \langle m(x,t)^\beta \rangle \leq C(\kappa + t)^d \mathrm{e}^{H(\beta t)}, \tag{4.9}$$

(ii) *For each $x \in U$, $t \geq 0$ and $\beta > 0$,*

$$\mathrm{e}^{H(\beta t) - 2d\kappa t} \leq \langle \widetilde{m}_U(x,t)^\beta \rangle \leq C(\kappa + t)^d \mathrm{e}^{H(\beta t)}, \tag{4.10}$$

where C is the constant of part (i).

(iii) *For each $\beta > 0$, $\gamma > 0$ and $a > 0$ we have that,*

$$\langle |m(x,t) - \widetilde{m}_{\gamma(\kappa t)^a}(x,t)|^\beta \rangle \leq C(\gamma(\kappa t)^a + 1)^d \mathrm{e}^{-2\beta\kappa t I(\gamma(\kappa t)^{a-1}/2)} \mathrm{e}^{H(\beta t)}, \tag{4.11}$$

for some constant $C > 0$, where $I\colon [0,\infty) \to [0,\infty)$ is defined in equation (3.2).

Proof. (i) The first inequality of equation (4.9) can be obtained from the Feynman–Kac representation (4.7), taking only into account the contribution of the path X_s which stays during the whole time interval $[0,t]$ at x ([12, p. 637]). To prove the second inequality of (4.9), let us note that by translation invariance it is enough to prove the estimate for $\langle m(0,t) \rangle$. On the other hand,

$$1(\tau_{\mathcal{G}(w)} > t)\mathrm{e}^{\int_0^t v(X_s)\,\mathrm{d}s} \leq \sum_{n=0}^{\infty} \mathrm{e}^{M_{R_n} t} 1(T_{n-1} \leq t < T_n),$$

where $T_{-1} = 0$, while for n natural T_n is the first exit time of the random walk $\{X_t : t \geq 0\}$ from the box Λ_{R_n}, with $R_n := R_0 2^n$ and $R_0 := \max\{\kappa t, 1\}$, while $M_{R_n} := \max_{x : \|x\| \leq R_n} |v(x)|$ as defined in equation (4.6). It follows that,

$$m(0,t,w) \leq \sum_{n=0}^{\infty} \mathrm{e}^{M_{R_n} t} P[T_{n-1} \leq t]. \tag{4.12}$$

Let us also remark that since $\beta > 0$, for each natural n we have the following inequality which will be used soon,

$$\langle \mathrm{e}^{\beta t M_{R_n}} \rangle \leq (2(R_n + 1))^d \exp\{H(\beta t)\}. \tag{4.13}$$

In fact, $\langle e^{\beta t M_{R_n}}\rangle \le \sum_{x\in[-R_n,R_n]^d}\langle e^{\beta t M_{R_n}} 1(v(x) = M_{R_n})\rangle$. Let us now consider the case $0 < \beta \le 1$. Then, by an application of inequality (4.5) to estimate (4.12), we conclude that,

$$m(0,t,w)^\beta \le e^{\beta t M_{R_0}} + \sum_{n=1}^{\infty} e^{\beta t M_{R_n}} P[T_{n-1} \le t]^\beta.$$

Taking expectations on both sides of this inequality and applying the estimate (4.13) and the second inequality of equation (3.3) for each of the d coordinates of the underlying random walk, we obtain,

$$\langle m(0,t)^\beta\rangle \le 2^d (R_0+1)^d e^{H(\beta t)} \left(1 + 4d\sum_{n=1}^{\infty} 2^{nd} e^{-\beta R_0 2^{n-1}\log(R_0 2^{n-1}/(2e\kappa t))}\right),$$

Now, since $e^{-\beta R_0 2^{n-1}\log(R_0 2^{n-1}/(2e\kappa t))} \le e^{-\beta 2^{n-1}\log(2^{n-1}/(2e))}$, and $R_0 + 1 \le 2R_0$, we see that there is a constant C such that inequality (4.9) is satisfied for $0 < \beta \le 1$.

Let us now consider the case $\beta > 1$. Let $\beta' > 1$ be defined by $1/\beta + 1/\beta' = 1$. Then, if we represent the left-hand side of (4.12) as

$$\sum_{k=0}^{\infty} e^{t M_{R_n}} P[T_{n-1} < t \le T_n]^{1/\beta'} P[T_{n-1} < t \le T_n]^{1/\beta},$$

by Hölder's inequality we get that,

$$m(0,t,w) \le \left(\sum_{n=0}^{\infty} e^{t\beta M_{R_n}} P[T_{n-1} < t \le T_n]\right)^{1/\beta}.$$

A computation similar to the case $0 < \beta \le 1$ finishes the proof of part (i).

(ii) The first inequality of equation (4.10) can be deduced by an argument analogous to the one leading to the first inequality of equation (4.9). Now, note from the representations (4.7) and (4.8) that $\widetilde{m}_U(x,t,w) \le m(x,t,w)$. Hence, the second inequality of equation (4.10) is a corollary of the second inequality of equation (4.9).

(iii) Let us remark from the Feynman–Kac representations (4.7) for $m(0,t)$ and (4.8) for $\widetilde{m}_{\gamma(\kappa t)^a}(0,t)$ that,

$$m(0,t) - \widetilde{m}_{\gamma(\kappa t)^a}(0,t) = \mathrm{E}_x\left[e^{\int_0^t v(X_s)\,ds} 1\left(\tau_{\Lambda^c_{\gamma(\kappa t)^a}} < t\right) 1\left(\tau_{\mathcal{G}(w)} > t\right)\right].$$

Hence, as in part (i) of the proof of this proposition,

$$|m(0,t) - \widetilde{m}_{\gamma(\kappa t)^a}(0,t)| \le \sum_{n=1}^{\infty} e^{M_{R'_n} t} P[T'_{n-1} \le t < T'_n], \tag{4.14}$$

where T'_n is the first exit time of the random walk $\{X_t : t \ge 0\}$ from the box $\Lambda_{R'_n}$, with $R'_n := R'_0 2^n$, $R'_0 := \max\{\gamma(\kappa t)^a, 1\}$ and $M_{R'_n} := \max_{x:\|x\|\le R'_n} |v(x)|$. Let us consider the case $0 < \beta \le 1$. By inequality (4.5) we obtain that,

$$|m(0,t) - \widetilde{m}_{\gamma(\kappa t)^a}(0,t)|^\beta \le \sum_{n=1}^{\infty} e^{t\beta M_{R'_n}} P[T'_{n-1} \le t]^\beta.$$

But, as in (4.13) we can conclude that $\langle \mathrm{e}^{\beta t M_{R'_n}} \rangle \leq 2^d(R'_n+1)^d \exp\{H(\beta t)\}$. Hence, from the first inequality of equation (3.3) we see that,

$$(4.15)\quad \langle |m(0,t) - \widetilde{m}_{\gamma(\kappa t)^a}(0,t)|^\beta \rangle \leq d4^\beta 2^d (R'_1+1)^d \mathrm{e}^{H(\beta t)} \mathrm{e}^{-2\beta\kappa t I(R'_0/(2\kappa t))} \times \sum_{n=1}^{\infty} \frac{(2(R'_0 2^n+1))^d}{(2(R'_0 2+1))^d} \mathrm{e}^{2\kappa t\beta(I(R'_0/(2\kappa t)) - I(R'_0 2^n/(2\kappa t)))}.$$

Now, using the fact that I is convex, nondecreasing and that $I(0)=0$, we see that,

$$I\left(\frac{R'_0}{2\kappa t}\right) - I\left(\frac{R'_0 2^n}{2\kappa t}\right) \leq -(2^n-1) I\left(\frac{\gamma(\kappa t)^{a-1}}{2}\right).$$

Substituting this back in (4.15), we obtain inequality (4.11) in the case $0<\beta\leq 1$. The case $\beta>1$ can be proved using similarly, using Hölder's inequality as in part (i). □

We end this section with important estimates involving the growth functions. Given a subset $U \subset \mathbb{Z}^d$, we define the discrete Laplacian operator with effective potential $v(0) = v_+(0) - v_-(0) \in [-\infty, \infty)$ by its action on functions $f : \mathbb{Z}^d \to \mathbb{R}$, which vanish outside $U_w := U \cap \mathcal{G}^c(w)$ ($f(x)=0$ for $x \notin U \cap \mathcal{G}^c(w)$) as,

$$(\Delta + v) f(x) = \sum_{e\in\mathcal{B}} (f(x+e) - f(x)) + v(x) f(x),$$

where $\mathcal{B}$ is the set formed by the elements of the basis of the free Abelian group $\mathbb{Z}^d$ and its inverses. Defining $L^2(U_w) := \{f : \sum_{x\in\mathbb{Z}^d} f(x)^2 < \infty, f(x) = 0 \text{ if } x \notin U_w\}$, we can check that $\Delta + v$ is a bounded self-adjoint operator on the Hilbert space $L^2(U_w)$ endowed with the inner product $(f,g) := \sum_{x\in\mathbb{Z}^d} f(x)g(x)$. We then define $\{\lambda_n(U,w) : 0 \leq n \leq \mathcal{U}-1\}$ as the set of eigenvalues of $\Delta+v$ in $L^2(U_w)$ in decreasing order, where $\mathcal{U}$ is the total number of eigenvalues. We will denote by $\psi_n^{U,w}$ the corresponding normalized eigenfunctions. Let $r\geq 0$. In the case in which $U = \Lambda(x,r)$, we will employ the notation $\{\lambda_n(x,r,w)\}$ instead of $\{\lambda_n(U,w)\}$ and $\psi_n^{x,r,w}$ instead of $\psi_n^{U,w}$.

We can now state the following important lemma.

Lemma 4.5. *Consider the quenched first moment $m(0,t,w)$. Then, for every $\beta>0$, $a>1$ and $t\geq 1$, there is a constant $k_1(d,a,\beta)$ such that,*

$$(4.16)\quad \frac{k_1^{-1}}{(\kappa t)^{da(\beta+1)}+1} \langle m(0,\beta t)\rangle \leq \langle m(0,t)^\beta \rangle \leq k_1\big((\kappa t)^{da(\beta+1)}+1\big)\langle m(0,\beta t)\rangle,$$

and

$$(4.17)\quad \frac{k_1^{-1}}{(\kappa t)^{da(\beta+1)}+1} \langle \widetilde{m}_{(\kappa t)^a}(0,\beta t)\rangle \leq \langle \widetilde{m}_{(\kappa t)^a}(0,t)^\beta\rangle \leq k_1\big((\kappa t)^{da(\beta+1)}+1\big)\langle \widetilde{m}_{(\kappa t)^a}(0,\beta t)\rangle.$$

PROOF. We will only prove (4.16), being the proof of equation (4.17) analogous. Let us first show that for every real $a>0$, $\beta>0$ and $t\geq 0$ there is a constant $c(d,\beta,a)$ such that,

$$(4.18)\quad m(x,t,w) \leq ([2(\beta\kappa t)^a]+1)^{d/2} \mathrm{e}^{t\lambda_0(x,(\beta\kappa t)^a,w)} + 4d\mathrm{e}^{-2\kappa t I((\beta\kappa t)^{a-1}/2)} \mathrm{e}^{tM_{(\beta\kappa t)^a}}.$$

Note that by inequality (4.14) with $\gamma = \beta^a$, we have that

$$m(x,t,w) \leq 4d\mathrm{e}^{-2\kappa t I((\beta\kappa t)^{a-1}/2)} \mathrm{e}^{tM_{(\beta\kappa t)^a}} + \widetilde{m}_U(x,t,w),$$

with $U = \Lambda(x, (\beta\kappa t)^a, w)$. We then need to estimate the truncated quenched first moment at scale $(\beta\kappa t)^a$, $\widetilde{m}_{(\beta\kappa t)^a}(x, t, w)$. First remark the following expansion in terms of the eigenvalues $\{\lambda_n(x, (\beta\kappa t)^a, w)\}$ and the corresponding eigenfunctions $\{\psi_n^{x,(\beta\kappa t)^a,w}\}$,

$$\widetilde{m}_{(\beta\kappa t)^a}(x, t, w) = \sum_{n=0}^{\mathcal{U}-1} \mathrm{e}^{t\lambda_n(x,(\beta\kappa t)^a,w)} \psi_n^{x,(\beta\kappa t)^a,w}(x) \big(\psi_n^{x,(\beta\kappa t)^a,w}, 1(A)\big), \tag{4.19}$$

where $A := \Lambda(x, (\beta\kappa t)^a, w)$. Now, by the Cauchy – Schwartz inequality we see that the right-hand side of equality (4.19) is upper-bounded by $\mathrm{e}^{t\lambda_0(x,(\beta\kappa t)^a,w)} \times \big(\sum_{n=0}^{\mathcal{U}-1} \big(\psi_n^{x,(\beta\kappa t)^a,w}, 1(\{x\})\big)^2 \sum_{n=0}^{\mathcal{U}-1} \big(\psi_n^{x,(\beta\kappa t)^a,w}, 1(A)\big)^2\big)^{1/2}$ which in turn is upper-bounded by $\mathrm{e}^{t\lambda_0(x,(\beta\kappa t)^a,w)}\sqrt{|A|}$, where $1(\{x\})(y)$ equals 1 if $y = x$ and 0 otherwise. Using the fact that $|A| = |\Lambda(x, (\beta\kappa t)^a, w)| \le ([2(\beta\kappa t)^a] + 1)^d$, finishes the proof of (4.18).

Let us now show that for every finite subset $U \subset \mathbb{Z}^d$, it is true that,

$$\frac{1}{|U|} \sum_{z\in U} m(z, t) \ge \frac{1}{|U|} \mathrm{e}^{t\lambda_0(U,w)}. \tag{4.20}$$

First note the trivial inequality $m(z, t, w) \ge \widetilde{m}_U(z, t, w)$. We also have the expansion, $\widetilde{m}_U(z, t, w) = \sum_{n=0}^{\mathcal{U}-1} \mathrm{e}^{\lambda_n(U,w)t} \psi_n^{U,w}(z) \big(\psi_n^{U,w}, 1(U)\big)$. Therefore we can see that,

$$\begin{aligned} \frac{1}{|U|} \sum_{z\in U} m(z, t) &\ge \frac{1}{|U|} \mathrm{e}^{\lambda_0(U,w)t} \big(\psi_0^{U,w}, 1(U)\big)^2 \\ &\ge \frac{1}{|U|} \mathrm{e}^{\lambda_0(U,w)t} \sum_{z\in U} (\psi_0^{U,w})^2(z) = \frac{1}{|U|} \mathrm{e}^{\lambda_0(U,w)t}, \end{aligned}$$

where we have used in the second inequality the fact that $\psi_0^{U,w}(x) \ge 0$ and in the last inequality the normalization condition $\sum_{z\in U} (\psi_0^{U,w})^2(z) = 1$.

Let us now prove the second inequality of (4.16). By Jensen's inequality (4.4), in the case $\beta \ge 1$, or inequality (4.5), in the case $0 < \beta < 1$, applied to (4.18), note that for some constant $c(a, d, \beta)$,

$$\langle m(x, t)^\beta \rangle \le c\big((\kappa t)^{da\beta/2} + 1\big) \big\langle \mathrm{e}^{\beta t\lambda_0(x,(\beta\kappa t)^a,w)} \big\rangle + 4d\mathrm{e}^{-2\beta\kappa t I((\beta\kappa t)^{a-1}/2)} \big\langle \mathrm{e}^{\beta t M_{(\beta\kappa t)^a}} \big\rangle.$$

Now, by the first inequality of equation (4.9) and a computation similar to the one leading to (4.13), the second term of the right hand side of the above inequality, is upper bounded by,

$$4d2^d \big((\beta\kappa t)^a + 1\big)^d \exp\Big\{ -2\beta\kappa t I\Big(\frac{(\beta\kappa t)^{a-1}}{2}\Big) + 2d\kappa t \Big\} \langle m(x, t)^\beta \rangle. \tag{4.21}$$

On the other hand, by inequality (4.20) with $U = \Lambda(0, (\kappa t)^a)$, we see that,

$$\langle \mathrm{e}^{\beta t\lambda_0(x,(\beta\kappa t)^a,w)} \rangle \le (2[(\beta\kappa t)^a] + 1)^d \langle m(x, \beta t) \rangle.$$

It follows that,

$$\langle m(x, t, w)^\beta \rangle \le c\big((\kappa t)^{da(\beta+1)} + 1\big) \langle m(x, \beta t) \rangle,$$

where we have used the fact that for $a > 1$, the term (4.21) is negligible with respect to $\langle m^\beta \rangle$. The second inequality of equation (4.16) now follows noting that for $a > 1$ the second term in the right-hand factor of the above inequality is negligible with respect to the first term.

Let us now prove the first inequality of equation (4.16). In the case $0 < \beta \leq 1$, by inequality (4.5) and the bound (4.20), we have that,

$$\sum_{x \in U} m^\beta(x,t,w) \geq e^{\beta t \lambda_0(U,w)}. \tag{4.22}$$

But when $\beta > 1$, by Jensen's inequality we have that

$$\sum_{x \in U} m^\beta(x,t,w) \geq |U|^{-(\beta-1)} \Big(\sum_{x \in U} m(x,t,w) \Big)^\beta,$$

and (4.22) is still satisfied. Choosing $U = \Lambda(0, (\beta\kappa t)^a)$, and using translation invariance we get,

$$\langle m(0,t)^\beta \rangle \geq (2[(\beta\kappa t)^a] + 1)^{-d} \langle e^{\beta t \lambda_0(0,(\beta\kappa t)^a, w)} \rangle.$$

Using again the bound (4.18), neglecting the second term, we finish the proof in the case $0 < \beta \leq 1$. □

An important consequence of Lemma 4.5, is that it shows that Assumption MI implies the so called *intermittency effect* [12]. Let us define for $\theta \neq 0$, the functions,

$$\bar{F}_\theta(t) := \frac{\log\langle m(0,t)^{1+\theta} \rangle - (1+\theta)\log\langle m(0,t)\rangle}{\theta}.$$

Note that by Jensen's inequality, $\bar{F}_\theta(t) \geq 0$.

Corollary 4.6. *Suppose that Assumptions* (E) *and* MI *are satisfied. Then, for every* $\theta \neq 0$,

$$\lim_{t \to \infty} \frac{\bar{F}_{2\theta}(t) - \bar{F}_\theta(t)}{\theta \log(\kappa t + e)} = \infty. \tag{4.23}$$

4.5. Correlation and variance estimates on the field of quenched first moments. In order to prove Theorem 2.3(ii), it will be important to have a control on the variance of the the quenched first moments. In the sequel of this paper, to avoid heavy notation, we will use m_a instead of $m_{(\kappa t)^a}$. Given $x, y \in \mathbb{Z}^d$ and $t \geq 0$, let us define,

$$c(x,y,t) := \langle m(x,t)m(y,t)\rangle - \langle m(x,t)\rangle\langle m(y,t)\rangle,$$

which we will call the *correlation between sites* x *and* y *at time* t of the field of quenched first moments. Similarly let us define for $a > 0$,

$$c_a(x,y,t) := \langle \widetilde{m}_a(x,t)\widetilde{m}_a(y,t)\rangle - \langle \widetilde{m}_a(x,t)\rangle\langle \widetilde{m}_a(y,t)\rangle,$$

the *correlation between sites* x *and* y *at time* t of the truncated field of quenched first moments at scale $(\kappa t)^a$. Let us begin with the following lemma.

Lemma 4.7. *Let* $t \geq 0$. *Consider the fields of quenched first moments* $\{m(x,t,w) : x \in \mathbb{Z}^d\}$ *and truncated quenched first moments at scale* $(\kappa t)^a$, $\{\widetilde{m}_a(x,t,w) : x \in \mathbb{Z}^d\}$. *Then the following statements are true.*

(i) *The sum of the correlations between site* 0 *and the other sites of the field of quenched first moments, behaves asymptotically as* $t \to \infty$ *like the sum of the correlations between site* 0 *and the other sites of the truncated field of quenched first moments at scale* $(\kappa t)^a$. *In other words,*

$$\sum_{y \in \mathbb{Z}^d} c(0,y,t) \sim \sum_{y \in \mathbb{Z}^d} c_a(0,y,t).$$

(ii) *Let $\{U_t : t > 0\}$ be a collection of subsets of the lattice $\mathbb{Z}^d$ indexed by $t > 0$. Assume that $|U_t| \sim |U_{t,(\kappa t)^a}|$ as $t \to \infty$, where $U_{t,r} := \{x \in U_t : \operatorname{dist}(x, U_t^c) \geq 2r\}$, for $r > 0$. Then,*

$$\operatorname{Var}_\mu \sum_{x \in \Lambda_{U_t}} m(x,t) \sim |U_t| \sum_{y \in \mathbb{Z}^d} c(0,y,t), \tag{4.24}$$

$$\operatorname{Var}_\mu \sum_{x \in \Lambda_{U_t}} \widetilde{m}_a(x,t) \sim |U_t| \sum_{y \in \mathbb{Z}^d} c_a(0,y,t), \tag{4.25}$$

and

$$\operatorname{Var}_\mu \sum_{x \in \Lambda_{U_t}} m(x,t) \sim \operatorname{Var}_\mu \sum_{x \in \Lambda_{U_t}} \widetilde{m}_a(x,t). \tag{4.26}$$

PROOF. From the Feynman–Kac representation (4.7) of Lemma 4.3, note that it is possible to write,

$$m(x,t,w) = \mathrm{E}_x\left[\mathrm{e}^{\sum_{z \in \mathbb{Z}^d} v(z)\mathcal{L}(z,t)} 1(A)\right],$$

where $A := \{\tau_{\mathcal{G}^c(w)} > t\}$, $\mathcal{L}(z,t) := \int_0^t \delta_z(X_s)\,\mathrm{d}s$ is the local time at the point z of the random walk $\{X_t : t \geq 0\}$ starting from x, and $\delta_z : \mathbb{Z}^d \to \{0,1\}$ is the indicator function of the set $\{z\}$. From here, using Fubini's theorem it follows that we have the following representation for the correlations between site x and y of the quenched field of first moments.

$$\begin{aligned} &c(x,y,t) \\ &\quad = \mathrm{E}_{x,y}\left[\left\langle \mathrm{e}^{\sum_{z \in \mathbb{Z}^d} v(\mathcal{L}+\tilde{\mathcal{L}})} 1(A \cap \tilde{A})\right\rangle - \left\langle \mathrm{e}^{\sum_{z \in \mathbb{Z}^d} v\mathcal{L}} 1(A)\right\rangle \left\langle \mathrm{e}^{\sum_{z \in \mathbb{Z}^d} v\tilde{\mathcal{L}}} 1(\tilde{A})\right\rangle\right] \end{aligned} \tag{4.27}$$

where $\tilde{\mathcal{L}}(z,t) := \int_0^t \delta_z(\widetilde{X}_s)\,\mathrm{d}s$ is the local time at the point z of the random walk $\{\widetilde{X}_t : t \geq 0\}$, independent of $\{X_t : t \geq 0\}$, starting from y, with law P_y, and $\tilde{A}$ is an identical copy of A, but defined in terms of the random walk $\{\widetilde{X}_t : t \geq 0\}$. Furthermore, $\mathrm{E}_{x,y} := \mathrm{E}_x \otimes \mathrm{E}_y$, denotes the expectation with respect to the law of the independent random walks $\{X_t\}$ and $\{\widetilde{X}_t\}$. Now, the expression (4.27) for the correlations can be written in terms of the cumulant generating function defined in equation (1.1),

$$c(x,y,t) = \mathrm{E}_{x,y}\left[\mathrm{e}^{\sum_{z \in \mathbb{Z}^d} H(\mathcal{L}+\tilde{\mathcal{L}})} - \mathrm{e}^{\sum_{z \in \mathbb{Z}^d} H(\mathcal{L})} \mathrm{e}^{\sum_{z \in \mathbb{Z}^d} H(\tilde{\mathcal{L}})}\right], \tag{4.28}$$

where we have used the independence of the coordinates of the effective potential $\{v(x) : x \in \mathbb{Z}^d\}$ under μ. Note that the super-additivity of H (Lemma 4.1(i)), implies that this expression is nonnegative. On the other hand, a reasoning similar to the one leading to the representation (4.28), this time based on the Feynman–Kac representation (4.8) of Lemma 4.3, enables us to deduce that,

$$\begin{aligned} &c_a(x,y,t) \\ &\quad = \mathrm{E}_{x,y}\left[1(\tau_a > t) 1(\tilde{\tau}_a > t)\left(\mathrm{e}^{\sum_{z \in \mathbb{Z}^d} H(\mathcal{L}+\tilde{\mathcal{L}})} - \mathrm{e}^{\sum_{z \in \mathbb{Z}^d} H(\mathcal{L})} \mathrm{e}^{\sum_{z \in \mathbb{Z}^d} H(\tilde{\mathcal{L}})}\right)\right], \end{aligned} \tag{4.29}$$

where $\tau_a := \tau_{\Lambda(x,(\kappa t)^a)} = \inf\{t \geq 0 : X_t \notin \Lambda(x,(\kappa t)^a)\}$ and $\tilde{\tau}_a := \tilde{\tau}_{\Lambda(x,(\kappa t)^a)} := \inf\{t \geq 0 : \widetilde{X}_t \notin \Lambda(y,(\kappa t)^a)\}$. From (4.28) and (4.29) it follows that,

$$\sum_{y \in \mathbb{Z}^d} c(0,y,t) \geq \sum_{y \in \mathbb{Z}^d} c_a(0,y,t). \tag{4.30}$$

But note that in reality, due to the independence between any pair of truncated quenched first moments at time t at two points at a distance larger than $2(\kappa t)^a$, we have $\sum_{y\in\mathbb{Z}^d} c_a(0,y,t) = \sum_{y\in\Lambda_{2(\kappa t)^a}} c_a(0,y,t)$. Furthermore,

$$\sum_{y\in\mathbb{Z}^d} c(0,y,t) = \sum_{y\in\Lambda_{2(\kappa t)^a}} c(0,y,t) + \sum_{y\notin\Lambda_{2(\kappa t)^a}} c(0,y,t). \tag{4.31}$$

Now, an application of the first inequality of equation (4.9) and Proposition 4.4(ii), shows that since $a>0$, it is true that $\sum_{y\in\Lambda_{2(\kappa t)^a}} c(0,y,t) \sim \sum_{y\in\Lambda_{2(\kappa t)^a}} c_a(0,y,t)$. And a second application of equation (4.9) and the first inequality in equation (3.3), shows that $\sum_{y\notin\Lambda_{2(\kappa t)^a}} c(0,y,t) \ll \sum_{y\in\Lambda_{2(\kappa t)^a}} c(0,y,t)$. This, together with inequality (4.30), ends up the proof of Lemma 4.7(i).

We will first prove (4.25). Remark that,

$$\mathrm{Var}_\mu \sum_{x\in U_t} \widetilde{m}_a(x,t) = \sum_{x,y\in U_t} c_a(x,y,t) = \sum_{\substack{x,y\in U_t\\ \|x-y\|\le 2(\kappa t)^a}} c_a(x,y,t),$$

where we used as in the proof of part (i), the fact that $\widetilde{m}_a(x,t,v)$ and $\widetilde{m}_a(y,t,v)$ are independent for $\|x-y\|\ge 2(\kappa t)^a$. Now, the right-most member of the above inequality is bounded by, $|U_t|\sum_{y:\|y\|\le 2t^a} c_a(0,y,t)$, which gives us the inequality,

$$\mathrm{Var}_\mu \sum_{x\in U_t} \widetilde{m}_a(x,t) \le |U_t| \sum_{y\in\mathbb{Z}^d} c_a(0,y,t).$$

Similarly we can conclude that,

$$\mathrm{Var}_\mu \sum_{x\in U_t} \widetilde{m}_a(x,t) \ge |U_{t,(\kappa t)^a}| \sum_{y\in\mathbb{Z}^d} c_a(0,y,t).$$

This finishes the proof of the statement of equation (4.25). The statement of equation (4.24) now follows from equation (4.25), Lemma 4.7(i) proved above, and the inequalities,

$$\mathrm{Var}_\mu \sum_{x\in U_t} \widetilde{m}_a(x,t) \le \mathrm{Var}_\mu \sum_{x\in U_t} m_a(x,t) \le |U_t| \sum_{y\in\mathbb{Z}^d} c(0,y,t).$$

Finally, note that (4.26) is a trivial consequence of (4.24) and (4.25). □

5. The annealed and Gaussian regimes

Here we will prove Theorem 2.3(i) and (ii), making use of the estimates obtained in the previous section, and Theorem 2.11. The main argument which will be used to prove Theorem 2.3 is a renormalization method, which we call *partition analysis*, as developed in [5]. In order to present a self-contained proof, in the first subsection we recall this technique. Then, in the second subsection, we derive some important estimates via the partition analysis technique, which will enable us to reduce the proofs to sums of independent random variables. In the third subsection we prove the law of large numbers, stated in equation (2.4). In the fourth subsection we will prove the negative part (absence of a law of large numbers) stated in equation (2.5). In Section 5.5, we will prove the central limit theorem stated in equation (2.6). Then, in Section 5.6 we prove the absence of a central limit behavior stated in equation (2.7). Finally, in Section 5.7, we prove Theorem 2.11.

5.1. Partition analysis. Here we shall follow closely Section 5.1 of Ben Arous et al. [5]. For a fixed natural L we consider the box $\Lambda_L = \{x \in \mathbb{Z}^d : \|x\| \leq L\}$. We will define two related but different kinds of partitions of Λ_L. The first one shows that Λ_L can be decomposed into disjoint *partition boxes* $\{\Lambda'_{\mathbf{i}} : \mathbf{i} \in \mathcal{I}\}$, indexed by some set $\mathcal{I}$, so that $\Lambda_L = \bigcup_i \Lambda'_i$. The second one defines a partition of Λ_L in a *strip set* and *main boxes* $\{\Lambda''_{\mathbf{i}} : \mathbf{i} \in \mathcal{I}\}$. In the first case, the index set $\mathcal{I}$ will be partitioned in disjoint subsets $\{\mathcal{I}_K : K \in \mathcal{K}\}$, where the cardinality of $\mathcal{K}$ is 2^d, in such a way that for each $K \in \mathcal{K}$ any pair of elements of the collection of partition boxes $\{\Lambda'_{\mathbf{i}} : \mathbf{i} \in \mathcal{I}_K\}$ is at a large Euclidean distance. In the second partition case, it turns out that the survival probabilities corresponding to sites in the strip set have a total sum which is negligible, while the main boxes happen to be essentially independent. To proceed we will need to introduce some notation defining the corresponding scales and subsets.

Our first parameter is a natural number L' smaller than or equal to L, which will be called the *mesoscopic scale*. By the division algorithm, we know that there exist natural numbers q and $\bar{q}$ such that $2L + 1 = qL' + \bar{q}$, with $0 \leq \bar{q} < q$. Note that this last equation can be written in the form

$$2L + 1 = \sum_{i=1}^{q} L'_i, \tag{5.1}$$

with $L'_i = L' + \theta_{\bar{q}}(i)$ and $\theta_{\bar{q}}(i) = 1$ for $i \leq \bar{q}$ and $\theta_{\bar{q}}(i) = 0$ for $i > \bar{q}$. For our purposes, the relevant fact is that $L' \leq L'_i \leq L' + 1$. In the sequel, for any given pair of real numbers a, b we will use the notation $[a,b]_l$ for $[a,b] \cap \mathbb{Z}$. We now will subdivide the box $[-L, L]_l$ in intervals according to equation (5.1). Thus, we define $I_1 := [-L, -L + L'_1 - 1]_l$ and for $1 < i \leq q$ we let $I_i := [-L + \sum_{j=1}^{i-1} L'_i, -L + \sum_{j=1}^{i} L'_i - 1]_l$. Note that $I_q = [L - L'_q + 1, L]_l$ and $|I_i| = L'_i$. Next, we introduce a second parameter r which is a natural number smaller than or equal to L'. We will call r the *fine scale*. Then, for each I_i we define an interval J_i such that $J_i \subset I_i$, $|J_i| = L' - 2r$ and the endpoints of J_i are at a distance larger than r to the endpoints of I_i. To do so, first let $r_i := r + \theta_{\bar{q}}(i)$. Then define $J_1 := [-L + r, -L + L'_1 - 1 - r_1]_l$ and for $1 < i \leq q$ we let $J_i := [-L + \sum_{j=1}^{i-1} L'_i + r, -L + \sum_{j=1}^{i} L'_i - 1 - r_i]_l$.

We now proceed to define the partition in Λ_L in partition boxes and define the corresponding decomposition of the index set. First we define the set $\mathcal{I} := \{1, 2, \ldots, q\}^d$, which will correspond to the indexes parametrizing the sub-boxes. For a given element $i \in \mathcal{I}$, of the form $\mathbf{i} = (i_1, \ldots, i_d)$ with $1 \leq i_k \leq q$, $1 \leq k \leq d$, we define

$$\Lambda'_{\mathbf{i}} := I_{i_1} \times I_{i_2} \times \cdots \times I_{i_d}.$$

We will call such a set a *partition box*. By definition the cardinality $|\Lambda'_{\mathbf{i}}|$ of a partition box satisfies,

$$(L')^d \leq |\Lambda'_{\mathbf{i}}| \leq (L' + 1)^d. \tag{5.2}$$

Note also that the partition boxes defines a partition of Λ_L so that $\Lambda_L = \bigcup_{\mathbf{i} \in \mathcal{I}} \Lambda'_{\mathbf{i}}$ where the union is disjoint.

Next we define a partition of the index set $\mathcal{I}$. Consider the collection $\mathcal{K}$ of subsets of $\{1, 2, \ldots, d\}$. Note that $|\mathcal{K}| = 2^d$. Now given $K \in \mathcal{K}$ we define $\mathcal{I}_K$ as the subset of $\mathcal{I}$ having coordinates which are even for $k \in K$ and odd for $k \notin K$. In other words, if we define $\mathbb{E}$ as the set of even natural numbers and $\mathbb{O}$ as the set of odd natural numbers then,

$$\mathcal{I}_K := \{\mathbf{i} = (i_1, \dots, i_d) \in \mathcal{I} : i_k \in \mathbb{E} \text{ if } k \in K, i_k \in \mathbb{O} \text{ if } k \notin K,\ 1 \le k \le d\}.$$

Note that $\{\mathcal{I}_K : K \in \mathcal{K}\}$ defines a partition of $\mathcal{I}$ so that $\mathcal{I} = \bigcup_{K\in\mathcal{K}} \mathcal{I}_K$ is a disjoint union. Hence,

$$\sum_{K\in\mathcal{K}} \sum_{\mathbf{i}\in\mathcal{I}_K} \sum_{x\in\Lambda'_{\mathbf{i}}} f(x) = \sum_{x\in\Lambda_L} f(x), \tag{5.3}$$

for every function $f\colon \Lambda_L \to \mathbb{R}$, which is a consequence of the fact that $\Lambda_L = \bigcup_{K\in\mathcal{K}} \bigcup_{\mathbf{i}\in\mathcal{I}_K} \Lambda'_{\mathbf{i}}$ is a disjoint union. We will refer to such a decomposition as the *parity partition at scale* L' of Λ_L. Furthermore, given $K \in \mathcal{K}$ and any pair of boxes $\Lambda'_{\mathbf{i}}$ and $\Lambda'_{\mathbf{j}}$ with $\mathbf{i}, \mathbf{j} \in \mathcal{I}_K$ and $\mathbf{i} \neq \mathbf{j}$ we have that,

$$\operatorname{dist}(\Lambda'_{\mathbf{i}}, \Lambda'_{\mathbf{j}}) \ge L'. \tag{5.4}$$

Here for any pair of subsets $A, B \subset \mathbb{Z}^d$ we define $\operatorname{dist}(A,B) := \inf_{x\in A, y\in B} |x-y|$. In other words (5.4) expresses the fact that the distance between any pair of partition boxes with different sub-indexes in $\mathcal{I}_K$ is larger than or equal to L'. This completes the description of the partition of Λ_L in partition boxes.

Next, we describe the partition of Λ_L into the strip set and main boxes. Given an $\mathbf{i} \in \mathcal{I}$ we let,

$$\Lambda''_{\mathbf{i}} := J_{i_1} \times J_{i_2} \times \cdots \times J_{i_d}.$$

Such a box will be called a *main box*. Its cardinality is $|\Lambda''_{\mathbf{i}}| = (L' - 2r)^d$. Now let,

$$S_L := \Lambda_L - \bigcup_{\mathbf{i}\in\mathcal{I}} \Lambda''_{\mathbf{i}}.$$

Such a set will be called the *strip set*. Note that S_L and $\{\Lambda''_L : \mathbf{i} \in \mathcal{I}\}$ define a partition of Λ_L. We will refer to such a partition as the *strip-box partition at scale* L' of Λ_L. We furthermore remark the following cardinality estimate for the strip set which will be useful later,

$$\frac{|S_L|}{(2L+1)^d} \le \frac{\left((L'+1)^d - (L'-2r)^d\right)}{(L')^d}, \tag{5.5}$$

where we have used the fact that $|\mathcal{I}| = q^d$.

5.2. Moment and decoupling inequalities. Following [5], we recall here some standard inequalities and derive inequalities involving sums of the quenched first moments that will be necessary to perform the partition analysis.

Let us first recall the well-known inequality of von Bahr and Esseen (p. 82, exercise 2.6.20, of Petrov [16]).

Lemma 5.1 (von Bahr – Esseen). *Let $X_1, \dots, X_n$ be mean zero independent random variables and $S_n := \sum_{k=1}^n X_i$. Then if* E *denotes the expectation with respect to the joint law of these random variables, and $1 \le r \le 2$, it is true that*

$$\mathrm{E}\,|S_n| \le 2 \sum_{k=1}^n \mathrm{E}\,|X_k|^r. \tag{5.6}$$

We continue with the following lemma (analogous to [5, Lemma 6]), which is a consequence of (5.6) and (4.5).

Lemma 5.2. *Let $a > 1$. Consider the field of truncated quenched first moments $\{\widetilde{m}_a(x,t,w) : x \in \mathbb{Z}^d\}$ at scale $(\kappa t)^a$. Let $L(t), L'(t) \colon [0,\infty) \to \mathbb{N}$ be functions such that $(\kappa t)^a \le L'(t) \le L(t)$. Then if $1 \le r \le 2$, it is true that,*

$$\left\langle \left| \sum_{x\in\Lambda_L} (\widetilde{m}_a - \langle \widetilde{m}_a \rangle) \right|^r \right\rangle \le 2(2L'+2)^{d(r-1)}(2L+1)^d \langle |\widetilde{m}_a - \langle \widetilde{m}_a\rangle|^r \rangle. \tag{5.7}$$

(i) PROOF. The proof of this result is analogous to that of [5, Lemma 6], requiring the use of the decomposition (5.3) with $f(x) = \widetilde{m}_1(x,t,w) - \langle \widetilde{m}_1\rangle(x,t)$, von Bahr – Esseen inequality (5.6) and Jensen's inequality (4.4). □

Next, we have the following estimate analogous to [5, Lemma 7].

Lemma 5.3. *Let $a > 1$. Consider the field of quenched first moments $\{m(x,t,w) : x \in \mathbb{Z}^d\}$ and of truncated quenched first moments $\{m_a(x,t,w) : x \in \mathbb{Z}^d\}$ at scale $(\kappa t)^a$. Let $L(t) \colon [0,\infty) \to \mathbb{N}$ be such that $(\kappa t)^a \ll L$. Then the following statements are true.*

(i) *Assume that,*

$$\left\langle \left| \frac{\sum_{x\in\Lambda_L} \widetilde{m}_a}{(2L+1)^d \langle \widetilde{m}_a \rangle} - 1 \right| \right\rangle \ll 1.$$

Then,

$$\left\langle \left| \frac{m^L}{\langle m \rangle} - 1 \right| \right\rangle \ll 1.$$

(ii) *Asymptotically as $t \to \infty$ we have,*

$$\left\langle \left| \frac{\sum_{x\in\Lambda_L}(m - \langle m\rangle)}{\sqrt{\mathrm{Var}_\mu \sum_{x\in\Lambda_L} m}} - \frac{\sum_{x\in\Lambda_L}(\widetilde{m}_a - \langle \widetilde{m}_a\rangle)}{\sqrt{\mathrm{Var}_\mu \sum_{x\in\Lambda_L} \widetilde{m}_a}} \right| \right\rangle \ll 1, \tag{5.8}$$

and

$$\left\langle \left| \frac{\sum_{x\in\Lambda_L}(m - \langle m\rangle)}{\sqrt{\mathrm{Var}_\mu \sum_{x\in\Lambda_L} m}} - \frac{\sum_{x\in\Lambda_L}(\widetilde{m}_a - \langle \widetilde{m}_a\rangle)}{\sqrt{\mathrm{Var}_\mu \sum_{x\in\Lambda_L} m}} \right| \right\rangle \ll 1. \tag{5.9}$$

PROOF. The case $\kappa = 0$ is trivial, so we will assume that $\kappa > 0$.

(i) A direct calculation shows us that,

$$\left| \frac{m^L}{\langle m\rangle} - \frac{\sum_{x\in\Lambda_L}\widetilde{m}_a}{(2L+1)^d\langle \widetilde{m}_a\rangle} \right| \le \frac{\sum_{x\in\Lambda_L}|m - \widetilde{m}_a|}{(2L+1)^d\langle m\rangle} + \frac{\langle |m - \widetilde{m}_a|\rangle}{\langle m\rangle} \left| \frac{\sum_{x\in\Lambda_L}\widetilde{m}_a}{(2L+1)^d\langle \widetilde{m}_a\rangle} \right|.$$

Now, an application of inequality (4.11) and of the first inequality in equation (4.9) shows us that, $\sum \langle |m - \widetilde{m}_a|\rangle / \big((2L+1)^d\langle m\rangle\big) \le \varepsilon_1(t)$, where $\varepsilon_1(t) := d42^d\big((\kappa t)^a + 1\big)^d \times e^{-2\kappa t (I((\kappa t)^{a-1}/2) - d)}$. Similarly we have that $\langle |m - \widetilde{m}_a|\rangle / \langle m\rangle \le \varepsilon_1(t)$. By the triangle inequality it follows that,

$$\left\langle \left| \frac{m^L}{\langle m\rangle} - \frac{\sum_{x\in\Lambda_L}\widetilde{m}_a}{(2L+1)^d\langle \widetilde{m}_a\rangle} \right| \right\rangle \le 2\varepsilon_1(t) + \left\langle \left| \frac{\sum_{x\in\Lambda_L}\widetilde{m}_a}{(2L+1)^d\langle \widetilde{m}_a\rangle} - 1 \right| \right\rangle. \tag{5.10}$$

Now, for $a > 1$, we have $I\big((\kappa t)^{a-1}/2\big) \to \infty$. Hence, $\varepsilon_1(t) \to 0$ as $t \to \infty$. This clearly implies the statement of part (i) of the lemma.

(ii) Let us first note that the left-hand side of equation (5.8) is upper-bounded by,

$$2\frac{\sum_{x\in\Lambda_L}\langle |m - \widetilde{m}_a|\rangle}{\sqrt{\mathrm{Var}_\mu \sum_{x\in\Lambda_L} m}} + \left\langle \frac{\sum_{x\in\Lambda_L}|\widetilde{m}_a - \langle \widetilde{m}_a\rangle|}{\sqrt{\mathrm{Var}_\mu \sum_{x\in\Lambda_L}\widetilde{m}_a}} \right\rangle \left(\frac{\sqrt{\mathrm{Var}_\mu \sum_{x\in\Lambda_L}\widetilde{m}_a}}{\sqrt{\mathrm{Var}_\mu \sum_{x\in\Lambda_L} m}} - 1 \right). \tag{5.11}$$

Now, by Cauchy – Schwartz inequality we have that

$$\left\langle \frac{\sum_{x\in\Lambda_L} |\widetilde{m}_a - \langle \widetilde{m}_a\rangle|}{\sqrt{\mathrm{Var}_\mu \sum_{x\in\Lambda_L} \widetilde{m}_a}} \right\rangle \leq 1.$$

On the other hand, by the assumption $t^a \ll L$ we know that the hypothesis of Lemma 4.7(ii) is satisfied for $U_t = \Lambda_{L(t)}$ so that the asymptotics (4.26) of this lemma holds, and hence the second term of (5.11) tends to 0 as $t \to \infty$. Furthermore, Cauchy – Schwartz inequality and Proposition 4.4(ii), imply that $\sum_{x\in\Lambda}\langle |m - \widetilde{m}_a|\rangle \leq (2L+1)^d \varepsilon_2(t)\sqrt{\langle m(0,t)^2\rangle}$, where

$$\varepsilon_2(t) := d42^d\big((\kappa t)^a + 1\big)^{d/2} \exp\big\{-2\kappa t I\big((\kappa t)^{a-1}/2\big) + 2d\kappa t\big\}.$$

In addition, by Lemma 4.7(ii) we have $\mathrm{Var}_\mu \sum_{x\in\Lambda} m \sim (2L+1)^d \sum_{x\in\mathbb{Z}^d} c(0,y,t)$. Now, $\sum_{x\in\mathbb{Z}^d} c(0,y,t) \geq \sqrt{\langle m^2\rangle - \langle m\rangle^2}$. Thus, the first term of (5.11) is upper-bounded by a quantity which asymptotically behaves as $t \to \infty$ like,

$$\frac{\varepsilon_2(t)}{\sqrt{1 - \langle m\rangle^2/\langle m^2\rangle}}.$$

But by Corollary 4.6, and the fact that $\overline{F}_1(t) \geq 0$, we conclude that $\langle m\rangle^2/\langle m^2\rangle \ll 1$. In brief, the first term of (5.11) is upper-bounded by a quantity which asymptotically as $t \to \infty$ behaves like, $\varepsilon_2(t)$. Obviously, when $a > 1$ we have $\varepsilon_2(t) \ll 1$. □

5.3. The annealed asymptotics. Let us now prove the law of large numbers stated in equation (2.4). To simplify the writing of the expressions in the calculations, we will redefine ϵ as 2ϵ, assuming that

$$L(t) \geq \exp\left\{\frac{1}{d} F_{2\epsilon}(t)\right\}, \tag{5.12}$$

for some $\epsilon > 0$, and prove that then in μ-probability it is true that,

$$\frac{m^L(0,t,w)}{\langle m(0,t)\rangle} \sim 1. \tag{5.13}$$

To do so we first remark that by inequality (5.10) of Lemma 5.3, it is enough to show that for $a = \frac{3}{2}$,

$$\left\langle \left| \frac{\sum_{x\in\Lambda_L} \widetilde{m}_a}{(2L+1)^d \langle \widetilde{m}_a\rangle} - 1 \right|^{1+\epsilon} \right\rangle \ll 1. \tag{5.14}$$

Remark that the right-hand side of equation (5.14) can be rewritten as,

$$\left\langle \left| \frac{\sum_{x\in\Lambda_L} \widetilde{m}_a}{(2L+1)^d \langle \widetilde{m}_a\rangle} - 1 \right|^{1+\epsilon} \right\rangle = \frac{\langle |\sum_{x\in\Lambda_L} (\widetilde{m}_a - \langle \widetilde{m}_a\rangle)|^{1+\epsilon}\rangle}{(2L+1)^{d(1+\epsilon)} \langle \widetilde{m}_a\rangle^{1+\epsilon}}. \tag{5.15}$$

At this point we make use of the parity partition decomposition for Λ_L previously defined to deal with the numerator of the right-hand side of equation (5.15) via inequality (5.7) with $r = 1 + \epsilon$. We will chose a time dependent mesoscopic scale $L'(t) = (\kappa t)^b$, where $0 < a < b$. Therefore, by Lemma 5.2, the right-hand side of equality (5.15) is upper-bounded by

$$\frac{2(2L'+2)^{d\epsilon} \langle |\widetilde{m}_a - \langle \widetilde{m}_a\rangle|^{1+\epsilon}\rangle}{(2L+1)^{d\epsilon} \langle \widetilde{m}_a\rangle^{1+\epsilon}}. \tag{5.16}$$

Now, since for any nonnegative reals x, y we have $|x-y|^{1+\epsilon} \leq |x|^{1+\epsilon} + |y|^{1+\epsilon}$, it follows that $\langle |\widetilde{m}_a - \langle \widetilde{m}_a\rangle|^{1+\epsilon}\rangle \leq \langle \widetilde{m}_a^{1+\epsilon}\rangle + \langle \widetilde{m}_a\rangle^{1+\epsilon} \leq 2 langle \widetilde{m}_a^{1+\epsilon}\rangle$, where in the last inequality we have used Jensen's inequality. Hence, since $\widetilde{m}_a \le m$, the expression of equation (5.16) is upper bounded by,

$$\frac{2(2L'+2)^{d\epsilon}\langle m^{1+\epsilon}\rangle}{(2L+1)^{d\epsilon}\langle \widetilde{m}_a\rangle^{1+\epsilon}}.$$

But by Proposition 4.4(ii) and (iii), we can replace in the denominator of the above expression the term $\langle \widetilde{m}_a\rangle$ by $\langle m\rangle$. Thus, (5.16) is upper-bounded by an expression which is asymptotically equivalent to,

$$\frac{2(2L'+2)^{d\epsilon+1}k_1((\kappa t)^{da(2+\epsilon)}+1)\mathrm{e}^{H_1((1+\epsilon)t)-(1+\epsilon)H_1(t)}}{(2L+1)^{d\epsilon}} \tag{5.17}$$
$$\le c((\kappa t)^{da(2+\epsilon)})+1)(L')^{d\epsilon}\mathrm{e}^{-\epsilon(F_{2\epsilon}(t)-F_\epsilon(t))},$$

where we have used the second inequality of equation (4.16) in the first inequality, and assumption (5.12) in the last inequality. Note that when $\kappa=0$ the last expression reduces to $\mathrm{e}^{-\epsilon(G_{2\epsilon}(t)-G_\epsilon(t))}$. Then, Assumption MI shows that the first and the second terms of the right-hand side of (5.17) tend to 0 as $t\to\infty$.

5.4. The nonannealed asymptotics. In this subsection we will prove the asymptotic behavior of equation (2.5). Again, we will redefine ϵ by 2ϵ, assuming that,

$$L(t) \le \exp\left\{\frac{1}{d}F_{-2\epsilon}(t)\right\}, \tag{5.18}$$

for some $\epsilon>0$. Note that it will be enough to show that,

$$\left\langle\left|\frac{m_L(0,t)}{\langle m(0,t)\rangle}\right|^{1-\epsilon}\right\rangle \ll 1. \tag{5.19}$$

To do so, note from inequality (4.5)) that the left-hand side of equation (5.19)) is upper-bounded by,

$$(2L+1)^{d\epsilon}\frac{\langle m^{1-\epsilon}(0,t)\rangle}{\langle m(0,t)\rangle^{1-\epsilon}}. \tag{5.20}$$

Now, by the second inequality of equation (4.16) applied with $\beta=1-\epsilon$ to the numerator of (5.20), we see that the left-hand side of equation (5.19) is upper bounded by,

$$(2L+1)^{d\epsilon}k_1((\kappa t)^{da(2-\epsilon)}+1)\mathrm{e}^{H_1((1-\epsilon)t)-(1-\epsilon)H_1(t)}.$$

Finally, by assumption (5.18), this expression is upper bounded by,

$$c((\kappa t)^{da(2-\epsilon)}+1)\mathrm{e}^{-\epsilon(F_{-\epsilon}(t)-F_{-2\epsilon}(t))}.$$

Now, the Assumption MI, implies that this expression converges to 0 as $t\to\infty$.

5.5. The Gaussian asymptotics. Here we prove the central limit theorem stated in equation (2.6). We will perform this time a strip-box partition of the box Λ_L into the strip set S_L and the main boxes. Let us fix $a>1$. Let us fix b and c such that $c>b>a$, and choose the mesoscopic scale $L'(t)=(\kappa t)^c$ and the fine scale $r=(\kappa t)^b$. Note that by Lemma 5.3(ii), it is enough to prove that,

$$\frac{\sum_{x\in\Lambda_L}(\widetilde{m}_a(x,t,w)-\langle\widetilde{m}_a(x,t)\rangle)}{\sqrt{\mathrm{Var}_\mu\sum_{x\in\Lambda_L}\widetilde{m}_a(x,t)}},$$

converges in distribution to the normal law $\mathcal{N}(0,1)$. To do so, we write,

$$\frac{\sum_{x\in\Lambda_L}(\widetilde{m}_a-\langle\widetilde{m}_a\rangle)}{\sqrt{\mathrm{Var}_\mu\sum_{x\in\Lambda_L}\widetilde{m}_a}}=\frac{\sum_{x\in S_L}(\widetilde{m}_a-\langle\widetilde{m}_a\rangle)}{\sqrt{\mathrm{Var}_\mu\sum_{x\in\Lambda_L}\widetilde{m}_a}}+\frac{\sum_{\mathbf{i}\in\mathcal{I}}\sum_{x\in\Lambda''_{\mathbf{i}}}(\widetilde{m}_a-\langle\widetilde{m}_a\rangle)}{\sqrt{\mathrm{Var}_\mu\sum_{x\in\Lambda_L}\widetilde{m}_a}}. \tag{5.21}$$

We will first show that the strip component of the decomposition (5.21) converges to 0 in probability. In fact, note that for t large enough, we have by statement (4.25) applied with $U_t=S_{L(t)}$ and $U_t=\Lambda_{L(t)}$, that

$$\frac{\mathrm{Var}_\mu\sum_{x\in S_L}\widetilde{m}_a}{\mathrm{Var}_\mu\sum_{x\in\Lambda_L}\widetilde{m}_a}\sim\frac{|S_L|}{(2L+1)^d}\le\min\{\kappa,(\kappa t)^{-(c-b)}\},$$

where for the last inequality we have used estimate (5.5). Since $c>b$, this tends to 0 as $t\to\infty$.

Therefore, it is enough to prove that the second term of the right-hand side of equality (5.21), tends in law to $\mathcal{N}(0,1)$. For this purpose, since the random variables $\{\sum_{x\in\Lambda''_{\mathbf{i}}}(\widetilde{m}_a-\langle\widetilde{m}_a\rangle):\mathbf{i}\in\mathcal{I}-\}$ are independent, it is enough to verify a version of the Lyapunov condition. Namely, we will show that,

$$\frac{\sum_{\mathbf{i}\in\mathcal{I}}\langle|\sum_{x\in\Lambda''_{\mathbf{i}}}(\widetilde{m}_a-\langle\widetilde{m}_a\rangle)|^{2+\epsilon}\rangle}{(\sum_{\mathbf{i}\in\mathcal{I}}\mathrm{Var}_\mu\sum_{x\in\Lambda''_{\mathbf{i}}}\widetilde{m}_a)^{1+\epsilon/2}}\ll 1, \tag{5.22}$$

and then apply again statement (4.25) to conclude that the variance of the denominator of the second term of equation can be substituted by $\sum_{\mathbf{i}\in\mathcal{I}}\mathrm{Var}_\mu\sum_{x\in\Lambda''_{\mathbf{i}}}\widetilde{m}_a$. Now, by the same token, we see that the denominator of the left-hand side of equation (5.22), behaves asymptotically as $t\to\infty$ like

$$(2L+1)^{d(1+\epsilon/2)}\Big(\sum_{x\in\mathbb{Z}^d}c_a(0,x,t)\Big)^{1+\epsilon/2}. \tag{5.23}$$

Thus, it is enough to prove the asymptotic behavior (5.22), with the denominator replaced by (5.23). Now, note that $\sum_{x\in\mathbb{Z}^d}c_a(0,x,t)$ is lower bounded by $\langle\widetilde{m}_a^2(0,t)\rangle-\langle\widetilde{m}_a(0,t)\rangle^2$. And by Proposition 4.4(i) and (iii) and Lemma 4.5, this variance is lower-bounded by an expression which is asymptotically equivalent to,

$$c\big((\kappa t)^{3da}+1\big)^{-1}\langle m(0,2t)\rangle-\langle m(0,t)\rangle^2\big). \tag{5.24}$$

Now, Assumption MI with $\theta=2$ and the inequality $\langle\widetilde{m}_a(0,t)\rangle\le\langle m(0,t)\rangle$, imply that, $(\log\langle m(0,2t)\rangle-2\log\langle m(0,t)\rangle\gg\log t$. This shows that the second term $c\langle m(0,t)\rangle^2$ of the expression (5.24), is negligible with respect to the first one $c\big((\kappa t)^{3da}+1\big)^{-1}\langle m(0,2t)\rangle$. Hence, we conclude that it is enough to prove that,

$$\big((\kappa t)^{3da}+1\big)\frac{\sum_{\mathbf{i}\in\mathcal{I}}\langle|\sum_{x\in\Lambda''_{\mathbf{i}}}(\widetilde{m}_a-\langle\widetilde{m}_a\rangle)|^{2+\epsilon}\rangle}{(2L+1)^{d(1+\epsilon/2)}\langle m(0,2t)\rangle^{(1+\epsilon/2)}}\ll 1. \tag{5.25}$$

By Jensen's inequality (4.4) and the upper-bound in equation (5.2), we see that, $\sum_{\mathbf{i}\in\mathcal{I}}\langle|\sum_{x\in\Lambda''_{\mathbf{i}}}(\widetilde{m}_a-\langle\widetilde{m}_a\rangle)|^{2+\epsilon}\rangle\le(L'+1)^{d(1+\epsilon)}(2L+1)^d\langle|\widetilde{m}_a-\langle\widetilde{m}_a\rangle|^{2+\epsilon}\rangle$. Using again the fact that for nonnegative reals x,y we have $|x-y|^{1+\epsilon}\le|x|^{1+\epsilon}+|y|^{1+\epsilon}$, it follows that $\langle|\widetilde{m}_a-\langle\widetilde{m}_a\rangle|^{2+\epsilon}\rangle\le\langle m^{2+\epsilon}\rangle+\langle m\rangle^{2+\epsilon}\le 2\langle m^{2+\epsilon}\rangle$, where we have used Jensen's inequality in the last inequality. We see that the left-hand side of (5.22) is upper-bounded by,

$$\big((\kappa t)^{3da}+1\big)(L'+1)^{d(1+\epsilon)}\frac{\langle m(0,t)^{2+\epsilon}\rangle}{(2L+1)^{d\epsilon/2}\langle m(0,2t)\rangle^{(1+\epsilon/2)}}.$$

Finally, by the hypothesis on the growth of L this can be bounded by,

$$\big((\kappa t)^{3da}+1\big)^2(L'+1)^{d(1+\epsilon)}\mathrm{e}^{\epsilon/2F_{\epsilon/2}(2t)-\epsilon/2F_\epsilon(2t)},$$

which by Condition MI, tends to 0 as $t\to\infty$.

5.6. The non-Gaussian asymptotics. Here we will prove the asymptotics of equation (2.7). It will be necessary to perform a parity partition of Λ_L with a mesoscopic scale $L'=(\kappa t)^b$ for some $b>1$. First note that by equation (5.9) it will be enough to show that for some $b>a>1$,

$$\left\langle\left|\frac{\sum_{x\in\Lambda_L}(\widetilde{m}_a-\langle\widetilde{m}_a\rangle)}{\sqrt{\mathrm{Var}_\mu\sum_{x\in\Lambda_L}m}}\right|^{2-\epsilon}\right\rangle\ll 1. \tag{5.26}$$

Now, by equation (4.24), the denominator $(\mathrm{Var}_\mu\sum_{x\in\Lambda_L}m)^{1-\epsilon/2}$, of the left-hand side of equation (5.26), can be lower-bounded by

$$\begin{aligned}(2L+1)^{d(1-\epsilon/2)}&(\langle m^2\rangle-\langle m\rangle^2)^{1-\epsilon/2}\\&\geq c(2L+1)^{d(1-\epsilon/2)}\langle m^2\rangle^{1-\epsilon/2}\\&\geq c\big((\kappa t)^{3da}+1\big)^{-1}(2L+1)^{d(1-\epsilon/2)}\langle m(0,2t)\rangle^{1-\epsilon/2},\end{aligned}$$

where in the second to last inequality we used the Assumption MI and Lemma 4.5 and in the last inequality assumption Lemma 4.5. On the other hand, inequality (5.7) of Lemma 5.2, applied with $r=2-\epsilon$, Jensen's inequality and $\widetilde{m}_a\leq m$, shows us that the numerator of the left-hand side of equation (5.26) is upper-bounded by $4(2L'+1)^{d(1-\epsilon)}(2L+1)^{d\epsilon/2}\langle m^{2-\epsilon}\rangle$. Using the upper-bound $\langle m^{2-\epsilon}(0,t)\rangle\leq k_1\big((\kappa t)^{3da}+1\big)\langle m(0,(2-\epsilon)t)\rangle$ of Lemma 4.5, the definition of the intermittency exponents $\{F_\theta\}$ in (1.3), and the assumption $L(t)\leq \mathrm{e}^{F_{-\epsilon}(2t)/2}$, we hence see that it is enough to show that,

$$(2L'+1)^{d(1-\epsilon)}\big((\kappa t)^{3da}+1)^2\mathrm{e}^{\epsilon/2F_{-\epsilon}(2t)}\mathrm{e}^{-\epsilon/2F_{-\epsilon/2}(2t)}\ll 1.$$

But this is a consequence of Assumption MI.

5.7. Proof of Theorem 2.11. Let us first prove part (i) for Weibull-type tails. Note that it is enough to find an $x>0$ such that,

$$\left\langle\left(\frac{m_L(0,t)}{\mathrm{e}^{(a(\gamma)+\delta)H(t)}}\right)^x\right\rangle\ll 1.$$

Now, this expression is upper bounded by,

$$\mathrm{e}^{-(x(a+\delta)H(t)-H(xt)-(1-x)\gamma H(t))+o(H(t))},$$

where we have used the asymptotics (1.5). Since $H\in R_{\rho'}$ for $\rho'=\frac{\rho}{\rho-1}$, we see that the above expression is upper bounded by,

$$\mathrm{e}^{-f_{\mathrm{W}}(x,a+\delta)H(t)+o(H(t))},$$

where $f_{\mathrm{W}}(x,b):=xb-x^{\rho'}-(1-x)\gamma$ for $x>0,b>0$. This function has a unique root at $x_0=\big((\rho-1)(b+\gamma)/\rho\big)^{\rho-1}$ when $b=a(\gamma)$. Choosing $x=x_0$, we get the upper bound,

$$\mathrm{e}^{-\delta x_0H(t)+o(H(t))}\ll 1.$$

This proves Theorem 2.11(i). The proof of part (iv) for Fréchet-type tails is completely analogous to the previous argument, so it will be omitted. To prove part (ii), note that in analogy to the proof of part (i), it is enough to show that,

$$e^{-(xH((a+\delta)t)/(a+\delta)-H(xt)-(1-x)\gamma t)+o(t)} \ll 1. \tag{5.27}$$

Now, by supposition (2.10) we have

$$\frac{H(xt) - xH\big((a+\delta)t\big)/(a+\delta)}{t} \sim -\rho x \log \frac{x}{a+\delta}.$$

Thus, the expression of equation (5.27), is upper-bounded by

$$e^{-f_{\mathrm{D}}(x,a+\delta)t+o(t)},$$

where $f_{\mathrm{D}}(x,b) := \rho x \log(x/b) - (1-x)\gamma$ for $x>0$, $b>0$. But this function has a single root at $x_0 = a e^{(\gamma-1)/\rho}$ when $b=a$. Hence, we obtain the upper bound,

$$e^{-x_0 \log(1+\delta/a)+o(t)} \ll 1.$$

This proves Theorem 2.11(ii). The proof of part (iii) for almost bounded potentials is analogous to the proof of part (ii) so it will be omitted. □

Acknowledgments. Alejandro Ramírez thanks the Courant Institute of Mathematical Science, New York City, for its kind hospitality, where part of this work was done.

References

1. P. Antal, *Trapping problems for the simple random walk*, Thesis, ETH, Zürich, 1994.
2. ———, *Enlargement of obstacles for the simple random walk*, Ann. Probab. **23** (1995), no. 1, 1061–1101.
3. K. B. Athreya and P. E. Ney, *Branching processes*, Grundlehren Math. Wiss., vol. 196, Springer, New York, 1972.
4. G. Ben Arous, L. V. Bogachev, and S. A. Molchanov, *Limit theorems for sums of random exponentials*, Probab. Theory Related Fields **132** (2005), no. 4, 579–612.
5. G. Ben Arous, S. Molchanov, and A. F. Ramírez, *Transition from the annealed to the quenched asymptotics for a random walk on random obstacles*, Ann. Probab. **33** (2005), no. 6, 2149–2187.
6. N. H. Bingham, C. M. Goldie, and J. L. Teugels, *Regular variation*, Encyclopedia Math. Appl., vol. 27, Cambridge Univ. Press, Cambridge, 1989.
7. M. Biskup and W. König, *Long-time tails in the parabolic Anderson model with bounded potential*, Ann. Probab. **29** (2001), no. 2, 636–682.
8. A. Bovier, I. Kurkova, and M. Löwe, *Fluctuations of the free energy in the REM and the p-spin SK models*, Ann. Probab. **30** (2002), no. 2, 605–651.
9. B. Derrida, *Random-energy model: limit of a family of disordered systems*, Phys. Rev. Lett. **45** (1980), no. 2, 79–82.
10. M. Donsker and S. R. S. Varadhan, *Asymptotics for the Wiener sausage*, Comm. Pure Appl. Math. **28** (1975), no. 4, 525–565.
11. ———, *On the number of distinct sites visited by a random walk*, Comm. Pure Appl. Math. **32** (1979), no. 6, 721–747.
12. J. Gärtner and S. A. Molchanov, *Parabolic problems for the Anderson model.* I: *Intermittency andrelated topics*, Comm. Math. Phys. **132** (1990), no. 3, 613–655.
13. ———, *Parabolic problems for the Anderson model.* II: *Second-order asymptotics and structure of high peaks*, Probab. Theory Related Fields **111** (1998), no. 1, 17–55.
14. R. van der Hofstad, W. König, and P. Mörters, *The universality classes in the parabolic Anderson model*, Comm. Math. Phys., to appear.
15. T. M. Liggett, *Interacting particle systems*, Grundlehren Math. Wiss., vol. 276, Springer, New York, 1985.

16. V. V. Petrov, *Limit theorems of probability theory: Sequences of independent random variables*, Oxford Stud. Probab., vol. 4, Oxford Univ. Press., New York, 1995.
17. A.-S. Sznitman, *Brownian motion obstacles and random media*, Springer Monogr. Math., Springer, Berlin, 1998.

École Polytechnique Fédérale de Lausanne, 1015 Lausanne, Switzerland & Courant Institute of Mathematical Sciences, New York University, 251 Mercer Street, New York, NY 10012-1185, USA
E-mail address: benarous@cims.nyu.edu

Department of Mathematics and Statistics, University of North Carolina — Charlotte, 376 Fretwell Bldg., 9201 University City Blvd., Charlotte, NC 28223-0001, USA
E-mail address: smolchan@uncc.edu

Facultad de Matemáticas, Pontificia Universidad Católica de Chile, Avenida Vicuña Mackenna 4860, Macul, Santiago 6904411, Chile
E-mail address: aramirez@mat.puc.cl
URL: http://www.mat.puc.cl/~aramirez

Centre de Recherches Mathématiques
CRM Proceedings and Lecture Notes
Volume **42**, 2007

Extreme Value Theory for Random Exponentials

Leonid V. Bogachev

ABSTRACT. We study the limit distribution of upper extreme values of i.i.d. exponential samples $\{e^{tX_i}\}_{i=1}^N$ as $t \to \infty$, $N \to \infty$. Two cases are considered: (A) $\operatorname{ess\,sup} X = 0$ and (B) $\operatorname{ess\,sup} X = \infty$. We assume that the function $h(x) = -\log P\{X > x\}$ (case B) or $h(x) = -\log P\{X > -1/x\}$ (case A) is (normalized) regularly varying at ∞ with index $1 < \varrho < \infty$ (case B) or $0 < \varrho < \infty$ (case A). The growth scale of N is chosen in the form $N \sim e^{\lambda H_0(t)}$ $(0 < \lambda < \infty)$, where $H_0(t)$ is a certain asymptotic version of the function $H(t) := \log E[e^{tX}]$ (case B) or $H(t) = -\log E[e^{tX}]$ (case A). As shown earlier by Ben Arous et al. [5], there are critical points $\lambda_1 < \lambda_2$, below which the LLN and CLT, respectively, break down, whereas for $0 < \lambda < \lambda_2$ the limit laws for the sum $S_N(t) = e^{tX_1} + \cdots + e^{tX_N}$ prove to be stable, with characteristic exponent $\alpha = \alpha(\varrho, \lambda) \in (0, 2)$. In this paper, we obtain the (joint) limit distribution of the upper order statistics of the exponential sample. In particular, $M_{1,N} = \max\{e^{tX_i}\}_{i=1}^N$ has asymptotically the Fréchet distribution with parameter α. We also show that the empirical extremal measure converges (in fdd) to a Poisson random measure with intensity $d(x^{-\alpha})$. These results are complemented by explicit representations of the joint limit distribution of $S_N(t)$ and $M_{1,N}(t)$ (and in particular of their ratio) in terms of i.i.d. random variables with standard exponential distribution.

1. Introduction

In this work, we are concerned with the asymptotic analysis of random samples of the form $\{e^{tX_i}, i = 1, \ldots, N\}$, where $X, X_1, X_2, \ldots$ are independent identically distributed (i.i.d.) random variables and both t and N tend to infinity. More specifically, our goal is to study the limiting distribution of the upper order statistics of the sample (in particular, of the maximum $M_{1,N}(t) = \max\{e^{tX_1}, \ldots, e^{tX_N}\}$) and also to explore the influence of the extreme values on the asymptotic behavior of the sums $S_N(t) = \sum_{i=1}^N e^{tX_i}$ depending on various rates of growth of the parameters t and N.

2000 *Mathematics Subject Classification.* Primary 60G50, 60G70; Secondary 60E07, 60F05.
Key words and phrases. Random exponentials, weak limit theorems, stable laws, regular variation, exponential Tauberian theorem, order statistics, Fréchet distribution, Poisson random measure.
This paper is based in part on joint work with G. Ben Arous and S. A. Molchanov [3].
The work of the author was partially supported by DFG Grant 436 RUS 113/722.
This is the final form of the paper.

It is natural to distinguish the following two cases:

Case A. X is *bounded* from above, $\operatorname{ess\,sup} X < \infty$,

Case B. X is *unbounded* from above, $\operatorname{ess\,sup} X = \infty$.

In case A, without loss in generality we may and will assume that the upper edge of the support of X is zero, $\operatorname{ess\,sup} X = 0$.

It is clear that the results will heavily depend on the structure of the upper tail of the distribution of X. In the present work, we focus on a fairly general class of distributions with the upper tail of the *Weibull/Fréchet* form

$$\mathrm{P}\{X > x\} \approx \begin{cases} \exp(-cx^{\varrho}) & \text{as } x \to \infty \text{ (case B)}, \\ \exp(-c(-x)^{-\varrho}) & \text{as } x \to 0- \text{ (case A)}, \end{cases} \tag{1.1}$$

where $1 < \varrho < \infty$ (case B) or $0 < \varrho < \infty$ (case A). In fact, cases B and A can be combined using the following $\pm$-convention: *in the symbols* $\pm$ *and* $\mp$, *the upper sign refers to case* B *and the lower sign to case* A; furthermore, the notation $x^{\pm}$ stands for the power $x^{\pm 1}$, and $f(x)^{\pm}$ is understood as $[f(x)]^{\pm}$. Setting

$$h(x) := -\log \mathrm{P}\{X > \pm x^{\pm}\}, \tag{1.2}$$

the precise meaning of (1.1) is then furnished by the assumption that the function h is regularly varying at infinity with index $\varrho \in (1, \infty)$ (case B) or $\varrho \in (0, \infty)$ (case A). For example, a normal distribution fits in case B with $\varrho = 2$.

Systematic study of this class of random exponentials has been initiated by Ben Arous et al. [4] (case B only), later on extended to both cases B and A in [5] (see also an earlier preprint [3]). (Let us also mention that the "limiting" value $\varrho = 0$ in case B, leading to random variables of the form X_i^t, has been recently considered by Bogachev [6].) Motivation comes from various areas in theoretical and applied probability, including the problem of a unified treatment of limit laws for sums and extreme values (see Schlather [22]), analysis of the free energy and its fluctuations in the Random Energy Model (REM) of spin glass (see Bovier et al. [7] and further references therein and in [5]), branching random walks in random environments (see [3, 5]), and risk theory (see [3, 5] and references therein).

One can expect that if the number of terms N in the sum $S_N(t)$ grows fast enough relative to the parameter t, then the Law of Large Numbers (LLN) and the Central Limit Theorem (CLT) should hold true in a conventional form. Ben Arous et al. [5] have shown that an appropriate growth scale of $N = N(t)$ is of the form

$$N \sim \mathrm{e}^{\lambda H_0(t)} \quad (t \to \infty), \tag{1.3}$$

where the rate function $H_0(t)$ is a certain asymptotic version of the cumulant generating function

$$H(t) = \pm \log \mathrm{E}[\mathrm{e}^{tX}]. \tag{1.4}$$

By a Tauberian theorem of Kasahara–de Bruijn (see Section 2 below), the function H is regularly varying with index ϱ' such that

$$\pm\frac{1}{\varrho} + \frac{1}{\varrho'} = 1. \tag{1.5}$$

The link between $h(x)$ and $H_0(t) \sim H(t)$ can be characterized explicitly, and in particular $H_0(t)$ can be found (for all t large enough) as the unique solution of the equation

$$\varrho' H_0 = \varrho h\big((\varrho' H_0/t)^{\pm}\big). \tag{1.6}$$

For example, if $h(x) = x^{\varrho}/\varrho$ ($x \ge 0$) then the solution H_0 is given by $H_0(t) = t^{\varrho'}/\varrho'$.

As shown in [5], the critical values of the parameter λ are given by

$$\lambda_1 = \frac{\varrho'}{\varrho}, \quad \lambda_2 = 2^{\varrho'}\frac{\varrho'}{\varrho}, \tag{1.7}$$

so that the LLN and CLT break down if $\lambda < \lambda_1$ and $\lambda < \lambda_2$, respectively. Moreover, for $0 < \lambda < \lambda_2$ one can prove (under a slightly more restrictive condition of *normalized regular variation* of h, see Section 2 below) that the distribution of $S_N(t)$, properly centered and normalized, converges to a stable law with characteristic exponent

$$\alpha \equiv \alpha(\varrho, \lambda) := \left(\frac{\lambda\varrho}{\varrho'}\right)^{1/\varrho'} \tag{1.8}$$

(note that the critical values (1.7) correspond to the canonical values $\alpha_1 = 1$, $\alpha_2 = 2$). Let us point out that the centering constant vanishes for $0 < \lambda < \lambda_1$, while the normalization constant is given for all $0 < \lambda < \lambda_2$ by $B(t) = \mathrm{e}^{\pm t\eta_1(t)}$, where η_1 is the (unique) root of the equation $h(\eta_1^{\pm}) = \lambda H_0(t)$ (note that the right-hand side of this equation is asymptotically equivalent to $\log N$, due to the scaling condition (1.3)). A more precise statement of these results will be given below in Section 3.

In the present paper, we complement this analysis by studying the asymptotic behavior of the upper extreme values of the exponential sample $\{\mathrm{e}^{tX_i}\}_{i=1}^N$. Surprisingly enough, the limiting picture here replicates the classical results in the i.i.d. extreme value theory, known in the case of attraction to the Fréchet distribution. In particular, the maximal term $M_{1,N}(t)$ normalized by $B(t) = \mathrm{e}^{\pm t\eta_1(t)}$ (for *all* $\lambda > 0$) converges in distribution to the Fréchet law Φ_α with the distribution function

$$\Phi_\alpha(x) = \exp(-x^{-\alpha}), \quad x > 0 \tag{1.9}$$

(Theorem 4.3). The proof was first sketched in [4] (in case B) and later on extended to both cases B and A in [3]. This result should be contrasted with the limit distribution of the maximum of "plain" exponentials, $\{\mathrm{e}^{X_i}\}_{i=1}^N$, which appears to be the Gumbel (double exponential) law Λ with the distribution function

$$\Lambda(x) = \exp(-\mathrm{e}^{-x}), \quad x \in \mathbb{R} \tag{1.10}$$

(Theorem 4.1).

Limit theorems for the (joint) distribution of individual order statistics confirm the Fréchet-type nature of exponential extremes (Theorems 4.8, 4.9). Of particular interest is Theorem 4.10, which reveals a Poisson limiting structure of the empirical extremal measure (in the sense of convergence of finite-dimensional distributions), with a "stable" intensity measure $d(x^{-\alpha})$. We also explore the role of the maximum in the sum. To this end, we use some representations of order statistics via uniform and exponential random variables, which allows us to work with a.s.-convergence. Using this approach, we obtain series representations for the joint limit distribution of $S_N(t)$ and $M_{1,N}(t)$.

The rest of the paper is laid out as follows. In Section 2 we specify our regularity assumptions and derive some consequences. Section 3 summarizes results about the limit laws for the sums $S_N(t)$ proved earlier. Section 4 deals with the limit (joint) laws for upper order statistics. In Section 5 we work out certain useful representations of order statistics, in order to obtain certain explicit characterizations of the limit laws of the sums $S_N(t)$. Finally, the Appendix contains a few direct proofs of a formula for the expected value of the limiting ratio $S_N(t)/M_{1,N}(t)$ in the case $0 < \alpha < 1$.

2. Regularity assumptions

Using the *log-tail distribution function* h defined in (1.2), the tail probability of X takes the form

$$\mathrm{P}\{X > x\} = \exp\{-h(\pm x^{\pm})\}. \tag{2.1}$$

We assume that *h is regularly varying at infinity with index ϱ* (we write $h \in \mathrm{R}_\varrho$), where $1 < \varrho < \infty$ (case B) or $0 < \varrho < \infty$ (case A). That is to say, for any $\kappa > 0$

$$\frac{h(\kappa x)}{h(x)} \to \kappa^{\varrho} \quad (x \to +\infty). \tag{2.2}$$

Moreover, by the Uniform Convergence Theorem (UCT) (see [2, Theorem 1.5.2], the limit (2.2) holds uniformly in κ on each interval $(0, K]$.

The *generalized inverse* of a function f can be defined as $f^{\leftarrow}(y) := \inf\{x : f(x) \geq y\}$ [20, Section 0.2]. Note [2, Theorem 1.5.12] that if $f \in \mathrm{R}_\varrho$ with $\varrho > 0$ then $f^{\leftarrow} \in \mathrm{R}_{1/\varrho}$ and

$$f^{\leftarrow}(f(x)) \sim f(f^{\leftarrow}(x)) \sim x \quad (x \to \infty). \tag{2.3}$$

Since the log-tail distribution function h is nondecreasing and right-continuous, its inverse $h^{\leftarrow}$ has the following useful property [20, eq. (0.6c)]:

$$h^{\leftarrow}(y) \leq x \iff y \leq h(x). \tag{2.4}$$

The asymptotic link between $h(x)$ and the cumulant generating function $H(t)$ (see (1.4)) is described by the Kasahara – de Bruijn exponential Tauberian theorem (see [2, Theorems 4.12.7 and 4.12.9]). Set

$$\varrho' := \frac{\varrho}{\varrho \mp 1} \in \begin{cases} (1, \infty) & \text{(case B)} \\ (0, 1) & \text{(case A)} \end{cases}$$

(cf. (1.5)), then $h \in \mathrm{R}_\varrho$ if and only if $H \in \mathrm{R}_{\varrho'}$. More precisely, let $\varphi \in \mathrm{R}_{1/\varrho}$ and put

$$\psi(u) := u\varphi(u)^{\mp} \in \mathrm{R}_{1/\varrho'}. \tag{2.5}$$

Then

$$h(x) \sim \frac{1}{\varrho}\varphi^{\leftarrow}(x) \ (x \to \infty) \iff H(t) \sim \frac{1}{\varrho'}\psi^{\leftarrow}(t) =: H_0(t) \ (t \to \infty). \tag{2.6}$$

In addition to regular variation, $h \in \mathrm{R}_\varrho$, we will we assume throughout the paper that h is *normalized regularly varying*, $h \in \mathrm{NR}_\varrho$ [2, p. 15], that is, h is absolutely continuous (and hence a.e.-differentiable) and satisfies

$$\frac{xh'(x)}{h(x)} \to \varrho \quad (x \to \infty). \tag{2.7}$$

Equivalently, for every $\delta > 0$ the function $h(x)/x^{\varrho-\delta}$ is ultimately increasing, whereas the function $h(x)/x^{\varrho+\delta}$ is ultimately decreasing [2, p. 24]. In particular, it follows that $h(x)$ itself is ultimately (strictly) increasing and therefore is invertible for x large enough.

As a benefit of the assumption $h \in \mathrm{NR}_\varrho$, the function $H_0(t) \sim H(t)$ in the Kasahara–de Bruijn theorem (see (2.6)) can be identified explicitly via equation (1.6) (see details in [5]). Furthermore, integration of (2.7) shows that h can be represented in the form

$$h(x) = h(0) + \int_0^x \frac{h(u)}{u}\bigl(\varrho + \varepsilon(u)\bigr)\,\mathrm{d}u, \tag{2.8}$$

where $\varepsilon(x) \to 0$ as $x \to \infty$.

The next lemma can be viewed as a refinement of the UCT.

Lemma 2.1. *If $h \in \mathrm{NR}_\varrho$ then, uniformly in κ on each interval $[\kappa_0, \kappa_1] \subset (0, \infty)$,*

$$\frac{h(\kappa x) - h(x)}{h(x)} = (\kappa^\varrho - 1)\bigl(1 + o(1)\bigr) \quad (x \to \infty).$$

PROOF. Using representation (2.8), after the substitution $u = xy$ we have

$$\frac{h(\kappa x) - h(x)}{h(x)} = \int_1^\kappa \frac{h(xy)}{h(x)y}\bigl(\varrho + \varepsilon(xy)\bigr)\,\mathrm{d}y. \tag{2.9}$$

By the UCT, the integrand in (2.9) tends to $\varrho y^{\varrho-1}$ as $x \to \infty$, uniformly in $y \in [\kappa_0, \kappa_1]$. Hence, the integral (2.9) converges, uniformly in $\kappa \in [\kappa_0, \kappa_1]$, to $\int_1^\kappa \varrho y^{\varrho-1}\,\mathrm{d}y = \kappa^\varrho - 1$. □

Let $\eta_1 \equiv \eta_1(t)$ be the unique solution of the equation

$$h(\eta_1^\pm) = \lambda H_0(t) \tag{2.10}$$

($\eta_1(t)$ is well defined for all t large enough). For $x > 0$, set

$$\eta_x \equiv \eta_x(t) := \eta_1(t) \pm \frac{\log x}{t}. \tag{2.11}$$

Combining equations (1.6) and (2.10) as $t \to \infty$ and using that $h \in \mathrm{R}_\varrho$, one can find

$$\lim_{t\to\infty} \frac{t\eta_1(t)}{H_0(t)} = \frac{\lambda\varrho}{\alpha}. \tag{2.12}$$

The next lemma plays the crucial role.

Lemma 2.2. *Uniformly in x on each interval $[x_0, x_1] \subset (0, \infty)$,*

$$\lim_{t\to\infty} \bigl[h(\eta_x(t)^\pm) - h(\eta_1(t)^\pm)\bigr] = \alpha \log x.$$

PROOF. Note that

$$\kappa_x(t) := \left(\frac{\eta_x(t)}{\eta_1(t)}\right)^\pm = \left(1 \pm \frac{\log x}{t\eta_1(t)}\right)^\pm \to 1 \quad (t \to \infty),$$

uniformly in $x \in [x_0, x_1] \subset (0, \infty)$. Therefore, for all large enough t the function $\kappa_x(t)$ is uniformly bounded, $0 < \kappa_0 \le \kappa_\tau(t) \le \kappa_1 < \infty$. Applying Lemma 2.1 and using (2.10), in the limit as $t \to \infty$ we obtain, uniformly in x,

$$h(\eta_x(t)^\pm) - h(\eta_1(t)^\pm) \sim h(\eta_1(t)^\pm)(\kappa_x(t)^\varrho - 1) = \lambda H_0(t)\left(\left(1 \pm \frac{\log x}{t\eta_1(t)}\right)^{\pm\varrho} - 1\right)$$
$$\sim \lambda H_0(t)\frac{\varrho \log x}{t\eta_1(t)} \sim \alpha \log x,$$

according to (2.12). □

3. Limit theorems for the sums

The main results of the paper [5] can be summarized in the following two theorems. We assume throughout that

$$\lim_{t\to\infty} N\mathrm{e}^{-\lambda H_0(t)} = 1, \tag{3.1}$$

where $\lambda > 0$ is a parameter (see (1.3)). Recall that α is given by equation (1.8).

Theorem 3.1. *Set*

$$A(t) := \begin{cases} N\,\mathrm{E}[\mathrm{e}^{tX}] = N\mathrm{e}^{\pm H(t)}, & \lambda > \lambda_1 \quad (\alpha > 1),\\ N\,\mathrm{E}\big[\mathrm{e}^{tX}\mathbf{1}_{\{X\le \pm\eta_1\}}\big], & \lambda = \lambda_1 \quad (\alpha = 1),\\ 0, & 0<\lambda<\lambda_1 \ (0<\alpha<1), \end{cases} \tag{3.2}$$

$$B(t) := \begin{cases} (N\,\mathrm{E}[\mathrm{e}^{2tX}])^{1/2} = N^{1/2}\mathrm{e}^{\pm H(2t)/2}, & \lambda > \lambda_2 \quad (\alpha > 2),\\ \big(N\,\mathrm{E}\big[\mathrm{e}^{2tX}\mathbf{1}_{\{X\le \pm\eta_1\}}\big]\big)^{1/2}, & \lambda = \lambda_2 \quad (\alpha = 2),\\ \mathrm{e}^{\pm t\eta_1(t)}, & 0<\lambda<\lambda_2 \ (0<\alpha<2). \end{cases} \tag{3.3}$$

Then, as $t\to\infty$,

$$\frac{S_N(t) - A(t)}{B(t)} \xrightarrow{\mathrm{d}} \begin{cases} \mathcal{N}(0,1), & \lambda \ge \lambda_2 \quad (\alpha \ge 2)\\ \mathcal{F}_\alpha, & 0<\lambda<\lambda_2 \ (0<\alpha<2), \end{cases} \tag{3.4}$$

where $\mathcal{N}(0,1)$ is the standard normal distribution (with zero mean and unit variance) and $\mathcal{F}_\alpha$ is a stable law with characteristic exponent α and skewness parameter $\beta \equiv 1$. The characteristic function of $\mathcal{F}_\alpha$ is given by[1]

$$\phi_\alpha(u) = \begin{cases} \exp\Big\{-\Gamma(1-\alpha)|u|^\alpha \exp\Big(-\dfrac{i\pi\alpha}{2}\,\mathrm{sgn}\, u\Big)\Big\}, & \alpha \ne 1,\\ \exp\Big\{iu(1-\gamma) - \dfrac{\pi}{2}|u|\Big(1 + i\,\mathrm{sgn}\, u \cdot \dfrac{2}{\pi}\log|u|\Big)\Big\}, & \alpha = 1, \end{cases} \tag{3.5}$$

where $\Gamma(s) = \int_0^\infty x^{s-1}\mathrm{e}^{-x}\,\mathrm{d}x$ is the gamma function, $\mathrm{sgn}\, u := u/|u|$ *and* $\gamma = 0.5772\ldots$ *is the Euler constant.*[2]

With some effort (see details in [3, 5]), one can check that for all $\lambda > 0$

$$\lim_{t\to\infty} \frac{B(t)}{A(t)} = 0.$$

Therefore, Theorem 3.1 implies the following LLN.

Theorem 3.2. *If $\lambda \ge \lambda_1$ then*

$$\frac{S_N(t)}{A(t)} \xrightarrow{\mathrm{p}} 1 \quad (t\to\infty), \tag{3.6}$$

where $A(t)$ is given by (3.2).

Remark 3.3. In fact, in the regions $\lambda > \lambda_1$ and $\lambda > \lambda_2$ the LLN and CLT, respectively, hold under the plain regularity condition $h \in \mathrm{R}_\varrho$, without any extra assumptions. Moreover, the function $H(t)$ can be used here instead of $H_0(t)$, and the scaling condition (3.1) can be relaxed to $\log N \gg \lambda_1 H(t)$ and $\log N \gg \lambda_2 H(t)$, respectively (see [3, 5]).

---

[1] For $1 < \alpha < 2$, we use an analytic continuation of the gamma function in (3.5), given by $\Gamma(1-\alpha) = \Gamma(2-\alpha)/(1-\alpha)$.

[2] See [12, #8.367].

Theorem 3.1 can be proved using the known asymptotic methods for sums of independent random variables (see, e.g., [17]). The proofs are technically quite involved since the condition of normalized regular variation guarantees only the limited smoothness of the distribution of X (i.e., ultimate monotonicity and a.e.-differentiability). However, this condition (in particular, Lemma 2.2), combined with the scaling condition (3.1) provide enough analytical control. For instance, in view of (2.1) the tail probability for e^{tX} is asymptotically given by

$$\begin{aligned}(3.7)\quad \mathrm{P}\{\mathrm{e}^{tX} > xB(t)\} = \mathrm{P}\{X > \pm\eta_x(t)\} = \mathrm{e}^{-h(\eta_x^{\pm})} &\sim N^{-1}\mathrm{e}^{h(\eta_1^{\pm})-h(\eta_x^{\pm})} \\ &\sim N^{-1}\mathrm{e}^{-\alpha\log x} = N^{-1}x^{-\alpha},\end{aligned}$$

where $B(t) = \mathrm{e}^{\pm t\eta_1(t)}$ (see (3.3)). Therefore, the Lévy–Khinchin spectral function, being the main ingredient of the limiting infinitely divisible law [17, pp. 81–82], is given for $x > 0$ by

$$(3.8)\qquad L(x) := -\lim_{t\to\infty} N\,\mathrm{P}\{\mathrm{e}^{tX} > xB(t)\} = -x^{-\alpha},$$

which indicates that α is indeed the characteristic exponent of the limit (stable) law (see [15, Theorem 2.2.1]).

Using (3.1), note that for $\lambda < \lambda_2$ we have

$$B(t) = \mathrm{e}^{\pm t\eta_1(t)} \sim N^{\pm t\eta_1(t)/(\lambda H_0(t))} \quad (t \to \infty),$$

and (2.12) implies that in case B, N is being raised to the power

$$\frac{t\eta_1(t)}{\lambda H_0(t)} \sim \frac{\varrho}{\alpha} > \frac{1}{\alpha}.$$

This should be compared to classical results in the i.i.d. case (see, e.g., [15, Theorem 2.1.1]), where the normalization is essentially of the form $N^{1/\alpha}$. As we see, in case B the sums $S_N(t)$ have a stable limit by virtue of a nonclassical (heavier) normalization. As for case A, we have $B(t) \sim N^{-t\eta_1(t)/(\lambda H_0(t))} \to 0$, which has no analogies in the classical theory.

However, another look at the tail probability reveals the mechanism of settling down to a stable law, analogous to that in the i.i.d. situation and acting in both cases A and B. Indeed, in order that i.i.d. random variables $\{Y_i\}$ belong to the domain of attraction of a stable law with characteristic exponent $\alpha > 0$, it is sufficient that for each $x > 0$

$$(3.9)\qquad \mathrm{P}\{Y > N^{1/\alpha}x\} \sim N^{-1}x^{-\alpha} \quad (N \to \infty)$$

(see [15, Theorem 2.6.1]). If we set $Y_i := \mathrm{e}^{tX_i}/B(t)N^{-1/\alpha}$ then, according to (3.7),

$$\mathrm{P}\{Y > N^{1/\alpha}x\} = \mathrm{P}\{\mathrm{e}^{tX} > xB(t)\} \sim N^{-1}x^{-\alpha} \quad (t \to \infty),$$

which mimics the condition (3.9). Thus, in the normalizing function represented in the form $B(t) = B(t)N^{-1/\alpha} \cdot N^{1/\alpha}$, the factor $B(t)N^{-1/\alpha}$ is responsible for the correct behavior of the distribution tail, whereas the conventional power $N^{1/\alpha}$ performs averaging towards a stable law with characteristic exponent α.

This observation helps explain heuristically the many similarities between the limit behavior of random exponentials e^{tX_i} and that of the usual i.i.d. random variables — from convergence to a stable law (Theorem 3.1) to the asymptotic properties of extreme values (Sections 4, 5 below).

We conclude this section with the following LLN for $\log S_N(t)$: as $t \to \infty$,

$$\frac{\log S_N(t)}{t\eta_1(t)} \xrightarrow{\mathrm{P}} \begin{cases} \dfrac{\alpha(\lambda \pm 1)}{\lambda\varrho}, & \lambda \ge \lambda_1, \\ \pm 1, & 0 < \lambda \le \lambda_1. \end{cases} \tag{3.10}$$

For $\lambda < \lambda_1$ this follows directly from Theorem 3.1. For $\lambda \ge \lambda_1$, Theorem 3.2 implies

$$\log S_N(t) = \log A(t) \cdot \theta(t), \tag{3.11}$$

where $\theta(t) \xrightarrow{\mathrm{P}} 1$ as $t \to \infty$. If $\lambda > \lambda_1$, then from (3.1) and (3.2)

$$\log A(t) = \log N \pm H(t) \sim (\lambda \pm 1) H_0(t),$$

and so (3.10) follows by (2.12). If $\lambda = \lambda_1 = \varrho'/\varrho$ (i.e., $\alpha = \alpha_1 = 1$), then $A(t) \le N\mathrm{e}^{\pm H(t)}$ (see (3.2)), so from (3.11) we get, according to (2.12) and (3.1),

$$\frac{\log S_N(t)}{t\eta_1(t)} \le \frac{\log N \pm H(t)}{t\eta_1(t)} \cdot \theta(t) \xrightarrow{\mathrm{P}} \frac{(\lambda_1 \pm 1)}{\lambda_1 \varrho} = \pm 1 \tag{3.12}$$

(cf. (1.5)). On the other hand, $S_N(t) \ge M_{1,N}(t) := \max\{\mathrm{e}^{tX_i}, i = 1, \dots, N\}$, hence

$$\frac{\log S_N(t)}{t\eta_1(t)} \ge \frac{\log M_{1,N}(t)}{t\eta_1(t)} \xrightarrow{\mathrm{P}} \pm 1 \quad (t \to \infty), \tag{3.13}$$

as we will show below (see (4.14)). Now, combination of (3.12) and (3.13) yields (3.10).

Remark 3.4. The limit (3.10) has the meaning of the *free energy* in the REM (see [3, Section 9; 5; 7] and further references therein). It is easy to check that (3.10), as a function of λ, is continuous and continuously differentiable everywhere including the critical point $\lambda = \lambda_1$, but its second derivative has a jump at this point. This corresponds to a "third order" phase transition [7].

4. Asymptotic behavior of extreme values

4.1. The plain exponential maximum. Let us first obtain the distribution of the maximum for plain exponentials, e^{X_i} (i.e., without the large scaling parameter t in the exponent). Namely, we will show that under our conditions on X_i, the random variables e^{X_i} belong to the domain of attraction of the Gumbel (double exponential) distribution Λ (see (1.10)).

Theorem 4.1. *Denote* $Y_i := \mathrm{e}^{X_i}$, $Y_{1,N} := \max\{Y_1, \dots, Y_N\}$, *and set*

$$a_N := \exp\bigl(\pm h^{\leftarrow}(\log N)^{\pm}\bigr), \quad b_N := \pm\frac{a_N \log a_N}{\varrho \log N}. \tag{4.1}$$

Then

$$\lim_{N\to\infty} \mathrm{P}\left\{\frac{Y_{1,N} - a_N}{b_N} \le x\right\} = \exp(-\mathrm{e}^{-x}), \quad x \in \mathbb{R}. \tag{4.2}$$

PROOF. It is not difficult to verify available sufficient conditions for convergence of the maximum's distribution to Λ (see, e.g., [11, Theorem 2.1.3]). However, it appears even simpler to prove (4.2) directly. Using (2.1), we can write

$$\mathrm{P}\{Y_{1,N} \le a_N + xb_N\} = (\mathrm{P}\{X \le \log(a_N + xb_N)\})^N = \bigl(1 - \mathrm{e}^{-h(L_N(x)^{\pm})}\bigr)^N, \tag{4.3}$$

where

$$L_N(x) := \pm \log(a_N + xb_N).$$

Note that, according to (4.1),

$$L_N(0)^{\pm} = \pm(\log a_N)^{\pm} = h^{\leftarrow}(\log N), \tag{4.4}$$

and since $h \in \mathrm{NR}_\varrho$, equation (4.4) can be inverted for all large enough N to yield

$$h(L_N(0)^{\pm}) = \log N. \tag{4.5}$$

Observe that, as $N \to \infty$,

$$\frac{b_N}{a_N \log a_N} = \pm\frac{1}{\varrho \log N} \to 0 \tag{4.6}$$

and, moreover,

$$\frac{b_N}{a_N} = \pm\frac{\log a_N}{\varrho \log N} = \frac{h^{\leftarrow}(\log N)^{\pm}}{\varrho \log N} \to 0, \tag{4.7}$$

since $h^{\leftarrow}(x) \in \mathrm{R}_{1/\varrho}$ and $h^{\leftarrow}(x)^{\pm}/x \in \mathrm{R}_{\pm 1/\varrho - 1}$ with $\pm 1/\varrho - 1 = -1/\varrho' < 0$. In view of (4.6) and (4.7), we obtain

$$\kappa_N(x) := \frac{L_N(x)^{\pm}}{L_N(0)^{\pm}} = \left(1 + \frac{\log(1 + x b_N/a_N)}{\log a_N}\right)^{\pm} \to 1 \quad (N \to \infty), \tag{4.8}$$

and hence $L_N(x)^{\pm} \sim L_N(0)^{\pm} \to +\infty$ (see (4.4)). Using (4.3) and (4.5), it follows that the limit (4.2) is reduced to showing that

$$\lim_{N\to\infty} \left[h(L_N(x)^{\pm}) - h(L_N(0)^{\pm})\right] = x.$$

To this end, we apply Lemma 2.1 and use (4.5), (4.8) to get

$$\begin{aligned} h(L_N(x)^{\pm}) - h(L_N(0)^{\pm}) &\sim h(L_N(0)^{\pm})(\kappa_N(x)^{\varrho} - 1) \\ &= \log N\left(\left(1 + \frac{\log(1 + x b_N/a_N)}{\log a_N}\right)^{\pm\varrho} - 1\right) \\ &\sim \log N \cdot \frac{(\pm\varrho) x b_N}{a_N \log a_N} = x, \end{aligned}$$

according to (4.6). □

Passing to logarithms, it is easy to show that the background random variables $X_i = \log \mathrm{e}^{X_i}$ also belong to the domain of attraction of the Gumbel distribution (see also [6, Proposition 10.1]).

Corollary 4.2. *Let* $X_{1,N} := \max\{X_1, \dots, X_N\}$ *and set*

$$\tilde{a}_N := \pm h^{\leftarrow}(\log N)^{\pm}, \quad \tilde{b}_N := \frac{h^{\leftarrow}(\log N)^{\pm}}{\varrho \log N}. \tag{4.9}$$

Then

$$\lim_{N\to\infty} \mathrm{P}\left\{\frac{X_{1,N} - \tilde{a}_N}{\tilde{b}_N} \le x\right\} = \exp(-\mathrm{e}^{-x}), \quad x \in \mathbb{R}. \tag{4.10}$$

Since $\tilde{b}_N/\tilde{a}_N \to 0$, the limit (4.10) readily implies the following LLN:

$$\frac{X_{1,N}}{h^{\leftarrow}(\log N)^{\pm}} \xrightarrow{\mathrm{p}} \pm 1 \quad (N \to \infty). \tag{4.11}$$

4.2. Limit distribution of the exponential maximum. We shall now see that when a nonlinear (power) scaling is switched on, $\mathrm{e}^X \mapsto \mathrm{e}^{tX}$, the limit law for the maximum changes dramatically. Denote by $M_{k,N} \equiv M_{k,N}(t) := \mathrm{e}^{tX_{k,N}}$ the nonincreasing order statistics of the sample $\{\mathrm{e}^{tX_i}\}_{i=1}^N$. In particular, $M_{1,N}(t) = \max\{\mathrm{e}^{tX_i}, i = 1, \dots, N\}$. For all $\lambda \in (0, \infty)$ denote

$$B(t) := \mathrm{e}^{\pm t\eta_1(t)}, \tag{4.12}$$

where $\eta_1(t)$ is defined by equation (2.10) (cf. (3.3)).

In this section, we obtain the limit distribution of $M_{1,N}(t)$.

Theorem 4.3. *For all $\lambda > 0$, as $t \to \infty$, $M_{1,N}(t)/B(t)$ converges in distribution to the Fréchet law Φ_α (see (1.9)).*

PROOF. Using (2.1), (2.11), and (4.12), for $x > 0$ we have

$$\begin{aligned} \mathrm{P}\{M_{1,N}(t) \le xB(t)\} &= \mathrm{P}\Big\{X_{1,N} \le \frac{\log x}{t} \pm t\eta_1(t)\Big\} \\ &= \mathrm{P}\{X_{1,N} \le \pm\eta_x\} = (\mathrm{P}\{X \le \pm\eta_x\})^N \\ &= \big(1 - \mathrm{e}^{-h(\eta_x^\pm)}\big)^N = \exp\Big(-N\mathrm{e}^{-h(\eta_x^\pm)}(1 + o(1))\Big). \end{aligned} \tag{4.13}$$

Furthermore, recalling (3.1) and (2.10) we obtain, as $t \to \infty$,

$$N\mathrm{e}^{-h(\eta_x^\pm)} \sim \mathrm{e}^{\lambda H_0(t) - h(\eta_x^\pm)} = \mathrm{e}^{h(\eta_1^\pm) - h(\eta_x^\pm)} \to \mathrm{e}^{-\alpha \log x} = x^{-\alpha},$$

according to Lemma 2.2. Returning to (4.13) we get

$$\lim_{t\to\infty} \mathrm{P}\{M_{1,N}(t) \le xB(t)\} = \exp(-x^{-\alpha}),$$

and the theorem is proved. □

Remark 4.4. The probability law Φ_α, known as the Fréchet distribution, represents one of the three types of possible weak limits for maxima of i.i.d. random variables (see [11, Theorem 2.4.1]). However, the known general theorems about convergence to Φ_α are not directly applicable in our case.

In view of (4.12), Theorem 4.3 implies the following LLN for $\log M_{1,N}$:

$$\frac{\log M_{1,N}(t)}{t\eta_1(t)} \xrightarrow{\mathrm{P}} \pm 1 \qquad (t \to \infty). \tag{4.14}$$

Remark 4.5. In contrast with a similar log-LLN for $S_N(t)$ (see (3.10)), the limit (4.14) does not involve any "phase transitions."

Remark 4.6. Comparing (4.14) and (3.10), we note that in the case $0 < \lambda \le \lambda_1$

$$\frac{\log M_{1,N}(t)}{\log S_N(t)} \xrightarrow{\mathrm{P}} 1 \quad (t \to \infty), \tag{4.15}$$

which indicates that the contribution of the maximal term $M_{1,N}(t)$ to the sum $S_N(t)$ is logarithmically equivalent to the whole sum. In the opposite case where $\lambda > \lambda_1$, the limit of the left-hand side in (4.15) can be shown to be strictly less than 1, so that $M_{1,N}(t)$ is negligible as compared to $S_N(t)$. This observation is supported by the LLN being valid for $\lambda \ge \lambda_1$ (see Theorem 3.2), and is further evidenced by Theorem 5.11 characterizing the limit distribution of the ratio $S_N(t)/M_{1,N}(t)$ in the case $\lambda < \lambda_1$.

Remark 4.7. Note that $\log M_{1,N} = tX_{1,N}$, so (4.14) amounts to

$$\frac{X_{1,N}(t)}{\eta_1(t)} \xrightarrow{\mathrm{P}} \pm 1 \quad (t \to \infty), \tag{4.16}$$

which is of course consistent with (4.11); indeed, by (2.3), (2.10), (3.1) and the UCT we have

$$\eta_1(t)^{\pm} \sim h^{\leftarrow}(h(\eta_1(t)^{\pm})) = h^{\leftarrow}\big(\lambda H_0(t)\big) \sim h^{\leftarrow}(\log N),$$

and so $\eta_1(t) \sim h^{\leftarrow}(\log N)^{\pm}$ (cf. (4.11) and (4.16)).

4.3. Joint asymptotic behavior of exponential order statistics. Recall that $M_{k,N} = \mathrm{e}^{tX_{k,N}}$ denotes the kth order statistic (arranged in the nonincreasing order) of the exponential sample $\{\mathrm{e}^{tX_i}\}_{i=1}^N$.

Theorem 4.8. *For all $\lambda > 0$ and each $k \in \mathbb{N}$,*

$$\lim_{t\to\infty} \mathrm{P}\left\{\frac{M_{k,N}(t)}{B(t)} \le x\right\} = \mathrm{e}^{-x^{-\alpha}} \sum_{j=0}^{k-1} \frac{x^{-\alpha j}}{j!} \quad (x > 0). \tag{4.17}$$

PROOF. We have

$$\mathrm{P}\{M_{k,N} \le xB(t)\} = \mathrm{P}\{X_{k,N} \le \pm\eta_x\} \tag{4.18}$$

(cf. (4.13)). The distribution of the kth order statistic $X_{k,N}$ is given by the following known formula (see [11, Section 2.8, eq. (135)]):

$$\mathrm{P}\{X_{k,N} \le u\} = \sum_{j=0}^{k-1} \binom{N}{j} [1 - F(u)]^j [F(u)]^{N-j},$$

where $F(u)$ is the common distribution function of the random variables X_i:

$$F(u) := \mathrm{P}\{X \le u\} = 1 - \mathrm{e}^{-h(\pm u^{\pm})}.$$

Setting $u = \pm\eta_x(t)$, similarly to the proof of Theorem 4.3 we obtain

$$\begin{aligned}
\mathrm{P}\{X_{k,N} \le \pm\eta_x(t)\} &= \sum_{j=0}^{k-1} \binom{N}{j} \mathrm{e}^{-jh(\eta_x^{\pm})} \big(1 - \mathrm{e}^{-h(\eta_x^{\pm})}\big)^{N-j} \\
&\sim \big(1 - \mathrm{e}^{-h(\eta_x^{\pm})}\big)^N \sum_{j=0}^{k-1} \frac{N^j}{j!} \mathrm{e}^{-jh(\eta_x^{\pm})} \\
&\sim \exp\big(-\mathrm{e}^{h(\eta_1^{\pm}) - h(\eta_x^{\pm})}\big) \sum_{j=0}^{k-1} \frac{1}{j!} \mathrm{e}^{j(h(\eta_1^{\pm}) - h(\eta_x^{\pm}))},
\end{aligned}$$

and (4.17) follows by Lemma 2.2. □

One can also derive the limiting form of the joint distribution for a finite number of upper extremes. For instance, let us prove the following assertion.

Theorem 4.9. *For any integers $r < s$ and all real $x, y > 0$,*

$$\lim_{t\to\infty} \mathrm{P}\left\{\frac{M_{r,N}}{B(t)} \le x,\ \frac{M_{s,N}}{B(t)} \le y)\right\} \tag{4.19}$$

$$= \begin{cases} \mathrm{e}^{-y^{-\alpha}} \displaystyle\sum_{j=0}^{r-1}\sum_{k=j}^{s-1} \frac{x^{-\alpha j}(y^{-\alpha} - x^{-\alpha})^{k-j}}{j!(k-j)!} & \text{if } x > y, \\ \mathrm{e}^{-x^{-\alpha}} \displaystyle\sum_{j=0}^{r-1} \frac{x^{-\alpha j}}{j!} & \text{if } x \le y. \end{cases}$$

PROOF. For $x \le y$ the inequality $M_{r,N} \le x$ implies $M_{s,N} \le y$, so that

$$\mathrm{P}\{M_{r,N} \le xB(t), M_{s,N} \le yB(t)\} = \mathrm{P}\{M_{r,N} \le xB(t)\},$$

and (4.19) follows by Theorem 4.17. For $x > y$, similarly to (4.18) we obtain

$$\mathrm{P}\{M_{r,N} \le xB(t), M_{s,N} \le yB(t)\} = \mathrm{P}\{X_{r,N} \le \pm\eta_x(t), X_{s,N} \le \pm\eta_y(t)\}.$$

The joint distribution of the order statistics $X_{r,N}$ and $X_{s,N}$ (with $r < s$, $u > v$) is given by (cf. [9, Section 2.2])

$$\mathrm{P}\{X_{r,N} \le u, X_{r,N} \le v\}$$
$$= \sum_{j=0}^{r-1}\sum_{k=j}^{s-1} \frac{N!}{j!(k-j)!(N-j)!} [1 - F(v)]^j [F(u) - F(v)]^{k-j} [F(u)]^{N-k}.$$

Taking $u = \pm\eta_x(t)$, $v = \pm\eta_y(t)$, similarly as above we arrive at (4.19). □

This theorem can be extended to the case of any given number of upper extremes. But it is more instructive to characterize the limit distribution of extreme values in a different way. Let us consider the random measure μ_N on $(0,\infty)$ (corresponding to the *empirical extremal process*) which counts the order statistics $M_{k,N}$ "from the right":

$$\mu_N(x,\infty) := \sum_{k=1}^{N} \mathbf{1}_{\{M_{k,N} > xB(t)\}}, \qquad x > 0. \tag{4.20}$$

The following theorem reveals a Poisson asymptotic structure of μ_N as $t \to \infty$.

Theorem 4.10. *In the sense of convergence of all finite-dimensional distributions, the measure μ_N converges to a Poisson random measure on $(0,\infty)$ with the mean measure $\nu(x,\infty) = x^{-\alpha}$. That is, for any $x_1 > \dots > x_n > 0$ and any integers $m_1, \dots, m_n \ge 0$*

$$\lim_{t\to\infty} \mathrm{P}\{\mu_N(\Delta_i) = m_i, i = 1,\dots,n\} = \prod_{i=1}^{n} \frac{[\nu(\Delta_i)]^{m_i}}{m_i!} \mathrm{e}^{-\nu(\Delta_i)}, \tag{4.21}$$

where $\Delta_i := (x_i, x_{i-1}]$, $x_0 := \infty$, and $\nu(\Delta_i) = x_i^{-\alpha} - x_{i-1}^{-\alpha}$.

Remark 4.11. The same answer was established by Weissman [23] for the counting measure of extremes in the case of i.i.d. random variables in the domain of attraction of the Fréchet distribution Φ_α. Analogous results were also obtained in [23] for the cases of attraction to the Weibull distribution Ψ_α and the Gumbel distribution Λ (see also [9, Section 9.4]).

PROOF OF THEOREM 4.10. By (4.20), the condition $\mu_N(\Delta_i) = m_i$ means that the interval $(\pm\eta_{x_i}, \pm\eta_{x_{i-1}}]$ contains exactly m_i terms from the sample $\{X_1, \dots, X_N\}$. Hence, putting $m_{n+1} := N - m_1 - \dots - m_n$, by the multinomial formula we obtain

$$\begin{aligned}
\mathrm{P}\{\mu_N(\Delta_i) = m_i, i = 1, \dots, n\} \\
&= N! \prod_{i=1}^{n} \frac{[F(\pm\eta_{x_{i-1}}) - F(\pm\eta_{x_i})]^{m_i}}{m_i!} \cdot \frac{[F(\pm\eta_{x_n})]^{m_{n+1}}}{m_{n+1}!} \\
&= N! \prod_{i=1}^{n} \frac{\left[\mathrm{e}^{-h(\eta_{x_i}^{\pm})} - \mathrm{e}^{-h(\eta_{x_{i-1}}^{\pm})}\right]^{m_i}}{m_i!} \cdot \frac{\left(1 - \mathrm{e}^{-h(\eta_{x_n}^{\pm})}\right)^{m_{n+1}}}{m_{n+1}!} \\
&\sim N^{m_1 + \dots + m_n} \prod_{i=1}^{n} \frac{\left[\mathrm{e}^{-h(\eta_{x_i}^{\pm})} - \mathrm{e}^{-h(\eta_{x_{i-1}}^{\pm})}\right]^{m_i}}{m_i!} \cdot \left(1 - \mathrm{e}^{-h(\eta_{x_n}^{\pm})}\right)^{N} \\
&\sim \prod_{i=1}^{n} \frac{\left[\mathrm{e}^{h(\eta_1^{\pm}) - h(\eta_{x_i}^{\pm})} - \mathrm{e}^{h(\eta_1^{\pm}) - h(\eta_{x_{i-1}}^{\pm})}\right]^{m_i}}{m_i!} \cdot \exp\left\{\mathrm{e}^{h(\eta_1^{\pm}) - h(\eta_{x_n}^{\pm})}\right\}.
\end{aligned}$$

Using that $h(\eta_1^{\pm}) - h(\eta_{x_i}^{\pm}) \to -\alpha \log x_i$ (see Lemma 2.2), we arrive at (4.21). □

5. Sums of random exponentials via order statistics

5.1. Some representations of extreme values. In this section, we record a few (basically well known) representations in distribution for extreme values $M_{k,N}$ and hence for the sum $S_N(t)$. These representations are expressed in terms of auxiliary sequences of i.i.d. random variables with either exponential or uniform distribution. In particular, they will be used to study the asymptotic behavior of $S_N(t)$ in the case $0 < \lambda < \lambda_1$. The advantage of such an approach (usually called the "method of common probability space") is that the random variables of interest will have a limit *with probability one*, rather than just in distribution.

Consider the random variable $\Xi := \pm h^{\leftarrow}(\xi)^{\pm}$, where ξ has the unit exponential distribution, that is, $\mathrm{P}\{\xi > x\} = \mathrm{e}^{-x}$ $(x > 0)$. A key observation is that $\Xi \stackrel{\mathrm{d}}{=} X$. Indeed, since h is right-continuous, we can use the property (2.4) to obtain

$$\begin{aligned}
\mathrm{P}\{\Xi \le x\} = \mathrm{P}\{\pm h^{\leftarrow}(\xi)^{\pm} \le x\} &= \mathrm{P}\{h^{\leftarrow}(\xi) \le \pm x^{\pm}\} \\
&= \mathrm{P}\{\xi \le h(\pm x^{\pm})\} = 1 - \mathrm{e}^{-h(\pm x^{\pm})} = \mathrm{P}\{X \le x\},
\end{aligned}$$

according to (2.1). Furthermore, if $\xi_1, \dots, \xi_N$ are i.i.d. random variables with unit exponential distribution, then the random variables $\Xi_i := \pm h^{\leftarrow}(\xi_i)^{\pm}$ are also independent and hence the random vector $(\Xi_i)_{i=1}^N$ has the same distribution as $(X_i)_{i=1}^N$. In particular, the joint distribution of the order statistics $\Xi_{1,N} \ge \dots \ge \Xi_{N,N}$ coincides with that of the order statistics $X_{1,N} \ge \dots \ge X_{N,N}$. We also note that since the function $\pm h^{\leftarrow}(x)^{\pm}$ is nondecreasing in its domain, the order statistics $\Xi_{k,N}$ can be represented through the underlying exponential order statistics $\xi_{1,N} \ge \dots \ge \xi_{N,N}$ as $\Xi_{k,N} = \pm h^{\leftarrow}(\xi_{k,N})^{\pm}$ $(k = 1, \dots, N)$.

Furthermore, let us recall the following representation in distribution of the exponential order statistics (see [9, Section 2.7]).

Lemma 5.1. *For each $k = 1, \dots, N$, the distribution of the kth order statistic $\xi_{k,N}$ coincides with the distribution of the random variable*

$$T_{k,N} := \sum_{i=k}^{N} \frac{\zeta_i}{i}, \tag{5.1}$$

where (ζ_i) is an auxiliary sequence of i.i.d. random variables, each having the unit exponential distribution. Moreover, $(\xi_{1,N}, \dots, \xi_{N,N}) \overset{\mathrm{d}}{=} (T_{1,N}, \dots, T_{N,N})$.

As a result, the joint distribution of the order statistics $M_{k,N} = \mathrm{e}^{tX_{k,N}}$ coincides with the joint distribution of the random variables $\mathrm{e}^{\pm th^{\leftarrow}(T_{k,N})^{\pm}}$. Hence, the sum $S_N(t) = \sum_{i=1}^{N} \mathrm{e}^{tX_i} = \sum_{k=1}^{N} \mathrm{e}^{tX_{k,N}}$ has the same distribution as the sum

$$\sum_{k=1}^{N} \mathrm{e}^{\pm th^{\leftarrow}(T_{k,N})^{\pm}} \tag{5.2}$$

with $T_{k,N}$ given by (5.1). It is convenient to rewrite $T_{k,N}$ as

$$T_{k,N} = \sum_{i=k}^{N} \left(\frac{\zeta_i}{i} - \log \frac{i+1}{i} \right) + \log \frac{N+1}{k}. \tag{5.3}$$

Lemma 5.2. *The series*

$$\sum_{i=1}^{\infty} \left(\frac{\zeta_i}{i} - \log \frac{i+1}{i} \right) \tag{5.4}$$

converges with probability one.

PROOF. Let us represent the series (5.4) as

$$\sum_{i=1}^{\infty} \left(\frac{\zeta_i}{i} - \log \frac{i+1}{i} \right) = \sum_{i=1}^{\infty} \frac{\zeta_i - 1}{i} + \sum_{i=1}^{\infty} \left(\frac{1}{i} - \log \frac{i+1}{i} \right) \tag{5.5}$$

and show that both series on the right are convergent. Recalling that ζ_i are exponentially distributed with mean 1, we have $\mathrm{E}[(\zeta_i - 1)/i] = 0$ and

$$\sum_{i=1}^{\infty} \mathrm{Var}\left(\frac{\zeta_i - 1}{i} \right) = \sum_{i=1}^{\infty} \frac{\mathrm{Var}[\zeta_i]}{i^2} = \sum_{i=1}^{\infty} \frac{1}{i^2} < \infty.$$

Since (ζ_i) are independent, this implies a.s.-convergence of the series $\sum_i (\zeta_i - 1)/i$ (see [17, Chapter IX, Section 2, Lemma 8]). Further, note that

$$\frac{1}{i} - \log\left(1 + \frac{1}{i}\right) = O(i^{-2}) \quad (i \to \infty),$$

hence the last series in (5.5) converges. □

Remark 5.3. A similar approach was used by Hall [14] to obtain a canonical representation for limiting extreme values in the i.i.d. scheme, following the idea suggested by Rényi [18] (see also [19, Chapter VIII, Section 9]]). LePage et al. [16] have used the same approach to study convergence to a stable law in the classical situation of i.i.d. random variables.

Lemma 5.2 implies that for each $k \ge 1$ the sum of the series

$$Z_k := \sum_{i=k}^{\infty} \left(\frac{\zeta_i}{i} - \log \frac{i+1}{i} \right) \tag{5.6}$$

is finite with probability one. It is not difficult to find the (joint) distribution of Z_k.

Lemma 5.4. *Let us set $\tau_k := k\mathrm{e}^{-Z_k}$ $(k = 1, 2, \dots)$. Then $(\tau_k) \stackrel{\mathrm{d}}{=} (\sigma_k)$, where $\sigma_k := \zeta_1' + \dots + \zeta_k'$ and (ζ_i') is a sequence of independent exponential random variables with mean 1. In particular, τ_k has the gamma distribution with mean k.*

Remark 5.5. The random variables (τ_k) are distributed as a sequence of arrival times of a Poisson process with unit rate.

Proof of Lemma 5.4. According to the definitions (5.6) and (5.1), we have

$$Z_k = \lim_{N\to\infty} \sum_{i=k}^{N} \left(\frac{\zeta_i}{i} - \log \frac{i+1}{i} \right) = \lim_{N\to\infty} (T_{k,N} - \log N) + \log k.$$

Hence,

$$\tau_k = \mathrm{e}^{-Z_k + \log k} = \lim_{N\to\infty} \left(N\mathrm{e}^{-T_{k,N}} \right). \tag{5.7}$$

Recall that by Lemma 5.1, the random variable $T_{k,N}$ has the same distribution as the kth (decreasing) order statistic $\xi_{k,N}$ of N independent exponential random variables $(\xi_i)_{i=1}^N$ (with mean 1). From this, one could derive the distribution of τ_k using the known limit results for the exponential order statistics (see [11, Example 1.3.1 for $k = 1$ and Theorem 2.8.1 for $k \ge 1$]).

However, a more neat proof is possible that requires almost no calculations and simultaneously allows one to establish independence of the successive differences $\tau_{k+1} - \tau_k$. Namely, observe that the random variables $U_i := \mathrm{e}^{-\xi_i}$ $(i = 1, \dots, N)$ are independent and uniformly distributed in $[0, 1]$. Therefore, $(\mathrm{e}^{-\xi_{k,N}})_{k=1}^N \stackrel{\mathrm{d}}{=} (U_{k,N})_{k=1}^N$, where $U_{1,N} \le \dots \le U_{N,N}$ are the order statistics of $(U_i)_{i=1}^N$. In turn, it is well known (see [10, Chapter III, §3]) that

$$(U_{1,N}, \dots, U_{N,N}) \stackrel{\mathrm{d}}{=} \left(\frac{\sigma_1}{\sigma_{N+1}}, \dots, \frac{\sigma_N}{\sigma_{N+1}} \right), \tag{5.8}$$

where σ_k are described in the lemma. Returning to (5.7), the distribution of the vector $(\tau_1, \dots, \tau_k)$ for each $k \ge 1$ can be computed as the weak limit

$$\left(\frac{N\sigma_k}{\sigma_{N+1}}, \dots, \frac{N\sigma_k}{\sigma_{N+1}} \right) \stackrel{\mathrm{d}}{\to} (\sigma_1, \dots, \sigma_k) \quad (N \to \infty),$$

where we used that, due to the Law of Large Numbers, $\sigma_{N+1}/N \stackrel{\mathrm{p}}{\to} 1$. □

5.2. Preparatory estimates. The main goal of this section is to establish a suitable uniform upper bound (with probability one) for the terms of the sum (5.2)) (see Lemma 5.7 below), which will allow us to pass to the limit in (5.2) as $t \to \infty$.

Lemma 5.6. *For each fixed $k \ge 1$, with probability one,*

$$\lim_{t\to\infty} \frac{\exp\{\pm t h^{\leftarrow}(T_{k,N})^{\pm}\}}{B(t)} = k^{-1/\alpha} \mathrm{e}^{Z_k/\alpha}, \tag{5.9}$$

where $B(t)$ is given by (3.3) and Z_k is defined in (5.6).

PROOF. Note that a number sequence (a_n) has a limit $a \in \mathbb{R}$ if and only if for each $c \neq a$,

$$\lim_{n\to\infty} \mathbf{1}_{\le c}(a_n) = \mathbf{1}_{\le c}(a), \tag{5.10}$$

where $\mathbf{1}_{\le c}(\cdot)$ denotes the indicator function of the interval $(-\infty, c]$. Let us fix $c > 0$ and consider the "level" inequality

$$\frac{\exp\{\pm t h^{\leftarrow}(T_{k,N})^{\pm}\}}{B(t)} \le c, \tag{5.11}$$

which, in view of notations (2.10), (3.3) amounts to $h^{\leftarrow}(T_{k,N}) \le \eta_c(t)^{\pm}$. Furthermore, due to the property (2.4) the last inequality can be rewritten as

$$T_{k,N} \le h(\eta_c(t)^{\pm}). \tag{5.12}$$

Our next step is to use equation (2.10) and represent inequality (5.12) as

$$T_{k,N} - \log N \le h(\eta_c(t)^{\pm}) - h(\eta_1(t)^{\pm}) + \lambda H_0(t) - \log N. \tag{5.13}$$

According to (5.3) and (5.6), the left-hand side of (5.13) amounts to

$$\sum_{i=k}^{N}\left(\frac{\zeta_i}{i} - \log\frac{i+1}{i}\right) + \log\frac{N+1}{N} - \log k \to Z_k - \log k \quad (N \to \infty), \tag{5.14}$$

while on the right, using Lemma 2.2 and the scaling relation (3.1), we have

$$h(\eta_c(t)^{\pm}) - h(\eta_1(t)^{\pm}) + \lambda H_0(t) - \log N \to \alpha \log c \quad (t \to \infty).$$

Therefore, in the limit $t \to \infty$, $N \to \infty$, inequality (5.13) takes the form $Z_k - \log k \le \alpha \log c$, or equivalently, $k^{-1/\alpha} \mathrm{e}^{Z_k/\alpha} \le c$. Comparing this inequality with (5.11) and applying (5.10), we obtain (5.9). □

Let us note that if $\sum_i a_i$ is a convergent series, then its partial sums $\sum_{i=k}^n a_i$ are uniformly bounded. Indeed, set $s_n := \sum_{i=1}^n a_i$, $s_0 := 0$, then $s^* := \sup_n s_n < +\infty$, $s_* := \inf_n s_n > -\infty$ and $\left|\sum_{i=k}^n a_i\right| = |s_n - s_{k-1}| \le s^* - s_* < \infty$ for all n and $k \le n$.

Therefore, since by Lemma 5.2 the series (5.4)) is convergent, there exists a proper random variable Z^* such that, with probability one,

$$\sum_{i=k}^{N}\left(\frac{\zeta_i}{i} - \log\frac{i+1}{i}\right) \le Z^* \quad \text{for all } N \text{ and } 1 \le k \le N. \tag{5.15}$$

Using (5.3) it follows that with probability one for all N and $1 \le k \le N$

$$T_{k,N} \le Z^* + \log\frac{N+1}{k}. \tag{5.16}$$

Lemma 5.7. *Let $B(t)$ be given by (3.3). Then for any $\tilde{\alpha} > \alpha$ and each $\varepsilon > 0$, with probability one for all large enough t and uniformly in $k \le N$*

$$\frac{\exp(\pm t h^{\leftarrow}(T_{k,N})^{\pm})}{B(t)} \le c_k := k^{-1/\tilde{\alpha}} \exp\left(\frac{Z^* + \varepsilon}{\alpha}\right). \tag{5.17}$$

PROOF. Similarly to (5.11), we can rewrite (5.17) in the form $T_{k,N} \le h(\eta_{c_k}(t)^{\pm})$ (cf. (5.12)). Furthermore, on account of (5.16) it suffices to check that

$$Z^* - \log k + \log(N+1) \le h(\eta_{c_k}(t)^{\pm}), \quad k = 1, \dots, N. \tag{5.18}$$

Note that with probability one, the ratio

$$\varkappa_k(t) := \frac{\eta_{c_k}(t)}{\eta_1(t)} = 1 \pm \frac{Z^* + \varepsilon}{\alpha t \eta_1(t)} \mp \frac{\log k}{\tilde{\alpha} t \eta_1(t)} \tag{5.19}$$

is ultimately bounded above and separated from 0, uniformly in $k \le N$. Indeed, $\varkappa_k(t)^{\pm} \le \varkappa_1(t)^{\pm} \to 1$ as $t \to \infty$. On the other hand, using (2.12) and (3.1), we get

$$\varkappa_k(t)^{\pm} \ge \varkappa_N(t)^{\pm} \to \left(1 \mp \frac{\alpha}{\tilde{\alpha}\varrho}\right)^{\pm} > \left(1 \mp \frac{1}{\varrho}\right)^{\pm} > 0,$$

since $\tilde{\alpha} > \alpha$. Hence, we can apply Lemma 2.1 and, using the elementary inequality

$$x^p - 1 \ge p(x-1) \quad (p > 1 \text{ or } p < 0)$$

(see [13, Theorem 41]) and also relations (2.10), (2.12) and (5.19), obtain

$$\begin{aligned} h(\eta_{c_k}^{\pm}) - h(\eta_1^{\pm}) &\sim h(\eta_1^{\pm})(\varkappa_k^{\pm\varrho} - 1) \ge \pm\varrho h(\eta_1^{\pm})(\varkappa_k - 1) \\ &= \frac{\varrho h(\eta_1^{\pm})}{\alpha t \eta_1}\left(Z^* + \varepsilon - \frac{\alpha \log k}{\tilde{\alpha}}\right) \sim Z^* + \varepsilon - \frac{\alpha \log k}{\tilde{\alpha}}, \end{aligned} \tag{5.20}$$

uniformly in $k \le N$. We also note that by (2.10) and (3.1),

$$\log(N+1) = h(\eta_1^{\pm}) + o(1) \quad (t \to \infty). \tag{5.21}$$

Estimates (5.20), (5.21) imply that inequality (5.18) will be proved once we check that

$$Z^* - \log k + o(1) \le \left(Z^* + \varepsilon - \frac{\alpha \log k}{\tilde{\alpha}}\right)(1 + o(1)),$$

where all $o(1)$ are uniform in $k \le N$. Rearranging, this is reduced to the inequality

$$0 \le \varepsilon + \left(1 - \frac{\alpha}{\tilde{\alpha}} + o(1)\right) \log k + o(1),$$

which holds as $t \to \infty$, since $\varepsilon > 0$, $1 - \alpha/\tilde{\alpha} > 0$ and $\log k \ge 0$. □

5.3. Convergence of the representations of sums. We are now in a position to prove the following result.

Theorem 5.8. *Assume that* $0 < \alpha < 1$. *Then, as* $t \to \infty$,

$$\frac{1}{B(t)} \sum_{k=1}^{N} \mathrm{e}^{\pm t h^{\leftarrow}(T_{k,N})^{\pm}} \xrightarrow{\text{a.s.}} V_\alpha := \sum_{k=1}^{\infty} k^{-1/\alpha} \mathrm{e}^{Z_k/\alpha}, \tag{5.22}$$

where the last series converges with probability one.

PROOF. Denote $v_k(t) := (1/B(t))\mathrm{e}^{\pm t h^{\leftarrow}(T_{k,N})^{\pm}}$ for $k \le N(t)$ and $v_k(t) := 0$ otherwise. By Lemma 5.6, $v_k(t) \xrightarrow{\text{a.s.}} k^{-1/\alpha}\mathrm{e}^{Z_k/\alpha}$ as $t \to \infty$. Let us pick some $\varepsilon > 0$ and for a given $\alpha \in (0,1)$ choose a number $\tilde{\alpha}$ such that $\alpha < \tilde{\alpha} < 1$. Then, according to Lemma 5.7 (see (5.17)), with probability one for all t large enough we have $v_k(t) \le v_k^* := k^{-1/\tilde{\alpha}} e^{(Z^*+\varepsilon)/\alpha}$, $k = 1, 2, \dots$ Since $1/\tilde{\alpha} > 1$, the series $\sum_k v_k^*$ is convergent, so the Lebesgue dominated convergence theorem yields

$$\frac{1}{B(t)} \sum_{k=1}^{N} e^{\pm t h^{\leftarrow}(T_{k,N})^{\pm}} = \sum_{k=1}^{\infty} v_k(t) \xrightarrow{\text{a.s.}} \sum_{k=1}^{\infty} k^{-1/\alpha} e^{Z_k/\alpha} \equiv V_\alpha \quad (t \to \infty),$$

also implying a.s.-convergence of the limiting series. □

Remark 5.9. The representation of the limit (5.22) can be rewritten as

$$V_\alpha = \sum_{k=1}^{\infty} \tau_k^{-1/\alpha}, \tag{5.23}$$

where (τ_k) are defined in Lemma 5.4. Note that convergence of the series (5.23) is obvious, since a.s. $\tau_k \sim k$ as $k \to \infty$ (by the strong LLN) and $\sum_k k^{-1/\alpha} < \infty$.

Recalling that $S_N(t)$ has the same distribution as the sum (5.2), from Theorem 5.8 it follows that the distribution of $S_N(t)/B(t)$ weakly converges, as $t \to \infty$, to the distribution of the random variable V_α. Comparing this result with Theorem 3.1, we arrive at the following assertion.

Theorem 5.10. *For $0 < \alpha < 1$, the random variable V_α defined in (5.22) has the stable distribution $\mathcal{F}_\alpha$ with characteristic function ϕ_α given by formula (3.5).*

This theorem can be viewed as a series representation of the stable distribution $\mathcal{F}_\alpha$ (with $0 < \alpha < 1$ and $\beta = 1$). However, being rewritten in the form (5.23) this representation amounts to one of the known formulas (cf. [21, Theorem 1.4.5]).

Theorem 5.11. *For $\alpha \in (0, 1)$, the ratio $S_N(t)/M_{1,N}(t)$ has a proper limit distribution, which can be represented via the random variable*

$$W_\alpha := e^{-Z_1/\alpha} V_\alpha \equiv 1 + \sum_{k=1}^{\infty} \exp\left(-\frac{1}{\alpha} \sum_{i=1}^{k} \frac{\zeta_i}{i}\right), \tag{5.24}$$

where (ζ_i) is a sequence of independent exponential random variables with mean 1 involved in the representation (5.6).

Proof. As stated after Lemma 5.1, the joint distribution of $M_{1,N}(t) = e^{tX_{1,N}}$ and $S_N(t) = \sum_{k=1}^{N} e^{tX_{k,N}}$ coincides with that of the pair

$$\exp\{\pm t h^{\leftarrow}(T_{1,N})^{\pm}\}, \quad \sum_{k=1}^{N} \exp\{\pm t h^{\leftarrow}(T_{k,N})^{\pm}\}.$$

In particular,

$$\frac{S_N(t)}{M_{1,N}(t)} \stackrel{\mathrm{d}}{=} \frac{\sum_{k=1}^{N} \exp\{\pm t h^{\leftarrow}(T_{k,N})^{\pm}\}}{\exp\{\pm t h^{\leftarrow}(T_{1,N})^{\pm}\}}. \tag{5.25}$$

Dividing both the numerator and denominator by the function $B(t)$ defined in (3.3) and applying Lemma 5.6 (with $k = 1$) and Theorem 5.8, we deduce that the right-hand side of (5.25) with probability one converges to

$$V_\alpha e^{-Z_1/\alpha} = 1 + \sum_{k=2}^{\infty} k^{-1/\alpha} e^{-(Z_1 - Z_k)/\alpha}. \tag{5.26}$$

From (5.6) it follows that for $k \geq 2$

$$Z_1 - Z_k = \sum_{i=1}^{k-1} \left(\frac{\zeta_i}{i} - \log \frac{i+1}{i} \right) = \sum_{i=1}^{k-1} \frac{\zeta_i}{i} - \log k. \tag{5.27}$$

Hence, (5.26) is reduced to the expression

$$1 + \sum_{k=2}^{\infty} \exp\left(-\frac{1}{\alpha} \sum_{i=1}^{k-1} \frac{\zeta_i}{i} \right),$$

which is the same as the right-hand part of (5.24). □

Remark 5.12. The random variable (5.24) can be represented as (cf. (5.23))

$$W_\alpha = \sum_{k=1}^{\infty} \left(\frac{\tau_1}{\tau_k} \right)^{1/\alpha}, \tag{5.28}$$

where (τ_k) are defined in Lemma 5.4.

Theorem 5.13. *The random variable W_α defined in* (5.24) *has the characteristic function given by*

$$f_\alpha(u) = \frac{e^{iu}}{1 - \alpha \int_0^1 (e^{iux} - 1)\, dx / x^{\alpha+1}}. \tag{5.29}$$

Remark 5.14. Remembering that W_α has emerged in Theorem 5.11 in relation to the limit of $S_N(t)/M_{1,N}(t)$, it is worthwhile to compare Theorem 5.13 with the analogous result by Darling [8, Theorem 5.1]; see also Arov and Bobrov [1, Corollary 4], stating that if (Y_i) is a sequence of i.i.d. random variables with distribution belonging to the domain of attraction of a stable law with exponent $0 < \alpha < 1$, then for $S_N := Y_1 + \cdots + Y_N$ and $M_{1,N} := \max\{Y_1, \dots, Y_N\}$ the ratio $S_N/M_{1,N}$ has the limit distribution with characteristic function (5.29).

Proof of Theorem 5.13. Using the observation in Remark 5.14, we will prove the theorem's statement indirectly, via establishing a representation analogous to (5.24) for the limit of the ratio $S_N/M_{1,N}$. Clearly, it suffices to do this with a suitable choice of random variables Y_i. Let us set $Y_i = e^{\xi_i/\alpha}$, where (ξ_i) is an i.i.d. sequence of exponentially distributed random variables with mean 1. Note that $Y_i > 1$ and

$$P\{Y_i > x\} = P\{\xi_i > \alpha \log x\} = e^{-\alpha \log x} = x^{-\alpha} \quad (x > 1).$$

Hence (see [15, Theorem 2.6.1]), the distribution of Y_i is in the domain of attraction of a stable law $\mathcal{F}_\alpha$ (with $\beta = 1$) and

$$\frac{S_N}{N^{1/\alpha}} = \frac{e^{\xi_1/\alpha} + \cdots + e^{\xi_N/\alpha}}{N^{1/\alpha}} \xrightarrow{d} \mathcal{F}_\alpha \quad (N \to \infty).$$

Passing to the order statistics $\xi_{1,N} \geq \cdots \geq \xi_{N,N}$, in a way similar to the above we represent the sum S_N as

$$S_N = \sum_{k=1}^{N} e^{\xi_{k,N}/\alpha} \stackrel{d}{=} \sum_{k=1}^{N} e^{T_{k,N}/\alpha},$$

where $T_{k,N}$ are given by (5.1). Analogously to (5.13) and (5.14), we obtain

$$\frac{S_N}{N^{1/\alpha}} \stackrel{\mathrm{d}}{=} \sum_{k=1}^{N} \mathrm{e}^{(T_{k,N}-\log N)/\alpha} \xrightarrow{\text{a.s.}} \sum_{k=1}^{\infty} k^{-1/\alpha}\mathrm{e}^{Z_k/\alpha} \equiv V_\alpha \quad (N \to \infty), \tag{5.30}$$

where a term-by-term passing to the limit can be justified as before, using (5.16).
Similarly, one shows that

$$\frac{M_{1,N}}{N^{1/\alpha}} \stackrel{\mathrm{d}}{=} \mathrm{e}^{(T_{1,N}-\log N)/\alpha} \xrightarrow{\text{a.s.}} \mathrm{e}^{Z_1/\alpha} \quad (N \to \infty). \tag{5.31}$$

Hence, dividing (5.30) by (5.31) we obtain (cf. (5.24))

$$\frac{S_N}{M_{1,N}} \xrightarrow{\mathrm{d}} \mathrm{e}^{-Z_1/\alpha} V_\alpha \equiv W_\alpha.$$

Comparing this with the result by Darling [8] mentioned above, we conclude that W_α has distribution with the characteristic function (5.29). □

Remark 5.15. It would be interesting to derive formula (5.29), or otherwise characterize the distribution of W_α directly from representation (5.24) (or (5.28)).

The following result, being an immediate consequence of Theorem 5.13, is of interest due to the striking simplicity of the answer.

Corollary 5.16. *For $0 < \alpha < 1$, the expected value of W_α is given by*

$$\mathrm{E}[W_\alpha] = \frac{1}{1-\alpha}. \tag{5.32}$$

PROOF. Differentiate formula (5.29) at $u = 0$. □

Remark 5.17. In the Appendix, we will give three alternative proofs of identity (5.32) based on representations (5.24) or (5.28).

Remark 5.18. Taking the expectation of W_α using (5.24) (see (A.1) below) and comparing with (5.32), we arrive at the following curious identity:

$$\sum_{k=1}^{\infty} \prod_{i=1}^{k} \left(1 + \frac{1}{i\alpha}\right)^{-1} = \frac{\alpha}{1-\alpha} \quad (0 < \alpha < 1).$$

The next assertion highlights an increasingly overwhelming role played by the maximal term $M_{1,N}(t)$ in the sum $S_N(t)$ as α tends to zero.

Proposition 5.19. *With probability one, $W_\alpha \to 1$ as $\alpha \to 0+$.*

PROOF. As we know, the series (5.24) is a.s.-convergent for all $0 < \alpha < 1$. Since $\zeta_i \geq 0$, its terms are nondecreasing functions of α, so that for $0 < \alpha \leq \alpha_0 < 1$ each one is dominated by the respective value with $\alpha = \alpha_0$. Moreover, with probability one each term is $o(1)$ as $\alpha \to 0+$, so the Lebesgue dominated convergence theorem implies that the series in (5.24) almost surely vanishes and hence $W_\alpha \to 1$. □

Using the series representation provided by Theorem 5.10, one can easily derive a limit theorem for the stable distribution $\mathcal{F}_\alpha$ as its parameter α tends to zero.

Proposition 5.20. *Let a random variable ζ_α have the stable distribution $\mathcal{F}_\alpha$ determined by (3.5), with parameters $0 < \alpha < 1$ and $\beta = 1$. Then, as $\alpha \to 0+$, the distribution of $\alpha \log \zeta_\alpha$ weakly converges to the double exponential distribution,*

$$\lim_{\alpha \to 0+} \mathrm{P}(\alpha \log \zeta_\alpha \leq x) = \exp(-\mathrm{e}^{-x}), \quad x \in \mathbb{R}. \tag{5.33}$$

PROOF. By Theorem 5.10, the random variable V_α has the distribution $\mathcal{F}_\alpha$, so it suffices to verify (5.33) for V_α. From (5.24) and using Proposition 5.19 we have $\alpha \log V_\alpha = Z_1 + \log W_\alpha \xrightarrow{\text{a.s.}} Z_1$ as $\alpha \to 0+$. To find the distribution of Z_1, note that according to the definition of τ_k in Lemma 5.4, we have $Z_1 = -\log \tau_1$, where τ_1 has the exponential distribution with mean 1. Hence, $\mathrm{P}\{Z_1 \le x\} = \mathrm{P}\{\tau_1 \ge \mathrm{e}^{-x}\} = \exp(-\mathrm{e}^{-x})$. □

Remark 5.21. The limit theorem (5.33) can be used to explore the limiting case $\alpha \to 0+$ of Shlather's conjecture concerning random norms (see [3, §9; 6; 22]). A general result of this kind was proved purely analytically by Zolotarev [24, Theorem 5; 25, Theorem 2.9.1].

We conclude this section by stating a series representation theorem for the limit in the case $\lambda_1 \le \lambda < \lambda_2$, analogous to Theorem 5.8.

Theorem 5.22. (a) *For* $\lambda_1 < \lambda < \lambda_2$,

$$\frac{1}{B(t)} \sum_{k=1}^{N} \left(\mathrm{e}^{\pm t h^{\leftarrow}(T_{k,N})^{\pm}} - \mathrm{E}\left[\mathrm{e}^{\pm t h^{\leftarrow}(T_{k,N})^{\pm}}\right]\right) \xrightarrow{\text{a.s.}} \sum_{k=1}^{\infty} \frac{\mathrm{e}^{Z_k/\alpha} - \mathrm{E}\left[\mathrm{e}^{Z_k/\alpha}\right]}{k^{1/\alpha}}.$$

(b) *For* $\lambda = \lambda_1$,

$$\frac{1}{B(t)} \sum_{k=1}^{N} \left(\mathrm{e}^{\pm t h^{\leftarrow}(T_{k,N})^{\pm}} - \mathrm{E}\left[\mathrm{e}^{\pm t h^{\leftarrow}(T_{k,N})^{\pm}} \mathbf{1}_{\{T_{k,N} \le h(\eta_1^{\pm})\}}\right]\right)$$

$$\xrightarrow{\text{a.s.}} \sum_{k=1}^{\infty} \frac{\mathrm{e}^{Z_k/\alpha} - \mathrm{E}\left[\mathrm{e}^{Z_k/\alpha} \mathbf{1}_{\{Z_k \le \log k\}}\right]}{k^{1/\alpha}}.$$

Here the limiting series converge with probability one.

This theorem can be proved along the same lines as in the case $0 < \lambda < \lambda_1$ above. However, the proof is technically more involved and will be presented elsewhere.

Appendix A. Direct proofs of Corollary 5.16

A.1. A proof using representation (5.24). From (5.24) it follows

$$\mathrm{E}\left[W_\alpha\right] = 1 + \sum_{k=1}^{\infty} \prod_{i=1}^{k} \mathrm{E}\left[\mathrm{e}^{-\zeta_i/(i\alpha)}\right] = 1 + \sum_{k=1}^{\infty} \prod_{i=1}^{k} \frac{i}{\alpha^{-1} + i}. \tag{A.1}$$

Note that the right-hand side of (A.1) coincides with the hypergeometric function [12, #9.100]

$$F(a, b; c; z) := 1 + \sum_{k=1}^{\infty} z^k \prod_{i=0}^{k-1} \frac{(a+i)(b+i)}{(c+i)(1+i)}$$

taken at the values $a = 1$, $b = 1$, $c = 1 + \alpha^{-1}$, $z = 1$. Furthermore, it is known [12, #9.122(1)] that for $z = 1$ one has

$$F(a, b; c; 1) = \frac{\Gamma(c)\Gamma(c-a-b)}{\Gamma(c-a)\Gamma(c-b)} \quad (c > a + b).$$

For the above specific values of the parameters this yields

$$F(1, 1; 1 + \alpha^{-1}; 1) = \frac{\Gamma(1 + \alpha^{-1})\Gamma(\alpha^{-1} - 1)}{\Gamma(\alpha^{-1})^2}. \tag{A.2}$$

Using that $\Gamma(1+x) = x\Gamma(x)$, we obtain

$$\Gamma(1+\alpha^{-1})\Gamma(\alpha^{-1}-1) = \alpha^{-1}\Gamma(\alpha^{-1}) \cdot \frac{\Gamma(\alpha^{-1})}{\alpha^{-1}-1} = \frac{\Gamma(\alpha^{-1})^2}{1-\alpha},$$

so substituting this into (A.2) we get (5.32).

A.2. A proof using representation (5.28). The proof above may not seem quite satisfactory, as it relies heavily on the 'external' analytic aid from tables of formulas. Here we give a simpler, self-contained proof based on another representation of W_α given by equation (5.28). Namely, using Lemma 5.4 and relation (5.8) we deduce from (5.28) that $W_\alpha \stackrel{\mathrm{d}}{=} 1 + \sum_{k=1}^{\infty} U_{1,k}^{1/\alpha}$, where $U_{1,k}$ is the minimum of independent random variables $U_1, \dots, U_k$ with uniform distribution on $[0,1]$. Therefore,

$$\mathrm{E}\bigl[W_\alpha\bigr] = 1 + \sum_{k=1}^{\infty} \mathrm{E}\bigl[U_{1,k}^{1/\alpha}\bigr]. \tag{A.3}$$

Note that $\mathrm{P}\{U_{1,k} > x\} = (1-x)^k$ $(0 \le x \le 1)$, and hence

$$\mathrm{E}\bigl[U_{1,k}^{1/\alpha}\bigr] = \int_0^1 x^{1/\alpha} k(1-x)^{k-1}\,\mathrm{d}x.$$

Substituting this expression into (A.3), we obtain

$$\mathrm{E}\bigl[W_\alpha\bigr] = 1 + \int_0^1 x^{1/\alpha} \sum_{k=1}^{\infty} k(1-x)^{k-1}\,\mathrm{d}x = 1 + \int_0^1 x^{1/\alpha-2}\,\mathrm{d}x = \frac{1}{1-\alpha},$$

and (5.32) follows.

A.3. Yet another proof. Finally, following an elegant idea of S. A. Molchanov (private communication), we give an elementary proof of identity (5.32) proceeding directly from representation (A.1). For $0 < \alpha < 1$, set

$$a_1 := \frac{1}{\alpha^{-1}-1}, \qquad a_i := \frac{i}{\alpha^{-1}+i-1} \quad (i \ge 2),$$

$$A_k := \prod_{i=1}^{k} a_i \quad (k \ge 1).$$

It is easy to check that for all $k \ge 1$,

$$\prod_{i=1}^{k} \frac{i}{\alpha^{-1}+i} = \prod_{i=1}^{k} a_i \cdot (1 - a_{k+1}) = A_k - A_{k+1}.$$

It follows that the series on the right-hand side of (A.1) is reduced to

$$1 + \sum_{k=1}^{\infty} (A_k - A_{k+1}) = 1 + \lim_{n\to\infty} (A_1 - A_n) = 1 + A_1,$$

since $A_n \to 0$ as $n \to \infty$. But

$$1 + A_1 = 1 + a_1 = \frac{1}{1-\alpha},$$

and formula (5.32) is proved.

Remark A.1. The different proofs of relation (5.32) given above, being of interest in their own right, may be useful for a direct proof of Theorem 5.13 (cf. Remark 5.14).

Acknowledgments. Financial support by the conference organizers and the CRM (Montréal) is gratefully acknowledged. This paper, as well as my talk at the conference, is based in part on joint work with G. Ben Arous and S. A. Molchanov.

My special thanks are due to Stanislav Alekseevich Molchanov, my former supervisor and subsequently a colleague, who has had a profound impact on my vision of Mathematics. For more than thirty years, I have been privileged — alongside the uncountably many other lucky ones — to witness his awesome creativity, benefit from his insightful ideas and enjoy his art as a champion story-teller. I am now very happy to add my voice to the choir of sincere congratulations on the occasion of his 65th birthday. Cheers!

References

1. D. Z. Arov and A. A. Bobrov, *The extreme terms of a sample and their role in the sum of independent random variables*, Theory Probab. Appl. **5** (1960), no. 4, 377–396.
2. N. H. Bingham, C. M. Goldie, and J. L. Teugels, *Regular variation*, Encyclopedia Math. Appl., vol. 27, Cambridge Univ. Press, Cambridge, 1989.
3. G. Ben Arous, L. V. Bogachev, and S. A. Molchanov, *Limit theorems for sums of random exponentials*, Technical Report NI03078-IGS, Isaac Newton Institute for Mathematical Sciences, Cambridge, 2003, `http://www.newton.cam.ac.uk/preprints/NI03078.pdf`.
4. ———, *Limit theorems for sums of random exponentials*, Recent Developments in Stochastic Analysis and Related Topics (Beijing, 2002) (S. Albeverio, Z.-M. Ma, and M. Roeckner, eds.), World Sci., Singapore, 2004, pp. 45 – 65.
5. ———, *Limit theorems for sums of random exponentials*, Probab. Theory Related Fields **132** (2005), no. 4, 579 – 612.
6. L. V. Bogachev, *Limit laws for norms of iid samples with Weibull tails*, J. Theoret. Probab. **19** (849 – 873), no. 4, 2006.
7. A. Bovier, I. Kurkova, and M. Löwe, *Fluctuations of the free energy in the REM and the p-spin SK models*, Ann. Probab. **30** (2002), no. 2, 605 – 651.
8. D. A. Darling, *The influence of the maximal term in the addition of independent random variables*, Trans. Amer. Math. Soc. **73** (1952), 95–107.
9. H. A. David, *Order statistics*, 2nd ed., Wiley Series in Probability and Mathematical Statistics, Wiley, New York, 1981.
10. W. Feller, *An introduction to probability theory and its applications: Vol.* II, 2nd ed., Wiley, New York, 1971.
11. J. Galambos, *The asymptotic theory of extreme order statistics*, Wiley Series in Probability and Mathematical Statistics, Wiley, New York, 1978.
12. I. S. Gradshteyn and I. M. Ryzhik, *Table of integrals, series, and products*, 5th ed., Academic Press, Boston, MA, 1994.
13. G. H. Hardy, J. E. Littlewood, and G. Pólya, *Inequalities*, 2nd ed., University Press, Cambridge, 1952.
14. P. Hall, *Representations and limit theorems for extreme value distributions*, J. Appl. Probab. **15** (1978), 639 – 644.
15. I. A. Ibragimov and Yu. V. Linnik, *Independent and stationary sequences of random variables*, Wolters-Noordhoff, Groningen, 1971.
16. R. LePage, M. Woodroofe, and J. Zinn, *Convergence to a stable distribution via order statistics*, Ann. Probab. **9** (1981), no. 2, 624 – 632.
17. V. V. Petrov, *Sums of independent random variables*, Ergeb. Math. Grenzgeb., vol. 82, Springer, Berlin, 1975.
18. A. Rényi, *On the theory of order statistics*, Acta Math. Acad. Sci. Hungar. **4** (1953), 191 – 227.
19. ———, *Probability theory*, North-Holland Ser. Appl. Math. and Mech., vol. 10, Elsevier, Amsterdam, 1970.

20. S. I. Resnick, *Extreme values, regular variation, and point processes*, Appl. Probab. Ser. Appl. Probab. Trust, vol. 4, Springer, New York, 1987.
21. G. Samorodnitsky and M. S. Taqqu, *Stable non-Gaussian random processes: Stochastic models with infinite variance*, Stochastic Model., Chapman & Hall, New York, 1994.
22. M. Schlather, *Limit distributions of norms of vectors of positive i.i.d. random variables*, Ann. Probab. **29** (2001), no. 2, 862–881.
23. I. Weissman, *Multivariate extremal processes generated by independent non-identically distributed random variables*, J. Appl. Probab. **12** (1975), 477–487.
24. V. M. Zolotarev, *The Mellin–Stieltjes transformation in probability theory*, Theory Probab. Appl. **2** (1957), no. 4, 433–460.
25. ———, *One-dimensional stable distributions*, Transl. Math. Monogr., vol. 65, Amer. Math. Soc., Providence, RI, 1986.

Department of Statistics, University of Leeds, Leeds LS2 9JT, UK
E-mail address: `bogachev@maths.leeds.ac.uk`

Centre de Recherches Mathématiques
CRM Proceedings and Lecture Notes
Volume **42**, 2007

Singular Continuous and Dense Point Spectrum for Sparse Trees with Finite Dimensions

Jonathan Breuer

ABSTRACT. Sparse trees are trees with sparse branchings. The Laplacian on some of these trees can be shown to have singular spectral measures. We focus on a simple family of sparse trees for which the dimensions can be naturally defined and shown to be finite. Generically, this family has singular spectral measures and eigenvalues that are dense in some interval.

1. Introduction

This paper extends and complements the paper [5] in which the notion of sparse trees was introduced. Sparse trees are trees which have arbitrarily long "one-dimensional" segments (by which we mean intervals of $\mathbb{Z}$), separated by occasional non-trivial branchings. It is shown in [5] that, when these trees are spherically symmetric, one may decompose the Laplacian as a direct sum of Jacobi matrices which have sparse "bumps" off the diagonal. The spectral theory of these matrices is similar to that of one-dimensional Schrödinger operators with sparse potentials (see [11] and references therein). In particular, matrices of this type exist for which the spectral measures are singular with respect to Lebesgue measure. These ideas make it possible to construct simple examples of trees for which the Laplacian has interesting spectral behavior. Several examples with singular continuous spectrum were presented in [5].

In this paper we will be concerned with a family of sparse trees that "interpolates" between $\mathbb{Z}^+$ and the Bethe lattice. These trees can be obtained from the Bethe lattice by replacing an edge, at a distance n from the root, by a segment of length $\sim \gamma^n$ for some fixed $\gamma > 1$. While the Bethe lattice is infinite-dimensional, a tree obtained in this manner can be shown to have dimensionality $= \log \gamma k / \log \gamma$, where k is the connectivity of the original Bethe lattice. (For our definition of dimension see Section 3). Thus, by letting γ vary from 1 to ∞, one gets a family of trees corresponding at one end ("$\gamma = 1$") to the Bethe lattice, and at the other end ("$\gamma = \infty$") to $\mathbb{Z}^+$.

2000 *Mathematics Subject Classification.* Primary 47B15; Secondary 81.

This research was supported in part by the Israel Science Foundation (grant no. 188/02) and by grant no. 2002068 from the United States – Israel Binational Science Foundation (BSF), Jerusalem, Israel.

This is the final form of the paper.

We shall analyze the spectral properties of the Laplacian on these trees with the help of the decomposition described above and some tools from the spectral theory of Schrödinger operators with sparse potentials. The constant branching, however, turns out to be a technical difficulty. We will bypass this difficulty by using an idea from [21], namely, we shall impose a certain probability measure on these trees and prove an "almost sure" result.

It turns out that the situation for these finite-dimensional structures is markedly different from the one for $\mathbb{Z}^d$. These trees (generically) have purely singular spectral measures and some dense point spectrum.

In addition to the new result described above, we also use this opportunity to expand the discussion on the basic setting and on some of the examples presented in [5]. Some basic facts that were briefly mentioned in that paper (such as the self-adjointness of the Laplacian on normal sparse trees), will be explained here in greater detail.

We remark that graphs with singular continuous [18] and pure point spectrum [13] are known to exist. In this context, the family of sparse trees is interesting in that, when varying two parameter sequences (namely, the branching size and the distances between branchings), one encounters a rich spectrum of phenomena. We note, in particular, the existence of examples with spectral measures of fractional Hausdorff dimensions (see [5] and Theorem 4.4 below).

This paper is structured as follows. The next section presents the notion of sparse trees and the decomposition theorem that is basic for all that follows. Some simple results concerning spectral measures for the Laplacian on sparse trees are given in Section 3. Section 4 describes the construction of the finite-dimensional trees mentioned above and our results for them. As mentioned above, this paper uses some ideas and tools from the spectral theory of discrete one-dimensional Schrödinger operators. Relevant notions and results are presented in the appendix.

We are grateful to Michael Aizenman, Nir Avni, Vojkan Jakšić, Yoram Last, Barry Simon, Simone Warzel and Andrej Zlatoš for useful discussions. We also wish to thank Michael Aizenman for the hospitality of Princeton, where some of this work was done.

2. Sparse trees

As noted in the introduction, basic to the analysis which follows is a certain decomposition of the Laplacian on a sparse tree. Since this is possible only when the tree has a certain spherical symmetry, we start with:

Definition 2.1 (Spherically homogeneous rooted tree). A rooted tree is called spherically homogeneous (SH) (see [3]) if any vertex v of generation j is connected with κ_j vertices of generation $j+1$. A locally finite spherically homogeneous tree is uniquely determined by the sequence $\{\kappa_j\}_{j=0}^{\infty}$. By locally finite we mean that the valence of any vertex is finite.

In the definition above, a vertex is said to be of generation j if it is at a distance j from the root O (where the distance between two vertices is defined as the number of edges of the unique path between them). Thus, for spherically homogeneous rooted trees, the valence of a vertex depends solely upon its location with respect to the root.

Let $\{L_n\}_{n=1}^{\infty}$ and $\{k_n\}_{n=1}^{\infty}$ be two sequences of natural numbers such that $k_n \geq 2$ for all n, and $\{L_n\}_{n=1}^{\infty}$ is strictly increasing. We say that Γ is a SH rooted

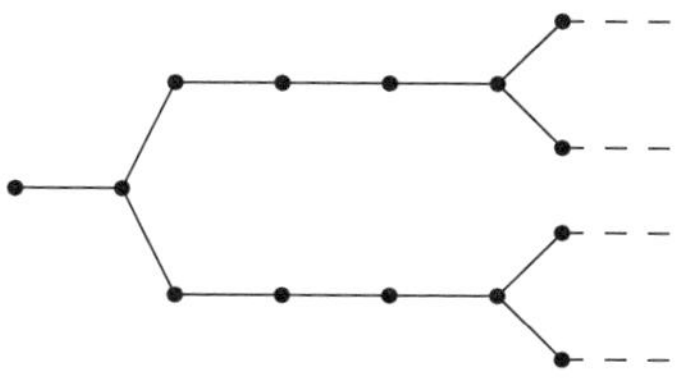

FIGURE 1. An example of a SH rooted tree with $\kappa_0 = 1$, $\kappa_1 = 2$, $\kappa_2 = \kappa_3 = \kappa_4 = 1$, $\kappa_5 = 2, \ldots$

tree of type $\{L_n, k_n\}_{n=1}^{\infty}$ if

$$\kappa_j = \begin{cases} k_n, & j = L_n \text{ for some } n, \\ 1, & \text{otherwise.} \end{cases} \tag{2.1}$$

We say that Γ is *sparse* if $(L_{n+1} - L_n) \to \infty$ as $n \to \infty$. Since sparse trees are not regular (the coordination number is not constant), there are two natural choices for the Laplacian:

$$(\Delta f)(x) = \sum_{y:d(x,y)=1} f(y), \tag{2.2}$$

and

$$(\tilde{\Delta} f)(x) = \sum_{y:d(x,y)=1} f(y) - \#\{y : d(x,y) = 1\} \cdot f(x) \tag{2.3}$$

where $\#A$ for a finite set A is the number of elements in A ($d(x,y)$ denotes the distance on the tree). For simplicity, we shall restrict our attention to Δ, though we note that all our results hold (when properly modified) for $\tilde{\Delta}$ as well.

It is clear that if $\{k_n\}_{n=1}^{\infty}$ is a bounded sequence then both Δ and $\tilde{\Delta}$, on the tree, are bounded and self-adjoint. For unbounded coordination number, the issue of self-adjointness has to be addressed.

Definition 2.2. We call a SH tree of type $\{L_n, k_n\}_{n=1}^{\infty}$–$\Gamma$ *normal* if $\{k_n\}$ unbounded implies that $\limsup_{n\to\infty}(L_{n+1} - L_n) > 1$.

The appendix has a proof that the Laplacians on normal SH rooted trees are self-adjoint. Clearly, any sparse tree is normal.

The main technical tool in the spectral analysis of sparse trees is the following theorem:

Theorem 2.3 ([5, Theorem 2.4])**.** *Let Γ be a normal rooted SH tree of type $\{L_n, k_n\}_{n=1}^{\infty}$. Let*

$$M_n = \begin{cases} \prod_{j=1}^{n} k_j - \prod_{j=1}^{n-1} k_j & n > 1 \\ k_1 - 1 & n = 1 \\ 1 & n = 0. \end{cases} \tag{2.4}$$

Furthermore, let $R_0 = 0$ and $R_n = L_n + 1$, for $n \geq 1$. Then Δ is unitarily equivalent to a direct sum of Jacobi matrices, each operating on a copy of $\ell^2(\mathbb{Z}^+)$:

$$\Delta \cong \bigoplus_{n=0}^{\infty} \underbrace{(J_n \oplus J_n \oplus \cdots \oplus J_n)}_{M_n \textit{ times}} \tag{2.5}$$

where $J_n = J(\{a_n(j)\}_{n=1}^{\infty}, \{b_n(j)\}_{n=1}^{\infty})$ *with*

$$a_n(j) = \begin{cases} \sqrt{k_m} & j = R_m - R_n \textit{ for some } m > n \\ 1 & \textit{otherwise} \end{cases} \tag{2.6}$$

and

$$b_n(j) \equiv 0. \tag{2.7}$$

Remarks. (1) The term Jacobi matrix stands for the semi-infinite matrix

$$J(\{a(j)\}_{j=1}^{\infty}, \{b(j)\}_{j=1}^{\infty}) = \begin{pmatrix} b(1) & a(1) & 0 & 0 & \cdots \\ a(1) & b(2) & a(2) & 0 & \cdots \\ 0 & a(2) & b(3) & a(3) & \ddots \\ \vdots & \ddots & \ddots & \ddots & \ddots \end{pmatrix} \tag{2.8}$$

with

$$b(j) \in \mathbb{R}, \quad a(j) > 0.$$

(2) For the case of a regular tree, a similar decomposition was discussed in [2, 8, 16] (see also [20] for a related result in the case of a metric tree).

(3) As noted above, this theorem holds for $\tilde{\Delta}$ as well, with the decomposition:

$$\tilde{\Delta} \cong \bigoplus_{n=0}^{\infty} \underbrace{(\tilde{J}_n \oplus \tilde{J}_n \oplus \cdots \oplus \tilde{J}_n)}_{M_n \text{ times}} \tag{2.9}$$

where $\tilde{J}_n = J(\{\tilde{a}_n(j)\}_{n=1}^{\infty}, \{\tilde{b}_n(j)\}_{n=1}^{\infty})$ with

$$\tilde{a}_n(j) = a_n(j) \tag{2.10}$$

and

$$\tilde{b}_n(j) = \begin{cases} -k_m - 1 & j = R_m - R_n \text{ for some } m > n \\ -2 & \text{otherwise.} \end{cases} \tag{2.11}$$

(4) Note that each J_n is a "tail" of J_{n-1} in the sense that one can get J_n by deleting a finite number of rows from the top, and the same number of columns from the left, of J_{n-1}.

3. Singular measures on sparse trees

In this and the next section we freely use terms (such as "transfer matrices") associated with the spectral theory of Jacobi matrices. The reader is referred to the appendix for their definitions, the notation we use, some basic results, and further references.

We start with a remark about the essential spectrum. Let Γ be a sparse tree of type $\{L_n, k_n\}_{n=1}^{\infty}$. If $k_n \to \infty$, perturbation theory arguments show that the essential spectrum of Δ on Γ is $[-2, 2]$. If $\{k_n\}$ is bounded, then from remark 4 after Theorem 2.3, it is easy to see that the essential spectrum of Δ is contained in the essential spectrum of J_0. Since the reverse inclusion is immediate, we have the following

Proposition 3.1. *Let Γ be a sparse tree of type $\{L_n, k_n\}_{n=1}^{\infty}$ and let $J_0 = J_0(\Gamma)$ be the corresponding Jacobi matrix appearing in Theorem 2.3. Let $\sigma_{\rm ess}(\Delta)$ be the essential spectrum of Δ on Γ and $\sigma_{\rm ess}(J_0)$ be the essential spectrum of J_0. Then, if either $k_n \to \infty$ or $\{k_n\}$ is bounded, then*

$$\sigma_{\rm ess}(\Delta) = \sigma_{\rm ess}(J_0). \tag{3.1}$$

Now, let H be a self-adjoint operator on a separable Hilbert space $\mathcal{H}$, and $\psi \in \mathcal{H}$. The *spectral measure* associated with ψ and H, μ_ψ, is the unique measure on $\mathbb{R}$ satisfying

$$\langle \psi, (H-z)^{-1}\psi \rangle = \int_{\mathbb{R}} \frac{\mathrm{d}\mu_\psi(x)}{x-z}, \qquad z \in \mathbb{C} \setminus \mathbb{R}$$

(see, e.g., [15]).

Theorem 2.3 reduces the spectral analysis of Δ on a sparse tree to the spectral analysis of Jacobi matrices with sparse "bumps" off the diagonal. (For $\tilde{\Delta}$, we get "bumps" on the diagonal as well). An application of (suitably adapted) methods from the spectral theory of one-dimensional Schrödinger operators with sparse potentials (see [11] for a review of the relevant theory) to the situation at hand, allows us to establish interesting spectral behavior for the Laplacian on certain sparse trees.

The following basic lemma was proved in the appendix of [1] for the case of the Bethe lattice. It holds for any SH rooted tree.

Lemma 3.2. *Let Γ be a normal rooted SH tree with root O. For any vertex v in Γ, let $\delta_v \in \ell^2(\Gamma)$ be the delta function at v, and μ_v the spectral measure associated with Δ and δ_v. Let (a,b) be an interval on which the absolutely continuous part of μ_O vanishes. Then, for any vertex v of Γ, the absolutely continuous part of μ_v vanishes on (a,b).*

Remark. Throughout this paper, "absolutely continuous" means absolutely continuous with respect to Lebesgue measure.

PROOF. The proof is a simple consequence of the identification of the essential support of the absolutely continuous spectrum with the set of energies for which the Green's function has positive imaginary part, combined with the recursion relation (see [1]) for the diagonal elements of the forward resolvents (these are the resolvents of Δ restricted to the various forward subtrees of Γ). □

For a normal rooted SH tree Γ, let $\{J_n(\Gamma)\}_{n=0}^{\infty}$ be the Jacobi matrices appearing in the decomposition of Δ on Γ, given by Theorem 2.3. Let μ_n be the spectral measure associated with $\delta_1 \in \ell^2(\mathbb{Z}^+)$ and J_n. Then Lemma 3.2 above, says that, if we want to prove that all spectral measures associated with the Laplacian on Γ are singular, it suffices to prove that μ_0 is singular. This will be useful later on.

The following lemma is another simple tool for proving singularity of all the spectral measures. Its proof features the "bump" transfer matrix which will prove itself useful throughout the rest of this paper. (See equations (B.10) – (B.13) for the definitions of the transfer matrices that we use below).

Lemma 3.3. *Let R_m be a strictly increasing sequence of natural numbers such that, for m large enough, $R_{m+1} - R_m \geq 2$. Let*

$$J = J(\{a(j)\}_{j=1}^{\infty}, \{b(j)\}_{j=1}^{\infty})$$

be a Jacobi matrix such that

$$a(j) = \begin{cases} \rho_m & j = R_m \text{ for some } m \\ 1 & \text{otherwise} \end{cases} \tag{3.2}$$

and

$$b(j) \equiv 0, \tag{3.3}$$

where $\rho_m > \delta > 0$ for all m. Then, if $\{\rho_m\}_{m=1}^{\infty}$ is unbounded, then the spectral measure, μ, associated with $\delta_1 \in \ell^2(\mathbb{Z}^+)$ and J, is singular with respect to Lebesgue measure.

PROOF. Assume $\lim_{l\to\infty} \rho_{m_l} = \infty$, $R_{m_l} - R_{m_l-1} \geq 2$, $R_{m_l+1} - R_{m_l} \geq 2$ and fix $E \in \mathbb{R}$. Then

$$\begin{aligned} S'_{m_l} \equiv T_{R_{m_l}+1, R_{m_l}-1}(E) &= S_{R_{m_l}+1}(E) S_{R_{m_l}}(E) \\ &= \begin{pmatrix} (E^2/\rho_{m_l} - \rho_{m_l}) & -E/\rho_{m_l} \\ E/\rho_{m_l} & -1/\rho_{m_l} \end{pmatrix}. \end{aligned} \tag{3.4}$$

It follows that

$$\max(1, \rho_{m_l} - E^2) \leq \|T_{R_{m_l}+1, R_{m_l}-1}(E)\|, \tag{3.5}$$

and so, applying Proposition B.3 with $m_j = R_{m_l} + 1$ and $l_j = R_{m_l} - 1$, (note that $a_{R_{m_l}-1} = 1$), we see that μ is singular on $\mathbb{R}$. □

Corollary 3.4. *Let Γ be a sparse tree of type $\{L_n, k_n\}_{n=1}^{\infty}$, with $\{k_n\}_{n=1}^{\infty}$ unbounded. Then all the spectral measures for Δ on Γ are singular with respect to Lebesgue measure.*

PROOF. The statement follows from Theorem 2.3, Lemma 3.3 and the fact that for a general Jacobi matrix, the vector δ_1 is a cyclic vector. □

On the other hand, a simple consequence of Proposition B.4 is the following

Lemma 3.5. *Let R_m be a strictly increasing sequence of natural numbers. Let*

$$J = J(\{a(j)\}_{j=1}^{\infty}, \{b(j)\}_{j=1}^{\infty})$$

be a Jacobi matrix such that

$$a(j) = \begin{cases} \rho_m & j = R_m \text{ for some } m \\ 1 & \text{otherwise} \end{cases} \tag{3.6}$$

and

$$b(j) \equiv 0, \tag{3.7}$$

where $\rho_m > 1$ for all m. Let $\{\beta_m\}_{m=1}^{\infty}$ be a sequence such that $\beta_m \geq \rho_m$ and $\lim_{m\to\infty} \beta_m = \infty$ and let $A_m = \prod_{l=1}^{m} \beta_l^2$. If for some $\varepsilon > 0$,

$$\limsup_{m\to\infty} \frac{(R_{m+1} - R_m)}{A_m^{(1+\varepsilon)}} > 0 \tag{3.8}$$

then the spectral measure, μ, associated with $\delta_1 \in \ell^2(\mathbb{Z}^+)$ and J, is continuous on $(-2, 2)$.

PROOF. Consider an arbitrary closed interval $I \subseteq (-2,2)$. We will show that $\mu(I \cap \cdot)$ is continuous. From this it will follow that $\mu\big((-2,2) \cap \cdot\big)$ is continuous. For $R_m + 1 < j < R_{m+1}$ we have that

$$S_j(E) = \begin{pmatrix} E & -1 \\ 1 & 0 \end{pmatrix}$$

so that $\det\big(S_j(E)\big) = 1$. Furthermore, define, as in the proof of Lemma 3.3

$$S'_m(E) \equiv T_{R_m+1,R_m-1}(E) = S_{R_m+1}(E)S_{R_m}(E) = \begin{pmatrix} (E^2/\rho_m - \rho_m) & -E/\rho_m \\ E/\rho_m & -1/\rho_m \end{pmatrix}. \tag{3.9}$$

Then we also have $\det\big(S'_m(E)\big) = 1$. Thus, if $j_2 \neq R_m$ and $j_1 \neq R_n$ for any m, n, we have that

$$\det\big(T_{j_1,j_2}(E)\big) = 1. \tag{3.10}$$

Note that if $R_m + 1 \leq j_2 < j_1 < R_{m+1}$, then $T_{j_1,j_2}(E)$ is just the transfer matrix for the free Laplacian, so that there is a constant C_I, depending only on the interval I, such that $1 \leq \|T_{j_1,j_2}(E)\| < C_I$ for any such j_1, j_2 and $E \in I$. In addition, we have from (3.9)

$$\|S'_{R_m(E)}\| \leq \rho_m + 5. \tag{3.11}$$

Thus, for $R_m < j < R_{m+1}$ we have

$$\|T_j(E)\| \leq C^m \prod_{j=0}^{m} \rho_j \tag{3.12}$$

for some constant C depending on I.

Let $m_l \to \infty$ be a sequence for which

$$\frac{R_{m_l+1} - R_{m_l}}{A_{m_l}^{(1+\varepsilon)}} > \delta \tag{3.13}$$

for some $\delta > 0$. Let M be chosen so that for all $m > M$, $\beta_m^\varepsilon > C$. Now, for sufficiently large $m_l > M$, we have, from (3.12),

$$\sum_{j=R_{m_l}+1}^{R_{m_l+1}-1} \|T_j(E)\|^{-2} \geq \frac{\delta}{2} A_{m_l}^{\varepsilon} C^{-2m_l} \geq \frac{\delta}{2} C^{-2M}. \tag{3.14}$$

Thus, the tail of the sum in (B.15) does not converge to zero. Therefore the sum is divergent and μ has no eigenvalues in I. This proves the lemma. □

Corollary 3.6. *Let Γ be a rooted SH tree of type $\{L_n, k_n\}_{n=1}^\infty$. Assume that, for some $\varepsilon > 0$,*

$$\limsup_{n\to\infty} \frac{(L_{n+1} - L_n)}{A_n^{(1+\varepsilon)}} > 0, \tag{3.15}$$

where $A_n = \prod_{l=1}^n \beta_l$ for some sequence $\beta_n \geq k_n$, such that $\beta_n \to \infty$. Then any spectral measure for Δ on Γ, is continuous on $(-2,2)$.

PROOF. The statement follows from Theorem 2.3, Lemma 3.5 and the fact that for a general Jacobi matrix, the vector δ_1 is a cyclic vector. □

A simple consequence of Corollaries 3.4, 3.6 and Proposition 3.1 is the following theorem from [5]:

Theorem 3.7 ([5, Theorem 4.1]). *Let $\{k_n\}_{n=1}^{\infty}$ be a sequence of natural numbers such that $k_n \to \infty$ as $n \to \infty$. Let $A_n = \prod_{j=1}^{n} k_j$. Assume that $(L_{n+1} - L_n) \to \infty$ and let Γ be a SH rooted tree of type $\{L_n, k_n\}_{n=1}^{\infty}$. Then the spectrum of Δ on Γ consists of the interval $[-2, 2]$ along with some discrete point spectrum outside this interval. If for some $\varepsilon > 0$,*

$$\limsup_{n \to \infty} \frac{(L_{n+1} - L_n)}{A_n^{(1+\varepsilon)}} > 0, \tag{3.16}$$

then any spectral measure for Δ is purely singular continuous on $(-2, 2)$.

Since the next section discusses trees with bounded k_n, we quote the corresponding result from [5]. We sketch its proof here since some of the ideas will appear in the sequel:

Theorem 3.8 ([5, Theorem 2.2]). *Let $k_0 \geq 2$ be a natural number and let $k_n \equiv k_0$. Assume that $(L_{n+1} - L_n) \to \infty$ and let Γ be a SH rooted tree, of type $\{L_n, k_n\}_{n=1}^{\infty}$. Then the essential spectrum of Δ on Γ contains the interval $[-2, 2]$ and, provided $(L_{n+1} - L_n)$ increase sufficiently rapidly, any spectral measure for Δ is purely singular continuous on $(-2, 2)$. By "sufficiently rapidly" we mean that $(L_{n+1} - L_n)$ has to be made sufficiently large with respect to $\{(L_{i+1} - L_i)\}_{i<n}$.*

Proof. The claim about the essential spectrum follows immediately from Proposition 3.1.

We want to show that if $(L_{n+1} - L_n)$ grow sufficiently rapidly (in the sense described in the theorem), then μ_{δ_O} (the spectral measure of the delta function at the root of Γ) is purely singular continuous on $(-2, 2)$. Lemmas 3.2 and 3.5 say that this suffices to imply that $(L_{n+1} - L_n)$ may be made to grow so fast as to make *all* spectral measures singular continuous.

Thus, our problem is reduced to the problem of studying a Jacobi matrix of the form $J(\{a(j)\}, \{b(j)\})$ with

$$a(j) = \begin{cases} \sqrt{k} & j = L_n + 1 \text{ for some } n \\ 1 & \text{otherwise} \end{cases} \tag{3.17}$$

and

$$b(j) \equiv 0. \tag{3.18}$$

The proof now follows closely Pearson's classical proof [14]. For any $E \in (-2, 2)$, let $\phi \in (0, \pi)$ be defined by $2\cos(\phi) = E$, and let $u_{1,2\cos(\phi)} \equiv u_{1,E}$ as defined in the appendix. Define the EFGP variables [10] $r_\phi(j)$ and $\theta_\phi(j)$ through:

$$r_\phi(j)\cos\big(\theta_\phi(j)\big) = u_{1,E}(j) - \cos(\phi)u_{1,E}(j-1) \tag{3.19}$$

$$r_\phi(j)\sin\big(\theta_\phi(j)\big) = \sin(\phi)u_{1,E}(j-1). \tag{3.20}$$

It is easy to see that for $L_n + 3 < j \leq L_{n+1} + 1$, $r_\phi(j) = r_\phi(j-1)$ and $\theta_\phi(j) = \theta_\phi(j-1) + \phi$ (since the evolution equations for $u_{1,E}$ there coincide with those of the usual Laplacian on $\mathbb{Z}^+$), so that for all n,

$$r_\phi(L_n + 1) = r_\phi(L_{n-1} + 3) \tag{3.21}$$

and

$$\theta_\phi(L_n+1)=\theta_\phi(L_{n-1}+3)+(L_n-L_{n-1}-2)\phi. \tag{3.22}$$

Since the map $g(\phi)=2\cos\phi$ is continuously invertible on $(0,\pi)$, it is clear that, instead of studying μ_{δ_O} on a given closed interval in $(-2,2)$, we may study its push-forward (via g^{-1}) ν on the corresponding closed interval in $(0,\pi)$. Now, since $a(j)$ is bounded, standard methods (see, e.g., [6, Section III.3]) show that ν on $(0,\pi)$ is equivalent to the measure defined by the limit

$$\lim_{n\to\infty}\int_I\frac{\mathrm{d}\phi}{\big(r_\phi(L_n+3)\big)^2} \tag{3.23}$$

where the integral is performed over closed subintervals $I\subseteq(0,\pi)$. Noting that

$$u_{1,2\cos(\phi)}(j-1)=\frac{\sin\big(\theta_\phi(j)\big)}{\sin(\phi)}r_\phi(j),\quad u_{1,2\cos(\phi)}(j)=\frac{\sin\big(\phi+\theta_\phi(j)\big)}{\sin(\phi)}r_\phi(j),$$

one can employ the transfer matrix

$$S'_n\big(2\cos(\phi)\big)\equiv S_{L_n+2}\big(2\cos(\phi)\big)S_{L_n+1}\big(2\cos(\phi)\big)$$

and (3.21) – (3.22) to express $\big(r_\phi(L_n+3)\big)^2$ as a function of $\big(r_\phi(L_{n-1}+3)\big)^2$.. Note that $S'_n\big(2\cos(\phi)\big)$ is unimodular. We get that

$$\int_I\frac{\mathrm{d}\phi}{r_\phi(L_n+3)^2}=\int_I\mathrm{d}\phi\prod_{i=1}^n f_i\big(\phi,N_i\phi,\theta_{i-1}(\phi)\big), \tag{3.24}$$

where $N_i=L_i-L_{i-1}-2$ and

$$f_i(\phi,y,\theta)^{-1}=A(\phi)+B(\phi)\cos 2(\theta+y)+C(\phi)\sin 2(\theta+y), \tag{3.25}$$

with $A^2-B^2-C^2=\det\big(S'_i(E)\big)=1$.

The situation described in (3.24) – (3.25) is exactly the same as that in [14, Section 3] (although the explicit expressions one gets for A, B and C are different). Note that the precise form of $S'_n(E)$ is of no importance. The unimodularity of this transfer matrix, together with the fact that $a(L_n+1)=\text{const}\neq 1$, suffice to imply that (as in [14]) the corollary to Theorem 1 from [14], applies in this situation. This shows that if the L_n are chosen to increase rapidly enough, ν, and thus μ_{δ_O}, is singular continuous. □

4. Finite-dimensional trees

As noted in the introduction, aside from providing interesting examples for the Laplacian, sparse trees are interesting as objects interpolating between the one-dimensional line ($\mathbb{Z}^+$) and the Bethe lattice (which is infinite-dimensional in a natural sense). By tuning the sequences $\{L_n\}$ and $\{k_n\}$, one may construct trees with dimensions having any real value between one and infinity. A simple example is obtained as follows: Let $k_n\equiv k\geq 2$ and take $L_n=[\gamma^n]$ for some $\gamma>2$. Denote the SH rooted tree of type $\{L_n,k_n\}_{n=1}^\infty$ by $\Gamma_{k,\gamma}$. A simple calculation gives:

Proposition 4.1. *Fix $\gamma>2$ and $\mathbb{N}\ni k\geq 2$. Let $\Gamma=\Gamma_{k,\gamma}$ and let $S_\Gamma(r)=\{v_i\in\mathcal{V}(\Gamma)\mid d(v_i,O)\leq r\}$, where $\mathcal{V}(\Gamma)$ is the set of vertices of Γ. Then*

$$\limsup_{r\to\infty}\frac{\log\#S_\Gamma(r)}{\log r}=\liminf_{r\to\infty}\frac{\log\#S_\Gamma(r)}{\log r}=\frac{\log\gamma k}{\log\gamma}.$$

Below, we shall refer to the quantity $\frac{\log \gamma k}{\log \gamma}$ as the *dimension* of $\Gamma_{k,\gamma}$.

In the context of the analogy between sparse trees and Schrödinger operators with sparse potentials, described in the previous section, $\Gamma_{k,\gamma}$ is analogous to a Schrödinger operator with bumps of fixed height placed at the sites $[\gamma^n]$ of $\mathbb{Z}^+$. Zlatoš deals with such operators in [21] and the analysis we present below is an adaptation of his methods (in particular [21, Section 6]) to the case at hand. While a large part of the argument translates word for word, there are a few significant changes, mainly having to do with the fact that the transfer matrices for our case are not, in general, unimodular. This is important for some of the arguments and, therefore, has to be bypassed to get the same results here. We discuss the changes below and give a sketch of the proof. However, we refer the reader to [21] for a more detailed discussion.

First,

Definition 4.2. Let Γ be a rooted tree. For any self-adjoint operator H on $\ell^2(\Gamma)$, and $-\pi/2 < \varrho < \pi/2$, let

$$H_\varrho = H - \tan(\varrho) P_O \tag{4.1}$$

where P_O is the orthogonal projection onto the subspace spanned by the delta function at O. We refer to H_ϱ as H with boundary condition ϱ.

We also need:

Definition 4.3. Let μ be a measure on $\mathbb{R}$. We say that μ has *exact local dimension* in $I \subseteq \mathbb{R}$ if for any $E \in I$ there is an $\alpha(E)$ and for any $\varepsilon > 0$ there is $\delta > 0$ for which $\mu\big((E-\delta, E+\delta) \cap \cdot\big)$ is both continuous with respect to $(\alpha(E)-\varepsilon)$-dimensional Hausdorff measure, and singular with respect to $(\alpha(E)+\varepsilon)$-dimensional Hausdorff measure. We call $\alpha(E)$ the *local dimension* of the measure μ.

Let ω_n be a random variable uniformly distributed over

$$[-n, -n+1, \ldots, n-1, n].$$

Let $(\Omega, \mathbb{P})$ be the product probability space for all ω_n, $n = 1, 2, \ldots$. Fix $1 < k \in \mathbb{N}$ and $\gamma > 2$ and for each $\omega \in \Omega$ let $\Gamma^\omega_{k,\gamma}$ be the SH rooted tree of type $\{L^\omega_n, k_n\}_{n=1}^\infty$ for $L^\omega_n = [\gamma^n] + \omega_n$ and $k_n \equiv k$. Clearly, Proposition 4.1 holds for any $\Gamma^\omega_{k,\gamma}$. The main result of this section is

Theorem 4.4. *For $\mathbb{P}$-a.e. ω, all the spectral measures for Δ on $\Gamma^\omega_{k,\gamma}$ are singular with respect to Lebesgue measure. Furthermore, let $V(k) = (1+k)^2/(4k)$ and let*

$$I = \left(-\sqrt{\frac{8(\gamma - V(k))}{2\gamma - 1}}, \sqrt{\frac{8(\gamma - V(k))}{2\gamma - 1}}\right) \tag{4.2}$$

if $\gamma > V$, and $I = \emptyset$ otherwise. Then for $\mathbb{P}$-a.e. ω and for Lebesgue a.e. $\varrho \in (-\pi/2, \pi/2)$, the spectral measure, μ_{δ_O}, associated with Δ_ϱ on $\Gamma^\omega_{k,\gamma}$, and with the delta function at the root, is purely singular continuous in I with exact local dimension

$$1 - \frac{\log\big((4V(k) - E^2/2)/(4 - E^2)\big)}{\log(\gamma)} \tag{4.3}$$

and it is dense pure point in the rest of $[-2, 2]$.

Corollary 4.5. *Assume $\gamma \geq 4$. Then if the dimension of $\Gamma^{\omega}_{k,\gamma}$ is at least 3, we have that $I = \emptyset$ and so, for $\mathbb{P}$-a.e. ω and for Lebesgue a.e. $\varrho \in (-\pi/2, \pi/2)$, the spectral measure μ_{δ_O} is dense pure point in $[-2, 2]$.*

PROOF OF THE COROLLARY. This is a simple computation. □

PROOF OF THEOREM 4.4. Theorem 2.3 and Lemma 3.2 imply that the theorem is an immediate consequence of Proposition 4.6 below. □

Proposition 4.6. *Fix $k \geq 2$ and $\gamma > 2$. For any $\omega \in \Omega$ and $-\pi/2 < \varrho < \pi/2$, let $J^{\omega}_{\varrho} = J(\{a^{\omega}(j)\}, \{b_{\varrho}(j)\})$ be a Jacobi matrix with*

$$a^{\omega}(j) = \begin{cases} \sqrt{k} & j = [\gamma^m] + \omega_m + 1 \text{ for some } m \\ 1 & \text{otherwise} \end{cases} \tag{4.4}$$

and

$$b_{\varrho}(j) = \begin{cases} -\tan(\varrho) & j = 1 \\ 0 & \text{otherwise.} \end{cases} \tag{4.5}$$

Then, for $\mathbb{P}$-a.e. ω, and for any $\varrho \in (-\pi/2, \pi/2)$, the spectral measure μ_{ϱ}, associated with the vector $\delta_1 \in \mathbb{Z}^+$ and with the Jacobi matrix J^{ω}_{ϱ}, is singular with respect to Lebesgue measure. Furthermore, for $\mathbb{P}$-a.e. ω and for Lebesgue a.e. $\varrho \in (-\pi/2, \pi/2)$, the spectral measure μ_{ϱ} is purely singular continuous in I with exact local dimension given by (4.3), where I is as defined in Theorem 4.4, and it is dense pure point in the rest of $[-2, 2]$.

A central role in the proof of the proposition will be played by the EFGP transform introduced in the proof of Theorem 3.8:

Fix $k \geq 2$ and let $J = J(\{a(j)\}, \{b(j)\})$ be a Jacobi matrix satisfying

$$a(j) = \begin{cases} \sqrt{k} & j = L_n + 1 \text{ for some } n \\ 1 & \text{otherwise} \end{cases} \tag{4.6}$$

for a sequence $\{L_n\}$ satisfying $L_{n+1} - L_n \geq 2$, and

$$b(j) \equiv 0. \tag{4.7}$$

For any $E \in (-2, 2)$, let $\phi \in (0, \pi)$ be defined by $2\cos(\phi) = E$, and let $u_{2\cos(\phi)} \equiv u_E$ solve (B.4) for J, namely

$$a(j)u(j+1) + a(j-1)u(j-1) = Eu(j), \quad j \geq 1 \tag{4.8}$$

with $a(0) = 1$. Recall that the EFGP variables [10] corresponding to u, $r_{\phi}(j)$ and $\theta_{\phi}(j)$, are defined through:

$$r_{\phi}(j)\cos(\theta_{\phi}(j)) = u_E(j) - \cos(\phi)u_E(j-1) \tag{4.9}$$

$$r_{\phi}(j)\sin(\theta_{\phi}(j)) = \sin(\phi)u_E(j-1). \tag{4.10}$$

First, note that there are positive constants, $C_1(\phi)$, $C_2(\phi)$, such that

$$C_1(\phi)(|u(j-1)|^2 + |u(j)|^2) \leq r_{\phi}(j)^2 \leq C_2(\phi)(|u(j-1)|^2 + |u(j)|^2). \tag{4.11}$$

We call $r(j)$ the *EFGP norm* of $u(j)$. Note, also, that for any j

$$C_1(\phi)\Big(\sum_{i=(j-1)}^{(j+2)} |u(i)|^2\Big) \leq r_{\phi}(j)^2 + r_{\phi}(j+2)^2 \leq C_2(\phi)\Big(\sum_{i=(j-1)}^{(j+2)} |u(i)|^2\Big). \tag{4.12}$$

For a function $f\colon \mathbb{Z}^+ \to \mathbb{C}$ and a sequence $\mathcal{L} = \{L_n\}_{n=1}^\infty$ of natural numbers, define

$$\|f\|_{L,\mathcal{L}} = \left(\sum_{\substack{1 \le j \le L \\ j \neq L_n + 2 \text{ for any } n}} |f(j)|^2 \right)^{1/2}$$

for any $L \in \mathbb{N}$, and extend to $L \in \mathbb{R}$ by linear interpolation. Then we have

Lemma 4.7. *Let $J = J(\{a(j)\}, \{b(j)\})$ be as defined in (4.6) – (4.7) and let $\mathcal{L} = \{L_n\}$. Assume that for some $1 \le \beta < 2$ and every E in some Borel set $A \subseteq (-2,2)$, the EFGP norm, r_u, of every solution u of (4.8) obeys*

$$\limsup_{L\to\infty} \frac{\|r_u\|^2_{L,\mathcal{L}}}{L^\beta} < \infty. \tag{4.13}$$

Then $\mu(A \cap \cdot)$ is continuous with respect to $(2-\beta)$-dimensional Hausdorff measure, where μ is the spectral measure associated with δ_1 and J.

Lemma 4.8. *Let $J = J(\{a(j)\}, \{b(j)\})$ be as defined in (4.6) – (4.7) and let $\mathcal{L} = \{L_n\}$. Let $u_{1,E}$ be the solution of (4.8) that satisfies*

$$u(0) = 0, \quad u(1) = 1,$$

and let $r_{1,E}$ denote the corresponding EFGP norm. If

$$\liminf_{L\to\infty} \frac{\|r_{1,E}\|^2_{L,\mathcal{L}}}{L^\alpha} = 0 \tag{4.14}$$

for every E in some Borel set $A \subseteq (-2,2)$, then $\mu(A \cap \cdot)$ is singular with respect to α-dimensional Hausdorff measure, where μ is the spectral measure associated with δ_1 and J.

Proof of Lemmas 4.7 and 4.8. The proofs follow immediately from (4.12), from the fact that $L_{n+1} - L_n \ge 2$, and from Propositions B.1 and B.2 respectively. □

Our aim, therefore, is to control the growth of the EFGP norms of generalized eigenfunctions. We recall from the previous section that for $L_n + 3 < j \le L_{n+1} + 1$, $r_\phi(j) = r_\phi(j-1)$ and $\theta_\phi(j) = \theta_\phi(j-1) + \phi$ and that, therefore,

$$r_\phi(L_n + 1) = r_\phi(L_{n-1} + 3) \tag{4.15}$$

and

$$\theta_\phi(L_n + 1) = \theta_\phi(L_{n-1} + 3) + (L_n - L_{n-1} - 2)\phi. \tag{4.16}$$

From Lemmas 4.7 and 4.8, we see that $r_\phi(L_n + 2)$ is irrelevant so all we need is to find the relation between $r_\phi(L_n + 1)$ and $r_\phi(L_n + 3)$. As before, using the relations

$$u_{2\cos(\phi)}(j-1) = \frac{\sin\big(\theta_\phi(j)\big)}{\sin(\phi)} r_\phi(j),$$

$$u_{2\cos(\phi)}(j) = \frac{\sin\big(\phi + \theta_\phi(j)\big)}{\sin(\phi)} r_\phi(j),$$

and the transfer matrix

$$\begin{aligned} S'_n\big(2\cos(\phi)\big) &\equiv S_{L_n+2}\big(2\cos(\phi)\big) S_{L_n+1}\big(2\cos(\phi)\big) \\ &= \begin{pmatrix} (4\cos^2(\phi)/\sqrt{k} - \sqrt{k}) & -2\cos(\phi)/\sqrt{k} \\ 2\cos(\phi)/\sqrt{k} & -1/\sqrt{k} \end{pmatrix}, \end{aligned}$$

we get that

$$\begin{aligned}(4.17)\quad & r_\phi(L_n+3)^2 \\ &= r_\phi(L_n+1)^2\cdot\Big(A(\phi)+B(\phi)\cos\big(2\theta_\phi(L_n+1)\big)+C(\phi)\sin\big(2\theta_\phi(L_n+1)\big)\Big),\end{aligned}$$

where

$$A(\phi)=\frac{1}{\sin^2(\phi)}\left(\frac{1+k^2}{2k}-\cos^2(\phi)\right)\geq 0 \tag{4.18}$$

and

$$A(\phi)^2-B(\phi)^2-C(\phi)^2=1. \tag{4.19}$$

Explicit formulas for $B(\phi)$ and $C(\phi)$ can be derived but they are of no consequence. The last relation follows from the fact that

$$\det\Big(S_n'\big(2\cos(\phi)\big)\Big)=1.$$

PROOF OF PROPOSITION 4.6. For a fixed $\phi\in(0,\pi)$ (and $E=2\cos(\phi)\in(-2,2)$), let

$$f(\theta)=\tfrac{1}{2}\log\big(A(\phi)+B(\phi)\cos(2\theta)+C(\phi)\sin(2\theta)\big). \tag{4.20}$$

Now, for a given $\omega\in\Omega$, let $\mathcal{L}^\omega=\{L_n^\omega\}_{n=1}^\infty$ (recall $L_n^\omega=[\gamma^n]+\omega_n$). Let $u_{1,E}^\omega$ be a generalized eigenfunction for J_0^ω such that $u_{1,E}^\omega(0)=0$ and $u_{1,E}^\omega(1)=1$, and let $r_{1,\phi}^\omega$ be the corresponding EFGP norm. Then, the above implies that

$$\begin{aligned}(4.21)\quad Y_n(\omega) &\equiv \log r_{1,\phi}^\omega(L_n^\omega+3)-\log r_{1,\phi}^\omega(L_{n-1}^\omega+3)\\ &= \tfrac{1}{2}\log\Big(A(\phi)+B(\phi)\cos\big(2\theta_\phi(L_n^\omega+1)\big)+C(\phi)\sin\big(2\theta_\phi(L_n^\omega+1)\big)\Big).\end{aligned}$$

Note that, if $\omega,\eta\in\Omega$ are such that $\omega_j=\eta_j$ for $j=1,2,\dots,n-1$, and $\eta_n=\omega_n+l$ for some l, and $Y_n(\omega)=f(\theta)$ for some θ, then $Y_n(\eta)=f(\theta+l\phi)$. Let

$$\begin{aligned}(4.22)\quad Z=\frac{1}{2\pi}\int_0^{2\pi}f(\theta)\,d\theta &= \frac{1}{2}\log\frac{A(\phi)+\sqrt{A(\phi)^2-B(\phi)^2-C(\phi)^2}}{2}\\ &= \frac{1}{2}\log\frac{A(\phi)+1}{2}=\frac{1}{2}\log\left(\frac{V(k)}{\sin^2(\phi)}-\frac{\cot^2(\phi)}{2}\right),\end{aligned}$$

and

$$\tilde f(\theta)=f(\theta)-Z. \tag{4.23}$$

Then $\tilde f(\theta)=\tilde f(\theta+2\pi)$, $\int_0^{2\pi}\tilde f(\theta)\,d\theta=0$ and $\tilde f$ is C^4, so $\tilde f$ satisfies the conditions of [21, Lemma 6.2]. Thus, it follows that, for any ϕ in a set of full Lebesgue measure $S\subseteq(0,\pi)$, and for $\mathbb{P}$-a.e. ω,

$$\frac{\sum_{n=1}^N(Y_n(\omega)-Z)}{N}\to 0,$$

(see [21, Section 6]). From this we get that for any $\phi\in S$ and for a.e. $\omega\in\Omega$, the EFGP norm $r_{1,\phi}^\omega$ satisfies

$$(L_n^\omega+3)^{d_1}\leq r_{1,\phi}^\omega(L_n^\omega+3)\leq(L_n^\omega+3)^{d_2} \tag{4.24}$$

for any

$$d_1<\frac{Z}{\log\gamma}<d_2$$

provided n is large enough. By Fubini's theorem, we get that (4.24) holds for a.e. ω and Lebesgue almost every ϕ. Furthermore, Lemma 4.9 below assures us that for each such (ω, ϕ), there exists a subordinate solution $u^{\omega}_{\mathrm{sub},\phi}$ to (4.8) whose EFGP norm $r^{\omega}_{\mathrm{sub},\phi}$ satisfies

$$r^{\omega}_{\mathrm{sub},\phi}(L^{\omega}_n + 3) \leq (L^{\omega}_n + 3)^{-d_1} \tag{4.25}$$

for large enough n. Since the absolutely continuous part of μ_ϱ (for any ϱ) is supported on the set of energies with no subordinate solution, it follows that, for any ϱ, μ_ϱ is purely singular on $(-2, 2)$. Furthermore, since the singular part of μ_ϱ is supported on the set of energies where the subordinate solutions obey the appropriate boundary conditions, rank-one perturbation arguments (see, e.g., [17]), imply that for Lebesgue almost any ϱ, μ_ϱ is supported on the set of energies where the subordinate solutions behave as in (4.25). Noting that, in this case,

$$\|r^{\omega}_{\mathrm{sub},\phi}\|^2_{L,\mathcal{L}^{\omega}} \leq CL^{1-2Z/\log\gamma}$$

if $Z/\log\gamma \leq \frac{1}{2}$ ($\Leftrightarrow E \in I$) and is square-summable otherwise, and that the EFGP norm r^{ω}_{ϕ} of any other solution to (4.8) satisfies

$$\|r^{\omega}_{\phi}\|^2_{L,\mathcal{L}^{\omega}} \leq CL^{1+2Z/\log\gamma},$$

we finish the proof with the help of Lemmas 4.7 and 4.8. □

The following lemma was used in the proof above. It is essentially the same as [21, Lemma 2.1], the only difference being in that not all transfer matrices for the general Jacobi case are unimodular. Since not all transfer matrices are relevant for the argument, this is inconsequential. In the proof below we only demonstrate this point. For the complete proof the reader is referred to [21].

Lemma 4.9. *Let $J = J(\{a(j)\}, \{b(j)\})$ be a Jacobi matrix satisfying*

$$a(j) = \begin{cases} \sqrt{k} & j = L_n + 1 \text{ for some } n \\ 1 & \text{otherwise} \end{cases} \tag{4.26}$$

for a sequence $\{L_n\}$ satisfying $L_{n+1} - L_n \geq 2$, and

$$b(j) \equiv 0. \tag{4.27}$$

Assume that for some $E \in (-2, 2)$, u is a solution of (4.8) whose EFGP norm satisfies

$$r(L_n + 3) = \mathrm{e}^{\sigma_n} \tag{4.28}$$

where $\sigma_n = \sum_{j=1}^{n}(Z_j + X_j)$ with $0 < d_1 \leq Z_j \leq d_2 < \infty$ and $\sum_{j=1}^{n} X_j = o(n)$. Then there exists a subordinate solution v of (4.8) for E, such that for any $d < d_1$ and for all sufficiently large n, the corresponding EFGP norm — p — satisfies

$$p(L_n + 3) \leq \mathrm{e}^{-dn}. \tag{4.29}$$

Proof. Let v be any solution of (4.8) different from u and let p be its EFGP norm. Since the transfer matrices $T_{L_n+2}(E)$ are unimodular, the argument of [10, Theorem 2.3] applies to show that there exist E-dependent constants, c_1, c_2 such that

$$c_1 \max(p_{L_n+3}, r_{L_n+3}) \leq \|T_{L_n+2}(E)\| \leq c_2 \max(p_{L_n+3}, r_{L_n+3}). \tag{4.30}$$

Note, further, that there exists a constant $B > 0$ such that

$$\|T_{n+1,n}(E)\| \equiv\| T_{L_{n+1}+2,L_n+2}(E) \|\leq B \tag{4.31}$$

and that $\det(T_{n+1,n}(E)) = 1$ as well. Thus, it follows that

$$\sum_{n=1}^{\infty} \frac{\|T_{n,n-1}(E)\|^2}{\|T_{L_n+2}(E)\|^2} < \infty \tag{4.32}$$

so one can apply [12, Theorem 8.1] to get a vector $\bar{v} \in \mathbb{R}^2$ such that

$$\frac{\|T_{L_n+2}(E)\bar{v}\|}{\|T_{L_n+2}(E)\bar{u}\|} \longrightarrow 0$$

for any other vector $\bar{u} \in \mathbb{R}^2$. From this point, the proof follows the proof of [21, Lemma 2.1], word for word, to show that $\bar{v}$ generates the claimed solution. □

Appendix A. Self-adjointness of the Laplacian on normal SH rooted trees

Proposition A.1. *Let Γ be a rooted SH tree of type $\{L_n, k_n\}_{n=1}^{\infty}$. Then the operator Δ_1 defined over*

$$D(\Delta_1) = \{u \in \ell^2(\Gamma) \mid u \text{ is of compact support}\} \tag{A.1}$$

via the equation

$$(\Delta_1 u)(x) = \sum_{y:d(x,y)=1} u(y), \tag{A.2}$$

is symmetric. If $\limsup_{n\to\infty}(L_{n+1} - L_n) > 1$ *then Δ — the closure of Δ_1 — is self-adjoint. The same statement holds for $\tilde{\Delta}_1$ and $\tilde{\Delta}$ (defined over the same domain), with equation* (A.2) *replaced by*

$$(\tilde{\Delta}_1 u)(x) = \sum_{y:d(x,y)=1} u(y) - \#\{y : d(x,y) = 1\} \cdot u(x). \tag{A.3}$$

PROOF. Since the proof for Δ and $\tilde{\Delta}$ is precisely the same, we use Δ.

It is trivial to see that Δ_1 is symmetric, so in order to show that Δ is self-adjoint, all we have to show is that $\ker(\Delta^* \pm \mathrm{i}) = \{0\}$.

Assume that $\Delta u = \mathrm{i}u$, then it follows that $\Delta\bar{u} = -\mathrm{i}\bar{u}$ (where $\bar{\ }$ for a complex number denotes complex conjugation). Let n_j be a subsequence for which $L_{n_j}+1 < L_{n_j+1}$. For a vertex v with $|v| \equiv d(v,O) = L_{n_j}+1$, let us denote its *unique* forward neighbor by $\hat{v}$. One can verify that an analogue of Green's formula (see, e.g., [4, Chapter VII, formula 1.4]) holds and we have:

$$2i \sum_{|v|\leq L_{n_j}+1} |u(v)|^2 = \sum_{|v|=L_{n_j}+1} u(\hat{v}) \cdot \overline{u(v)} - \sum_{|v|=L_{n_j}+1} \overline{u(\hat{v})} \cdot u(v)$$

so

$$\sum_{|v|\leq L_{n_j}+1} |u(v)|^2 \leq \sum_{|v|=L_{n_j}+1} |u(\hat{v})| \cdot |u(v)|.$$

From this it follows that if $u \neq 0$, then the RHS above does not converge to zero and therefore $u \notin \ell^2(\Gamma)$. This proves the proposition. □

Appendix B. Eigenfunctions and transfer matrices for Jacobi matrices

Let $J = J(\{a(j)\}, \{b(j)\})$ be a Jacobi matrix with $b(j) \in \mathbb{R}$ and $a(j) > 0$ satisfying $\sum_{j=1}^{\infty} 1/a(j) = \infty$ (which suffices for J to be self-adjoint [4]). A basic idea in the spectral theory of Jacobi matrices is to relate spectral properties of J as reflected by $\mu = \mu_{\delta_1}$ — the spectral measure of δ_1 — to properties of formal eigenfunctions

(B.1) $$Ju = Eu.$$

By this term we mean functions $u\colon \mathbb{Z}^+ \to \mathbb{C}$ which satisfy

(B.2) $$a(j)u(j+1) + a(j-1)u(j-1) + b(j)u(j) = Eu(j), \quad j > 1$$

(B.3) $$a(1)u(2) + b(1)u(1) = Eu(1).$$

Since, for a given $E \in \mathbb{R}$, all solutions to (B.2) – (B.3) are linearly dependent (determined by $u(1)$), it suffices to study $u_{1,E}$ which is the solution satisfying $u_{1,E}(1) = 1$. It is convenient to define $a(0) = 1$ and to extend (B.2) to $j = 1$ by demanding $u_{1,E}(0) = 0$. Thus $u_{1,E}$ is the unique solution to

(B.4) $$a(j)u(j+1) + a(j-1)u(j-1) + b(j)u(j) = Eu(j), \quad j \geq 1$$

with

(B.5) $$u_{1,E}(0) = 0, \quad u_{1,E}(1) = 1.$$

We further define $u_{2,E}$ as the unique solution to (B.4) satisfying

(B.6) $$u_{2,E}(0) = 1, \quad u_{2,E}(1) = 0.$$

Note that any solution, u, to (B.4) with $u(1) \neq 0$ can be viewed as $u_{1,E}$ for a slightly modified Jacobi matrix. Namely, u solves (B.2) – (B.3) for the same set of parameters except with $b(1)$ changed to $\big(b(1) + u(0)/u(1)\big)$. This remark is basic for the analysis of Section 4. We say that $u_{1,E}$ is subordinate if

(B.7) $$\lim_{L\to\infty} \frac{\|u_{1,E}\|_L}{\|u_{2,E}\|_L} = 0$$

where

$$\|f\|_L = \left(\sum_{j=1}^{[L]} |f(j)|^2 + (L - [L])|f([L]+1)|^2 \right)^{1/2}.$$

The Gilbert – Pearson theory of subordinacy [7] says that the singular part of μ is supported on the set of energies where $u_{1,E}$ is subordinate, and that the absolutely continuous part of μ is supported off this set. The Jitomirskaya – Last extension of this theory [9] analyzes further the singular part of μ according to its singularity/continuity with respect to dimensional Hausdorff measures (see [9] for the concept of Hausdorff measures and dimensions). In Section 4 of the paper we use the following results from [9]:

Proposition B.1 ([9]). *Assume that $1 \leq a(j) \leq M$ for some $M > 1$. Assume that for some $1 \leq \beta < 2$ and every E in some Borel set A, every solution u of* (B.4) *obeys*

(B.8) $$\limsup_{L\to\infty} \frac{\|u\|_L^2}{L^\beta} < \infty.$$

Then $\mu(A \cap \cdot)$ is continuous with respect to $(2-\beta)$-dimensional Hausdorff measure.

Proposition B.2 ([9]). *Assume that* $1 \le a(j) \le M$ *for some* $M > 1$. *If*

$$\liminf_{L\to\infty} \frac{\|u_{1,E}\|_L^2}{L^\alpha} = 0 \tag{B.9}$$

for every E *in some Borel set* A, *then* $\mu(A \cap \cdot)$ *is singular with respect to* α-*dimensional Hausdorff measure.*

The next results we quote relate the properties of μ to the properties of the transfer matrices corresponding to J. These are the 2×2 matrices

$$T_j(E) = S_j(E)S_{j-1}(E)\cdots S_1(E), \tag{B.10}$$

where

$$S_j(E) = \begin{pmatrix} (E - b(j))/a(j) & -a(j-1)/a(j) \\ 1 & 0 \end{pmatrix}. \tag{B.11}$$

It isn't hard to see that

$$T_j(E) = \begin{pmatrix} u_{1,E}(j+1) & u_{2,E}(j+1) \\ u_{1,E}(j) & u_{2,E}(j) \end{pmatrix} \tag{B.12}$$

so it is not surprising to find that the behavior of $T_j(E)$ is related to the behavior of the eigenfunctions. For any j_1, j_2, we use the shorthand

$$T_{j_1,j_2}(E) \equiv T_{j_1}(E)T_{j_2}(E)^{-1}. \tag{B.13}$$

The following is a generalization of [12, Theorem 1.2] relating the behavior of $T_j(E)$ with the existence of absolutely continuous spectrum.

Proposition B.3. *Let* m_j, l_j *be arbitrary sequences of natural numbers and let*

$$A_1 = \left\{ E \;\middle|\; \liminf_{j\to\infty} \frac{1}{a_{l_j}} \|T_{m_j,l_j}(E)\| < \infty \right\}. \tag{B.14}$$

Then A_1 *supports the a.c. part of* μ *in that* $\mu_{\mathrm{ac}}(\mathbb{R} \setminus A_1) = 0$.

Proof. Note that $\det(T_{l_j}(E)) = 1/a_{l_j}$ so that

$$\|T_{l_j}(E)^{-1}\| = a_{l_j}\|T_{l_j}(E)\|.$$

Thus, we have that

$$\begin{aligned} \frac{1}{a_{l_j}}\|T_{m_j}(E)T_{l_j}(E)^{-1}\| &\le \frac{1}{a_{l_j}}\|T_{m_j}(E)\|\,\|T_{l_j}(E)^{-1}\| \\ &= \|T_{m_j}(E)\|\,\|T_{l_j}(E)\|. \end{aligned}$$

From here one proceeds exactly as in the proof of [12, Theorem 3.4D] to get the conclusion: We know that $u_{1,E}(j)$, viewed as functions of E, are orthonormal polynomials with respect to $d\mu$, and that $u_{2,E}(j)$ are orthonormal polynomials with respect to another measure $d\tilde{\mu}$, such that the measure $d\nu = \min(d\mu, d\tilde{\mu})$ is purely absolutely continuous and equivalent to the absolutely continuous part of $d\mu$ (see [12]). Thus, from the characterization (B.12), we get that

$$\int_{\mathbb{R}} \|T_j(E)\|^2 \, \mathrm{d}\nu(E) \le 4.$$

The theorem now follows from an application of the Cauchy–Schwarz inequality and Fatou's lemma. □

In order to show absence of eigenvalues, the following idea of Simon–Stolz [19] is useful:

Proposition B.4 (See [19]). *For a given $E \in \mathbb{R}$, if*

$$\sum_{j,\det(T_j(E))=1} \|T_j(E)\|^{-2} = \infty, \tag{B.15}$$

then there can be no eigenvalue at E.

PROOF. The proposition follows from the observation that if

$$\det(T_m(E)) = 1$$

and u solves (B.2) then

$$|u(m+1)|^2 + |u(m)|^2 \geq \frac{|u(1)|^2 + |u(0)|^2}{\|T_m(E)\|^2}.$$

□

References

1. M. Aizenman, R. Sims, and S. Warzel, *Stability of the absolutely continuous spectrum of random Schrödinger operators on tree graphs*, Probab. Theory Related Fields **136** (2006), no. 3, 363–394.
2. C. Allard and R. Froese, *A Mourre estimate for a Schrödinger operator on a binary tree*, Rev. Math. Phys. **12** (2000), no. 12, 1655–1667.
3. H. Bass, M. V. Otero-Espinar, D. Rockmore, and C. Tresser, *Cyclic renormalization and automorphism groups of rooted trees*, Lecture Notes in Math., vol. 1621, Springer, Berlin, 1996.
4. Ju. M. Berezanskii, *Expansions in eigenfunctions of selfadjoint operators*, Transl. Math. Monogr., vol. 17, Amer. Math. Soc., Providence, RI, 1968.
5. J. Breuer, *Singular continuous spectrum for the Laplacian on certain sparse trees*, Comm. Math. Phys. **269** (2007), no. 3, 851–857.
6. R. Carmona and J. Lacroix, *Spectral theory of random Schrödinger operators*, Probab. Appl., Birkhäuser, Boston, MA, 1990.
7. D. J. Gilbert and D. B. Pearson, *On subordinacy and analysis of the spectrum of one-dimensional Schrödinger operators*, J. Math. Anal. Appl. **128** (1987), no. 1, 30–56.
8. V. Georgescu and S. Golénia, *Isometries, Fock spaces and spectral analysis of Schrödinger operators on trees*, J. Funct. Anal. **227** (2005), no. 2, 389–429.
9. S. Ya. Jitomirskaya and Y. Last, *Power-law subordinacy and singular spectra.* I: *Half-line operators*, Acta Math. **183** (1999), no. 2, 171–189.
10. A. Kiselev, Y. Last, and B. Simon, *Modified Prüfer and EFGP transforms and the spectral analysis of one-dimensional Schrödinger operators*, Comm. Math. Phys. **194** (1998), no. 1, 1–45.
11. Y. Last, *Spectral theory of Sturm–Liouville operators on infinite intervals: A review of recent developments*, Sturm–Liouville Theory: Past and Present (W. O. Amrein, A. M. Hinz, and D. B. Pearson, eds.), Birkhäuser, Basel, 2005, pp. 99–120.
12. Y. Last and B. Simon, *Eigenfunctions, transfer matrices, and absolutely continuous spectrum of one-dimensional Schrödinger operators*, Invent. Math. **135** (1999), no. 2, 329–367.
13. L. Malozemov and A. Teplyaev, *Pure point spectrum of the Laplacians on fractal graphs*, J. Funct. Anal. **129** (1995), no. 2, 390–405.
14. D. B. Pearson, *Singular continuous measures in scattering theory*, Comm. Math. Phys. **60** (1978), no. 1, 13–36.
15. M. Reed and B. Simon, *Methods of modern mathematical physics.* I: *Functional analysis*, 2nd ed., Academic Press, New York, 1980.
16. R. V. Romanov and G. E. Rudin, *Scattering on the Bruhat–Tits tree.* I, Phys. Lett. A **198** (1995), no. 2, 113–118.

17. B. Simon, *Spectral analysis of rank one perturbations and applications*, Mathematical Quantum Theory. II: Schrödinger Operators (J. Feldman, R. Froese, and L. Rosen, eds.), CRM Proc. Lecture Notes, vol. 8, Amer. Math. Soc., Providence, RI, 1995, pp. 109–149.
18. ———, *Operators with singular continuous spectrum.* VI: *Graph Laplacians and Laplace–Beltrami operators*, Proc. Amer. Math. Soc. **124** (1996), no. 4, 1177–1182.
19. B. Simon and G. Stolz, *Operators with singular continuous spectrum.* V: *Sparse potentials*, Proc. Amer. Math. Soc. **124** (1996), no. 7, 2073–2080.
20. M. Solomyak, *On the spectrum of the Laplacian on regular metric trees*, Waves Random Media **14** (2004), no. 1, S155–S171.
21. A. Zlatoš, *Sparse potentials with fractional Hausdorff dimensions*, J. Funct. Anal. **207** (2004), no. 1, 216–252.

Institute of Mathematics, The Hebrew University of Jerusalem, Jerusalem, 91904, Israel

E-mail address: jbreuer@math.huji.ac.il

Centre de Recherches Mathématiques
CRM Proceedings and Lecture Notes
Volume **42**, 2007

Some New Estimates on the Spectral Shift Function Associated with Random Schrödinger Operators

Jean-Michel Combes, Peter D. Hislop, and Frédéric Klopp

ABSTRACT. We prove some new pointwise-in-energy bounds on the expectations of various spectral shift functions (SSF) associated with random Schrödinger operators in the continuum having Anderson-type random potentials in both finite-volume and infinite-volume. These estimates are a consequence of our new Wegner estimate for finite-volume random Schrödinger operators [5]. For lattice models, we also obtain a representation of the infinite-volume density of states in terms of the expectation of a SSF for a single-site perturbation. For continuum models, the corresponding measure, whose density is given by this SSF, is absolutely continuous with respect to the density of states and agrees with it in certain cases. As an application of one-parameter spectral averaging, we give a short proof of the classical pointwise upper bound on the SSF for finite-rank perturbations.

1. Introduction: The Wegner estimate and the spectral shift function

Some recent analyses of random Schrödinger operators have involved three related concepts: the Wegner estimate for the finite-volume Hamiltonians, the spectral shift function (SSF), and the integrated density of states (IDS). In this note, we prove some new pointwise bounds on the expectation of some SSFs that occur in the theory of random Schrödinger operators in the continuum. These bounds result from an improved version of the Wegner estimate [5]. In earlier work [4, 8], we used L^p-bounds on the SSF in order to obtain better estimates on the IDS. In our most recent work, we obtain an optimal Wegner estimate directly without using the SSF and found, as a consequence, new pointwise bounds on the expectation of the SSF. It has often been conjectured that in the case of ergodic, random, Schrödinger operators of the form considered here the SSF for a local single-site perturbation should be in $L^\infty_{\text{loc}}(\mathbb{R})$ once it is averaged over the random variables on which the disordered potential depends. We prove this in this note. We mention that these types of bounds on the SSF also play a motivating role in the fractional moment method for proving localization in the continuum [1]. For lattice models,

2000 *Mathematics Subject Classification.* Primary: 35P25, 81Q10; Secondary: 47A40, 81U20.

The work of the second author is supported in part by NSF grant DMS-0503784.

This is the final form of the paper.

the pointwise bounds on the SSF are a simple consequence of the fact that the corresponding perturbations are finite-rank (cf. [2,22] and Section 4.3).

We first recall a special case of the Wegner estimate proved in [5] that will be used for the bounds on the SSF. We refer to [5] for the general statement, valid for arbitrary bounded processes on $\mathbb{Z}^d$, and the proofs. The family of Schrödinger operators $H_\omega = H_0 + V_\omega$, on $L^2(\mathbb{R}^d)$, is constructed from a deterministic, $\mathbb{Z}^d$-periodic, background operator $H_0 = (-i\nabla - A_0)^2 + V_0$. We consider an Anderson-type random potential V_ω constructed from the single-site potential u as

$$V_\omega(x) = \sum_{j\in\mathbb{Z}^d} \omega_j u(x-j). \tag{1.1}$$

The family of random variables is assumed to be independent, and identically distributed (iid). The results are independent of the disorder provided it is nonzero.

We define local versions of the Hamiltonians and potentials associated with bounded regions in $\mathbb{R}^d$. By $\Lambda_L(x)$, we mean the open cube of side length L centered at $x \in \mathbb{R}^d$. For $\Lambda \subset \mathbb{R}^d$, we denote the lattice points in Λ by $\tilde{\Lambda} = \Lambda \cap \mathbb{Z}^d$. For a cube Λ, we take H_0^Λ and H_Λ (omitting the index ω) to be the restrictions of H_0 and H_ω, respectively, to the cube Λ, with periodic boundary conditions on the boundary $\partial\Lambda$ of Λ. We denote by $E_0^\Lambda(\cdot)$ and $E_\Lambda(\cdot)$ the spectral families for H_0^Λ and H_Λ, respectively. Furthermore, for $\Lambda \subset \mathbb{R}^d$, let χ_Λ be the characteristic function for Λ. The local potential V_Λ is defined by

$$V_\Lambda(x) = V_\omega(x)\chi_\Lambda(x), \tag{1.2}$$

and we assume this can be written as

$$V_\Lambda(x) = \sum_{j\in\tilde{\Lambda}} \omega_j u(x-j). \tag{1.3}$$

For example, if the support of u is contained in a single unit cube, the formula (1.3) holds. We refer to the discussion in [4] when the support of u is compact, but not necessarily contained inside one cube. In this case, V_Λ can be written as in (1.3) plus a boundary term of order $|\partial\Lambda|$ and hence it does not contribute to the large $|\Lambda|$ limit. Hence, we may assume (1.3) without any loss of generality. We will also use the local potential obtained from (1.3) by setting all the random variables to one, that is,

$$\widetilde{V}_\Lambda(x) = \sum_{j\in\tilde{\Lambda}} u_j(x), \tag{1.4}$$

where we will write $u_j(x) = u(x-j)$.

We will always make the following four assumptions:

(H1) The background operator $H_0 = (-i\nabla - A_0)^2 + V_0$ is a lower semi-bounded, $\mathbb{Z}^d$-periodic Schrödinger operator with a real-valued, $\mathbb{Z}^d$-periodic, potential V_0, and a $\mathbb{Z}^d$-periodic vector potential A_0. We assume that V_0 and A_0 are sufficiently regular so that H_0 is essentially self-adjoint on $C_0^\infty(\mathbb{R}^d)$.

(H2) The periodic operator H_0 has the unique continuation property, that is, for any $E \in \mathbb{R}$ and for any function $\phi \in H^2_{\mathrm{loc}}(\mathbb{R}^d)$, if $(H_0 - E)\phi = 0$, and if ϕ vanishes on an open set, then $\phi \equiv 0$.

(H3) The nonzero, non negative, compactly supported, bounded single-site potential $u \in L_0^\infty(\mathbb{R}^d)$, and it is strictly positive on a nonempty open set.

(H4) The random coupling constants $\{\omega_j \mid j \in \mathbb{Z}^d\}$, are independent and identically distributed. The probability distribution μ_0 of ω_0 is compactly supported with a bounded density $h_0 \in L_0^\infty(\mathbb{R})$.

These imply that the infinite-volume random Schrödinger operator H_ω is ergodic with respect to the group of $\mathbb{Z}^d$-translations.

Our results also apply to the randomly perturbed Landau Hamiltonian $H_\omega(\lambda) = H_L(B) + \lambda V_\omega$, for $\lambda \neq 0$, where V_ω is an Anderson-type potential as in (1.1). The Landau Hamiltonian $H_L(B)$ on $L^2(\mathbb{R}^2)$ is given by

$$H_L(B) = (-\mathrm{i}\nabla - A_0)^2, \quad \text{with } A_0(x,y) = \frac{B}{2}(-y,x). \tag{1.5}$$

The constant $B \neq 0$ is the magnetic field strength.

Under these assumptions, the Wegner estimate necessary for our purposes has the following form.

Theorem 1.1. *We assume that the family of random Schrödinger operators $H_\omega = H_0 + V_\omega$ on $L^2(\mathbb{R}^d)$ satisfies hypotheses* (H1) – (H4). *Then, there exists a locally uniform constant $C_W > 0$ such that for any $E_0 \in \mathbb{R}$, and $\epsilon \in (0,1]$, the local Hamiltonians H_Λ satisfy the following Wegner estimate*

$$\mathbb{P}\{\mathrm{dist}(\sigma(H_\Lambda), E_0) < \epsilon\} \leq \mathbb{E}\{\mathrm{Tr}\, E_\Lambda([E_0 - \epsilon, E_0 + \epsilon])\} \leq C_W \epsilon |\Lambda|. \tag{1.6}$$

A similar estimate holds for randomly perturbed Landau Hamiltonians.

This theorem immediately implies the Lipschitz continuity of the integrated density of states [5]. As a consequence, the density of states (DOS) exists and is a locally bounded function. In this note, we use Theorem 1.1 to prove new pointwise bounds on the expectation of the SSF for both finite-volume and infinite-volume random Schrödinger operators. We comment on the relation of these results to various results concerning the SSF for random Schrödinger operators in Section 4. We also show that one-parameter spectral averaging can be used to recover the classical pointwise bound on the SSF for finite-rank perturbations.

2. Bounds on the spectral shift function for finite-volume Hamiltonians

We use the result of Theorem 1.1 to bound the expectation of the SSF for a single-site perturbation of a finite-volume Hamiltonian H_Λ. Since the size of the support of the perturbation u is of order one relative to $|\Lambda|$, we expect the SSF to be of order one also. For a discussion of the relation between the IDS and the SSF, we refer the reader to [8] and references therein. A nice review of results concerning the SSF may be found in [2]. We recall that for a pair of self-adjoint operators $(H(1), H(0))$, such that $f(H(1)) - f(H(0))$ is trace-class, the SSF $\xi(E; H(1), H(0))$ is defined through the trace formula. For example, for any $f \in C_0^1(\mathbb{R})$, we have

$$\mathrm{Tr}[f(H(1)) - f(H(0))] = \int_{\mathbb{R}} f'(E)\xi(E; H(1), H(0))\, \mathrm{d}E. \tag{2.1}$$

We first consider a one-parameter family of self-adjoint operators $H(\lambda) = H_0 + \lambda V$, with $V \geq 0$, and λ uniformly distributed on $[0,1]$. Birman and Solomyak proved a relation (cf. [19]) between the averaged, weighted, trace of the spectral projector $E_\lambda(\cdot)$ of $H(\lambda)$, and the SSF for the pair $H(1) \equiv H(\lambda = 1)$ and $H(0) \equiv H(\lambda = 0) = H_0$. For any measurable $\Delta \subset \mathbb{R}$, this formula has the form

$$\int_0^1 d\lambda \,\mathrm{Tr}\, V^{1/2} E_\lambda(\Delta) V^{1/2} = \int_\Delta dE\, \xi(E; H_0 + V, H_0), \tag{2.2}$$

whenever all the terms exist. For example, if V is relatively H_0-trace class, then all the terms are well-defined.

We apply this formula as follows. First, we must also make a stronger hypothesis on the probability distribution than (H4). We will assume:

(H4′) The random coupling constants $\{\omega_j \mid j \in \mathbb{Z}^d\}$, are independent and identically distributed. The probability distribution μ_0 of ω_0 is the uniform distribution on $[0, 1]$.

As, for some $C > 0$, one has $0 \le \sum_j u_j \le C$, it follows trivially that

$$\sum_{j \in \tilde{\Lambda}} \mathrm{Tr}\, u_j^{1/2} E_\Lambda(\Delta) u_j^{1/2} \le C\, \mathrm{Tr}\, E_\Lambda(\Delta). \tag{2.3}$$

We now consider the effect of the variation of one random variable ω_j, for $j \in \tilde{\Lambda}$, on the local Hamiltonian. In formula (2.2), we take $H(0) = H_\Lambda(\omega_j = 0)$, $H(1) = H_\Lambda(\omega_j = 1)$, so that $\lambda = \omega_j$, and $V = u_j \ge 0$. We write $H_{j\perp}^\Lambda$ for H_Λ with $\omega_j = 0$. The Birman-Solomyak formula (2.2) then has the form

$$\int_0^1 d\omega_j \;\mathrm{Tr}\, u_j^{1/2} E_\Lambda(\Delta) u_j^{1/2} = \int_\Delta dE\, \xi(E; H_{j\perp}^\Lambda + u_j, H_{j\perp}^\Lambda). \tag{2.4}$$

Taking the expectation of (2.3) and using formula (2.4), we obtain

$$\begin{aligned} \mathbb{E}\{\sum_{j \in \tilde{\Lambda}} \mathrm{Tr}\, u_j^{1/2} E_\Lambda(\Delta) u_j^{1/2}\} &= \sum_{j \in \tilde{\Lambda}} \mathbb{E}\Big\{\int_\Delta dE\, \xi(E; H_{j\perp}^\Lambda + u_j, H_{j\perp}^\Lambda)\Big\} \\ &\le C\mathbb{E}\{\mathrm{Tr}\, E_\Lambda(\Delta)\} \\ &\le C_0 |\Delta|\, |\Lambda|, \end{aligned} \tag{2.5}$$

where we used the result of the proof of Theorem 1.1 on the last line. We conclude from (2.5) that

$$\frac{1}{|\Delta|} \int_\Delta dE \left\{ \frac{1}{|\Lambda|} \sum_{j \in \tilde{\Lambda}} \mathbb{E}\{\xi(E; H_{j\perp}^\Lambda + u_j, H_{j\perp}^\Lambda)\} \right\} \le C_0. \tag{2.6}$$

If the spatially averaged expectation of the SSF is $L^1_{\mathrm{loc}}(\mathbb{R})$ in E, we can conclude a pointwise bound from (2.6), for Lebesgue almost every energy E, of the form

$$\mathbb{E}\left\{ \frac{1}{|\Lambda|} \sum_{j \in \tilde{\Lambda}} \xi(E; H_{j\perp}^\Lambda + u_j, H_{j\perp}^\Lambda) \right\} \le C_0. \tag{2.7}$$

In [8], we proved that the SSF for local Schrödinger operators with compactly-supported perturbations is locally-L^1, so this pointwise bound (2.7) holds. Finally, we observe that due to the periodic boundary conditions on $\partial\Lambda$ and the $\mathbb{Z}^d$-periodicity of H_0, we have that for any $j, k \in \tilde{\Lambda}$

$$\mathbb{E}\{\xi(E; H_{j\perp}^\Lambda + u_j, H_{j\perp}^\Lambda)\} = \mathbb{E}\{\xi(E; H_{k\perp}^\Lambda + u_k, H_{k\perp}^\Lambda)\}, \tag{2.8}$$

and consequently it follows from (2.7) that for any $j \in \tilde{\Lambda}$,

$$\mathbb{E}\{\xi(E; H_{j\perp}^\Lambda + u_j, H_{j\perp}^\Lambda)\} \le C_0. \tag{2.9}$$

Theorem 2.1. *Under the hypotheses* (H1) – (H4′), *the expectation of the spectral shift function, corresponding to the variation of a single site of the finite-volume Hamiltonian, is uniformly locally bounded in energy. That is, for any bounded energy interval, there is a constant* $C_I > 0$, *independent of* Λ, *so that for Lebesgue almost every* $E \in I$, *we have*

$$\mathbb{E}\{\xi(E; H_{j\perp}^{\Lambda} + u_j, H_{j\perp}^{\Lambda})\} \leq C_I, \tag{2.10}$$

for any $j \in \tilde{\Lambda}$.

In the lattice case, the perturbation u_j is rank-one, so by the general theory (cf. [2, 22], or see Section 4.3), we have the bound

$$\xi(E; H_{j\perp}^{\Lambda} + u_j, H_{j\perp}^{\Lambda}) \leq 1, \tag{2.11}$$

for any $j \in \tilde{\Lambda}$, uniformly in $E \in \mathbb{R}$.

3. Bounds on the spectral shift function for infinite-volume Hamiltonians

We consider the thermodynamic limit of the SSF in (2.7). The Birkhoff Ergodic Theorem implies that the limit of the right side of (2.7) is the expectation of the SSF corresponding to the pair of infinite-volume Hamiltonians $(H_{0\perp}, H_{0\perp} + u_0)$ if we replace $\xi(E; H_{j\perp}^{\Lambda} + u_j, H_{j\perp}^{\Lambda})$ by $\xi(E; H_{j\perp} + u_j, H_{j\perp})$, where $H_{j\perp}$ is the infinite-volume Hamiltonian with $\omega_j = 0$. This is the content of the next proposition.

Theorem 3.1. *Let* $H_{0\perp}$ *be the infinite-volume random Hamiltonian* H_ω *with* $\omega_0 = 0$ *and assume hypotheses* (H1) – (H4′). *Then the SSF* $\xi(E; H_{0\perp} + u_0, H_{0\perp})$ *is well-defined and* $\mathbb{E}\{\xi(E; H_{0\perp} + u_0, H_{0\perp})\} \in L^\infty_{\text{loc}}(\mathbb{R})$.

PROOF. (1) We begin with the integrated expression (2.6) and write

$$\begin{aligned} \frac{1}{|\Delta|} \int_\Delta \mathrm{d}E\, \mathbb{E}\Big\{ \frac{1}{|\Lambda|} \sum_{j \in \tilde{\Lambda}} \xi(E; H_{j\perp}^{\Lambda} + u_j, H_{j\perp}^{\Lambda}) \Big\} \\ = \frac{1}{|\Delta|} \int_\Delta \mathrm{d}E\, \mathbb{E}\Big\{ \frac{1}{|\Lambda|} \sum_{j \in \tilde{\Lambda}} \xi(E; H_{j\perp} + u_j, H_{j\perp}) \Big\} + \frac{\mathcal{E}_\Lambda(\Delta)}{|\Delta|}, \end{aligned} \tag{3.1}$$

where the error term is

$$\mathcal{E}_\Lambda(\Delta) \equiv \int_\Delta \mathrm{d}E\, \mathbb{E}\Big\{ \frac{1}{|\Lambda|} \sum_{j \in \tilde{\Lambda}} \big(\xi(E; H_{j\perp}^{\Lambda} + u_j, H_{j\perp}^{\Lambda}) - \xi(E; H_{j\perp} + u_j, H_{j\perp})\big) \Big\}. \tag{3.2}$$

We will prove below that $\mathcal{E}_\Lambda(\Delta) \to 0$ as $|\Lambda| \to \infty$. Assuming this for the moment, it follows from the Birkhoff Ergodic Theorem and (3.1) that

$$\begin{aligned} \lim_{|\Lambda| \to \infty} \frac{1}{|\Delta|} \int_\Delta \mathrm{d}E\, \mathbb{E}\Big\{ \frac{1}{|\Lambda|} \sum_{j \in \tilde{\Lambda}} \xi(E; H_{j\perp} + u_j, H_{j\perp}) \Big\} \\ = \frac{1}{|\Delta|} \int_\Delta \mathrm{d}E\, \mathbb{E}\{\xi(E; H_{0\perp} + u_0, H_{0\perp})\} \leq C_I < \infty. \end{aligned} \tag{3.3}$$

In order to justify the interchange of the expectation and the infinite-volume limit, we note that the nonnegative series in brackets on the first line of (3.3) converges pointwise a.e. to the integrand on the second line of (3.3). As the SSF $\xi(E; H_{j\perp} + u_j, H_{j\perp}) \in L^1_{\text{loc}}(\mathbb{R})$, the term in the brackets on the right of the first line of (3.3) is uniformly bounded, so the exchange is justified by the Lebesgue Dominated Convergence Theorem. We apply the Lebesgue Differentiation Theorem to the second line of (3.3), since the SSF is in $L^1_{\mathrm{loc}}(\mathbb{R})$, and obtain the pointwise bound in Theorem 3.1.

(2) It remains to prove the vanishing of the error term in (3.2) in the infinite-volume limit. Using the identity on the first line of (2.5), we obtain

$$\mathcal{E}_\Lambda(\Delta) = \mathbb{E}\Big\{ \frac{1}{|\Lambda|} \sum_{j \in \tilde{\Lambda}} [\mathrm{Tr}\, u_j^{1/2} E_\Lambda(\Delta) u_j^{1/2} - \mathrm{Tr}\, u_j^{1/2} E(\Delta) u_j^{1/2}] \Big\}. \tag{3.4}$$

We define a local nonnegative measure κ_Λ by

$$\kappa_\Lambda(\Delta) \equiv \frac{1}{|\Lambda|} \mathbb{E}\Big\{ \sum_{j \in \tilde{\Lambda}} \mathrm{Tr}\, u_j^{1/2} E_\Lambda(\Delta) u_j^{1/2} \Big\}, \tag{3.5}$$

and the nonnegative measure $\tilde{\kappa}_\Lambda$, defined similarly but with the spectral projection $E(\cdot)$ for the infinite-volume Hamiltonian H_ω. In terms of these local measures, we can express the right side of (3.4) as

$$\mathcal{E}_\Lambda(\Delta) = \frac{1}{|\Lambda|} [\kappa_\Lambda(\Delta) - \tilde{\kappa}_\Lambda(\Delta)]. \tag{3.6}$$

We first prove that the measure $\mathcal{E}_\Lambda(\cdot)$ converges vaguely to zero by computing the Laplace transform of the measure. The Laplace transform $\mathcal{L}(\mathcal{E}_\Lambda)(t)$ is easily seen to be given by

$$\mathcal{L}(\mathcal{E}_\Lambda)(t) = \frac{1}{|\Lambda|} \mathbb{E}\{\mathrm{Tr}\, \widetilde{V}_\Lambda (e^{-tH_\omega} - e^{-tH_\Lambda})\}. \tag{3.7}$$

Using the Feynman – Kac formula for the heat semigroups, for example, one easily shows, as in [12], that

$$\lim_{|\Lambda| \to \infty} \mathcal{L}(\mathcal{E}_\Lambda)(t) = 0, \tag{3.8}$$

for $t > 0$ pointwise, for a reasonable expanding family of regions Λ. This implies the measure $\mathcal{E}_\Lambda(\cdot)$ converges vaguely to zero which, in turn, implies that the right side of (3.6) converges to zero. □

A consequence of this result is an apparently new relationship between the infinite-volume SSF and the DOS for lattice models. The analogous relation for continuum models defines a new measure absolutely continuous with respect to Lebesgue measure and to the DOS measure. These results follow easily from the proof of Theorem 3.1.

Corollary 3.1. *Let ν be the DOS measure for the random Hamiltonian H_ω. For lattice models, for any Borel set $A \subset \mathbb{R}$, we have*

$$\nu(A) = \int_A \mathrm{d}E\, \mathbb{E}\{\xi(E; H_{0^\perp} + u_0, H_{0^\perp})\}. \tag{3.9}$$

For continuum models, there is a nonnegative measure κ, absolutely continuous with respect to the DOS measure and Lebesgue measure, with distribution given in (3.13), *so that*

$$\kappa(A) = \int_A \mathrm{d}E\, \mathbb{E}\{\xi(E; H_{0^\perp} + u_0, H_{0^\perp})\}. \tag{3.10}$$

For any closed bounded interval $I \subset \mathbb{R}$, there are constants $0 < c_I \leq C_I < \infty$, so that for any Borel set $A \subset I$, we have

$$0 \leq \kappa(A) \leq c_I |A| \quad \text{and} \quad 0 \leq \kappa(A) \leq C_I \nu(A). \tag{3.11}$$

PROOF. From the Birkhoff Ergodic Theorem, and expression (2.4), we can express the integral on the right in (3.3) in terms of a positive measure κ as follows

$$\begin{aligned} \int_\Delta \mathrm{d}E\, \mathbb{E}\{\xi(E; H_{0^\perp} + u_0, H_{0^\perp})\} &= \mathbb{E}\Big(\lim_{|\Lambda| \to \infty} \frac{1}{|\Lambda|} \sum_{j \in \tilde{\Lambda}} \operatorname{Tr} u_j^{1/2} E(\Delta) u_j^{1/2} \Big) \\ &= \mathbb{E}\{\operatorname{Tr} u_0^{1/2} E(\Delta) u_0^{1/2}\} \\ &\equiv \kappa(\Delta), \end{aligned} \tag{3.12}$$

where $\kappa(\cdot)$ is the nonnegative measure with distribution function given by

$$K(E) \equiv \mathbb{E}\{\operatorname{Tr} u_0^{1/2} P(E) u_0^{1/2}\}, \tag{3.13}$$

where $P(E)$ is the spectral family for H_ω. For the lattice case, this measure is just the DOS measure, since $u_0 = \delta_0$, so that $\mathbb{E}\{\xi(E; H_{0^\perp} + u_0, H_{0^\perp})\}$ is a representation of the DOS. It follows immediately from (3.12) and Theorem 3.1 that for any closed bounded interval $I \subset \mathbb{R}$, there exists a finite constant $0 < C_I < \infty$, so that for any Lebesgue measurable set $A \subset I$, we have

$$0 \leq \kappa(A) = \int_A \mathrm{d}E\, \mathbb{E}\{\xi(E; H_{0^\perp} + u_0, H_{0^\perp})\} \leq C_I |A|. \tag{3.14}$$

It remains to prove that κ is bounded above by the DOS measure. This implies the absolute continuity with respect to ν. We simply note that there exists a constant $0 < C_0 < \infty$, depending only on u, so that

$$0 \leq \sum_{j \in \tilde{\Lambda}} \operatorname{Tr} u_j^{1/2} E_\Lambda(\Delta) u_j^{1/2} \leq C_0 \operatorname{Tr} E_\Lambda(\Delta), \tag{3.15}$$

and recall the definition of the DOS measure. This implies that $0 \leq \kappa(A) \leq C_I \nu(A)$, for $A \subset I \subset \mathbb{R}$. □

This measure κ is similar to the DOS measure for continuum models. The distribution function for the DOS for continuum models is given by $N(E) = \mathbb{E}\{\operatorname{Tr} \chi_{\Lambda_1(0)} P(E) \chi_{\Lambda_1(0)}\}$. The measure κ is equivalent to the DOS measure ν if the single-site potential satisfies $c_0 \chi_{\Lambda_1(0)} \leq u$, for some $c_0 > 0$, and it is equal to ν in the special case that $u = \chi_{\Lambda_1(0)}$. The equivalence of measures means that there are constants $C_0, c_0 > 0$ so that

$$c_0 \nu(A) \leq \kappa(A) \leq C_0 \nu(A), \tag{3.16}$$

for all Borel subsets $A \subset \mathbb{R}$.

4. Comments

We make three comments on various other results concerning the SSF associated with random Schrödinger operators that have recently occurred in the literature related to random Schrödinger operators. For deterministic Schrödinger operators, *pointwise bounds* are known only in a few specific cases, such as finite-rank perturbations (cf. [2,22] and Theorem 4.2 below) or perturbations of the Laplacian on $L^2(\mathbb{R}^d)$ by sufficiently smooth potentials [20].

4.1. Related results on the averaged SSF. Bounds on the L^p-norm of the SSF, for $0 < p \leq 1$, were proved in [8] and improved in [11]. More recently, Hundertmark et al.[10], obtained some new integral bounds on the SSF that indicate that one cannot expect that, in general, the SSF is locally bounded. Indeed, Kirsch [13, 14] proved that if the Dirichlet Laplacian in Λ_L, a cube of side length L centered at the origin, is perturbed by a nonnegative, bounded potential supported inside the unit cube Λ_1, then the corresponding finite-volume SSF, at any positive energy, diverges as $L \to \infty$. Raikov and Warzel [18] considered the SSF for the Landau Hamiltonian (1.5) and a perturbation by a compactly-supported potential. They showed that the SSF diverges at the Landau energies.

The averaged SSF is expected to be better behaved. In addition to the pointwise bounds of Theorems 2.1 and 3.1, Aizenman et al.[1] proved an interesting bound on a spectral shift function related to the ones treated here. They consider the SSF $\xi(t, E)$ for a pair of Hamiltonians $H_t = H_0 + tV$ and $H_t + U$, where V and U are nonnegative bounded potentials such that V is strictly positive on a neighborhood of the support of U. Specifically, for any $\delta > 0$, we define the set $Q_\delta \equiv \{x \in \mathbb{R}^d \mid \operatorname{dist}\big(x, \operatorname{supp}(U)\big) < \delta\}$. We then require that V be strictly positive on Q_δ.

Theorem 4.1. *For any $0 < s < \min(2/d, 1/2)$, there is a finite positive constant $C_{s,\delta} > 0$ so that the SSF $\xi(t, E)$ satisfies the bound*

$$\int_0^1 |\xi(t, E)|^s \mathrm{d}t \leq C_{s,\delta} \|U\|_\infty (1 + |E - E_0| + \|V\|_\infty)^{2s(d+1)}, \tag{4.1}$$

where $E \geq E_0 \equiv \inf \sigma(H_0)$.

4.2. Spectral shift density. Kostrykin and Schrader [15, 16] introduced the *spectral shift density* (SSD) that is closely related to the integrated density of states. The SSD is the density of a measure Ξ obtained by the thermodynamic limit

$$\int_{\mathbb{R}} g(\lambda)\, \mathrm{d}\Xi(\lambda) = \lim_{|\Lambda| \to \infty} \int_{\mathbb{R}} g(\lambda) \frac{\xi(E; H_0 + \chi_\Lambda V_\omega)}{|\Lambda|}. \tag{4.2}$$

Note that the size of the perturbation $\chi_\Lambda V_\omega$ is of order $|\Lambda|$. They prove that the SSD $\tilde{\xi}(E)$ is given as

$$\tilde{\xi}(E) = N_0(E) - N(E), \quad \text{a.e. } E \in \mathbb{R}, \tag{4.3}$$

where $N_0(E)$ is the IDS of H_0 and $N(E)$ is the IDS of H_ω.

4.3. A pointwise bound on the SSF for finite-rank perturbations. We consider the SSF for a finite-rank perturbation. Let $B \geq 0$ be a nonnegative finite-rank operator with rank N. Let $H_s = H_0 + sB$ be the one-parameter perturbation of a self-adjoint, lower-semibounded operator H_0. The variable $s \in [0, 1]$ is uniformly distributed. We consider the SSF $\xi(E; H_1, H_0)$ and recover the classical pointwise upper bound N usually obtained by other methods cf. [2, 22].

Theorem 4.2. *The spectral shift function for the pair of self-adjoint operators (H_1, H_0), where $0 \leq B \equiv H_1 - H_0$ is a finite-rank operator of rank $N < \infty$, satisfies the bound*

$$0 \leq \xi(E; H_1, H_0) \leq N. \tag{4.4}$$

PROOF. Let $f \in C_0^1(\mathbb{R})$ and consider the formula for the SSF:

$$\begin{aligned} \operatorname{Tr}\big(f(H_1) - f(H_0)\big) &= -\int_{\mathbb{R}} f'(E)\xi(E; H_1, H_0)\,\mathrm{d}E \\ &= \int_0^1 \frac{d}{ds} \operatorname{Tr} f(H_s)\,\mathrm{d}s \\ &= \int_0^1 \mathrm{d}s \operatorname{Tr} B^{1/2} f'(H_s) B^{1/2} \\ &= \sum_{j=1}^N \int_0^1 \mathrm{d}s\, \langle \phi_j, B^{1/2} f'(H_s) B^{1/2} \phi_j \rangle \end{aligned} \tag{4.5}$$

Let $E_s(\cdot)$ be the spectral family for H_s. The matrix element in (4.5) is written as

$$\begin{aligned} \langle \phi_j, B^{1/2} f'(H_s) B^{1/2} \phi_j \rangle &= \int_{\mathbb{R}} f'(\lambda)\,\mathrm{d}\langle B^{1/2}\phi_j, E_s(\lambda) B^{1/2}\phi_j \rangle \\ &= \int_{\mathbb{R}} f'(\lambda)\,\mathrm{d}\mu_{H_s}^{\psi_j}(\lambda), \end{aligned} \tag{4.6}$$

where $\psi_j \equiv B^{1/2}\phi_j$ and $\mu_{H_s}^{\psi_j}$ is the corresponding spectral measure for H_s and ψ_j. We divide the support of f' into p subintervals Δ_k and bound the absolute value of the integral over λ in (4.6) from above as

$$\left| \int_{\mathbb{R}} f'(\lambda)\,\mathrm{d}\mu_{H_s}^{\psi_j}(\lambda) \right| \le \sum_{k=1}^p |f'(x_k)| \mu_{H_s}^{\psi_j}(\Delta_k), \tag{4.7}$$

where $x_k \in \Delta_k$ is such that $|f(x_k)| = \sup_{x\in\Delta_k} |f'(x)|$. Inserting this into the integral over s in (4.5), we see that it remains to estimate

$$\int_0^1 \mu_{H_s}^{\psi_j}(\Delta_k)\,\mathrm{d}s = \int_0^1 \langle B^{1/2}\phi_j, E_s(\Delta_k) B^{1/2}\phi_j \rangle\,\mathrm{d}s. \tag{4.8}$$

For bounded probability distributions with compact support, the integral on the right in (4.8) was estimated in [3]. The result is

$$\int_0^1 \langle B^{1/2}\phi_j, E_s(\Delta_k) B^{1/2}\phi_j \rangle\,\mathrm{d}s \le |\Delta_k|, \tag{4.9}$$

since $\|\psi_j\| \le 1$. Combining this bound with (4.5) – (4.6) and recalling the approximation (4.7), we obtain

$$\left| \int_{\mathbb{R}} f'(E)\xi(E; H_1, H_0)\,\mathrm{d}E \right| \le \sum_{j=1}^N \sum_{k=1}^p |f'(x_k)|\,|\Delta_k| \le N\|f'\|_1, \tag{4.10}$$

which, extending the estimate to any $f \in L^1(\mathbb{R})$, we conclude that

$$|\xi(E; H_1, H_0)| \le N, \tag{4.11}$$

proving the result. □

References

1. M. Aizenman, A. Elgart, S. Naboko, J. Schenker, and G. Stolz, *Moment analysis for localization in random Schrödinger operators*, Invent. Math. **163** (2006), no. 2, 343 – 413.
2. M. Sh. Birman and D. R. Yafaev, *The spectral shift function. The papers of M. G. Kreĭn and their further development*, Algebra i Analiz **4** (1992), no. 5, 1 – 44 (Russian); English transl., St. Petersburg Math. J. **4** (1993), no. 5, 833 – 870.
3. J.-M. Combes and P. D. Hislop, *Localization for some continuous, random Hamiltonians in d-dimensions*, J. Funct. Anal. **124** (1994), no. 1, 149 – 180.
4. J.-M. Combes, P. D. Hislop, and F. Klopp, *Hölder continuity of the integrated density of states for some random operators at all energies*, Int. Math. Res. Not. **2003**, no. 4, 179 – 209.
5. ______, *An optimal Wegner estimate and its application to the global continuity of the integrated density of states for random Schrödinger operators*, preprint.
6. J.-M. Combes, P. D. Hislop, F. Klopp, and G. Raikov, *Global continuity of the integrated density of states for random Landau Hamiltonians*, Comm. Partial Differential Equations **29** (2004), no. 7-8, 1187 – 1213.
7. J.-M. Combes, P. D. Hislop, and E. Mourre, *Spectral averaging, perturbation of singular spectra, and localization*, Trans. Amer. Math. Soc. **348** (1996), no. 12, 4883 – 4894.
8. J.-M. Combes, P. D. Hislop, and S. Nakamura, *The* L^p*-theory of the spectral shift function, the Wegner estimate and the integrated density of states for some random operators*, Comm. Math. Phys. **218** (2001), no. 1, 113 – 130.
9. P. D. Hislop and F. Klopp, *The integrated density of states for some random operators with nonsign definite potentials*, J. Funct. Anal. **195** (2002), no. 1, 12 – 47.
10. D. Hundertmark, R. Killip, S. Nakamura, P. Stollmann, and I. Veselić, *Bounds on the spectral shift function and the density of states*, Comm. Math. Phys. **262** (2005), no. 2, 489 – 503.
11. D. Hundertmark and B. Simon, *An optimal* L^p*-bound on the Krein spectral shift function*, J. Anal. Math. **87** (2002), 199 – 208.
12. W. Kirsch, *Random Schrödinger operators. A course*, Schrödinger Operators (Sønderborg, 1988) (H. Holden and A. Jensen, eds.), Lecture Notes in Phys., vol. 345, Springer, Berlin, 1989, pp. 264 – 370.
13. ______, *Small perturbations and the eigenvalues of the Laplacian on large bounded domains*, Proc. Amer. Math. Soc. **101** (1987), no. 3, 509 – 512.
14. ______, *The stability of the density of states of Schrödinger operator under very small perturbations*, Integral Equations Operator Theory **12** (1989), no. 3, 383 – 391.
15. V. Kostrykin and R. Schrader, *Scattering theory approach to random Schrödinger operators in one dimension*, Rev. Math. Phys. **11** (1999), no. 2, 187 – 242.
16. ______, *The density of states and the spectral shift density of random Schrödinger operators*, Rev. Math. Phys. **12** (2000), no. 6, 807 – 847.
17. S. N. Naboko, *The structure of singularities of operator functions with a positive imaginary part*, Funktsional. Anal. i Prilozhen. **25** (1991), 1 – 13 (Russian); English transl., Funct. Anal. Appl. **25** (1991), no. 4, 243 – 253.
18. G. Raikov and S. Warzel, *Quasi-classical versus non-classical spectral asymptotics for magnetic Schrödinger operators with decreasing electric potentials*, Rev. Math. Phys. **14** (2002), no. 10, 1051 – 1072.
19. B. Simon, *Spectral averaging and the Krein spectral shift*, Proc. Amer. Math. Soc. **126** (1998), no. 5, 1409 – 413.
20. A. Sobolev, *Efficient bounds for the spectral shift functions*, Ann. Inst. H. Poincaré Phys. Théor. **58** (1993), no. 1, 55 – 83.
21. B. Sz.-Nagy and C. Foiaş, *Harmonic analysis of operators on Hilbert space*, North-Holland Publishing Co., Amsterdam, 1970.
22. D. R. Yafaev, *Mathematical scattering theory: General theory*, Transl. Math. Monogr., vol. 105, Amer. Math. Soc., Providence, RI, 1992.

Département de Mathématiques, Université du Sud, Toulon-Var, 83130 La Garde, France
E-mail address: combes@cpt.univ-mrs.fr

Department of Mathematics, University of Kentucky, Lexington, KY 40506-0027, USA
E-mail address: hislop@ms.uky.edu

Laboratoire Analyse Géométrie et Applications, Institut Galilée, Université Paris-Nord, 93430 Villetaneuse, France and Institut Universitaire de France
E-mail address: klopp@math.univ-paris13.fr

Centre de Recherches Mathématiques
CRM Proceedings and Lecture Notes
Volume **42**, 2007

On Phase Transitions and Limit Theorems for Homopolymers

M. Cranston and S. Molchanov

ABSTRACT. Given the measure on random walk paths P_0 and a Hamiltonian H the Gibbs perturbation of H defined by

$$\frac{\mathrm{d}P_{\beta,t}}{\mathrm{d}P_0} = Z_{\beta,t}^{-1} \exp\{\beta H(t,x)\}$$

with

$$Z_{\beta,t} = \int \exp\{\beta H(t,x)\}\,\mathrm{d}P_0(x)$$

gives a new measure on paths x which can be viewed as polymers. In the case $H(t,x) = \int_0^t \delta_0(x_s)\,\mathrm{d}s$ we say the resulting measure is concentrated on "homopolymers" and are interested in the influence of dimension and β on their behavior.

1. Introduction

We consider a model for so-called homopolymers which is given in the form of a Gibb's distribution on the space of paths in $\mathbb{Z}^d$. Let Ω be the set of right continuous, left limit trajectories $\omega\colon [0,\infty) \to \mathbb{Z}^d$. We shall use x to denote the coordinate process, $x_t(\omega) = \omega(t)$ and, as is customary, drop the ω writing simply x_t. Accordingly, each $\omega \in \Omega$ has jump times $0 < t_1 < t_2 < \cdots < t_k < \cdots$, finitely many on each bounded interval, depending on ω, satisfying $|x_{t_i} - x_{t_{i-1}}| = 1$ and $x_{t_i} = x_t$ for $t_i \le t < t_{i+1}$. Let P_0 denote the measure on these paths generated by continuous time, simple, symmetric random walk with jump rate 1. The 0 in P_0 denotes the infinite temperature regime, $\beta = 0$, in the family of Gibb's probability measures $P_{\beta,t}$ defined by

$$dP_{\beta,t}(x) = Z_{\beta,t}^{-1} \exp\{\beta H(t,x)\}\,\mathrm{d}P_0(x), \tag{1.1}$$

$$Z_{\beta,t} = \int_{\Omega_t} \exp\{\beta H(t,x)\}\,\mathrm{d}P_0(x). \tag{1.2}$$

The Hamiltonian $H(t,x)$ describes the energy of the configuration x, which captures the interaction of x with the external field and the energy of self-interaction of x.

2000 *Mathematics Subject Classification.* Primary 82D60; Secondary 60G50.

Key words and phrases. homopolymer, free energy Gibb's measure, randomwalk.

Research of both authors supported by a grant from NSF.
This is the final form of the paper.

The normalization $Z_{\beta,t}$ is a statistical sum and is the usual partition function. As $t \to \infty$, physicists expect that the distribution of $P_{\beta,t}$ has, in an appropriate sense, a limit. This limit would be the Gibb's measure for the infinite system. The existence of the limit usually follows from general compactness arguments. The limit may not be unique. A classical example is the Ising model of ferromagnetism on the lattice $\mathbb{Z}^d$, $d \geq 2$, in the case of low temperature, i.e., $\beta > \beta_{\rm cr}(d)$. There are situations where the limit is unique, but dependence on β is not analytic. Non-uniqueness or non-analyticity of the limiting Gibb's measure are manifestations of phase transitions with respect to temperature in large systems. Usually, such transitions imply that the free energy

$$F(\beta) = \lim_{t\to\infty} \frac{1}{t} \ln Z_{\beta,t} \tag{1.3}$$

is also non-analytic. The limit (1.3) exists in all physically interesting situations and can be considered as the definition of the "size" t. The function $F(\beta)$ typically has a different analytic expression (branches) for $\beta > \beta_{\rm cr}$, $\beta < \beta_{\rm cr}$, corresponding to different phases of the large physical system. In this paper we'll discuss the problem of phase transition of long polymers, similar to the transition of albumen from the liquid phase to the solid state under the action of high temperature. Another example is that of the transition of caoutchouc to rubber in the process of vulcanization. In both situations, we have in the background a transition from the "globular" state of the polymer, when it is coiled up into a compact ball, to the diffusive state, when the polymer is similar to the trajectory of a Brownian motion. For more on physical theories of the polymer's phase transitions, see [4, 5]. Typically, these works focus mainly on the analysis of the free energy $F(\beta)$. We will study the structure of the polymer chains probabilistically, deriving limit theorems for example.

2. Simple homopolymer model, phase transitions

Recall that P_0 is the distribution of the symmetric, continuous time, random walk on $\mathbb{Z}^d$ which has the discrete Laplacian, Δ, as its generator. Here for a function $\psi\colon \mathbb{Z}^d \to \mathbb{Z}^d$, we define $\Delta\psi(x) = \sum_{|y-x|=1} (\psi(y) - \psi(x))$. The transition probabilities associated to P_0, namely, $p_0(s,x,y) = P(x_{r+s} = y | x_r = x)$ are given by the fundamental solution of the heat equation,

$$\begin{aligned} \frac{\partial p_0}{\partial t}(t,x,y) &= \Delta_x p_0(t,x,y) = \Delta_y p_0(t,x,y), \\ p_0(0,x,y) &= \delta_x(y). \end{aligned} \tag{2.1}$$

As usual, the finite dimensional distributions are given, with the convention $x_0 = 0$, $t_0 = 0$, by

$$P_0(x_{t_1} = x_1, \ldots, x_{t_k} = x_k) = \Pi_{j=1}^k p_0(t_j - t_{j-1}, x_{t_{j-1}}, x_{t_j}).$$

We think of the measure P_0 as being supported on the so-called free polymers of any length t. We now construct a measure on polymers of length t by specifying a probability measure on the paths Ω_t which consists of the restriction of paths in Ω to the interval $[0,t]$. This measure takes into account the interaction of the polymer chain with an external field and its dependence on the reciprocal temperature, β. The case $\beta = 0$ will correspond to the free polymer measure P_0. The measure $P_{\beta,t}$

is the Gibb's perturbation of P_0 given by specifying its density with respect to P_0 as

$$P_{\beta,t}(d\omega) = Z_{\beta,t}^{-1} \exp\left\{\beta \int_0^t \delta_0(x_s)\,\mathrm{d}s\right\} P_0(\mathrm{d}\omega), \tag{2.2}$$

where $Z_{\beta,t}$ is the partition function (normalizing constant) given by

$$Z_{\beta,t} = E_0\left[\exp\left\{\beta \int_0^t \delta_0(x_s)\,\mathrm{d}s\right\}\right].$$

The Hamiltonian $H(t,x) = \int_0^t \delta_0(x_s)\,\mathrm{d}s$ describes the interaction between the polymer chain x_s, $0 \le s \le t$, and the attracting potential $V(x) = \delta_0(x)$, which provides the external, time independent field. This model is called a homopolymer. The theory of homopolymers is simple and well-known (in the physical literature.) Our goal is to give a probabilistic treatment and eventually to go on to more complicated models.

3. Main results

We now present our theorems

Theorem 3.1. *Let*

$$H^\beta = \Delta + \beta\delta_0$$

and denote the spectrum of H^β by Σ^β. Then

$$\lambda_0(\beta) = \sup \Sigma^\beta.$$

Then the free energy

$$F(\beta) = \lim_{t\to\infty} \frac{1}{t} \ln Z_{\beta,t}$$

exists and

$$F(\beta) = \lambda_0(\beta).$$

In the dimensions $d = 1,2$, we have $\lambda_0(\beta) > 0$ for $\beta > 0$. In dimensions $d \ge 3$, there is a phase transition for the free energy, i.e., there is a $\beta_{\mathrm{cr}}(d) > 0$ such that $\lambda_0(\beta) > 0$ for $\beta > \beta_{\mathrm{cr}}(d)$ and $\lambda_0(\beta) = 0$ for $\beta \le \beta_{\mathrm{cr}}(d)$. Moreover, F is analytic for $\beta > \beta_{\mathrm{cr}}(d)$ and non-analytic at $\beta = \beta_{\mathrm{cr}}(d)$.

The proof depends on the following lemma.

Lemma 3.2. *The spectrum of H^β is*

$$\Sigma^\beta = \Sigma^0 \cup \{\lambda_0(\beta)\},$$

where $\lambda_0(\beta) > 0$ if $d = 1,2$ for $\beta > 0$, and there is a $\beta_{\mathrm{cr}}(d) > 0$ such that for $d \ge 3$, we have $\lambda_0(\beta) > 0$ for $\beta > \beta_{\mathrm{cr}}(d)$. If $d \ge 3$ and $\beta \le \beta_{\mathrm{cr}}(d)$ then

$$\Sigma^\beta = \Sigma^0.$$

If $\Sigma^\beta = \Sigma^0$ then the spectrum of H^β is purely absolutely continuous. In the case when the positive eigenvalue exists, it corresponds to a unique eigenfunction ψ_β, which is strictly positive and exponentially decaying.

PROOF. The result is well known, but we include the proof for the reader's convenience. Introduce the fundamental solution by

$$\begin{aligned}\frac{\partial p_\beta}{\partial t}(t,x,y) &= \Delta_x p_\beta(t,x,y) + \beta\delta_0(x)p_\beta(t,x,y),\\ p_\beta(0,x,y) &= \delta_y(x).\end{aligned} \tag{3.1}$$

and consider the resolvent kernel, $R_\lambda(x,y) = \int_0^\infty p_\beta(t,x,y)\mathrm{e}^{-\lambda t}\,\mathrm{d}t$, which solves the equation

$$\Delta_x R_\lambda(x,y) + \beta\delta_0(x)R_\lambda(x,y) = \lambda R_\lambda(x,y) - \delta_y(x).$$

Here λ is large enough to guarantee the analyticity of $R_\lambda(x,y)$ in λ. Taking Fourier transforms, we obtain

$$\hat{\Delta}(\phi)\hat{R}_\lambda(\phi,y) + \beta R_\lambda(0,y) = \lambda\hat{R}_\lambda(\phi,y) - \mathrm{e}^{-i(\phi,y)}$$

Selecting the standard orthonormal basis $\{e_1,\dots,e_d\}$ for $\mathbb{R}^d$ and setting $\phi_j = (\phi, e_j)$ we write $-\Phi(\phi) \equiv \hat{\Delta}(\phi,y) = -2\sum_{j=1}^d(1-\cos\phi_j)$. Then solve for $\hat{R}_\lambda(\phi,y)$ obtaining

$$\hat{R}_\lambda(\phi,y) = \frac{\beta R_\lambda(0,y) + \mathrm{e}^{-i(\phi,y)}}{\lambda+\Phi(\phi)}.$$

Integrating in ϕ over the d-dimensional torus, $\mathbb{T}^d$, yields,

$$\begin{aligned}R_\lambda(0,y) &= \frac{1}{(2\pi)^d}\int_{\mathbb{T}^d}\hat{R}_\lambda(\phi,y)\,\mathrm{d}\phi\\ &= \beta R_\lambda(0,y)\frac{1}{(2\pi)^d}\int_{\mathbb{T}^d}\frac{1}{\lambda+\Phi(\phi)}\,\mathrm{d}\phi + \frac{1}{(2\pi)^d}\int_{\mathbb{T}^d}\frac{\mathrm{e}^{-i(\phi,y)}}{\lambda+\Phi(\phi)}\,\mathrm{d}\phi.\end{aligned} \tag{3.2}$$

Set

$$I(\lambda,y) = \frac{1}{(2\pi)^d}\int_{\mathbb{T}^d}\frac{\mathrm{e}^{-i(\phi,y)}}{\lambda+\Phi(\phi)}\,\mathrm{d}\phi.$$

and

$$I(\lambda,0) = I(\lambda)$$

It's easy to see that for any $\lambda > 0$, the function $I(\lambda,y)$ is exponentially decreasing in $|y|$ and strictly positive. The function $I(\lambda)$ is analytic for $\operatorname{Re}\lambda > 0$ and for real λ it is increasing monotonically as $\lambda \searrow 0$. Using the asymptotic $\Phi(\phi) \sim |\phi|^2$, for $|\phi| \searrow 0$, we see that,

$$\lim_{\lambda\to 0} I(\lambda) = +\infty, \quad d = 1,2. \tag{3.3}$$

Whereas,

$$\lim_{\lambda\to 0} I(\lambda) = I(0) = \frac{1}{(2\pi)^d}\int_{\mathbb{T}^d}\frac{1}{\Phi(\phi)}\,\mathrm{d}\phi < \infty, \quad d \geq 3. \tag{3.4}$$

Since

$$R_\lambda(0,y) = \frac{I(\lambda,y)}{(1-\beta I(\lambda))},$$

it follows that

$$\hat{R}_\lambda(\phi,y) = \frac{\beta I(\lambda,y)}{(1-\beta I(\lambda))(\lambda+\Phi(\phi))} + \frac{\mathrm{e}^{-i(\phi,y)}}{(\lambda+\Phi(\phi))}. \tag{3.5}$$

Taking the inverse Fourier transform of $\hat{R}_\lambda(\phi, y)$ yields the following formula for the resolvent,

$$R_\lambda(x,y) = \frac{\beta I(\lambda,x)I(\lambda,y)}{(1-\beta I(\lambda))} + I(\lambda, y-x). \tag{3.6}$$

Thus, for fixed $x, y \in \mathbb{Z}^d$, the function $R_\lambda(x,y)$ is not analytic in λ if and only if $\lambda \in \Sigma^0 \cup \{\lambda \in (0,\infty) : 1 - \beta I(\lambda) = 0\}$. In the case $\lambda \in (0,\infty)$ with $1 - \beta I(\lambda) = 0$, the resolvent kernel has a pole at the point $\lambda = \lambda_0(\beta) > 0$, where $I(\lambda_0(\beta)) = 1/\beta$. According to (3.3) this positive pole exists if $d = 1, 2$ for all $\beta > 0$ but by (3.4), for $d \geq 3$, it exists for $\beta > \beta_{\mathrm{cr}}(d) = 1/I(0+)$. By the Spectral Theorem, if $\lambda_0(\beta)$ and ψ_β are the principal eigenvalue and its normalized eigenfunction, respectively, for H^β, then

$$R_\lambda(x,y) = \frac{\psi_\beta(x)\psi_\beta(y)}{\lambda - \lambda_0(\beta)} + \int_{-4d}^{0} \frac{d(E_z\delta_y, \delta_x)}{\lambda - z} = \frac{\psi_\beta(x)\psi_\beta(y)}{\lambda - \lambda_0(\beta)} + \int_{-4d}^{0} \frac{\mu_{x,y}(dz)}{\lambda - z}. \tag{3.7}$$

where $\mu_{x,y}(dz) = d(E_z\delta_y, \delta_x)$ is the spectral measure. In particular, for $x = y$, the spectral measure $\mu_{y,y}(dz)$ is positive and finite. Comparison of (3.6) and (3.7) gives

$$\psi_\beta(x) = cI(\lambda_0(\beta), x),$$

with a normalizing constant c such that $\|\psi_0\|_2 = 1$. (One can also prove that $\lim_{s\to 0} \operatorname{Im} R_{\lambda+is}(x,y) = \operatorname{Im} R_\lambda(x,y)$ is a bounded continuous function of λ on $[-4d, 0]$ which means that the spectral measure is a.c. on $[-4d, 0]$. This fact won't be used however.) □

From now on we shall write β_{cr} instead of $\beta_{\mathrm{cr}}(d)$, suppressing its dependence on d. We now turn to the proof of Theorem 3.1.

PROOF OF THEOREM 3.1. Defining

$$\psi(t,x) = E^x\left[\exp\left\{\beta \int_0^t \delta_0(x_s)\, ds\right\}\delta_0(x_t))\right],$$

we see that

$$Z_{\beta,t} = \sum_{x\in\mathbb{Z}^d} \psi(t,x).$$

Notice that the Feynman – Kac formula gives

$$\begin{aligned} \frac{\partial\psi}{\partial t}(t,x) &= \Delta\psi(t,x) + \beta\delta_0(x)\psi(t,x), \\ \psi(0,x) &= \delta_0(x) \end{aligned} \tag{3.8}$$

Consequently $\psi(t,x) = p_\beta(t,0,x)$ is the fundamental solution of the heat equation for the operator H^β given at (3.1). If $\mathbf{1}$ denotes the vector with all components equal to 1, it then follows that

$$Z_{\beta,t} = \langle p_\beta(t,0,\cdot), \mathbf{1}\rangle = \sum_{x\in\mathbb{Z}^d} p_\beta(t,0,x). \tag{3.9}$$

From (3.9) we see that the asymptotic behavior of $Z_{\beta,t}$ as $t \to \infty$ is closely related to the spectral properties of the operator H^β. Notice H^β is a bounded operator with $\|H^\beta\| \leq 2d + \beta$. The spectrum of H^0 is $\Sigma_0 = [-4d, 0]$. The Fourier representation of H^0 is given by multiplication by its symbol. Namely, select an orthonormal basis $e_1, e_2, \ldots, e_d$ of $\mathbb{R}^d$ and setting $\phi_j = \langle \phi, e_j \rangle$,

$$\widehat{\Delta f}(\phi) = 2\sum_{j=1}^{d}(\cos\phi_j - 1)\hat{f}(\phi). \tag{3.10}$$

Let's return now to (3.8). Taking the Laplace transform in time of both sides, and writing $M_\lambda(x) = \int_0^\infty p_\beta(t, 0, x)\mathrm{e}^{-\lambda t}\,\mathrm{d}t$, we obtain

$$\Delta_x M_\lambda(x) + \beta\delta_0(x)M_\lambda(x) = \lambda M_\lambda(x) - \delta_0(x).$$

Solving gives,

$$M_\lambda(x) = R_\lambda(x, 0). \tag{3.11}$$

Since

$$Z_{\beta,t} = \sum_{x\in\mathbb{Z}^d} p_\beta(t, 0, x),$$

we have by Plancherel,

(3.12)

$$\begin{aligned}\int_0^\infty Z_{\beta,t}\mathrm{e}^{-\lambda t}\,\mathrm{d}t &= \langle M_\lambda, \mathbf{1}\rangle = (2\pi)^d \int_{\mathbb{T}^d} \hat{M}_\lambda(\phi)\delta_0(\phi)\,\mathrm{d}\phi \\ &= (2\pi)^d\hat{M}_\lambda(0) = (2\pi)^d\left(\frac{\beta I(\lambda)}{1-\beta I(\lambda)}\frac{1}{\lambda} + \frac{1}{\lambda}\right) = \frac{(2\pi)^d}{\lambda(1-\beta I(\lambda))}.\end{aligned}$$

Inversion of the Laplace transform now gives the formula

$$Z_{\beta,t} = \frac{1}{(2\pi)^d}\int_{-\infty}^{\infty}(< M_{\sigma+is}, \nVdash > \mathrm{e}^{(\sigma+is)t}\,\mathrm{d}s$$

which is valid for any $\sigma > \lambda_0(\beta)$. By deformation of contour, (3.12) yields the asymptotic formula

$$Z_{\beta,t} = \frac{\mathrm{e}^{\lambda_0(\beta)t}(2\pi)^d}{\beta\lambda_0(\beta)|I'(\lambda_0(\beta))|} + O(\mathrm{e}^{\delta t}),$$

for any $\delta > 0$. Consequently, if $\lambda_0(\beta) > 0$, then

$$\lim_{t\to\infty}\frac{1}{t}\ln Z_{\beta,t} = \lambda_0(\beta).$$

If $\lambda_0(\beta) = 0$, then

$$\lim_{t\to\infty}\frac{1}{t}\ln Z_{\beta,t} \le \delta,$$

for any $\delta > 0$. On the other hand, since $Z_{\beta,t} \ge p_\beta(t, 0, 0) \ge p_0(t, 0, 0) \sim ct^{-\frac{d}{2}}$, it follows that

$$\lim_{t\to\infty}\frac{1}{t}\ln Z_{\beta,t} \ge 0.$$

Thus, in the case $\lambda_0(\beta) = 0$, we have

$$\lim_{t\to\infty}\frac{1}{t}\ln Z_{\beta,t} = 0.$$

Finally, regardless of the value of $\lambda_0(\beta)$, we have proved

$$\lim_{t\to\infty}\frac{1}{t}\ln Z_{\beta,t} = \lambda_0(\beta).$$

The analyticity of $F(\beta)$ for $\beta > \beta_{\rm cr}$ follows from the analyticity of $I(\lambda)$ and the relation $\beta I(\lambda_0(\beta)) = 1$. □

4. Markov description of polymer in globular phase

We now give a description, in the globular phase $\beta > \beta_{\rm cr}$, not just of the distribution of the endpoint $x(t)$, but of the initial part of the path x_r, $0 \le r \ll t$ as $t \to \infty$. We remind the reader that finite-dimensional distributions under $P_{\beta,t}$ change with t. To be more precise, if $s < t$ and $0 < t_1 < t_2 < \cdots < t_k < s$ and $A_1, A_2, \dots, A_k \subset \mathbb{Z}^d$, then in general, $P_{\beta,t}(x_{t_1} \in A_1, x_{t_2} \in A_2, \dots, x_{t_k} \in A_k) \neq P_{\beta,s}(x_{t_1} \in A_1, x_{t_2} \in A_2, \dots, x_{t_k} \in A_k)$. In the next theorem we show that as the polymer gets very long, its initial segment converges to a positive recurrent Markov process and we give the characteristics of that process.

Theorem 4.1. *Let $p_\beta(t,x,y)$ be the fundamental solution of the parabolic problem at* (3.1). *For $\beta > \beta_{\rm cr}$, define the Doob transform kernel,*

$$r_\beta(t,x,y) = \frac{p_\beta(t,x,y)\psi_\beta(y)}{\psi_\beta(x)} \mathrm{e}^{-\lambda_0(\beta)t}. \tag{4.1}$$

Then $r_\beta(t,x,y)$ is the transition density for an ergodic, pure jump, Markov process on $\mathbb{Z}^d$ with invariant probability distribution

$$\pi(x) = \psi_\beta^2(x),$$

and generator

$$A_\beta f(x) = \sum_{|y-x|=1} a_\beta(x,y)\big(f(y) - f(x)\big),$$

where

$$a_\beta(x,y) = \begin{cases} 0, & \text{if } |x-y| > 1, \\ \dfrac{\psi_\beta(y)}{\psi_\beta(x)}, & \text{if } |x-y| = 1, \\ \lambda_0(\beta) + 2d - \beta\delta_0(x), & \text{if } y = x. \end{cases}$$

The endpoint of the polymer, x_t, with respect to $P_{\beta,t}$ satisfies

$$\lim_{t\to\infty} P_{\beta,t}(x_t = x) = \frac{\psi_\beta(x)}{\langle \psi_\beta, \mathbb{1} \rangle}. \tag{4.2}$$

For any finite T, the distribution of the process x_s, $s \in [0,T]$, with respect to the measure $P_{\beta,t}$ converges weakly to the Markov process with transition density r_β as $t \to \infty$. The asymptotic invariant distribution of any initial segment of the polymer in the globular phase is thus given by $\pi(x) = \psi_\beta^2(x)$, $x \in \mathbb{Z}^d$, in the sense that,

$$\lim_{s\to\infty} \lim_{t\to\infty} P_{\beta,t}(x_s = x) = \psi_\beta^2(x). \tag{4.3}$$

Also,

$$Z_{\beta,t} = \mathrm{e}^{\lambda_0(\beta)t}\psi_\beta(0) \sum_{x\in\mathbb{Z}^d} \psi_\beta(x) + O(\mathrm{e}^{\delta t}), \quad \textit{for all } 0 < \delta < \lambda_0(\beta). \tag{4.4}$$

Proof. We've already established (4.4). Using

$$p_\beta(t,x,y) = \psi_\beta(x)\psi_\beta(y)\mathrm{e}^{\lambda_0(\beta)t} + O(\mathrm{e}^{\delta t}), \quad \text{for all } 0 < \delta < \lambda_0(\beta) \tag{4.5}$$

we get (4.2) since

$$P_{\beta,t}(x_t = x) = \frac{p_\beta(t,0,x)}{Z_{\beta,t}} = \frac{e^{\lambda_0(\beta)t}\psi_\beta(0)\psi_\beta(x) + O(e^{\delta t})}{e^{\lambda_0(\beta)t}\psi_\beta(0)\sum_{x\in\mathbb{Z}^d}\psi_\beta(x) + O(e^{\delta t})}, \tag{4.6}$$

since $\sum_{x\in\mathbb{Z}^d}\psi_\beta(x) = \langle\psi_\beta, \mathbf{1}\rangle$. Write the semigroup generated by p_β as

$$Q_t f(x) = \sum_{y\in\mathbb{Z}^d} p_\beta(t,x,y)f(y) = e^{tH^\beta}f(x)$$

where the reader will recall that $H^\beta = \Delta + \beta\delta_0$ and Q_t acts on the space of bounded functions. Since ψ_β is the eigenfunction corresponding to $\lambda_0(\beta)$,

$$Q_t\psi_\beta(x) = e^{t\lambda_0(\beta)}\psi_\beta(x).$$

Now the kernel

$$r_\beta(t,x.y) = \frac{p_\beta(t,x,y)\psi_\beta(y)}{\psi_\beta(x)}e^{-\lambda_0(\beta)t}$$

generates a semigroup, which we denote by R_t with $R_t\mathbf{1} = \mathbf{1}$. The generator A_β of R_t as calculated by the formula

$$A_\beta f(x) = \lim_{h\searrow 0}\frac{R_h f(x) - f(x)}{h}$$

results in the expression

$$A_\beta f(x) = \sum_{|y-x|=1} a_\beta(x,y)(f(y) - f(x)),$$

claimed in the theorem. Using the asymptotic formula

$$p_\beta(t,x,y) \sim \psi_\beta(x)\psi_\beta(y)e^{\lambda_0(\beta)t},$$

together with the definition of $r_\beta(t,x,y)$ we see that both

$$\lim_{t\to\infty} r_\beta(t,x,y) = \psi_\beta^2(y).$$

and

$$\sum_{x\in\mathbb{Z}^d}\psi_\beta^2(x)r_\beta(t,x,y) = \psi_\beta^2(y).$$

In addition,

$$\begin{aligned}\sum_z r_\beta(t,x,z)\psi_\beta^{-1}(z) &= \sum_z p_\beta(t,x,z)\psi_\beta^{-1}(x)e^{-\lambda_0(\beta)t} \\ &= \sum_z e^{\lambda_0(\beta)t}\psi_\beta(x)\psi_\beta(z)\psi_\beta^{-1}(x)e^{-\lambda_0(\beta)t} + o(1) \\ &= \langle\psi_\beta, \mathbb{1}\rangle + o(1).\end{aligned} \tag{4.7}$$

This yields the asymptotic finite dimensional distributions of the initial part of the polymer trajectory in the globular ($\beta > \beta_{\mathrm{cr}}$) phase. Namely, if $0 = \tau_0 < \tau_1 < \cdots <$

τ_n: $x_0 = 0, x_1, \ldots, x_n \in \mathbb{Z}^d$, then using (4.7),

$$(4.8)\quad P_{\beta,\tau_n+t}\Big(\bigcap_{j=0}^{n}\{x_{\tau_j} = x_j\}\Big)$$
$$= \frac{\sum_{z\in\mathbb{Z}^d}\prod_{j=1}^{n} p_\beta(\tau_j - \tau_{j-1}, x_{j-1}, x_j)p_\beta(t, x_n, z)}{\sum_{z\in\mathbb{Z}^d} p_\beta(\tau_n + t, x_0, z)}$$
$$= \frac{\sum_{z\in\mathbb{Z}^d}\prod_{j=1}^{n} \mathrm{e}^{\lambda_0(\beta)(\tau_j-\tau_{j-1})}r_\beta(\tau_j - \tau_{j-1}, x_{j-1}, x_j)\mathrm{e}^{\lambda_0(\beta)t}r_\beta(t, x_n, z)\psi_\beta^{-1}(z)}{\sum_{z\in\mathbb{Z}^d} \mathrm{e}^{\lambda_0(\beta)(\tau_n+t)}r_\beta(\tau_n + t, x_0, z)\psi_\beta^{-1}(z)}$$
$$= \prod_{j=1}^{n} r_\beta(\tau_j - \tau_{j-1}, x_{j-1}, x_j)\frac{\sum_z r_\beta(t, x_n, z)\psi_\beta^{-1}(z)}{\sum_z r_\beta(\tau_n + t, x_0, z)\psi_\beta^{-1}(z)}$$
$$\to \prod_{j=1}^{n} r_\beta(\tau_j - \tau_{j-1}, x_{j-1}, x_j), \quad \text{as } t \to \infty.$$

In order to obtain weak converge of the measures $P_{\beta,t}$ on the space of trajectories up to time T we need to establish tightness. According to [2], this requires control of the oscillations. Set

$$\omega_T(x, [t_{i-1}, t_i)) \equiv \sup_{s,t\in[t_{i-1},t_i)} |x_t - x_s|$$

and

$$\omega_T'(x, \delta) \equiv \inf_{\{t_i\}} \max_{1\le i\le r} \omega_T(x, [t_{i-1}, t_i)),$$

where the inf is taken over finite sets $\{t_i\}$ such that

$$(4.9)\qquad 0 = t_0 < t_1 < \cdots < t_r = T, \quad t_i - t_{i-1} > \delta.$$

Then, since our paths all start at 0 under $P_{\beta,t}$ tightness will follow provided for each positive ϵ and η there exists a $\delta \in (0, 1)$and T_0 such that

$$(4.10)\qquad P_{\beta,t}(\omega_T'(x, \delta) \ge \epsilon) \le \eta, \quad t \ge T_0.$$

But, we have

$$(4.11)\quad P_{\beta,t}(\omega_T'(x, \delta) \ge \epsilon) = Z_{\beta,t}^{-1}E_0[\mathrm{e}^{\beta\int_0^n \delta_0(x_s)\,\mathrm{d}s}1_{\{\omega_T'(x,\delta)\ge\epsilon\}}]$$
$$= Z_{\beta,t}^{-1}E_0\big[\mathrm{e}^{\beta\int_0^n \delta_0(x_s)\,\mathrm{d}s}1_{\{\omega_T'(x,\delta)\ge\epsilon\}}E_{x_T}[\mathrm{e}^{\beta\int_0^{t-T} \delta_0(x_s)\,\mathrm{d}s}]\big]$$
$$\le E_0[\mathrm{e}^{\beta T}1_{\{\omega_T'(x,\delta)\ge\epsilon\}}]\frac{Z_{\beta,t-T}}{Z_{\beta,t}}$$
$$\le \mathrm{e}^{\beta T}P_0(\omega_T'(x, \delta) \ge \epsilon) \to 0, \quad t \to \infty.$$

This proves the convergence in distribution of the process on $[0, T]$ under $P_{\beta,t}$ as $t \to \infty$. It also provides the invariant distribution statement (4.3) since

$$\lim_{t\to\infty} P_{\beta,t}\,(x_s = x) = r_\beta(s, 0, x).$$

and

$$\lim_{s\to\infty} r_\beta(s, 0, x) = \psi_\beta^2(x).$$

This proves the theorem. □

An immediate consequence is a relative of a result of Alexander and Sidoravicius [1]. In that paper the notion of pinning is defined in terms of the local time, $L_{x,t} = \int_0^t \delta_x(x_s)\,\mathrm{d}s$. The polymer is said to be pinned if for some $\delta > 0$,

$$\lim_{t\to\infty} P_{\beta,t}\left(\frac{L_{0,t}}{t} > \delta\right) = 1.$$

Corollary 4.2. *Let $L_{x,s} = \int_0^s \delta_x(x_r)\,\mathrm{d}r$, be the local time of the polymer at x up to time s. Then, if $\beta > \beta_{\mathrm{cr}}$, for every $\epsilon > 0$,*

$$\lim_{s\to\infty}\lim_{t\to\infty} P_{\beta,t}\left(\left|\frac{L_{x,s}}{s} - \psi_\beta^2(x)\right| > \epsilon\right) = 0.$$

We now turn attention to the behavior of the polymer trajectory in the diffusive phase, namely $\beta < \beta_{\mathrm{cr}}(d)$, $d \geq 3$. Since this should behave like the free polymer, we consider briefly the random walk with transition probability $p_0(t,x,y)$. Since p_0 satisfies (2.1), the Fourier transform, $\hat{p}_0(t,\phi,y) = \sum_{z\in\mathbb{Z}^d} p_0(t,z,y)\mathrm{e}^{i(\phi,z)}$ satisfies

$$\begin{aligned}\frac{\partial \hat{p}_0}{\partial t}(t,\phi,y) &= -\Phi(\phi)\hat{p}_0(t,\phi,y),\\ \hat{p}_0(0,\phi,y) &= \mathrm{e}^{i(\phi,y)}.\end{aligned} \tag{4.12}$$

So, $\hat{p}_0(t,\phi,y) = \mathrm{e}^{i(\phi,y)}\mathrm{e}^{-t\Phi(\phi)}$. Setting $y = 0$ and inverting gives

$$p_0(t,x,0) = p_0(t,0,x) = \mathrm{e}^{-2dt}\frac{1}{(2\pi)^d}\int_{\mathrm{T}^d} \mathrm{e}^{2t\sum_{j=1}^d \cos\phi_j - i(\phi,x)}\,\mathrm{d}\phi.$$

An explicit formula may be given in terms of Bessel functions which leads to the following results,

$$\begin{aligned} E[\langle\xi,x_t\rangle^2] &= |\xi|^2 t, \quad \xi\in\mathbb{R}^d\\ \operatorname{Cov} x_t &= tI\\ P_0\left(\frac{x_t}{\sqrt{t}}\in A\right) &\to \frac{1}{\sqrt{4\pi}^d}\int_A \mathrm{e}^{-|y|^2}\,\mathrm{d}y.\end{aligned} \tag{4.13}$$

We expect similar results hold for the polymer in the non-globular phase $\beta < \beta_{\mathrm{cr}}$. This expectation arises from the "depinning" of the polymer, in this regime since the attraction of the potential isn't sufficient to make typically transient paths repeatedly visit the origin. As a result, the paths under the polymer measure are very much like the free paths since they eventually stop visiting the origin.

Theorem 4.3. *In the high temperature regime $\beta < \beta_{\mathrm{cr}}$, $d \geq 3$, as $t \to \infty$*

$$\begin{aligned} Z_{\beta,t} &\to \frac{1}{1-\beta I(0)}\\ E_{\beta,t}[\langle\xi,x_t\rangle^2] &\sim |\xi|^2 t, \quad \xi\in\mathbb{R}^d\\ \operatorname{Cov}_{\beta,t} x_t &\sim tI\\ P_{\beta,t}\left(\frac{x_t}{\sqrt{t}}\in A\right) &\to \frac{1}{\sqrt{4\pi}^d}\int_A \mathrm{e}^{-|y|^2}\,\mathrm{d}y\end{aligned} \tag{4.14}$$

PROOF. Assuming $\beta < \beta_{\text{cr}}$, the Fourier transform of the fundamental solution $p_\beta(t,0,x)$ satisfies

$$\begin{aligned}\frac{\partial \hat{p}_\beta}{\partial t}(t,0,\phi) &= -\Phi(\phi)\hat{p}_\beta(t,0,\phi) + \beta p_\beta(t,0,0),\\ \hat{p}_\beta(0,0,\phi) &= 1.\end{aligned} \tag{4.15}$$

Thus,

$$\hat{p}_\beta(t,0,\phi) = \mathrm{e}^{-t\Phi(\phi)} + \beta \int_0^t \mathrm{e}^{-\Phi(\phi)(t-s)} p_\beta(s,0,0)\,\mathrm{d}s. \tag{4.16}$$

Inverting this equation we have

$$p_\beta(t,0,x) = p_0(t,0,x) + \beta \int_0^t p_0(t-s,0,x)p_\beta(s,0,0)\,\mathrm{d}s. \tag{4.17}$$

Introduce the Laplace transforms of $p_\beta(t,0,0)$ and $p_0(t,0,0)$ using the notation

$$\mathcal{P}_\beta(\lambda) = \int_0^\infty \mathrm{e}^{-\lambda t} p_\beta(t,0,0)\,\mathrm{d}t$$

and

$$\mathcal{P}_0(\lambda) = \int_0^\infty \mathrm{e}^{-\lambda t} p_0(t,0,0)\,\mathrm{d}t.$$

Then using $\hat{p}_0(t,0,\phi) = \mathrm{e}^{-t\Phi(\phi)}$ and the Fourier inversion formula

$$p_0(t,0,0) = \frac{1}{(2\pi)^d}\int_{\mathbb{T}} \hat{p}_0(t,0,\phi)\,\mathrm{d}\phi$$

one obtains

$$\mathcal{P}_0(\lambda) = \frac{1}{(2\pi)^d}\int_{\mathbb{T}} \frac{d\phi}{\lambda + \Phi(\phi)} \equiv I(\lambda).$$

It follows from this and (4.17) that

$$\mathcal{P}_\beta(\lambda) = \frac{\mathcal{P}_0(\lambda)}{1 - \beta\mathcal{P}_0(\lambda)}.$$

Now, as we're in the case $\beta < \beta_{\text{cr}}(d)$, it follows that

$$\int_0^\infty p_\beta(t,0,0)\,\mathrm{d}t = \mathcal{P}_\beta(0) = \frac{I(0)}{1-\beta I(0)} < \infty. \tag{4.18}$$

By (4.18), the Gibb's normalization factor converges, namely

$$\begin{aligned} Z_{\beta,t} &= \sum_{x\in\mathbb{Z}^d} p_\beta(t,0,x) = \hat{p}_\beta(t,0,0) = 1 + \beta\int_0^t p_\beta(s,0,0)\,\mathrm{d}s\\ &\to 1 + \beta\int_0^\infty p_\beta(s,0,0)\,\mathrm{d}s, \quad \text{as } t\to\infty\\ &= \frac{1}{1-\beta I(0)}.\end{aligned} \tag{4.19}$$

The asymptotic variance is easy to handle,

$$E_{\beta,t}[\langle\xi,x_t\rangle^2] = Z_{\beta,t}^{-1}\sum_{x\in\mathbb{Z}^d} p_\beta(t,0,x)\langle\xi,x\rangle^2 \equiv D_\beta(\xi,t). \tag{4.20}$$

Then, by (4.17) we have

$$D_\beta(t) = Z_{\beta,t}^{-1}|\xi|^2\left(t + \beta\int_0^t (t-s)p_\beta(s,0,0)\,\mathrm{d}s\right),$$

Since $Z_{\beta,t} \to 1/(1-\beta I(0))$, this implies

$$D_\beta(t) \sim (1-\beta I(0))|\xi|^2 t\left(1 + \beta\int_0^t p_\beta(s,0,0)\,\mathrm{d}s\right) \sim |\xi|^2 t.$$

Thus, in the case $\beta < \beta_{\text{cr}}$, the polymer has linear variance (like the free polymer.) The statement about the covariance is handled in an entirely analogous manner. In order to derive the Central Limit Theorem, evaluate the Fourier transform $\hat{p}_\beta(t,0,\phi)$ at $\phi = \tau/\sqrt{t}$, $\phi \in \mathbb{R}$. Since $\Phi(\phi) \sim |\phi|^2$ as $\phi \to 0$, and using (4.16) we see that

$$\hat{p}_\beta(t,0,\phi) = (1+o(1))\mathrm{e}^{-|\phi|^2} + (1+o(1))\beta\int_0^t \mathrm{e}^{-|\phi|^2(1-\frac{s}{t})}p_\beta(s,0,0)\,\mathrm{d}s \tag{4.21}$$

$$\to \mathrm{e}^{-|\phi|^2}\frac{1}{1-\beta I(0)}, \qquad \text{as } t \to \infty.$$

From (4.21) and (4.19) follows immediately the Central Limit Theorem for the distribution of $x_t/\sqrt{t}$ with respect to the Gibb's measure $P_{\beta,t}$ as $t \to \infty$.. □

5. Behavior of the polymer at $\beta = \beta_{\text{cr}}(d)$

Our analysis in this case rests on the asymptotic behavior of

$$I(\lambda) = \frac{1}{(2\pi)^d}\int_{\mathbb{T}^d}\frac{d\phi}{\lambda + \Phi(\phi)},$$

as $\lambda \to 0$. We examine this in the next lemma for $d = 3$. Higher dimensions can be treated in an entirely analogous manner.

Lemma 5.1. *If $d = 3$, there is a constant $c_3 > 0$ such that as $\lambda \to 0$,*

$$I(\lambda) = I(0) - c_3\sqrt{\lambda}(1+o(1)).$$

PROOF. This was established in [3]. □

According to the lemma, the Laplace transform satisfies

$$\mathcal{P}_{\beta_{\text{cr}}}(\lambda) = \frac{I(\lambda)}{1-\beta_{\text{cr}}I(\lambda)} \sim \frac{I(0)}{c\beta_{\text{cr}}\sqrt{\lambda}} = \frac{c}{\sqrt{\lambda}}, \quad \lambda \to 0. \tag{5.1}$$

This last asymptotic expression is valid not only for real λ but also for $\lambda \in \mathbb{C}$. Inverting gives

$$p_{\beta_{\text{cr}}}(t,0,0) = \frac{1}{2\pi}\int_{-\infty}^{\infty}\mathcal{P}_{\beta_{\text{cr}}}(\lambda + ix)\mathrm{e}^{t(\lambda+ix)}\,\mathrm{d}x,$$

and by deformation of contour, we get

$$p_{\beta_{\text{cr}}}(t,0,0) = \frac{1}{2\pi}\int_\gamma \mathcal{P}_{\beta_{\text{cr}}}(z)\mathrm{e}^{tz}\,\mathrm{d}z,$$

where γ is the curve with image the imaginary axis above i and below $-i$ and the semicircle $x^2+y^2 = \delta^2$, $x > 0$. By the smoothness of $\mathcal{P}_{\beta_{\text{cr}}}(z)$ for $|\operatorname{Im} z| > \delta$, we obtain

Lemma 5.2. *For $\beta = \beta_{\rm cr}$, $d = 3$, there is a positive constant c_2 such that*

$$p_{\beta_{\rm cr}}(t,0,0) \sim \frac{c_2}{\sqrt{t}}, \qquad t \to \infty.$$

As a consequence,we can derive the limiting distribution and variance of x_t with respect to $P_{\beta_{\rm cr},t}$ as $t \to \infty$.

Theorem 5.3. *For $d = 3$ and $\beta = \beta_{\rm cr}$ as $t \to \infty$,*

$$\begin{aligned} Z_{\beta_{\rm cr},t} &\sim 2c_2\beta_{\rm cr}\sqrt{t}, \\ E_{\beta_{\rm cr},t}[\langle \xi, x_t\rangle^2] &\sim \tfrac{2}{3}|\xi|^2 t, \\ \operatorname{Cov}_{\beta_{\rm cr},t}(x_t) &\sim \tfrac{2}{3} tI, \\ P_{\beta_{\rm cr},t}\Big(\frac{x_t}{\sqrt{t}} \in \cdot\Big) &\to P(\xi \in \cdot) \end{aligned} \tag{5.2}$$

where ξ is a random vector with characteristic function

$$f(\phi) = \frac{1}{2}\int_0^1 \mathrm{e}^{-|\phi|^2(1-u)} \frac{1}{\sqrt{u}}\,\mathrm{d}u, \qquad \phi \in \mathbb{R}^d.$$

PROOF. First substitute $\phi = 0$ into (4.16) and use Lemma 5.2 to get

$$Z_{\beta_{\rm cr},t} = \hat{p}_{\beta_{\rm cr}}(t,0,0) = 1 + \beta_{\rm cr}\int_0^t p_{\beta_{\rm cr}}(s,0,0)\,\mathrm{d}s \sim 2c_2\beta_{\rm cr}\sqrt{t}, \quad t \to \infty. \tag{5.3}$$

As before,

$$\begin{aligned} E_{\beta_{\rm cr},t}&[\langle \xi, x_t\rangle^2] \\ &= Z_{\beta_{\rm cr},t}^{-1} \sum_{x\in\mathbb{Z}^d} p_{\beta_{\rm cr}}(t,0,x)\langle \xi, x\rangle^2 \\ &= Z_{\beta_{\rm cr},t}^{-1}\Big(|\xi|^2 t + \beta_{\rm cr}|\xi|^2\int_0^t (t-s)p_{\beta_{\rm cr}}(s,0,0)\,\mathrm{d}s\Big) \\ &\sim (2c_2\beta_{\rm cr}\sqrt{t})^{-1}|\xi|^2 t\Big(1 + \beta_{\rm cr}c_2 t^{1/2}\int_0^1 (1-u)u^{-1/2}\,\mathrm{d}u\Big), \quad t\to\infty \\ &\sim \tfrac{2}{3}|\xi|^2 t. \end{aligned} \tag{5.4}$$

A similar argument handles the covariance claim. As a result, we see that the proper normalization of x_t for a limiting law would be $x_t/\sqrt{t}$. Interestingly, the limiting distribution in the critical case $\beta = \beta_{\rm cr}$ is a mixture of Gaussians. Evaluate (4.16) at $\phi/\sqrt{t}$ with $\phi \in \mathbb{R}$ to get

$$\hat{p}_{\beta_{\rm cr}}\Big(t,0,\frac{\phi}{\sqrt{t}}\Big) = \mathrm{e}^{-|\phi|^2}(1+o(1)) + c_2\beta_{\rm cr}\int_0^t (1+o(1))\mathrm{e}^{-|\phi|^2(1-s/t)}s^{-1/2}\,\mathrm{d}s. \tag{5.5}$$

After normalization by $Z_{\beta_{\rm cr},t} \sim 2c_2\beta_{\rm cr}\sqrt{t}$ it follows that

$$E_{\beta_{\rm cr},t}[\mathrm{e}^{-\langle x_t,\phi/\sqrt{t}\rangle}] \to \frac{1}{2}\int_0^1 \mathrm{e}^{-|\phi|^2(1-u)}\frac{1}{\sqrt{u}}\,\mathrm{d}u.$$

That completes the proof. □

We now examine the approach to the mixed Gaussian phase at $\beta = \beta_{\rm cr}$. That is, given a large t, we give a value $\beta(t) < \beta_{\rm cr}$ close to $\beta_{\rm cr}$ such that the distribution of $x_t/\sqrt{t}$ under $P_{\beta(t),t}$ is approximately the mixed Gaussian appearing at the critical parameter $\beta_{\rm cr}$. The result is the following

Theorem 5.4. *If $\beta_{\mathrm{cr}}-\beta(t)=o(t^{-1/2})$ as $t\to\infty$ then the distribution of $x_t/\sqrt{t}$ approaches the mixed Gaussian with characteristic function $\frac{1}{2}\int_0^1 \mathrm{e}^{-|\phi|^2(1-u)}\,\mathrm{d}u/\sqrt{u}$.*

PROOF. For $0\le\beta<\beta_{\mathrm{cr}}$ define for $0\le s\le t$,

$$\Delta_\beta(s)=p_{\beta_{\mathrm{cr}}}(s,0,0)-p_\beta(s,0,0).$$

Now fix a large value of t and let $\beta=\beta(t)$ depend on t in such a way that $\beta_{\mathrm{cr}}-\beta(t)=o(t^{-1/2})$. From (4.17) we derive the formula

$$\Delta_\beta(s)=(\beta_{\mathrm{cr}}-\beta)\Delta_0(s)+\beta\int_0^s p_0(s-r,0,0)\Delta_\beta(r)\,\mathrm{d}r,\quad 0\le s\le t. \tag{5.6}$$

Evaluating at $s=t$ and iterating, we get with $t_0=0$,

$$\begin{aligned}\Delta_\beta(t)&=(\beta_{\mathrm{cr}}-\beta)\\ &\times\left(\Delta_0(t)+\sum_{n=1}^\infty\beta^n\int_0^t\int_0^{t_1}\cdots\int_0^{t_{n-1}}\prod_{i=1}^n p_0(t_i-t_{i-1},0,0)\Delta_0(t_n)\,\mathrm{d}t_n\cdots\mathrm{d}t_1\right).\end{aligned} \tag{5.7}$$

And using the simple bound $|\Delta_0(s)|\le 1$, it follows that

$$\begin{aligned}&|\Delta_\beta(t)|\\ &\le(\beta_{\mathrm{cr}}-\beta)\left(1+\sum_{n=1}^\infty\beta^n\int_0^t\int_0^{t_1}\cdots\int_0^{t_{n-1}}\prod_{i=1}^n p_0(t_i-t_{i-1},0,0)\,\mathrm{d}t_n\cdots\mathrm{d}t_1\right).\end{aligned} \tag{5.8}$$

But,

$$\begin{aligned}&\int_0^t\int_0^{t_1}\cdots\int_0^{t_{n-1}}\prod_{i=1}^n p_0(t_i-t_{i-1},0,0)\,\mathrm{d}t_n\cdots\mathrm{d}t_1\\ &\quad=\int_0^t\int_0^{t_1}\cdots\int_0^{t_{n-1}}\prod_{i=1}^n p_0(t_i,0,0)\,\mathrm{d}t_n\cdots\mathrm{d}t_1\\ &\quad=\frac{\left(\int_0^t p_0(s,0,0)\,\mathrm{d}s\right)^n}{n!}\le\frac{\left(\int_0^\infty p_0(s,0,0)\,\mathrm{d}s\right)^n}{n!}=\frac{I^n(0)}{n!}=\frac{1}{\beta_{\mathrm{cr}}^n n!}.\end{aligned} \tag{5.9}$$

Inserting (5.9) in (5.8) gives

$$|\Delta_\beta(t)|\le(\beta_{\mathrm{cr}}-\beta)\mathrm{e}^{\beta/\beta_{\mathrm{cr}}}. \tag{5.10}$$

Now the characteristic function of $x_t/\sqrt{t}$ with respect to $P_{\beta,t}$ is

$$\frac{\hat{p}_\beta(t,0,\phi/\sqrt{t})}{Z_{\beta,t}}.$$

Analyzing the denominator, by Lemma 5.2,

$$p_{\beta_{\mathrm{cr}}}(t,0,0)\sim\frac{c_2}{\sqrt{t}}$$

and

$$\left|\int_0^t\Delta_\beta(s)\,\mathrm{d}s\right|\le(\beta_{\mathrm{cr}}-\beta)\mathrm{e}^{\beta/\beta_{\mathrm{cr}}}t$$

so that in view of

$$(\beta_{\mathrm{cr}}-\beta)=o(t^{-1/2})$$

we have

$$
\begin{aligned}
(5.11)\qquad Z_{\beta,t} &= 1-\beta\int_0^t \Delta_\beta(s)\,\mathrm{d}s+\beta\int_0^t p_{\beta_{\mathrm{cr}}}(s,0,0)\,\mathrm{d}s\\
&= 1-\beta\int_0^t \Delta_\beta(s)\,\mathrm{d}s+(\beta-\beta_{\mathrm{cr}})\int_0^t p_{\beta_{\mathrm{cr}}}(s,0,0)\,\mathrm{d}s\\
&\qquad\qquad +\beta_{\mathrm{cr}}\int_0^t p_{\beta_{\mathrm{cr}}}(s,0,0)\,\mathrm{d}s\\
&= \beta_{\mathrm{cr}}c_2\sqrt{t}\bigl(1+o(1)\bigr).
\end{aligned}
$$

For the numerator, if t is large enough so that $\mathrm{e}^{-\Phi(\phi/\sqrt{t})}\sim \mathrm{e}^{-|\phi|^2}$,

$$
\begin{aligned}
(5.12)\quad &\hat{p}_\beta\Bigl(t,0,\frac{\phi}{\sqrt{t}}\Bigr)\\
&\sim \mathrm{e}^{-|\phi|^2}-\beta\int_0^t \mathrm{e}^{-|\phi|^2(1-s/t)}\Delta_\beta(s)\,\mathrm{d}s+\beta\int_0^t \mathrm{e}^{-|\phi|^2(1-s/t)}p_{\beta_{\mathrm{cr}}}(s,0,0)\,\mathrm{d}s\\
&= \mathrm{e}^{-|\phi|^2}+o(1)\sqrt{t}-(\beta_{\mathrm{cr}}-\beta)\int_0^t \mathrm{e}^{-|\phi|^2(1-s/t)}p_{\beta_{\mathrm{cr}}}(s,0,0)\,\mathrm{d}s\\
&\qquad\qquad +\beta_{\mathrm{cr}}\int_0^t \mathrm{e}^{-|\phi|^2(1-s/t)}p_{\beta_{\mathrm{cr}}}(s,0,0)\,\mathrm{d}s\\
&\sim \mathrm{e}^{-|\phi|^2}+o(1)\sqrt{t}+(\beta_{\mathrm{cr}}-\beta)\sqrt{t}+\beta_{\mathrm{cr}}\int_0^t \mathrm{e}^{-|\phi|^2(1-s/t)}p_{\beta_{\mathrm{cr}}}(s,0,0)\,\mathrm{d}s.
\end{aligned}
$$

Thus,

$$
\begin{aligned}
(5.13)\quad &\frac{\hat{p}_\beta(t,0,\phi/\sqrt{t})}{Z_{\beta,t}}\\
&\sim \frac{1}{c_2\beta_{\mathrm{cr}}\sqrt{t}}\biggl(O(1)+o(1)\sqrt{t}+(\beta_{\mathrm{cr}}-\beta)\sqrt{t}+\beta_{\mathrm{cr}}\int_0^t \mathrm{e}^{-|\phi|^2(1-s/t)}p_{\beta_{\mathrm{cr}}}(s,0,0)\,\mathrm{d}s\biggr)\\
&\sim \frac{\beta_{\mathrm{cr}}\int_0^t \mathrm{e}^{-|\phi|^2(1-s/t)}p_{\beta_{\mathrm{cr}}}(s,0,0)\,ds}{c_2\;\beta_{\mathrm{cr}}\sqrt{t}}\\
&\sim \frac{1}{2}\int_0^1 \mathrm{e}^{-|\phi|^2(1-u)}\frac{\mathrm{d}u}{\sqrt{u}}
\end{aligned}
$$

as was to be shown. □

References

1. K. S. Alexander and V. Sidoravicius, *Pinning of polymers and interfaces by random potentials*, Ann. Appl. Probab. **16** (2006), no. 2, 636–669, available at arXiv:math.PR/0501028.
2. P. Billingsley, *Convergence of probability measures*, 2nd ed., Wiley Ser. Probab. Statist. Probab. Statist., Wiley, New York, 1999.
3. R. A. Carmona and S. A. Molchanov, *Parabolic Anderson problem and intermittency*, Mem. Amer. Math. Soc. **108** (1994), no. 518.
4. A. M. Gutin, A. Yu. Grosberg, and E. I. Shakhnovich, *Globular state of branched random heteropolymers*, J. Phys. A **26** (1993), no. 5, 1037–1049.
5. I. M. Lifshitz, A. Yu. Grosberg, and A. R. Khokhlov, *Some problems of the statistical physics of polymer chains with volume interaction*, Rev. Modern Phys. **50** (1978), no. 3, 683–713.

Department of Mathematics, University of California — Irvine, 281 Multipurpose Science & Technology Bldg, Irvine, CA 92697-3875, USA
E-mail address: mcransto@math.uci.edu

Department of Mathematics and Statistics, University of North Carolina — Charlotte, 376 Fretwell Bldg., 9201 University City Blvd., Charlotte, NC 28223-0001, USA
E-mail address: smolchan@uncc.edu

Centre de Recherches Mathématiques
CRM Proceedings and Lecture Notes
Volume **42**, 2007

Asymptotics of the Poincaré Functions

Gregory Derfel, Peter J. Grabner, and Fritz Vogl

Abstract. The asymptotic behavior of the solutions of of Poincaré's functional equation $f(\lambda z) = P(f(z))$ ($\lambda \in \mathbb{C}$, $|\lambda| > 1$) for P a polynomial of degree ≥ 2 is studied in different regions of the complex plane.

1. Introduction

In 1890 H. Poincaré [12] has studied the equation

$$f(\lambda z) = R(f(z)), \quad z \in \mathbb{C}, \tag{1.1}$$

where $R(z)$ is a rational function and $\lambda \in \mathbb{C}$. He proved that, if $R(0) = 0$, $R'(0) = \lambda$ and $|\lambda| > 1$, then there exists a meromorphic or entire solution of (1.1). After Poincaré, solutions of (1.1) are called *the Poincaré functions admitting a multiplication theorem* (cf. [16]). Later on, G. Valiron [15, 16] elaborated the case, where $R(z) = P(z)$ is a polynomial, i.e.,

$$f(\lambda z) = P(f(z)), \quad z \in \mathbb{C}, \tag{1.2}$$

and obtained conditions for the existence of an entire solution $f(z)$. Furthermore, he derived the following asymptotic formula for $M(r) = \max_{|z|=r} |f(z)|$:

$$\log M(r) \sim r^{\rho} Q\left(\frac{\log r}{\log|\lambda|}\right), \quad r \to \infty. \tag{1.3}$$

Here Q is a 1-periodic function bounded between two positive constants, $\rho = \log m / \log|\lambda|$ and $m = \deg P(z)$.

An interesting example of such equations, which stems from the description of Brownian motion on Sierpinski's gasket, has been studied in [1, 7, 8]. This is the functional equation

$$f(5z) = 4f^2(z) - 3f(z). \tag{1.4}$$

Similar equations arise also in the theory of branching processes (cf. [2, 9]).

Different aspects of the Poincaré functions are discussed in papers [3, 5, 6, 13] closely related to the present one.

2000 *Mathematics Subject Classification.* Primary 30D05; Secondary 39B32, 30D15, 37F10.

The second author is supported by the Austrian Science Foundation FWF, project S9605, part of the Austrian National Research Network "Analytic Combinatorics and Probabilistic Number Theory".

This is the final form of the paper.

In our paper we derive further results of Valiron's type. Namely, in addition to (1.3) we find asymptotics of entire solutions $f(z)$ on various rays $\arg z = \vartheta$ of the complex plane. It turns out that this heavily depends on the arithmetic nature of λ.

For instance, the following statement (Theorem 4.5, below) is proved:

If $\arg\lambda = 2\pi\beta$ *and* β *is irrational, then* $f(z)$ *is unbounded along any ray* $\arg z = \vartheta$. *Moreover, if we denote* $\varphi(z) = \log|f(z)|$ *(where the main branch of logarithm is taken) then there exists a sequence* $r_n \to \infty$, *such that the limit*

$$\lim_{n\to\infty} \frac{\varphi(r_n e^{i\vartheta})}{r_n^{\rho}} = L \tag{1.5}$$

exists and $L > 0$.

On the other hand, if β is rational (and, in particular, if $\beta = 0$, i.e., λ is real) $f(z)$ may be bounded on some rays and even in whole sectors. Nevertheless, for rational β, the limit (1.5) still exists under some additional assumptions. Denote $\beta = t/s$ and suppose that t, s are relatively prime. Put $q = \lambda^s$ (note that $1 < q \in \mathbb{R}$). Here we have the following result (see Theorems 5.2 and 5.3 below):

Suppose that either $|\lambda| > m^2$ *or* $s > 2\rho$. *Then* $f(z)$ *is unbounded on any ray, and one can find a geometric progression* $r_n = q^n r_0$, $(r_0 > 0)$, *for which the limit* (1.5) *exists and* $L > 0$.

Further refinements are possible when $\lambda > 1$ is real and $P(z) = p_m z^m + \cdots + p_1 z + p_0$ is a polynomial with real coefficients. Namely, we prove the the following statement (Theorem 6.3):

Assume that: (a) $\lambda > m$; (b) $p_i \geq 0$, *for* $i \geq 2$; (c) *All preimages of* 0 *under* P, *i.e.,* $P^{-n}(\{0\})$ *are real. Let* $f(z)$ *be an entire solution of* (1.2) *such that* $f(0)$ *and* $f'(0)$ *are real and* $f(0) \geq 0$ *and* $f'(0) > 0$. *Then* $f(z)$ *tends to infinity along any ray in the sector* $0 < |\vartheta| < \pi/2$. *Moreover,*

$$\varphi(re^{i\vartheta}) \sim r^{\rho} Q_{\vartheta}\left(\frac{\log r}{\log\lambda}\right), \qquad r \to \infty \tag{1.6}$$

where $Q_{\vartheta}(z)$ *is* 1*-periodic and bounded between two positive constants.*

Condition (c) plays an important role in the last statement. If $P(z)$ is quadratic polynomial (a case, arising in some applications) it is possible to give an exact criterion for reality of $P^{-n}(\{0\})$ (see Lemma 6.7 below):

Let

$$P(z) = az(z-\omega), \quad 0 \neq \omega \in \mathbb{R} \tag{1.7}$$

All preimages of 0 *under* P *are real, if and only if the following condition is fulfilled*

$$a|\omega| \geq \begin{cases} 2 & \text{for } \omega > 0 \\ 4 & \text{for } \omega < 0. \end{cases} \tag{1.8}$$

The paper is organized as follows. In Section 2 we discuss some elementary but important properties of the Poincaré equation (1.2) and its (formal) power series solutions. Section 3 is devoted to a study of the asymptotics of solutions for (1.2) along *spirals* (i.e., geometric progression of the form $z_n = \lambda^n z_0$, $\lambda \in \mathbb{C}$). As a consequence of these results, in Section 4, we obtain asymptotics of solutions on different *rays* of the complex plane. We split the presentation into two different cases:

(a) $\beta = \frac{1}{2\pi}\arg\lambda$ is rational (Section 4.1)

(b) β is irrational (Section 4.2)

Sufficient conditions for unboundedness of solutions along all rays emanating from the origin are given in Section 5.

Further refinements are possible, when $\lambda > 1$ is real and $P(z)$ is a polynomial with real and "almost positive" coefficients — these are derived in Section 6. Also, we specialize some of our results for quadratic polynomials.

In the final Section 7 we present some results on the asymptotics of solutions $f(z)$ (rather than $|f(z)|$) in some angular regions of the complex plane.

2. Preliminary remarks, simplifications of the equation, formal power series solution

Consider the equation

$$f(\lambda z) = P[f(z)], \tag{2.1}$$

where

$$P(z) = p_m z^m + p_{m-1} z^{m-1} + \cdots + p_0. \tag{2.2}$$

Throughout this paper we assume that $m \geq 2$ and $|\lambda| > 1$.

2.1. Simplifications of the equation. The following Proposition 2.1 shows that, without loss of generality, we may always assume that $p_m = 1$.

Proposition 2.1. *Let $f(z)$ be a solution for* (2.1). *Then*

$$g(z) = p_m^{-1/(m-1)} f(z) \tag{2.3}$$

is a solution of the equation

$$g(\lambda z) = \widetilde{P}[g(z)], \tag{2.4}$$

where

$$\widetilde{P}(z) = z^m + \tilde{p}_{m-1} z^{m-1} + \cdots + \tilde{p}_0,$$

is a polynomial with the leading coefficient equal to 1.

(*More precisely,*

$$\tilde{p}_k = p_k c^{k-1}; \quad \text{for } k = 0, \ldots, m \text{ with } c = p_m^{-1/(m-1)}, \tag{2.5}$$

In particular, $\tilde{p}_0 = p_0 c^{-1} = p_0 p_m^{1/(m-1)}$.)

Proof. Put $f(z) = cg(z)$ and insert this in (2.1).

Then we have

$$cg(\lambda z) = p_m c^m [g(z)]^m + p_{m-1} c^{m-1} [g(z)]^{m-1} + \cdots + p_0$$

or

$$g(\lambda z) = p_{m-1} c^{m-1} [g(z)]^m + p_{m-1} c^{m-2} [g(z)]^{m-2} + \cdots + c^{-1} p_0$$

Next, choose c so that $p_m c^{m-1} = 1$, i.e., $c = p_m^{-1/(m-1)}$. This completes the proof. □

Example 1. By virtue of the formula $\cos 2z = 2\cos^2 z - 1$ the function $f(z) = \cos z$ satisfies the equation

$$f(2z) = 2[f(z)]^2 - 1$$

Here $p_2 = 2, c = p_2^{-1} = 2^{-1}$ and therefore $\tilde{p}_2 = 1$, $\tilde{p}_0 = -2$.

Therefrom, according to Proposition 2.1, the function $g(z) = 2\cos z = e^{iz} + e^{-iz}$ satisfies the equation

$$g(2z) = g(z)^2 - 2$$

Proposition 2.2. *Let $f(z)$ be a solution of (2.1). Then*

$$g(z) = f(z^\gamma), \quad \text{for } \gamma > 0 \tag{2.6}$$

is a solution of the equation

$$g(\lambda^{1/\gamma} z) = P[g(z)] \tag{2.7}$$

PROOF.

$$g(\lambda^{1/\gamma} z) = f[(\lambda^{1/\gamma} z)^\gamma] = f(\lambda z^\gamma) = P[f(z^\gamma)] = P[g(z)]$$ □

Example 2. As it was mentioned already the function $f(z) = \cos z$ satisfies equation

$$f(2z) = P[f(z)],$$

where $P(z) = 2z^2 - 1$. Put $g(z) = \cos\sqrt{z}$, i.e., choose $\gamma = 1/2$ in (2.6). Then, in accordance with Proposition 2.2, the (entire) function $g(z) = \cos\sqrt{z}$ satisfies the equation

$$g(4z) = \cos\sqrt{4z} = \cos 2\sqrt{z} = 2(\cos\sqrt{z})^2 - 1 = P[g(z)].$$

H. Poincaré in [11, 12] and G. Valiron in [16] assumed that $P(0) = 0$. There is no loss of generality in this assumption, as is shown in the following Proposition 2.3. The condition $P(0) = 0$ can be always achieved by simple change of variables.

Proposition 2.3. *Suppose that $f(z)$ satisfies equation (2.1) and initial condition*

$$f(0) = c_0.$$

(In accordance with Lemma 2.5 below, it is necessary that $c_0 = P(c_0)$, i.e., c_0 is fixed point of $P(z)$.) Then the function $g(z) = f(z) - c_0$ satisfies the equation

$$g(\lambda z) = \widetilde{P}[g(z)],$$

where

$$\widetilde{P}(z) = P(z + c_0) - c_0$$

and the initial condition

$$g(0) = 0$$

In addition,

$$\widetilde{P}(0) = 0,$$

i.e., 0 is a fixed point of $\widetilde{P}(z)$ (which corresponds to the fixed point c_0 of $P(z)$). Furthermore, if c_i $(i = 0, 1, \ldots, m-1)$ are fixed points of $P(z)$ then $\tilde{c}_i = c_i - c_0$ $(i = 0, \ldots, \ell)$ are fixed points of $\widetilde{P}(z)$ and $\widetilde{P}'(\tilde{c}_i) = P'(c_i)$

PROOF. Firstly

$$g(\lambda z) = f(\lambda z) - c_0 = P[f(z)] - c_0 = P[g(z) + c_0] - c_0 \equiv \widetilde{P}[g(z)]$$

Also, clearly

$$g(0) = f(0) - c_0 = c_0 - c_0 = 0; \quad \widetilde{P}(0) = P(c_0) - c_0 = 0$$

Next, suppose that c_i $(i = 0, \ldots, \ell)$ is a fixed point of $P(z)$. Then

$$\widetilde{P}(c_i - c_0) = P[(c_i - c_0) + c_0] - c_0 = P(c_i) - c_0 = c_i - c_0$$

i.e. $\widetilde{c} = c_i - c_0$ is a fixed point of $\widetilde{P}(z)$. Finally, $\widetilde{P}'(z) = P'(z + c_0)$ for any z and, in particular, $\widetilde{P}'(\widetilde{c}_i) = \widetilde{P}'(c_i - c_0) = P'(c_i)$. □

2.2. Formal power series solutions of (2.1). Let

$$f(z) = \sum_{k=0}^{\infty} f_k z^k \tag{2.8}$$

be a formal power series solution of (2.1). It is easily verified that $f(z)$ satisfies (2.1) if and only if the coefficients f_n satisfy the following recurrence

$$P(f_0) = f_0 \tag{2.9}$$

$$(\lambda - \alpha) f_1 = 0, \quad \text{with } \alpha := P'(f_0) = \sum_{k=1}^{m} k p_k f_0^{k-1} \tag{2.10}$$

$$\begin{aligned}(\lambda^n - \alpha) f_n &= \sum_{k=2}^{m} p_k \sum_{\substack{\ell_1 + \cdots + \ell_k = n \\ \ell_1, \ldots, \ell_k < n}} f_{\ell_1} f_{\ell_2} \cdots f_{\ell_K} \\ &=: S_n(p_2, \ldots, p_m, f_0, f_1, \ldots f_{n-1})\end{aligned} \tag{2.11}$$

Remark 2.4. We call a power series solution *trivial*, if it degenerates to the constant f_0. Note, that in view of (2.1), a nontrivial solution is nonpolynomial.

The following lemma immediately follows from (2.9) – (2.11).

Lemma 2.5. *The following assertions hold:*

(1) *The only admissible values for f_0 (at most m) are the fixed points of $P(z)$.*
(2) *A nontrivial formal power solution $f(z)$ exists if and only if there is an $n_0 \in \mathbb{N}$ such that*

$$\lambda^{n_0} = \alpha = P'(f_0) \tag{2.12}$$

(3) *If $n_0 \geq 2$ and $f(z)$ is a nontrivial solution, then*

$$f_{kn_0+1} = f_{kn_0+2} = \cdots = f_{(k+1)n_0-1} = 0 \quad \textit{for } k = 0, 1, \ldots \tag{2.13}$$

and there are infinitely many f_i, such that $f_i \neq 0$.
(4) *$S_n(p_2, \ldots, p_m, f_0, f_1, \ldots, f_{n-1})$ is a polynomial of its arguments with positive coefficients.*

Remark 2.6. Let $f(z)$ be a nontrivial formal power solution of (2.1), which satisfies (2.12) with $n_0 \geq 2$. Then, from Lemma 2.5(3) it follows that:

(1) $f(z)$ is even, if n_0 is even
(2) the function

$$g(z) = f(\sqrt[n_0]{z})$$

can be developed into a formal power series and satisfies the equation

$$g(\lambda^{n_0} z) = P[g(z)]$$

in a line with Proposition 2.2. In other words, $g(z)$ is formal power series solution of equation

$$g(\mu z) = P[g(z)],$$

where

$$\mu := \lambda^{n_0} = \alpha = P'(g_0) = P'(f_0)$$

i.e. $g(z)$ satisfies (2.12), with $n_0 = 1$

2.3. Entire solution versus formal solution. We conclude this section with the following classical theorem of Valiron.

Theorem 2.7 (Valiron [16]). *Any formal power series solution $f(z)$ of (2.1) is an entire function of order*

$$\rho = \frac{\log m}{\log |\lambda|}.$$

Furthermore, denote $M(r) = \max_{|z|\leq r} |f(z)|$. Then

$$\log M(r) \sim r^\rho Q\left(\frac{\log r}{\log |\lambda|}\right), \qquad r \to \infty,$$

where Q is a 1-periodic function bounded between two positive constants.

3. Asymptotics of solutions along spirals

As it was mentioned already in the introduction the main objective of this paper is to derive asymptotics for the solutions of (2.1) along different rays $\arg z = \vartheta$ of the complex plane $\mathbb{C}$.

Latter asymptotics can be derived from the asymptotics of $f(z)$ along spirals (geometric progressions) of the form $z_n = \lambda^n z_0$ where $z_0 = r_0 \mathrm{e}^{\mathrm{i}\vartheta_0} \in \mathbb{C}$ will be specified below. Corresponding results are given in Theorem 3.1, below.

Consider the Poincaré equation

$$f(\lambda z) = P[f(z)], \quad |\lambda| > 1 \tag{3.1}$$

where (as indicated in Section 2.1, we can always assume that P is normalized)

$$P(z) = z^m + p_{m-1} z^{m-1} + \cdots + p_1 z + p_0, \quad m \geq 2. \tag{3.2}$$

Denote

$$K = \max\{|p_0|, \ldots, |p_{m-1}|\}. \tag{3.3}$$

Suppose that $z_0 = r_0 \mathrm{e}^{\mathrm{i}\vartheta_0} \in \mathbb{C}$ is such a point, that

$$|f(z_0)| > \max\{\mathrm{e}, 2mK\}. \tag{3.4}$$

Denote

$$\varphi(z) = \log |f(z)|; \quad z_n = \lambda^n z_0; \quad \rho = \frac{\log m}{\log |\lambda|} \tag{3.5}$$

(It follows from (3.4) that $\varphi(z_0) > 1$.)

Theorem 3.1. *Suppose that $f(z)$ is a solution of (3.1) and (3.4) is satisfied. Then the limit along the spiral $z_n = \lambda^n z_0$*

$$\lim_{n\to\infty} \frac{\varphi(z_n)}{|z_n|^\rho} = L(z_0) \tag{3.6}$$

exists and

$$\frac{\varphi(z_0) - 3Km/(2|f(z_0)|)}{|z_0|^\rho} < L(z_0) < \frac{\varphi(z_0) + 3Km/(2|f(z_0)|)}{|z_0|^\rho}. \tag{3.7}$$

In particular, in view of (3.4), $L(z_0)$ is bounded between two positive constants:

$$\frac{\varphi(z_0) - 3/4}{|z_0|^\rho} < L < \frac{\varphi(z_0) + 3/4}{|z_0|^\rho}. \tag{3.8}$$

Furthermore, if $f(z)$ is an entire solution of (3.1), then L is a continuous function on the domain $\{z \in \mathbb{C} \mid |f(z)| > \max(\mathrm{e}, 2mK)\}$.

Remark 3.2. Throughout this paper we deal mainly with the asymptotics of entire solutions of (3.1). However, some statements of the present section and Section 4 are valid for arbitrary solutions of (3.1). In particular, for validity of (3.6) and (3.7) no assumptions on the smoothness of $f(z)$ are needed.

The proof is similar to the one of [4, Lemma 1]. It may be divided into two main steps. We shall state these two steps as Lemmas 3.3 and 3.8 below.

Lemma 3.3. *Suppose that* (3.4) *is fulfilled. Then*

$$\log f(\lambda z_0) = m \log f(z_0) + r(z_0), \tag{3.9}$$

where

$$r(z) = \log\left(1 + \frac{p_{m-1}}{f(z)} + \cdots + \frac{p_0}{f(z)^m}\right)$$

and

$$|r(z_0)| \leq \frac{3mK}{2|f(z_0)|}. \tag{3.10}$$

Denote $\Re r(z_0) = r_1(z_0)$ and equate real parts in (3.9).
As an immediate consequence of of Lemma 3.3 we obtain

Corollary 3.4. *If* (3.4) *is satisfied, then*

$$\log |f(\lambda z_0)| = m \log |f(z_0)| + r_1(z_0) \tag{3.11}$$

or

$$\varphi(\lambda z_0) = m\varphi(z_0) + r_1(z_0), \tag{3.12}$$

where

$$|r_1(z_0)| \leq \frac{3mK}{2|f(z_0)|}. \tag{3.13}$$

PROOF OF LEMMA 3.3. Let us write $P(z)$ in the form

$$P(z) = z^m\left(1 + \frac{p_m}{z} + \cdots + \frac{p_0}{z^m}\right) =: z^m\big(1 + R(z)\big). \tag{3.14}$$

Then we have for $|z| \geq 1$

$$|R(z)| \leq \frac{mK}{|z|}.$$

From (3.14) it follows that

$$\log P(z) = m \log z + \tilde{r}(z), \tag{3.15}$$

where $\tilde{r}(z) = \log\big(1 + R(z)\big)$. Moreover, if (3.4) is satisfied, then

$$|\tilde{r}(z_0)| \leq \frac{3}{2}|R(z_0)| \leq \frac{3}{2}\frac{mK}{|z_0|} \tag{3.16}$$

(in view of the inequality $|\log(1+z)| \leq \frac{3}{2}|z|$, which is valid for $|z| \leq \frac{1}{2}$).

Next, combining (3.1) and (3.15), we obtain (3.9), where $r(z_0) = \tilde{r}\big(f(z_0)\big)$. Furthermore, it follows from (3.16) that $r(z_0)$ satisfies (3.10). □

Remark 3.5. From the proof of Lemma 3.3 it follows that $r(z)$ is an analytic function and $r_1(z)$ is continuous in the domain $\{z \in \mathbb{C} \mid |f(z)| > \max(\mathrm{e}, 2mK)\}$, if $f(z)$ is an entire solution of (3.1).

Now, from Corollary 3.4 we derive the two simple additional Corollaries 3.6 and 3.7.

Corollary 3.6. *The sequence* $\varphi(\lambda^n z_0)$ *is monotonically increasing and tends to infinity, as* $n \to \infty$. *Moreover* $\varphi(\lambda^n z_0) > (\frac{5}{4})^n \varphi(z_0)$.

PROOF. It follows from (3.4) that

$$\varphi(z_0) > 1 \quad \text{and} \quad |r_1(z_0)| < \tfrac{3}{4}. \tag{3.17}$$

Combining (3.12), (3.13), and (3.17) we obtain

$$\begin{aligned}\varphi(\lambda z_0) &= m\left(\varphi(z_0) + \frac{r_1(z_0)}{m}\right) \geq m\left(\varphi(z_0) - \frac{3}{4m}\right) \\ &\geq m\left(\varphi(z_0) - \frac{3}{4m}\varphi(z_0)\right) = \frac{4m-3}{4}\varphi(z_0) \geq \frac{5}{4}\varphi(z_0).\end{aligned} \tag{3.18}$$

This implies monotonicity as well as the lower bound for $\varphi(\lambda^n z_0)$. □

Combining Corollary 3.6 with the estimate (3.13) yields

Corollary 3.7. *For* $n = 0, 1, \ldots$ *we have*

$$|r_1(\lambda^n z_0)| < \frac{3}{2}\frac{mK}{|f(z_0)|} = \frac{3}{2} mK e^{-\varphi(z_0)}. \tag{3.19}$$

Lemma 3.8. *Suppose that* (3.4) *is satisfied. Then the limit*

$$\lim_{n\to\infty} \frac{\varphi(\lambda^n z_0)}{m^n} = L_1(z_0) \tag{3.20}$$

exists and

$$|L_1(z_0) - \varphi(z_0)| \leq \frac{3mK}{2|f(z_0)|}. \tag{3.21}$$

Furthermore, if f *is an entire solution, then* L *is a continuous function on some neighborhood of* z_0.

PROOF. From (3.12) we get by iteration

$$\frac{\varphi(\lambda^n z_0)}{m^n} = \varphi(z_0) + \frac{1}{m} r_1(z_0) + \cdots + \frac{1}{m^n} r_1(\lambda^{n-1} z_0). \tag{3.22}$$

Denote

$$S(z_0) = \sum_{k=1}^{\infty} \frac{r_1(\lambda^{k-1} z_0)}{m^k}, \tag{3.23}$$

where the series converges uniformly by (3.19) and

$$|S(z_0)| \leq \frac{3mK}{2|f(z_0)|}. \tag{3.24}$$

Therefore the limit (3.20) exists and equals $\varphi(z_0) + S(z_0)$. The inequality (3.21) follows from (3.24). Continuity of L_1 is a consequence of uniform convergence of the series (3.23) and the continuity of φ and r_1. □

Now Theorem 3.1 is an immediate consequence of Lemma 3.3.

PROOF OF THEOREM 3.1. Since $z_n = \lambda^n z_0$ and $\rho = \log m / \log |\lambda|$ then $|\lambda|^n = |z_n/z_0|$ and $m = |\lambda|^\rho$. Therefore,

$$m^n = (|\lambda|^\rho)^n = \left|\frac{z_n}{z_0}\right|^\rho.$$

Inserting this into (3.20) we get

$$\lim_{n\to\infty} \frac{\varphi(z_n)}{|z_n|^\rho} = \lim_{n\to\infty} \frac{\varphi(z_n)}{|z_0|^\rho m^n} = L_1(z_0)|z_0|^{-\rho} = L(z_0). \tag{3.25}$$

Combining this with (3.21) we obtain

$$\left|L(z_0) - \frac{\varphi(z_0)}{|z_0|^\rho}\right| \le \frac{3mK}{2|f(z_0)||z_0|^\rho}, \tag{3.26}$$

which gives (3.7). □

As can be seen from the proof, it is possible to relax the conditions of Theorem 3.1 slightly and to restate it in the following modified form.

Theorem 3.9. *Suppose that $f(z)$ satisfies* (3.11) (*or, what is the same, $\varphi(z)$ satisfies* (3.12)). *Suppose further that* (3.13) *is satisfied. Then the conclusion of Theorem* 3.1 *holds, i.e., $\varphi(z)$ satisfies* (3.6) *and* (3.7).

Remark 3.10. One can see from the proof (see (3.22) – (3.24)) that not only (3.26) (or (3.7)) is true, but the slightly stronger statement

$$\left|\frac{\varphi(z_n)}{|z_n|^\rho} - \frac{\varphi(z_0)}{|z_0|^\rho}\right| \le \frac{3mK}{2|f(z_0)||z_0|^\rho} \quad \text{for } n = 1, 2, \dots \tag{3.27}$$

holds.

4. Asymptotics along rays

As a consequence of Theorem 3.1 on the asymptotic behavior of $f(z)$ along spirals we shall derive the asymptotics along rays $re^{i\vartheta}$ in this section. It turns out that the latter heavily depends on the arithmetic nature of $\arg\lambda$.

Denote $\arg\lambda = 2\pi\beta$. Then it is natural to distinguish two cases: β is rational and β is irrational.

4.1. Rational angle.

Theorem 4.1. *Suppose that the conditions of Theorem* 3.1 (*or Theorem* 3.9) *are satisfied. Suppose further that $\beta = t/s$ is rational* (*with s and t relatively prime*). *Denote $q = \lambda^s$* (*note that q is a real number > 1*); *$r_n = q^n r_0$* (*and recall the notation $z_0 = r_0 e^{i\vartheta_0}$*). *Then the limit*

$$\lim_{n\to\infty} \frac{\varphi(r_n e^{i\vartheta})}{r_n^\rho} = L(z_0) \tag{4.1}$$

exists and satisfies (3.7).

Remark 4.2. Similar asymptotics occur on the rays

$$\arg z = \vartheta_0 + \frac{2k\pi}{s}, \quad k = 0, 1, \dots, s-1.$$

PROOF. If $f(z)$ satisfies (3.1) then (by iteration of (3.1)) it also satisfies

$$f(\lambda^s z) = P^{(s)}(f(z)), \tag{4.2}$$

where $P^{(s)}$ denotes the sth iteration of P. By our assumptions $q = \lambda^s$ is real and > 1. The assertion follows by an application of Theorem 3.1 to $P^{(s)}$. □

If in addition to the assumptions of Theorem 4.1 it is known that $f(z) \to \infty$ as $z \to \infty$ along the ray $\arg z = \vartheta_0$, then one can obtain stronger results. Suppose that there exists an interval $I = [r_0, r_0 q]$ such that condition (3.4) is satisfied for all $z \in \mathrm{e}^{\mathrm{i}\vartheta_0} I$, i.e.,

$$\forall z \in \mathrm{e}^{\mathrm{i}\vartheta_0} I\colon \quad |f(z)| > \max\{\mathrm{e}, 2mK\}. \tag{4.3}$$

Denote

$$M_0 = \max_{z \in \mathrm{e}^{\mathrm{i}\vartheta_0} I} \varphi(z) \quad \text{and} \quad m_0 = \min_{z \in \mathrm{e}^{\mathrm{i}\vartheta_0} I} \varphi(z).$$

Then the following is an immediate consequence of Theorem 4.1.

Theorem 4.3. *Suppose that the assumptions of Theorem* 3.1 (*or Theorem* 3.9) *are satisfied,* $\beta = t/s$ *is rational and* (4.3) *holds. Then*

$$\varphi(r\mathrm{e}^{\mathrm{i}\vartheta_0}) \sim r^\rho Q\left(\frac{\log r}{\log q}\right), \tag{4.4}$$

where Q *is a continuous* 1*-periodic real function of a real variable bounded between two positive constants:*

$$\forall t \in \mathbb{R}\colon \quad \frac{m_0 - 3/4}{(q r_0)^\rho} \le Q(t) \le \frac{M_0 + 3/4}{r_0^\rho}. \tag{4.5}$$

Remark 4.4. The theorem of Valiron [16] cited in Section 1 follows from Theorem 4.3.

PROOF. Suppose that $f(z)$ is an entire solution of (3.1). Denote

$$M(r) = \max_{|z|=r} |f(z)| \quad \text{and} \quad \psi(r) = \log M(r).$$

Then arguments analogous to those which led us to (3.12) and (3.13) yield

$$\psi(|\lambda| r) = m\psi(r) + R(r) \tag{4.6}$$

with

$$|R(r)| < \frac{3mK}{2M(r)}. \tag{4.7}$$

Thus the conditions of Theorem 3.9 are satisfied for the *real* scaling factor $|\lambda|$ in (4.6). Clearly, also $M(r) > \max\{\mathrm{e}, 2mK\}$ for r large enough, which implies (4.3). From this we obtain

$$\log M(r) \sim r^\rho Q\left(\frac{\log r}{\log |\lambda|}\right) \tag{4.8}$$

which implies Valiron's theorem. □

4.2. Irrational angle. In this case we cannot claim the existence of geometric progressions along a given ray, for which (4.1) holds. Instead we will prove that for any ray $re^{i\vartheta}$ there exists a sequence of positive real numbers (s_n) with $s_n \to \infty$ such that

$$\lim_{n\to\infty} \frac{\varphi(s_n e^{i\vartheta})}{s_n^\rho} = L > 0$$

exists. In particular, this implies that in the case of irrational β solutions of the Poincaré equation are unbounded on any ray. As far as we know this phenomenon has not been mentioned in the literature before. An alternative proof of the latter statement based on the Phragmén – Lindelöf principle will be given in Section 5.

Theorem 4.5. *Suppose that the assumptions of Theorem* 3.1 (*or Theorem* 3.9) *are satisfied and* $\beta = 1/(2\pi)\arg\lambda$ *is irrational. Then on any ray* $\arg z = \vartheta$ *there exists a sequence* $(s_n e^{i\vartheta})$ *with* $s_n \to \infty$ *such that the limit*

$$\lim_{n\to\infty} \frac{\varphi(s_n e^{i\vartheta})}{s_n^\rho} = L(z_0) \tag{4.9}$$

exists and satisfies (3.7).

PROOF. Let z_0 satisfy (3.4). Then there exists an $\varepsilon > 0$ such that $|f(z)| > \max(e, 2mK)$ holds for all $z \in B(z_0, \varepsilon)$. By Lemma 3.8

$$L_1(z) = \lim_{n\to\infty} \frac{\varphi(\lambda^n z)}{m^n} \tag{4.10}$$

is a continuous function of z on $B(z_0, \varepsilon)$ and convergence is uniform in $B(z_0, \varepsilon)$ as indicated in the proof of Lemma 3.8.

By the density of the set $\{n\arg\lambda \mid n \in \mathbb{N}\}$ on the unit circle there exists a sequence of integers $(n_k)_{k\in\mathbb{N}}$ such that

$$\lim_{k\to\infty} \arg\lambda^{n_k} = \vartheta - \arg z_0 \mod 2\pi.$$

Let

$$\zeta_k = |z_0| e^{i\vartheta} \left(\frac{\lambda}{|\lambda|}\right)^{-n_k}.$$

Then clearly $\lim_{k\to\infty} \zeta_k = z_0$ and we have

$$\lim_{k\to\infty} \frac{\varphi(\lambda^{n_k}\zeta_k)}{m^{n_k}} = \lim_{k\to\infty} \frac{\varphi(|z_0| e^{i\vartheta} |\lambda|^{n_k})}{m^{n_k}} = L_1(z_0)$$

by uniform convergence. Denoting $s_k = |z_0||\lambda|^{n_k}$ we have then

$$\lim_{k\to\infty} \frac{\varphi(s_k e^{i\vartheta})}{s_k^\rho} = \lim_{k\to\infty} \frac{\varphi(s_k e^{i\vartheta})}{(|z_0|\,|\lambda|^{n_k})^\rho} = \frac{L_1(z_0)}{|z_0|^\rho} = L(z_0). \qquad \square$$

5. Conditions for the unboundedness of solutions along rays

Note that Theorem 3.1 and Corollary 4.1 are results of conditional type. In general, we do not know whether condition (3.4) is satisfied in a specific point $z_0 = r_0 e^{i\vartheta_0}$. In this section we give some conditions under which any nontrivial entire solution $f(z)$ of (3.1) is unbounded on any ray $\arg z = \vartheta$. Denote as before $\lambda = |\lambda| e^{2\pi i\beta}$.

We can assure the unboundedness of $f(z)$ along all rays under the following conditions either on $|\lambda|$ or on $\arg\lambda$:

(a) $|\lambda|$ is large compared to $m = \deg P(z)$

(b) β is rational, $\beta = t/s$ (in lowest terms) and s is large compared to $\rho = \log m / \log |\lambda|$
(c) β is irrational.

The proofs of (b) and (c) are based on the Phragmén–Lindelöf principle, or more precisely on the following corollary.

Corollary 5.1 (Phragmén–Lindelöf, [10]). *Assume that $f(z)$ is a nonconstant entire function of order $\rho \geq \frac{1}{2}$. Suppose that n rays emanate from the origin and split the complex plane $\mathbb{C}$ into n angular regions each of which has angle less than π/ρ. Then $f(z)$ is unbounded on at least one of these rays.*

5.1. Large $|\lambda|$.

Theorem 5.2. *Suppose that $|\lambda| > m^2$, i.e., $\rho < \frac{1}{2}$. Then*

(1) *$f(z)$ is unbounded along any ray $\arg z = \vartheta$*
(2) *For any ϑ there exists a sequence $r_n = r_n(\vartheta) \to \infty$ such that the limit*

$$\lim_{n\to\infty} \frac{\varphi(r_n e^{i\vartheta})}{r_n^\rho} = L(\vartheta) \tag{5.1}$$

exists.
(3) *These limits $L(\vartheta)$ are bounded between two positive constants for all $\vartheta \in [0, 2\pi]$.*
(4) *Under the additional assumption that $\beta = 1/(2\pi)\arg\lambda$ is rational ($\beta = t/s$ in lowest terms) further refinements are possible. Denote $q = \lambda^s$ (and note that $q > 1$). Then r_n can be chosen as a geometric progression $r_n = r_0 q^n$ independent of the direction ϑ.*

PROOF. Unboundedness of $f(z)$ along any ray follows from the assumption $\rho < \frac{1}{2}$ (see [14, 8.73]). Moreover there exists a sequence $\tilde{r}_n \to \infty$ such that

$$m(\tilde{r}_n) = \min_{|z|=\tilde{r}_n} |f(z)| \to \infty.$$

In particular, one can find a circle $C_0 = \{z \in \mathbb{C} \mid |z| = r_0\}$ such that condition (3.4) is satisfied for all $z \in C_0$. Then (2) and (3) follow either from Corollary 4.1 (if β is rational) or from Theorem 4.5 (if β is irrational). Applying Corollary 4.1 once again yields (4). □

5.2. Rational β with large denominator.

Theorem 5.3. *Suppose that $\beta = 1/(2\pi)\arg\lambda$ is rational ($\beta = t/s$ in lowest terms). Suppose further that $s > 2\rho$. Then*

(1) *$f(z)$ is unbounded on any ray $\arg z = \vartheta$.*
(2) *Furthermore, for any ϑ one can find a geometric progression $r_n = r_n(\vartheta) = q^n r_0(\vartheta)$ (with $q = \lambda^s > 1$) such that the limit (5.1) exists and $L(\vartheta) > 0$.*

PROOF. Suppose that there exists a ray $\arg z = \vartheta_1$ ($r > 0$) such that $f(z)$ is bounded along this ray. Then in view of equation (3.1) $f(z)$ is bounded along all rays $\arg z = \vartheta_1 + 2k\pi t/s \mod 2\pi$ ($k = 0, \ldots, s-1$). The angle between two consecutive rays is $2\pi/s < \pi/\rho$. From Corollary 5.1 it follows that f is constant. Assertion (2) follows by applying Corollary 4.1 as in the proof of Theorem 5.2. □

5.3. Irrational β. Suppose that $\beta = 1/(2\pi)\arg\lambda$ is irrational. It was already proved in Section 4.2 as an immediate consequence of Theorem 4.5 that every nontrivial entire solution $f(z)$ of (3.1) is unbounded along any ray $\arg z = \vartheta$. Here we present an alternative proof of this statement using again the Phragmén–Lindelöf principle. However the existence of the limit (4.9) cannot be obtained in this way.

Theorem 5.4. *If β is irrational, then every nontrivial entire solution $f(z)$ of (3.1) is unbounded along any ray $\arg z = \vartheta$.*

PROOF. Suppose that $f(z)$ is bounded on a ray $\arg z = \vartheta_1$. Then again by (3.1) $f(z)$ is bounded along all rays $\arg z = \vartheta_1 + 2k\pi\beta \mod 2\pi$ for $k \in \mathbb{N}$. Since the angles $\vartheta_1 + 2k\pi\beta \mod 2\pi$ are dense in $[0, 2\pi]$, one can find n such rays with angles between consecutive rays less than π/ρ. Corollary 5.1 implies then that $f(z)$ is constant. □

6. Real case: real $\lambda > 1$ and polynomial $P(z)$ with real coefficients

Throughout this section we will assume that $\lambda > 1$ is real and all coefficients of $P(z)$ are real. Also we use the notations of Section 2 above.

Immediately, from Lemma 2.5 (points (3) and (4)) we derive

Corollary 6.1. *Suppose that*

(1) $p_i \geq 0$ *for* $i \geq 2$
(2) $f_0 \geq 0$ *and* $f_{n_0} > 0$.

Then

(a) $f_i \geq 0$ *for all* i
(b) *there are infinitely many* i *for which* $f_i > 0$
(c) $f(z) > 0$ *for all* $z \in \mathbb{R}^+$.

From this we obtain

Corollary 6.2. *Under the assumptions of Corollary 6.1 $f(z) \to \infty$ in $\mathbb{R}^+$. Moreover,*

$$f(r) = M(r) = \max_{|z|=r} |f(z)|$$

and

$$\log f(r) \sim r^{\rho} Q\left(\frac{\log r}{\log \lambda}\right) \quad \text{for } r \to \infty, \tag{6.1}$$

where Q is a continuous 1-periodic function bounded between two positive constants.

Further refinements are possible under additional assumptions (cf. also [3, Theorem 1]).

Theorem 6.3. *Assume that*

(1) $\lambda > m$
(2) $p_i \geq 0$ *for* $i \geq 2$
(3) *all preimages of* 0 *under* P *are real.*

Let $f(z)$ be an entire solution of (3.1) *such that $f(0)$ and $f'(0)$ are real and $f(0) \geq 0$ and $f'(0) > 0$. Then $f(z)$ tends to infinity along any ray in the sector $|\arg z| < \frac{\pi}{2}$. Moreover,*

$$\log|f(re^{i\vartheta})| \sim r^\rho Q_\vartheta\left(\frac{\log r}{\log\lambda}\right) \quad \textit{for } r \to \infty, \tag{6.2}$$

where Q_ϑ is a 1-periodic function bounded between two positive constants.

The proof is based on the following three lemmas.

Lemma 6.4. *Suppose that all preimages of* 0 *under P are real. Let $f(z)$ be an entire solution of* (3.1) *such that $f(0)$ and $f'(0) \neq 0$ are real. Then $f(z)$ has only real zeros.*

PROOF. Suppose that $f(z) = 0$ has a nonreal solution z_0. Then

$$0 = f(z_0) = P(f(z_0\lambda^{-1})) = \cdots = P^{(n)}(f(z_0\lambda^{-n})),$$

which implies that $f(z_0\lambda^{-n})$ is an nth preimage of 0 under P and is therefore real by our assumptions. Using Taylor's formula we obtain

$$\lim_{n\to\infty} \lambda^n(f(z\lambda^{-n}) - f(0)) = f'(0)z.$$

Here the left-hand side is real and the right-hand side is not, which gives a contradiction. □

Lemma 6.5. *Suppose that the assumptions of Lemma* 6.4 *are satisfied. Assume further that*

(1) $p_i \geq 0$ *for* $i \geq 2$
(2) $f(0) \geq 0$ *and* $f'(0) > 0$.

Then $f(z)$ has nonpositive real zeros only.

PROOF. The assumption $f'(0) \neq 0$ implies that $n_0 = 1$. Furthermore, we have $f_{n_0} > 0$. Thus by Corollary 6.1 $f(z) > 0$ on $\mathbb{R}^+$. Thus the zeros have to be nonpositive. □

Lemma 6.6. *Assume that $f(z)$ is an entire function of order $0 < \rho < 1$ with only negative real zeros. Then $f(z)$ tends to infinity faster than any power of z along any ray in the sector $|\arg z| < \pi/2$.*

PROOF. Because $0 < \rho < 1$ according to Hadamard's theorem we have

$$f(z) = \prod_{n=1}^{\infty}\left(1 + \frac{z}{r_n}\right), \tag{6.3}$$

where $(-r_n)_{n\in\mathbb{N}}$ with $r_n > 0$ are the negative zeros of f. Write $z = x + iy$ and $S_\vartheta = \{z \in \mathbb{C} \mid |\arg z| \leq \vartheta\}$ for $\vartheta < \pi/2$. Then we have for $z \in S_\vartheta$

$$\left|1 + \frac{z}{r_n}\right| > 1 + \frac{x}{r_n} \geq 1 + \frac{|z|\cos\vartheta}{r_n} \geq \max\left(1, \frac{|z|\cos\vartheta}{r_n}\right).$$

Inserting this estimate into (6.3) yields

$$|f(z)| = \prod_{n=1}^{\infty}\left|1 + \frac{z}{r_n}\right| > \prod_{n=1}^{N}\left|1 + \frac{z}{r_n}\right| > C|z|^N,$$

where $C = \cos^N\vartheta/(r_1\cdots r_N)$ and N is an arbitrary positive integer. □

PROOF OF THEOREM 6.3. Combining Lemma 6.6 and Corollary 4.3 proves the theorem. □

To conclude this section we present a lemma, which gives a condition for the reality of the preimages of 0 in the case of quadratic polynomials.

Lemma 6.7. *Let*

$$P(z) = az(z-\omega), \quad 0 \neq \omega \in \mathbb{R} \tag{6.4}$$

All preimages of 0 *under* P *are real, if and only if the following condition is fulfilled*

$$a|\omega| \geq \begin{cases} 2 & \textit{for } \omega > 0 \\ 4 & \textit{for } \omega < 0 \end{cases} \tag{6.5}$$

PROOF. We will give a short geometric proof of this lemma using the plots of the functions $az(z-\omega)$ for $\omega > 0$ and $\omega < 0$ in Figure 1.

We will give the proof for $\omega > 0$; the proof for $\omega < 0$ is similar.

The fixed points of $P(z)$ are 0 and $b = \omega + 1/a$. Next, $P^{-1}(\{0\}) = \{0, \omega\}$, $P^{-1}(\{b\}) = \{-1/a, b\}$, and $P^{-1}(\omega)$ consists of two points from the interval $I = [-1/a, b]$.

Now, to prove sufficiency of (6.5) it is enough to prove that under condition (6.5) every preimage of a point of I lies in I. As can be seen from Figure 1, this is true, if

$$\frac{1}{a} \leq -\min_{z \in \mathbb{R}} P(z) = -P\left(\frac{\omega}{2}\right) = \frac{a}{4}\omega^2,$$

which is satisfied if $a\omega \geq 2$ and $\omega > 0$. This proves sufficiency of (6.5).

On the other hand, suppose that $a\omega < 2$. Then

$$\frac{1}{a} > -\min_{z \in \mathbb{R}} P(z). \tag{6.6}$$

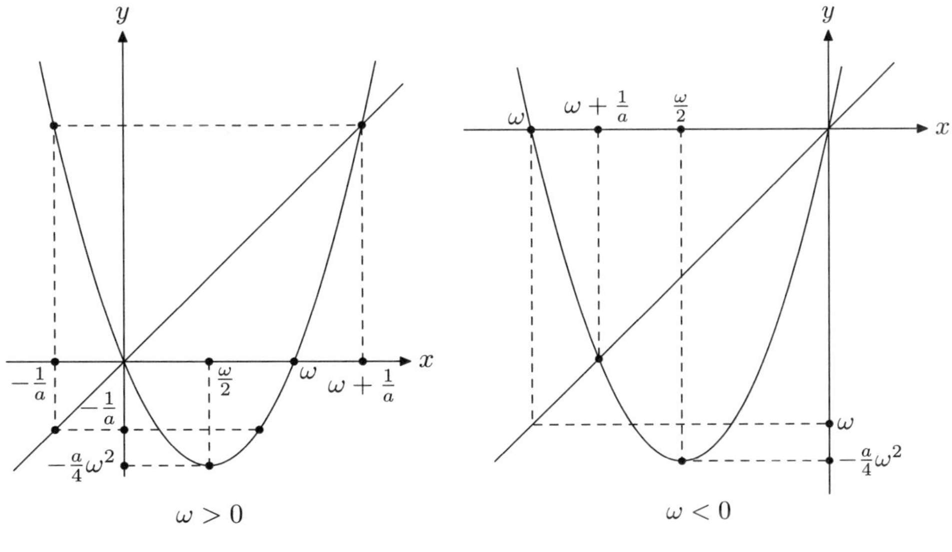

FIGURE 1. The graphs of $az(z-\omega)$ for $\omega > 0$ and $\omega < 0$.

Then there exist points $c \in P^{-n}(\{0\})$ arbitrarily close to the fixed point b. Thus there exists a point $d \in P^{-1}(\{c\}) \subset P^{-n-1}(\{0\})$ arbitrarily close to $-1/a \in P^{-1}(\{b\})$. By (6.6) the preimages of $-1/a$ and therefore of d are not real. Thus $P^{-n-1}(\{0\})$ contains nonreal points. This proves necessity of (6.5). □

7. Asymptotic behavior in angular regions

In this section, as well as in Section 6, we assume that $\lambda > 1$ is real and all coefficients of $P(z) = z^m + p_{m-1}z^{m-1} + \cdots + p_0$ (see (3.2)) are real. Moreover, (by virtue of Proposition 2.3 and Remark 2.6) without loss of generality, we assume that $P(0) = 0$, $P'(0) = \lambda$, $f(0) = 0$, and $f'(0) > 0$.

Let $f(z)$ be an entire solution of (2.1). In contrast with the previous sections (where asymptotics of $|f(z)|$ is studied) here we collect some results on the asymptotics of the solution $f(z)$ itself in some angular regions of the complex plane.

As a starting point we present a theorem, which we proved in [3].

Theorem 7.1 ([3, Theorem 1]). *Let f be an entire solution of the functional equation (2.1). Furthermore, suppose that $\mathcal{F}_\infty$, the Fatou component of ∞ of P, contains an angular region of the form*

$$W_\beta = \{z \in \mathbb{C} \setminus \{0\} \mid |\arg z| < \beta\}$$

for some $\beta > 0$. Then for any $\varepsilon > 0$ and any $M > 0$ the asymptotic relation

$$f(z) = \exp\left(z^\rho Q\left(\frac{\log z}{\log \lambda}\right) + o(|z|^{-M})\right) \tag{7.1}$$

holds uniformly for $z \in W_{\beta-\varepsilon}$, where Q is a periodic holomorphic function of period 1 on the strip $\{w \in \mathbb{C} \mid |\Im w| < \beta/\log \lambda\}$. The real part of $z^\rho Q(\log z/\log \lambda)$ is bounded between two positive constants; Q takes real values on the real axis.

Remark 7.2. Notice that the condition on the Fatou component $\mathcal{F}_\infty$ is used in the proof of this theorem to ensure that $f(z)$ tends to infinity in the angular region W_β. Therefore, this condition could be replaced by

$$\lim_{z\to\infty} f(z) = \infty \quad \text{for } |\arg z| < \beta.$$

Remark 7.3. E. Romanenko and A. Sharkovsky have studied equation (2.1) on $\mathbb{R}$ (rather than on $\mathbb{C}$). Applying Sharkovsky's method of "first integrals" ("invariant graphs") they obtained not only an asymptotic formula of type (7.1) but a full asymptotic expansion for all solutions $f(x)$, such that $f(x) \to \infty$ for $x \to \infty$.

We will present a number of consequences of this theorem.

From Theorem 7.1 and Theorem 4.5 we obtain

Corollary 7.4. *Let f be a solution of (2.1). Assume that there exists $r > 0$ such that*

$$\forall x \in [r, \lambda r] : |f(x)| > \max(\mathrm{e}, 2mK).$$

Then there exists $\gamma > 0$ such that for any $0 < \varepsilon < \gamma$ and any $M > 0$ the asymptotic relation (7.1) holds uniformly for $z \in W_{\gamma-\varepsilon}$.

As an immediate corollary of Theorem 7.1 we get

Corollary 7.5. *Let f be a solution of (2.1). If the Julia set of P is a subset of $\mathbb{R}^- \cup \{0\}$, then for any $\varepsilon > 0$ and any $M > 0$ the asymptotic relation (7.1) holds uniformly for $z \in W_{\pi-\varepsilon}$.*

Acknowledgments. We are grateful to E. Romanenko and A. Sharkovsky for having sent their preprint and for stimulating discussions. We thank M. Sodin for useful remarks.

References

1. M. T. Barlow and E. A. Perkins, *Brownian motion on the Sierpiński gasket*, Probab. Theory Related Fields **79** (1988), no. 4, 543 – 623.
2. J. D. Biggins and N. H. Bingham, *Near-constancy phenomena in branching processes*, Math. Proc. Cambridge Philos. Soc. **110** (1991), no. 3, 545–558.
3. G. Derfel, P. J. Grabner, and F. Vogl, *The zeta function of the Laplacian on certain fractals*, Trans. Amer. Math. Soc., to appear.
4. G. Derfel and F. Vogl, *Divide-and-conquer recurrences — classification of asymptotics*, Aequationes Math. **60** (2000), no. 3, 243 – 257.
5. A Eremenko and G. Levin, *Periodic points of polynomials*, Ukrainian Math. J. **41** (1989), no. 11, 1258 – 1262.
6. A. Eremenko and M. Sodin, *Iterations of rational functions and the distributions of the values of Poincaré functions*, J. Soviet Math. **58** (1992), no. 6, 504 – 509.
7. P. J. Grabner, *Functional iterations and Brownian motion on the Sierpiński gasket*, Mathematika **44** (1997), no. 2, 374–400.
8. P. J. Grabner and W. Woess, *Functional iterations and periodic oscillations for the simple random walk on the Sierpiński graph*, Stochastic Process. Appl. **69** (1997), no. 1, 127–138.
9. S. Karlin and J. McGregor, *Embeddability of discrete time simple branching processes into continuous time branching processes*, Trans. Amer. Math. Soc. **132** (1968), 115–136.
10. A. I. Markushevich, *Theory of functions of a complex variable, Vol.* I, II, III, 2nd English ed., translated by R. A. Silverman, Chelsea Publishing Co., New York, 1977.
11. H. Poincaré, *Sur une classe étendue de transcendantes uniformes*, C. R. Acad. Sci. Paris **103** (1886), 862–864.
12. ———, *Sur une classe nouvelle de transcendantes uniformes*, J. Math. Pures Appl. (4) **6** (1890), 316 – 365.
13. E. Romanenko and A. Sharkovsky, *Long time properties of solutions of simplest q-difference equations* (2000), preprint (Russian).
14. E. C. Titchmarsh, *The theory of functions*, 2nd corr. ed., Oxford Univ. Press., Oxford, 1979.
15. G. Valiron, *Lectures on the general theory of integral functions*, Édouard Privat, Toulouse, 1923.
16. ———, *Fonctions analytiques*, Presses Universitaires de France, Paris, 1954.

DEPARTMENT OF MATHEMATICS AND COMPUTER SCIENCE, BEN GURION UNIVERSITY OF THE NEGEV, BEER SHEVA 84105, ISRAEL
E-mail address: derfel@math.bgu.ac.il

INSTITUT FÜR MATHEMATIK A, TECHNISCHE UNIVERSITÄT GRAZ, STEYRERGASSE 30, 8010 GRAZ, AUSTRIA
E-mail address: peter.grabner@tugraz.at

INSTITUT FÜR ANALYSIS UND SCIENTIFIC COMPUTING, TECHNISCHE UNIVERSITÄT WIEN, WIEDNER HAUPTSTRASSE 8–10, 1040 WIEN, AUSTRIA
E-mail address: fvogl@osiris.tuwien.ac.at

Centre de Recherches Mathématiques
CRM Proceedings and Lecture Notes
Volume **42**, 2007

Hamiltonian Extension and Eigenfunctions for a Time Dispersive Dissipative String

Alexander Figotin and Jeffrey Schenker

Abstract. We carry out a detailed analysis of a time dispersive and dissipative (TDD) string, using our recently developed theories of conservative and Hamiltonian extensions of TDD systems. This analysis of the TDD string includes, in particular: (i) an explicit construction of its conservative Hamiltonian extension, consisting of the original string coupled to "hidden strings"; (ii) explicit formulas for energy and momentum densities in the extended system, providing a transparent physical picture accounting precisely for the dispersion and dissipation; (iii) the eigenmodes for the extended string system, which provide an eigenmode expansion for solutions to the TDD wave equation governing the solution to the TDD string. The obtained results provide a solid basis for the rigorous treatment of the long standing problem of scattering by a TDD scatterer, illustrated here by the computation of scattering states for a string with dissipation restricted to a half line.

1. Introduction

The need for a Hamiltonian description of dissipative systems has long been known, having been emphasized by Morse and Feshbach [5, Chapter 3.2] forty years ago. Recently we have introduced conservative [1] and then Hamiltonian [2, 3] theories of time dispersive and dissipative (TDD) systems addressing that need. The extended Hamiltonian is constructed by coupling the given TDD system to a system of "hidden strings" in a canonical way so that it has a transparent interpretation as the system energy. It turns out that such a system of hidden strings is, in fact, a canonical heat bath as described in [4, Section 2; 6, Section 2].

This work is intended as a self contained companion to the papers [1–3], illustrating the phenomena described there through explicit calculations for the typical example of a TDD string.

To keep everything elementary, let us consider a scalar wave equation in one spatial dimension, such as might be used to describe wave propagation along a

2000 *Mathematics Subject Classification.* Primary 37L50; Secondary 35P10.

The effort of the first author was sponsored by the Air Force Office of Scientific Research, Air Force Materials Command, USAF, under grant number FA9550-04-1-0359.

The second author received travel and materials support under the aforementioned USAF grant and was supported by a National Science Foundation post-doctoral fellowship.

This is the final form of the paper.

homogeneous string. In the absence of dispersion, the displacement $\phi(x,t)$ of the string at position $x \in \mathbb{R}$ and time $t \in \mathbb{R}$ evolves by the 1D wave equation

$$\partial_t^2 \phi(x,t) - \gamma \partial_x^2 \phi(x,t) = f(x,t), \quad x,t \in \mathbb{R}, \tag{1.1}$$

where

- γ is the tension of the string.
- we have taken units in which the mass per unit length is 1.
- $f(x,t)$ is an external *driving force* per unit length, usually supposed bounded and compactly supported.

For the most part, we consider (1.1) with the string *at rest* at $t = -\infty$,

$$\lim_{t \to -\infty} \phi(x,t) = \lim_{t \to \infty} \partial_t \phi(x,t) = 0. \tag{1.2}$$

Thus the solution $\phi(x,t)$ is a function of the driving force $f(x,t)$, indeed

$$\phi(x,t) = \frac{1}{2v} \int_{-\infty}^{t} \int_{x-v(t-t')}^{x+v(t-t')} f(y,t')\, dy\, dt' \tag{1.3}$$

with $v = 1/\sqrt{\gamma}$ the speed of propagation on the string.

In this paper we consider a modification of (1.1) incorporating friction in the form of a dissipative term with time dispersion. Specifically, we consider the following equation

$$\partial_t \left\{ \partial_t \phi(x,t) + \int_{-\infty}^{t} \chi(x,t-t') \partial_{t'} \phi(x,t')\, dt' \right\} - \gamma \partial_x^2 \phi(x,t) = f(x,t), \tag{1.4}$$

with $\chi(x,\tau)$ a given function, called the susceptibility, satisfying a power dissipation condition — (1.23) below. The physical idea behind (1.4) is as follows. The wave equation (1.1) can be expressed in terms of the string momentum $\pi(x,t) = \partial_t \phi(x,t)$ as

$$\partial_t \pi(x,t) = \text{sum of all forces at } x = \underbrace{f(x,t)}_{\text{driving force}} + \underbrace{\gamma \partial_x^2 \phi(x,t)}_{\text{string tensile force}}. \tag{1.5}$$

In the damped string (1.4), the basic relation $\partial_t \pi =$ "sum of all forces" still holds, but the simple relationship $\pi = \partial_t \phi$ between the string momentum and velocity is replaced by the *material relation*

$$\pi(x,t) = \partial_t \phi(x,t) + \int_{-\infty}^{t} \chi(x,t-t') \partial_{t'} \phi(x,t')\, dt'. \tag{1.6}$$

Equation (1.6) is supposed to be a phenomenological relation describing the interaction of the string with a surrounding medium, expressing the fact that some of the string momentum is absorbed by the medium and then partially retransmitted to the string with delay.

The wave equation (1.1) may be expressed as a Hamiltonian system. As described in [2, 3], the procedure of going from (1.1) to (1.4) is naturally understood in this context and indeed (1.4) *can be derived from a larger Hamiltonian system, with additional variables.* The additional variables of this extension may be interpreted as describing a "hidden string," with internal coordinate s, attached to each point x of the physical string (see Figure 1). These "hidden strings" are coupled to the physical string via a coupling function $\varsigma(x,s)$, but do not interact directly with

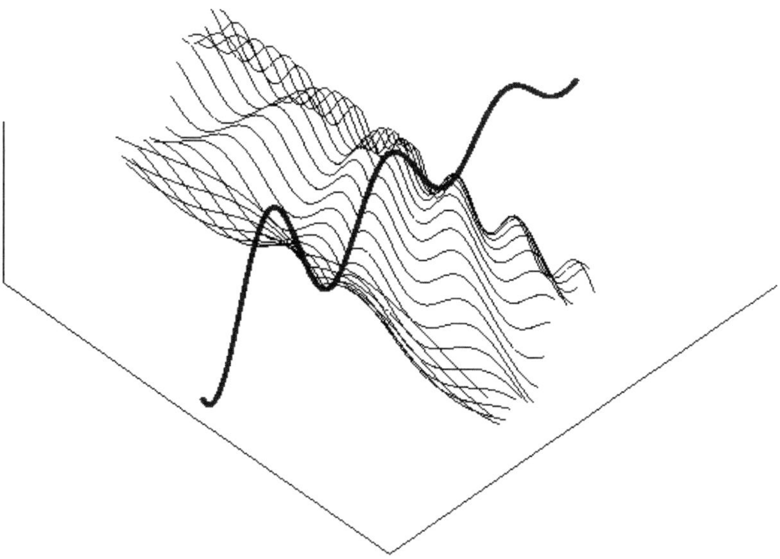

FIGURE 1. The extended system consists of the physical string (thick) coupled to an independent "hidden string" (thin) at every point of dissipation. This Hamiltonian system models exactly the dissipation and dispersion in the physical string via the exchange of energy with the hidden strings.

one another.[1] Upon "integrating out" the hidden strings, the result is an effective dynamics of the form (1.4) for the physical string, with susceptibility

$$\chi(x,\tau) = \frac{1}{2}\int_{\mathbb{R}} \varsigma(s,x) \int_{s-\tau}^{s+\tau} \varsigma(x,r)\,\mathrm{d}r\,\mathrm{d}s. \tag{1.7}$$

Different choices of the coupling function $\varsigma(x,s)$ produce different susceptibilities $\chi(x,\tau)$, and furthermore [2, 3] any susceptibility satisfying the power dissipation condition — (1.23) below — can be obtained in this way.

Given the role of the TDD wave equation (1.4) as a phenomenological description for the interaction of the string with its environment, it is not surprising that an extended system exists which produces the effective dynamics (1.4), at least approximately. However, the main point of [2, 3], which we explain below, is that it is possible to construct explicitly an extended system which *exactly* reproduces (1.4). Further, using that system we can derive simple expressions for quantities like the energy and wave momentum densities for the damped string in terms of the susceptibility χ.

In addition, as illustrated below, the extended system clarifies the nature and role of eigenfunctions in a TDD system by providing a natural eigenmode expansion

[1] The lack of interaction between hidden strings is not essential. Indeed the methods of [1–3] allow to consider systems with "spatial dispersion" in the material relation

$$\pi(x,t) = \partial_t\phi(x,t) + \int_{\mathbb{R}}\int_0^\infty \chi(x',\tau)\partial_t\phi(x-x',t-\tau)\,\mathrm{d}\tau\,\mathrm{d}x'.$$

The extensions for such systems involve interaction between the hidden strings, but for simplicity we do not pursue this here.

for solutions to (1.4). In this context, an eigenfunction $\exp(-\mathrm{i}\omega t)\phi_\omega(x)$ is a time periodic steady state solution to (1.4). For the simple nondissipative string (1.1), an eigenmode solves

$$\gamma\partial_x^2\phi_\omega(x) = -\omega^2\phi_\omega(x), \tag{1.8}$$

which is to say it is a linear superposition of the plane waves $\exp(\pm\mathrm{i}\omega x/\sqrt{\gamma})$. For the TDD string (1.4), the dissipation induced by the dispersion in (1.6) may preclude the existence of a steady state solution without a source term. However, the extended system is without dissipation, and it is no trouble to construct its eigenmodes. We find below that the resulting eigenmode equation for ϕ_ω

$$\gamma\partial_x^2\phi_\omega(x) = -\omega^2\big(1 + \operatorname{Re}\widehat{\chi}(x,\omega)\big)\phi_\omega(x) - \mathrm{i}\omega^2\operatorname{Im}\widehat{\chi}(x,\omega)g_\omega(x), \tag{1.9}$$

has a source term involving an *arbitrary* function g_ω, which depends on the configuration of the hidden strings. Here $\widehat{\chi}$ is the τ-Fourier transform of χ,

$$\widehat{\chi}(x,\omega) = \int_0^\infty \mathrm{e}^{\mathrm{i}\omega\tau}\chi(x,\tau)\,\mathrm{d}\tau. \tag{1.10}$$

Let us now sketch the construction of the Hamiltonian extension. To start we write (1.1) as a first order system, introducing the momentum density $\pi(x,t) = \partial_t\phi(x,t)$, so that

$$\begin{aligned}\partial_t\phi(x,t) &= \pi(x,t)\\ \partial_t\pi(x,t) &= \gamma\partial_x^2\phi(x,t) + f(x,t).\end{aligned} \tag{1.11}$$

This first order system is Hamiltonian, with symplectic form

$$J(\phi_1,\pi_1;\phi_2,\pi_2) = \int_{\mathbb{R}} (\phi_1(x)\pi_2(x) - \phi_2(x)\pi_1(x))\,\mathrm{d}x \tag{1.12}$$

and time dependent Hamilton function

$$H_f(\phi,\pi,t) = \frac{1}{2}\int_{\mathbb{R}} (\pi(x)^2 + \gamma|\partial_x\phi(x)|^2)\,\mathrm{d}x - \int_{\mathbb{R}} f(x,t)\phi(x)\,\mathrm{d}x. \tag{1.13}$$

As the system is nonautonomous, $H_f(\pi(\cdot,t),\phi(\cdot,t),t)$ is not conserved. Instead along a trajectory to (1.11), we have

$$\frac{\mathrm{d}H_f}{\mathrm{d}t} = \partial_t H_f(\phi(\cdot,t),\pi(\cdot,t),t) = -\int_{\mathbb{R}} \partial_t f(x,t)\phi(x,t)\,\mathrm{d}x. \tag{1.14}$$

We take the internal energy of the (undamped) string to be the Hamilton function $H_0(\phi,\pi)$ of the system with $f \equiv 0$. Thus, using (1.14) we obtain

$$\frac{\mathrm{d}H_0}{\mathrm{d}t} = \frac{\mathrm{d}H_f}{\mathrm{d}t} + \frac{\mathrm{d}}{\mathrm{d}t}\int_{\mathbb{R}} f(x,t)\phi(x,t)\,\mathrm{d}x = \int_{\mathbb{R}} f(x,t)\partial_t\phi(x,t)\,\mathrm{d}x. \tag{1.15}$$

In other words, the rate of work done on the system is the integral of the external force times the velocity — "power = force × velocity."

Observe that the Hamiltonian H_0 can be written

$$H_0(\phi,\pi) = \tfrac{1}{2}\big(\|K_\pi\pi\|^2_{L^2(\mathbb{R})} + \|K_\phi\phi\|^2_{L^2(\mathbb{R})}\big) \tag{1.16}$$

where

$$K_\phi = \sqrt{\gamma}\partial_x, \quad K_\pi = \text{Identity}. \tag{1.17}$$

Based on this expression, we separate the equations of motion (1.11), into "material relations"

$$f_\phi(x) = K_\phi \phi(x) = \sqrt{\gamma}\partial_x \phi(x) \tag{1.18}$$

$$f_\pi(x) = K_\pi \pi(x) = \pi(x), \tag{1.19}$$

and dynamical equations,

$$\begin{aligned} \partial_t \phi(x,t) &= K_\pi^\dagger f_\pi(x,t) \\ \partial_t \pi(x,t) &= -K_\phi^\dagger f_\phi(x,t) + f(x,t). \end{aligned} \tag{1.20}$$

To obtain a first order TDD system leading to (1.4), we follow [2,3] and replace the identity (1.19) with the "material relation,"

$$f_\pi(x,t) + \int_0^\infty \chi(x,\tau) f_\pi(x,t-\tau)\,\mathrm{d}\tau = \pi(x,t), \tag{1.21}$$

but maintain (1.18) and the equations of motion (1.20), i.e.,

$$\begin{aligned} \partial_t \phi(x,t) &= f_\pi(x,t) \\ \partial_t \pi(x,t) &= \gamma \partial_x^2 \phi(x,t) + f(x,t). \end{aligned} \tag{1.22}$$

(One could also introduce dispersion in the relation between ϕ and f_ϕ, or even dispersion mixing f_ϕ and f_π, but for simplicity we consider here only TDD in the momentum.) The key requirements on the *susceptibility* χ are

(1) Causality, manifest in (1.21) in that the integral on the l.h.s. depends only on the past.
(2) Power dissipation, which is the requirement that for every x

$$\int_{-\infty}^\infty g(t) \int_0^\infty \chi(x,\tau)\partial_t g(t-\tau)\,\mathrm{d}\tau\,\mathrm{d}t \geq 0 \tag{1.23}$$

for an arbitrary function $g(t)$, say compactly supported.

The significance of the power dissipation condition is as follows. We suppose that the internal energy of the TDD system at time t is given by

$$H_0 = \frac{1}{2}\int_{\mathbb{R}} (f_\pi(x,t)^2 + \gamma \partial_x \phi(x,t)^2)\,\mathrm{d}x, \tag{1.24}$$

with $f_\pi(x,t)$ as in (1.21). Then

$$\begin{aligned} \frac{\mathrm{d}}{\mathrm{d}t} H_0(t) &\\ = \int_{\mathbb{R}} &\big(\gamma \partial_x \phi(x,t) \partial_t \partial_x \phi(x,t) + f_\pi(x,t)\partial_t f_\pi(x,t)\big)\,\mathrm{d}x \\ = \int_{\mathbb{R}} &\bigg(\gamma \partial_x \phi(x,t)\partial_x f_\pi(x,t) + f_\pi(x,t)\partial_t \pi(x,t) \\ &\qquad - f_\pi(x,t)\int_0^\infty \chi(x,\tau)\partial_t f_\pi(x,t-\tau)\,\mathrm{d}\tau\bigg)\,\mathrm{d}x \\ = \int_{\mathbb{R}} &\bigg(\gamma \partial_x \phi(x,t)\partial_x f_\pi(x,t) + \gamma f_\pi(x,t)\partial_x^2 \phi(x,t) \\ &\qquad + f_\pi(x,t) f(x,t) - f_\pi(x,t)\int_0^\infty \chi(x,\tau)\partial_t f_\pi(x,t-\tau)\,\mathrm{d}\tau\bigg)\,\mathrm{d}x \\ = \int_{\mathbb{R}} &f(x,t)\partial_t\phi(x,t)\,\mathrm{d}x - \int_{\mathbb{R}}\bigg[\int_0^\infty \chi(x,\tau)\partial_t^2 \phi(x,t-\tau)\,\mathrm{d}\tau\bigg]\partial_t \phi(x,t)\,\mathrm{d}x, \end{aligned} \tag{1.25}$$

where in going from the third to the final line we have used integration by parts and the assumption that $\partial_x \phi$ and f_π vanish at spatial infinity. The first term on the r.h.s. is the rate of work done on the system by the external force f. Similarly, we interpret the second term as the rate of work done by the dissipative force, $-\int \chi(\tau)\partial_t^2\phi(x,t-\tau)\,\mathrm{d}\tau$. Thus, power dissipation amounts to the requirement that, for any trajectory, the *total work done by the dissipative force is negative.*

Let us note that the right hand side of (1.23) is equal to

$$\begin{aligned}\chi(x,0)\int_{-\infty}^{\infty} g(t)^2\,\mathrm{d}t + \int_{-\infty}^{\infty} g(t)\int_0^{\infty} \partial_\tau\chi(x,\tau)g(t-\tau)\,\mathrm{d}\tau\,\mathrm{d}t \\ = \int_{\mathbb{R}^2} \{\chi(x,0)\delta(t_1-t_2) + \tfrac{1}{2}[\partial_\tau\chi](x,|t_1-t_2|)\}g(t_1)g(t_2)\,\mathrm{d}^2t,\end{aligned} \tag{1.26}$$

which follows from integration by parts after noting that $\partial_t g(t-\tau) = -\partial_\tau g(t-\tau)$. Here $\delta(t)$ is the Dirac delta function. Thus the power dissipation condition is equivalent to the statement that for every x the (generalized) function

$$D_\chi(x,t) = \chi(x,0)\delta(t) + \tfrac{1}{2}[\partial_\tau\chi](x,|t|) \tag{1.27}$$

is *positive definite* in the sense of the classical Bochner's Theorem, see [7, Theorem IX.9]. Thus, by Bochner's Theorem, the time Fourier transform of D is a nonnegative measure,

$$\begin{aligned}\widehat{D}_\chi(x,\omega) &= \chi(x,0) + \frac{1}{2}\int_{-\infty}^{\infty} [\partial_\tau\chi](x,|t|)\mathrm{e}^{\mathrm{i}\omega t}\,\mathrm{d}t \\ &= \chi(x,0) + \int_0^{\infty} \partial_\tau\chi(x,\tau)\cos(\omega\tau)\,\mathrm{d}\tau \\ &= \omega\int_0^{\infty} \chi(x,\tau)\sin(\omega\tau)\,\mathrm{d}\tau = \omega\,\mathrm{Im}\,\widehat{\chi}(x,\omega) \geq 0.\end{aligned} \tag{1.28}$$

Since χ is real, $\widehat{D}_\chi(x,\omega)$ is symmetric under $\omega \leftrightarrow -\omega$, and indeed Bochner's Theorem shows that a symmetric measure of suitably bounded growth is nonnegative if and only if it is the Fourier transform of a real positive definite distribution.

The simplest physically relevant example of a susceptibility satisfying the power dissipation condition is obtained with $\chi(x,\tau) = \alpha > 0$, a positive constant. Then

$$f_\pi(x,t) + \alpha\int_0^{\infty} f_\pi(x,t-\tau)\,\mathrm{d}\tau \;=\; \pi(x,t). \tag{1.29}$$

Using the equation of motion $f_\pi = \partial_t\phi$, we find that

$$\pi(x,t) = \partial_t\phi(x,t) + \alpha\int_0^{\infty} \partial_t\phi(x,t-\tau)\,\mathrm{d}\tau = \partial_t\phi(x,t) + \alpha\phi(x,t), \tag{1.30}$$

where we have applied the boundary condition $\lim_{t\to-\infty}\phi(x,t) = 0$. Combined with the equation of motion $\partial_t\pi = \gamma\partial_x^2\phi + f$ we obtain

$$\partial_t^2\phi(x,t) + \alpha\partial_t\phi(x,t) - \gamma\partial_x^2\phi(x,t) = f(x,t), \tag{1.31}$$

which is the dynamical equation for a driven damped string, with damping force per unit length $-\alpha\partial_t\phi(x,t)$. Note that α is dimensionally an inverse time and $1/\alpha$ is the characteristic time for the damping of oscillations.

A more realistic model for damping is obtained by supposing $\chi(x,\tau)$ to be a nontrivial function of τ as in (1.21). To allow for damping restricted to only a part of the string, we suppose that χ depends on x as well. For instance, we could take the Debye susceptibility

$$\chi(x,\tau) = \alpha(x)\mathrm{e}^{-\nu(x)\tau}, \tag{1.32}$$

with $\alpha(x) \geq 0$ and $\nu(x) \geq 0$ nonnegative functions of x. This results in the following integral-differential equation for ϕ

$$\partial_t^2\phi(x,t) + \alpha(x)\int_0^\infty \mathrm{e}^{-\nu(x)\tau}\partial_t^2\phi(x,t-\tau)\,\mathrm{d}\tau - \gamma\partial_x^2\phi(x,t) = f(x,t), \tag{1.33}$$

or after integration by parts

$$\begin{aligned} \partial_t^2\phi(x,t) &+ \alpha(x)\partial_t\phi(x,t) - \alpha(x)\nu(x)\phi(x,t) \\ &+ \alpha(x)\nu(x)^2\int_0^\infty \mathrm{e}^{-\nu(x)\tau}\phi(x,t-\tau)\,\mathrm{d}\tau - \gamma\partial_x^2\phi(x,t) = f(x,t). \end{aligned} \tag{1.34}$$

Then

$$\widehat{D}_\chi(x,\omega) = \alpha(x)\int_0^\infty \mathrm{e}^{-\nu(x)\tau}\omega\sin(\omega\tau)\,\mathrm{d}\tau = \alpha(x)\frac{\omega^2}{\nu(x)^2+\omega^2} \geq 0, \tag{1.35}$$

so the Debye susceptibility satisfies the power dissipation condition.

The results of [2,3] show that under the above conditions it is possible to find a coupling function $\varsigma(x,s)$, $s \in \mathbb{R}$, such that solutions to the TDD equations are generated by solutions to the following extended system

$$\begin{aligned} \partial_t\phi(x,t) &= f_\pi(x,t) \\ \partial_t\pi(x,t) &= \gamma\partial_x^2\phi(x,t) + f(x,t) \\ \partial_t\psi(x,s,t) &= \theta(x,s,t) \\ \partial_t\theta(x,s,t) &= \partial_s^2\psi(x,s,t) + \varsigma(x,s)f_\pi(x,t), \end{aligned} \tag{1.36}$$

with

$$f_\pi(x,t) = \pi(x,t) - \int_{-\infty}^\infty \varsigma(x,s)\psi(x,s,t)\,\mathrm{d}s. \tag{1.37}$$

That is, given a solution (ϕ,π,ψ,θ) to (1.36), at rest at $t=-\infty$ and with f_π given by (1.37), the first two coordinates (ϕ,π) obey (1.22) with f_π given by (1.21), and conversely any solution to (1.21) – (1.22) with the string at rest at $t=-\infty$ arises in this way. Note that for each fixed x the additional variables θ,ψ may be interpreted as describing the oscillations of a "hidden string" with coordinate s, displacement $\psi(x,s,t)$, momentum $\theta(x,s,t)$ and driven by external force $\varsigma(x,s)f_\pi(x,t)$. Thus (1.36) is precisely the system of coupled strings described above and illustrated in Figure 1.

The extended system, consisting of the physical and hidden strings is Hamiltonian with symplectic form

$$\begin{aligned} \mathcal{J}(\phi_1,\pi_1,\psi_1,\theta_1;\phi_2,\pi_2,\psi_2,\theta_2) &= \int_{\mathbb{R}}\{\phi_1(x)\pi_2(x) - \phi_2(x)\pi_1(x)\}\,\mathrm{d}x \\ &+ \int_{\mathbb{R}^2}\{\psi_1(x,s)\theta_2(x,s) - \psi_2(x,s)\theta_1(x,s)\}\,\mathrm{d}s\,\mathrm{d}x \end{aligned} \tag{1.38}$$

and Hamilton function

$$(1.39)\quad \mathcal{H}_f(\phi,\pi,\psi,\theta,t) = \frac{1}{2}\int_{\mathbb{R}}\{f_\pi(x)^2 + \gamma\big(\partial_x\phi(x)\big)^2\}\,\mathrm{d}x - \int_{\mathbb{R}} f(x,t)\phi(x)\,\mathrm{d}x + \frac{1}{2}\int_{\mathbb{R}^2}\{\theta(x,s)^2 + \big(\partial_s\psi(x,s)\big)^2\}\,\mathrm{d}s\,\mathrm{d}x,$$

where f_π is given by (1.37), that is $f_\pi(x) = \pi(x) - \int \varsigma(x,s)\psi(x,s)\,\mathrm{d}s$.

The dynamical equation for an excitation of the hidden string at x is a driven wave equation

$$(1.40)\qquad \partial_t^2\psi(x,s,t) = \partial_s^2\psi(x,s,t) + \varsigma(x,s)f_\pi(x,t).$$

The solution to (1.40) with the hidden string at rest at $t = -\infty$ is easily written down, see (1.3):

$$(1.41)\qquad \psi(x,s,t) = \frac{1}{2}\int_0^\infty \left[\int_{s-\tau}^{s+\tau}\varsigma(x,r)\,\mathrm{d}r\right] f_\pi(x,t-\tau)\,\mathrm{d}\tau.$$

Thus

$$(1.42)\quad \int_{\mathbb{R}}\varsigma(x,s)\psi(x,s,t)\,\mathrm{d}s = \int_0^\infty\left[\frac{1}{2}\int_{\mathbb{R}}\varsigma(x,s)\int_{s-\tau}^{s+\tau}\varsigma(x,r)\,\mathrm{d}r\,\mathrm{d}s\right] f_\pi(x,t-\tau)\,\mathrm{d}\tau.$$

Comparing (1.21) and (1.37) we see that for the extension to reproduce the TDD system upon elimination of ψ and θ it is necessary and sufficient that the coupling $\varsigma(x,s)$ satisfy (1.7), that is

$$(1.43)\qquad \chi(x,\tau) = \frac{1}{2}\int_{\mathbb{R}}\varsigma(s,x)\int_{s-\tau}^{s+\tau}\varsigma(x,r)\,\mathrm{d}r\,\mathrm{d}s.$$

The existence of such a function, which is unique under a natural symmetry assumption, is guaranteed by the power dissipation condition [2,3].

Indeed, if we let $\widehat{\varsigma}(x,\sigma)$ denote the s-Fourier transform of ς,

$$(1.44)\qquad \widehat{\varsigma}(x,\sigma) = \int_{\mathbb{R}}\varsigma(x,s)\mathrm{e}^{\mathrm{i}\sigma s}\,\mathrm{d}s,$$

then (1.43) is equivalent to

$$(1.45)\qquad \widehat{D}_\chi(x,\omega) = \widehat{\varsigma}(x,\omega)\widehat{\varsigma}(x,-\omega),$$

so it suffices to take [2]

$$(1.46)\qquad \widehat{\varsigma}(x,\sigma) = \sqrt{2\widehat{D}_\chi(x,\sigma)}.$$

Furthermore, this choice of $\widehat{\varsigma}(x,\sigma)$ is unique if we ask further that $\widehat{\varsigma}(x,\sigma)\geq 0$ and that $\sigma\mapsto\widehat{\varsigma}(x,\sigma)$ be symmetric. Then,

$$(1.47)\quad \varsigma(x,s) = \frac{1}{2\pi}\int_{\mathbb{R}}\sqrt{2\widehat{D}_\chi(x,\sigma)}\mathrm{e}^{-\mathrm{i}\sigma s}\,\mathrm{d}\sigma = \frac{1}{2\pi}\int_{\mathbb{R}}\sqrt{2\widehat{D}_\chi(x,\sigma)}\cos(\sigma s)\,\mathrm{d}\sigma,$$

is symmetric and positive definite.

For example, in view of (1.28), (1.35), and (1.47), the following coupling produces the Debye susceptibility,

$$(1.48)\qquad \varsigma(x,s) = \sqrt{2\alpha(x)}\partial_s\Psi(\nu(x)s).$$

where

$$\Psi(s) = \frac{1}{2\pi \mathrm{i}} \int_{-\infty}^{\infty} \frac{1}{\omega} \sqrt{\frac{\omega^2}{1+\omega^2}} \mathrm{e}^{\mathrm{i}\omega s} \, \mathrm{d}\omega = \frac{s}{|s|} \frac{1}{\pi} \int_0^1 \frac{1}{\sqrt{1-u^2}} \mathrm{e}^{-u|s|} \, \mathrm{d}u = \frac{s}{|s|} \frac{1}{\pi} \int_0^{\pi/2} \mathrm{e}^{-|s|\sin(\phi)} \, \mathrm{d}\phi. \tag{1.49}$$

That is,

$$\varsigma(x,s) = \sqrt{2\alpha(x)} \left(\delta(s) - \frac{\nu(x)}{\pi} \int_0^{\pi/2} \sin(\phi) \mathrm{e}^{-\nu(x)|s|\sin(\phi)} \, \mathrm{d}\phi \right). \tag{1.50}$$

2. Local energy and momentum conservation in the extended system

We interpret the Hamiltonian $\mathcal{H}_0$ with $f \equiv 0$ as the internal energy of the damped string system consisting of the coupled physical and hidden strings. We have conservation of energy in the extended system, in the form

$$\frac{\mathrm{d}}{\mathrm{d}t} \mathcal{H}_0 = \int_{\mathbb{R}} f(x,t) \partial_t \phi(x,t) \, \mathrm{d}x, \tag{2.1}$$

i.e., the rate of change of $\mathcal{H}_0$ is the rate of work done on the system by the external force.

A significant advantage of working with the extend system is a transparent interpretation of the energy of the dissipative string as a sum of contributions from the physical and hidden strings. That is, it is natural to break the $\mathcal{H}_0$ into two pieces

$$\mathcal{H}_0(\phi, \pi, \psi, \theta, t) = H_0(\phi, \pi, \psi) + H_{\mathrm{hs}}(\psi, \theta), \tag{2.2}$$

the energy of the physical string and hidden strings respectively,

$$H_0(\phi, \pi, \psi) = \frac{1}{2} \int_{\mathbb{R}} \left\{ f_\pi(x)^2 + \gamma \big(\partial_x \phi(x) \big)^2 \right\} \mathrm{d}x \tag{2.3}$$

$$H_{\mathrm{hs}}(\psi, \theta) = \frac{1}{2} \int_{\mathbb{R}^2} \left\{ \theta(x,s)^2 + \big(\partial_s \psi(x,s) \big)^2 \right\} \mathrm{d}s \, \mathrm{d}x, \tag{2.4}$$

with f_π given by (1.37).

The internal energy $\mathcal{H}_0$ can be written as the integral of a local energy density

$$\mathcal{E}(x,t) = E_0(x,t) + E_{\mathrm{hs}}(x,t) \tag{2.5}$$

with

$$\begin{aligned} E_0(x,t) &= \tfrac{1}{2} \left\{ \big(\partial_t \phi(x,t) \big)^2 + \gamma \big(\partial_x \phi(x,t) \big)^2 \right\} \\ E_{\mathrm{hs}}(x,t) &= \frac{1}{2} \int_{\mathbb{R}} \left\{ \big(\partial_t \psi(x,s,t) \big)^2 + \big(\partial_s \psi(x,s,t) \big)^2 \right\} \mathrm{d}s. \end{aligned} \tag{2.6}$$

The energy conservation law (2.1) has the following local expression

$$\partial_t \mathcal{E}(x,t) + \partial_x J(x,t) = f(x,t) \partial_t \phi(x,t) \tag{2.7}$$

with the energy current

$$J(x,t) = -\gamma f_\pi(x,t) \partial_x \phi(x,t) = -\gamma \partial_t \phi(x,t) \partial_x \phi(x,t). \tag{2.8}$$

It is interesting to compute the time derivatives of E_0 and E_{hs} alone:

$$\partial_t E_0(x,t) + \partial_x J(x,t) = f(x,t) \partial_t \phi(x,t) - \partial_t \Delta(x,t) \partial_t \phi(x,t) \tag{2.9}$$

$$\partial_t E_{\mathrm{hs}}(x,t) = \partial_t \Delta(x,t) \partial_t \phi(x,t) \tag{2.10}$$

with

$$\begin{aligned}\Delta(x,t) = \pi(x,t) - f_\pi(x,t) &= \int_{\mathbb{R}} \varsigma(x,s)\psi(x,s,t)\,\mathrm{d}s \\ &= \int_0^\infty \chi(x,\tau)\partial_t\phi(x,t-\tau)\,\mathrm{d}\tau,\end{aligned} \tag{2.11}$$

where we have used (1.42). The first of these equations (2.9) is simply the local version of the energy law for the TDD string (1.25). From the second (2.10), we see that the energy of the hidden strings, which is the energy lost to dissipation up to time t, is

$$\begin{aligned}E_{\mathrm{hs}}(x,t) &= \int_{-\infty}^t \partial_{t'}\phi(x,t') \int_0^\infty \chi(x,\tau)\partial_{t'}^2\phi(x,t'-\tau)\,\mathrm{d}\tau\,\mathrm{d}t'. \\ &= \int_{-\infty}^t \int_{-\infty}^t D_\chi(x,t_1-t_2)\partial_{t_1}\phi(x,t_1)\partial_{t_2}\phi(x,t_2)\,\mathrm{d}t_1\,\mathrm{d}t_2.\end{aligned} \tag{2.12}$$

If the susceptibility χ — and hence the coupling ς — is independent of x, then the extended system is invariant under spatial translations. Associated to this symmetry is a local conservation law

$$\partial_t p(x,t) + \partial_x T(x,t) = -f(x,t)\partial_x\phi(x,t), \tag{2.13}$$

for the wave momentum density

$$\begin{aligned}p(x,t) &= -\pi(x,t)\partial_x\phi(x,t) - \int_{\mathbb{R}} \theta(x,s,t)\partial_x\psi(x,s,t)\,\mathrm{d}s \\ &= -\{\partial_t\phi(x,t) + \Delta(x,t)\}\partial_x\phi(x,t) - \int_{\mathbb{R}} \partial_t\psi(x,s,t)\partial_x\psi(x,s,t)\,\mathrm{d}s,\end{aligned} \tag{2.14}$$

with wave momentum flux, called *stress*,

$$\begin{aligned}T(x,t) = E_0(x,t) &+ \Delta(x,t)\partial_t\phi(x,t) \\ &+ \int \left\{(\partial_t\psi(x,s,t))^2 - (\partial_s\psi(x,s,t))^2\right\}\mathrm{d}s.\end{aligned} \tag{2.15}$$

When the driving force vanishes, $f = 0$, the total wave momentum

$$P = \int_{\mathbb{R}} p(x,t)\,\mathrm{d}x \tag{2.16}$$

is a conserved quantity.

The wave momentum density $p = p_0 + p_{\mathrm{hs}}$ is again a sum of contributions

$$p_0(x,t) = -\{\partial_t\phi(x,t) + \Delta(x,t)\}\partial_x\phi(x,t) \tag{2.17}$$

$$p_{\mathrm{hs}}(x,t) = -\int_{\mathbb{R}} \partial_t\psi(x,s,t)\partial_x\psi(x,s,t)\,\mathrm{d}s \tag{2.18}$$

from the physical and hidden strings. Likewise we separate the stress

$$T(x,t) = T_0(x,t) + T_{\mathrm{hs}}(x,t) \tag{2.19}$$

into two pieces,

$$T_0(x,t) = E_0(x,t) + \Delta(x,t)\partial_t\phi(x,t) \tag{2.20}$$

$$T_{\mathrm{hs}}(x,t) = \frac{1}{2}\int \{(\partial_t\psi(x,s,t))^2 - (\partial_s\psi(x,s,t))^2\}\,\mathrm{d}s. \tag{2.21}$$

Then

$$\partial_t p_0(x,t) + \partial_x T_0(x,t) = -f(x,t)\partial_x\phi(x,t) + \partial_x\Delta(x,t)\partial_t\phi(x,t) \tag{2.22}$$

$$\partial_t p_{\rm hs}(x,t) + \partial_x T_{\rm hs}(x,t) = -\partial_x\Delta(x,t)\partial_t\phi(x,t). \tag{2.23}$$

When we study eigenfunctions below, it will be convenient to work with complex valued solutions. In the above expressions, terms which are quadratic in the field variables should be modified in the complex case by the replacement $ab \to \operatorname{Re}\bar{a}b$. That is

$$\begin{aligned}
E_0(x,t) &= \tfrac{1}{2}\{|\partial_t\phi(x,t)|^2 + \gamma|\partial_x\phi(x,t)|^2\}\\
E_{\rm hs}(x,t) &= \frac{1}{2}\int_{\mathbb{R}}\{|\partial_t\psi(x,s,t)|^2 + |\partial_s\psi(x,s,t)|^2\}\,\mathrm{d}s\\
J(x,t) &= -\gamma\operatorname{Re}\overline{\partial_t\phi(x,t)}\partial_x\phi(x,t)\\
p_0(x,t) &= -\operatorname{Re}\overline{\partial_t\phi(x,t)}\partial_x\phi(x,t) - \operatorname{Re}\overline{\Delta(x,t)}\partial_x\phi(x,t),
\end{aligned} \tag{2.24}$$

etc.

3. The eigenfunction equation

It is useful and interesting to study steady state solutions to the extended system (1.36), for example solutions which are periodic in time $\mathrm{e}^{-\mathrm{i}\omega t}\big(\phi_\omega(x), \pi_\omega(x), \psi_\omega(x,s), \theta_\omega(x,s)\big)$. We refer to the spatial component $\Phi_\omega(x,s) = \big(\phi_\omega(x), \pi_\omega(x), \psi_\omega(x,s), \theta_\omega(x,s)\big)$ of such a time periodic solution as an eigenfunction for the linear system (1.36) with eigenvalue ω. Thus an eigenfunction satisfies

$$\begin{aligned}
-\mathrm{i}\omega\phi_\omega(x) &= f_\pi(x)\\
-\mathrm{i}\omega\pi_\omega(x) &= \gamma\partial_x^2\phi_\omega(x)\\
-\mathrm{i}\omega\psi_\omega(x,s) &= \theta_\omega(x,s)\\
-\mathrm{i}\omega\theta_\omega(x,s) &= \partial_s^2\psi_\omega(x,s) + \varsigma(x,s)f_\pi(x),
\end{aligned} \tag{3.1}$$

with

$$f_\pi(x) + \int_{\mathbb{R}}\varsigma(x,s)\psi_\omega(x,s)\,\mathrm{d}s - \pi_\omega(x) = 0. \tag{3.2}$$

We see that the displacement of the hidden string at position x satisfies

$$-\partial_s^2\psi_\omega(x,s) - \omega^2\psi_\omega(x,s) = -\mathrm{i}\omega\varsigma(x,s)\phi_\omega(x), \tag{3.3}$$

so

$$\begin{aligned}
&\psi_\omega(x,s)\\
&= a(x)\cos(\omega s) + b(x)\sin(\omega s) - \frac{\mathrm{i}\omega\phi_\omega(x)}{2\pi}\,\mathrm{P.V.}\int_{-\infty}^{\infty}\mathrm{e}^{-\mathrm{i}\sigma s}\frac{1}{\sigma^2-\omega^2}\widehat{\varsigma}(x,\sigma)\,\mathrm{d}\sigma\\
&= a(x)\cos(\omega s) + b(x)\sin(\omega s) + \frac{\mathrm{i}\phi_\omega(x)}{2}\int_{-\infty}^{\infty}\sin(\omega|s'-s|)\varsigma(x,s')\,\mathrm{d}s',
\end{aligned} \tag{3.4}$$

with $a(x)$ and $b(x)$ undetermined functions of x. Here, P. V. denotes the "principle value" integral,

$$\text{(3.5)}\quad \text{P. V.}\int_{-\infty}^{\infty} \mathrm{e}^{-i\sigma s}\frac{1}{\sigma^2-\omega^2}\widehat{\varsigma}(x,\sigma)\,\mathrm{d}\sigma$$
$$= \lim_{\delta\downarrow 0}\int_{\{\sigma:|\sigma^2-\omega^2|\geq\delta\}} \mathrm{e}^{-i\sigma s}\frac{1}{\sigma^2-\omega^2}\widehat{\varsigma}(x,\sigma)\,\mathrm{d}\sigma$$
$$= \lim_{\delta\downarrow 0}\frac{1}{2}\int_{\mathbb{R}} \mathrm{e}^{-i\sigma s}\left[\frac{1}{\sigma^2-(\omega+\mathrm{i}\delta)^2}+\frac{1}{\sigma^2-(\omega-\mathrm{i}\delta)^2}\right]\widehat{\varsigma}(x,\sigma)\,\mathrm{d}\sigma.$$

By (3.4), we see that

$$\text{(3.6)}\quad \int_{\mathbb{R}}\varsigma(x,s)\psi_\omega(x,s)\,\mathrm{d}s = a(x)\widehat{\varsigma}(x,\omega)$$
$$-\frac{\phi_\omega(x)}{2\pi}\,\text{P. V.}\int_{-\infty}^{\infty}\frac{\mathrm{i}\omega}{\sigma^2-\omega^2}\widehat{\varsigma}(x,\sigma)\widehat{\varsigma}(x,-\sigma)\,\mathrm{d}\sigma.$$

Recalling (1.45), that $\widehat{\varsigma}(x,\sigma)\widehat{\varsigma}(x,-\sigma) = \widehat{D}_\chi(x,\sigma) = 2\sigma\,\mathrm{Im}\,\widehat{\chi}(x,\sigma)$, we see that the final term on the r.h.s. can be expressed

$$\text{(3.7)}\quad -\frac{\phi_\omega(x)}{2\pi}\,\text{P. V.}\int_{-\infty}^{\infty}\frac{2\mathrm{i}\omega}{\sigma^2-\omega^2}\sigma\,\mathrm{Im}\,\widehat{\chi}(x,\sigma)\,\mathrm{d}\sigma$$
$$= -\frac{\mathrm{i}\phi_\omega(x)}{2\pi}\,\text{P. V.}\int_{-\infty}^{\infty}\left\{\frac{1}{\sigma-\omega}+\frac{1}{-\sigma-\omega}\right\}\sigma\,\mathrm{Im}\,\widehat{\chi}(x,\sigma)\,\mathrm{d}\sigma$$
$$= -\mathrm{i}\omega\,\mathrm{Re}\,\widehat{\chi}(x,\omega)\phi_\omega(x),$$

where we have used the symmetry $\sigma\,\mathrm{Im}\,\widehat{\chi}(x,\sigma) = -\sigma\,\mathrm{Im}\,\widehat{\chi}(x,-\sigma)$ and the "Kramers–Krönig relation"

$$\text{(3.8)}\quad \omega\,\mathrm{Re}\,\widehat{\chi}(x,\omega) = \text{P. V.}\frac{1}{\pi}\int_{-\infty}^{\infty}\frac{1}{\sigma-\omega}\sigma\,\mathrm{Im}\,\widehat{\chi}(x,\sigma)\,\mathrm{d}\sigma.$$

Plugging this result into (3.2) and combining the first two equations of (3.1) we find that the string displacement ϕ solves the following elliptic problem

$$\text{(3.9)}\quad \gamma\partial_x^2\phi_\omega(x) = -\omega^2(1+\mathrm{Re}\,\widehat{\chi}(x,\omega))\phi_\omega(x) - \mathrm{i}\omega a(x)\widehat{\varsigma}(x,\omega).$$

In analyzing (3.9) we should distinguish two cases:

(1) $\varsigma(x,\omega) = 0$ for all x.
(2) $\varsigma(x,\omega)$ nonzero for x in an open set.

In the first case, which is nongeneric, the medium described by the hidden strings is not absorbing at the given frequency ω, that is $\widehat{\chi}(x,\omega)$ is real for every x. Thus (3.9) is a very strong restriction on ϕ_ω, namely that it should satisfy the Schrödinger equation

$$\text{(3.10)}\quad (-\gamma\partial_x^2+V(x))\phi_\omega(x) = \lambda\phi_\omega(x),$$

with potential $V(x) = -\omega^2\widehat{\chi}(x,\omega)$ and spectral parameter $\lambda = \omega^2$. In the second case, the physical string displacement may be decomposed as follows

$$\text{(3.11)}\quad \phi_\omega(x) = \phi_\omega^{(1)}(x) + \phi_\omega^{(2)}(x),$$

where $\phi_\omega^{(1)}(x)$ is any solution to the nondissipative eigenfunction equation

$$\text{(3.12)}\quad \gamma\partial_x^2\phi_\omega^{(1)}(x) = -\omega^2(1+\mathrm{Re}\,\widehat{\chi}(x,\omega))\phi_\omega^{(1)}(x),$$

and $\phi_\omega^{(2)}$ is an *arbitrary function with support in* $\{x : \widehat{\varsigma}(x,\omega) \neq 0\}$. Indeed, we need only choose a to be

$$a(x) = \mathrm{i}\frac{\gamma\partial_x^2\phi_\omega^{(2)}(x) + \omega^2\big(1 + \operatorname{Re}\widehat{\chi}(x,\omega)\big)\phi_\omega^{(2)}(x)}{\omega\widehat{\varsigma}(x,\omega)}. \tag{3.13}$$

Thus, the physical string displacement for an eigenmode $\phi_\omega(x)$ may be chosen arbitrarily within that part of the string which is absorbing at the given frequency ω. We want to emphasize the significance of this fact, since the formal eigenvalue problem

$$\gamma\partial_x^2\phi_\omega(x) = -\omega^2(1 + \widehat{\chi}(x,\omega))\phi_\omega(x) \tag{3.14}$$

does not allow for such arbitrariness in the choice of $\phi_\omega(x)$. The resolution to this apparent contradiction lies in recognizing that a TDD system is an open part of a larger conservative Hamiltonian system. It is only for the extended Hamiltonian system that the eigenmodes Φ_ω are unambiguously defined with $\mathrm{e}^{\mathrm{i}\omega t}\Phi_\omega$ a stationary solution to the canonical Hamiltonian evolution equation. The "legitimate" eigenmodes for the original TDD string consist of projections ϕ_ω of the eigenmodes Φ_ω onto the subspace of the physical string. Thus a TDD string, being an open system has as many eigenmodes, i.e stationary solutions, as its the minimal conservative extension introduced and described in [1–3]. In particular, a finite-dimensional TDD system typically has infinitely many stationary solutions and, hence, infinitely many eigenmodes. Another view on the construction of eigenmodes follows.

The eigenfunctions written down above, involving as they do arbitrary excitations of the hidden strings, may not be relevant to the dynamics (1.36) with the external force acting on the physical string. Indeed, note that the coefficient $b(x)$ does not appear in the effective equation (3.9) for ϕ_ω. To see which eigenfunctions appear in the expansion of a general solution to (1.36) with a compactly supported external force, it is convenient to introduce the Fourier-Laplace transform

$$\widetilde{h}(\zeta) = \int_{\mathbb{R}} \mathrm{e}^{\mathrm{i}\zeta t} h(t)\,\mathrm{d}t, \tag{3.15}$$

defined for

(1) $\zeta \in \mathbb{C}$ if $h \to 0$ super exponentially fast as $|t| \to \infty$, for instance if h is compactly supported in time.
(2) $\operatorname{Im}\zeta > 0$ if $h \to 0$ super exponentially fast as $t \to -\infty$.
(3) $\operatorname{Im}\zeta < 0$ if $h \to 0$ super exponentially fast as $t \to \infty$.

Taking the Fourier – Laplace transform of (1.36) results in the following equations

$$\begin{aligned}
-\mathrm{i}\zeta\tilde{\phi}(x,\zeta) &= \tilde{f}_\pi(x,\zeta)\\
-\mathrm{i}\zeta\tilde{\pi}(x,\zeta) &= \gamma\partial_x^2\tilde{\phi}(x,\zeta) + \tilde{f}(x,\zeta)\\
-\mathrm{i}\zeta\tilde{\psi}(x,s,\zeta) &= \tilde{\theta}(x,s,\zeta)\\
-\mathrm{i}\zeta\tilde{\theta}(x,s,\zeta) &= \partial_s^2\tilde{\psi}(x,s,\zeta) + \varsigma(x,s)\tilde{f}_\pi(x,\zeta),
\end{aligned} \tag{3.16}$$

with

$$\tilde{f}_\pi(x,\zeta) = \tilde{\pi}(x,\zeta) - \int_{-\infty}^{\infty} \varsigma(x,s)\tilde{\psi}(x,s,\zeta)\,\mathrm{d}s. \tag{3.17}$$

Due to our consideration of solutions with the strings at rest at $t = -\infty$, we expect the quantities in (3.16) to be well defined only for $\operatorname{Im}\zeta > 0$. However, if the driving force is compactly supported in time then $\widetilde{f}(x,\zeta)$ is defined for all $\zeta \in \mathbb{C}$, and we may extend $\tilde{\phi}(x,\zeta)$ to $\zeta \in \mathbb{C}\setminus\mathbb{R}$ by solving (3.16). Following the eigenfunction analysis, we obtain, with $\varepsilon =$ sign of $\operatorname{Im}\zeta$,

$$\begin{aligned}(3.18)\qquad \tilde{\psi}(x,s,\zeta) &= -\frac{\mathrm{i}\zeta\tilde{\phi}(x,\zeta)}{2\pi}\int_{-\infty}^{\infty} \mathrm{e}^{-\mathrm{i}\sigma s}\frac{1}{\sigma^2-\zeta^2}\widehat{\varsigma}(x,\sigma)\,\mathrm{d}\sigma \\ &= \varepsilon\frac{\tilde{\phi}(x,\zeta)}{2}\int_{-\infty}^{\infty} \mathrm{e}^{\mathrm{i}\varepsilon\zeta|s'-s|}\varsigma(x,s')\,\mathrm{d}s' \\ (3.19)\qquad \gamma\partial_x^2\tilde{\phi}(x,\zeta) &= -\zeta^2\big(1+\widehat{\chi}(x,\varepsilon\zeta)\big)\tilde{\phi}(x,\zeta)+\tilde{f}(x,\zeta),\end{aligned}$$

where $\widehat{\chi}(x,\zeta)$ is defined for $\operatorname{Im}\zeta > 0$ as

$$(3.20)\qquad \widehat{\chi}(x,\zeta) = \int_0^{\infty} \mathrm{e}^{\mathrm{i}\tau\zeta}\chi(x,\tau)\,\mathrm{d}\tau.$$

For a function h vanishing at $+\infty$ or $-\infty$ we have the Fourier inversion formula

$$(3.21)\qquad h(t) = \frac{1}{2\pi}\int_{\mathbb{R}} \mathrm{e}^{-\mathrm{i}\omega t\mp\eta t}\widetilde{h}(\omega\mp\mathrm{i}\eta)\,\mathrm{d}\omega, \quad \Big(\lim_{t=\pm\infty} h(t) = 0\Big),$$

with $\eta > 0$ arbitrary. Inverting the solution to (3.16) with $\eta > 0$ or < 0 produces distinct solutions to (1.36): for $\eta > 0$ we obtain the desired *causal* solution with the strings at rest at $t = -\infty$, while for $\eta < 0$ we obtain the *anti-causal* solution with the strings at rest at $t = +\infty$.

In a certain sense we are only interested in the causal solution to (1.36) and thus to the solution to (3.16) only for ζ in the upper half-plane. However, since (3.16) involves a source term, this solution, often called the *resolvent*, is not directly expressed as a superposition of eigenfunctions. However, there is a general procedure for decomposing the solution to (3.16) into a superposition of eigenfunctions. Namely for $\omega \in \mathbb{R}$ we define

$$\begin{aligned}(3.22)\quad &(\phi_\omega(x),\pi_\omega(x),\psi_\omega(x,s),\theta_\omega(x,s)) \\ &= \lim_{\delta\to 0}\big(\tilde{\phi}(x,\omega+\mathrm{i}\delta)-\tilde{\phi}(x,\omega-\mathrm{i}\delta),\tilde{\pi}(x,\omega+\mathrm{i}\delta)-\tilde{\pi}(x,\omega-\mathrm{i}\delta), \\ &\qquad \tilde{\psi}(x,s,\omega+\mathrm{i}\delta)-\tilde{\psi}(x,s,\omega-\mathrm{i}\delta),\tilde{\theta}(x,s,\omega+\mathrm{i}\delta)-\tilde{\phi}(x,s,\omega-\mathrm{i}\delta)\big).\end{aligned}$$

Since $\tilde{f}(x,\zeta)$ is continuous at each $\zeta = \omega \in \mathbb{R}$, it follows from (3.16) that $(\phi_\omega, \pi_\omega, \psi_\omega, \theta_\omega)$ is an eigenfunction for each ω. To recover the resolvent for ζ in the upper half-plane from the eigenfunctions (3.22), suppose that the external force is supported in the set $\{t : t > t_0\}$. Then

$$(3.23)\qquad |\widetilde{f}(x,\zeta)| \le C\mathrm{e}^{-\operatorname{Im}\zeta t_0}, \quad (\operatorname{Im}\zeta > 0).$$

Thus $\exp(-\mathrm{i}\zeta t_0)f(x,\zeta)$ is bounded in the upper half-plane, and by analyticity we have,[2]

$$(3.24)\quad \big(\tilde{\phi}(x,\zeta),\tilde{\pi}(x,\zeta),\tilde{\psi}(x,s,\zeta),\tilde{\theta}(x,s,\zeta)\big)$$
$$= \frac{1}{2\pi\mathrm{i}}\int_{\mathbb{R}} \frac{\mathrm{e}^{\mathrm{i}(\zeta-\omega)t_0}}{\omega-\zeta}\big(\phi_\omega(x),\pi_\omega(x),\psi_\omega(x,s),\theta_\omega(x,s)\big)\,\mathrm{d}\omega, \quad (\operatorname{Im}\zeta>0)$$

expressing the solution to (3.16) as a superposition of eigenfunctions. (There is a similar formula for the advanced solution with $\operatorname{Im}\zeta<0$, involving an upper bound t_1 on the support of the external force.)

Now let us fix ω and suppose that we have an eigenfunction of the form (3.22). Then by (3.18) and (3.19),

$$(3.25)\quad \psi_\omega(x,s) = \frac{\mathrm{i}\phi_\omega(x)}{2}\int_{-\infty}^{\infty}\sin(\omega|s'-s|)\varsigma(x,s')\,\mathrm{d}s' + \frac{g_\omega(x)}{2}\cos(\omega s)\widehat{\varsigma}(x,\omega)$$

$$(3.26)\quad \gamma\partial_x^2\phi_\omega(x) = -\omega^2\big(1+\operatorname{Re}\widehat{\chi}(x,\omega)\big)\phi_\omega(x) - \mathrm{i}\omega^2\operatorname{Im}\widehat{\chi}(x,\omega)g_\omega(x),$$

with

$$(3.27)\quad g_\omega(x) = \lim_{\delta\to 0}\{\tilde{\phi}(x,\omega+\mathrm{i}\delta)+\tilde{\phi}(x,\omega-\mathrm{i}\delta)\}.$$

In other words, the eigenfunctions which appear in the expansion (3.24) are of the form (3.9) with $b(x)=0$. We refer to such eigenfunctions as *spectral* eigenfunctions. By (3.24), the spectral eigenfunctions form a complete set for the description of the dynamics of the extended system (1.36). Thus, the freedom to choose g_ω in (3.26) provides us with a rich enough family of solutions to describe the dynamics of the TDD string.

Note that a solution to (3.16) satisfies the eigenvalue equation away from the spatial support of the driving force. Thus, one may try to produce an eigenfunction as the limit of a sequence or solutions to (3.16) with a sequence of driving forces supported farther and farther from the origin. When this procedure works to produce a nontrivial eigenmode, the result is of the form (3.26) with a special choice of the arbitrary function g_ω. We can obtain in this way,

(1) The *causal eigenfunctions* with $g_\omega(x)=\phi_\omega(x)$,

$$(3.28)\quad \psi_\omega(x,s) = \frac{\phi_\omega(x)}{2}\int_{-\infty}^{\infty}\mathrm{e}^{\mathrm{i}\omega|s'-s|}\varsigma(s',x)\,\mathrm{d}s'$$

$$(3.29)\quad \gamma\partial_x^2\phi_\omega(x) = -\omega^2(1+\widehat{\chi}(x,\omega))\phi_\omega(x),$$

corresponding to a driving force in the distant past and the causal boundary condition with the strings at rest at $t=-\infty$.

(2) The *anti-causal eigenfunctions* with $g_\omega(x)=-\phi_\omega(x)$,

$$(3.30)\quad \psi_\omega(x,s) = \frac{\phi_\omega(x)}{2}\int_{-\infty}^{\infty}\mathrm{e}^{-\mathrm{i}\omega|s'-s|}\varsigma(s',x)\,\mathrm{d}s'$$

$$(3.31)\quad \gamma\partial_x^2\phi_\omega(x) = -\omega^2(1+\overline{\widehat{\chi}(x,\omega)})\phi_\omega(x),$$

[2]For a general linear system the limit on the right hand side of (3.22) is a defined only as a vector valued measure and the integral on the right hand side of (3.24) should be interpreted as the integral of $1/(\omega-\zeta)$ against this measure. For the systems of coupled strings considered here, however, these boundary measures are always absolutely continuous, so (3.24) holds with $(\phi_\omega,\pi_\omega,\psi_\omega,\theta_\omega)$ defined for almost every ω by (3.22).

corresponding to a driving force in the distant future and the anti-causal, or advanced, boundary condition with the strings at rest at $t = +\infty$.

4. Energy flux in an eigenfunction

The energy density $\mathcal{E}$ in an eigenfunction, at a point x with $\widehat{\varsigma}(x,\omega) \neq 0$, is typically infinite due to the contribution from the hidden strings. This result is to be expected physically as the eigenfunction represents the idealized steady propagation of a monochromatic wave through an absorbing medium with infinite heat capacity. However, due to the decoupling between the hidden strings, energy can flow only through the physical string and we expect the energy flux J to be finite. Furthermore as we shall see $\partial_x J$, which by (2.7) is formally equal to $-\partial_t \mathcal{E}$, can be nonzero at a point x, in which case the eigenfunction is a steady state in which energy is dissipated to or absorbed from the medium at x at a constant rate.

By (2.8), the energy flux for an eigenfunction is

$$J(x,t) = -\gamma \operatorname{Re}\Big\{ \mathrm{i}\omega \overline{\phi_\omega(x)} \partial_x \phi_\omega(x) \Big\} = \gamma\omega \operatorname{Im} \overline{\phi_\omega(x)} \partial_x \phi_\omega(x). \tag{4.1}$$

Thus, by (3.26),

$$-\partial_x J(x) = -\omega \operatorname{Im} \overline{\phi_\omega(x)} \gamma \partial_x^2 \phi_\omega(x) = \omega^3 \operatorname{Im} \widehat{\chi}(x,\omega) \operatorname{Re}\Big\{ \overline{\phi_\omega(x)} g_\omega(x) \Big\}. \tag{4.2}$$

for a spectral eigenfunction, where $g_\omega(x)$ is the arbitrary function describing the excitation of the medium, as represented by the hidden strings. The energy density of the physical string

$$E_0(x) = \tfrac{1}{2}\{\omega^2 |\phi_\omega(x)|^2 + \gamma |\partial_x \phi_\omega(x)|^2\}, \tag{4.3}$$

is constant in time. We have a constant rate of dissipation at each x with $-\partial_x J(x) \neq 0$,

$$\partial_t \mathcal{E}(x) = \partial_t E_{\mathrm{hs}}(x) = \omega^3 \operatorname{Im} \widehat{\chi}(x,\omega) \operatorname{Re}\Big\{ \overline{\phi_\omega(x)} g_\omega(x) \Big\}. \tag{4.4}$$

Thus the eigenfunction represents a steady state situation in which energy is flowing into or out of the hidden strings at a constant rate at each point x with $\operatorname{Im} \widehat{\chi}(x,\omega) \neq 0$.

For a general spectral eigenfunction (3.26), the dissipation $\partial_t \mathcal{E}(x)$ may be positive or negative, however for a causal eigenfunction (3.29), with $g_\omega = \phi_\omega$, we have

$$\partial_t \mathcal{E}(x) = \partial_t E_{\mathrm{hs}}(x) = \omega^3 \operatorname{Im} \widehat{\chi}(x,\omega) |\phi_\omega(x)|^2 > 0, \qquad \text{(causal eigenfunction)} \tag{4.5}$$

corresponding to a steady dissipation of energy from the physical string into the medium, represented by the hidden strings. Similarly, for an anti-causal eigenfunction (3.31) there is a steady flux of energy *out of the medium and into the physical string*

$$\partial_t E_{\mathrm{hs}}(x) = -\omega^3 \operatorname{Im} \widehat{\chi}(x,\omega) |\phi_\omega(x)|^2 < 0, \qquad \text{(anti-causal eigenfunction).} \tag{4.6}$$

5. Plane waves and momentum flux in an homogeneous system

Suppose that the susceptibility $\chi(x,\tau) = \chi(\tau)$ is constant for the whole range of $x \in \mathbb{R}$. Then it is interesting to look for plane wave solutions $\mathrm{e}^{\mathrm{i}(kx-\omega t)}\big(\phi_0, \pi_0, \psi_0(s), \theta_0(s)\big)$ with ϕ_0, π_0 constants and $\psi_0(s)$, $\theta_0(s)$ independent of x, i.e., eigenfunctions $\propto \mathrm{e}^{\mathrm{i}kx}$.

A first observation is that there are no causal or anti-causal eigenfunctions of this form, at least at frequencies with $\operatorname{Im}\widehat{\chi}(\omega) > 0$. Indeed a causal solution would satisfy

$$-\gamma k^2 \phi_0 = -\omega^2 \big(1 + \widehat{\chi}(\omega)\big)\phi_0, \tag{5.1}$$

and the only nontrivial solutions to this equation are exponentially growing as $x \to \pm\infty$ unless $\widehat{\chi}(\omega)$ is real. Such evanescent waves play a role in constructing scattering states for in-homogeneous systems but are physically irrelevant in the homogeneous system.

Thus to find plane wave solutions it is necessary to look beyond causal eigenfunctions. For a plane wave spectral eigenfunction, see (3.26), we have

$$-\gamma k^2 \phi_0 = -\mathrm{i}\omega^2 g_0 \operatorname{Im}\widehat{\chi}(\omega) - \omega^2\big(1 + \operatorname{Re}\widehat{\chi}(\omega)\big)\phi_0, \tag{5.2}$$

$$\psi_0(s) = \frac{g_0}{2}\cos(\omega s)\widehat{\varsigma}(\omega) + \frac{\mathrm{i}\phi_0}{2}\int_{-\infty}^{\infty} \sin(\omega|s' - s|)\varsigma(s')\,\mathrm{d}s', \tag{5.3}$$

with g_0 an arbitrary constant describing the excitation of the hidden strings. A bounded solution results only for k real, so

$$g_0 = \mathrm{i}\alpha\phi_0, \quad \alpha \in \mathbb{R}, \tag{5.4}$$

must be a pure imaginary multiple of ϕ_0. Furthermore, we must have

$$1 + \operatorname{Re}\widehat{\chi}(\omega) - \alpha \operatorname{Im}\widehat{\chi}(\omega) \geq 0. \tag{5.5}$$

Then setting

$$\gamma k^2 = \omega^2\big(1 + \operatorname{Re}\widehat{\chi}(\omega) - \alpha\operatorname{Im}\widehat{\chi}(\omega)\big), \tag{5.6}$$

we obtain a nontrivial plane wave solution.

If the medium is not dissipative at frequency ω, so $\operatorname{Im}\widehat{\chi}(\omega) = 0$, there is a *dispersion* relation between k and ω

$$\gamma k^2 = \omega^2\big(1 + \operatorname{Re}\widehat{\chi}(\omega)\big), \quad (\operatorname{Im}\widehat{\chi}(\omega) = 0). \tag{5.7}$$

At frequencies ω with nontrivial dissipation, that is $\operatorname{Im}\widehat{\chi}(\omega) \neq 0$, there is *no relation between k and ω*. Indeed k may be chosen arbitrarily provided we take

$$\alpha = \frac{1 + \operatorname{Re}\widehat{\chi}(\omega)}{\operatorname{Im}\widehat{\chi}(\omega)} - \frac{\gamma k^2}{\omega^2 \operatorname{Im}\widehat{\chi}(\omega)}. \tag{5.8}$$

By (4.1), the energy flux for a plane wave

$$J = \gamma\omega k|\phi_0|^2 \tag{5.9}$$

is constant. Thus $\partial_x J = 0$ and plane wave solutions represent a steady state flow without dissipation of energy. More precisely, the dissipation of energy to the medium, as described by the hidden strings, is exactly balanced by the energy emitted from the medium. The energy density of the physical string is

$$E_0 = (\omega^2 + \gamma k^2)|\phi_0|^2 \tag{5.10}$$

and the energy density of the medium, described by the hidden strings, is of course infinite.

As the system is homogeneous, we can also consider the wave momentum density and stress in a plane wave eigenfunction. By (2.17) the wave momentum density of the physical string is

$$p_0 = \omega k|\phi_0|^2 - k\operatorname{Im}\overline{\Delta_0}\phi_0, \tag{5.11}$$

where

$$\Delta_0 = \int_{\mathbb{R}} \varsigma(s)\psi_0(s)\, \mathrm{d}s = \mathrm{i}\omega\phi_0\big(\alpha \operatorname{Im}\widehat{\chi}(\omega) - \operatorname{Re}\widehat{\chi}(\omega)\big). \tag{5.12}$$

Thus

$$p_0 = \omega k\{1 + \operatorname{Re}\widehat{\chi}(\omega) - \alpha \operatorname{Im}\widehat{\chi}(\omega)\}|\phi|^2 = \frac{\gamma k^3}{\omega}|\phi_0|^2. \tag{5.13}$$

The wave momentum density of the hidden strings is infinite, since

$$p_{\mathrm{hs}} = \omega k \int_{\mathbb{R}} |\psi_0(s)|^2\, \mathrm{d}s \tag{5.14}$$

by (2.18).

Of more interest is the stress, which by (2.19) – (2.20) is

$$T = E_0 + \omega \operatorname{Im} \overline{\Delta_0}\phi_0 + T_{\mathrm{hs}}, \tag{5.15}$$

where E_0 given by (5.10) and T_{hs} is formally

$$T_{\mathrm{hs}} = \frac{1}{2}\int_{-\infty}^{\infty} \{\omega^2|\psi_0(s)|^2 - |\partial_s\psi_0(s)|^2\}\, \mathrm{d}s. \tag{5.16}$$

The integral on the r.h.s. of (5.16) does not converge absolutely, but we can regularize it by defining $T_{\mathrm{hs}} = \lim_{\delta\to 0} T_{\mathrm{hs}}(\delta)$ with

$$T_{\mathrm{hs}}(\delta) = \frac{1}{2}\int_{-\infty}^{\infty} \mathrm{e}^{-\delta|s|}\{\omega^2|\psi_0(s)|^2 - |\partial_s\psi_0(s)|^2\}\, \mathrm{d}s. \tag{5.17}$$

To evaluate the integral in (5.17) it is useful to write

$$\begin{aligned} |\partial_s\psi_0(s)|^2 &= \frac{1}{2}\partial_s^2|\psi_0(s)|^2 - \operatorname{Re}\overline{\psi_0(s)}\partial_s^2\psi_0(s) \\ &= \frac{1}{2}\partial_s^2|\psi_0(s)|^2 + \omega^2|\psi_0(s)|^2 + \omega\varsigma(s)\operatorname{Im}\overline{\psi_0(s)}\phi_0, \end{aligned} \tag{5.18}$$

where we have used (3.3). Thus

$$\begin{aligned} T_{\mathrm{hs}}(\delta) &= -\frac{1}{2}\int_{-\infty}^{\infty} \mathrm{e}^{-\delta|s|}\{\tfrac{1}{2}\partial_s^2|\psi_0(s)|^2 + \omega\varsigma(s)\operatorname{Im}\overline{\psi_0(s)}\phi_0\}\, \mathrm{d}s \\ &= \frac{\delta}{2}|\psi_0(0)|^2 - \frac{\delta^2}{4}\int_{-\infty}^{\infty} \mathrm{e}^{-\delta|s|}|\psi_0(s)|^2\, \mathrm{d}s \\ &\qquad - \frac{1}{2}\omega\int_{-\infty}^{\infty} \mathrm{e}^{-\delta|s|}\varsigma(s)\operatorname{Im}\overline{\psi_0(s)}\phi_0\, \mathrm{d}s. \end{aligned} \tag{5.19}$$

It follows that

$$T_{\mathrm{hs}} = \lim_{\delta\to 0} T_{\mathrm{hs}}(\delta) = -\frac{1}{2}\omega\int_{-\infty}^{\infty} \varsigma(s)\operatorname{Im}\overline{\psi_0(s)}\phi_0\, \mathrm{d}s = -\frac{1}{2}\omega \operatorname{Im}\overline{\Delta_0}\phi_0, \tag{5.20}$$

if, for instance, $|\varsigma(s)|$ is integrable. Thus

$$\begin{aligned} T &= E_0 + \frac{1}{2}\omega\operatorname{Im}\overline{\Delta_0}\phi_0 \\ &= \frac{1}{2}\left(\omega^2(1 + \operatorname{Re}\widehat{\chi}(\omega) - \alpha\operatorname{Im}\widehat{\chi}(\omega)) + \gamma k^2\right)|\phi_0|^2 = \gamma k^2|\phi_0|^2, \end{aligned} \tag{5.21}$$

which is identical to the result for the undamped string!

6. Scattering eigenfunctions for a semi-infinite medium

To close, we would like to illustrate the power of the above method by unambiguously constructing the scattering states representing an idealized description of the following experiment. Imagine that we have an long, i.e., infinite, string whose right half ($x > 0$) is subject to dispersion and dissipation with susceptibility $\chi(\tau)$. That is, the the susceptibility for the whole string is

$$\chi(x, \tau) = \begin{cases} 0 & x < 0 \\ \chi(\tau) & x > 0 \end{cases}. \tag{6.1}$$

Suppose we drive the left end of the string, at $x = -\infty$, periodically with frequency ω, sending an incoming wave toward the TDD right half of the string. After some time a steady state is reached in which a certain fraction of the incoming wave is absorbed by the dissipative half of the string and a certain fraction is reflected.

The steady state eigenfunction describing the above experiment is a causal eigenfunction since the source is at $x = -\infty$ and started in the distant past. Thus by (3.29), the string displacement satisfies

$$\gamma \partial_x^2 \phi_\omega(x) = -\omega^2 \big(1 + I[x > 0]\widehat{\chi}(\omega)\big)\phi_\omega(x), \tag{6.2}$$

where $I[x > 0] = 1$ for $x > 0$ and 0 otherwise. Indeed, this is the naive equation that one might right down for the scattering states of a TDD string. However, it is only within the context of the extended Hamiltonian system that we understand this to be only one of many possible eigenfunction equations, the choice of which is dictated by the physics under consideration, namely a source at spatial infinity in the distant past.

By (6.2), we have

$$\phi_\omega(x) = \mathrm{e}^{\mathrm{i}k_< x} + r\mathrm{e}^{-\mathrm{i}k_< x}, \quad k_< = \frac{\omega}{\sqrt{\gamma}}, \quad (x < 0) \tag{6.3}$$

on the left half of the string, up to an over all multiple which we fix to be 1 without loss. In (6.3) the term $\mathrm{e}^{\mathrm{i}k_< x}$ is the incoming wave and $r\mathrm{e}^{-\mathrm{i}k_> x}$ is the reflected wave.

To find the reflection coefficient r we need to solve for the form of the eigenfunction on the right half of the string. Again by (6.2), we have

$$\gamma \partial_x^2 \phi_\omega(x) = -\omega^2 \big(1 + \widehat{\chi}(\omega)\big)\phi_\omega(x), \quad (x > 0) \tag{6.4}$$

for $x > 0$. Furthermore, since we require the solution to be bounded this determines ϕ uniquely up to an over all multiple

$$\phi_\omega(x) = v\mathrm{e}^{\mathrm{i}k_> x}, \tag{6.5}$$

where v is an, as yet, undetermined transmission coefficient and

$$k_> = \sqrt[+]{\frac{\omega^2}{\gamma}\big(1 + \widehat{\chi}(\omega)\big)}. \tag{6.6}$$

Here $\sqrt[+]{z}$ denotes the unique square root of z in the upper half-plane.[3] Since $\operatorname{Im} k_> > 0$, at least if $\operatorname{Im}\widehat{\chi}(\omega) \neq 0$, we see that $\phi_\omega(x)$ decays exponentially in the dissipative half of the physical string $(x > 0)$. The excitation of the hidden strings, which are restricted to $x > 0$, is given by (3.28):

$$\psi_\omega(x,s) = \frac{v}{2}\mathrm{e}^{\mathrm{i}k_> x}\int_{-\infty}^{\infty} \mathrm{e}^{\mathrm{i}\omega|s-s'|}\varsigma(s')\,\mathrm{d}s', \quad (x>0). \tag{6.7}$$

To determine v and r we note that the eigenfunction equation (6.2) forces ϕ_ω and $\partial_x\phi_\omega$ to be continuous functions of x, in particular at $x = 0$. Thus,

$$\lim_{x\uparrow 0}\phi_\omega(x) = 1 + r = v = \lim_{x\downarrow 0}\phi_\omega(x) \tag{6.8}$$

$$\lim_{x\uparrow 0}\partial_x\phi_\omega(x) = k_<(1-r) = k_> v = \lim_{x\downarrow 0}\partial_x\phi_\omega(x). \tag{6.9}$$

We conclude that

$$v = \frac{2}{1+\rho(\omega)}, \quad r = \frac{1-\rho(\omega)}{1+\rho(\omega)}, \tag{6.10}$$

where

$$\rho(\omega) = \frac{k_>}{k_<} = \begin{cases} \sqrt[+]{1+\widehat{\chi}(\omega)} & \omega > 0, \\ -\sqrt[+]{1+\widehat{\chi}(\omega)} & \omega < 0. \end{cases} \tag{6.11}$$

It is useful to consider some general properties of $\rho(\omega)$ and the reflection and transmission coefficients. To begin note that since $\omega \operatorname{Im}\widehat{\chi}(\omega) \geq 0$ we have

$$\omega \operatorname{Re} \sqrt[+]{1+\widehat{\chi}(\omega)} \geq 0, \tag{6.12}$$

and thus

$$\operatorname{Re}\rho(\omega) \geq 0. \tag{6.13}$$

This implies that

$$1 - |r|^2 = \frac{4}{1+|\rho(\omega)|^2}\operatorname{Re}\rho(\omega) = |v|^2 \operatorname{Re}\rho(\omega) = |v|^2\frac{\operatorname{Re} k_>}{k_<} \geq 0, \tag{6.14}$$

so, in particular,

$$|r| \leq 1. \tag{6.15}$$

Equation (6.14) expresses the continuity of the energy flux $J(x)$ at $x = 0$, since by (4.1)

$$\begin{aligned} J(x) &= \gamma\omega \operatorname{Im}\overline{\phi_\omega(x)}\partial_x\phi_\omega(x) \\ &= \gamma\omega\begin{cases} k_<(1-|r|^2) & x<0 \\ \operatorname{Re} k_>|v|^2\mathrm{e}^{-2\operatorname{Im} k_> x} & x>0 \end{cases} \\ &= \sqrt{\gamma}\omega^2(1-|r|^2)\begin{cases} 1 & x<0, \\ \mathrm{e}^{-2\operatorname{Im} k_> x} & x>0. \end{cases} \end{aligned} \tag{6.16}$$

[3] If $\widehat{\chi}(\omega)$ is real and $(1+\widehat{\chi}(\omega)) > 0$ then the medium transmits at frequency ω and $k_>$ should be determined as the limit

$$k_> = \lim_{\delta\downarrow 0}\sqrt[+]{\frac{(\omega+\mathrm{i}\delta)^2}{\gamma}(1+\widehat{\chi}(\omega+\mathrm{i}\delta))}.$$

Furthermore, we see that the energy flux is nonnegative, representing a flow of energy from the source at $x = -\infty$, and the rate of dissipation, $-\partial_x J(x) \neq 0$, is nonzero for every $x > 0$,

$$-\partial_x J(x) = 2\sqrt{\gamma}\omega^2(1-|r|^2)\operatorname{Im} k_> \mathrm{e}^{-2\operatorname{Im} k_> x}, \quad (x>0). \tag{6.17}$$

Finally, we note that the energy density of the physical string is

$$\begin{aligned} E_0(x) &= \frac{1}{2}\{\omega^2|\phi_\omega(x)|^2 + \gamma|\partial_x\phi_\omega(x)|^2\} \\ &= \begin{cases} \omega^2(1+|r|^2) & x<0, \\ \omega^2\frac{1}{2}(1+|1+\widehat{\chi}(\omega)|)|v|^2\mathrm{e}^{-2\operatorname{Im} k_> x} & x>0, \end{cases} \\ &= \omega^2(1+|r|^2)\begin{cases} 1 & x<0, \\ \mathrm{e}^{-2\operatorname{Im} k_> x} & x>0, \end{cases} \end{aligned} \tag{6.18}$$

where we have noted that by (6.10)

$$1+|r|^2 = \frac{1+|\rho(\omega)|^2}{2}|v|^2 = \frac{1+|1+\widehat{\chi}(\omega)|}{2}|v|^2. \tag{6.19}$$

In summary, the scattering eigenmode

$$\phi(x,t) = \begin{cases} \mathrm{e}^{\mathrm{i}\omega(x/\sqrt{\gamma}-t)} + r\mathrm{e}^{-\mathrm{i}\omega(x/\sqrt{\gamma}+t)} & x<0 \\ v\mathrm{e}^{\mathrm{i}\omega(\rho(\omega)x/\sqrt{\gamma}-t)} & x>0, \end{cases} \tag{6.20}$$

$$\psi(x,s,t) = \frac{v}{2}\mathrm{e}^{\mathrm{i}\omega(\rho(\omega)x/\sqrt{\gamma}-t)}\int_{\mathbb{R}} \mathrm{e}^{\mathrm{i}\omega|s-s'|}\varsigma(s')\,\mathrm{d}s', \quad (x>0), \tag{6.21}$$

describes a steady state in which the incoming wave is partially reflected, with the remainder an evanescent transmitted wave that penetrates the dissipative part of the string with an exponential profile resulting in an excitation of the hidden strings accounting for dispersion and dissipation. The total rate of dissipation in the TDD portion of the string is

$$-\int_0^\infty \partial_x J(x)\,\mathrm{d}x = J(0) = \sqrt{\gamma}\omega^2(1-|r|^2). \tag{6.22}$$

References

1. A. Figotin and J. H. Schenker, *Spectral theory of time dispersive and dissipative systems*, J. Stat. Phys. **118** (2005), no. 1-2, 199–263.
2. ———, *Hamiltonian treatment of time dispersive and dissipative media within the linear response theory*, J. Comput. Appl. Math., to appear, available at `arXiv:physics/0410127`.
3. ———, *Hamiltonian structure for dispersive and dissipative dynamical system*, submitted, available at `arXiv:math-ph/0608003`.
4. V. Jakšić and C.-A. Pillet, *Ergodic properties of classical dissipative systems*. I, Acta Math. **181** (1998), no. 2, 245 – 282.
5. P. M. Morse and H. Feshbach, *Methods of theoretical physics*, McGraw-Hill, New York, 1953.
6. L. Rey-Bellet, *Open classical systems*, (Grenoble, 2003), Open Quantum Systems. II: The Markovian Approach (S. Attal, A. Joyce, and C.-A. Pillet, eds.), Lecture Notes in Math., vol. 181, Springer, Berlin, 2006, pp. 41 – 78.
7. M. Reed and B. Simon, *Methods of modern mathematical physics*. I: *Functional analysis*, 2nd ed., Academic Press, New York, 1980.

Department of Mathematics, University of California — Irvine, Irvine, CA 92697-3875, USA
E-mail address: afigotin@uci.edu

Department of Mathematics, Michigan State University, East Lansing, MI 48824, USA
Current address: School of Mathematics, Institute for Advanced Study, Princeton, NJ 08540, USA
E-mail address: jeffrey@math.msu.edu

Centre de Recherches Mathématiques
CRM Proceedings and Lecture Notes
Volume **42**, 2007

Localization at Low Energies for Attractive Poisson Random Schrödinger Operators

François Germinet, Peter D. Hislop, and Abel Klein

ABSTRACT. We prove exponential and dynamical localization at low energies for the Schrödinger operator with an attractive Poisson random potential in any dimension. We also conclude that the eigenvalues in that spectral region of localization have finite multiplicity.

1. Introduction and main results

The motion of an electron moving in an amorphous medium where identical impurities have been randomly scattered, each impurity creating a local attractive potential, is described by a Schrödinger equation with Hamiltonian

$$H_X := -\Delta + V_X \quad \text{on } \mathrm{L}^2(\mathbb{R}^d), \tag{1.1}$$

where the potential is given by

$$V_X(x) := -\sum_{\zeta \in X} u(x - \zeta), \tag{1.2}$$

with X being the location of the impurities and $-u(x - \zeta) \leq 0$ the attractive potential created by the impurity placed at ζ. Since the impurities are randomly distributed, it is natural to model the configurations of the impurities by a Poisson process on $\mathbb{R}^d$ [35, 37].

The *attractive Poisson Hamiltonian* is the random Schrödinger operator $H_{\mathbf{X}}$ in (1.1) with $\mathbf{X}$ a Poisson process on $\mathbb{R}^d$ with density $\varrho > 0$; $V_{\mathbf{X}}$ being then an *attractive Poisson random potential.* The attractive Poisson Hamiltonian $H_{\mathbf{X}}$ is an $\mathbb{R}^d$-ergodic family of random self-adjoint operators; it follows from standard results (cf. [27, 37]) that there exists fixed subsets of $\mathbb{R}$ so that the spectrum of $H_{\mathbf{X}}$, as well as the pure point, absolutely continuous, and singular continuous components, are equal to these fixed sets with probability one.

Poisson Hamiltonians have been known to have Lifshitz tails, a strong indication of localization, for quite a long time [6, 12, 31, 33, 37, 41, 43]. In particular, the

2000 *Mathematics Subject Classification.* Primary 35P05, 81Q10; Secondary 35J10, 47A10.
The second author was supported in part by NSF Grant DMS-0503784.
The third author was supported in part by NSF Grant DMS-0457474.
This is the final form of the paper.

existence of Lifshitz tails for attractive Poisson Hamiltonians is proved in [33]. But up to recently localization was known only in one dimension [42]; the multidimensional case remaining an open question (cf. [34]).

We have recently proved localization at the bottom of the spectrum for Schrödinger operators with positive Poisson random potentials in arbitrary dimension [16, 17]. We obtained both exponential (or Anderson) localization and dynamical localization, as well as finite multiplicity of eigenvalues. In this article we extend these results to attractive Poisson Hamiltonians, proving localization at low energies.

Localization has been known for Anderson-type Hamiltonians [1, 7, 20, 28, 30, 32, 36]. In random amorphous media, localization was known for some Gaussian random potentials [13, 34, 44]. In all these case there is an a priori Wegner estimate in all scales (e.g., [7–11, 13, 26, 30, 36]).

Bourgain and Kenig's proved localization for the Bernoulli – Anderson Hamiltonian, an Anderson-type Hamiltonian where the coefficients of the single-site potentials are Bernoulli random variables [3]. They established a Wegner estimate by a multiscale analysis using "free sites" and a new quantitative version of unique continuation which gives a lower bound on eigenfunctions. Since they obtained weak probability estimates and had discrete random variables, they also introduced a new method to prove Anderson localization from estimates on the finite-volume resolvents given by a single-energy multiscale analysis. The new method does not use the perturbation of singular spectra method nor Kotani's trick as in [7, 40], which requires random variables with bounded densities. It is also not an energy-interval multiscale analysis as in [14, 29, 45], which requires better probability estimates.

To prove localization for Poisson Hamiltonians [17], we exploited the probabilistic properties of Poisson point processes to use the new ideas introduced by Bourgain and Kenig [2, 3].

Here we study attractive single-site potentials, which we write as $-u$, where u is a nonnegative, nonzero L^∞-function on $\mathbb{R}^d$ with compact support, with

$$(1.3) \qquad u_- \chi_{\Lambda_{\delta_-}(0)} \le u \le u_+ \chi_{\Lambda_{\delta_+}(0)} \quad \text{for some constants } u_\pm, \delta_\pm \in \left]0, \infty\right[.$$

($\Lambda_L(x)$ denotes the box of side L centered at $x \in \mathbb{R}^d$.) It follows that H_X is essentially self-adjoint on $C_c^\infty(\mathbb{R}^d)$ and $\sigma(H_{\mathbf{X}}) = \mathbb{R}$ with probability one [6, 37]. We show that the conclusions of [17] hold in an interval of negative energies of the form $]-\infty, E_0(\varrho)]$ for some $E_0(\varrho) < 0$. We obtain both exponential (or Anderson) localization and dynamical localization, as well as finite multiplicity of eigenvalues.

For a given set B, we let χ_B be its characteristic function, $\mathcal{P}_0(B)$ the collection of its countable subsets, and $\#B$ its cardinality. Given $X \in \mathcal{P}_0(A)$ and $A \subset B$, we set $X_A := X \cap A$ and $N_X(A) := \#X_A$. We write $|A|$ for the Lebesgue measure of a Borel set $A \subset \mathbb{R}^d$. We let $\Lambda_L(x) := x + (-L/2, L/2)^d$ be the box of side L centered at $x \in \mathbb{R}^d$. By Λ we will always denote some box $\Lambda_L(x)$, with Λ_L denoting a box of side L. We set $\chi_x := \chi_{\Lambda_1}(x)$, the characteristic function of the box of side 1 centered at $x \in \mathbb{R}^d$. We write $\langle x \rangle := \sqrt{1 + |x|^2}$, $T(x) := \langle x \rangle^\nu$ for some fixed $\nu > d/2$. By $C_{a,b,\ldots}$, $c_{a,b,\ldots}$, $K_{a,b,\ldots}$, etc., will always denote some finite constant depending only on $a, b, \ldots$.

A Poisson process on a Borel set $B \subset \mathbb{R}^d$ with density $\varrho > 0$ is a map $\mathbf{X}$ from a probability space $(\Omega, \mathbb{P})$ to $\mathcal{P}_0(B)$, such that for each Borel set $A \subset B$ with $|A| < \infty$ the random variable $N_{\mathbf{X}}(A)$ has Poisson distribution with mean $\varrho|A|$, i.e.,

$$(1.4)\qquad \mathbb{P}\{N_{\mathbf{X}}(A)=k\}=\frac{(\varrho|A|)^k}{k!}\mathrm{e}^{-\varrho|A|}\quad \text{for } k=0,1,2,\ldots,$$

and the random variables $\{N_{\mathbf{X}}(A_j)\}_{j=1}^n$ are independent for disjoint Borel subsets $\{A_j\}_{j=1}^n$ (e.g., [25, 39]).

Theorem 1.1. *Let $H_{\mathbf{X}}$ be an attractive Poisson Hamiltonian on $\mathrm{L}^2(\mathbb{R}^d)$ with density $\varrho>0$. Then there exists an energy $E_0=E_0(\varrho)<0$ for which the following holds $\mathbb{P}$-a.e.: The operator $H_{\mathbf{X}}$ has pure point spectrum in $]-\infty,E_0]$ with exponentially localized eigenfunctions, and, if ϕ is an eigenfunction of $H_{\mathbf{X}}$ with eigenvalue $E\in\,]-\infty,E_0]$ we have, with $m_E:=\frac{1}{8}\sqrt{\frac{1}{2}E_0-E}\le m_{E_0}:=\frac{1}{8}\sqrt{-\frac{1}{2}E_0}$, that*

$$(1.5)\qquad \|\chi_x\phi\|\le C_{\mathbf{X},\phi}\mathrm{e}^{-m_E|x|}\quad \textit{for all } x\in\mathbb{R}^d.$$

Moreover, there exist $\tau>1$ and $s\in\,]0,1[$ such that for all eigenfunctions ψ, ϕ (possibly equal) with the same eigenvalue $E\in\,]-\infty,E_0]$ we have

$$(1.6)\qquad \|\chi_x\psi\|\,\|\chi_y\phi\|\le C_{\mathbf{X}}\|T^{-1}\psi\|\|T^{-1}\phi\|\mathrm{e}^{\langle y\rangle^\tau}\mathrm{e}^{-|x-y|^s}\quad \textit{for all } x,y\in\mathbb{Z}^d.$$

In particular, the eigenvalues of $H_{\mathbf{X}}$ in $]-\infty,E_0]$ have finite multiplicity, and $H_{\mathbf{X}}$ exhibits dynamical localization in $]-\infty,E_0]$, that is, for any $p>0$ we have

$$(1.7)\qquad \sup_t \|\langle x\rangle^p e^{-itH_{\mathbf{X}}}\chi_{]-\infty,E_0]}(H_{\mathbf{X}})\chi_0\|_2^2<\infty.$$

The proof of Theorem 1.1 relies on the construction introduced in [17] for Poisson Hamiltonians, based on the new multiscale analysis of Bourgain [2] and Bourgain – Kenig [3] for Bernoulli – Anderson Hamiltonians. Exponential localization follows as in [3]. The decay of eigenfunction correlations given in (1.6) then follows from [23] as in [17]. Dynamical localization and finite multiplicity of eigenvalues are consequences of (1.6).

The Bourgain – Kenig multiscale analysis requires some detailed knowledge about the location of the impurities, as well as information on "free sites," and relies on conditional probabilities. To deal with these issues and also handle the measurability questions that appear for the Poisson process, in [17] we performed a finite volume reduction in each scale as part of the multiscale analysis.

In this note we review the basic construction of [17], and apply it to attractive Poisson Hamiltonians. But since these are unbounded from below, we need to modify the finite volume reduction and the a priori finite volume estimates.

2. Attractive Poisson Hamiltonians

The Poisson process $\mathbf{X}$ on $\mathbb{R}^d$ with density ϱ is constructed from a marked Poisson process as follows: Let $\mathbf{Y}$ be a Poisson process on $\mathbb{R}^d$ with density 2ϱ, and to each $\zeta\in Y$ associate a Bernoulli random variable ε_ζ, either 0 or 1 with equal probability, with $\boldsymbol{\varepsilon}_{\mathbf{Y}}=\{\varepsilon_\zeta\}_{\zeta\in\mathbf{Y}}$ independent random variables. Then $(\mathbf{Y},\boldsymbol{\varepsilon}_{\mathbf{Y}})$ is a Poisson process with density 2ρ on the product space $\mathbb{R}^d\times\{0,1\}$, the *marked Poisson process*; its underlying probability space will still be denoted by $(\Omega,\mathbb{P})$. (We use the notation $(Y,\varepsilon_Y):=\{(\zeta,\varepsilon_\zeta);\zeta\in Y\}\in\mathcal{P}_0(\mathbb{R}^d\times\{0,1\})$.) Define maps $\mathcal{X},\mathcal{X}'\colon\mathcal{P}_0(\mathbb{R}^d\times\{0,1\})\to\mathcal{P}_0(\mathbb{R}^d)$ by

$$(2.1)\qquad \mathcal{X}(\widetilde{Z}):=\{\zeta\in\mathbb{R}^d;(\zeta,1)\in\widetilde{Z}\},\quad \mathcal{X}'(\widetilde{Z}):=\{\zeta\in\mathbb{R}^d;(\zeta,0)\in\widetilde{Z}\},$$

for all $\widetilde{Z}\in\mathcal{P}_0(\mathbb{R}^d\times\{0,1\})$. Then the maps $\mathbf{X},\mathbf{X}'\colon\Omega\to\mathcal{P}_0(\mathbb{R}^d)$, given by

$$(2.2)\qquad \mathbf{X}:=\mathcal{X}(\mathbf{Y},\boldsymbol{\varepsilon}_{\mathbf{Y}}),\quad \mathbf{X}':=\mathcal{X}'(\mathbf{Y},\boldsymbol{\varepsilon}_{\mathbf{Y}}),$$

i.e., $\mathbf{X}(\omega) = \mathcal{X}(\mathbf{Y}(\omega), \varepsilon_{\mathbf{Y}(\omega)}(\omega))$, $\mathbf{X}'(\omega) = \mathcal{X}'(\mathbf{Y}(\omega), \varepsilon_{\mathbf{Y}(\omega)}(\omega))$, are Poisson processes on $\mathbb{R}^d$ with density ϱ (cf. [25, Section 5.2; 39, Example 2.4.2]), and we have

$$N_{\mathbf{X}}(A) + N_{\mathbf{X}'}(A) = N_{\mathbf{Y}}(A) \quad \text{for all Borel sets } A \subset \mathbb{R}^d. \tag{2.3}$$

If $\mathbf{X}$ is a Poisson process on $\mathbb{R}^d$ with density ϱ, then $\mathbf{X}_A$ is a Poisson process on A with density ϱ for each Borel set $A \subset \mathbb{R}^d$, with $\{\mathbf{X}_{A_j}\}_{j=1}^n$ being independent Poisson processes for disjoint Borel subsets $\{A_j\}_{j=1}^n$. Similar considerations apply to $\mathbf{X}'$ and to the marked Poisson process $(\mathbf{Y}, \varepsilon_{\mathbf{Y}})$, with $\mathbf{X}_A, \mathbf{X}'_A, \mathbf{Y}_A, \varepsilon_{\mathbf{Y}_A}$ satisfying (2.2).

From now on we fix a probability space $(\Omega, \mathbb{P})$ on which the Poisson processes $\mathbf{X}$ and $\mathbf{X}'$, with density ϱ, and $\mathbf{Y}$, with density 2ϱ, are defined, as well as the Bernoulli random variables $\varepsilon_{\mathbf{Y}}$, and we have (2.2). All events will be defined with respect to this probability space. $H_{\mathbf{X}}$ (and $H_{\mathbf{Y}}$) will always denote an attractive Poisson Hamiltonian on $\mathrm{L}^2(\mathbb{R}^d)$ with density $\varrho > 0$ (2ϱ), as in (1.1)–(1.3).

We start by showing that the attractive Poisson Hamiltonian is self-adjoint and we have trace estimates needed in the multiscale analysis.

Proposition 2.1. *The attractive Poisson Hamiltonians $H_{\mathbf{X}}$ and $H_{\mathbf{Y}}$ are essentially self-adjoint on $C_c^\infty(\mathbb{R}^d)$ with probability one. In addition, we have*

$$\operatorname{tr}\{T^{-1}\mathrm{e}^{-tH_{\mathbf{X}}}T^{-1}\} < \infty \quad \textit{for all } t > 0 \ \mathbb{P}\textit{-a.e.}, \tag{2.4}$$

and

$$\operatorname{tr}(T^{-1}\chi_{]-\infty,E]}(H_{\mathbf{X}})T^{-1}) < \infty \quad \textit{for all } E \in \mathbb{R} \ \mathbb{P}\textit{-a.e.} \tag{2.5}$$

Since the potential is attractive, it may create infinitely deep wells. This is controlled by the following estimate.

Lemma 2.2. *Given a box Λ we set*

$$\mathbb{J}_{\delta_+,\Lambda} := (\tfrac{1}{2}\delta_+\mathbb{Z}^d) \cap \Lambda. \tag{2.6}$$

There exists $L^* = L^*(d, \varrho, \delta_+)$, such that for any $L \geq L^*$ we have

$$\mathbb{P}\{N_{\mathbf{Y}}(\Lambda_{2\delta_+}(j)) \leq \varrho \log L \ \forall j \in \mathbb{J}_{\delta_+,\Lambda_L}\} \geq 1 - L^{-2(\log\log L)/3} \tag{2.7}$$

and

$$\mathbb{P}\{\|\chi_{\Lambda_L} V_{\mathbf{Y}}\|_\infty \leq u_+ \varrho \log L\} \geq 1 - L^{-2(\log\log L)/3}. \tag{2.8}$$

It follows that for $\mathbb{P}$-a.e. ω we have

$$V_{\mathbf{X}(\omega)}(x) \geq V_{\mathbf{Y}(\omega)}(x) \geq -c_\omega \log(1 + |x|) \quad \textit{for all } x \in \mathbb{R}^d, \tag{2.9}$$

where $c_\omega > 0$ (depending also on $p, \delta_+, \varrho, u_+$).

Proof. We may assume $\varrho \log L \geq 1$. Standard bounds on Poisson random variables (cf. [17, Eq. (2.7)]) give

$$\mathbb{P}\{N_{\mathbf{Y}}(\Lambda_{2\delta_+}(x)) > \varrho \log L\} \leq \left(\frac{2\varrho(2\delta_+)^d}{\varrho \log L}\right)^{\varrho \log L} \leq L^{-3(\log\log L)/4} \tag{2.10}$$

for any $x \in \mathbb{R}^d$, if $L \geq L_1^*(d, \varrho, \delta_+)$. It follows that for $L \geq L^*(d, \varrho, \delta_+)$ we have

$$\begin{aligned} \mathbb{P}\{N_{\mathbf{Y}}(\Lambda_{2\delta_+}(j)) \leq \varrho \log L \ \forall j \in \mathbb{J}_{\delta_+,\Lambda_L}\} &\geq 1 - \left(\frac{2L}{\delta_+}\right)^d L^{-3(\log\log L)/4} \\ &\geq 1 - L^{-2(\log\log L)/3}. \end{aligned} \tag{2.11}$$

Now, for any $x \in \Lambda_L$ there exists $j \in \mathbb{J}_{\delta_+,\Lambda_L}$ s.t. $\Lambda_{\delta+}(x) \subset \Lambda_{2\delta+}(j)$. Hence, we also have

$$\mathbb{P}\{N_{\mathbf{Y}}(\Lambda_{\delta_+}(x) \cap \Lambda_L) \le \varrho \log L \; \forall x \in \Lambda_L\} \ge 1 - L^{-2(\log\log L)/3}. \tag{2.12}$$

But if the event in (2.12) occurs, it follows from (1.3) that

$$|V_{\mathbf{Y}}(x)| \le u_+ \varrho \log L \quad \forall x \in \Lambda_L, \tag{2.13}$$

and (2.8) follows, since

$$0 \ge V_{\mathbf{X}}(x) \ge V_{\mathbf{Y}}(x) \tag{2.14}$$

because of (2.2). The Borel – Cantelli Lemma now gives (2.9). □

In the one-dimensional case an estimate similar to (2.9) can be found in [24].

PROOF OF PROPOSITION 2.1. In view of (2.9), it follows from the Faris – Levine Theorem [38, Theorem X.38] that $H_{\mathbf{X}}$ and $H_{\mathbf{Y}}$ are essentially self-adjoint on $\mathcal{C}_c^\infty(\mathbb{R}^d)$ with probability one.

The trace estimate (2.4) follows from Gaussian bounds on heat kernels [4, Lemma 1.7], which hold $\mathbb{P}$-a.e. in view of (2.9). As a consequence, we have

$$\begin{aligned} \operatorname{tr}(T^{-1}\chi_{]-\infty,E]}(H_{\mathbf{X}})T^{-1}) &= \operatorname{tr}(T^{-1}\mathrm{e}^{-H_{\mathbf{X}}}\mathrm{e}^{2H_{\mathbf{X}}}\chi_{]-\infty,E]}(H_{\mathbf{X}})\mathrm{e}^{-H_{\mathbf{X}}}T^{-1}) \\ &\le \mathrm{e}^{2E}\operatorname{tr}(T^{-1}\mathrm{e}^{-2H_{\mathbf{X}}}T^{-1}) < \infty \quad \text{for all } E \in \mathbb{R} \ \mathbb{P}\text{-a.e.} \end{aligned} \tag{2.15}$$ □

Given two disjoint configurations $X, Y \in \mathcal{P}_0(\mathbb{R}^d)$ and $t_Y = \{t_\zeta\}_{\zeta\in Y} \in [0,1]^Y$, we set

$$H_{X,(Y,t_Y)} := -\Delta + V_{X,(Y,t_Y)}, \quad \text{where } V_{X,(Y,t_Y)}(x) := V_X(x) + \sum_{\zeta\in Y} t_\zeta u(x-\zeta). \tag{2.16}$$

In particular, given $\varepsilon_Y \in \{0,1\}^Y$ we have, recalling (2.1), that

$$H_{X,(Y,\varepsilon_Y)} = H_{X \sqcup \mathcal{X}(Y,\varepsilon_Y)}. \tag{2.17}$$

We also write $H_{(Y,t_Y)} := H_{\varnothing,(Y,t_Y)}$ and

$$H_\omega := H_{\mathbf{X}(\omega)} = H_{(\mathbf{Y}(\omega),\boldsymbol{\varepsilon}_{\mathbf{Y}(\omega)}(\omega))}. \tag{2.18}$$

3. Finite volume

The multiscale analysis requires finite volume operators,which are defined as follows. Given a box $\Lambda = \Lambda_L(x)$ in $\mathbb{R}^d$ and a configuration X, we set

$$H_{X,\Lambda} := -\Delta_\Lambda + V_{X,\Lambda} \quad \text{on } \mathrm{L}^2(\Lambda), \tag{3.1}$$

where Δ_Λ is the the Laplacian on Λ with Dirichlet boundary condition, and

$$V_{X,\Lambda} := \chi_\Lambda V_{X_\Lambda} \quad \text{with } V_{X_\Lambda} \text{ as in (1.2).} \tag{3.2}$$

The finite volume resolvent is $R_{X,\Lambda}(z) := (H_{X,\Lambda} - z)^{-1}$.

We have $\Delta_\Lambda = \nabla_\Lambda \cdot \nabla_\Lambda$, where ∇_Λ is the gradient with Dirichlet boundary condition. We sometimes identify $\mathrm{L}^2(\Lambda)$ with $\chi_\Lambda \mathrm{L}^2(\mathbb{R}^d)$ and, when necessary, will use subscripts Λ and $\mathbb{R}^d$ to distinguish between the norms and inner products of $\mathrm{L}^2(\Lambda)$ and $\mathrm{L}^2(\mathbb{R}^d)$. Note that we always have

$$\chi_{\hat\Lambda} V_{X,\Lambda} = \chi_{\hat\Lambda} V_{X,\Lambda'}, \tag{3.3}$$

where

$$\hat{\Lambda} = \hat{\Lambda}_L(x) := \Lambda_{L-\delta_+}(x) \quad \text{with } \delta_+ \text{ as in (1.3)}, \tag{3.4}$$

which suffices for the multiscale analysis.

The multiscale analysis estimates probabilities of desired properties of finite volume resolvents at energies $E \in \mathbb{R}$. ($L^{p\pm}$ means $L^{p\pm\delta}$ for some small $\delta > 0$. We will write $\sqcup$ for disjoint unions: $C = A \sqcup B$ means $C = A \cup B$ with $A \cap B = \varnothing$.)

Definition 3.1. Consider an energy $E \in \mathbb{R}$ and a rate of decay $m > 0$. A box Λ_L is said to be (X, E, m)-good if

$$\|R_{X,\Lambda_L}(E)\| \le \mathrm{e}^{L^{1-}} \tag{3.5}$$

and

$$\|\chi_x R_{X,\Lambda_L}(E)\chi_y\| \le \mathrm{e}^{-m|x-y|}, \quad \text{for } x, y \in \Lambda_L \text{ with } |x-y| \ge \frac{L}{10}. \tag{3.6}$$

We say that the box Λ_L is (ω, E, m)-good if it is $(\mathbf{X}(\omega), E, m)$-good.

Condition (3.6) is the standard notion of a regular box in the multiscale analysis [14,15,18,29,45]. Condition (3.5) plays the role of a Wegner estimate. This control on the resolvent is just good enough so that it does not destroy the exponential decay obtained with (3.6). Since Wegner is proved scale by scale (as in [2,3,5]), it is incorporated in the definition of the goodness of a given box.

But *goodness* of boxes does not suffice for the induction step in the multiscale analysis given in [2,3], which also needs an adequate supply of *free sites* to obtain a Wegner estimate at each scale. Given two disjoint configurations $X, Y \in \mathcal{P}_0(\mathbb{R}^d)$ and $t_Y = \{t_\zeta\}_{\zeta \in Y} \in [0,1]^Y$, we recall (2.16) and define the corresponding finite volume operators $H_{X,(Y,t_Y),\Lambda}$ as in (3.1) and (3.2) using X_Λ, Y_Λ and t_{Y_Λ}, i.e.,

$$H_{X,(Y,t_Y),\Lambda} := -\Delta_\Lambda + V_{X,(Y,t_Y),\Lambda} \quad \text{with} \quad V_{X,(Y,t_Y),\Lambda} := \chi_\Lambda V_{X_\Lambda,(Y_\Lambda,t_{Y_\Lambda})}, \tag{3.7}$$

with $R_{X,(Y,t_Y),\Lambda}(z)$ being the corresponding finite volume resolvent.

Definition 3.2. Consider two configurations $X, Y \in \mathcal{P}_0(\mathbb{R}^d)$ and an energy E. A box Λ_L is said to be (X, Y, E, m)-good if $X \cap Y = \varnothing$ and we have (3.5) and (3.6) with $R_{X,(Y,t_Y),\Lambda_L}(E)$ for all $t_Y \in [0,1]^Y$. In this case Y consists of (X, E)-free sites for the box Λ_L. (*In particular, the box* Λ_L *is* $(X \sqcup \mathcal{X}(Y, \varepsilon_Y), E, m)$*-good for all* $\varepsilon_Y \in \{0,1\}^Y$.)

The multiscale analysis requires some detailed knowledge about the location of the impurities, that is, about the Poisson process configuration, as well as information on "free sites." To deal with this and also handle the measurability questions that appear for the Poisson process, a finite volume reduction was performed in [17] as part of the multiscale analysis. The key is that a Poisson point can be moved a little bit without spoiling the goodness of boxes [17, Lemma 3.3]. We now recall the construction of [17], with some slight adaptations to our present setting.

Definition 3.3. Let $\eta_L := \mathrm{e}^{-L^{10^6 d}}$ for $L > 0$. Given a box $\Lambda = \Lambda_L(x)$, set

$$\mathbb{J}_\Lambda := \{j \in x + \eta_L \mathbb{Z}^d;\ \Lambda_{\eta_L}(j) \subset \Lambda\}. \tag{3.8}$$

A configuration $X \in \mathcal{P}_0(\mathbb{R}^d)$ is said to be Λ-acceptable if (recall (2.6))

$$N_X\big(\Lambda_{2\delta_+}(j)\big) < \varrho \log L \quad \forall j \in \mathbb{J}_{\delta_+,\Lambda}, \tag{3.9}$$

$$N_X\big(\Lambda_{\eta_L}(j)\big) \le 1 \qquad \text{for all } j \in \mathbb{J}_\Lambda, \tag{3.10}$$

and

$$N_X\Big(\Lambda \setminus \bigcup_{j\in\mathbb{J}_\Lambda} \Lambda_{\eta_L(1-\eta_L)}(j)\Big) = 0; \tag{3.11}$$

it is Λ-acceptable$'$ if it satisfies (3.9),(3.10), and the weaker

$$N_X\Big(\Lambda \setminus \bigcup_{j\in\mathbb{J}_\Lambda} \Lambda_{\eta_L}(j)\Big) = 0. \tag{3.12}$$

We set

$$\mathcal{Q}_\Lambda^{(0)} := \{X \in \mathcal{P}_0(\mathbb{R}^d); X \text{ is } \Lambda\text{-acceptable}\}, \tag{3.13}$$

$$\mathcal{Q}_\Lambda^{(0\prime)} := \{X \in \mathcal{P}_0(\mathbb{R}^d); X \text{ is } \Lambda\text{-acceptable}'\}, \tag{3.14}$$

and consider the event (recall that $\mathbf{Y}$ is the Poisson process with density 2ϱ)

$$\Omega_\Lambda^{(0)} := \{\mathbf{Y} \in \mathcal{Q}_\Lambda^{(0)}\}. \tag{3.15}$$

Note that $\Omega_\Lambda^{(0)} \subset \{\mathbf{X} \in \mathcal{Q}_\Lambda^{(0)}\}$ and $\mathcal{Q}_\Lambda^{(0)} \subset \mathcal{Q}_\Lambda^{(0\prime)}$. Condition (3.11) is put in to avoid ambiguities in the multiscale analysis.

Remark 3.4. Note that (3.9) is not the same as [17, Eq. (3.19)], and hence the above definitions of Λ-acceptable and Λ-acceptable$'$ configurations are slightly different from the ones given in [17, Definition 3.4]. The reason for (3.9) is to ensure the estimates (3.17) and (3.18) in the next lemma.

Lemma 3.5. *Let $\Lambda = \Lambda_L$. Then for all Λ-acceptable$'$ configurations Y we have*

$$N_Y(\Lambda) \le \varrho(2\delta_+)^{-d} L^d \log L, \tag{3.16}$$

$$V_Y(x) \ge -u_+\varrho \log L \quad \textit{for all } x \in \Lambda, \tag{3.17}$$

and

$$H_{(Y,t_Y),\Lambda} \ge -u_+\varrho \log L \quad \textit{for all } t_Y \in [0,1]^Y. \tag{3.18}$$

PROOF. The estimate (3.16) is an immediate consequence of (3.9). It also follows from (3.9), by the same argument used for (2.13), that for all Λ-acceptable$'$ configurations Y we have the lower bound (3.17), from which we get (3.18) since $H_{(Y,t_Y),\Lambda} \ge H_{Y,\Lambda} \ge V_{Y,\Lambda}$. ☐

Lemma 3.6. *There exists a scale $\bar{L} = \bar{L}(d,\varrho,\delta_+) < \infty$, such that if $L \ge \bar{L}$ we have*

$$\mathbb{P}\{\Omega_{\Lambda_L}^{(0)}\} \ge 1 - L^{-(\log\log L)/2}. \tag{3.19}$$

PROOF. Recalling (2.7), we have

$$\mathbb{P}\{\Omega_{\Lambda_L}^{(0)}\} \ge 1 - L^{-2(\log\log L)/3} - 4d\varrho(L^{d-1} + L^d)\eta_L - 2\varrho^2 L^d \eta_L^d, \tag{3.20}$$

and hence (3.19) follows for large L. ☐

Lemma 3.6 tells us that inside the box Λ, outside an event of negligible probability in the multiscale analysis, we only need to consider Λ-acceptable configurations of the Poisson process Y.

Fix a box $\Lambda = \Lambda_L(x)$, then

$$X \overset{\Lambda}{\sim} Y \iff N_X(\Lambda_{\eta_L}(j)) = N_Y(\Lambda_{\eta_L}(j)) \quad \text{for all } j \in \mathbb{J}_\Lambda \tag{3.21}$$

introduces an equivalence relation in both $\mathcal{Q}_\Lambda^{(0\prime)}$ and $\mathcal{Q}_\Lambda^{(0)}$; the equivalence class of X in $\mathcal{Q}_\Lambda^{(0\prime)}$ will be denoted by $[X]'_\Lambda$. If $X \in \mathcal{Q}_\Lambda^{(0)}$, then $[X]_\Lambda = [X]'_\Lambda \cap \mathcal{Q}_\Lambda^{(0)}$ is its equivalence class in $\mathcal{Q}_\Lambda^{(0)}$. Note that $[X]'_\Lambda = [X_\Lambda]'_\Lambda$. We also write

$$[A]_\Lambda := \bigcup_{X \in A} [X]_\Lambda \quad \text{for subsets } A \subset \mathcal{Q}_\Lambda^{(0)}. \tag{3.22}$$

The following lemma [17, Lemma 3.6] tells us that "goodness" of boxes is a property of equivalence classes of acceptable′ configurations: changing configurations inside an equivalence class takes good boxes into just-as-good (jgood) boxes. Proceeding as in the lemma, we find that changing configurations inside an equivalence class takes jgood boxes into what we may call just-as-just-as-good (jjgood) boxes, and so on. Since we will only carry this procedure a bounded number of times, the bound independent of the scale, we will simply call them all jgood boxes.

Lemma 3.7 ([17]). *Fix $E_0 > 0$ and consider an energy $E \in [0, E_0]$. Suppose the box $\Lambda = \Lambda_L$ (with L large) is (X, E, m)-good for some $X \in \mathcal{Q}_{\Lambda_L}^{(0\prime)}$. Then for all $Y \in [X]'_\Lambda$ the box Λ is (Y, E, m)-jgood, that is,*

$$\|R_{Y,\Lambda}(E)\| \le \mathrm{e}^{L^{1-}+\eta_L^{1/4}} \sim \mathrm{e}^{L^{1-}} \tag{3.23}$$

and

$$\|\chi_x R_{Y,\Lambda}(E)\chi_y\| \le \mathrm{e}^{-m|x-y|} + \eta_L^{1/4} \sim \mathrm{e}^{-m|x-y|}, \quad \text{for } x, y \in \Lambda \text{ with } |x-y| \ge \frac{L}{10}. \tag{3.24}$$

Moreover, if $X, Y, X \sqcup Y \in \mathcal{Q}_\Lambda^{(0\prime)}$ and the box Λ is (X, Y, E, m)-good, then for all $X_1 \in [X]'_\Lambda$ and $Y_1 \in [Y]'_\Lambda$ we have $X_1 \sqcup Y_1 \in [X \sqcup Y]'_\Lambda$, and the box Λ is (X_1, Y_1, E, m)-jgood as in (3.23) and (3.24).

We also have a lemma [17, Lemma 3.8] about the distance to the spectrum inside equivalence classes.

Lemma 3.8 ([17]). *Fix $E_0 > 0$ and consider an energy $E \in [0, E_0]$ and a box $\Lambda = \Lambda_L$ (with L large). Suppose $\operatorname{dist}(E, \sigma(H_{X,\Lambda})) \le \tau_L$ for some $X \in \mathcal{Q}_{\Lambda_L}^{(0\prime)}$, where $\sqrt{\eta_L} \ll \tau_L < \frac{1}{2}$. Then*

$$\operatorname{dist}(E, \sigma(H_{Y,\Lambda})) \le \mathrm{e}^{\eta_L^{1/4}} \tau_L \quad \textit{for all } Y \in [X]'_\Lambda. \tag{3.25}$$

In view of (3.16) – (3.10) we have

$$\mathcal{Q}_\Lambda^{(0)} / \overset{\Lambda}{\sim} = \{[J]_\Lambda; J \in \mathcal{J}_\Lambda\}, \quad \text{where } \mathcal{J}_\Lambda := \{J \subset \mathbb{J}_\Lambda; (3.9) \text{ holds}\}, \tag{3.26}$$

and we can write $\mathcal{Q}_\Lambda^{(0)}$ and $\Omega_\Lambda^{(0)}$ as

$$\mathcal{Q}_\Lambda^{(0)} = \bigsqcup_{J \in \mathcal{J}_\Lambda} [J]_\Lambda \quad \text{and} \quad \Omega_\Lambda^{(0)} = \bigsqcup_{J \in \mathcal{J}_\Lambda} \{\mathbf{Y} \in [J]_\Lambda\}. \tag{3.27}$$

We now introduce the basic Poisson configurations and basic events with which we will construct all the relevant probabilistic events. These basic events have to take into account in their very structure the presence of free sites and the finite volume reduction. The following definitions are borrowed from [17].

Definition 3.9. Given $\Lambda = \Lambda_L(x)$, a Λ-bconfset (basic configuration set) is a subset of $\mathcal{Q}_\Lambda^{(0)}$ of the form

$$C_{\Lambda,B,S} := \bigsqcup_{\varepsilon_S \in \{0,1\}^S} [B \cup \mathcal{X}(S,\varepsilon_S)]_\Lambda = \bigsqcup_{S' \subset S} [B \cup S']_\Lambda, \tag{3.28}$$

where we always implicitly assume $B \sqcup S \in \mathcal{J}_\Lambda$. $C_{\Lambda,B,S}$ is a Λ-dense bconfset if S satisfies the density condition (cf. (3.4))

$$\#(S \cap \hat{\Lambda}_{L^{1-}}) \geq L^{d-} \quad \text{for all boxes } \Lambda_{L^{1-}} \subset \Lambda_L. \tag{3.29}$$

We also set

$$C_{\Lambda,B} := C_{\Lambda,B,\varnothing} = [B]_\Lambda. \tag{3.30}$$

Definition 3.10. Given $\Lambda = \Lambda_L(x)$, a Λ-bevent (basic event) is a subset of $\Omega_\Lambda^{(0)}$ of the form

$$\mathcal{C}_{\Lambda,B,B',S} := \{\mathbf{Y} \in [B \sqcup B' \sqcup S]_\Lambda\} \cap \{\mathbf{X} \in C_{\Lambda,B,S}\} \cap \{\mathbf{X}' \in C_{\Lambda,B',S}\}, \tag{3.31}$$

where we always implicitly assume $B \sqcup B' \sqcup S \in \mathcal{J}_\Lambda$. In other words, the Λ-bevent $\mathcal{C}_{\Lambda,B,B',S}$ consists of all $\omega \in \Omega_\Lambda^{(0)}$ satisfying

$$\begin{aligned} N_{\mathbf{X}(\omega)}(\Lambda_{\eta_L}(j)) &= 1 && \text{if } j \in B, \\ N_{\mathbf{X}'(\omega)}(\Lambda_{\eta_L}(j)) &= 1 && \text{if } j \in B', \\ N_{\mathbf{Y}(\omega)}(\Lambda_{\eta_L}(j)) &= 1 && \text{if } j \in S, \\ N_{\mathbf{Y}(\omega)}(\Lambda_{\eta_L}(j)) &= 0 && \text{if } j \in \mathbb{J}_\Lambda \setminus (B \sqcup B' \sqcup S). \end{aligned} \tag{3.32}$$

$\mathcal{C}_{\Lambda,B,B',S}$ is a Λ-dense bevent if S satisfies the density condition (3.29). In addition, we set

$$\mathcal{C}_{\Lambda,B,B'} := \mathcal{C}_{\Lambda,B,B',\varnothing} = \{\mathbf{X} \in C_{\Lambda,B}\} \cap \{\mathbf{X}' \in C_{\Lambda,B'}\}. \tag{3.33}$$

The number of possible bconfsets and bevents in a given box is always finite. We always have

$$\mathcal{C}_{\Lambda,B,B',S} \subset \{\mathbf{X} \in C_{\Lambda,B,S}\} \cap \Omega_\Lambda^{(0)}, \tag{3.34}$$

$$\mathcal{C}_{\Lambda,B,B',S} \subset \mathcal{C}_{\Lambda,\varnothing,\varnothing,B \sqcup B' \sqcup S} = \{\mathbf{Y} \in [B \sqcup B' \sqcup S]_\Lambda\}. \tag{3.35}$$

Note also that it follows from (3.15), (3.26) and (3.33) that

$$\Omega_\Lambda^{(0)} = \bigsqcup_{\{(B,B');B \sqcup B' \in \mathcal{J}_\Lambda\}} \mathcal{C}_{\Lambda,B,B'} \tag{3.36}$$

Moreover, for each $S_1 \subset S$ we have

$$C_{\Lambda,B,S} = \bigsqcup_{S_2 \subset S_1} C_{\Lambda,B \sqcup S_2, S \setminus S_1}, \tag{3.37}$$

$$\mathcal{C}_{\Lambda,B,B',S} = \bigsqcup_{S_2 \subset S_1} \mathcal{C}_{\Lambda,B \sqcup S_2, B' \sqcup (S_1 \setminus S_2), S \setminus S_1}. \tag{3.38}$$

Lemma 3.7 leads to the following definition.

Definition 3.11. Consider an energy $E \in \mathbb{R}$, $m > 0$, and a box $\Lambda = \Lambda_L(x)$. The Λ-bevent $\mathcal{C}_{\Lambda,B,B',S}$ and the Λ-bconfset $C_{\Lambda,B,S}$ are (Λ, E, m)-good if the box Λ is (B, S, E, m)-good. (*Note that Λ is then (ω, E, m)-jgood for every $\omega \in \mathcal{C}_{\Lambda,B,B',S}$*). Those (Λ, E, m)-good bevents and bconfsets that are also Λ-dense will be called (Λ, E, m)-adapted.

Definition 3.12. Consider an energy $E \in \mathbb{R}$, a rate of decay $m > 0$, and a box Λ. We call Ω_Λ a (Λ, E, m)-localized event if there exist disjoint (Λ, E, m)-adapted bevents $\{\mathcal{C}_{\Lambda,B_i,B'_i,S_i}\}_{i=1,2,\ldots,I}$ such that

$$\Omega_\Lambda = \bigsqcup_{i=1}^{I} \mathcal{C}_{\Lambda,B_i,B'_i,S_i}. \tag{3.39}$$

If Ω_Λ is a (Λ, E, m)-localized event, note that $\Omega_\Lambda \subset \Omega_\Lambda^{(0)}$ by its definition, and hence, recalling (3.38) and (3.33) , we can rewrite Ω_Λ in the form

$$\Omega_\Lambda = \bigsqcup_{j=1}^{J} \mathcal{C}_{\Lambda,A_j,A'_j}, \tag{3.40}$$

where the $\{\mathcal{C}_{\Lambda,A_j,A'_j}\}_{j=1,2,\ldots,J}$ are disjoint (Λ, E, m)-good bevents.

We will need (Λ, E, m)-localized events of scale appropriate probability.

Definition 3.13. Fix $p > 0$. Given an energy $E \in \mathbb{R}$ and a rate of decay $m > 0$, a scale L is (E, m)-localizing if for some box $\Lambda = \Lambda_L$ (and hence for all) we have a (Λ, E, m)-localized event Ω_Λ such that

$$\mathbb{P}\{\Omega_\Lambda\} > 1 - L^{-p}. \tag{3.41}$$

4. A priori finite volume estimates

Given an energy E, to start the multiscale analysis we will need, as in [2, 3], an *a priori* estimate on the probability that a box Λ_L is good with an adequate supply of free sites, for some sufficiently large scale L. The multiscale analysis will then show that such a probabilistic estimate also holds at all large scales.

Proposition 4.1. *Let $H_{\mathbf{X}}$ be an attractive Poisson Hamiltonian on $\mathrm{L}^2(\mathbb{R}^d)$ with density $\varrho > 0$, and fix $p > 0$. Then there exists a scale $\bar{L}_0 = \bar{L}_0(d, u, \varrho, p) < \infty$, such that for all scales $L \geq \bar{L}_0$, setting*

$$\delta_L := (\varrho^{-1}(p + d + 1) \log L)^{1/d}, \quad E_L := -2u_+ \varrho \log L, \tag{4.1}$$

and

$$m_{L,E} := \tfrac{1}{2}\sqrt{\tfrac{1}{2}E_L - E} \leq m_L := \tfrac{1}{2}\sqrt{-\tfrac{1}{2}E_L} \quad \text{for all } E \in \,]-\infty, E_L], \tag{4.2}$$

the scale L is $(E, m_{L,E})$-localizing for all energies $E \in \,]-\infty, E_L]$.

Proof. Let $\Lambda = \Lambda_L(x)$, and let δ_L and E_L be as in (4.1). If $Y \in \mathcal{Q}_\Lambda^{(0)}$, it follows from (3.18) and the Combes–Thomas estimate (e.g., [19, Eq. (19)]) that for all $E \in \,]-\infty, E_L]$ and all $t_Y \in [0, 1]^Y$ we have, with $m_L(E)$ as in (4.2), that

$$\|R_{(Y,t_Y),\Lambda}(E)\| \leq (2m_{L,E})^{-2} \tag{4.3}$$

and

$$\|\chi_y R_{(Y,t_Y),\Lambda}(E)\chi_{y'}\| \leq \tfrac{1}{2} m_{L,E}^{-2}\, \mathrm{e}^{-2m_{L,E}|y-y'|} \quad \text{for } y, y' \in \Lambda \text{ with } |y - y'| \geq 4\sqrt{d}. \tag{4.4}$$

We now require $L > \delta_L + \delta_+$, and set

$$J := \{j \in x + \delta_L \mathbb{Z}^d \cap \Lambda;\ \Lambda_{\delta_L}(j) \subset \hat{\Lambda}\}, \tag{4.5}$$

$$\widehat{\mathcal{J}}_\Lambda := \{S \in \mathcal{J}_\Lambda;\ N_S(\Lambda_{\delta_L}(j)) \geq 1 \text{ for all } j \in J\}. \tag{4.6}$$

If $S \in \widehat{\mathcal{J}}_\Lambda$, the density condition (3.29) for S in Λ follows from (4.6), and it follows from (4.3) and (4.4) that $\mathcal{C}_{\Lambda,\emptyset,\emptyset,S}$ is a $(\Lambda, E, m_{L,E})$-adapted bevent for all $E \in]-\infty, E_L]$ if $L \geq \bar{L}_1(d,u,\varrho,p)$. We conclude that

$$\Omega_\Lambda = \bigsqcup_{S \in \widehat{\mathcal{J}}_\Lambda} \mathcal{C}_{\Lambda,\emptyset,\emptyset,S} = \bigsqcup_{S \in \widehat{\mathcal{J}}_\Lambda} \{\mathbf{Y} \in [S]_\Lambda\} \tag{4.7}$$

is a $(\Lambda, E, m_{L,E})$-localizing event for all $E \in]-\infty, E_L]$.

To establish (3.41), let $\delta'_L := \delta_L/2^d$ and consider the event

$$\Omega_\Lambda^{(\ddagger)} := \{N_{\mathbf{Y}}(\Lambda_{\delta'_L}(j)) \geq 1 \quad \text{for all} \quad j \in J\}. \tag{4.8}$$

We have

$$\mathbb{P}\{\Omega_\Lambda^{(\ddagger)}\} \geq 1 - \left(\frac{L}{\delta_L}\right)^d \mathrm{e}^{-2\varrho(\delta'_L)^d} = 1 - \frac{\varrho}{(p+d+1)L^{p+1}\log L} \geq 1 - \frac{1}{L^{p+1}}, \tag{4.9}$$

if $L \geq \mathrm{e}^{\varrho/(p+d+1)}$. Since $\Omega_\Lambda \supset \Omega_\Lambda^{(\ddagger)} \cap \Omega_\Lambda^{(0)}$, (3.41) follows from (4.9) and (3.19) for $L \geq \bar{L}_0(d,u,\varrho,p)$. □

5. The multiscale analysis and the proof of localization

The Bourgain–Kenig multiscale analysis, namely [3, Proposition A′], was adapted to Poisson Hamiltonians in [17, Proposition 5.1]. To apply the latter to attractive Poisson Hamiltonians we must show that the requirements of this multiscale analysis are satisfied. More precisely, we must show that attractive Poisson Hamiltonians satisfy appropriate versions of Properties SLI (Simon–Lieb inequality), EDI (eigenfunction decay inequality), IAD (independence at a distance), NE (average number of eigenvalues), and GEE (generalized eigenfunction expansion); see [18]. The Wegner estimate is proved by the multiscale analysis; it is not an "a priori requirement."

Since events based on disjoint boxes are independent, we have Property IAD. Property GEE is satisfied in view of (2.4), and we also have (2.5), which is needed in the multiscale analysis.

But Properties SLI, EDI and NE require some care and modification. In a box Λ_L we always work with Λ_L-acceptable configurations X, whence the potential V_{X,Λ_L} satisfies the lower bound (3.17). Inside the box Λ_L, Properties SLI and EDI (see [3, Section 2; 21, Theorem A.1]) are governed by the same constant $\gamma_{E,L}$ given in [21, Eq. (A.2)], and hence for Λ_L-acceptable configurations we have

$$\gamma_{]-\infty,0],L} := \sup_{E \in]-\infty,0],L} \gamma_{E,L} \leq C_d\sqrt{u_+ \varrho \log L}. \tag{5.1}$$

For Property NE, it follows by the argument in [21, Eqs (A.6)–(A.7)] that for all Λ_L-acceptable configurations X and energies $E \in]-\infty, 0]$ we have

(5.2)
$$\operatorname{tr}\{\chi_{(-\infty,E)}(H_{X,\Lambda_L})\} \leq \operatorname{tr}\{\chi_{(-\infty,E+u_+\varrho\log L)}(-\Delta_{\Lambda_L})\} \leq C_d(u_+\varrho\log L)^{d/2}L^d.$$

The extra factors of $\sqrt{\log L}$ in (5.1) and (5.2) are acceptable in the multiscale analysis.

The Wegner estimate is proved at each scale using [3, Lemma 5.1′]. The sign of the single-site potential does not matter in this argument, as long as the single-site potential has a definite sign, positive or negative, to ensure the monotonicity of the eigenvalues in the free sites couplings.

Thus the following proposition follows from Proposition 4.1 and [17, Proposition 5.1].

Proposition 5.1. *Let* $H_{\mathbf{X}}$ *be an attractive Poisson Hamiltonian on* $\mathrm{L}^2(\mathbb{R}^d)$ *with density* $\varrho > 0$ *and* $p = \frac{3}{8}d-$. *Then there exist an energy* $E_0 = E_0(\varrho) < 0$ *and a scale* $L_0 = L_0(\varrho)$, *such that setting* $m_E := \frac{1}{4}\sqrt{\frac{1}{2}E_0 - E} \leq m_{E_0} := \frac{1}{4}\sqrt{-\frac{1}{2}E_0}$, *the scale* L *is* (E, m_E)*-localizing for all* $L \geq L_0$ *and* $E \in]-\infty, E_0]$.

Theorem 1.1 now follows from Proposition 5.1 as in [17, Proposition 6.1].

References

1. M. Aizenman, A. Elgart, S. Naboko, J. Schenker, and G. Stolz, *Moment analysis for localization in random Schrödinger operators*, Invent. Math. **163** (2006), no. 2, 343–413.
2. J. Bourgain, *On localization for lattice Schrödinger operators involving Bernoulli variables*, Geometric aspects of functional analysis, Lecture Notes in Math., vol. 1850, Springer, Berlin, 2004, pp. 77–99.
3. J. Bourgain and C. E. Kenig, *On localization in the continuous Anderson–Bernoulli model in higher dimension*, Invent. Math. **161** (2005), no. 2, 389–426.
4. K. Broderix, H. Leschke, and P. Müller, *Continuous integral kernels for unbounded Schrödinger semigroups and their spectral projections*, J. Funct. Anal. **212** (2004), no. 2, 287–323.
5. R. Carmona, A. Klein, and F. Martinelli, *Anderson localization for Bernoulli and other singular potentials*, Comm. Math. Phys. **108** (1987), no. 1, 41–66.
6. R. Carmona and J. Lacroix, *Spectral theory of random Schrödinger operators*, Probab. Appl., Birkhäuser, Boston, MA, 1990.
7. J.-M. Combes and P. D. Hislop, *Localization for some continuous, random Hamiltonians in d-dimension*, J. Funct. Anal. **124** (1994), no. 1, 149–180.
8. J.-M. Combes, P. D. Hislop, and F. Klopp, *Hölder continuity of the integrated density of states for some random operators at all energies*, Int. Math. Res. Not. **2003**, no. 4, 179–209.
9. J.-M. Combes, P. D. Hislop, F. Klopp, and S. Nakamura, *The Wegner estimate and the integrated density of states for some random operators*, Proc. Indian Acad. Sci. Math. Sci. **112** (2002), no. 1, 31–53.
10. J.-M. Combes, P. D. Hislop, and E. Mourre, *Spectral averaging, perturbation of singular spectra, and localization*, Trans. Amer. Math. Soc. **348** (1996), no. 12, 4883–4894.
11. J.-M. Combes, P. D. Hislop, and S. Nakamura, *The* L^p*-theory of the spectral shift function, the Wegner estimate and the integrated density of states for some random operators*, Comm. Math. Phys. **218** (2001), no. 1, 113–130.
12. M. Donsker and S. R. S. Varadhan, *Asymptotics for the Wiener sausage*, Comm. Pure Appl. Math. **28** (1975), no. 4, 525–565.
13. W. Fischer, H. Leschke, and P. Müller, *Spectral localization by Gaussian random potentials in multi-dimensional continuous space*, J. Statist. Phys. **101** (2000), no. 5-6, 935–985.
14. J. Fröhlich, F. Martinelli, E. Scoppola, and T. Spencer, *Constructive proof of localization in the Anderson tight binding model*, Comm. Math. Phys. **101** (1985), no. 1, 21–46.
15. J. Fröhlich and T. Spencer, *Absence of diffusion with Anderson tight binding model for large disorder or low energy*, Comm. Math. Phys. **88** (1983), no. 2, 151–184.
16. F. Germinet, P. Hislop, and A. Klein, *On localization for the Schrödinger operator with a Poisson random potential*, C. R. Acad. Sci. Paris Sér. I Math. **341** (2005), no. 8, 525–528.
17. ———, *Localization for Schrödinger operators with Piosson random potential*, preprint.
18. F. Germinet and A. Klein, *Bootstrap multiscale analysis and localization in random media*, Comm. Math. Phys. **222** (2001), no. 2, 415–448.
19. ———, *Operator kernel estimates for functions of generalized Schrödinger operators*, Proc. Amer. Math. Soc. **131** (2003), no. 3, 911–920.
20. ———, *Explicit finite volume criteria for localization in random media and applications*, Geom. Funct. Anal. **13** (2003), no. 6, 1201–1238.
21. ———, *A characterization of the Anderson metal-insulator transport transition*, Duke Math. J. **124** (2004), no. 2, 309–351.
22. ———, *New characterizations of the region of complete localization for random Schrödinger operators*, J. Stat. Phys. **122** (2006), no. 1, 73–94.

23. ———, in preparation.
24. I. W. Herbst and J. S. Howland, *The Stark ladder and other one-dimensional external field problems*, Comm. Math. Phys. **80** (1981), no. 1, 23 – 42.
25. J. F. C. Kingman, *Poisson processes*, Oxford Stud. Probab., vol. 3, Oxford Univ. Press., New York, 1993.
26. W. Kirsch, *Wegner estimates and Anderson localization for alloy-type potentials*, Math. Z. **221** (1996), no. 3, 507 – 512.
27. W. Kirsch and F. Martinelli, *On the ergodic properties of the spectrum of general random operators*, J. Reine Angew. Math. **334** (1982), 141 – 156.
28. W. Kirsch, P. Stollmann, and G. Stolz, *Localization for random perturbations of periodic Schrödinger operators*, Random Oper. Stochastic Equations **6** (1998), no. 3, 241 – 268.
29. A. Klein, *Multiscale analysis and localization of random operators*, Random Schrödinger Operators: Methods, Results, and Perspectives, Panor. Synthèse, Soc. Math. France, Paris, to appear.
30. F. Klopp, *Localization for continuous random Schrödinger operators*, Comm. Math. Phys. **167** (1995), no. 3, 553 – 569.
31. ———, *A low concentration asymptotic expansion for the density of states of a random Schrödinger operator with Poisson disorder*, J. Funct. Anal. **145** (1997), no. 2, 267 – 295.
32. ———, *Weak disorder localization and Lifshitz tails: continuous Hamiltonians*, Ann. Henri Poincaré **3** (2002), no. 4, 711 – 737.
33. F. Klopp and L. Pastur, *Lifshitz tails for random Schrödinger operators with negative singular Poisson potential*, Comm. Math. Phys. **206** (1999), no. 1, 57 – 103.
34. H. Leschke, P. Müller, and S. Warzel, *A survey of rigorous results on random Schrödinger operators for amorphous solids*, Markov Process. Related Fields **9** (2003), no. 4, 729 – 760.
35. I. M. Lifshits, S. A. Gredeskul, and L. A. Pastur, *Introduction to the theory of disordered systems*, Wiley-Intersci. Publ., Wiley, New York, 1988.
36. F. Martinelli and H. Holden, *On absence of diffusion near the bottom of the spectrum for a random Schrödinger operator on* $L^2(\mathbb{R}^\nu)$, Comm. Math. Phys. **93** (1984), no. 2, 197 – 217.
37. L. Pastur and A. Figotin, *Spectra of random and almost-periodic operators*, Grundlehren Math. Wiss., vol. 297, Springer, New York, 1992.
38. M. Reed and B. Simon, *Methods of modern mathematical physics.* II: *Fourier analysis, self-adjointness*, Academic Press, Boston, MA, 1975.
39. R.-D. Reiss, *A course on point processes*, Springer Ser. Statist., Springer, New York, 1993.
40. B. Simon and T. Wolff, *Singular continuum spectrum under rank one perturbations and localization for random Hamiltonians*, Comm. Pure Appl. Math. **39** (1986), no. 1, 75 – 90.
41. P. Stollmann, *Lifshitz asymptotics via linear coupling of disorder*, Math. Phys. Anal. Geom. **2** (1999), no. 3, 279 – 289.
42. G. Stolz, *Localization for random Schrödinger operators with Poisson potential*, Ann. Inst. H. Poincaré Phys. Théor. **63** (1995), no. 3, 297 – 314.
43. A.-S. Sznitman, *Brownian motion, obstacles and random media*, Springer Monogr. Math., Springer, Berlin, 1998.
44. N. Ueki, *Wegner estimates and localization for Gaussian random potentials*, Publ. Res. Inst. Math. Sci. **40** (2004), no. 1, 29 – 90.
45. H. von Dreifus and A. Klein, *A new proof of localization in the Anderson tight binding model*, Comm. Math. Phys. **124** (1989), 285 – 299.

Département de Mathématiques, Université de Cergy-Pontoise, Site de Saint Martin, 2 avenue Adolphe Chauvin, 95302 Cergy-Pontoise Cedex, France
E-mail address: germinet@math.u-cergy.fr

Department of Mathematics, University of Kentucky, Lexington, KY 40506-0027, USA
E-mail address: hislop@ms.uky.edu

Department of Mathematics, University of California — Irvine, Irvine, CA 92697-3875, USA
E-mail address: aklein@uci.edu

Centre de Recherches Mathématiques
CRM Proceedings and Lecture Notes
Volume **42**, 2007

On the Influence of Random Perturbations on the Propagation of Waves Described by a Periodic Schrödinger Operator

Yuri A. Godin, Stanislav Molchanov, and Boris Vainberg

Abstract. We study the variance of the propagation matrix of a periodic Schrödinger operator assuming that the monodromy matrix undergoes small random perturbations over each period. It is shown that if the frequency of propagation lies inside the band, then the total variance is proportional to the sum of the variances of single perturbations, i.e., obeys the probability law. However, if the frequency of propagation is close to the band edge, the resulting variance is proportional to the sum of cubic roots of the variances that make propagation highly sensitive to small perturbations.

1. Introduction

The product $\mathbf{\Pi}_N$ of $N \gg 1$ random unimodular (or symplectic) matrices is a typical object in the theory of random media (localization, propagation of electromagnetic waves through periodically stratified media, evolution of the magnetic field in turbulence flows of a conducting fluid, and more). One can find some general facts on $\mathbf{\Pi}_N$ in [4, 7]. There are two approaches to the asymptotic analysis of $\mathbf{\Pi}_N$, $N \to \infty$. One can study the a.s. behavior of $\|\mathbf{\Pi}_N\|_2 = [\operatorname{tr}(\mathbf{\Pi}_N \mathbf{\Pi}_N^{\mathrm{T}})]^{1/2}$. It is known (the Furstenberg theorem) that with the minimum of assumptions there exists so called quenched (or a.s.) Lyapunov exponent which is the (positive) limit

$$\gamma_q = \lim_{N\to\infty} \frac{\ln \|\mathbf{\Pi}_N\|_2}{N} > 0.$$

Another approach requires the prior averaging over the ensemble of realizations and the study of the norm moments of $\mathbf{\Pi}_N$:

$$\gamma_a^{(p)} = \lim_{N\to\infty} \frac{\ln \mathrm{E}\, \|\mathbf{\Pi}_N\|_2^p}{N},$$

These limits are called the annealed Lyapunov exponents of order p, and due to Jensen's inequality

$$(p+1)^{-1}\gamma_a^{(p+1)} \geq p^{-1}\gamma_a^{(p)} \geq \gamma_q.$$

2000 *Mathematics Subject Classification.* Primary 35R60, 34F05; Secondary 35Q60, 35B27.
This is the final form of the paper.

The goal of this paper is to study how a small random perturbation may affect transmission through a long slab of a periodic media, in particular, near the band edge of the spectrum.

Let H be a Schrödinger operator on $L^2(\mathbb{R})$ with a potential $q = q(x)$ supported on an interval $(0, \ell)$:

$$H\psi = -\frac{\mathrm{d}^2\psi}{\mathrm{d}x^2} + q(x)\psi(x), \tag{1.1}$$

and let ψ be the scattering solution for H, i.e., (1.1) holds and

$$\psi(x) = \begin{cases} \mathrm{e}^{\mathrm{i}\omega x} + r(\omega)\mathrm{e}^{-\mathrm{i}\omega x}, & x < 0; \\ t(\omega)\mathrm{e}^{\mathrm{i}\omega x}, & x > \ell. \end{cases}$$

Then the transmission coefficient $t(\omega)$ satisfies the following relation (see, for example, [3]):

$$|t(\omega)|^2 = \frac{4}{\|\mathbf{T}_\omega(0,\ell)\|_2^2 + 2}, \tag{1.2}$$

where $\mathbf{T}_\omega(a, b)$ is the Prüffer transfer matrix for the interval (a, b)

$$\mathbf{T}_\omega(a,b)\begin{bmatrix} \psi_\omega(a) \\ \omega^{-1}\psi'_\omega(a) \end{bmatrix} = \begin{bmatrix} \psi_\omega(b) \\ \omega^{-1}\psi'_\omega(b) \end{bmatrix}$$

for any solution $\psi = \psi_\omega(x)$ of $H\psi_\omega = \omega^2\psi_\omega$, and $\|\mathbf{T}\|_2$ is the Hilbert–Schmidt norm of the matrix $\mathbf{T} = [t_{i,j}]$

$$\|\mathbf{T}\|_2^2 = \operatorname{tr} \mathbf{T}\mathbf{T}^{\mathrm{T}} = \sum t_{i,j}^2.$$

We will also consider 1D Schrödinger operator with a periodic potential $q(x)$:

$$H\psi = -\frac{\mathrm{d}^2\psi}{\mathrm{d}x^2} + q(x)\psi(x), \quad q(x+L) \equiv q(x), \tag{1.3}$$

on $L^2(\mathbb{R})$ and operator H_N with truncated potential q_N containing N periods:

$$H_N\psi = -\frac{\mathrm{d}^2}{\mathrm{d}x^2} + q_N(x)\psi, \quad q_N(x) = I_{[0,NL)}(x)q(x). \tag{1.4}$$

Spectral properties of operators (1.3) or (1.4) depend on the monodromy operator $\mathbf{M}_\omega$(transfer matrix over the period L)

$$\mathbf{M}_\omega \begin{bmatrix} \psi_\omega(0) \\ \omega^{-1}\psi'_\omega(0) \end{bmatrix} = \begin{bmatrix} \psi_\omega(L) \\ \omega^{-1}\psi'_\omega(L) \end{bmatrix} \tag{1.5}$$

(in the Prüffer form). Here ψ_ω is an arbitrary solution of $H\psi = \omega^2\psi$.

If $t_N(\omega)$ is the transmission coefficient for the operator H_N then (1.2) implies

$$|t_N(\omega)|^2 = \frac{4}{\|\mathbf{M}_\omega^N\|_2^2 + 2}. \tag{1.6}$$

Thus, transmission through the truncated periodic medium depends on the upper bound of $\|\mathbf{M}_\omega^N\|_2$.

Let us suppose now that the monodromy operators through each period $[(n-1)L, nL]$, $n = 1, 2, \ldots, N$, contains some noise (defects, absorption, etc.), i.e., the corresponding propagator through the period $\mathbf{M}_{\omega,n} = \mathbf{T}_\omega([(n-1)L, nL])$ is a small random perturbation of the monodromy operator $\mathbf{M}_\omega$. Formula (1.6) takes the following form in this case

$$|t_N(\omega)|^2 = \frac{4}{\|\prod_{n=1}^N \mathbf{M}_{\omega,n}\|_2^2 + 2} \tag{1.7}$$

We will consider two models where the perturbation is either additive or multiplicative. In the first case, the propagator $\mathbf{M}_{\omega,n}$ has the form

$$\mathbf{M}_{\omega,n} = \mathbf{M}_\omega + \sigma \mathbf{W}_n, \tag{1.8}$$

where σ is a small parameter, and $\mathbf{W}_n = \begin{bmatrix} \xi_n & \eta_n \\ \zeta_n & \mu_n \end{bmatrix}$. Here ξ_n, η_n, ζ_n, μ_n are independent (for all n) $\mathcal{N}(0,1)$ random variables depending only on ω. In the multiplicative case we assume that

$$\mathbf{M}_{\omega,n} = \mathbf{M}_\omega(\mathbf{I} + \sigma \mathbf{V}_n), \tag{1.9}$$

where σ is a small parameter, and $\mathbf{V}_n = \begin{bmatrix} \xi_n & \eta_n \\ \zeta_n & -\xi_n \end{bmatrix}$. Here ξ_n, η_n, ζ_n are random variables which are not necessary independent, but have a joint Gaussian distribution with the covariance

$$\mathbf{B} = [b_{i,j}] = \mathrm{E} \begin{bmatrix} \xi_n^2 & \xi_n\eta_n & \xi_n\zeta_n \\ \xi_n\eta_n & \eta_n^2 & \eta_n\zeta_n \\ \xi_n\zeta_n & \eta_n\zeta_n & \zeta_n^2 \end{bmatrix}. \tag{1.10}$$

Clearly matrix (1.8) can be written in the form (1.9) and vice versa. The difference between the additive and multiplicative perturbations is in the assumptions on the random variables. In the first case we assume that the entries of the perturbation matrix are independent identically distributed random variables (isotropic case). This assumption makes the problem simpler. However, the disadvantage of this model is that $\det \mathbf{M}_{\omega,n} - 1$ for small σ has the order $O(\sigma)$. Since the determinant of the transfer matrix for operator (1.1) is equal to one, the additive model is applicable only to the situation when $q(x)$ is complex valued (lossy medium) or the operator (1.3) contains a term with first derivative. In the multiplicative case, the entries of the matrix $\mathbf{V}_n$ are not independent and the trace of $\mathbf{V}_n$ is zero. This leads to the fact that

$$\det \mathbf{M}_{\omega,n} = 1 + O(\sigma^2), \quad \sigma \to 0,$$

which makes the multiplicative perturbations more realistic than the additive ones.

We are going to show now that the multiplicative model appears as a first order (in σ) approximation in the case of a small random perturbation of a deterministic potential.

Consider the perturbation of periodic Schrödinger equation by the white noise $\xi(x)$

$$-y'' + [V(x) + \sigma\xi(x)]y = \omega^2 y, \quad V(x+L) = V(x), \tag{1.11}$$

where σ is a small parameter. We represent two linear independent solutions $y_i(x)$, $i = 1, 2$, of (1.11) in the form

$$y_i(x) = \psi_i(x) + \sigma \int_0^x G(x,s)\xi(s)y_i(s)\,\mathrm{d}s, \tag{1.12}$$

where $G(x,s)$ is the Green's function

$$G(x,s) = \begin{cases} \psi_2(x)\psi_1(s) - \psi_1(x)\psi_2(s), & s < x, \\ 0, & s > x, \end{cases} \tag{1.13}$$

and $\psi_1(x)$, $\psi_2(x)$ is the fundamental set of solutions of the unperturbed equation (1.11). Within the linear approximation in σ, solution of (1.11) is given by

$$y_i(x) = \psi_i(x) + \sigma \int_0^x G(x,s)\xi(s)\psi_i(s)\,\mathrm{d}s + O(\sigma^2), \quad i = 1,2. \tag{1.14}$$

If we denote

$$\eta_{ij} = \int_0^L \psi_i(s)\psi_j(s)\xi(s)\,\mathrm{d}s, \quad i = 1,2, \tag{1.15}$$

then the values of $y(L)$ and $y'(L)$ in the linear approximation can be express from (1.14) as follows

$$\begin{aligned} y_i(L) &= \psi_i(L) + \sigma[\eta_{1i}\psi_2(L) - \eta_{i2}\psi_1(L)], \\ y_i'(L) &= \psi_i'(L) + \sigma[\eta_{1i}\psi_2'(L) - \eta_{i2}\psi_1'(L)], \quad i = 1,2. \end{aligned}$$

Therefore, the monodromy matrix $\mathbf{M}$ of the perturbed equation (1.11) has the form

$$\begin{aligned} \mathbf{M} &= \begin{bmatrix} y_1(L) & \omega y_2(L) \\ \omega^{-1} y_1'(L) & y_2'(L) \end{bmatrix} \\ &= \begin{bmatrix} \psi_1(L) & \omega\psi_2(L) \\ \omega^{-1}\psi_1'(L) & \psi_2'(L) \end{bmatrix} \\ &\quad + \sigma \begin{bmatrix} \eta_{11}\psi_2(L) - \eta_{12}\psi_1(L) & \omega\eta_{12}\psi_2(L) - \omega\eta_{22}\psi_1(L) \\ \omega^{-1}\eta_{11}\psi_2'(L) - \omega^{-1}\eta_{12}\psi_1'(L) & \eta_{12}\psi_2'(L) - \eta_{22}\psi_1'(L) \end{bmatrix} \\ &= \mathbf{M}_\omega \left(\mathbf{I} + \sigma \begin{bmatrix} -\eta_{12} & -\omega\eta_{22} \\ \omega^{-1}\eta_{11} & \eta_{12} \end{bmatrix} \right), \end{aligned} \tag{1.16}$$

where $\mathbf{M}_\omega$ is the unperturbed monodromy matrix

$$\mathbf{M}_\omega = \begin{bmatrix} \psi_1(L) & \omega\psi_2(L) \\ \omega^{-1}\psi_1'(L) & \psi_2'(L) \end{bmatrix}. \tag{1.17}$$

The problem under investigation depends on two parameters: $\sigma \to 0$ and $N \to \infty$. Our goal is to find the error in the transfer matrix over N periods,

$$\mathbf{\Pi}_N = \prod_{n=1}^N \mathbf{M}_{\omega,n}, \tag{1.18}$$

due to a random perturbation of the monodromy matrix on each period.

We deal with the second moment, or to be more precise, we estimate the scalar variance of $\mathbf{\Pi}_N$,

$$D_N = \mathrm{E}\,\mathrm{tr}(\mathbf{\Pi}_N\mathbf{\Pi}_N^{\mathrm{T}}) - \mathrm{tr}(\mathbf{\Pi}_N\mathbf{\Pi}_N^{\mathrm{T}})|_{\sigma=0}, \tag{1.19}$$

as $\sigma \to 0$, $N \to \infty$. If D_N is small, then the variance of any smooth function of $\mathbf{\Pi}_N$ (and the covariance of any smooth vector or matrix function of $\mathbf{\Pi}_N$) is small. In particular, the variance of the transmission coefficient has the same order as D_N.

In the linear theory, the error in the process of summation is increasing slowly (as the function of the sample volume). If $S_N = \sigma\zeta_1 + \ldots \sigma\zeta_N$ where ζ_j are independent $\mathcal{N}(0,1)$ random variables, then the variance of S_N is equal to $\sigma^2 N$, i.e., the error due to summation of N small random perturbations is negligible if

$\sigma^2 N \ll 1$. The main conclusion of the present paper is that in both additive and multiplicative cases $D_N = O(\sigma^2 N)$ when ω belongs to the band of the absolutely continuous spectrum and $D_N = O(\sigma^{2/3} N)$ when ω is close to the band edge. Thus, the transmission is much more sensitive to perturbations and admissible values of $\sigma = o(N^{-3/2})$ are much smaller in the case when the frequency is close enough to the spectral band edge. On the other hand, spectral band edges provide important features of the dispersion function. These points are surrounded by bands, and transmission is possible for all frequencies in their neighborhoods, while the dispersion relation is flat and the group velocity vanishes at those points. However, transmission in this case is very sensitive to random perturbations. As we will see, this difference in the order of D_N is related to the existence of the Jordan block of the monodromy matrix $\mathbf{M}_\omega$ when ω is at the band edge.

2. Some facts on periodic Schrödinger operators

The monodromy matrix $\mathbf{M}_\omega = \left[\begin{smallmatrix} \alpha & \beta \\ \gamma & \delta \end{smallmatrix}\right]$ of the periodic Schrödinger operator (1.3) is an entire function of ω, $\det \mathbf{M}_\omega \equiv 1$, and the band-gap spectral structure of H is determined by Hill's discriminant

$$\Delta(\omega) = \tfrac{1}{2} \operatorname{tr} \mathbf{M}_\omega = \tfrac{1}{2}(\alpha + \delta)(\omega, L).$$

Namely, $|\Delta(\omega)| < 1$ inside the bands (the intervals of the ω-axis where $\lambda = \omega^2$ belongs to the a.c. spectrum of H), $|\Delta(\omega)| > 1$ in the gaps (open intervals separating the bands), equations $\Delta(\omega) = \pm 1$ give the edges of the bands. The dispersion relation $k = k(\omega)$ (ω belongs to a band) is defined by the formula

$$\Delta(\omega) = \cos k(\omega). \tag{2.1}$$

The points where the edges of two neighboring bands coincide are called degenerate gaps. It is well-known (see [2] or [6]) that the monodromy operator has a Jordan block with a superdiagonal and a multiple eigenvalue ± 1 when the frequency ω coincides with a nondegenerate bandgap edge.

There is a simple expression for the Hilbert – Schmidt norm of powers of the monodromy operator through the dispersion (2.1) (see [5]):

$$\|\mathbf{M}_\omega^N\|_2^2 = (\|\mathbf{M}_\omega\|_2^2 - 2)\frac{\sin^2 Nk(\omega)}{\sin^2 k(\omega)} + 2. \tag{2.2}$$

Note that $\det \mathbf{M}_\omega \equiv 1$ implies that

$$\varkappa = \|\mathbf{M}_\omega\|_2^2 - 2 \geq 0, \tag{2.3}$$

and

$$\varkappa > 0 \quad \text{at the band edges.} \tag{2.4}$$

In fact,

$$\|\mathbf{M}_\omega\|_2^2 = \alpha^2 + \beta^2 + \gamma^2 + \delta^2 = (\alpha - \delta)^2 + (\beta + \gamma)^2 + 2 \det \mathbf{M}_\omega \geq 2,$$

and the equality is possible only if $\alpha = \delta$, $\beta = -\gamma$. One can easily check that the latter cannot occur at the band edges since $\mathbf{M}_\omega$ has a nontrivial Jordan cell structure there.

3. Error in the transmission due to additive perturbations

Theorem 3.1. *The variance D_N has the following asymptotic behavior when ω is fixed, $\sigma \to 0$, and N is bounded or growing not very fast.*

If ω is an interior point of a band and $\sigma^2 N < C < \infty$, then

$$D_N = c(\omega)\sigma^2 N\big(1+o(1)\big). \tag{3.1}$$

If ω is a spectral band edge and $\sigma^{2/3} N < C < \infty$, then

$$D_N = c(\omega)\sigma^{2/3} N\big(1+o(1)\big). \tag{3.2}$$

The proof of this statement is based on the following two simple propositions. Let

$$\mathbf{A}_N = \mathrm{E}(\mathbf{\Pi}_N \mathbf{\Pi}_N^{\mathrm{T}}), \quad x_N = x_N(\omega,\sigma) = \operatorname{tr}\mathbf{A}_N, \quad N \geq 1, \tag{3.3}$$

and $\mathbf{A}_0 = \mathbf{I}$ (where $\mathbf{I}$ is the identity matrix), $x_0 = \operatorname{tr}\mathbf{I} = 2$. Then

$$D_N = x_N(\omega,\sigma) - x_N(\omega,0), \quad N \geq 1. \tag{3.4}$$

Proposition 3.2. *Let $N = 1$ and therefore,*

$$\mathbf{A}_1 = \mathrm{E}(\mathbf{M}_{\omega,1}\mathbf{M}_{\omega,1}^{\mathrm{T}}).$$

Then $\mathbf{A}_1 = \mathbf{M}_\omega \mathbf{M}_\omega^{\mathrm{T}} + \sigma^2 x_0 \mathbf{I}$.

PROOF. In fact, for $\mathbf{W}_1 = \begin{bmatrix} \xi & \eta \\ \zeta & \mu \end{bmatrix}$ we have

$$\mathbf{A}_1 = \mathbf{M}_\omega \mathbf{M}_\omega^{\mathrm{T}} + \sigma^2\,\mathrm{E}(\mathbf{W}_1\mathbf{W}_1^{\mathrm{T}}) = \mathbf{M}_\omega \mathbf{M}_\omega^{\mathrm{T}} + \sigma^2\,\mathrm{E}\begin{bmatrix} cc\xi^2+\eta^2 & \xi\zeta+\eta\mu \\ \xi\zeta+\eta\mu & \zeta^2+\mu^2 \end{bmatrix}$$
$$= \mathbf{M}_\omega \mathbf{M}_\omega^{\mathrm{T}} + 2\sigma^2\mathbf{I}. \qquad \square$$

Proposition 3.3. *If $\mathbf{A} = \mathbf{A}^{\mathrm{T}}$ is a real nonrandom symmetric matrix then*

$$\mathrm{E}[\mathbf{M}_\omega + \sigma\mathbf{W}_n]\mathbf{A}[\mathbf{M}_\omega^{\mathrm{T}} + \sigma\mathbf{W}_n^{\mathrm{T}}] = \mathbf{M}_\omega\mathbf{A}\mathbf{M}_\omega^{\mathrm{T}} + \sigma^2(\operatorname{tr}\mathbf{A})\mathbf{I}.$$

PROOF. In fact,

$$\mathrm{E}[\mathbf{M}_\omega + \sigma\mathbf{W}_n]\mathbf{A}[\mathbf{M}_\omega^{\mathrm{T}} + \sigma\mathbf{W}_n^{\mathrm{T}}] = \mathbf{M}_\omega\mathbf{A}\mathbf{M}_\omega^{\mathrm{T}} + \sigma^2\,\mathrm{E}(\mathbf{W}_n\mathbf{A}\mathbf{W}_n^{\mathrm{T}}).$$

It remains to note that for $\mathbf{W} = \begin{bmatrix} \xi & \eta \\ \zeta & \mu \end{bmatrix}$, $\mathbf{W}^{\mathrm{T}} = \begin{bmatrix} \xi & \zeta \\ \eta & \mu \end{bmatrix}$, $\mathbf{A} = \begin{bmatrix} a & b \\ b & d \end{bmatrix}$, we have

$$\mathrm{E}(\mathbf{W}\mathbf{A}\mathbf{W}^{\mathrm{T}}) = \mathrm{E}\begin{bmatrix} a\xi+b\eta & b\xi+d\eta \\ a\zeta+b\mu & b\zeta+d\mu \end{bmatrix}\begin{bmatrix} \xi & \eta \\ \zeta & \mu \end{bmatrix}$$
$$= \begin{bmatrix} a+d & 0 \\ 0 & a+d \end{bmatrix} = \begin{bmatrix} \operatorname{tr}\mathbf{A} & 0 \\ 0 & \operatorname{tr}\mathbf{A} \end{bmatrix}. \qquad \square$$

PROOF OF THEOREM 3.1. Applying successively Proposition 3.3 and using Proposition 3.2, we get

$$\begin{aligned}
\mathbf{A}_n &= \mathbf{M}\mathbf{A}_{n-1}\mathbf{M}^{\mathrm{T}} + \sigma^2(\operatorname{tr}\mathbf{A}_{n-1})\mathbf{I} \\
&= \mathbf{M}[\mathbf{M}\mathbf{A}_{n-2}\mathbf{M}^{\mathrm{T}} + \sigma^2(\operatorname{tr}\mathbf{A}_{n-2})\mathbf{I}]\mathbf{M}^{\mathrm{T}} + \sigma^2(\operatorname{tr}\mathbf{A}_{n-1})\mathbf{I} \\
&= \mathbf{M}^2\mathbf{A}_{n-2}(\mathbf{M}^{\mathrm{T}})^2 + \sigma^2(\operatorname{tr}\mathbf{A}_{n-2})\mathbf{M}\mathbf{M}^{\mathrm{T}} + \sigma^2(\operatorname{tr}\mathbf{A}_{n-1})\mathbf{I} \\
&\ \ \vdots \\
&= \mathbf{M}^n(\mathbf{M}^{\mathrm{T}})^n + \sigma^2\sum_{s=0}^{n-1}(\operatorname{tr}\mathbf{A}_s)\mathbf{M}^{n-s-1}(\mathbf{M}^{\mathrm{T}})^{n-s-1},
\end{aligned}$$

where $n \geq 1$, and $\mathbf{A}_0 = \mathbf{I}$. We apply the operator tr to both sides of the equation above and put

$$\mu_s = \operatorname{tr} \mathbf{M}^s (\mathbf{M}^{\mathrm{T}})^s.$$

This implies

$$x_n = \mu_n + \sigma^2 (x_0 \mu_{n-1} + \cdots + x_{n-1} \mu_0). \tag{3.5}$$

In particular,

$$x_0 = \mu_0 = 2,$$
$$x_1 = \mu_1 + \sigma^2 x_0 \mu_0,$$
$$x_2 = \mu_2 + \sigma^2 (x_0 \mu_1 + x_1 \mu_0).$$

Let us introduce generating functions of sequences $\{x_n\}$ and $\{\mu_n\}$:

$$X(z) = \sum_{n=0}^{\infty} x_n z^n, \quad \mu(z) = \sum_{n=0}^{\infty} \mu_n z^n.$$

The sequence $\{\mu_n\}$ is bounded for each fixed value of ω_0 (see (2.2)). Then (3.5) implies that $|x_n| \leq C(n+1)|x_{n-1}|$. Thus, the generating functions are well defined. It follows from the recursive relation (3.5) that

$$X(z) = \mu(z) + z\sigma^2 X(z) \mu(z),$$

and therefore,

$$X(z) = \frac{\mu(z)}{1 - z\sigma^2 \mu(z)}.$$

Now we calculate $\mu(z)$. We rewrite (2.2) in the form

$$\mu_s = \varkappa \frac{\sin^2 sk}{\sin^2 k} + 2.$$

where $\varkappa \geq 0$. Hence

$$\mu(z) = \sum_{s=0}^{\infty} \varkappa \frac{\sin^2 sk}{\sin^2 k} z^s + \frac{2}{1-z}.$$

Calculating the sum of geometric series

$$\sum_{n=0}^{\infty} p^n \cos nx = \operatorname{Re} \sum_{n=0}^{\infty} p^n e^{inx} = \frac{1 - p\cos x}{1 - 2p\cos x + p^2},$$

we get

$$\begin{aligned} \mu(z) &= \frac{2}{1-z} + \frac{\varkappa z(1+z)}{(1-z)(1 - 2z\cos 2k + z^2)} \\ &= \frac{2(1 - 2z\cos 2k + z^2) + \varkappa z(1+z)}{(1-z)(1 - 2z\cos 2k + z^2)}. \end{aligned}$$

The function $X(z) = \mu(z)/(1 - \sigma^2 z \mu(z)) = P_2(z)/Q_3(z)$ is a rational function. Here

$$P_2(z) = 2(1 - 2z\cos 2k(\omega) + z^2) + \varkappa z(1+z), \tag{3.6}$$

$$Q_3(z) = (1-z)(1 - 2z\cos 2k(\omega) + z^2) - \sigma^2 z P_2(z). \tag{3.7}$$

Assume that the denominator $Q_3(z)$ has three simple roots 0, z_1, $\bar{z}_1$. Then

$$\frac{P_2(z)}{Q_3(z)} = \frac{c_0}{z_0 - z} + \frac{c_1}{z_1 - z} + \frac{\bar{c}_1}{\bar{z}_1 - z}$$

where

$$c_0 = \frac{P_2(z_0)}{(1+2\sigma^2)(z_0-z_1)(z_0-\bar{z}_1)},$$
$$c_1 = \frac{P_2(z_1)}{(1+2\sigma^2)(z_0-z_1)(\bar{z}_1-z_1)}.$$

Expansion of each simple fraction into geometric series implies

$$x_N = \frac{c_0}{z_0^{N+1}} + \frac{c_1}{z_1^{N+1}} + \frac{\bar{c}_1}{\bar{z}_1^{N+1}}. \tag{3.8}$$

This explicit formula allows to find the asymptotic behavior of x_N as $\sigma \to 0$, $N \to \infty$.

Consider now two cases when ω belongs to a band and when ω is a spectral band edge. If ω belongs to a band then $|\cos 2k(\omega)| < 1$ and therefore the roots of the polynomial Q_3 with $\sigma = 0$ are $z = 1$ and $z = \mathrm{e}^{\pm 2ik(\omega)}$. Clearly, when $\sigma \to 0$, the roots have the form

$$z_0 = 1 + \alpha_0(\omega)\sigma^2 + O(\sigma^4), \quad z_1 = \mathrm{e}^{2ik} + \alpha_1(\omega)\sigma^2 + O(\sigma^4), \quad z_2 = \bar{z}_1.$$

Thus

$$x_N(\omega,\sigma) = c_0(\omega)e^{-\alpha_0(\omega)\sigma^2 N}\big(1+o(1)\big) + 2\,\mathrm{Re}\, c_1(\omega)e^{-\alpha_1(\omega)\sigma^2 N}\big(1+o(1)\big),$$
$$\sigma \to 0, \ \sigma^2 N < C.$$

Together with (3.4) this proves the first statement of Theorem 3.1.

Assume now that ω is a spectral band edge. Then $\cos 2k(\omega) = 1$, and (3.6)–(3.7) imply that

$$Q_3 = (1-z)^3 + \sigma^2 z P_2(z),$$

where $P_2(1) = 2\varkappa > 0$ due to (2.4). Thus, the roots of Q_3 can be found from the equation

$$1 - z + \mu\varphi(z) = 0, \quad \mu = \sigma^{2/3}\mathrm{e}^{\mathrm{i}2j\pi/3}, \quad \varphi = \big(zP_2(z)\big)^{1/3},$$

where φ is analytic in a neighborhood of $z = 1$, and $\varphi(1) = (2\varkappa)^{1/3} > 0$. Hence the roots of Q_3have the form

$$z_j = 1 + \alpha_j(\omega)\sigma^{2/3} + O(\sigma^{4/3}),\ j = 0,1,2, \quad \alpha_j(\omega) = \alpha_0(\omega)\mathrm{e}^{\mathrm{i}2j\pi/3},$$

when $\sigma \to 0$. Here z_0 is real, $z_2 = \bar{z}_1$. Now (3.8) implies that for $\sigma \to 0$, $\sigma^{2/3}N < C$,

$$x_N(\omega,\sigma) = c_0(\omega)e^{-\alpha_0(\omega)\sigma^{2/3}N}\big(1+o(1)\big) + 2\,\mathrm{Re}\, c_1(\omega)e^{-\alpha_1(\omega)\sigma^{2/3}N}\big(1+o(1)\big), \tag{3.9}$$

and this leads to the second statement of the theorem. □

4. Multiplicative perturbation

Theorem 4.1. *If ω is an interior point of a spectral band, then the first statement* (3.1) *of Theorem* 3.1 *holds.*

If ω is a spectral band edge, then the second statement of Theorem 3.1 (3.2) *is valid provided that the correlation matrix $\mathbf{B}(\omega)$ is nondegenerate at that frequency.*

PROOF. In the multiplicative case, perturbation of the monodromy matrix over the period has the form (1.9) and the variance is defined by (1.19), where $\mathbf{\Pi}_n$ is given by (1.18). Similar to the additive case, $\mathbf{A}_n$ is evaluated by induction.

The following formula plays the role of Proposition 3.2

$$\begin{aligned}\mathbf{A}_1 = \mathrm{E}(\mathbf{M}_{\omega,1}\mathbf{M}_{\omega,1}^{\mathrm{T}}) &= \mathrm{E}(\mathbf{M}_\omega(\mathbf{I}+\sigma\mathbf{V}_1)(\mathbf{I}+\sigma\mathbf{V}_1^{\mathrm{T}})\mathbf{M}_\omega^{\mathrm{T}}) \\ &= \mathbf{M}_\omega\big(\mathbf{I}+\sigma^2\,\mathrm{E}(\mathbf{V}_1\mathbf{V}_1^{\mathrm{T}})\big)\mathbf{M}_\omega^{\mathrm{T}}.\end{aligned} \tag{4.1}$$

Formula (3.3), with $N = n$, after evaluating the expectations with respect to random variables with subindexes $s < n$, becomes

$$\mathbf{A}_n = \mathrm{E}(\mathbf{M}_\omega(\mathbf{I}+\sigma\mathbf{V}_n)\mathbf{A}_{n-1}(\mathbf{I}+\sigma\mathbf{V}_n^{\mathrm{T}})\mathbf{M}_\omega^{\mathrm{T}}), \quad n > 1. \tag{4.2}$$

Now (4.1) – (4.2) can be written as follows

$$\mathbf{A}_n = \mathbf{T}_0(\mathbf{I}+\sigma^2\mathbf{T}_1)\mathbf{A}_{n-1}, \quad n \geq 1, \quad \mathbf{A}_0 = \mathbf{I}, \tag{4.3}$$

where $\mathbf{T}_0$ and $\mathbf{T}_1$ are linear operators in the 3D space $\mathbb{S}$ of symmetric matrices. In fact,

$$\mathbf{T}_0\mathbf{A} = \mathbf{M}_\omega\mathbf{A}\mathbf{M}_\omega^{\mathrm{T}}, \tag{4.4}$$

$$\mathbf{T}_1\mathbf{A} = \mathrm{E}(\mathbf{V}_n\mathbf{A}\mathbf{V}_n^{\mathrm{T}}). \tag{4.5}$$

Thus,

$$\mathbf{A}_n = [\mathbf{T}_0(\mathbf{I}+\sigma^2\mathbf{T}_1)]^n\mathbf{A}_0 = [\mathbf{T}_0(\mathbf{I}+\sigma^2\mathbf{T}_1)]^n. \tag{4.6}$$

Let $\lambda_j = \lambda_j(\omega,\sigma)$, $j = 1,2,3$, be eigenvalues of the operator

$$\mathbf{T} = \mathbf{T}_0(\mathbf{I}+\sigma^2\mathbf{T}_1). \tag{4.7}$$

Then

$$x_n(\omega,\sigma) = \operatorname{tr}\mathbf{A}_n = \sum_{j=1}^{3}\lambda_j^n(\omega,\sigma). \tag{4.8}$$

To proceed the proof, we need

Lemma 4.2. *If the matrix $\mathbf{M}_\omega$ has the eigenvalues λ_1 and λ_2 then the eigenvalues of $\mathbf{T}_0$ in (4.4) are λ_1^2, λ_2^2, and $\lambda_1\lambda_2$. In particular, if ω is inside a spectral band, then $\lambda_{1,2} = \mathrm{e}^{\pm ik(\omega)}$, and the eigenvalues of $\mathbf{T}_0$ are $\mathrm{e}^{\pm 2ik(\omega)}$ and 1.*

If ω is a nondegenerate band edge, then $\lambda_1 = \lambda_2 = \pm 1$, and $\mathbf{T}_0$ has the eigenvalue $\lambda = 1$ of multiplicity three with one-dimensional eigenspace.

PROOF. Let $\mathbf{M}_\omega\mathbf{x}_j = \lambda_j\mathbf{x}_j$ and $\mathbf{A} = \mathbf{x}_i\mathbf{x}_j^{\mathrm{T}}$, $i,j = 1,2$. Then

$$\mathbf{M}_\omega\mathbf{A}\mathbf{M}_\omega^{\mathrm{T}} = \mathbf{M}_\omega\mathbf{x}_i\mathbf{x}_j^{\mathrm{T}}\mathbf{M}_\omega^{\mathrm{T}} = \mathbf{M}_\omega\mathbf{x}_i(\mathbf{M}_\omega\mathbf{x}_j)^{\mathrm{T}} = \lambda_i\lambda_j\mathbf{x}_i\mathbf{x}_j^{\mathrm{T}} = \lambda_i\lambda_j\mathbf{A}. \tag{4.9}$$

Similarly, $\mathbf{A}^{\mathrm{T}} = \mathbf{x}_j\mathbf{x}_i^{\mathrm{T}}$ also satisfies (4.9). Then matrix $\mathbf{A}+\mathbf{A}^{\mathrm{T}}$ is symmetric and is an eigenmatrix of $\mathbf{T}_0$ with the eigenvalue $\lambda_i\lambda_j$.

Consider now ω being a nondegenerate band edge. Then matrix $\mathbf{M}_\omega$ is similar to a Jordan block [2]

$$\mathbf{M}_\omega = \mathbf{C}^{-1}\mathbf{J}\mathbf{C}, \quad \mathbf{J} = \begin{bmatrix}\mu & 1\\ 0 & \mu\end{bmatrix}, \ \det\mathbf{C} \neq 0, \ \mu = \pm 1. \tag{4.10}$$

The first statement of Lemma 4.2 provides existence of eigenmatrix $\mathbf{A} = \mathbf{x}_1\mathbf{x}_1^{\mathrm{T}}$ for $\mathbf{T}_0$ with the eigenvalue equal to one. Let us show that the eigenspace of $\mathbf{T}_0$ is one-dimensional. To this end, it suffices to compute the first and the second adjoint eigenmatrices. The monodromy matrix $\mathbf{M}_\omega$ has an eigenvector $\mathbf{x}_1$ and an adjoint vector $\mathbf{y}_1$ such that

$$\mathbf{M}_\omega \mathbf{x}_1 = \mu \mathbf{x}_1, \tag{4.11}$$

$$\mathbf{M}_\omega \mathbf{y}_1 = \mu \mathbf{y}_1 + \mathbf{x}_1. \tag{4.12}$$

Then one can easily check that

$$\mathbf{X}_2 = \tfrac{1}{2}(\mathbf{x}_1\mathbf{x}_1^{\mathrm{T}} + \mathbf{x}_1\mathbf{y}_1^{\mathrm{T}} + \mathbf{y}_1\mathbf{x}_1^{\mathrm{T}}). \tag{4.13}$$

is the first adjoint eigenmatrix. Indeed,

$$\begin{aligned} \mathbf{M}_\omega \mathbf{X}_2 \mathbf{M}_\omega^{\mathrm{T}} &= \tfrac{1}{2}\mathbf{M}_\omega(\mathbf{x}_1\mathbf{x}_1^{\mathrm{T}} + \mathbf{x}_1\mathbf{y}_1^{\mathrm{T}} + \mathbf{y}_1\mathbf{x}_1^{\mathrm{T}})\mathbf{M}_\omega^{\mathrm{T}} \\ &= \tfrac{1}{2}(\mathbf{M}_\omega\mathbf{x}_1(\mathbf{M}_\omega\mathbf{x}_1)^{\mathrm{T}} + \mathbf{M}_\omega\mathbf{x}_1(\mathbf{M}_\omega\mathbf{y}_1)^{\mathrm{T}} + \mathbf{M}_\omega\mathbf{y}_1(\mathbf{M}_\omega\mathbf{x}_1)^{\mathrm{T}}) \\ &= \tfrac{1}{2}(\mathbf{x}_1\mathbf{x}_1^{\mathrm{T}} + \mathbf{x}_1(\mathbf{y}_1^{\mathrm{T}} + \mathbf{x}_1^{\mathrm{T}}) + (\mathbf{y}_1 + \mathbf{x}_1^{\mathrm{T}})\mathbf{x}_1^{\mathrm{T}}) = \mathbf{X}_2 + \mathbf{X}_1. \end{aligned} \tag{4.14}$$

Similar to (4.14), one can show that $\mathbf{X}_3$ given by

$$\mathbf{X}_3 = \tfrac{1}{2}(\mathbf{x}_1\mathbf{x}_1^{\mathrm{T}} + \mathbf{y}_1\mathbf{y}_1^{\mathrm{T}} + \mathbf{x}_1\mathbf{y}_1^{\mathrm{T}} + \mathbf{y}_1\mathbf{x}_1^{\mathrm{T}}), \tag{4.15}$$

satisfies the equation

$$\mathbf{M}_\omega \mathbf{X}_3 \mathbf{M}_\omega^{\mathrm{T}} = \mathbf{X}_3 + \mathbf{X}_2. \tag{4.16}$$

□

Now we complete the proof of Theorem 4.1. If ω is inside the spectral band, then due to Lemma 4.2 eigenvalues of $\mathbf{T}_0$ are distinct and equal to 1, $\mathrm{e}^{\pm 2ik(\omega)}$. Hence, the eigenvalues of the operator $\mathbf{T}$ have the form

$$\lambda_1 = \mathrm{e}^{2ik(\omega)} + g(\omega)\sigma^2 + O(\sigma^4), \tag{4.17}$$

$$\lambda_2 = \bar{\lambda}_1, \tag{4.18}$$

$$\lambda_3 = 1 + h(\omega)\sigma^2 + O(\sigma^4). \tag{4.19}$$

This, together with (4.8) and (3.4), leads to (3.1). It remains to show that if $\det \mathbf{B} \neq 0$ (the covariance matrix is not degenerate) then

$$\lambda_i(\omega,\sigma) = 1 + c_i(\omega)\sigma^{2/3} + O(\sigma^{4/3}), \quad \sigma \to 0,\ c_i(\omega) \neq 0. \tag{4.20}$$

Let us calculate the characteristic polynomial of $\mathbf{T}$ (4.7). Observe that due to Lemma 4.2

$$\det(\mathbf{T}_0 - \lambda \mathbf{I}) = (1-\lambda)^3. \tag{4.21}$$

Expansion (4.20) is apparent if we show that

$$\det(\mathbf{T} - \mathbf{I}) = \det(\mathbf{T}_0(\mathbf{I} + \sigma^2\mathbf{T}_1) - \mathbf{I}) = c(\omega)\sigma^2 + O(\sigma^4), \quad \sigma \to 0, \text{ with } c(\omega) \neq 0. \tag{4.22}$$

In an orthonormal Jordan basis the monodromy matrix $\mathbf{M}_\omega$ has the form

$$\tilde{\mathbf{J}} = \begin{bmatrix} \mu & d \\ 0 & \mu \end{bmatrix}, \quad d \neq 0, \tag{4.23}$$

to which it can be reduced by an orthonormal matrix $\mathbf{O}$.

Operator $\mathbf{T}$ defined by (4.7), has the form

$$\mathbf{T}\mathbf{X} = \mathbf{M}\mathbf{X}\mathbf{M}^{\mathrm{T}} + \sigma^2\mathbf{M}\,\mathrm{E}(\mathbf{V}\mathbf{X}\mathbf{V}^{\mathrm{T}})\mathbf{M}^{\mathrm{T}}. \tag{4.24}$$

In the Jordan basis it has the following representation

$$\mathbf{TX} = \mathbf{O}(\mathbf{M}\mathbf{O}^{\mathrm{T}}\mathbf{X}\mathbf{O}\mathbf{M}^{\mathrm{T}} + \sigma^2\mathbf{M}\,\mathrm{E}(\mathbf{V}\mathbf{O}^{\mathrm{T}}\mathbf{X}\mathbf{O}\mathbf{V}^{\mathrm{T}})\mathbf{M}^{\mathrm{T}})\mathbf{O}^{\mathrm{T}} = \tilde{\mathbf{J}}\mathbf{X}\tilde{\mathbf{J}}^{\mathrm{T}} + \sigma^2\tilde{\mathbf{J}}\,\mathrm{E}(\widetilde{\mathbf{V}}\mathbf{X}\widetilde{\mathbf{V}}^{\mathrm{T}})\tilde{\mathbf{J}}^{\mathrm{T}}, \tag{4.25}$$

where $\widetilde{\mathbf{V}}$ has the form

$$\widetilde{\mathbf{V}} = \mathbf{O}\mathbf{V}\mathbf{O}^{\mathrm{T}} = \begin{bmatrix} \tilde{\xi}_1 & \tilde{\xi}_2 \\ \tilde{\xi}_3 & -\tilde{\xi}_1 \end{bmatrix}, \tag{4.26}$$

and $\tilde{\xi}_i$ are linear combinations of random variables ξ_i (entries of random potential $\mathbf{V}$) such that $\operatorname{tr}\widetilde{\mathbf{V}} = \operatorname{tr}\mathbf{V} = 0$.

Let

$$\widetilde{\mathbf{B}} = \begin{bmatrix} \tilde{b}_{11} & \tilde{b}_{12} & \tilde{b}_{13} \\ \tilde{b}_{21} & \tilde{b}_{22} & \tilde{b}_{23} \\ \tilde{b}_{31} & \tilde{b}_{32} & \tilde{b}_{33} \end{bmatrix} \tag{4.27}$$

be the covariance matrix of random variables $\tilde{\xi}_i$, $i = 1, 2, 3$. Since the linear mapping $\xi \to \tilde{\xi}$ is nondegenerate, $\det\widetilde{\mathbf{B}} > 0$.

Now we represent symmetric matrix $\mathbf{X} = \left[\begin{smallmatrix} x & y \\ y & z \end{smallmatrix}\right]$ as a vector $\left[\begin{smallmatrix} x & y & z \end{smallmatrix}\right]^{\mathrm{T}}$ and evaluate $\det(\mathbf{T} - \mathbf{I})$ (4.22). In this notation, operator $\mathbf{T}$ has the form of a 3×3 matrix $\widehat{\mathbf{T}}$, and one can easily find that operator $\mathbf{T}_0$ is given by the matrix

$$\widehat{\mathbf{T}}_0 = \begin{bmatrix} 1 & 2d & d^2 \\ 0 & 1 & d \\ 0 & 0 & 1 \end{bmatrix}, \tag{4.28}$$

so that

$$\widehat{\mathbf{T}} = \widehat{\mathbf{T}}_0(\mathbf{I} + \sigma^2\widehat{\mathbf{T}}_1), \tag{4.29}$$

where $\widehat{\mathbf{T}}_1$ is given by (4.27).

Finally

$$\det(\mathbf{T} - \mathbf{I}) = \det\left(\begin{bmatrix} 0 & 2d & d^2 \\ 0 & 0 & d \\ 0 & 0 & 0 \end{bmatrix} + \sigma^2 \begin{bmatrix} 1 & 2d & d^2 \\ 0 & 1 & d \\ 0 & 0 & 1 \end{bmatrix} \begin{bmatrix} \tilde{b}_{11} & \tilde{b}_{12} & \tilde{b}_{13} \\ \tilde{b}_{21} & \tilde{b}_{22} & \tilde{b}_{23} \\ \tilde{b}_{31} & \tilde{b}_{32} & \tilde{b}_{33} \end{bmatrix}\right) = 2\tilde{b}_{33}\sigma^2 + O(\sigma^4), \quad \sigma \to 0. \tag{4.30}$$

Here $\tilde{b}_{33} > 0$ since $\det\widetilde{\mathbf{B}} > 0$. This implies (4.20) and completes the proof of Theorem 4.1. □

5. Numerical example

As an example we consider perturbation of periodic delta-function potential [5]

$$V(x) = A\sum_n \delta(x - nL), \tag{5.1}$$

where $A > 0$ is a constant. The monodromy matrix in this case has the form

$$\mathbf{M}_\omega = \begin{bmatrix} \alpha & \beta \\ \gamma & \delta \end{bmatrix} = \begin{bmatrix} \cos\omega L & \sin\omega L \\ -\sin\omega L & \cos\omega L \end{bmatrix} \begin{bmatrix} 1 & 0 \\ A/\omega & 1 \end{bmatrix} = \begin{bmatrix} \cos\omega L + (A/\omega)\sin\omega L & \sin\omega L \\ (A/\omega)\cos\omega L - \sin\omega L & \cos\omega L \end{bmatrix}, \tag{5.2}$$

so that in (5.8) – (5.9)

$$\alpha = \cos(\omega L) + \frac{A}{\omega}\sin(\omega L), \tag{5.3}$$

$$\beta = \sin(\omega L), \tag{5.4}$$

$$\gamma = \frac{A}{\omega}\cos(\omega L) - \sin(\omega L), \tag{5.5}$$

$$\delta = \cos(\omega L). \tag{5.6}$$

Calculation of the covariance matrix $\mathbf{B} = \mathrm{E}[\xi_{i,j}]$ is reduced to evaluation of the following integrals

$$\mathbf{B} = \begin{bmatrix} \int_0^L \psi_1^2(s)\psi_2^2(s)\,\mathrm{d}s & \int_0^L \psi_1(s)\psi_2^3(s)\,\mathrm{d}s & -\int_0^L \psi_1^3(s)\psi_2(s)\,\mathrm{d}s \\ \int_0^L \psi_1(s)\psi_2^3(s)\,\mathrm{d}s & \int_0^L \psi_2^4(s)\,\mathrm{d}s & -\int_0^L \psi_1(s)\psi_2^3(s)\,\mathrm{d}s \\ -\int_0^L \psi_1^3(s)\psi_2(s)\,\mathrm{d}s & -\int_0^L \psi_1(s)\psi_2^3(s)\,\mathrm{d}s & \int_0^L \psi_1^4(s)\,\mathrm{d}s \end{bmatrix}, \tag{5.7}$$

where $\psi_1(s) = \cos(\omega s)$ and $\psi_2 = \sin(\omega s)$.

For numerical calculations it is more convenient to work with vectors rather than two-dimensional matrices. Therefore, we will represent symmetric matrices $\mathbf{A}_n$ by vectors. Corresponding matrices of the linear operators in $\mathbb{R}^3$ defined by (4.3) are denoted by $\widehat{\mathbf{T}}_0$ and $\widehat{\mathbf{T}}_1$

$$\widehat{\mathbf{T}}_0 = \begin{bmatrix} \alpha^2 & 2\alpha\beta & \beta^2 \\ \alpha\gamma & 2\alpha\delta - 1 & \beta\delta \\ \gamma^2 & 2\gamma\delta & \delta^2 \end{bmatrix}. \tag{5.8}$$

The elements of the perturbation matrix $\widehat{\mathbf{T}}_1$ are expresses through the covariance matrix (5.7) as follows

$$\widehat{\mathbf{T}}_1 = \begin{bmatrix} b_{11} & 2b_{12} & b_{22} \\ b_{13} & -b_{11} + b_{23} & -b_{12} \\ b_{33} & -2b_{13} & b_{11} \end{bmatrix}. \tag{5.9}$$

Figure 1 illustrates the sensitivity of the scalar variance of the transfer matrix over one period on the propagating frequency ω for $L = 1$ and $A = 6$. If $\omega = 2.6$ is

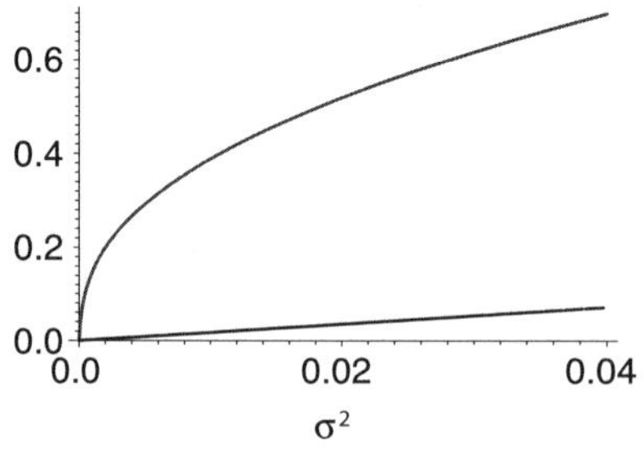

FIGURE 1. Dependence of the scalar variance of the transfer matrix over one period on the perturbation variance σ^2 for $L = 1$ and $A = 6$. The lower curve corresponds to the frequency $\omega = 2.6$ in the middle of the first band, while the upper one matches $\omega = 1.977$ near the left edge of the first band.

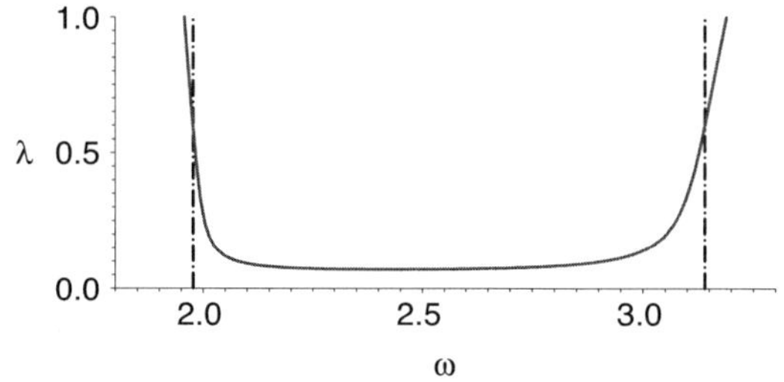

FIGURE 2. Dependence of the scalar variance of the transfer matrix over one period on the frequency ω for the perturbation variance $\sigma^2 = 0.04$ and $L = 1$, $A = 6$. Band edges are denoted by vertical lines.

close to the middle of the band, then the variance of the transfer matrix is increasing linearly with σ^2 having the same order (lower curve). If, however, $\omega = 1.977$ is close to the band edge then the eigenvalue undergoes drastic changes with σ^2 (cf. (3.1)–(3.2)).

Dependence of the variance of the transfer matrix on the frequency ω is shown in Figure 2 for the perturbation variance $\sigma^2 = 0.04$ and $L = 1$, $A = 6$. As the frequency approaches the band edges, the variance of the transfer matrix increases dramatically.

References

1. A. Figotin and I. Vitebskiy, *Oblique frozen modes in periodic layered media*, Phys. Rev. E (3) **68** (2003), no. 3, 036609.
2. B. M. Levitan and I. S. Sargsjan, *Introduction to spectral theory: selfadjoint ordinary differential operators*, translated by A. Feinstein, Transl. Math. Monogr., vol. 39, Amer. Math. Soc., Providence, RI, 1975.
3. I. M. Lifshits, S. A. Gredeskul, and L. A. Pastur, *Introduction to the theory of disordered systems*, Wiley-Intersci. Publ., Wiley, New York, 1988.
4. S. Molchanov, *Lectures on random media*, Lectures on Probability Theory (Saint-Flour, 1992) (P. Bernard, ed.), Lectures Notes in Math., vol. 1581, Springer, New York, 1994, pp. 242–411.
5. S. Molchanov and B. Vainberg, *Slowing down and reflection of waves in truncated periodic media*, J. Funct. Anal. **231** (2006), no. 2, 287–311.
6. S. Novikov, S. V. Manakov, L. P. Pitaevskiĭ, and V. E. Zakharov, *Theory of solitons: The inverse scattering method*, Plenum, New York, 1984.
7. L. Pastur and A. Figotin, *Spectra of random and almost-periodic operators*, Grundlehren Math. Wiss., vol. 297, Springer, New York, 1992.

DEPARTMENT OF MATHEMATICS AND STATISTICS, UNIVERSITY OF NORTH CAROLINA — CHARLOTTE, CHARLOTTE, NC 28223, USA

E-mail address, Yu. A. Godin: `ygodin@uncc.edu`

E-mail address, S. Molchanov: `smolchan@uncc.edu`

E-mail address, B. Vainberg: `brvainbe@email.uncc.edu`

Centre de Recherches Mathématiques
CRM Proceedings and Lecture Notes
Volume **42**, 2007

Spectral Theory of 1-D Schrödinger Operators with Unbounded Potentials

A. Gordon, J. Holt, and S. Molchanov

ABSTRACT. This paper presents the spectral theory of the Schrödinger operator with piecewise constant random and deterministic potentials that are unbounded from above and below, and in several cases strongly fluctuating. In particular, we will present models exhibiting a type of transition from discrete spectrum, to pure point spectrum, to singular continuous spectrum.

1. Introduction

Many years ago two of the authors along with B. Tsagani published in the Russian journal *Functional Analysis and its Applications* a brief paper [6] about random and deterministic operators with unbounded potentials. We were interested mainly in bifurcations from dense point spectrum to discrete spectrum. This paper did not contain proofs, but the subject was interesting and as a result, it attracted certain interest and generated responses. One of the models from this paper (that of the sparse rectangular barriers) was studied in a more general form in the paper [3] by one of the authors. We decided to publish a complete analysis of the three remaining models from [6]. Such analysis is important due to the growing interest in spectral bifurcations and borderline examples. Some of the ideas, which we had in mind in the early 1990s, are probably still of some interest. The paper is dedicated to the memory of Professor B. Tsagani, who proposed the return to this subject and started to work on two of the models before suddenly passing away in 2004.

Model I is related to the classical criterion of Weyl (whose final version was given by A. Molchanov, see details in [2, 9]) of the discreteness of the spectrum. Specifically, if $V(x) \geq c_0 > -\infty$ for all x, then $\Sigma(H)$ is discrete for any boundary phase θ_0 if and only if for all $h > 0$

$$\lim_{x\to\infty} \int_x^{x+h} V_+(z)\,\mathrm{d}z = \infty. \tag{1.1}$$

2000 *Mathematics Subject Classification.* Primary 34L40, 47E05; Secondary 34B24, 34C10.
The research of the two last authors is partially supported by NSF grant DMS-0405927.
This is the final form of the paper.

Here, as usual, the operator $H = H_\theta : L^2(0,\infty) \to L^2(0,\infty)$ is given by

$$H\psi = -\psi'' + V(x)\psi \tag{1.2}$$

$$\psi(0)\cos\theta - \psi'(0)\sin\theta = 0 \tag{1.3}$$

where $\theta \in [0,\pi)$. A similar result is true on the full axis, but we'll consider, in the sake of simplicity, only the half axis case.

We will give examples of operators H with exotic essential spectrum and with singular continuous spectral measure. Model I presents piecewise constant nonnegative potentials increasing in mean. Model II presents piecewise constant potentials unbounded from above and below. In both cases we describe transitions from dense pure point spectrum to discrete spectrum on the entire energy axis. Finally, model III describes a class of potentials which are non-positive and tend to $-\infty$ very fast. We'll discuss the problem of the essential self-adjointness of such operators (initiated by the classical example of E. Nelson, see [11]) and the transition from discrete spectrum (unbounded from above and below) to dense pure point spectrum.

2. Model I. Potentials increasing in mean

Consider a partition of $\mathbb{R}^+$ into intervals $I_n = [n, n+1)$ and define

$$V(x) = \begin{cases} 0 & \text{if } x \in I_{2n} \\ A_n & \text{if } x \in I_{2n+1} \end{cases} \tag{2.1}$$

for $n \geq 0$ where $A_n > 0$ and $A_n \to \infty$.

Potential V is bounded from below, but (1.1) does not hold. In particular $\int_x^{x+2} V_+(z)\,dz \to \infty$, while for $h < 1$, $\liminf_{x\to\infty} \int_x^{x+h} V_+(z)\,dz = 0$. Hence $\Sigma(H)$ is not pure discrete for any $\theta_0 \in [0,\pi)$. It means that $\Sigma_{\text{ess}}(H)$ is non-empty.

Theorem 2.1. *With $V(x)$ as above, $\Sigma_{\text{ess}}(H) = \{\pi^2 k^2 : k = 1, 2, \dots\}$.*

Operator H (as any 1-d Schrödinger operator on the half axis) has simple spectrum (recall that a self-adjoint operator on $\mathcal{H}$ has simple spectrum if there exists $g \in \mathcal{H}$ such that the $\{E([\alpha,\beta])g : \alpha,\beta \in R\}$ is dense in $\mathcal{H}$). From this fact Theorem 2.1 implies

Corollary 2.2. *For any $\theta \in [0,\pi)$ operator H of model* I *has pure point spectrum with accumulation points $\lambda_k = \pi^2 k^2$.*

To prove Theorem 2.1 we will first show that for each $k \geq 1$, $\lambda_k \in \Sigma_{\text{ess}}(H)$ using the classical Weyl's theorem for essential spectrum, i.e., $\lambda \in \Sigma_{\text{ess}}(H) <$ if there exists a sequence ψ_n with $\|\psi_n\| = 1$ for which ψ_n converges weakly to 0 and $\|H\psi_n - \lambda\psi_n\| \to 0$ as $n \to \infty$. To prove that for every $\lambda \neq \lambda_k$, $\lambda \notin \Sigma_{\text{ess}}(H)$ we'll apply the Sturm oscillation theory. The following lemma will play a central role in the current model I as well as model II.

Lemma 2.3. *For $A' \geq A > 0$ let $V_{A,A'}(x)$ be the "test potential" defined by*

$$V_{A,A'}(x) = \begin{cases} A & \text{if } x < -\frac{1}{2} \\ -B(A) & \text{if } |x| \leq \frac{1}{2} \\ A' & \text{if } x > \frac{1}{2}. \end{cases} \tag{2.2}$$

where $B(A) > 0$. For some fixed $\delta_0 \in (0,1]$ let $\lambda(A) = A^{1-\delta_0}$ and $B = O\big(\lambda(A)\big)$. Then for $A \to \infty$ and $n \in \mathbb{Z}^+$ such that $n^2\pi^2 \le \lambda(A)$ the operator $H = -\mathrm{d}^2/\mathrm{d}x^2 + V_{A,A'}(x)$ has an eigenvalue $\lambda_n(A, A')$ with eigenfunction $\psi_n(x; A, A')$ such that $\lambda_n + B(A) = \omega_n^2$ where

$$\omega_n = n\pi - n\pi\left(\frac{1}{\sqrt{A+B}} + \frac{1}{\sqrt{A'+B}}\right) + O\left(\left(\frac{n}{\sqrt{A+B}}\right)^3\right) \tag{2.3}$$

while for $|x| \le \frac{1}{2}$

$$\psi_n(x) = \alpha_n \cos\sqrt{\lambda_n}(x + \tfrac{1}{2}) + \beta_n \sin\sqrt{\lambda_n}(x + \tfrac{1}{2}) \tag{2.4}$$

where $\alpha_n^2, \beta_n^2 \in R^+$ and for $|x| > \frac{1}{2}$

$$|\psi_n(x)| \le c_1 \frac{1}{\sqrt{A}} \mathrm{e}^{-c_0\sqrt{A}|x|} \tag{2.5}$$

for some $c_0 > \frac{1}{2} - \epsilon_0$, $c_1 > 0$ and $\epsilon_0 = \epsilon_0(\delta_0)$ uniformly in n for $n \le \sqrt{\lambda(A)}/\pi$.

PROOF. Let $-B(A) < \lambda \le \lambda(A)$. The equation $H\psi = \lambda\psi$ has on $(-\infty, -\frac{1}{2})$ the decreasing solution

$$\psi_\lambda(x) = \frac{1}{\sqrt{A-\lambda}} \mathrm{e}^{\sqrt{A-\lambda}(x+1/2)}.$$

Similarly, on $[\frac{1}{2}, \infty)$ the equation $H\psi = \lambda\psi$ has a solution of the form

$$\tilde{\psi}_\lambda(x) = Ce^{-\sqrt{A'-\lambda}(x-1/2)}$$

where C is constant. The solution $\psi_\lambda(x)$ has a simple extension onto $[-\frac{1}{2}, \frac{1}{2}]$;

$$\psi_\lambda(x) = \frac{1}{\sqrt{A-\lambda}} \cos\sqrt{\lambda+B}(x + \tfrac{1}{2}) + \frac{\sin\sqrt{\lambda+B}(x+\frac{1}{2})}{\sqrt{\lambda+B}}$$

$$\psi_\lambda'(x) = -\sqrt{\frac{\lambda+B}{A-\lambda}} \sin\sqrt{\lambda+B}(x+\tfrac{1}{2}) + \cos\sqrt{\lambda+B}(x+\tfrac{1}{2}).$$

The eigenvalue equation for λ_n reflects the continuity of the functions ψ_λ, ψ_λ' for $x \le \frac{1}{2}$ and $\tilde{\psi}_\lambda, \tilde{\psi}_\lambda'$ for $x > \frac{1}{2}$ in that

$$\frac{\psi_\lambda}{\psi_\lambda'}(\tfrac{1}{2} - 0) = \frac{\tilde{\psi}_\lambda}{\tilde{\psi}_\lambda'}(\tfrac{1}{2} + 0)$$

which becomes

$$\frac{\cos\sqrt{\lambda+B} + \sqrt{(A-\lambda)/(\lambda+B)}\sin\sqrt{\lambda+B}}{-\sin\sqrt{\lambda+B} + \sqrt{(A-\lambda)/(\lambda+B)}\cos\sqrt{\lambda+B}} = -\sqrt{\frac{\lambda+B}{A'-\lambda}}. \tag{2.6}$$

Put $a = (A+B)/(\lambda+B)$, $b = (A'+B)/(\lambda+B)$ and let $\omega = \sqrt{\lambda+B}$. Then it follows easily that $b > a \ge K(A)A^{\delta_0}$ where, uniformly in A, $K(A) \ge K$ for some fixed $K > 0$. In the new notations (2.6) can be can be rewritten as

$$\tan\omega = -\frac{\sqrt{b-1} + \sqrt{a-1}}{\sqrt{b-1}\sqrt{a-1} - 1} = O\left(\frac{1}{\sqrt{a-1}}\right).$$

More precisely,

$$\tan\omega = -\omega\left(\frac{1}{\sqrt{A+B}} + \frac{1}{\sqrt{A'+B}}\right) + O\left(\left(\frac{\omega}{\sqrt{A+B}}\right)^3\right). \tag{2.7}$$

Iterations $\omega_n = \pi n - \delta_n$ now lead to formula (2.3) for the eigenvalues. Then

$$\begin{aligned}\psi_n(\tfrac{1}{2}) &= \frac{1}{\sqrt{A-\lambda_n}}\cos\sqrt{\lambda_n+B} + \frac{1}{\sqrt{\lambda_n+B}}\sin\sqrt{\lambda_n+B} \\ &= \frac{1}{\sqrt{A-\lambda_n}}\cos\omega_n + \frac{1}{\omega_n}\sin\omega_n \\ &= \frac{(-1)^n}{\sqrt{A-\lambda_n}} + O(\delta_n^2).\end{aligned}$$

This proves (2.5). □

Remark 2.4. We'll use this lemma in the proof of Theorem 2.1 in the case when $B(A) = 0$ and $\lambda_i = O(1)$. In its full generality this lemma will be essential for model II. We also note that if $\lambda(A) \sim \theta A$ where $\theta \in (0,1)$ then the result of Lemma 2.3 is wrong: the deviation of λ_n from $\pi^2 n^2$ for $n \sim \sqrt{\theta A}$ is of the order 1, not $o(1)$.

Lemma 2.5. *Let $\lambda_n(A, A')$ be the eigenvalues of Lemma 2.3 and $\psi_n(x; A, A')$ the corresponding eigenfunctions. For $B(A) = 0$ and $\lambda_n \le \lambda(A)$ uniformly in n,*

$$\int_{-1}^{1} |H\psi_n(x;A) - \pi^2 n^2 \psi_n(x;A)|^2 \,\mathrm{d}x = o(1)$$

as $A \to \infty$. In fact

$$\int_{-1}^{1} |H\psi_n(x;A) - n^2\pi^2 \psi_n(x;A)|^2 \,\mathrm{d}x \sim \frac{2\lambda_n}{A}.$$

PROOF. Since $\psi_n(x; A)$ is an eigenfunction for λ_n it follows that

$$\int_{-1}^{1} |H\psi_n(x;A) - n^2\pi^2\psi_n(x;A)|^2 \,\mathrm{d}x = (\lambda_n - n^2\pi^2)^2 \int_{-1}^{1} |\psi_n(x;A)|^2 \,\mathrm{d}x.$$

Brief calculations reveal that

$$\int_{-1}^{1} |\psi_n(x;A)|^2 \,\mathrm{d}x \sim \frac{1}{2\lambda_n}$$

and that

$$\frac{(\lambda_n - n^2\pi^2)^2}{\lambda_n} \sim \frac{2\lambda_n}{A}$$

for each fixed n as $A \to \infty$. □

PROOF OF THEOREM 2.1. To show that in model I for each l, $\pi^2 l^2 \in \Sigma_{\mathrm{ess}}(H)$, we define a sequence of almost eigenfunctions. Let $\psi_l(x; A_n, A_{n+1})$ be the eigenfunctions of Lemma 2.3 corresponding to λ_l. By Lemma 2.5 it follows that $\|H\psi_{l,n} - l^2\pi^2\psi_{l,n}\|_{L^2}^2 = o(1)$ as $n \to \infty$. Let $a_n' = 2n - \frac{1}{2}$, $a_n'' = 2n - \frac{1}{4}$, $a_{n+1}'' = 2n + \frac{5}{4}$ and $a_{n+1}' = 2n + \frac{3}{2}$. Define the sequence $\varphi_{l,n}(x)$ by

$$\varphi_{l,n}(x) = v(x)\psi_l(x - (2n + \tfrac{1}{2}); A_n, A_{n+1})$$

where $v(x) \in C_0^\infty$ is chosen such that $v(x) \equiv 0$ for $|x - (2n + \frac{1}{2})| > 1$, $v(x) \equiv 1$ for $|x - (2n + \frac{1}{2})| < \frac{3}{4}$ and $0 \le v(x) \le 1$ for all x. Then,

$$\|H\varphi_{l,n} - l^2\pi^2\varphi_{l,n}\|_{L^2}^2 = \int_{2n-1/2}^{2n+3/2} |H\varphi_{l,n}(x) - l^2\pi^2\varphi_{l,n}(x)|^2 \,\mathrm{d}x.$$

Lemma 2.6. *For fixed $l \in \mathbb{Z}^+$ and $n \to \infty$*

$$\|H\varphi_{l,n} - \pi^2 l^2 \varphi_{l,n}\|_2 \sim \frac{2\lambda_l}{A_n}. \tag{2.8}$$

The proof is a direct consequence of Lemma 2.5. Lemma 2.6 shows (due to Weyl's criterion) that $\lambda_l = \pi^2 l^2 \in \Sigma_{\text{ess}}(H)$ for integers $l \geq 1$ and we have proved the first part of Theorem 2.1. The proof of the second part will be based on the phase formalism and the Sturm oscillation theory. For $\lambda = k^2$ one can introduce the monodromy operators for even and odd integer intervals. We use the "modified" Prüfer's form of the monodromy operators:

$$M_k(a,b)\colon \begin{pmatrix} \psi_k(a) \\ \psi_k'(a)/k \end{pmatrix} \longrightarrow \begin{pmatrix} \psi_k(b) \\ \psi_k'(b)/k \end{pmatrix} \tag{2.9}$$

where $\psi_k(x)$ is a solution of the problem $H\psi_k = k^2\psi$ satisfying the boundary condition (1.3). As easy to see, on the even intervals $I_{2n} = [2n, 2n+1]$,

$$O_n(k) \equiv M_k(I_{2n}) = \begin{pmatrix} \cos k & \sin k \\ -\sin k & \cos k \end{pmatrix}. \tag{2.10}$$

This simple formula is the main advantage of the modified Prüfer's monodromy matrices. Similarly,

$$M_k(I_{2n+1}) = \begin{pmatrix} \cosh\sqrt{A_n - k^2} & (k/\sqrt{A_n - k^2})\sinh\sqrt{A_n - k^2} \\ (\sqrt{A_n - k^2}/k)\sinh\sqrt{A_n - k^2} & \cosh\sqrt{A_n - k^2} \end{pmatrix}. \tag{2.11}$$

on the odd intervals $I_{2n+1} = [2n+1, 2n+2]$. If $A_n > k^2$ (which is true for all sufficiently large n if k is fixed or bounded) then the matrices $M_k(I_{2n+1})$ have real eigenvalues

$$\lambda_n^{\pm}(k) = e^{\pm\mu_n}$$

with corresponding eigenvectors

$$e_n^+(k) = \begin{pmatrix} a_n \\ 1 \end{pmatrix}, \quad e_n^-(k) = \begin{pmatrix} -a_n \\ 1 \end{pmatrix}$$

where $\mu_n = \mu_n(k) = \sqrt{A_n - k^2}$ and $a_n = a_n(k) = k/\mu_n$. Vector $e_n^+(k)$ is an attractor for the linear mapping $M_k(I_{2n+1})$ while vector $e_n^-(k)$ is an attractor for $M_k^{-1}(I_{2n+1})$. For any vector $\vec{x} = (x_1, x_2)^T$ we have

$$M_k(I_{2n+1})\vec{x} = \frac{e^{\mu_n}}{2a_n}(x_1 + a_n x_2)\begin{pmatrix} a_n \\ 1 \end{pmatrix} + \frac{e^{-\mu_n}}{2a_n}(-x_1 + a_n x_2)\begin{pmatrix} -a_n \\ 1 \end{pmatrix}. \tag{2.12}$$

Along with the monodromy operators we can introduce the modified Prüfer's phase $\theta_k(t)$ and amplitude $r_k(t)$. They are given by the standard relations

$$\psi_k(t) = r_k(t)\sin\theta_k(t) \tag{2.13}$$

$$\frac{\psi_k'(t)}{k} = r_k(t)\cos\theta_k(t). \tag{2.14}$$

It is well known that for H_θ

$$\theta_k'(t) = k - \frac{V(t)\sin^2\theta_k(t)}{k}, \quad \theta_k(0) = \theta \tag{2.15}$$

$$r_k'(t) = r_k(t)\frac{V(t)\sin 2\theta_k(t)}{2k}, \quad r_k(0) = 1. \tag{2.16}$$

To show that $\Sigma_{\text{ess}}(H) \subset \{l^2\pi^2 : l \in \mathbb{Z}^+\}$ we will show that for each fixed $n \in \mathbb{Z}^+$ and arbitrarily small $\zeta > 0$

$$\limsup_{t\to\infty} \frac{1}{\pi}\big(\theta_{(n+1)^2\pi^2-\zeta}(t) - \theta_{n^2\pi^2+\zeta}(t)\big) < \infty.$$

Then it follows (see [9,12]) that $\Sigma_{\text{ess}}(H) \cap (n^2\pi^2+\zeta, (n+1)^2\pi^2-\zeta) = \varnothing$. Without loss of generality consider the solutions $\theta_{(n+1)^2\pi^2-\zeta}(t)$ and $\theta_{n^2\pi^2+\zeta}(t)$ of (2.15) such that $\theta_{(n+1)^2\pi^2-\zeta}(2M_0) = \alpha$ and $\theta_{n^2\pi^2+\zeta}(2M_0) = \beta$ where $\alpha, \beta \in [\zeta, \pi-\zeta]$ and where $M_0 = M_0(\zeta)$ is chosen so that $A_m > n^2\pi^2$ and $1/\sqrt{A_m} < \zeta^2$ for all $m > 2M_0$. Then on the even intervals I_{2n}

$$\theta_k(t) = \theta_k(2n) + k(t-2n), \quad r_k(t) = r_k(2n)$$

which is equivalent to (2.9). In particular, for

$$k_1 = \sqrt{n^2\pi^2+\zeta} \quad \text{and} \quad k_2 = \sqrt{(n+1)^2\pi^2-\zeta}$$

$$\theta_{k_2}(2M_0+1) - \theta_{k_1}(2M_0+1) = (\alpha-\beta) + \pi - \frac{2n+1}{2\pi(n^2+n)}\zeta + O(\zeta^2).$$

On the intervals I_{2n+1} the behavior of the phase is completely different. For one, there is no rotation of the phase $\theta_k(t)$ for $t \in I_{2n+1}$. Let's consider $\theta_k(t) \bmod \pi$ and introduce the angles $\gamma_m = \tan^{-1}(a_m)$, and $\pi - \gamma_m$. Without loss we take M_0 so large that $\gamma_m \in (0, \zeta^2/2)$ for all $m > M_0$ (hence $\pi - \gamma_m \in (\pi - \zeta^2/2, \pi)$) and in this case then for all $m > M_0$ we have $A_m^{-1} < \zeta^4/\big(2\pi(n+1)\big)^2 + O(\zeta^7)$. Of course, $\gamma_m \to 0$ and $\pi - \gamma_m \to \pi$ if $m \to \infty$. Using (2.11) and (2.12) it follows that both $\theta_{k_2}(2M_0+2)$ and $\theta_{k_1}(2M_0+2)$ are exponentially close to the angles $\gamma_{2M_0+2}(k)$ for $k = k_1 = k_2$, i.e.,

$$\theta_k(2M_0+2) = \gamma_{2M_0+2}(k) + O\big(\mathrm{e}^{-2\mu_{2M_0+2}(k)}\big)$$

and therefore

$$|\theta_{k_2}(2M_0+2) - \theta_{k_1}(2M_0+2)| = \frac{1}{\sqrt{A_{2M_0+1}}}\Big(\pi - \frac{1}{2\pi}\frac{2n+1}{n(n+1)}\zeta\Big) + O\Big(\frac{\zeta^2}{\sqrt{A_{2M_0+1}}}\Big).$$

Since $1/\sqrt{A_m} < \text{const}\cdot\zeta^2 + O(\zeta^3)$ for all $m > M_0$ we have

$$|\theta_{k_2}(2M_0+2) - \theta_{k_1}(2M_0+2)| = O(\zeta^2).$$

Hence, we have shown that for fixed $l \in \mathbb{Z}^+$ and arbitrarily small $\zeta > 0$ that for any given $\alpha, \beta \in [0, \pi)$ there exists $M_0(\zeta)$ large enough so that the phases $\theta_k(t)$ satisfying (2.15) with initial conditions $\theta_{k_2}(2M_0) = \alpha$, $\theta_{k_1}(2M_0) = \beta$ satisfy the relation

$$|\theta_{k_2}(2M_0+2) - \theta_{k_1}(2M_0+2)| = O(\zeta^2).$$

Define now $\theta_{k_2}(2M_0+2) = \alpha$ and $\theta_{k_1}(2M_0+2) = \beta$. Then similar steps (using either formula (2.12) or direct integration of the Ricatti equation (2.15) with initial conditions $\theta_k(2M_0+2)$ where $k = k_1, k_2$) give by induction that for any $\alpha, \beta \in [\zeta, \pi-\zeta]$ and $m \geq M_0(\zeta)$ that $\theta_k(2m) \in [-\zeta^2/2, \zeta^2/2]$ (and even exponentially small neighborhood of the point $\gamma_m(k)$ (of order $\exp(-2\mu_m)$) shown in Figure 1 below by shading). The last fact (together with the description of the phase on the even intervals I_{2m}, $m \geq 0$) implies that for fixed l and $\zeta > 0$ the difference

$$\theta_{(l+1)^2\pi^2-\zeta}(t) - \theta_{l^2\pi^2+\zeta}(t), \quad \text{for } t \geq 2M_0(\zeta)$$

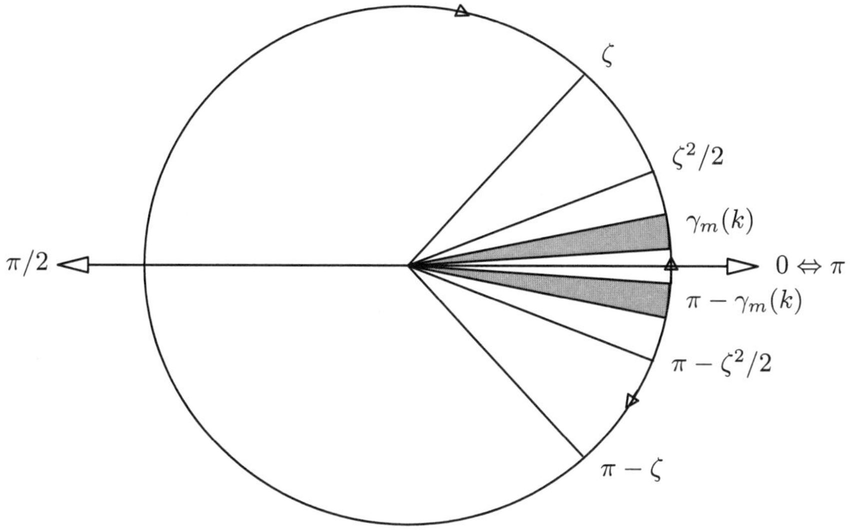

FIGURE 1. Arrows on the circle indicate the direction of the phase $\theta_k(t)$ for $t \in [2m+1, 2m+2)$

(both phases started from the same initial value $\theta_0 = 0$ for $t = 2M_0(\zeta)$) cannot leave the interval $(\zeta, \pi - \zeta)$. It means that the difference above is bounded on the full half axis, i.e,. the Hamiltonian H is this model has at most finitely many discrete eigenvalues inside the interval $[\pi^2 l^2 + \zeta/2, \pi^2(l+1) - \zeta/2]$. □

Let's note that the last fact implies that

$$\ln\|M_\lambda([0,2n))\| \sim \sum_{k=1}^{n} \sqrt{A_k - \lambda} \sim \sum_{k=1}^{n} \sqrt{A_k}.$$

The last estimation guarantees that eigenfunctions of our Hamiltonian of model I decay super-exponentially.

We can generalize model I in several directions. We will consider only one such generalization. Let $\vartheta > 0$ be fixed and consider the sequence $\{B_n\} \subset [0, \vartheta]$. Define

$$V(x) = \begin{cases} B_n & \text{if } x \in I_{2n+1} \\ A_n & \text{if } x \in I_{2n} \end{cases} \tag{2.17}$$

where as before $A_n > 0$ and $A_n \to \infty$. In addition, let Γ be the set of limit points of the sequence $\{B_n\}$, i.e., $\Gamma = \{B : B = \lim_{k\to\infty} B_{n_k}\}$

Theorem 2.7. *With $V(x)$ as above, $\Sigma_{\text{ess}}(H_{\theta_0}) = \bigcup_{n=1}^{\infty}(\{n^2\pi^2\} \oplus \Gamma)$ for all $\theta_0 \in [0, \pi)$. If Γ is discrete, i.e.,* card$(\Gamma \cap \Delta) < \infty$ *for any $\Delta \subset [0, \vartheta]$, the spectral measure $\mu(H_{\theta_0})$ of H_{θ_0} is pure point.*

PROOF. The proof that

$$\Sigma_{\text{ess}}(H_{\theta_0}) \supset \bigcup_{n=1}^{\infty}(\{n^2\pi^2\} \oplus \Gamma)$$

is exactly the same as in Theorem 2.1. Using Lemmas 2.3 and 2.5 we construct a sequence of almost eigenfunctions $\varphi_{l,n}(x)$. They have exactly the same form as before. As for the opposite inclusion,

$$\Sigma_{\text{ess}}(H_{\theta_0}) \subset \bigcup_{n=1}^{\infty}(\{n^2\pi^2\} \oplus \Gamma)$$

the proof is slightly different in that on the intervals I_{2n} the transfer matrix $M_k(I_{2n})$ is not orthogonal under the modified Prüfer variables (2.13) and (2.14). In this case it is better to define the standard Prüfer variables by

$$\psi_\lambda(t) = r_\lambda(t)\sin\theta_\lambda(t) \tag{2.18}$$

$$\psi'_\lambda(t) = r_\lambda(t)\cos\theta_\lambda(t). \tag{2.19}$$

It is well-known that the phase $\theta_\lambda(t)$ and amplitude $r_\lambda(t)$ obey the equations

$$\theta'_\lambda(t) = \cos^2\theta_\lambda(t) + (\lambda - V(t))\sin^2\theta_\lambda(t) \tag{2.20}$$

$$r'_\lambda(t) = r_\lambda(t)\frac{\sin 2\theta_\lambda(t)}{2\sqrt{\lambda}}(1 + V(t) - \lambda) \tag{2.21}$$

The transfer matrices on even and odd intervals (for $B_n < \lambda < A_n$) have the usual forms:

$$M_\lambda(I_{2n}) = \begin{pmatrix} \cos\sqrt{\lambda - B_n} & (\sin\sqrt{\lambda - B_n})/\sqrt{\lambda - B_n} \\ -\sqrt{\lambda - B_n}\sin\sqrt{\lambda - B_n} & \cos\sqrt{\lambda - B_n} \end{pmatrix}$$

$$M_\lambda(I_{2n+1}) = \begin{pmatrix} \cosh\sqrt{A_n - \lambda} & (\sinh\sqrt{A_n - \lambda})/\sqrt{A_n - \lambda} \\ \sqrt{A_n - \lambda}\sinh\sqrt{A_n - \lambda} & \cosh\sqrt{A_n - \lambda} \end{pmatrix}$$

and from this it follows that

$$\begin{aligned} M_\lambda([0,2n]) &= M_\lambda(I_{2n-1})M_\lambda(I_{2n-2})\cdots M_\lambda(I_1)M_\lambda(I_0) \\ &\equiv \prod_{j=0}^{2n-2} M_\lambda(I_{j+1})M_\lambda(I_j). \end{aligned} \tag{2.22}$$

For $n \geq 0$ let R_n be the matrix defined by

$$R_n = \begin{pmatrix} \sqrt[4]{\lambda - B_n} & 0 \\ 0 & \sqrt[-4]{\lambda - B_n} \end{pmatrix}.$$

Then it follows easily that

$$M_\lambda([0,2n]) = R_{n+1}\left(\prod_{j=0}^{2n-2} \widetilde{M}_\lambda(I_{j+1})\widetilde{M}_\lambda(I_j)\right)R_0^{-1}$$

where

$$\widetilde{M}_\lambda(I_{2n}) = R_n^{-1}M_\lambda(I_{2n})R_n = \begin{pmatrix} \cos\sqrt{\lambda - B_n} & \sin\sqrt{\lambda - B_n} \\ -\sin\sqrt{\lambda - B_n} & \cos\sqrt{\lambda - B_n} \end{pmatrix}$$

and

$$\widetilde{M}_\lambda(I_{2n+1}) = R_{n+2}^{-1}M_\lambda(I_{2n+1})R_{n+1}. \tag{2.23}$$

These formulas admit the following representation. The appropriate phase (selected differently for different even intervals I_{2n}, $n \geq 0$) rotates uniformly with speed $v_n = \sqrt{\lambda - B_n}$ on each even interval I_{2n} after which it is subjected to the strong hyperbolic transformation $\widetilde{M}_\lambda(I_{2n-1})$, $n \geq 1$ with eigenvalues of the order $\exp(\pm\sqrt{A_n - \lambda})$. Now the proof of the second part of Theorem 2.7 is similar to the proof of the corresponding part of Theorem 2.1. Assume that $[\lambda_1, \lambda_2] \cap$

$(\Gamma \oplus \{n^2\pi^2 : n \geq 1\}) = \varnothing$. Then, for sufficiently small $\zeta > 0$ and $n > n_0(\zeta) \gg 1$ and condition $\psi_\lambda(2n_0) = 0$ for $\lambda \in [\lambda_1, \lambda_2]$, we have for $\lambda_1 < \lambda' < \lambda < \lambda_2$ that

$$\tilde{\theta}_\lambda(2n_0 + 1) - \tilde{\theta}_{\lambda'}(2n_0 + 1) \in [\zeta/2, \pi - \zeta/2].$$

The hyperbolic transformation $\widetilde{M}_\lambda(I_{2n_0+1})$ will return both phases $\tilde{\theta}_{\lambda'}$ and θ_λ (with exponentially high accuracy) to an angle of order $O(A_n^{-1/2})$ and after this forever the difference between $\tilde{\theta}_{\lambda'}$ and $\tilde{\theta}_\lambda$ will be less than $\pi - 2\zeta$. □

Remark 2.8. If Γ contains an interval $[a, b]$ then the well-known result by Gordon [4, 5] and Del Rio et al.[1] gives that $\mu(H_{\theta_0}) = \mu_{\text{pp}}(H_{\theta_0})$ only for a.e. $\theta_0 \in [0, \pi)$ and that there exists a set $\mathcal{G} \subset [0, \pi)$ of Lebesgue measure 0 and of second category, such that $\mu(H_{\theta_0}) = \mu_{\text{sc}}(H_{\theta_0})$ on $\bigcup_{n=1}^{\infty}(\{n^2\pi^2\} \oplus [a, b])$.

Remark 2.9. We have now seen in model I examples where, as a set, $\Sigma_{\text{ess}}(H)$ is discrete and unbounded from above. One might ask if it is possible to construct nonnegative potentials for which the essential spectrum is discrete and finite? The answer (see [10]) is no. In this case $\Sigma_{\text{ess}}(H)$ is either empty or unbounded from above. This fact will be one of the main differences between model I and model II below.

3. Model II. Potentials unbounded from above and below.

Let $A_n, B_n > 0$ $A_n, B_n \to \infty$ where we assume $B_n < A_n^{1-\delta_0}$ for some $\delta_0 > 0$. Define

$$V(x) = \begin{cases} -B_n & \text{if } x \in I_{2n-1} \\ A_n & \text{if } x \in I_{2n} \end{cases} \tag{3.1}$$

Then $V(x)$ is unbounded from above and below. If $A_n = \infty$ for all n, then the corresponding Hamiltonian $H_{\theta_0}\psi(x) = -\psi''(x) + V(x)\psi(x)$, $\psi(0)\cos\theta_0 - \psi'(0)\sin\theta_0 = 0$ on $L_2(0, \infty)$ has the system of non-interacting potential wells of unite length with depths $\{-B_n\}$ surrounded by infinitely high walls. The spectrum in this case is the set

$$\Lambda = \{\lambda_{m,n} = \pi^2 m^2 - B_n : m, n \geq 0\}.$$

If the barriers A_n are finite then the operator H_{θ_0} is essentially self-adjoint independently of B_n due to the following result by R. Ismagilov (see [8, 11]). Recall that a differential expression of the form $L := \mathrm{d}^2/dx^2 + V(x)$ is in the limit point case at ∞ if for every λ there exists a solution of $Lu = \lambda u$ which is not square integrable at ∞.

Theorem 3.1. *Suppose that $V \in L^1_{\text{loc}}$ and let there exist a sequence of intervals (x_n, y_n) for which $V(x) \geq -(y_n - x_n)^2$ for all $x \in (x_n, y_n)$ and $\sum(y_n - x_n)^{-2} = \infty$. Then L is in the limit point case at ∞ (without any assumptions on V outside of the intervals (x_n, y_n)).*

Therefore the spectral measure $\mu(d\lambda)$ for H_{θ_0} is unique, and the spectral properties of H_{θ_0} depend mainly on the structure of the set $\{\lambda_{m,n} : n \geq 0, m \geq 1\}$.

Theorem 3.2. *Consider the set of limit points of the set Λ, $L_\Lambda = \{\lambda : \lambda = \lim_{k\to\infty} \lambda_{m_k,n_k}\}$. Then $\Sigma_{\text{ess}}(H_{\theta_0}) = L_\Lambda$ and if L_Λ is discrete (i.e., $L_\Lambda \cap \Delta$ is finite for any $\Delta \subset R$) then $\mu(H_{\theta_0}) = \mu_{\text{pp}}(H_{\theta_0})$ for all $\theta_0 \in [0, \pi)$.*

Remark 3.3. As in model I it follows by Gordon's theorem that if L_λ contains an interval $[a,b]$ then the most we can say concerning the nature of the spectral measure $\mu(H_{\theta_0})$ is that $\mu(H_{\theta_0}) = \mu_{\text{pp}}(H_{\theta_0})$ for a.e. $\theta \in [0,\pi)$ and there exists a dense $\mathcal{G}_\delta$ for which $\mu(H_{\tilde{\theta}_0}) = \mu_{sc}(H_{\tilde{\theta}_0})$ on $[a,b]$ for all $\tilde{\theta}_0 \in \mathcal{G}_\delta$.

Corollary 3.4. *Suppose the set Λ is discrete so that $L_\Lambda = \varnothing$. Then $\Sigma(H) = \Sigma_d(H)$ on $(-\infty,\infty)$ for all $\theta_0 \in [0,\pi)$.*

While the proof of Theorem 3.2 is analogous to Theorem 2.7, there are significant differences between models I and II. For one, while the essential spectrum (if nonempty) in model I must always be unbounded from above (see [10]), whereas in model II the essential spectrum can be finite (see the example below). Secondly, the spectrum of model II can be pure discrete while in model I the spectrum must always have nonempty essential spectrum.

Example 3.5. Let $B_n = \pi^2 n^4$ and $A_n = \pi^2 n^5$. Then $\lambda_{m,n} = m^2\pi^2 - B_n = \pi^2(m-n^2)(m+n^2)$. Obviously $0 \in L_\Lambda$ since $0 = \lambda_{n^2,n}$ for all $n \geq 1$. Also, 0 is the only limit point Λ. Therefore by Theorem 3.1 we have $\Sigma_{\text{ess}}(H) = \{0\}$ and $\mu(H_{\theta_0})$ is pure point for all $\theta_0 \in [0,\pi)$.

4. Model III. Negative potentials decreasing to $-\infty$

Let $V(x) = -v_n^2$ for $x \in [n-1,n)$ where for all n, $v_n = 2\pi M_n + \delta_n$ and $M_n \in \mathbb{Z}^+$ and $\delta_n \in [-\pi/2,\pi/2]$. Because V can decrease arbitrarily fast at ∞ it is not clear that L is in the limit point case at ∞ and indeed as we shall see below there are important classes of potentials V giving rise to L in the limit point and limit circle cases at ∞.

We will first consider a class of potentials $V(x)$ for which L is in the limit point case at ∞. Let $V(x)$ be as above where $\delta_n = 0$ for all n. We will call this type of potential a "Nelson" potential after E. Nelson who presented results for potentials having a similar structure to the one above. Our first result is very much in the spirit of his known example.

Lemma 4.1. *For any sequence $M_n \in \mathbb{Z}^+$ the operator H_{θ_0} corresponding to a Nelson potential is essentially self-adjoint.*

PROOF. The proof is straight forward. We simply construct two solutions $\psi_0(x)$ and $\psi_1(x)$ of the equation $L\psi = 0$ satisfying $\psi_0(0) = 0 = \psi_1'(0)$ and $\psi_0'(0) = 1 = \psi_1(0)$. From their structure it follows that both are not from $L^2(0,\infty)$. Therefore L is in the limit point case and hence H_{θ_0} is self-adjoint. □

Again motivated by E. Nelson we define the class of potentials $V(x)$ with $\delta_n = \pi/2$ for all n. We call this class of potentials "semi-Nelson" type potentials. The next several results show that not only is there a wide class of semi-Nelson potentials giving rise to L not in the limit point case at ∞ , but that even certain "small" perturbations of such potentials do not destroy this property.

Lemma 4.2. *Let $V(x)$ be as above with $\delta_n = \pi/2$ for all n and define the sequence $\{h_n\}$ by*

$$h_0 = 1, \qquad h_n = \frac{1}{v_n h_{n-1}} \quad \textit{for } n \geq 1.$$

Then there exist linearly independent solutions $\psi_0(x)$ and $\psi_1(x)$ of the equation $L\psi = 0$ in $D(H)$ such that

$$\int_0^{2m} |\psi_0(x)|^2\,dx = \sum_{n=1}^{m} h_{2n-1}^2, \qquad \int_1^{2m+1} |\psi_1(x)|^2\,dx = \sum_{n=1}^{m} h_{2n}^2$$

for any $m \in \mathbb{Z}^+$.

Hence, if $\{h_n\} \in l^2$ then $L\psi = 0$ has two linearly independent $L^2(0,\infty)$ solutions. Thus L is in the limit circle case at ∞ so that H_{θ_0} is non-essentially self-adjoint.

As an illustrative example, let us consider the particular case when v_n have the for m $v_n = 2\pi\lfloor n^\alpha\rfloor + \pi/2$ where $\alpha > 0$ and $\lfloor x\rfloor = \max\{n \in \mathbb{Z}^+ : n \le x\}$. As usual for any real number x, we can write $x = \lfloor x\rfloor + \{x\}$ where $\{x\}$ is the fractional part.

Lemma 4.3. *Let $V(x) = -v_n^2$ on $[n-1,n)$ where $v_n = 2\pi\lfloor n^\alpha\rfloor + \pi/2$. Then, if $\alpha > 1$ the solutions $\psi_0(x)$ and $\psi_1(x)$ constructed above are from $L^2(0,\infty)$. Therefore the operator H_{θ_0} is non-essentially self-adjoint.*

PROOF. Let's note that h_{2n} can be presented as

$$h_{2n} = \frac{v_1 v_3 \cdots v_{2n-1}}{v_2 v_4 \cdots v_{2n}} = \exp\left(\sum_{k=1}^{n} \ln\left(\frac{v_{2k-1}}{v_{2k}}\right)\right)$$

where

$$\ln\left(\frac{v_{2k-1}}{v_{2k}}\right) = \ln\left(\frac{2\pi\lfloor(2k-1)^\alpha\rfloor + \pi/2}{2\pi\lfloor(2k)^\alpha\rfloor + \pi/2}\right) = \ln\left(\frac{\lfloor(2k-1)^\alpha\rfloor}{\lfloor(2k)^\alpha\rfloor}\right) + O\left(\frac{1}{\lfloor k^\alpha\rfloor}\right).$$

Now with the fact that $\lfloor n^\alpha\rfloor \sim n^\alpha$ it follows

$$\begin{aligned}\ln\left(\frac{v_{2k-1}}{v_{2k}}\right) &= \ln\left(\frac{2k-1}{2k}\right)^\alpha + O\left(\frac{1}{k^\alpha}\right)\\ &= -\frac{\alpha}{2k} + O\left(\frac{1}{k^2}\right) + O\left(\frac{1}{k^\alpha}\right) = -\frac{\alpha}{2k} + O\left(\frac{1}{k^2}\right).\end{aligned}$$

So then,

$$\exp\left(\sum_{k=1}^{n} \ln\left(\frac{v_{2k-1}}{v_{2k}}\right)\right) = \exp\left(\sum_{k=1}^{n}\left(-\frac{\alpha}{2k} + O\left(\frac{1}{k^2}\right)\right)\right)$$

and therefore

$$h_{2n}^2 \sim K_1 \exp\left(\sum_{k=1}^{n} -\frac{\alpha}{k}\right) \sim K_2 \exp\left(-\alpha \ln n\right) = \frac{K_2}{n^\alpha}.$$

This shows that $\sum_{n=0}^{\infty} h_{2n}^2$ must converge and it shows that $\psi_1(x) \in L^2(0,\infty)$. Similar procedures show $\psi_0(x) \in L^2(0,\infty)$. □

Small perturbations of the potential V preserve the property of non-essential self-adjointness as the next lemma shows.

Lemma 4.4. *Let $\{\varepsilon_n\}$ be a sequence for which $\varepsilon_n \in 2\pi Z$, and*

$$\sum_{k=0}^{\infty} \frac{|\varepsilon_n|}{a_n} < \infty.$$

Let $V(x) = -a_n^2$ *for* $x \in [n-1, n)$ *and set* $V_\varepsilon(x) = -(a_n + \varepsilon_n)^2$. *Then if the operator* $H_{\theta_0}\psi = -\psi'' + V\psi$ *is non-essentially self-adjoint, the the operator* $H_{\theta_0}^\varepsilon \psi = -\psi'' + V_\varepsilon \psi$ *is non-essentially self-adjoint.*

Remark 4.5. The two cases considered above show that the bifurcation between the essentially self-adjoint operators and the non-essentially self-adjoint operators is quite complex.

Theorem 4.6. *Suppose that* $V(x)$ *is a Nelson potential. Suppose the sequence* $\{\upsilon_n\}$ *satisfies the following:*

$$\lim_{n\to\infty} \frac{\upsilon_{n+1}}{\upsilon_n} = 1 \quad \sum_{k=1}^{\infty} \frac{1}{\upsilon_k} < \infty.$$

Then the spectrum $\sum(H_{\theta_0})$ *is discrete.*

Before proving this result we first discuss a few technical details. Consider the transfer matrix for our operators, i.e., the matrix $M_\lambda(0,x)$ for which

$$M_\lambda(0,x)\colon \begin{bmatrix} \psi_k(0) \\ \psi_k'(0) \end{bmatrix} \longrightarrow \begin{bmatrix} \psi_k(x) \\ \psi_k'(x) \end{bmatrix}$$

where $\psi_k(x)$ satisfies $-\psi_k'' + V\psi_k = k^2\psi_k$ with initial conditions $\psi_k(0) = \sin\theta_0$, $\psi_k'(0) = \cos\theta_0$. Due to the type of potentials we are considering the matrices have the nice form

$$M_k(0,n) = M_k(n,n-1)M_k(n-1,n-2)\cdots M_k(0,1) \equiv \prod_{j=1}^{n} M_k(j-1,j)$$

for $x = n \in \mathbb{Z}^+$ where

$$M_k(j-1,j) = \begin{bmatrix} \cos\sqrt{k^2+\upsilon_j^2} & (1/\sqrt{k^2+\upsilon_j^2})\sin\sqrt{k^2+\upsilon_j^2} \\ -\sqrt{k^2+\upsilon_j^2}\sin\sqrt{k^2+\upsilon_j^2} & \cos\sqrt{k^2+\upsilon_j^2} \end{bmatrix}$$

with $\vec{\psi}_k(j) = M_k(j-1,j)\vec{\psi}_k(j-1)$ and $\vec{\psi}_k(x) = (\psi_k(x), \psi_k'(x))^T$. Let

$$I_j(k) = \begin{bmatrix} \sqrt[4]{k^2+\upsilon_j^2} & 0 \\ 0 & \sqrt[-4]{k^2+\upsilon_j^2} \end{bmatrix}$$

and

$$O_j(k) = \begin{bmatrix} \cos\sqrt{k^2+\upsilon_j^2} & \sin\sqrt{k^2+\upsilon_j^2} \\ -\sin\sqrt{k^2+\upsilon_j^2} & \cos\sqrt{k^2+\upsilon_j^2} \end{bmatrix}.$$

Then $M_k(j-1,j)$ has the decomposition

$$M_k(j-1,j) = I_j^{-1}(k)O_j(k)I_j(k).$$

Let also

$$\begin{aligned} D_{j-1,j}(k) &= I_j(k)I_{j-1}^{-1}(k) \\ &= \begin{bmatrix} \sqrt[4]{(k^2+a_j^2)/(k^2+a_{j-1}^2)} & 0 \\ 0 & \sqrt[-4]{(k^2+a_{j-1}^2)/(k^2+a_j^2)} \end{bmatrix}. \end{aligned}$$

Then we can write

$$M_k(0,n) = I_{n+1}^{-1}(k) \cdot \prod_{j=1}^{n+1} D_{j,j+1}(k)O_{j+1}(k) \cdot I_1(k).$$

It is the middle matrix $\widetilde{M}_k(0,n) = \prod_{j=1}^{n+1} D_{j,j+1}(k)O_{j+1}(k)$ that contains the information concerning the spectrum of the operator H_{θ_0}. We will first show that under minimal assumptions on the sequence $\{v_n\}$ the spectrum in the Nelson case is always discrete. As is usual one needs to show that

$$\dim(E_{H^{\theta_0}}(\lambda) - E_{H_{\theta_0}}(0)) < \infty$$

for any λ where $E_{H_{\theta_0}}$ is a spectral projection of H_{θ_0}. Let $N_s(E,\theta_0)$ be the number of zeros of the solution $u(x,E)$ to the problem $(L-E)\psi = 0$ on $(0,s)$ satisfying the boundary conditions $u(0) = \sin\theta_0$, $u'(0) = \cos\theta_0$. Let also

$$n_{\theta_0}(0,\lambda) = \liminf_{s\to\infty}\big(N_s(\lambda,\theta_0) - N_s(0,\theta_0)\big).$$

Then (see [12]) it follows

$$\dim(E_{H^{\theta_0}}(\lambda) - E_{H^{\theta_0}}(0)) < n_{\theta_0}(0,\lambda) + 1.$$

At this point, we can introduce the Prüffer variables (2.18) and (2.19). Recall though that now $u(x,\lambda) = 0$ if and only if $\theta_\lambda(x) = 0 \pmod{\pi}$. Therefore

$$N_L(\lambda,\theta_0) = \left\lfloor \frac{\theta_\lambda(L)}{\pi} \right\rfloor.$$

Therefore we seek estimation of

$$\frac{\theta_\lambda(L) - \theta_0(L)}{\pi}.$$

Using the Prüffer relationship the transfer matrix relationship for $m < n$ becomes

$$\begin{bmatrix}\rho_\lambda(n)\sin\theta_\lambda(n)\\ \rho_\lambda(n)\cos\theta_\lambda(n)\end{bmatrix} = M_\lambda(m,n)\begin{bmatrix}\rho_\lambda(m)\sin\theta_\lambda(m)\\ \rho_\lambda(m)\cos\theta_\lambda(m)\end{bmatrix}$$

where for $N \geq 1$ the transfer matrix $M_\lambda(m,m+N)$ can be presented as above in the form

$$\begin{aligned} M_k(m,m+N) &= I_{m+N+1}^{-1}\cdot\prod_{j=1}^{N} D_{m+j,m+j+1}(k)O_{m+j}(k)\cdot I_{m+1}(k)\\ &= I_{m+N+1}^{-1}\widetilde{M}_k(m,m+N)I_{m+1}(k). \end{aligned}$$

This expression can be replaced by the equivalent expression:

$$\begin{bmatrix}\rho_\lambda(L)\cos\theta_\lambda(L)\\ \rho_\lambda(L)\sin\theta_\lambda(L)\end{bmatrix} = \widehat{M}_\lambda(m,L)\begin{bmatrix}\rho_\lambda(m)\cos\theta_\lambda(m)\\ \rho_\lambda(m)\sin\theta_\lambda(m)\end{bmatrix}$$

where

$$\widehat{M}_k(m,m+N) = I_{m+N+1}(k)\cdot\prod_{j=1}^{N} D_{m+j,m+j+1}(k)O_{m+j}(k)\cdot I_{m+1}(k). \tag{4.1}$$

It is this form that we prefer and this form of the transfer matrix will be used in the proof of Theorem 4.6.

PROOF OF THEOREM 4.6. Using the transfer matrix $\widehat{M}_k(m,L)$ we can track the behavior of the phases $\theta_\lambda(x)$ and $\theta_0(x)$ for the solutions $\psi_\lambda(x)$ and $\psi_0(x)$. In particular, we will show that the relative differences (i.e., modulo π) can in fact, never exceed π, i.e., $(\theta_\lambda(x) - \theta_0(x)) \bmod \pi < \pi$. Now since $v_{n+1}/v_n \to 1$ we can write

$$\frac{v_{n+1}}{v_n} = 1 + \epsilon_n$$

where $\epsilon_n \to 0$ as $n \to \infty$. Now choose m so large that

$$\sum_{k \geq m} \frac{\lambda}{2v_k} < \epsilon$$

where ϵ is arbitrarily small. Let $\theta_\lambda(m) = \delta_\lambda$, $\theta_0(m) = \delta_0$ where $\delta_\lambda, \delta_0 \in [0, \pi)$. The structure of v_n then gives $\theta_\lambda(m+1) = \delta_\lambda + \sqrt{\lambda + v_m^2}$ and $\theta_0(m+1) = \delta_0 + v_m$ so that $\theta_\lambda(m+1) - \theta_0(m+1) = \delta_\lambda - \delta_0 + \lambda/(2v_m) + O(v_m^{-3})$. Now, if this difference exceeds π then it cannot exceed by much. In particular it could exceed by at most $\lambda/(2v_m) + O(v_m^{-3})$ and in this case after considering this difference modulo π we would have $\theta_\lambda(m+1) - \theta_0(m+1) < \lambda/(2v_m)$ which is arbitrarily close to 0. So then, let us assume, without loss of generality, that at m, $\delta_\lambda = \delta_0 = \theta_0$ where $\theta_0 \in [0, \pi)$. Assume also that the amplitudes have been normalized, $\rho_\lambda(m) = \rho_0(m) = 1$. Now, multiplication of the rotation matrices $O_m^{-1}(s)$ (for $s = 0, \lambda$) from the left gives

$$\begin{bmatrix} \cos\tilde{\theta}_s \\ \sin\tilde{\theta}_s \end{bmatrix} = O_m^{-1}(s) \begin{bmatrix} \cos\theta_0 \\ \sin\theta_0 \end{bmatrix} = \begin{bmatrix} \cos(\theta_0 + \sqrt{s + v_m^2}) \\ \cos(\theta_0 + \sqrt{s + v_m^2}) \end{bmatrix}.$$

Therefore $\theta_0 \to \tilde{\theta}_{s,1} \equiv \theta_0 + \sqrt{s + v_m^2}$ and the difference becomes

$$\tilde{\theta}_{\lambda,1} - \tilde{\theta}_{0,1} = \frac{\lambda}{2v_m} + O(v_m^{-3}).$$

Multiplication from the left by the diagonal matrices $D_{m,m+1}^{-1}(s)$ then give the relation

$$\begin{bmatrix} \cos\tilde{\theta}_{s,1} \\ \sin\tilde{\theta}_{s,1} \end{bmatrix} \to \rho_s^1 \begin{bmatrix} \cos\theta_{s,1} \\ \sin\theta_{s,1} \end{bmatrix} = \begin{bmatrix} \gamma_m^{-1}(s)\cos\tilde{\theta}_s \\ \gamma_m(s)\sin\tilde{\theta}_s \end{bmatrix}$$

so that $\tan\theta_{s,1} = \gamma_m^2 \tan\tilde{\theta}_{s,1}$ where $\gamma_m(s) > 1$ is as before. Now either $\tilde{\theta}_{s,1} \in (\pi/2, \pi)$ or $\tilde{\theta}_{s,1} \in [0, \pi/2]$. If $\tilde{\theta}_{s,1} \in (\pi/2, \pi)$ then $\theta_{s,1} < \tilde{\theta}_{s,1}$. Multiplication by the next rotation $O_{m+1}^{-1}(s)$ followed by the diagonal $D_{m+1,m+2}^{-1}(s)$ produces a similar relationship between angles on the next interval $[m+1, m+2)$ in that

$$D_{m+1,m+2}^{-1}(s) O_{m+1}^{-1}(s) \colon \theta_{s,1} \to \tilde{\theta}_{s,2} \to \theta_{s,2}$$

and $\theta_{s,2} < \tilde{\theta}_{s,2} < \theta_0 + \sqrt{s + a_m^2} + \sqrt{s + a_{m+1}^2}$ and in general (following in a similar pattern), for any $k > 1$

$$\tilde{\theta}_{s,k} < \theta_0 + \sum_{j=m}^{m+k-1} \sqrt{s + a_j^2}.$$

Brief estimations on the differences $\theta_{\lambda,k} - \theta_{0,k}$ using the fact that $v_{n+1} \sim v_n$ give also that

$$\tilde{\theta}_{\lambda,k} - \tilde{\theta}_{0,k} < \sum_{j=m}^{m+k} \frac{\lambda}{2a_j}$$

for any $k > 1$. The convergence of the series on the right implies then that

$$\lim_{k \to \infty} (\tilde{\theta}_{\lambda,k} - \tilde{\theta}_{0,k}) < \epsilon(\lambda)$$

i.e., the difference is bounded. Similar ideas and estimations show the same conclusion if $\tilde{\theta}_1 \in [0, \pi/2]$. Hence the spectrum on the interval $[0, \lambda]$ is at most discrete. □

Using similar ideas one can generalize Theorem 4.6 slightly:

Theorem 4.7 (Discrete Spectrum). *Let* $V(x) = -(2\pi M_n + \delta_n)^2$ *for* $x \in [n, n+1)$ *where* $M_n \in Z^+$ *and* $\delta_n \in (-\pi/2, \pi/2)$. *If* $\sum |\delta_n| < \infty$ *then the spectrum* $\Sigma(H_{\theta_0})$ *is discrete.*

We would like to note that an additional topic of interest related to Theorem 4.7 are the quasi-classical formulas for the counting functions $N([0,\lambda])$ and $N([-\lambda,0])$ for the number of eigenvalues in the intervals $[0,\lambda]$ and $[-\lambda,0]$ respectively, as $\lambda \to \infty$ (see [7]).

Theorem 4.8 (Essential Spectrum). *Let* $\{M_n\}$ *be any sequence such that* $M_n \nearrow \infty$ *and* $M_{n+1}/M_n \to C \geq 1$ *where* $C = \infty$ *is allowed. Then for any* $|\lambda_0| > 0$ *there exists a sequence* $\{\delta_n\}$ *with* $\delta_n \in (-\pi/2, \pi/2)$ *such that for any* $\theta_0 \in [-\pi/2, \pi/2)$ *the operator* $H_{\theta_0} = H_{\theta_0}(V)$, *corresponding to the potential* V *defined by* $V(x) = a_n^2$ *and* $a_n = 2\pi M_n + \delta_n$ *for* $x \in [n, n+1)$, *has nonempty essential spectrum on the interval* $(0, \lambda_0)$, *that is,* $\Sigma_{\text{ess}}(H_{\theta_0}) \cap (0,\lambda_0) \neq \varnothing$.

PROOF. We'll show the result only for $\lambda_0 > 0$ for the case $\lambda_0 < 0$ follows similarly. We take $n_0 \in Z^+$ with $n_0 \gg \lambda_0$. Just as before in the Nelson case above we may assume that $\theta_{\lambda_0}(n_0) = 0 = \theta_0(n_0) \equiv \theta_0$ and as in (4.1) the transfer matrix has the form

$$M_\lambda(n_0, n_0+k) = \tilde{I}_{n_0+k+1} \cdot \prod_{j=1}^{k+1} \widetilde{D}_{n_0+j} \widetilde{O}_{n_0+j} \cdot \tilde{I}_{n_0+1}$$

where

$$\tilde{I}_{n_0+1}(\lambda) = \begin{bmatrix} \sqrt[-4]{\lambda + a_{n_0+1}^2} & 0 \\ 0 & \sqrt[4]{\lambda + a_{n_0+1}^2} \end{bmatrix}.$$

This matrix has the attracting eigenvector $\begin{bmatrix} 0 & 1 \end{bmatrix}^T$. Since λ_0 is fixed, the matrix $\tilde{I}_{n_0+1}(\lambda_0)$ has an eigenvalue asymptotic to $\sqrt{2\pi M_n}$ as $n \to \infty$, and it means that for given fixed θ_0, the matrix $\tilde{I}_{n_0+1}$ maps the angle $\theta_0 \in [-\pi/2, \pi/2)$ into

$$\tilde{\theta}_0(\lambda) = \tan^{-1}\left(\sqrt{\lambda + a_{n_0+1}^2} \tan \theta_0\right)$$

which is strongly attracted to $\pm\pi/2$ for any $\theta_0 \neq 0$. It is clear that with $\theta_0 = 0$ there is no deviation from this angle after the mapping $\tilde{I}_{n_0+1}$ so that $\tilde{\theta}_0(\lambda) = 0$. The rotation $O_{n_0+1}(\lambda)$ then maps $\tilde{\theta}_0(\lambda)$ to the angle

$$\tilde{\theta}_1(\lambda) = 2\pi M_{n_0+1} + \delta_{n_0+1} + \frac{\lambda}{2a_{n_0+1}} + r_{n_0+1}(\lambda)$$

where $r_{n_0+1}(\lambda) = O(a_{n_0+1}^{-2})$ for all λ in a bounded region. Let

$$\delta_{n_0+1} = -\frac{\lambda_0}{4a_{n_0+1}} - \frac{r_{n_0+1}(\lambda_0)}{2}.$$

Then

$$\tilde{\theta}_1(\lambda_0) = \frac{\lambda_0}{4a_{n_0+1}} + \frac{r_{n_0+1}(\lambda_0)}{2} \quad \text{and} \quad \tilde{\theta}_1(0) = -\frac{\lambda_0}{4a_{n_0+1}} - \frac{r_{n_0+1}(\lambda_0)}{2}$$

modulo π of course. Now the angles $\tilde{\theta}_1(0)$, $\tilde{\theta}_1(\lambda_0)$ are symmetrically located across the zero phase. The diagonal matrix $D_{n_0+1}(\lambda)$ maps $\tilde{\theta}_1(0)$, $\tilde{\theta}_1(\lambda_0)$ respectively to

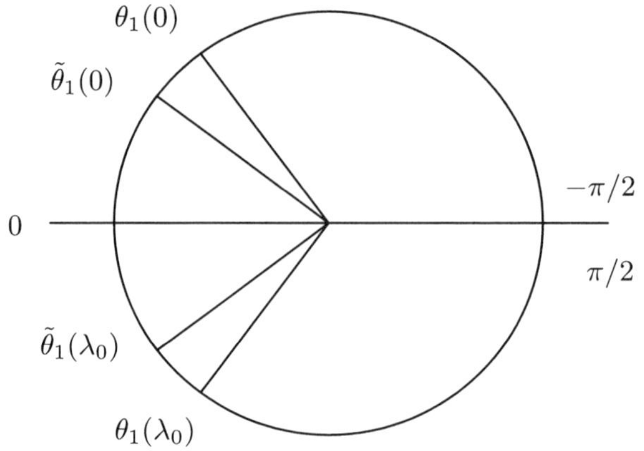

FIGURE 2. Separation of the phases $\theta_1(\lambda_0)$ and $\theta_1(0)$

the angles

$$\theta_1(\lambda_0) = \tan^{-1}\big(\gamma^2_{n_0+1}(\lambda_0)\tan\tilde\theta_1(\lambda_0)\big)$$
$$\theta_1(0) = -\tan^{-1}\big(\gamma^2_{n_0+1}(0)\tan\tilde\theta_1(0)\big)$$

where

$$\gamma_n^2(\lambda) = \sqrt{\frac{\lambda + a^2_{n+1}}{\lambda + a_n^2}}.$$

Note that the angles are wider after multiplication by the diagonal matrix but still practically symmetric across the zero phase for large n_0 since $\gamma_n^2(\lambda_0) \sim \gamma_n^2(0)$ as $n \to \infty$ (see Figure 2).

It is clear then that we can select $r_{n_0+2}(\lambda) = O(a^{-2}_{n_0+2})$ and

$$\delta_{n_0+2} = -\frac{\lambda_0}{4a_{n_0+2}} - \frac{r_{n_0+2}(\lambda_0)}{2}$$

so that the rotation matrix $O_{n_0+2}(\lambda)$ takes $\theta_1(0)$, $\theta_1(\lambda_0)$ to $\tilde\theta_2(0)$, $\tilde\theta_2(\lambda_0)$ given by

$$\tilde\theta_2(0) = -\tan^{-1}\left(\gamma^2_{n_0+1}(0)\tan\left(\frac{\lambda_0}{4a_{n_0+1}}\right)\right) - \frac{\lambda_0}{4a_{n_0+2}} - \frac{r_{n_0+2}(\lambda_0)}{2}$$
$$\tilde\theta_2(\lambda_0) = \tan^{-1}\left(\gamma^2_{n_0+1}(\lambda_0)\tan\left(\frac{\lambda_0}{4a_{n_0+1}}\right)\right) + \frac{\lambda_0}{4a_{n_0+2}} + \frac{r_{n_0+2}(\lambda_0)}{2}$$

so that these angles are symmetric about the zero phase. The divergence of the product $\prod \gamma_n^2$ and the choice of δ_k imply that we can find some L_0 large enough so that $\theta_{L_0}(0) = -\pi/2 + \epsilon/2$ and $\theta_{L_0}(\lambda_0) = \pi/2 - \epsilon/2$ where ϵ is arbitrarily small. At this moment set $\delta_{n_0+L_0} = 0$. Then, the next rotation will not change the position of $\theta_{L_0}(0)$, i.e., modulo π we have $\tilde\theta_{L_0+1}(0) = \theta_{L_0}(0)$. Also like before the diagonal matrix maps $\tilde\theta_{L_0+1}(0)$ to $\theta_{L_0+1}(0) = \tan^{-1}\big(\gamma^2_{L_0+1}(0)\tan\tilde\theta_{L_0+1}(0)\big)$ and brief calculations give

$$\theta_{L_0+1}(0) = -\frac{\pi}{2} + \frac{a_{n_0+L_0}}{a_{n_0+L_0+1}}\cdot\frac{\epsilon}{2} + O(\epsilon^2).$$

Continuing with this pattern gives for any $k > 0$

$$\theta_{L_0+k}(0) = -\frac{\pi}{2} + \frac{a_{n_0+L_0}}{a_{n_0+L_0+k}} \cdot \frac{\epsilon}{2} + O(\epsilon^2).$$

For λ_0 the procedure is the same only this time we must take into account that the rotations are not exactly multiples of 2π, but instead multiples of 2π plus some small extra turns. As we have said L_0 is chosen so that $\theta_{L_0}(\lambda_0) = \pi/2 - \epsilon/2$. Denote by $\alpha(u, v)$ the difference between angles u and v. Then $\alpha(\theta_{L_0}(\lambda_0), \pi/2) = \epsilon/2$. The counter-clockwise rotation $O_{L_0+1}(\lambda_0)$ gives

$$\alpha(\tilde{\theta}_{L_0+1}(\lambda_0), \pi/2) = \frac{\epsilon}{2} - \frac{\lambda_0}{2a_{n_0+L_0+1}} + O\Big(\frac{1}{a^2_{n_0+L_0+1}}\Big)$$

which we choose to rewrite, using the fact that a_{n_0} is large in comparison with λ_0, as

$$\alpha(\tilde{\theta}_{L_0+1}(\lambda_0), \pi/2) = \frac{\epsilon}{2} - \frac{\lambda_0}{2\sqrt{\lambda_0 + a^2_{n_0+L_0+1}}} + O\Big(\frac{1}{a^2_{n_0+L_0+1}}\Big)$$

(this will simplify the calculations that follow). Now, after applying the diagonal matrix we get (again after brief calculations and dropping the smaller order terms from here on to reduce space)

$$\alpha(\theta_{L_0+1}(\lambda_0), \pi/2) = \sqrt{\frac{\lambda_0 + a^2_{n_0+L_0+1}}{\lambda_0 + a^2_{n_0+L_0+2}}}\Big(\frac{\epsilon}{2} - \frac{\lambda_0}{2\sqrt{\lambda_0 + a^2_{n_0+L_0+1}}}\Big).$$

Now the next rotation gives

$$\begin{aligned}\alpha(\tilde{\theta}_{L_0+2}(\lambda_0), \pi/2)&\\ &= \sqrt{\frac{\lambda_0 + a^2_{n_0+L_0+1}}{\lambda_0 + a^2_{n_0+L_0+2}}}\Big(\frac{\epsilon}{2} - \frac{\lambda_0}{2\sqrt{\lambda_0 + a^2_{n_0+L_0+1}}}\Big) - \frac{\lambda_0}{2\sqrt{\lambda_0 + a^2_{n_0+L_0+2}}}.\end{aligned}$$

For brevity in notation let $K_0 = n_0 + L_0$. Then multiplication by the next diagonal and brief calculations give

$$\alpha(\theta_{L_0+2}(\lambda_0), \pi/2) = \frac{1}{\sqrt{\lambda_0 + a^2_{K_0+3}}}\Big(\sqrt{\lambda_0 + a^2_{K_0+1}} \cdot \frac{\epsilon}{2} - 3 \cdot \frac{\lambda_0}{2}\Big).$$

Repeating we have in general then, for any $j > 1$

$$\alpha(\theta_{L_0+j-1}(\lambda_0), \pi/2) = \frac{1}{\sqrt{\lambda_0 + a^2_{K_0+j}}}\Big(\sqrt{\lambda_0 + a^2_{K_0+1}} \cdot \frac{\epsilon}{2} - j \cdot \frac{\lambda_0}{2}\Big).$$

Now combining the above results with these we see that the difference between $\theta_{L_0+j}(0)$ and $\theta_{L_0+j}(\lambda_0)$ is

$$\frac{1}{\sqrt{\lambda_0 + a^2_{K_0+j}}}\Big(\sqrt{\lambda_0 + a^2_{K_0+1}} \cdot \frac{\epsilon}{2} - j \cdot \frac{\lambda_0}{2}\Big) + \frac{\sqrt{\lambda_0 + a^2_{K_0+1}}}{\sqrt{\lambda_0 + a^2_{K_0+j}}} \cdot \frac{\epsilon}{2}$$

which simplifies

$$\frac{1}{\sqrt{\lambda_0 + a^2_{K_0+j}}}\Big(\sqrt{\lambda_0 + a^2_{K_0+1}} \cdot \epsilon - j \cdot \frac{\lambda_0}{2}\Big).$$

From this expression it follows that this difference can be made negative and that this would give one rotation in the relative difference of the phase $\theta_{\lambda_0} - \theta_0$. From here we can repeat the above construction so that by induction there exist infinitely many rotations in the phase. □

References

1. R. Del Rio, N. Makarov, and B. Simon, *Operators with singular continuous spectrum.* II: *Rank one operators*, Comm. Math. Phys. **165** (1994), no. 1, 59–67.
2. I. M. Glazman, *Direct methods of qualitative spectral analysis of singular differential operators*, Israel Program for Scientific Translation, Jerusalem, 1965.
3. A. Ya. Gordon, *A deterministic potential with a purely point spectrum*, Mat. Zametki **48** (1990), no. 6, 38–46 (Russian); English transl., Math. Notes **48** (1990), no. 5-6, 1197–1203.
4. ———, *Exceptional value of the boundary phase for the Schrödinger equation on the semi-axis*, Uspekhi Mat. Nauk **47** (1992), no. 1(283), 211–212 (Russian); English transl., Russian Math. Surveys **47** (1992), 260–261.
5. ———, *Pure point spectrum under* 1*-parameter perturbations and instability of Anderson localization*, Comm. Math. Phys. **164** (1994), no. 3, 489–505.
6. A. Ya. Gordon, S. A. Molchanov, and B. Tsagani, *Spectral theory for one-dimensional Schrödinger operators with strongly fluctuating potentials*, Funktsional. Anal. i Prilozhen. **25** (1991), no. 3, 89–92 (Russian); English transl., Funct. Anal. Appl. **25** (1991), 236–238.
7. J. Holt and S. Molchanov, *On the Bohr formula for the one-dimensional Schrödinger operator with increasing potetial*, Appl. Anal. **84** (2005), no. 6, 555–569.
8. R. S. Ismagilov, *The spectrum of the Sturm–Liouville equation with oscillating potential*, Mat. Zametki **37** (1985), no. 6, 869–879 (Russian); English transl., Math. Notes **37** (1985), no. 5-6, 476–482.
9. B. M. Levitan and I. S. Sargsjan, *Introduction to spectral theory: selfadjoint ordinary differential operators*, Transl. Math. Monogr., vol. 39, Amer. Math. Soc., Providence, RI, 1975.
10. C. R. Putnam, *On the unboundedness of the essential spectrum*, Amer. J. Math. **74** (1952), no. 3, 578–586.
11. M. Reed and B. Simon, *Methods of modern mathematical physics.* II: *Fourier analysis, self-adjointness*, Academic Press, Boston, MA, 1975.
12. J. Weidmann, *Spectral theory of ordinary differential operators*, Lecture Notes in Math., vol. 1258, Springer, Berlin, 1987.

Department of Mathematics and Statistics, University of North Carolina — Charlotte, 376 Fretwell Bldg., 9201 University City Blvd., Charlotte, NC 28223-0001, USA

E-mail address, A. Gordon: `aygordon@uncc.edu`

E-mail address, J. Holt: `jholt@uncc.edu`

E-mail address, S. Molchanov: `smolchan@uncc.edu`

Centre de Recherches Mathématiques
CRM Proceedings and Lecture Notes
Volume **42**, 2007

Fermi – Dirac Generators and Tests for Randomness

Alexander Gordon, Stanislav Molchanov, and Joseph Quinn

ABSTRACT. A Fermi – Dirac experiment FDO(n, N) is an ordered random sampling of size n without replacement from the integers from 1 to N. A complete sampling FDO(N, N) yields a string (random permutation of size N) that can be used to simulate Fermi – Dirac experiments for any n less than or equal to N. We refer to such strings, regardless of their origin, as Ordered Fermi – Dirac Generators, (FDO) generators. Because of their lack of collisions, such generators cannot hope to simulate Maxwell – Boltzmann (MB) experiments (in a previous publication, the authors showed this to still be the case even if truncation is used to overcome the lack of any collisions). All pseudo-random number generators (PRNGs) are, at some level, FDO generators. They are routinely combined with some sort of projection to provide simulations of MB generators.

Another approach is possible. Instead of trying to produce a good MB PRNG directly, one can focus on producing a high quality PRNG that does a good job of simulating a complete FDO experiment, where the absence of collisions (Pauli exclusion principle) is an intrinsic property. There is a natural duality between unordered FD(n, N) experiments and Bose – Einstein (BE$(N - n, n + 1)$) experiments (consisting of distributing $N - n$ indistinguishable balls to $n + 1$ cells). We develop the asymptotics of occupation statistics for BE experiments which have many analogues to known results for MB experiments. However, some of the results are completely new with no analogue in the MB theory. We show that one of these has considerable ability to distinguish between the performances of a fairly large number of standard PRNGs. Other tests give an indication of the degree to which algorithmic chaos has destroyed local correlations inherent in dynamical systems. Finally, we several ways by which a high quality FDO generator can be transformed into an MB generator of essentially the same quality.

1. Introduction

In our previous publication, [2], we explored the fact that, at some level, all dynamical systems on a finite phase space cannot repeat values and that this can result in detectable differences between the sequences produced by a pseudo random number generator, PRNG, and those that would be produced by a true Maxwell – Boltzmann experiment. In this paper, we want to focus our attention on Fermi – Dirac sequences (no repetitions). There are three reasons for doing this:

2000 *Mathematics Subject Classification.* Primary 60A99; Secondary 60P99, 60K99.
This is the final form of the paper.

(1) As already mentioned, all PRNGs produce, at some level, Fermi – Dirac sequences. We will make this clear below.

(2) In some sense, Fermi – Dirac sequences and the random models associated with them are the most basic and simplest sources and might, therefore, be easier to validate or test. Furthermore, as we shall see, if one has a "good" Fermi – Dirac sequence, one can obtain from it a "good" Maxwell – Boltzmann sequence.

(3) Historically, much of modern interest in probability and notions of randomness stems from the attempts begun by Maxwell and Boltzmann, and later by Gibbs, to understand how macroscopic thermodynamic properties could result from microscopic dynamical laws governing the evolution of the molecules of a gas. The explanation offered for how a nonequilibrium system could evolve to one whose energy distribution is, asymptotically, that of the appropriate stationary distribution requires one to view the system on two levels: a coarse-grained level, and a trajectory level. We are interested in seeing that the time average spent by the trajectory in any given coarse grained element is equal to the "size" of that element with respect to the appropriate invariant measure. Under the assumption of metric indecomposability, this will happen for almost all trajectories and, assuming we are on one of those trajectories, we can imagine choosing a ΔT and noting the position of the system at each ΔT time interval. At the level of the dynamical system, the resulting sequence of points X_0, $X_{\Delta T}$, $X_{2\Delta T}, \ldots$, $X_{n\Delta T}, \ldots$, will never repeat itself. However, at the coarse grained level, repetitions can occur. For any finite sequence thus extracted, the coarse-grained boxes can be refined into finer grained boxes so that each $X_{i\Delta T}$ will be in only one of them. Now, for that sequence, the "positions" can be labeled by a pair of integers (i, j) where the first coordinate is the label of the coarse-grained box and the second coordinate is the label of the refined box. The resulting "trajectory," at the coarse-grained level, will then be a projection of the full nonrepeating position (i, j) onto its first coordinate. Models like this will be referred to as generalized Fermi – Dirac models. But, we emphasize, at some level these are Fermi – Dirac models although this may not be explicit to the observer.

We return to reason (1) above and recall some of the arguments from [2] that support this assertion. In the background of any algorithmical random number generator (PRNG), there is a discrete dynamical system, i.e., a one-to-one mapping $F\colon X \to X$ of a suitable finite set X into itself (preserving therefore the uniform measure on X). We can restrict our mapping to one of the ergodic components $X' \subset X$. An $X' \subset X$ is an ergodic component if it consists of a point of X and all of its iterates under the mapping F. If $x_1 \in X$, then $X' = \{x_1, x_2 = F(x_1), \ldots, x_t = F(x_{t-1})\}$, where t is the first time that $x_1 = F(x_t)$. Note, $t = |X'|$. In many cases, $X' = X = W_k$, where W_k is the set of binary words of length k. In this case, $|W_k| = 2^k$. Some typical examples can be found in [5]. The mapping F may be followed by an additional mapping $\varphi\colon X \to Y$. Then the output of the PRNG will be $y_1 = \varphi(x_1)$, $y_2 = \varphi(x_2), \ldots$, $y_t = \varphi(x_t)$, which may not be distinct. In many cases, φ is a truncation of some of the higher order bits in the full word which are often considered to be flawed. Let $N = |\{y_1, y_2, \ldots, y_t\}|$, and let $z_1, z_2, \ldots, z_N$ be a labeling of the distinct values of the y_i, $1 \le i \le t$. The X' fibers over each z_i will be of the form $\varphi^{-1}(z_i) \cap X'$ and will form a partition of the ergodic component X'. If $M_i = |\varphi^{-1}(z_i) \cap X'|$, then $t = \sum_{i=1}^{N} M_i$. A straightforward model for such a cycle might consist of distributing t balls to N boxes in some way so that the number of balls ending up in the ith box is equal to M_i. Important examples include the lagged Fibonacci generators under modular addition or multiplication (for historical reasons, the xor lagged Fibonacci generators should also be mentioned). In these cases, for a typical cycle, the fibers over the output words have varying heights. There are any number of ways that the ordered outcomes of a typical cycle for such a generator could have been realized from a genuinely random experiment. One way is that the assignment could have resulted from a run of a true Maxwell–Boltzmann experiment in which the N outcomes are associated with the same number, N, of boxes and the t balls are placed in the boxes sequentially in such a manner that each ball has an equal probability of going into any one of the boxes. A generator F with such an ergodic component could then be defined by taking $F\colon \{1,2,\ldots,N\} \times \{1,2,\ldots,t\} \to \{1,2,\ldots,N\} \times \{1,2,\ldots,t\}$ to be any one-to-one, onto extension of the map $f\colon \{(N(1),1),(N(2),2),\ldots,(N(t),t)\} \to \{(N(1),1),(N(2),2),\ldots,(N(t),t)\}$ where $f(N(t),t) = (N(1),1)$, for $1 \leq i < t$, $f\big((N(i),i)\big) = (N(i+1),i+1)$, and $N(i)$ is the label of the box that the ith ball enters. This same construction (which is, by the way, a "copy" algorithm) will work for any other way of assigning the t balls to the N boxes. When we think of a PRNG as providing uniform numbers over some range, we are considering the sequence it produces as being somehow representative or similar to that which would have been produced by a Maxwell–Boltzmann experiment. From the discussion above, it is clear that all generators can be identified with a collection of finite ordered sequences of elements from some word space. Thus, for the rest of this paper, by a simple generator, we will mean a sequence $y_1, y_2, \ldots, y_n$, $y_i \in Y$, $1 \leq i \leq n$. A *seed* for such a generator consists of an initial index and a *sample of size* k, $k \geq 1$ will consist of the next k elements of the sequence following the element indexed by the seed. A *multi-generator*, will consist of a collection of simple generators. The length of the string, n, is the *period of the simple generator.*

Another random way that the ordered outcomes of a generator could have been produced would be to arrange the possible outcomes in towers of boxes labeled (i,j), $0 \leq i \leq N-1$, $1 \leq j \leq M_i$. The experiment would then consist of placing $t = \sum_{i=1}^{N} M_i$ balls into the t boxes in such a way that every unoccupied box has an equal probability of receiving the next ball. Of course, such experiments could be run with any number, n, $n \leq t$, of balls. We refer to such an experiment as an ordered generalized Fermi–Dirac experiment with parameters n, N, $\{M_i\}$, $\mathrm{GFDO}(n,N,\{M_i\})$. If, for each i, $M_i = 1$, the experiment is an ordered Fermi–Dirac experiment, $\mathrm{FDO}(n,N)$. Note that a $\mathrm{GFDO}(t,N,\{M_i = 1\})$ is equivalent to an $\mathrm{FDO}(t,t)$ experiment. If we forget the order, the experiments are referred to as generalized Fermi–Dirac, $\mathrm{GFD}(n,N,\{M_i\})$, and Fermi–Dirac, $\mathrm{FD}(n,N)$. A generator $y_1,\ldots,y_N$ that has been produced from an $\mathrm{FDO}(N,N)$ experiment will be referred to as an $\mathrm{FDO}(N,N)$ generator or an FDO generator of order N, or, sometimes, just an FDO generator.

In this paper, we will focus on Fermi–Dirac generators, $y_1,\ldots,y_N$, for some integer $N \geq 1$ over an appropriate symbol space and both the ordered and unordered samples that can be extracted from them. There are two kinds of samples that we will be interested in when discussing these generators: (1) the contiguous kind, already mentioned above, which consists of a block $y_{i_0+1},\ldots,y_{i_0+k}$ (for a sample of size k and seed i_0); and (2) the noncontiguous kind based on considering successors of intervals of integers. To illustrate, suppose we are interested in a noncontiguous sample based on the interval of integers $[1, m]$. Define the corresponding sequence of indices $i(1) =$ that i, $1 \leq i \leq n$, such that $y_i = 1$, $i(2) =$ that index i such that $y_i = 2$, and so on. The noncontiguous sample then consists of the $y_{i(1)+1}, y_{i(2)+1}, \ldots, y_{i(m)+1}$. Note that, if the generator is the result of a PRNG, F, then $y_{i(1)+1} = F(1), \ldots, y_{i(m)+1} = F(m)$. The samples we extract from Fermi – Dirac generators are naturally ordered, however, if we forget the order we are merely extracting a subset of the integers in the label space.

Contiguous samples are the kind that are often extracted from a generator. Statistics based on them, therefore, are of principle interest in efforts to evaluate the "quality" or performance of a generator. Noncontiguous samples are involved in assessing generators based on dynamical systems. In order to destroy the correlations inherent in dynamical systems, even chaotic ones, the sampling interval ΔT must be large in comparison to the time scale on which the dynamical system evolves. As we will see, noncontiguous samples provide an indication of whether a PRNG has been built in such a way as to break the natural ("short-term") correlations inherent in dynamical systems.

In order to evaluate the performance of a Fermi – Dirac generator with respect to statistics based on contiguous samples, one needs a battery of statistics whose theoretical distributions are well understood and for which relevant probabilities can be effectively approximated. For Fermi – Dirac generators, there is a duality between statistics associated with Fermi – Dirac models and corresponding occupation statistics for Bose – Einstein models. The asymptotic theory of these occupation statistics parallel the well known analogous theory for Maxwell – Boltzmann models.

In Section 2, we review the Maxwell – Boltzmann asymptotic theory for occupation statistics. In Section 3, we develop the asymptotic theory of occupation statistics associated with Fermi – Dirac/Bose – Einstein models. The main value of these asymptotic results appears to be the development of intuition into what might represent useful relationships to impose on the parameters of the models. To obtain hard estimates on tail probabilities and sometimes on the entire distribution, when needed, one can utilize a technique developed in [2] that utilizes factorial moments and classical results due to Chebyshev. In Section 4, we discuss some testing results for a number of generators based on occupation statistics for contiguous samples. We also discuss results for noncontiguous samples for several PRNGs. Finally, in Section 5, we discuss several ways that a Fermi – Dirac generator can be used to produce (with varying efficiency) a Maxwell – Boltzmann generator.

2. Review of limit theorems for MB statistics

Following [2], we denote by $\mathrm{MBO}(n, N)$ the result of an ordered Maxwell – Boltzmann experiment in which n balls are placed sequentially in N boxes, labeled from 0 to $N - 1$, and the order is kept track of. Each ball has an equal probability of being placed in each of the boxes. Each possible outcome $i_1, \ldots, i_n$ of such an experiment has probability $(1/N^n)$. The classical Maxwell – Boltzmann experiment, $\mathrm{MB}(n, N)$, consists of ignoring the order and only keeping track of the cell occupation statistics: $x_0, \ldots, x_i, \ldots, x_{N-1}$ where $x_i = \#\{\text{balls in cell } i\}$. For these, one obtains the classical multinomial probability distribution $P_{\mathrm{MB}}(x_0 = k_0, \ldots, x_{N-1} = k_{N-1}) = n!/(k_0! \cdots k_{N-1}!)(1/N^n)$, here $k_0 + \cdots + k_{N-1} = n$. So, the elementary outcomes are the ways in which the integer n can be written as the sum of N integers $k_0, \dots, k_{N-1}$. From the point of view of serving as a source of uniformly distributed numbers, an MBO(n, N) is much more useful. If one starts anywhere in the outcome sequence $i_1, \dots, i_n$, say at the position j_0, and takes the next m, $m \le n - j_0$, results, according to assumption, one will have a theoretically valid, ordered sample of size m from the numbers ranging from 0 to $N-1$ under the uniform distribution. Of course, one cannot rule out that the entire sequence obtained by the experiment was not a wildly improbable one and might not be representative in any useful way of the "typical" MBO(n, N) experiment. However, if n is large and much larger than m, then with exception as unlikely as one might wish, one can expect the collection of all such samples to be representative of what one would get in sampling from a uniform distribution on the numbers from 0 to $N-1$. In this section, we will recall some of the results from [2], with some additional details, on the limit theorems for MB experiments. Suppose we have available an MB(n, N) experiment. Let $\pi_0 = \pi_0(n, N)$, $\pi_1 = \pi_1(n, N)$,..., $\pi_{N-1} = \pi_{N-1}(n, N)$ be the corresponding occupation numbers of these cells. These random variables satisfy a multinomial distribution: for any $x_0, x_1, \dots, x_{N-1}$; $\sum_{i=0}^{N-1} x_i = n$, $P(\pi_0 = x_0, \dots, \pi_{N-1} = x_{N-1}) = n!/(x_0! \cdots x_{N-1}!)(1/N)^n$. An important fact is that for any $\lambda > 0$ and independent Poissonian random variables $\nu_0, \nu_2, \dots, \nu_{N-1}$ with parameter λ,

$$P(\pi_0 = x_0, \dots, \pi_{N-1} = x_{N-1}) \\ = P(\nu_0 = x_0, \dots, \nu_{N-1} = x_{N-1} | x_0 + \dots + x_{N-1} = n).$$

In analytical applications, the optimal choice of λ is $\lambda = n/N$ (the density of the particles).

Let's also introduce the random variables $\gamma_0, \gamma_1, \dots$ depending on the parameters n, N:

$$\gamma_s(n, N) = \#\{\pi_i = s, i = 0, \dots, N-1\}, \quad s = 0, 1, \dots, n.$$

Of course,

$$\begin{aligned} \gamma_1 + 2\gamma_2 + \dots + n\gamma_n &= n \\ \gamma_0 + \gamma_1 + \dots + \gamma_n &= N. \end{aligned} \tag{2.1}$$

In statistical physics, formulas (2.1) are equivalent to the conservation laws for energy and momentum.

The rth-factorial moment, $\mu_r(s)$, for γ_s is

$$\mu_r(s) = r!\binom{N}{r}\frac{n!}{(s!)^r(n-sr)!}\frac{(N-r)^{n-rs}}{N^n}.$$

Based on this, it follows that

$$\left[\frac{N}{s!}\left(\frac{n}{N}\right)^s e^{-n/N}\right]^r e^{-nr/(N(N-1))-r(r-1)/(N-(r-1))-sr(sr-1)/(n-(sr-1))+r^2s/(N-1)} \\ \le \mu_r(s) \le \left[\frac{N}{s!}\left(\frac{n}{N}\right)^s e^{-n/N}\right]^r e^{-r(r-1)/N-sr(sr-1)/n+r^2s/N}.$$

The asymptotics of the $\mu_r(s)$ are determined by the $[(N/s!)(n/N)^s e^{-n/N}]^r$ factor that appears in both the lower and upper bounds for $\mu_r(s)$. There are two different situations: the low density case, $n << N$, and the high density case, $n >> N$. In the first case, $\gamma_0 = o(N)$, in the second case, $\gamma_0 = O(N)$. The limit theorem results we now formulate were used in [2], those concerning the γ_s statistics are essentially due to von Mises, see [9]. We begin with the low density case.

Proposition 2.1. *Assume that* $(N/s!)(n/N)^s \to a_s$, $n, N \to \infty$. *Then* $\mu_r(s) \approx (a_s)^r$ *and, therefore,* γ_s *is asymptotically Poisson with parameter* a_s.

In particular, if $n^2/(2N) \to a_2$, $n, N \to \infty$ then $\gamma_2(n,N)$ has asymptotically a Poisson distribution with parameter a_2, and $\gamma_s(n,N) \to 0$ in probability for any $s \geq 3$. Furthermore, if we let $M_n = \max_{i\leq N} \pi_i$. Then, asymptotically, $\mathrm{P}\{M_n = 1\} \to \mathrm{e}^{-a_2}$, $\mathrm{P}\{M_n > 1\} \to 1 - \mathrm{e}^{-a_2}$.

Joint asymptotic distribution behavior is modeled by

Proposition 2.2. *Let* $n^3/(6N^2) \to a_3$. *Then, for fixed* z *and* $x_3 \geq 0$, *we have*

$$\mathrm{P}\left\{\gamma_2 = x_2 = \frac{3a_3N}{n} + z\sqrt{\frac{3a_3N}{n}}, \gamma_3 = x_3\right\} \approx \frac{\mathrm{e}^{-z^2/2}}{\sqrt{2\pi(3a_3N/n)}}\mathrm{e}^{-a_3}\frac{(a_3)^{x_3}}{x_3!}.$$

Proposition 2.3 (Dynamical limit theorem). *Let*

$$\tau_2^1 = \min\{t : \max_{i\leq N-1}\{\pi_i(t,N)\} = 2\},$$

i.e., τ_2^1 *is the first collision time. Then for fixed* $a \geq 0$,

$$\mathrm{P}\left\{\frac{\tau_2^1}{\sqrt{N}} > a\right\} \to \mathrm{e}^{-a^2/2}.$$

Thus, the random variable $(\tau_2^1)^2/(2N)$ has the standard exponential limiting distribution. One can introduce the second, third, etc. collision times and find the joint limiting distribution, see [2].

Let now $n \gg N$ (the high density case) and more specifically,

$$\frac{N}{s!}\left(\frac{n}{N}\right)^s \mathrm{e}^{-n/N} \to a_s > 0,$$

that is

$$n = N\big[\ln N - s\ln(\ln N)\big(1 + o(1)\big)\big].$$

then

Proposition 2.4. *The number of s-occupancy cells* $\gamma_s(n,N)$ *has asymptotically a Poisson distribution with parameter* a_s.

Proposition 2.5. *Let* $T_0 = \min\{t : \gamma_0(t,N) = 0\}$. *Then* $\mathrm{P}\{T_0 \leq t\} = \mathrm{P}\{\gamma_0(t,N) = 0\} \approx \mathrm{e}^{-a_0}$, $a_0 = N\mathrm{e}^{-t/N}$, *or*

$$\mathrm{P}\left\{T_0^* = \frac{T_0 - N\ln N}{N} \leq -\ln a_0\right\} = \mathrm{e}^{-a_0}.$$

This gives, for $-\ln a_0 = x$,

$$\mathrm{P}\{T_0^* \leq x\} \to \mathrm{e}^{-\mathrm{e}^{-x}}, \quad x \in R;$$

the classical double exponential distribution.

Proposition 2.6. *If* $m(t) = \min_{i\leq N-1}\{\pi_i(t,N)\}$, *then*

$$\mathrm{P}\{m(t,N) = 0\} = \mathrm{P}\{\gamma_0(t) > 0\} \to 1 - \mathrm{e}^{-a_0}.$$

The theory for the limiting distribution of $M(t,N) = \max_{i \leq N-1}\{\pi_i(t,N)\}$ is fairly complicated and depends on the behavior of the limit parameter $t/(N \ln N)$ as $t, N \to \infty$. We will not go into details, but we summarize in broad fashion the results that can be found in [4, Chapter 2, Section 6]. We want to include them for comparison with the corresponding results for Bose–Einstein models.

Proposition 2.7. *The asymptotic distribution for $M(t,N)$, $t, N \to \infty$ depends on the behavior of the limit parameter $t/(N \ln N)$, $t, N \to \infty$ as follows:* (1) *when $t/(N \ln N) \to 0$, the limiting distribution of $M(t,N)$ is concentrated in one or two points*; (2) *when $t/(N \ln N) \to x > 0$, the limiting distribution of $M(t,N)$ is concentrated on countably many points and takes the form $\mathrm{P}\{M(t,N) \leq r+k\} \to \exp\{-\lambda\gamma^{k+1}/(1-\gamma)\}$ where γ is the root of the $\gamma + x(\ln\gamma - \gamma + 1) = 0$ in an interval $0 < \gamma < 1$*; (3) *when $t/(N \ln N) \to \infty$, $t, N \to \infty$, then $M(t,N)$, when suitably centered and scaled has, asymptotically, the classical double exponential distribution.*

3. FD(n,N)/BE$(N-n, n+1)$ duality and limit theorems

Each run of a true FD(n,N) experiment chooses a partition of the remaining (unchosen) $N-n$ cells into $n+1$ groups so that each partition is equally likely. This implies that each run of an FD(n,N) experiment is equivalent to running a BE$(N-n, n+1)$ experiment. An extensive family of tests can be based on occupation statistics for various patterns associated with FD experiments and corresponding patterns associated with their equivalent BE experiments. These are discussed in [8]. In this paper, we are going to limit ourselves to BE analogs of the statistics for MB experiments that were recalled in the previous section. Before proceeding with this agenda, however, there are a couple of technical issues that we need to discuss. We have identified an FDO-generator with a sequence $y_1, y_2, \ldots, y_N$ where $y_i \in \{0, 1, \ldots, N-1\}$ and each integer from 0 to $N-1$ occurs exactly once in the sequence. An FDO-generator could be the result of a run of an FDO(N,N) experiment and we are interested in testing/validating whether that is the case for a given FDO-generator. If a generator is the result of an FDO(N,N) experiment, then we can use the first n, $n \leq N$, terms in the sequence, $y_1, y_2, \ldots, y_n$ to simulate/represent the outcome of an FDO(n,N) experiment. Running such an experiment can be accomplished in two steps:

(1) First running an FD(n,N) experiment (equivalent to uniformly picking a subset of the N cells of size n).
(2) Then uniformly picking a permutation on the n cells to determine the order in which they are filled.

If we use an FDO-generator to simulate an FDO(n,N) experiment, $n < N$, the result $y_1, y_2, \ldots, y_n$ generates both a candidate FD(n,N) experiment and an experiment in choosing a random permutation of order n. The generator can be evaluated on the basis of how well such simulations produce representative results for both types of experiments. However, we have found that most decent FD-PRNGs do not produce obviously flawed results on such experiments. Of course, just because a statistic based on a single sample doesn't supply enough evidence to reject a null hypothesis, doesn't mean that, for instance, the empirical distribution of the statistic based on all possible samples or a large collection of "independent" samples might not indicate that there is a problem. This leads to a desire to repeat samples. If one takes the next n elements in the sequence, $y_{n+1}, y_2, \ldots, y_{2n}$, the result should produce a valid random permutation of size n, however, it cannot be claimed to be a run of an $\mathrm{FD}(n, N)$ experiment, since the elements of the subset $\{y_1, y_2, \ldots, y_n\}$ are not available. With modification, of course, it can be considered to be an $\mathrm{FD}(n, N-n)$ experiment. Thus, tests based on repeated $\mathrm{FD}(n, \cdot)$ experiments cannot be considered to be identically distributed, which leads to additional complications. We are not going to deal with these complications in this paper. Rather, we will focus here on some tests that avoid this problem but still illustrate some of the issues involved in building tests that are appropriate for FDO-generators. There is one last technical issue to mention. An FDO-generator has a beginning. If we choose a seed $i_0 < N - n$ and then extract a contiguous sample of size n, $y_{i_0+1}, \ldots, y_{i_0+n}$, we have to view this as having first chosen a sample $y_1, \ldots, y_{i_0}$ (or equivalently, $R(i_0) = \{1, \ldots, N\} - \{y_1, \ldots, y_{i_0}\}$) and then selecting a random nonrepeating string of length n from $R(i_0)$. So, if we are considering an FDPRNG to be a simulation of an FDO-experiment, we need to choose a seed to identify which FDO-experiment. Thus, a given FDPRNG can be considered to be a simulation of N different FDO-experiments.

3.1. Limit theorems for BE experiments. By definition, this experiment consists of distributing n indistinguishable particles over N cells. Let $\pi_i(n)$, $i = 0, 2, \ldots, N-1$, be the corresponding occupation numbers. As before, $\gamma_s(n) = \#\{i : 0 \le i \le N-1, \pi_i(n) = s\}$, $s = 0, 1, 2, \ldots$. Of course, $\pi_0(n) + \cdots + \pi_{N-1}(n) = n$, $\gamma_1(n) + 2\gamma_2(n) + \cdots = n$, $\gamma_0(n) + \gamma_1(n) + \cdots = N$, and there is a one-to-one correspondence between nonnegative solutions of the equation $x_0 + \cdots + x_{N-1} = n$ and the elementary events $\{\pi_0 = x_0, \ldots, \pi_{N-1} = x_{N-1}\}$. It is well known that the number of such solutions is equal to $\binom{N+n-1}{n} = \binom{N+n-1}{N-1}$.

The random variables π_i are conditionally independent and identically geometrically distributed with an arbitrary parameter p. That is $\mathrm{P}\{\pi = l\} = qp^l$, $l \ge 0$, $\mathrm{E}\,\pi = p/q = p/(1-p)$. One can optimally select p from the equation $\mathrm{E}\,\pi = n/N$. So, take p so that $p/(1-p) = n/N$, or

$$p = \frac{n}{N+n}.$$

Combinatorial calculations in the BE case are simpler than in the MB case of Section 2 above. As before, we will discuss two cases: the low density case, $n/N \ll 1$, and the high density case, $n/N \gg 1$.

The easiest way to establish the asymptotics of the γ_s statistics is to use techniques developed in [9] to derive formulas for their factorial moments of all order, recognize them asymptotically as those of a Poisson distribution and then use a result due to von Mises to imply that they are then asymptotically Poisson. The technique in [9] is based on inclusion exclusion methods that establish that the kth-factorial moment of γ_s, $\mu_k(s) = k!S(s,k)$, where $S(s,k) = \sum \mathrm{P}\{\pi_{i_1} = s, \ldots, \pi_{i_k} = s\}$ and the sum is taken over all distinct subsets of $\{0, \ldots, N-1\}$ of size k.

We now establish an expression for $\mu_k(s)$ that will be specialized to handle the low and high density cases.

Proposition 3.1. *The kth factorial moment of* γ_s, $\mu_k(s)$ *is given by*

$$\mu_k(s) = (N^2n^s)^k \times \frac{(1-1/N)^2\cdots(1-(k-1)/N)^2(1-k/N)(1-1/n)\cdots(1-(ks-1)/n)}{(N+n)^{k(s+1)}\bigl(1-1/(N+n)\bigr)\cdots\bigl(1-k(s+1)/(N+n)\bigr)}$$

PROOF.

$$\begin{aligned}
S(s,k) &= \binom{N}{k}\frac{|\mathrm{BE}(n-ks,N-k)|}{|\mathrm{BE}(n,N)|} = \binom{N}{k}\frac{\binom{n-ks+N-k-1}{N-k-1}}{\binom{N+n-1}{N-1}} \\
&= \frac{N!}{k!(N-k)!}\frac{(n+N-k(s+1)-1)!/\bigl((N-k-1)!(n-ks)!\bigr)}{(N+n-1)!/\bigl((N-1)!n!\bigr)} \\
&= \frac{1}{k!}\frac{N(N-1)\cdots\bigl(N-(k-1)\bigr)(N-1)\cdots(N-k)n(n-1)\cdots\bigl(n-(ks-1)\bigr)}{(N+n-1)\cdots\bigl(N+n-k(s+1)\bigr)} \\
&= \frac{1}{k!}\frac{N(N-1)\cdots\bigl(N-(k-1)\bigr)(N-1)\cdots(N-k)n(n-1)\cdots\bigl(n-(ks-1)\bigr)}{(N+n)^{k(s+1)}\bigl(1-1/(N+n)\bigr)\cdots\bigl(1-k(s+1)/(N+n)\bigr)} \\
&= \frac{1}{k!}(N^2n^s)^k\frac{(1-1/N)^2\cdots(1-(k-1)/N)^2(1-k/N)(1-1/n)\cdots(1-(ks-1)/n)}{(N+n)^{k(s+1)}\bigl(1-1/(N+n)\bigr)\cdots\bigl(1-k(s+1)/(N+n)\bigr)}.
\end{aligned}$$

Since $\mu_k(s) = k!S(s,k)$, the result follows. □

3.1.1. *The low density case.* This first result is a specialization of Proposition 3.1 to the case when $\frac{n}{N} \to 0, n, N \to \infty$.

Proposition 3.2. *Let* $N(n/N)^s \to a_s > 0$, $n, N \to \infty$. *Then the random variable* $\gamma_s(n)$ *is asymptotically Poisson with parameter* a_s *and*

$$\mu_k(s) = \left(N\left(\frac{n}{N}\right)^s\right)^k \times \frac{(1-1/N)^2\cdots(1-(k-1)/N)^2(1-k/N)(1-1/n)\cdots(1-(ks-1)/n)}{(1+n/N)^{k(s+1)}\bigl(1-1/(N+n)\bigr)\cdots\bigl(1-k(s+1)/(N+n)\bigr)}.$$

We can also consider BE-random variables such as $\gamma_{\geq s} = \#\{i : 1 \leq i \leq N, \pi_i(n) \geq s\}$ but, except for the $s = 2$ case, we defer that analysis to [8]. Here we deal only with obtaining explicit estimates for $\mathrm{P}\{\gamma_{\geq 2} = 0\}$ which we have found useful below and in the testing section.

Proposition 3.3. *Let* $n^2/N \to a_2 > 0$, $n, N \to \infty$. *Then* $\mathrm{P}\{\gamma_{\geq 2} = 0\} \to \mathrm{e}^{-a_2}$ *and, more explicitly*:

(1) $\mathrm{P}\{\gamma_{\geq 2} = 0\} = \frac{(1-1/N)(1-2/N)\cdots(1-(n-1)/N)}{(1+1/N)(1+2/N)\cdots(1+(n-1)/N)}$;

(2) $\mathrm{e}^{-n(n-1)/(N-A)} \leq \mathrm{P}\{\gamma_{\geq 2} = 0\} \leq \mathrm{e}^{-n(n-1)/N}$, *where* $A = \max_{n,N}\{n^2/N\}$ $(n^2/N \to a_2)$.

PROOF. Let's begin by considering the probability $\mathrm{P}\{\gamma_{\geq 2} = 0\}$. Clearly,

$$\mathrm{P}\{\gamma_{\geq 2} = 0\} = \frac{\binom{N}{n}}{\binom{N+n-1}{n}} = \frac{N(N-1)\cdots(N-n+1)}{(N+n-1)\cdots(N+1)N} = \frac{(1-1/N)(1-2/N)\cdots(1-(n-1)/N)}{(1+1/N)(1+2/N)\cdots(1+(n-1)/N)}.$$

But, $\ln\big((1-x)/(1+x)\big) = -2x - 2x^3/3 - 2x^5/5 - \cdots$, so $-2x/(1-x^2) < \ln\big((1-x)/(1+x)\big) < -2x$. It follows that for $(i/N)^2 \le n^2/N^2 \le A/N$, where $A = \max_{n,N}\{n^2/N\}$ $(n^2/N \to a_2)$,

$$\mathrm{e}^{-(2\sum_{i=1}^{n-1} i/N)/(1-A/N)} \le \mathrm{P}\{\gamma_{\ge 2} = 0\} \le \mathrm{e}^{-2\sum_{i=1}^{n-1} i/N} = \mathrm{e}^{-n(n-1)/N}$$

or

$$\mathrm{e}^{-n(n-1)/(N(1-A/N))} \le \mathrm{P}\{\gamma_{\ge 2} = 0\} \le \mathrm{e}^{-n(n-1)/N}$$

From this, it is clear that

$$\mathrm{P}\{\gamma_{\ge 2} = 0\} \to \mathrm{e}^{-a_2}$$

and

$$\mathrm{e}^{-n(n-1)/(N-A)} \le \mathrm{P}\{\gamma_{\ge 2} = 0\} \le \mathrm{e}^{-n(n-1)/N}.$$

provides effective estimation in terms of the asymptotics. □

It is not straightforward to introduce the "time" $\tau_{\ge 2}(t) = \min\{t : \gamma_{\ge 2}(t) > 0\}$, nonetheless, one can formally put

$$\mathrm{P}\{\tau_{\ge 2} < a\sqrt{N}\} = 1 - \mathrm{P}\{\gamma_{\ge 2}(a\sqrt{N}) = 0\} \to 1 - \mathrm{e}^{-a_2}.$$

Proposition 3.4. *If $M_{N,n} = \max_{i \le N-1}\{\pi_i(n)\}$, $m_{N,n} = \min_{i \le N-1}\{\pi_i(n)\}$. Then, under the condition $\frac{n^2}{N} \to a_2$,*

$$\mathrm{P}\{m_{N,n} = 0\} = 1, \qquad \mathrm{P}\{M_{N,n} = 0\} = 0,$$
$$\mathrm{P}\{M_{N,n} = 1\} \to \mathrm{e}^{-a_2}, \qquad \mathrm{P}\{M_{N,n} \ge 2\} \to 1 - \mathrm{e}^{-a_2}.$$

3.1.2. *The high density case.* The results for the high density case are completely different both from those for the low density case above and from those for the high density MB case. Assume that

$$N^2/n \to a > 0. \tag{3.1}$$

We begin with the factorial moments. This result follows from Proposition 3.1 as did Proposition 3.2 with only one minor and evident modification.

Proposition 3.5. *Suppose $N^2/n \to a > 0$, $n, N \to \infty$. Then, for all s, the random variable γ_s is asymptotically Poisson with parameter a and*

$$\mu_k(s) = \left(\frac{N^2}{n}\right)^k \times \frac{(1-1/N)^2 \cdots (1-(k-1)/N)^2(1-k/N)(1-1/n)\cdots(1-(ks-1)/n)}{(1+N/n)^{k(s+1)}\big(1-1/(N+n)\big)\cdots\big(1-k(s+1)/(N+n)\big)}.$$

Actually, a stronger result is available:

Proposition 3.6. *Under condition (3.1) and fixed $s \ge 0$, the random variables $\gamma_0(n), \gamma_1(n), \ldots, \gamma_s(n)$ are asymptotically independent and Poissonian with the same parameter $\lambda = a$.*

We will only offer a sketch of the proof of this result, which would be quite tedious to present fully. We begin by proving this result for $\gamma_0(n)$ and $\gamma_1(n)$. For

$N > m_0 + m_1$, $n > N$, $\widetilde{N} = N - (m_0 + m_1)$, $\tilde{n} = n - 2\widetilde{N} - m_1$ we have

$$\begin{aligned}
&\mathrm{P}\{\gamma_0(n) = m_0, \gamma_1(n) = m_1\} \\
&= \frac{N!/(m_0!m_1!(N - m_0 - m_1)!)\binom{\widetilde{N}+\tilde{n}-1}{\tilde{n}}}{\binom{N+n-1}{n}} \\
&= \frac{N!/(m_0!m_1!(N - m_0 - m_1)!)\binom{n-N+m_0-1}{N-m_0-m_1-1}}{\binom{N+n-1}{n}} \\
&= \frac{N!}{m_0!m_1!(N - m_0 - m_1)!} \frac{(N-1)!}{(N - 1 - m_0 - m_1)!} \\
&\qquad \times \frac{(n - N + m_0 - 1)!}{(n - 2N + 2m_0 + m_1)!} \frac{n!}{(n + N - 1)!} \\
&= \left[\frac{(N^2)^{m_0+m_1}}{m_0!m_1!}\left(1 - \frac{1}{N}\right)^2 \cdots \left(1 - \frac{m_0 + m_1 - 1}{N}\right)^2 \left(1 - \frac{m_0 + m_1}{N}\right)\right] \\
&\qquad \times \frac{n^{N-(m_0+m_1)-1}}{n^{N-1}} \frac{(1 - (N - m_0)/n)^{N-(m_0+m_1)-1}}{(1 + 1/n)\cdots(1 + (N-1)/n)} \\
&\qquad \times \left(1 - \frac{1}{n - N + m_0}\right)\cdots\left(1 - \frac{N - m_0 - m_1 - 1}{n - N + m_0}\right) \\
&= \left[\frac{1}{m_0!m_1!}\left(\frac{N^2}{n}\right)^{m_0+m_1} \mathrm{e}^{o(1/n)}\right]\mathrm{e}^{-2N^2/n+o(1/n)} \\
&\sim \left[\frac{1}{m_0!}a^{m_0}\mathrm{e}^{-a}\right]\left[\frac{1}{m_1!}a^{m_1}\mathrm{e}^{-a}\right].
\end{aligned}$$

To deal with the general situation involving $\{\gamma_p = m_p, \gamma_s = m_s\}$, $p < s$, $s \geq 1$, one writes the $\mathrm{P}\{\gamma_p = m_p, \gamma_s = m_s\}$ as a (possibly multiple) sum of probabilities of the form

$$\mathrm{P}\{\gamma_0 = m_0, \ldots, \gamma_s = m_s\} = \frac{N!}{m_0!\cdots m_s!(N - m_0 - \cdots - m_s)!} \frac{\binom{\widetilde{N}+\tilde{n}-1}{\widetilde{N}-1}}{\binom{N+n-1}{N-1}}$$

with $\widetilde{N} = N - m_0 - \cdots - m_s$, $\tilde{n} = n - \widetilde{N}(s+1) - (m_1 + 2m_2 + \cdots + sm_s)$. Calculations similar to those above allow one to separate the summands which, except for the appropriate factor, can be seen to iteratively sum (asymptotically) to one.

There is no similar result available for MB experiments.

We introduce a "fictional" collision time by the formula

$$T_0 = \min\Big\{t : \min_{i \leq N-1} \pi_i(t) \geq 1\Big\} = \min\{t : \gamma_0(t) = 0\}.$$

Then

$$\mathrm{P}\{\frac{T_0}{N^2} \leq a\} = \mathrm{P}\{\gamma_0(aN^2) = 0\} = \mathrm{e}^{-N^2/(aN^2)} = \mathrm{e}^{-1/a}.$$

3.2. Limit theorems for extremes. Put $m_{N,n} = \min_{0 \leq i \leq N-1}\{\pi_i(n)\}$ and $M_{N,n} = \max_{0 \leq i \leq N-1}\{\pi_i(n)\}$. We want to prove limit theorems for $m_{N,n}$ and $M_{N,n}$ under different assumptions on the relationships between the number of particles n and the number of boxes N. We'll concentrate on the BE situation. Let $X_1, \ldots, X_N$ be i.i.d. random variables with a geometric distribution $\mathrm{P}\{X_i = k\} = \mathrm{P}\{X_1 = k\} =$

qp^k, $k \geq 0$, $p = n/(N+n)$, $q = N/(N+n)$. As is well known,

$$\mathrm{P}\{\pi_0 = x_0, \dots, \pi_{N-1} = x_{N-1}\} \\ = \mathrm{P}\{X_0 = x_0, \dots, X_{N-1} = x_{N-1} \mid X_0 + \cdots + X_{N-1} = n\}.$$

The following formula reduces calculations for the distributions of $m_{N,n}$ and $M_{N,n}$ to the case of extremes for the r.v.s X_i, $i = 0, 1, \dots, N-1$ and the local CLT for sums of truncated i.i.d. r.v.s:

Lemma 3.7.

$$\text{(1)} \qquad \mathrm{P}\{M_{N,n} < a\} = \mathrm{P}\{\pi_i < a, i = 0, 1, \dots, N-1\} \\ = \frac{\mathrm{P}\{X_0 < a\}^N \, \mathrm{P}\{\widetilde{X}_0 + \cdots + \widetilde{X}_{N-1} = n\}}{\mathrm{P}\{X_0 + \cdots + X_{N-1} = n\}},$$

where the $\widetilde{X}_i$ *are i.i.d. r.v.s with the truncated law* $\mathrm{P}\{\widetilde{X}_0 = s\} = qp^s/(\sum_{s=0}^{a-1} qp^s) = qp^s/(1-p^a)$, $s = 0, 1, \dots, a-1$.

$$\text{(2)} \qquad \mathrm{P}\{m_{N,n} \geq a\} = \mathrm{P}\{\pi_i \geq a, i = 0, 1, \dots, N-1\} \\ = \frac{\mathrm{P}\{X_0 \geq a\}^N \, \mathrm{P}\Big\{\widetilde{\widetilde{X}}_0 + \cdots + \widetilde{\widetilde{X}}_{N-1} = n\Big\}}{\mathrm{P}\{X_0 + \cdots + X_{N-1} = n\}},$$

where the $\widetilde{\widetilde{X}}_i$ *are i.i.d. r.v.s with values greater than or equal to* a *and* $\mathrm{P}\Big\{\widetilde{\widetilde{X}}_0 = s\Big\} = qp^s/(\sum_{i=a}^{\infty} qp^i) = qp^{s-a}$ *(i.e.,* $\widetilde{\widetilde{X}} = X + a$*).*

We will prove that, in all natural situations, the ratio of the probabilities of sums to truncated sums being equal to n tends to 1 and, therefore, the limit theorems for extremes of the i.i.dř.v.s X_i, $i = 0, \dots, N-1$, are the same as for the (dependent) occupation numbers π_i, $0 \leq i \leq N-1$.

For given n, N, let's consider, first, the limit theorems for the extremes of N independent geometric random variables with parameter $p = p(n, N) = n/(N+n)$. For this situation, the random variables of interest are just the maximum and minimum values, respectively, obtained on the N trials and we are no longer conditioning on the sum being equal to n. We will denote these maximum and minimum values by $M'_{N,n}$ and $m'_{N,n}$, respectively. This topic is a classical one (cf. [3]). However, the results for integer-valued random variables have some new features.

For $M'_{N,n}$, there are three situations with different answers: $n = o(N)$, $n \sim \alpha N$, and $n \gg N$, i.e., $N = o(n)$. We begin with the case $n = o(N)$. In this case with $N, n \to \infty$, $p = n/(N+n) \to 0$, we have on the basis of standard calculations that

$$\mathrm{e}^{-p^k/(1-p^k)N} \leq (1-p^k)^N \leq \mathrm{e}^{-p^k N}$$

and hence the asymptotics of $(1-p^k)^N$ will be the same as those of $\mathrm{e}^{-p^k N}$. Furthermore, for each integer $k \geq 1$, $\mathrm{P}\{M'_{N,n} < k\} = (1-p^k)^N \sim \mathrm{e}^{-p^k N}$ and there will be a largest value for k, $K = K(n, N)$ such that $p^K N = \big(n/(N+n)\big)^K N \geq 1$. To find K, we consider the inequality $-K \log\big((N+n)/n\big) + \log(N) \geq 0$, which leads to

$$K = \left[\frac{\log N}{\log N + \log(1 + n/N) - \log n}\right] = \left[\frac{1}{1 + (\log(1 + n/N)/\log N) - \log n/\log N}\right].$$

The asymptotics of $\log n/\log N$ will essentially determine how $M'_{N,n}$ will distribute itself as $n, N \to \infty$.

Proposition 3.8. *If $n = o(N)$, $N, n \to \infty$, then one can find integers $K = K(n, N) = \left[\ln N/\ln\left((N+n)/n\right)\right]$ such that*

$$\mathrm{P}\{M'_{N,n} > K + 1\} \to 0 \quad \textit{and} \quad \mathrm{P}\{M'_{N,n} < K - 1\} \to 0.$$

Furthermore, each of the following possibilities can occur:

- $\mathrm{P}\{M'_{N,n} = K\} \to \alpha$ *and* $\mathrm{P}\{M'_{N,n} = K - 1\} \to 1 - \alpha$, $0 < \alpha < 1$;
- $\mathrm{P}\{M'_{N,n} = K\} \to \alpha$ *and* $\mathrm{P}\{M'_{N,n} = K + 1\} \to 1 - \alpha$, $0 < \alpha < 1$;
- $\mathrm{P}\{M'_{N,n} = K\} \to 1$.

Proof. As pointed out above, $\mathrm{P}\{M'_{N,n} < K\} \sim \mathrm{e}^{-p^K N} = \mathrm{e}^{-p^{-\beta}}$ if $K = \left[\log N/\left(\log((N+n)/n)\right)\right]$ and $\beta = \log N/\log\left((N+n)/n\right) - K$. Similarly, $\mathrm{P}\{M'_{N,n} < K+1\} \sim \mathrm{e}^{-p^{1-\beta}}$, $\mathrm{P}\{M'_{N,n} < K+2\} \sim \mathrm{e}^{-p^{2-\beta}}$, and $\mathrm{P}\{M'_{N,n} < K - 1\} \sim \mathrm{e}^{-p^{-(1+\beta)}}$. Now, the only conditions on β are $0 \le \beta < 1$, but for a given n, N it can be anywhere in this range. For the purpose of discussion, we will assume that $n, N \to \infty$ in such a way that β approaches some fixed number in the interval $[0, 1]$. Since $p \to 0$ as $n, N \to \infty$, $p^{-(1+\beta)} \to \infty$ and $p^{2-\beta} \to 0$. It follows, therefore, that $\mathrm{P}\{M'_{N,n} < K - 1\} \to 0$ and $\mathrm{P}\{M'_{N,n} < K + 2\} \to 1$. Thus, asymptotically, the only values that can carry mass are $K - 1$, K, $K + 1$. If $p^{-\beta} \to \infty$, which is usually going to be the case, then $\mathrm{P}\{M'_{N,n} < K\} \to 0$ and, if $p^{(1-\beta)} \to 0$ then $\mathrm{P}\{M'_{N,n} < K+1\} \to 1$, so $\mathrm{P}\{M'_{N,n} = K\} \to 1$. On the other hand, β may approach 1 in such a way that $p^{(1-\beta)} \to c \in (0, 1]$ and then $\mathrm{P}\{M'_{N,n} = K\} \to \mathrm{e}^{-c}$ and $\mathrm{P}\{M'_{N,n} = K+1\} \to 1 - \mathrm{e}^{-c}$ (this merely requires that $\beta \sim 1 - \log(1/c)/\log(N/n)$ since then $p^{1-\beta} \sim \mathrm{e}^{-(1-\beta)\log((N+n)/n)} \sim \mathrm{e}^{-\log(1/c)} = c$). Finally, β may approach 0 in such a way that $p^{-\beta} \to c \in [1, \infty)$, in which case, $\mathrm{P}\{M'_{N,n} < K\} \to \mathrm{e}^{-c}$ and so $\mathrm{P}\{M'_{N,n} = K - 1\} \to \mathrm{e}^{-c}$ and $\mathrm{P}\{M'_{N,n} = K\} \to 1 - \mathrm{e}^{-c}$ (this requires that $\beta \sim \log c/\log(N/n)$). □

Let P_1, P_2 be two probability measures defined on the nonnegative integers. We will say that P_1 is ϵ-equivalent to the P_2 on the set $L \subset \{0, 1, 2, \dots\}$ if for each $j \in L$, $1 - \epsilon \le \mathrm{P}_1(j)/\mathrm{P}_2(j) \le 1 + \epsilon$.

Proposition 3.9. *Suppose $n \sim aN$, a a fixed positive real number. Let $J = J(N, n) = [\log N/(\log(a+1) - \log a)]$, $\beta = \beta(N, n) = \log N/(\log(a+1) - \log a) - J$, $d = d(N, n) = p^{-\beta}$, and $c_0 = -\log(p_0)$, $p_0 = a/(a+1)$. Then for any integer z, $\mathrm{P}\{M'_{N,n} < z + J\} \sim \mathrm{e}^{-de^{-c_0 z}}$.*

Proof. Since $n \sim aN$, $p = n/(N+n) \sim aN/(N+aN) = a/(a+1) < 1$. Given $\epsilon > 0$, there is an integer J_1 such that $\mathrm{e}^{-p^{2J_1}/(1-p^{J_1})} > 1 - \epsilon$. It will follow from this that, for $z \ge J_1$, $1 - \epsilon \le \mathrm{e}^{-p^z N/(1-p^z)}/ee^{-p^z N} \le \mathrm{P}\{M'_{N,n} < z\}/\mathrm{e}^{-p^z N} \le 1$. By choosing an even bigger J_1, we may replace p by p_0 and still have the inequalities

$1-\epsilon \le \mathrm{P}\{M'_{N,n} < z\}/\mathrm{e}^{-p_0^z N} \le 1+\epsilon$. Since $N \to \infty$, we can choose an N large enough so that $\log N/(\log(a+1) - \log a) \gg J_1$ and $[\log N/(\log(N+n) - \log n)] = [\log N/(\log(a+1) - \log a)]$. For any such N, let $J_0 = [\log N/(\log(a+1) - \log a)]$ and let $\beta = \log N/(\log(a+1) - \log a) - [\log N/(\log(a+1) - \log a)]$. We have that, for any integer z such that $z + J \ge J_1$, $\mathrm{P}\{M'_{N,n} < z + J\} \sim \mathrm{e}^{-p_0^{z+J} N} = \mathrm{e}^{-p_0^{J+\beta} N p_0^{z-\beta}} = \mathrm{e}^{-p_0^{-\beta} e^{-c_0 z}} = \mathrm{e}^{-de^{-c_0 z}}$, where $d = p^{-\beta}$, $c_0 = -\log(p_0)$. The assertion follows. □

Remark 3.10. The inequality $\mathrm{P}\{M'_{N,n} < z + J\} \le \mathrm{e}^{-p_0^{z+J} N}$ holds for all z such that $z + J \ge 0$. Thus, the asymptotic distribution can be used to give upper bounds on the initial segment $[0, J_1]$ but not effective lower bounds there.

Proposition 3.11. *For any real z and $a_N = N^2/n \to a > 0$,*

$$\mathrm{P}\left\{ M'_{N,n} < \frac{N \ln N}{a_N} + zN \right\} \xrightarrow[N\to\infty]{} \exp(-\mathrm{e}^{-az})$$

PROOF. If h is an integer, $h > N \log N/2a_N$,

$$\mathrm{P}\{M'_{N,n} < h\} = (1 - p^h)^N = \left(1 - \left(\frac{1}{1 + a_N/N}\right)^h\right)^N$$
$$\sim \left(1 - e^{-(a_N/N)h}\right)^N \sim \mathrm{e}^{-N\mathrm{e}^{-(a_N/N)h}}.$$

If now $h = (N \ln N)/a_N + zN$, then $\mathrm{P}\{M'_{N,n} < (N \ln N)/a_N + zN\} \sim \mathrm{e}^{-N\mathrm{e}^{-\log N - z}} = \mathrm{e}^{-\mathrm{e}^{-z}}$. The result follows from these observations. □

The asymptotics for the minimum are essentially simpler and follow from the observation that $\mathrm{P}\{m'_{N,n} \ge a\} = p^{aN} = (n/(n+N))^{aN} = (1/(1+N/n))^{aN} \sim \mathrm{e}^{-aN^2/n}$, when $N/n \to 0$, a any nonnegative integer (note $(1/(1+N/n))^{aN} < e^{-aN^2/n}$ even when $N/n \to r \le 1$). Of course, if $n < N$, then $\mathrm{P}\{m'_{N,n} = 0\} = 1$. We state the following proposition without formal proof.

Proposition 3.12. *Let $n \ge N \to \infty$, then*
- $\mathrm{P}\{m'_{N,n} = 0\} = 1 + o(1)$ *if* $n = o(N^2)$;
- $\mathrm{P}\{m'_{N,n} \ge a\} \sim (\mathrm{e}^{-1/c_0})^a$ *if* $n/N^2 \to c_0$;
- $\mathrm{P}\{m'_{N,n}/A_N \ge a\} \to \mathrm{e}^{-a}$ *if* $n = N^2 A_N$, $A_N \to \infty$.

We now want to establish the corresponding results for random variables $M_{N,n}$ and $m_{N,n}$. As indicated above, this depends on the asymptotics of the ratios

$$\frac{\mathrm{P}\{\widetilde{X}_0 + \cdots + \widetilde{X}_{N-1} = n\}}{\mathrm{P}\{X_0 + \cdots + X_{N-1} = n\}} \quad \text{and} \quad \frac{\mathrm{P}\{\widetilde{\widetilde{X}}_0 + \cdots + \widetilde{\widetilde{X}}_{N-1} = n\}}{\mathrm{P}\{X_0 + \cdots + X_{N-1} = n\}},$$

respectively. We deal first with the ratio

$$\frac{\mathrm{P}\{\widetilde{X}_0 + \cdots + \widetilde{X}_{N-1} = n\}}{\mathrm{P}\{X_0 + \cdots + X_{N-1} = n\}}.$$

Our results require the application of the local CLT for nonnegative integer valued random variables the application of which requires the calculation of the means and standard deviations of the variables X_0 and $\widetilde{X}_0$. The generating functions for these random variables are $\varphi(z) = \sum_{n=0}^{\infty} qp^n z^n = q/(1-pz)$ and

$$\varphi_k(z) = \frac{\sum_{i=0}^{k} qp^i z^i}{\sum_{i=0}^{k} qp^i} = \frac{q(1-p^{k+1}z^{k+1})/(1-pz)}{q(1-p^{k+1})/(1-p)} = q\frac{(1-p^{k+1}z^{k+1})}{(1-pz)(1-p^{k+1})}$$

for X_0 and $\widetilde{X}_0$, respectively. From these, the first and second factorial moments can be calculated and used to determine the means and variances of these distributions. For X_0 these are

$$\mathrm{E}(X_0) = \frac{p}{q}, \quad \mathrm{Var}(X_0) = \frac{p}{q^2}.$$

For $\widetilde{X}_0$, they are

$$\begin{aligned}\mathrm{E}(\widetilde{X}_0) &= \frac{p}{q} - \frac{p^{k+1}(k+1)}{1-p^{k+1}},\\ \mathrm{Var}(\widetilde{X}_0) &= \frac{p}{q^2} - \frac{p^{k+1}(k+1)}{1-p^{k+1}}\left(1 - \frac{p^{k+1}(k+1)}{1-p^{k+1}}\right).\end{aligned}$$

Propositions 3.8 and 3.11 will go through with essentially no modifications, but the situation for Proposition 3.9 is not so straightforward.

Proposition 3.13. *Let $n = o(N)$, $N, n \to \infty$, then one can find integers $K = K(n, N) = \lfloor \ln N / \ln((N+n)/n) \rfloor$ such that*

$$\mathrm{P}\{M_{N,n} > K+1\} \to 0 \quad \textit{and} \quad \mathrm{P}\{M_{N,n} < K-1\} \to 0.$$

Furthermore, each of the following possibilities can occur:

- $\mathrm{P}\{M_{N,n} = K\} \to \alpha$ *and* $\mathrm{P}\{M_{N,n} = K-1\} \to 1-\alpha$, $0 < \alpha < 1$;
- $\mathrm{P}\{M_{N,n} = K\} \to \alpha$ *and* $\mathrm{P}\{M_{N,n} = K+1\} \to 1-\alpha$, $0 < \alpha < 1$;
- $\mathrm{P}\{M_{N,n} = K\} \to 1$.

PROOF. Utilizing the formulas above for the variances and expectations of X_0 and $\widetilde{X}_0$, and keeping in mind that $n/N \to 0$ as $n, N \to \infty$, the result will follow if we can show that

$$\frac{\mathrm{P}\{\widetilde{X}_0 + \cdots + \widetilde{X}_{N-1} = n\}}{\mathrm{P}\{X_0 + \cdots + X_{N-1} = n\}} \to 1.$$

From the local CLT, see [3] or [7], this will follow if we can show that

$$\begin{gathered}\mathrm{Var}(\widetilde{X}_0)/\mathrm{Var}(X_0) \to 1,\\ \left(n - N\,\mathrm{E}(X_0)\right)^2/(\mathrm{Var}(X_0)N) \to 0,\\ (n - N\,\mathrm{E}(\widetilde{X}_0))^2/(\mathrm{Var}(\widetilde{X}_0)N) \to 0.\end{gathered}$$

To see that $\mathrm{Var}(\widetilde{X}_0)/\mathrm{Var}(X_0) \to 1$, we note that

$$\begin{aligned}\frac{\mathrm{Var}(\widetilde{X}_0)}{\mathrm{Var}(X_0)} &= \frac{p/q^2 - p^{k+1}/(1-p^{k+1})\left(1 - p^{k+1}(k+1)/(1-p^{k+1})\right)}{p/q^2}\\ &= 1 - \frac{q}{p^2}\frac{p^{k+1}}{1-p^{k+1}}\left(1 - \frac{p^{k+1}(k+1)}{1-p^{k+1}}\right)\\ &= 1 - q\frac{p^{k-1}}{1-p^{k+1}}\left(1 - \frac{p^{k+1}(k+1)}{1-p^{k+1}}\right) \to 1\end{aligned}$$

since $p \to 0$ and $k-1 \geq 1$. Since $\left(n - N\,\mathrm{E}(X_0)\right)^2/(\mathrm{Var}(X_0)N) = 0$, it remains only to show that $\left(n - N\,\mathrm{E}(\widetilde{X}_0)\right)^2/(\mathrm{Var}(\widetilde{X}_0)N) \to 0$. But,

$$\begin{aligned}
&\frac{\left(n - N\,\mathrm{E}(\widetilde{X}_0)\right)^2}{\mathrm{Var}(\widetilde{X}_0)N} \\
&\quad = \frac{(Np)^2 p^{2k}(k+1)^2/(1-p^{k+1})^2}{Np\left(1/q^2 - p^k(k+1)/(1-p^{k+1})\left(1 - p^{k+1}(k+1)/(1-p^{k+1})\right)\right)} \\
&\quad = \frac{Np^{2k+1}(k+1)^2 q^2}{(1-p^{k+1})^2\left(1 - q^2 p^k(k+1)/(1-p^{k+1})(1 - p^{k+1}(k+1)/(1-p^{k+1}))\right)}.
\end{aligned}$$

Now, we are only concerned with those $k \geq 1$ that are members of the set $\{K = [\ln N/\ln((N+n)/n)], K+1, K+2\}$ and for such k, the final term above is asymptotic to $p^{K+(1-\alpha)}p^{\beta} \to 0$, $\beta \in \{0,2,4\}$ since $Np^{K+\alpha} = 1$, $\alpha = \ln N/\ln((N+n)/n) - K$. The result is established. □

Proposition 3.14. *Suppose $n \sim aN$, a a fixed positive real number. Let $J = J(N,n) = [\log N/(\log(a+1) - \log a)]$, $\beta = \beta(N,n) = \log N/(\log(a+1) - \log a) - J$, $d = d(N,n) = p^{-\beta}$, and $c_0 = -\log(p_0)$, $p_0 = a/(a+1)$. Then for any integer z, $\mathrm{P}\{M_{N,n} < z + J\} \sim \mathrm{e}^{-de^{-c_0 z}}$.*

Proof. Referring to the proof of Proposition 3.9, assume that $J/2 > J_1$ and suppose that $k \geq \gamma J$, $\frac{1}{2} < \gamma < 1$. Then $p^{k+1} < p^{\gamma J} \to 0$ and $Np^{2k+1} < Np^{2\gamma J+1} \leq p^{(2\gamma-1)J} \to 0$ (recall that $p^{J+1} < 1$). These observations all us to modify the two crucial points in the proof of Proposition 3.13 to complete the proof of this result. □

Proposition 3.15. *For any real z and $a_N = N^2/n \to a > 0$,*

$$\mathrm{P}\{M_{N,n} < \frac{N\ln N}{a_N} + zN\} \xrightarrow[N\to\infty]{} \exp(-e^{-az})$$

Proof. If we take $k > \gamma(N\log N)/a_N$, $\frac{1}{2} < \gamma < 1$, then for any such k,

$$\begin{aligned}
p^{k+1} < p^{\gamma(N\log N)/a_N} &= \left(\frac{1}{1 + a_N/N}\right)^{\gamma(N\log N)/a_N} \\
&\sim \mathrm{e}^{-(a_N/N)\gamma((N\log N)/a_N)} = N^{-\gamma} \to 0.
\end{aligned}$$

Furthermore, we have

$$Np^{2k+1} < N\left(\frac{1}{1 + a_N/N}\right)^{2\gamma(N\log N)/a_N} \sim N\mathrm{e}^{-(a_N/N)2\gamma(N\log N/a_N)} = N^{1-2\gamma} \to 0.$$

These observations allow us to modify the two crucial points in the proof of Proposition 3.13 to complete the proof of this result. □

Remark 3.16. Both Proposition 3.13 and Proposition 3.14 involve parameters that depend on N, n. Depending on how $n, N \to \infty$, the support of the asymptotic distribution in Proposition 3.13 can vary over the entire set of positive integers or it could oscillate between two pairs $\{K-1, K\}$ and $\{K, K+1\}$. Actually, almost any pattern of oscillation on sequences of pairs could be achieved. Because the remainder $\beta(N,n)$ in Proposition 3.14 can range over the interval $[0,1)$, the parameter $d = p_0^{-\beta}$ can also vary over the interval $(1/p_0, 1]$. Again, virtually any pattern could be realized.

The result for the random variables $m_{N,n}$ is straightforward.

Proposition 3.17. *Let $n \geq N \to \infty$, then*

- $\mathrm{P}\{m_{N,n} = 0\} = 1 + o(1)$ *if* $n = o(N^2)$;
- $\mathrm{P}\{m_{N,n} \geq a\} \sim (\mathrm{e}^{-1/c_0})^a$ *if* $n/N^2 \to c_0$;
- $\mathrm{P}\{m_{N,n}/A_N \geq a\} \to \mathrm{e}^{-a}$ *if* $n = N^2 A_N$, $A_N \to \infty$.

Proof. For each $a \geq 0$, let $\tilde{\tilde{X}}^a = a + X$. Then $\mathrm{E}(\tilde{\tilde{X}}^a) = a + p/q$ and $\mathrm{Var}(\tilde{\tilde{X}}^a) = p/q^2$. Since the $\mathrm{Var}(X_0) = \mathrm{Var}(\tilde{\tilde{X}}^a_0)$, we trivially have that $\mathrm{Var}(\tilde{\tilde{X}}^a_0)/\mathrm{Var}(X_0) \to 1$. Furthermore, we still have that $\big(n - NE(X_0)\big)^2/(Var(X_0)N) = 0$. However, $\big(n - N\,\mathrm{E}(\tilde{\tilde{X}}^a_0)\big)^2/(\mathrm{Var}(X_0)N) = a/\big(N(1 + N/n)\big)(N^2/n)^2$. If $n = o(N^2)$, this does not have to converge to 0. Take $n = N^{3/2}$ for example. Then, for $a = 1$, $a(1/N)(N^2/n)^2\big(1/(1 + N/n)\big) \to 1$ and

$$\frac{\mathrm{P}\{\tilde{\tilde{X}}^a_0 + \cdots + \tilde{\tilde{X}}^a_{N-1} = n\}}{\mathrm{P}\{X_0 + \cdots + X_{N-1} = n\}} \to \frac{1}{\mathrm{e}}.$$

This doesn't matter, however, because $\mathrm{P}\{m'_{N,n} > 0\} \to 0$, so the first part of the proposition still holds. In the other two cases it is easy to check that

$$a\left(\frac{1}{N}\right)\left(\frac{N^2}{n}\right)^2\left(\frac{1}{1 + N/n}\right) \to 0,$$

just what is needed to establish the other two results. The proposition is established. □

For testing purposes, it will be convenient to have the following specialized precise result:

Proposition 3.18.

$$\mathrm{P}\{m_{N,n} > k\} = \frac{(1 - (kN + 1)/n) \cdots \big(1 - \big(kN + (N - 1)\big)/n\big)}{(1 + 1/n) \cdots (1 + (N - 1)/n)}.$$

4. Testing

The tests we include in this section are not meant to provide any sort of validation for generators. It can be argued that validation of generators really needs to be conducted in terms of some particular application or applications, which is not our focus right now. First, let's describe the generators that are considered.

Excelrand: Excel's `rand()` generator is widely used and well thought of. It seeds from the clock and we have not seen a description of its algorithm. In particular, we do not know how it relates to the linear congruential generator `RAND`, the Unix system `rand()`, or the MAPLE generator "`rand()`."

Excelgen: Excel provides another way to generate random numbers through its Tools drop down menu under Data Analysis/Random Number Generation. It is possible to seed the uniform generator provided under this option, although it is laborious. It seems likely that this generator is `rand()`, but we do not know that for sure. It can be used to extract a relatively small noncontiguous sample.

Unixrand: Unix provides two standard generators that are described in the man pages. This one is evoked by the command `rand()` and can be seeded.

`random()`: The other generator provided by Unix. The algorithm is not disclosed, except to say that it uses a nonlinear additive feedback random number generator. It is considered to be a good generator. It can be seeded.

GGL: is a multiplicative linear congruential PRNG that uses the algorithm

$$X_n = 16807X_{n-1} \bmod (2^{31} - 1).$$

Lagged-Fibonacci generators:: Lagged-Fibonacci generators use an initial set of seed values $x_1, x_2, \ldots, x_p$ and two lags p and q, $1 \leq q < p$, to generate successive elements x_i for $i > p$ by means of the recursion $x_i = x_{i-p} \circ x_{i-q}$. Here, $\circ$ is some modular binary operation which might be $+, -, *$, or (xor) (exclusive or). Following Marsaglia [5], we designate such a generator by $F(p, q, \circ)$. Lagged-Fibonacci generators of maximal period are obtained when an associated polynomial is primitive over appropriate modular rings, see [1]. This is true for example of $F(17, 5, +)$ on integers $\bmod 2^m$ (period is $(2^{17}-1)2^{m-1}$, $m \leq p$). Other "good" choices are $F(31, 13, +)$, $F(55, 24, +)$, and $F(250, 103, +)$. The initial seeds of these generators can be represented by $m \times p$ binary matrices (and to get the maximal orbit, you need this matrix to have maximal rank). The $+$ and $*$ versions have massive numbers of disjoint ergodic components, which has made them candidates for massively parallel applications, see [6]. Even though these generators are not FDO-generators, certain tests can be applied to them anyway. We include them in part to illustrate this and because of their importance. We will deal only with $+$ versions.

fdo-3: We have been involved in an effort to produce true random numbers from alpha decay. We used data from a prototype to produce an FDO-generator on the integers from 0 to $2^{18} - 1$.

randomhalf: This is an FDO-generator on $[0, 2^{18} - 1]$ that was produced from the Unix generator `random()`.

mrand: This is the MAPLE random number generator. It can be seeded.

As indicated in the introduction, we are interested in conducting analyses for two different types of samples. Given a generator $y_1, \ldots, y_n$ a contiguous sample of size k and seed i_0 consists of the next k sequence elements after $y_{i_0}, y_{i_0+1}, \ldots, y_{i_0+k}$. Unless otherwise specified, we will consider the index sums to be modulo n. A noncontiguous sample consists of the successors to the integers in an interval assuming that the generator assumes all the values in the interval. Thus to extract the noncontiguous sample associated with the integer interval $[1, 2]$, we would determine i_1, i_2 so that $y_{i_1} = 1$, $y_{i_2} = 2$. The noncontiguous sample would then consist of y_{i_1+1}, y_{i_2+1} (and order does matter). For PRNGs, $y_{i_1+1} = F(1)$, $y_{i_2+1} = F(2)$ if F is the underlying dynamical mapping. Noncontiguous samples give some indication of whether the supposedly chaotic system being sampled by the PRNG involves time intervals long enough to wipe out short-term correlations between the nearest phase points in the sample. The assumption that the generator sequence is a good FDO or MBO generator would certainly require this. There should be no clear correlations between successors of integer intervals and a noncontiguous sample should pass all randomness tests, those for independence between its elements in particular. If this does not turn out to be the case, then one can expect that contiguous samples which have "close" points in common will be likely to suffer other undesirable correlations. There are some generators, such as the lagged Fibonacci generators, for which these concepts are a bit ambiguous. The seed of one of these generators is a

TABLE 1. Tests of generators based on noncontiguous samples

Generator	Sample Size	Statistic	p-value	Pattern
GGL	400/(200 bits)	S_{200}	2^{-199}	All ones
mrand	200/(100 bits)	S_{100}	10^{-12}	1000000...
Unixrand	200/(100 bits)	S_{100}	10^{-4}	No obvious
random()	200/(100 bits)	S_{100}	10^{-6}	Alt. with runs[a]
Excelgen	40/(20 bits)	S_{20}	2^{-19}	All ones
randomhalf	800/(400 bits)	S_{400}	$> .64$	No discernible
fdo-3	800/(400 bits	S_{400}	~ 1	No discernible

[a]The pattern is alternating 0, 1 punctuated with runs of 1s.

vector of numbers which has full rank (in a certain sense) and the number 1 is not going to be in the cycle at the level of the vectors. We can still look at the sequence generated by any given seed and extract noncontiguous samples, but there is no reason to believe that they will be correlated or not correlated. In fact, one can easily dream up seeds for particular lagged Fibonacci generators where there will be a strong correlation and others where there is no reason to believe there will be any obvious correlations. For example, the cycles corresponding to the seeds $[0, 2, 0, 3, 2, 2, 0, 3, 2, 1, 1, 3, 3, 1, 3, 2, 1]$ and $[0, 2, 0, 3, 2, 2, 0, 3, 2, 1, 1, 3, 3, 1, 3, 2, 0]$ for $F(17, 5, +)$ ($+$ mod 4) are disjoint but have significantly correlated substrings much longer than those involved in just the two seeds themselves. Since disjoint cycles for such generators are proposed as independent sources for massively parallel machines, this may be something to consider further.

We will deal with the noncontiguous samples first.

4.1. Noncontiguous samples. The generators that we have extracted noncontiguous samples from and analyzed are: GGL, mrand, Excelgen, Unixrand, random(), and randomhalf. Table 1 presents the results.

The statistic S_n is calculated as follows: a noncontiguous sample starting at 1 of size $2n$ is extracted from the generator and n successive pairs are then compared. A 1 is assigned if the first term is less than the second, a 0 is assigned otherwise. S_n is then the number of 1s. S_n should be binomial if the generator is FDO. If the generator is MBO, S_n should still be binomial on samples for which the same value is never assigned to two consecutive integers. In an MBO experiment, the likelihood of two consecutive balls going into the same cell is vanishingly small and is easy to check for.

4.2. Contiguous samples. We looked at the performance of a number of generators on a very large number of individual occupation statistics γ_s. On single sample tests, all generators appear to perform reasonably well. If one wants then to consider the empirical distributions of these statistics, one runs into issues with repeated samples. Repeat sampling of permutations is straightforward and requires no adjustments. However, once a BE sample has been produced, the next sample will be from a different phase space. Thus, the statistics one gathers are not identically distributed. Adjusting for this takes some effort and will be handled elsewhere. Still, there are two statistics that use the BE statistics derived here but are based on permutations and can therefore be repeat sampled, namely, the

TABLE 2. generators based on contiguous samples

Generator	Sample Size	$Min > 0$	p-value	$Min > 1$	p-value
	1008	.4226	$> .30$	.1736	.03
Excelrand	1013	.4452	.011	.1737	.03
fdo-3	1043	.4497	.003	.1829	.002
randomhalf	1018	.4676	.00007	.2014	.000003
$F(31, 13, +)$	1440	.4632	.00001	.1903	.00002

$M_{N,n} = \max_{i \le N-1}\{\pi_i(n)\}$ (in the low-density case) and $m_{N,n} = \min_{i \le N-1}\{\pi_i(n)\}$ (in the high-density case). The tests included here are run by repeat sampling contiguous samples of size 109 from the generator (except for Excelrand, where we cannot control the seeding). These samples determine a random permutation of the integers from 0 to 108 (if the generator is FDO, this is automatic but is still the case when the generator is MBO and the sample contains no repeats). The resulting permutation is treated as an FDO(109, 109) experiment and based on the corresponding FDO generator, taking the first $m = 99$ of them, we obtain a run of a BE(10, 100) experiment with $n^2/N = 1$ and the $\mathrm{P}\{M_{100,10} = 1\}$ can be calculated precisely (using MAPLE) from the formula for $\mathrm{P}\{\gamma_{\ge 2} = 0\}$ in Proposition 3.3. If instead, we take the first $m = 9$ terms, we obtain a run of a BE(100, 10) experiment with $N^2/n = 1$ and the $\mathrm{P}\{m_{10,100} > k\}$ can be calculated precisely using Proposition 3.18 (which we did for $k \in \{0, 1\}$). At the sample sizes that we ran, $M_{N,n}$ did not show anything interesting. However, the high-density $m_{N,n}$ statistic shows considerable promise to actually differentiate between generators. We tested GGL, Excelrand, fdo-3, randomhalf and $F(31, 13, +)$. The results for tests based on contiguous samples are included in Table 2

The true value for $\mathrm{P}\{m_{10,100} > 0\} = .4060$ and for $\mathrm{P}\{m_{10,100} > 1\} = .1491$. It is interesting that *GGL* appears to perform the best on this test among generators that are usually thought to be better than it is. The relatively poor performance of fdo-3 is significant, since this generator was derived from alpha decay data, furthermore, the output of the device from which it was derived passed standard batteries of tests that are used to "certify" generators (including the NIST tests).

5. Transforming FDO generators to MBO generators

Let's compare the entropy of both the ideal MBO(n, N) and FDO(n, N) experiments. The distribution of n distinguishable particles over N boxes contains entropy

$$H\big(\mathrm{MBO}(n, N)\big) = n \log_2(N).$$

The entropy for the corresponding FDO(n, N) experiment is slightly smaller, namely, if n/N is bounded away from 1, then

$$\begin{aligned} H\big(\mathrm{FDO}(n, N)\big) &= \log_2(N) + \cdots + \log_2(N - n + 1) \\ &\sim \frac{1}{\ln(2)}\left[n \ln N - \frac{1 + 2 + \cdots + (n-1)}{N}\right] \sim n \log_2 N - \frac{n^2}{(2\ln 2)N}. \end{aligned}$$

If $n/N \to 1$, then

$$\begin{aligned} H(\mathrm{FDO}(N,N)) &= \log_2(N!) - \log_2((N-n)!) \\ &= N\log_2 N - N\log_2 e + O(\log_2 N) - (N-n)\log_2(N-n) \\ &\qquad + (N-n)\log_2 e - O(\log_2(N-n)) \end{aligned}$$

From these considerations, an exhaustion argument can be constructed that shows that the per bit entropy of both experiments are asymptotically the same as $N \to \infty$. This indicates that if one has, for each N, an ideal FDO generator (equivalently, an ideal $\mathrm{FDO}(N,N)$) then one should be able to use it to simulate an ideal $\mathrm{MBO}(n,M)$ experiment for some suitable range of values for n and M with as little loss of per bit entropy as desired. We will discuss three methods for simulating $\mathrm{MBO}(n,M)$ from appropriate FDO generators. The first method will be the most efficient in the sense that the asymptotic average per bit entropy will converge to 1. However, a price will be paid in terms of computations and memory required. Another method, called the "adding boxes" method, is attractive from some points of view, is less complicated to program, but still requires memory and sacrifices a significant portion of the per bit entropy. The third method only uses the permutation data. It is the most straightforward to implement and requires no memory, however, for fixed n, asymptotically the average per bit entropy approaches 0.

Before proceeding with the first method, let's consider a specific example to illustrate some of the difficulties that will have to be dealt with.

Example 5.1. Let's take $N = 10$, $n = 3$, $M = 5$. We want to use an $\mathrm{FDO}(10,10)$ generator to simulate an $\mathrm{MBO}(3,5)$ experiment. Suppose the FDO generator is $y_1,\ldots,y_{10}$, $0 \le y_i \le 9$. From y_1, we can determine which of the equally likely equivalence classes $\{0,1\}$, $\{2,3\}$, $\{4,5\}$, $\{6,7\}$, $\{8,9\}$ that y_1 is a member of and thus determine a unique, uniformly distributed number from 0 to 4, which we denote by z_1. Note, some entropy has been lost in doing this. Now, y_2 can be considered to be a value from a uniformly distributed random variable on the $N-1=9$ numbers from 0 to 8 (by determining its rank among the numbers $\{0,1,\ldots,9\}-\{y_1\}$). If $0 \le y_2 \le 4$, then y_2 does represent an equally likely selection from these five integers and we can use it to determine z_2. The probability that z_2 is determined on this second draw is $\frac{5}{9}$. But, with probability $\frac{4}{9}$, $5 \le y_2 \le 8$ and conditioned on this, the value chosen is for a uniformly distributed distribution on four values 0 to 3, not five. We may keep this value in mind and look at y_3. With probability $\frac{5}{8}$, z_2 will be determined, if it was not determined by the y_2. But, with probability $\frac{3}{8}$, y_3 will represent a value for a uniformly distributed random variable on the three values 0 to 2. We could be forced to continue to y_4 without z_2 yet determined. If y_4 is not between 0 and 4 (which would happen with probability $\frac{5}{7}$), then it represents a value from a uniformly distributed random variable on the two values $0,1$. We could thus be forced to continue. On the next draw, the outcome is either between 0 and 4 or is uniquely determined, so z_2 may still not be determined. If it is not, the next draw will determine z_2 for certain, but we will not be able to determine z_3 except in some approximate sense. Thus, an $\mathrm{FDO}(10,10)$ generator can determine a perfect $\mathrm{MBO}(2,5)$ experiment (although it might need six draws to do it) but it cannot, with probability one, determine a perfect $\mathrm{MBO}(3,5)$ experiment. If $N \ge 15$, however, we would always be able to determine a perfect MBO(3, 5) experiment (in general, if $N \geq nM$, an FDO(N, N) generator will be able to simulate a perfect MBO(n, M) experiment).

We are going to do three things to simplify the rest of our discussion of this case: (1) we will fix n, (2) we will seek samples of size n which are uniform on $[0, M-1]$ where M is a power of 2 and $M \geq M_0 = 2^{m_0}$ for some positive integer m_0,and (3) we will be allowing N to always be large enough so that an MBO(n, M_0) experiment can always be successfully completed (or, at least, we aren't going to worry about that kind of complication). Let's first consider how we can get on average as many bits as possible out of a draw from a uniform distribution on the numbers from 0 to $N-1$ (this would correspond to taking the case that $M = 2$). If $N = 2^k$ for some positive integer k, then all the numbers from 0 to $N - 1 = 2^k - 1$ are equally probable and so we can extract k lower bits with probability 1 (that is, the dyadic expansion for y_1 may be extracted). If $2^k < N < 2^{k+1}$, then we have to do something different. Let $N = 2^{k_l} + 2^{k_{l-1}} + \cdots + 2^{k_1}$ be the nonzero terms of a dyadic expansion for N. Here, $0 \leq k_1 < k_2 < \cdots < k_l = k$. As in the example above, if $0 \leq y_1 \leq 2^{k_l} - 1$, then all these numbers are equally likely, so we may extract the first $k_l = k = [\log_2 N]$ bits in the dyadic expansion for y_1 (using $k+1$ terms), this will occur with probability $2^{k_l}/N$. If, on the other hand, $2^{k_l} \leq y_1 \leq 2^{k_l} + 2^{k_{l-1}} - 1$, then we may extract the first k_{l-1} bits of the dyadic expansion for y_1 with equal probability on all configurations. This will occur with probability $2^{k_{l-1}}/N$. Proceeding in this fashion, we see that the expected number of bits that can be extracted is given by the expression

$$
\begin{aligned}
(5.1) \qquad \mathrm{E}(\#\mathrm{bits}) &= k_l \frac{2^{k_l}}{N} + k_{l-1}\frac{2^{k_{l-1}}}{N} + \cdots + k_1 \frac{2^{k_1}}{N} \\
&= \frac{k_l 2^{k_l} + k_{l-1}2^{k_{l-1}} + \cdots + k_1 2^{k_1}}{2^{k_l} + 2^{k_{l-1}} + \cdots + 2^{k_1}} \\
&= k_l \frac{N}{N} - \left[\frac{(k_l - k_{l-1})2^{k_{l-1}} + \cdots + (k_l - k_1)2^{k_1}}{2^{k_l} + 2^{k_{l-1}} + \cdots + 2^{k_1}}\right] \\
&= k_l - \left[\frac{(k_l - k_{l-1})2^{k_{l-1}-k_l} + \cdots + (k_l - k_1)2^{k_1 - k_l}}{1 + 2^{k_{l-1}-k_l} + \cdots + 2^{k_1 - k_l}}\right] \\
&\geq k_l - [(k_l - k_{l-1})2^{k_{l-1}-k_l} + \cdots + (k_l - k_1)2^{k_1 - k_l}] \\
&\geq k_l - 2 = [\log_2 N] - 2 \geq \log_2 N - 3.
\end{aligned}
$$

It follows from this that the average per bit entropy of this extraction scheme is asymptotic to 1. For a given $M_0 = 2^{m_0}$, we need m_0 bits for each sampling. For a given N, we can extract bits from each draw according to the above scheme and keep going until we have at least m_0 bits. We then have a sample point from a uniform distribution on $[0, M-1]$ where $M = 2^m$, $m = \#$bits extracted. Again, as $N \to \infty$, such an extraction method will yield an average per bit entropy of 1. If we want to extract a sample of size $n \geq 1$, then we just continue this process until we have at least nm_0 bits. We then extract our sample by taking the largest block of bits, say k bits, such that $k \geq m_0$ and $nk \leq \#$bits extracted. Taking $M = 2^k$, we have a sample of size n by taking the first n consecutive k-blocks. The reader should not find it difficult to see that the average per bit entropy of such a procedure will still approach 1 as $N \to \infty$. We formalize the algorithm below. Assume that we have an FDO generator, $y_1, \ldots, y_N$.

Algorithm 5.2. Let m_0, $N \gg 2^{m_0}$ and n be positive integers. Let B be a binary array of length $|B| = m(\log_2 N + 1)$. Below, for any integers $z \geq 0$ and $k \geq 0$, z_k means the kth bit in the binary representation of z, and $r(y_j)$, $1 \leq j \leq N$, is the rank of y_j in the set $\{0, 1, \ldots, N-1\} - \{y_1, \ldots, y_{j-1}\}$.

```
    s ← 0
    j ← 1
L:  R ← N − j + 1
    l ← max{k: x_k < R_k}
    (B_s, B_{s+1} ..., B_{s+l−1}) ← (x_0, x_1, ..., x_{l−1})
    s ← s + l
    if s ≥ nm_0 then
        return (B_0, B_1, ..., B_{s−1})
    else
        j ← j + 1
        if j > N then stop
        else go to L
        end if
    end if
```

The random sample is then extracted from B by extracting n blocks of size $k = [s/n]$, where $[a]$ denotes the integer part of a real number a.

The following proposition summarizes the results above:

Proposition 5.3. *Let m_0, n, $N \gg 2^{m_0}$ be positive integers. If Algorithm 5.2 is used to extract a sample of n uniformly distributed r.v.s on the integers 0 to $M-1$, where $M = 2^k$ and k is determined as indicated, then the average per bit entropy of the method approaches 1 as $N \to \infty$.*

A similar result is valid for an arbitrary positive integer M. An $M-$*bit* is an integer from 0 to $M-1$. One has to modify the argument in (5.1). We expand N in terms of powers of M obtaining $N = \alpha_l M^{k_l} + \alpha_{l-1} M^{k_{l-1}} + \cdots + \alpha_1 M^{k_1}$. A natural modification of the bit extraction method to extract $M-$ bits instead and a little thought leads to the expression

$$\begin{aligned} \mathrm{E}(\#M - \text{bits}) &= k_l \frac{\alpha_l M^{k_l}}{N} + k_{l-1} \frac{\alpha_{l-1} M^{k_{l-1}}}{N} + \cdots + k_1 \frac{\alpha_1 M^{k_1}}{N} \\ &= \frac{k_l \alpha_l M^{k_l} + k_{l-1} \alpha_{l-1} M^{k_{l-1}} + \cdots + k_1 \alpha_1 M^{k_1}}{\alpha_l M^{k_l} + \alpha_{l-1} M^{k_{l-1}} + \cdots + \alpha_1 M^{k_1}}. \end{aligned}$$

From this a lower bound on the average number of $M-$ bits that can be extracted per sample point can be established which can be used to show that the average per bit entropy again approaches 1 as $N \to \infty$.

Remark 5.4. The above method does allow one to extract MBO samples from an FDO(N, N) generator with practically no loss of entropy provided N is large enough and we are willing to allow the exact size of the interval $[0, M-1]$ being sampled to be determined by the process. However, it requires significant computations and may also involve significant memory. If we wish to fix $M = M_0$ ahead of time and if $Mn \ll N$, then it is highly probable that more than enough $M-$ bits will be extracted from the first draw. This brings up some interesting questions concerning efficiencies, since, if the extra bits are thrown away, they will represent a significant loss of entropy. Exactly how M, n, nM should relate to

N, under these circumstances, is something we have not looked into here. In most cases, one will be given a fixed N, say for example in the case of the generator GGL. In that case, determining the conditions on M and n that will guarantee that the sample will be extractable are of interest (as seen in Example 5.1, there will be values of M, n for which one will not be certain that the sample can be extracted). There are also interesting efficiency questions such as, for fixed N, which combinations of M, n are most efficient (say, from the entropy point of view). We will not pursue these issues in this paper, our purpose here has been to show that arbitrarily efficient extraction of MBO sample points from FDO generators is possible and to indicate some of the issues that might be involved in dealing with real generators of a fixed size. An alternative approach to the one taken above is to just convert the entire FDO generator into a Bernoulli sequence, which can be done without loss of average per bit entropy (asymptotically) as indicated by Proposition 5.3. However, depending on the size of the generator, this may be computationally expensive, the result would have to be stored in memory and this would destroy the advantages of working with a PRNG, if that happens to be the source. Still, there are interesting questions in this direction as well, such as in determining the statistics of the number of bits that would be extracted from a given FDO(N, N).

If we are willing to give up some of the entropy, we can extract MBO samples from a given, good FDO generator in a manner that is computationally a little more straightforward than the above algorithm but still requires some memory. This method is appealing in other ways, which we will mention a little later. Rather than write down the tedious details of an algorithm for the method, we will describe it more intuitively in terms of running a perfect FDO(N, N) experiment and generating an MBO(n, M) sample in the process.

Let $M < N$. The possible outcomes of the FDO(N, N) experiment are identified with boxes labeled $0, 1, \ldots, N-1$. At any moment of the algorithm's lifetime, some M out of N boxes are *target boxes* with *target numbers* from 0 to $M-1$ (identified with the possible outcomes of the MBO experiment). Initially, the target boxes are those labeled 0 to $M-1$, their target numbers being identical to their labels. In running the FDO(N, N) experiment, we are distributing balls one at a time to the boxes $[0, N-1]$. If the ball goes to a box that is currently not a target box, then the box is discarded and no output of the MBO experiment is recorded. If, on the other hand, the ball goes to a target box whose target number is i say, then i is recorded as the current outcome of the MBO experiment, the target box is discarded, and the next available (not discarded) box becomes a target box with the same target number i. This process is continued until either the MBO(n, M) experiment has been completed or it terminates because no box is available to replace the target box just discarded (the latter occurs after $N - M + 1$ steps of the FDO(N, N) experiment).

Clearly, for given $M < N$, there will be sample sizes n for which this algorithm will fail because we will run out of assignment boxes before the sample can be generated. On the other hand, if M is somewhat smaller than N, say $M \leq N/2$, and $n \ll N - M$, then the sample will almost certainly be generated but with a significant loss of entropy. For example, if N is even and $M = N/2$, $n \ll N/2$ then the per bit average entropy of such a scheme can be expected to be around $\frac{1}{2}$. If one wants the sample size $n > M$, then M will generally have to be a smaller fraction of N in order to have a high probability of completing the sample and one can expect this to cost more in terms of average per bit entropy.

While the algorithm for this "adding boxes" technique is tedious to encode, it is not complex. The scheme does require a possibly significant amount of memory. The main advantage is that the details of the algorithm do not change with M or N, so, if perfect FDO experiments were available for arbitrarily large N, the same code could be used to generate MBO(n, M) experiments for arbitrary (n, M).

There are many interesting questions left open by this discussion. Determining the exact relationship between the average per bit entropy of the scheme for given n, M, N is of interest. Issues concerning optimal relationships between n, M, N that take into account the probability that the scheme may not complete are also of interest. We leave these to the future, for now.

The final scheme rests on simply using the permutations that can be associated with the ordered outcomes of the generator. To illustrate, suppose we have an FDO generator $y_1, \ldots, y_N$, $N \geq 2$, then we can produce a symmetric Bernoulli string of size $[N/2]$ by associating with the ordered pairs of outcomes $(y_1, y_2), \ldots, (y_{2[N/2]-1}, y_{2[N/2]})$ the values 0 or 1 depending on whether the first coordinate is less than the second or not. More generally, suppose $M_0 \geq 2$ and $n \geq 1$ are given integers and we want to simulate an MBO(M, n) experiment with $M \geq M_0$. Take the first integer K, such that $K! \geq M_0$, and then use the outcomes $(y_1, \ldots, y_K), (y_{K+1}, \ldots y_{2K})$, etc., to determine permutations on K elements $(1, 2, \ldots, K)$ each of which is equally likely and can be associated uniquely with an integer from 0 to $K! - 1$ thus providing sample points from the uniform distribution on the integers $[0, K! - 1]$. The only programming issues here are with the encoding of the permutations and associating the basic permutations with the FDO outcomes $(y_{sK+1}, \ldots, y_{(s+1)K})$. This second issue is resolved by sorting the $y_{sK+1}, \ldots, y_{(s+1)K}$ $(s \geq 1)$ and keeping track of final positions of the indices. Once we have a standard permutation on the integers from 1 to K, it is clear what we want to do: we map each permutation to its rank in the set of all permutation sorted lexicographically reduced by 1, so that the identity permutation $(1, 2, 3, \ldots, K-1, K)$ goes to 0, $(1, 2, 3, \ldots, K-2, K, K-1)$ goes to 1, $(1, 2, 3, \ldots, K-1, K-2, K)$ goes to 2, and so on. We have implemented this scheme in MAPLE.

As N goes to ∞, such a scheme's average per bit entropy will approach 0. For a given situation, this still may be the most practical scheme to use. It has many advantages. As long as $N \geq nK$, the scheme is guaranteed to produce the desired sample. It requires no memory and is readily adapted to methods that involve random seeding of the FDO generator (it can be started at any point in the string without having to worry about outcomes that may have preceded the ones before that term). It can even accommodate within-sample random reseeding if a simple algorithm is employed to check that no overlaps are occurring between the permutation samples (this would require memory, however).

As an example, consider GGL as an FDO$(2^{31} - 1, 2^{31} - 1)$ generator. If we were interested in a uniform distribution on the integers from 0 to $M - 1$, where $M \geq 2^{21}$, then taking $K = 10$ will suffice, since $10! = 3628800 > 2097152 = 2^{21}$. Now, there are $[N/10] = [(2^{31} - 1)/10] = 214748364$ permutations that can be sampled from GGL, which is between 2^{27} and 2^{28}, which is quite a respectable MBO generator size. Indeed, this may very well be the best way to convert GGL to an MBO generator that has the possibility of collisions. In [2], we projected GGL onto its most significant 21 bits and rejected that it was an MBO source based on the fact that the X statistic considered in [2] was too small, this required a sample of size 8,000,000. We have not run the experiment yet, but we conjecture that, if we use the permutations instead, the resulting generator will no longer have that problem.

References

1. R. P. Brent, *On the periods of generalized Fibonacci recurrences*, Math Comp. **63** (1994), no. 207, 389–401.
2. A. Figotin, A. Gordon, S. Molchanov, J. Quinn, and N. Stavrakas, *Occupancy numbers in testing random number generators*, SIAM J. Appl. Math. **62** (2002), no. 6, 1980–2011.
3. B. V. Gnedenko and A. N. Kolmogorov, *Limit distributions for sums of independent random variables*, Addison-Wesley Co., Cambridge, MA, 1954.
4. V. F. Kolchin, B. A. Sevast′yanov, and V. P. Chistyakov, *Random allocations*, Scripta Series in Mathematics, V. H. Winston & Sons, Washington DC, 1978.
5. G. Marsaglia, *A current view of random number generators*, Computer Science and Statistics: The Interface (Atlanta, GA, 1984) (L. Billard, ed.), North-Holland, Amsterdam, 1985, pp. 3–10.
6. M. Mascagni, M. L. Robinson, D. V. Pryor, and S. A. Cuccaro, *Parallel pseudorandom number generation using additive lagged-Fibonacci recursions.*
7. V. V. Petrov, *Sums of independent random variables*, Ergeb. Math. Grenzgeb., vol. 82, Springer, New York–Heidelberg, 1975.
8. J. Quinn, *Occupation statistics revisited: Factorial moment formulas for general patterns*, in preparation.
9. R. von Mises, *Mathematical theory of probability and statistics*, Academic Press, New York–London, 1964.

Department of Mathematics and Statistics, University of North Carolina — Charlotte, 376 Fretwell Bldg., 9201 University City Blvd., Charlotte, NC 28223-0001, USA

E-mail address, A. Gordon: `aygordon@unc.edu`
E-mail address, S. Molchanov: `smolchan@uncc.edu`
E-mail address, J. Quinn: `jequinn@uncc.edu`

Centre de Recherches Mathématiques
CRM Proceedings and Lecture Notes
Volume **42**, 2007

The Spectral Problem, Substitutions and Iterated Monodromy

Rostislav Grigorchuk, Dmytro Savchuk, and Zoran Šunić

ABSTRACT. We provide a self-similar measure for the self-similar group $\mathcal{G}$ acting faithfully on the binary rooted tree, defined as the iterated monodromy group of the quadratic polynomial $z^2 + \mathrm{i}$. We also provide an L-presentation for $\mathcal{G}$ and calculations related to the spectrum of the Markov operator on the Schreier graph of the action of G on the orbit of a point on the boundary of the binary rooted tree.

Introduction

It was observed recently that the class of self-similar groups naturally appears in mathematics. The most recent examples come from combinatorics and are related to one of the most famous combinatorial problems known as Hanoi Towers Game [14]. Slightly older examples, due first of all to Nekrashevych [19] (see also [2]), are related to holomorphic dynamics and random walks.

Self-similar groups can be defined as groups generated by all the states of a (not necessarily finite) Mealey automaton [12]. Of particular interest and importance is the case when the automaton is finite, since in that case the obtained group is finitely generated.

Self-similar groups have many interesting and important properties. The class of self-similar groups contains exotic examples, such as groups of Burnside type or groups of intermediate growth, as well as familiar examples, such as free groups or free products of finite groups, that are well known and are regular objects of study in combinatorial group theory.

One of the most remarkable discoveries in the recent years is the observation, due to Nekrashevych, that the so-called iterated monodromy groups (IMG), which can be related to any self-covering map, belong to the class of self-similar groups and that, in the most natural situations, there is an explicit procedure representing them by finite automata.

Even in the case of quadratic maps over $\mathbb{C}$ one gets a rich theory with wonderful applications both to holomorphic dynamics and to group theory [4, 19].

2000 *Mathematics Subject Classification.* Primary 20E08; Secondary 37F10, 60B15.
The first author is partially supported by NSF grants DMS-0308985 and DMS-0456185.
The third author is partially supported by NSF grant DMS-0600975.
This is the final form of the paper.

Already the simplest examples of quadratic polynomials, such as $z^2 - 1$ or $z^2 + \mathrm{i}$, show that the corresponding groups can be quite complicated and can have extraordinary properties.

The group $\mathrm{IMG}(z^2 - 1)$ is called the Basilica group after the Julia set of $z^2 - 1$ which (mildly) resembles the roof of the San Marco Basilica in Venice (the top part of the Julia set is the roof and the bottom part is its reflection in the water). Basilica group is torsion free, of exponential growth, amenable but not elementary (and not even subexponentially) amenable [6, 16], has trivial Poisson boundary, is weakly branch, and has many other interesting properties.

The main object of this article is the group $\mathrm{IMG}(z^2 + \mathrm{i})$ (denoted by $\mathcal{G}$ in the rest of the text), introduced in [2] and later studied by Bux and Perez [7], who proved that $\mathcal{G}$ has intermediate growth. This is not the first example of a self-similar group of intermediate growth (the first examples were constructed in [8, 9]), but it is the first example of a group of intermediate growth that naturally appears in the area of applications of group theory.

We start with a quick introduction to the theory of self-similar groups and, in particular, to iterated monodromy groups. We are aiming for a self-contained treatment, which would make it possible for the reader to understand the context of the paper completely without reading other sources. In particular we give very detailed calculation of the action of $\mathcal{G} = \mathrm{IMG}(z^2 + \mathrm{i})$ on the binary rooted tree. Then we show that the group $\mathcal{G}$ is a regular branch group, thus presenting an example of a branch group which naturally appears in holomorphic dynamics. The main body of the article is devoted to the calculation of an L-presentation for $\mathcal{G}$ (i.e., a presentation of a group by generators and relations which involves a finite set of relators and their iterations by substitution). Although it is known that L-presentations are quite common for groups of branch type the number of examples in which explicit computation is possible is rather small.

The presence of L-presentations is important from different points of view. Such presentations are at the first level of complexity after the finite presentations and quite often provide the simplest way to describe a group that is not finitely presented ($\mathcal{G}$ is not finitely presented [20]). Further, such presentations can be used to embed a group into a finitely presented group in a way that preserves many properties of the original group. We use the obtained L-presentation of $\mathcal{G}$ to embed $\mathcal{G}$ into a finitely presented group with 4 generators and 10 relators, which is amenable but not elementary amenable (this approach has been used for the first time in [10]).

During the last stage of the preparation of the manuscript we were informed by Bartholdi and Nekrashevych of their work [5] providing calculation of an L-presentation for $\mathrm{IMG}(z^2 + c)$ for every post-critically finite quadratic polynomial $z^2 + c$. It would be interesting to compare the obtained presentations.

The rest of the article deals with finding a self-similar measure on $\mathcal{G}$. The notion of a self-similar measure was introduced by Kaimanovich in [18], who extends some ideas (in particular the idea of self-similarity of a random walk) that appeared in the work of Bartholdi and Virág [6].

The self-similar measure is closely related to the problem of computation of the spectrum of a Hecke type operator that can be related to any group acting on a rooted tree and to the problem of computation of the spectrum of the discrete

Laplace operator (or, what is almost the same, the Markov operator) on the boundary Schreier graph of a group (i.e., the graph of the action of the group on the orbit of a point of the boundary). A general approach to the spectral problem (which extends the ideas outlined in [1,15]) based on a renormalization principle and leading to questions on amenability, multidimensional dynamics and multiparametric self-similarity of operators is described in [13]. Unfortunately, the spectral problem is not solved yet in our situation. What we are able to construct is a rational map on $\mathbb{R}^3$ whose proper invariant set (shaped as a "strange attractor") gives the spectrum of the Markov operator after intersection by a corresponding line. Here we have a situation analogous to the case of Basilica group [17]. Further efforts in the description of the shape of the attractor (and hence of the spectrum) are needed.

The Schreier graph in this case, viewed through a macroscope, has a form of a dendrite and this is a reflection of a general fact relating the geometry of Schreier graphs and Julia sets proved by Nekrashevych [19].

In any case, our computations allow us to find a self-similar nondegenerate measure on $\mathcal{G}$, which gives a self-similar random walk on the group. The study of asymptotic properties of such random walks is a promising direction and will be one of our subsequent subjects of investigation.

1. Iterated monodromy groups

The theory of iterated monodromy groups was developed mostly by Nekrashevych. A very detailed exposition can be found in his monograph [19]. Here we give a definition and some basic properties of these groups.

Consider a path connected and locally path connected topological space M. Let M_1 be an open and path connected subset of M and $f\colon M_1 \to M$ be a d-fold covering map. Fix a base point $t \in M$ and let $\pi_1(M,t)$ be the corresponding fundamental group. The set of iterated preimages of t under f has a natural structure of a d-ary rooted tree T. Namely, each point s from this set has exactly d preimages $s_1, \dots, s_d$ and these preimages are declared to be adjacent to s in T. The nth level of the tree T consists of the d^n points in the set $f^{-n}(t)$. Note that although the intersection of $f^{-n}(t)$ and $f^{-m}(t)$ may be nonempty for $m \neq n$, we formally consider the set of vertices of T to be a disjoint union of the sets $f^{-n}(t)$, $n \geq 0$.

There is a natural action of $\pi_1(M,t)$ on the tree T. Let $\gamma \in \pi_1(M,t)$ be a loop based at t. For any point s of $f^{-n}(t)$, there is a unique preimage $\gamma_{[s]}$ of γ under f^n which starts at s and ends at a point s', which also belongs to $f^{-n}(t)$. We define an action of γ on T by setting $\gamma(s) = s'$. This action induces a permutation of $f^{-n}(t)$ because the preimages of γ^{-1} starting at the points of $f^{-n}(t)$ are defined uniquely as well. The group of all permutations of $f^{-n}(t)$ induced by all elements of $\pi_1(M,t)$ is called the *nth monodromy group* of f. If $\gamma(s) = s'$ then $\gamma(f(s)) = f(s')$ since $f(\gamma_{[s]}) = \gamma_{[f(s)]}$, so γ acts on T by a tree automorphism.

The action of $\pi_1(M,t)$ on T is not necessary faithful. Let N be the kernel of this action.

Definition 1.1. The group $\mathrm{IMG}(f) = \pi_1(M,t)/N$ is called the *iterated monodromy group* of f.

It can be shown (see [19] for details) that, up to isomorphism, $\mathrm{IMG}(f)$ does not depend on the choice of the base point t.

In order to describe the automorphisms induced on T by the loops from $\pi_1(M,t)$ we need to come up with a "coordinate system" on T. Let $X = \{0, 1, \dots, d-1\}$ be a standard alphabet of cardinality d. Then the set X^* of all finite words over X also has the structure of a d-ary rooted tree, where v is adjacent to vx, for any $v \in X^*$ and $x \in X$.

The group $\operatorname{Aut} X^*$ of all automorphisms of X^* has the structure of an infinite iterated permutational wreath product $\wr_{i\geq 1} \operatorname{Sym}(d)$ (because $\operatorname{Aut} X^* \cong \operatorname{Aut} X^* \wr_X \operatorname{Sym}(d)$, where $\operatorname{Sym}(d)$ acts naturally on X by permutations). This gives a convenient way to express automorphisms from $\operatorname{Aut} X^*$ in the form

$$g = (g|_0, g|_1, \dots, g|_{d-1})\sigma_g, \tag{1.1}$$

where $g|_0, g|_1, \dots, g|_{d-1}$ are automorphisms of the subtrees of X^* with roots at the vertices $0, 1, \dots, d-1$ (these subtrees are canonically identified with X^*) induced by g, and σ_g is the permutation of X induced by g (i.e., $\sigma_g(x) = g(x)$ — the action of g on $x \in X$). More generally, for every $u \in X^*$ we define $g|_u$ to be the automorphism of the subtree of X^* rooted at u (identified with X^*) induced by g. The automorphism $g|_u$ is called the *section* of g at u and is uniquely determined by $g(uw) = g(u)g|_u(w)$, for all $w \in X^*$.

The product of automorphisms written in form (1.1) is performed in the following way. If $h = (h|_0, h|_1, \dots, h|_{d-1})\sigma_h$ then

$$gh = (g|_0 h|_{\sigma_g(0)}, \dots, g|_{d-1} h|_{\sigma_g(d-1)})\sigma_g\sigma_h.$$

By definition $gh(w) = h\big(g(w)\big)$.

Definition 1.2. A group $G \leq \operatorname{Aut} X^*$ is called *self-similar* if $g|_u \in G$ for all $g \in G$ and $u \in X^*$.

A convenient way to describe a particular finitely generated self-similar group G generated by automorphisms $g_1, g_2, \dots, g_n$ is through a, so-called, *wreath recursion.* In this presentation we simply write the action of each g_i in the form

$$g_i = \big(w_1(g_1, \dots, g_n), \dots, w_d(g_1, \dots, g_n)\big)\sigma_{g_i},$$

where w_i, $i = 1, \dots, n$, are words in the free group of rank n.

Another language which describes self-similar groups is the language of automaton groups (see the survey paper [12] for details).

Definition 1.3. A Mealy *automaton* is a tuple (Q, X, π, λ), where Q is a set (a set of states), X is a finite alphabet, $\pi\colon Q \times X \to Q$ is a transition function and $\lambda\colon Q \times X \to X$ is an output function. If the set of states Q is finite the automaton is called *finite.*

One can think of an automaton as a sequential machine which, at each moment of time, is in one of its states. Given a word $w \in X^*$ the automaton acts on it as follows. It "eats" the first letter x in w and depending on this letter and on the current state q it "spits out" a new letter $\lambda(q,x) \in X$ and changes its state to $\pi(q,x)$. The new state then handles the rest of word w in the same fashion. Thus the map λ can be extended to $\lambda\colon Q \times X^* \to X^*$ — we just feed the automaton with letters of $u \in X^*$ one by one. Each state q of the automaton defines a map, also denoted by q, from X^* to itself defined by $q(w) = \lambda(q, w)$. In the special case when, for all $q \in Q$, the map $\lambda(q, \cdot)$ is a permutation of X the map $q\colon X^* \to X^*$ is invertible and hence, an automorphism of the tree X^*. In this case the automaton is called *invertible.*

Definition 1.4. A group of automorphisms of X^* generated by all the states of an invertible automaton $\mathcal{A}$ is called the *automaton group* generated by $\mathcal{A}$.

The class of automaton groups coincides with the class of self-similar groups. Indeed, the action on X^* of every element g of a self-similar group can be encoded by an automaton whose states are the sections of g on the words from X^*, transition and output functions are derived from the representation (1.1). Namely, for each $u \in X^*$, set $\pi(g|_u, x) = g|_{ux}$ and $\lambda(g|_u, x) = g|_u(x)$.

Important subclass of automaton groups consists of groups generated by *finite* automata. For example, we know that for groups in this class the word problem is solvable.

A standard way to visualize automata is by so-called *Moore diagrams.* Such a diagram is an oriented graph where the set of vertices is Q and for every $q \in Q$, $x \in X$, there is an edge from q to $\pi(q, x)$ labeled by $\big(x, \lambda(q, x)\big)$. In case of invertible automata it is common to label states by the corresponding permutations of X and leave only the first coordinate on the edge labels. An example of a Moore diagram is presented in Figure 5.

We go back now to iterated monodromy groups and construct an isomorphism $\Lambda\colon X^* \to T$ such that the induced action of $\pi_1(M, t)$ on X^* becomes particularly nice (self-similar).

We construct Λ level by level. Set $\Lambda(\varnothing) = t$. For each vertex v in X^n we will construct a path l_v in M joining t to one of its preimages s_v from $f^{-n}(t)$ and define $\Lambda(v) = s_v$. Choose arbitrarily d paths $l_0, \dots, l_{d-1}$ in M connecting t to its d preimages in $f^{-1}(t)$ and, for $x \in X$, define $\Lambda(x)$ to be the end of the path l_x. Now assume we have already defined $\Lambda(v)$ and corresponding paths l_v for all $v \in X^m$, $m \le n$ and Λ is an isomorphism between the first n levels of X^* and T such that, for all vertices v on the first n levels, $\Lambda(v)$ is the endpoint of ℓ_v. For any word $xu \in X^{n+1}$ with $x \in X$ and $u \in X^n$ define

$$l_{xu} = l_u f^{-n}_{[\Lambda(u)]}(l_x),$$

where $f^{-n}_{[\Lambda(u)]}(l_x)$ is the unique preimage of the path l_x under f^n starting at the vertex $\Lambda(u)$ (composition of paths is performed from left to right, i.e., the path on the left is traversed first). Define $\Lambda(xu)$ to be the end of the path l_{xu}.

In order to prove that Λ is an isomorphism of trees we need to show that $f\big(\Lambda(xvy)\big) = \Lambda(xv)$, for all $x, y \in X$ and $v \in X^*$. Indeed,

$$f(l_{xvy}) = f(l_{vy}) f(f^{-n}_{[\Lambda(vy)]}(l_x)) = f(l_{vy}) f^{-(n-1)}_{[\Lambda(v)]}(l_x).$$

By definition, $f^{-(n-1)}_{[\Lambda(v)]}(l_x)$ is a path going from $\Lambda(v)$ to $\Lambda(xv)$, so the end $\Lambda(xvy)$ of the path l_{xvy} is mapped to $\Lambda(xv)$ under f. Abusing the notation, we often identify the trees T and X^* and write v for $\Lambda(v)$ (see Figure 1, where solid lines represent edges in the tree T and dashed lines represent paths in M).

Definition 1.5. The action of $\mathrm{IMG}(f)$ on X^* induced by the isomorphism Λ is called the *standard action* of $\mathrm{IMG}(f)$.

The tree isomorphism Λ allows us to compute iterated monodromy groups using the language of self-similar groups [19]. We provide the details here in order to keep the paper relatively self-contained and to help the understanding of the computations that follow. Recall, that for any loop γ based at t and any $u \in f^{-n}(t)$

we denote by $\gamma_{[u]}$ the unique preimage of γ under f^n that starts at the point u. Similarly, $f_{[u]}^{-n}(l_x)$ denotes the unique preimage of the path l_x starting at u.

Theorem 1.6. *The standard action of* IMG(f) *is self-similar. More precisely, the section* $\gamma|_x$ *of* $\gamma \in \mathrm{IMG}(f)$ *at* $x \in X$ *is given by*

$$\gamma|_x = l_x \gamma_{[x]} (l_{\gamma(x)})^{-1}. \tag{1.2}$$

PROOF. Let $v \in X^n$ be an arbitrary word and suppose $\gamma(xv) = yu$, for $y \in X$ and $u \in X^n$. Then vertices v and u are connected by the path

$$p = f_{[v]}^{-n}(l_x) \cdot \gamma_{[xv]} \cdot \left(f_{[u]}^{-n}(l_y)\right)^{-1},$$

which goes through the vertices $v \to xv \to yu \to u$ (see Figure 2, where solid curves represent paths in M and dashed lines represent paths in the tree X^*)). We have

$$f^n(p) = l_x \gamma_{[x]} l_y^{-1}.$$

Thus the loop $\ell = l_x \gamma_{[x]} l_y^{-1}$ based at t represents the element of IMG(f) which moves v to u. The loop ℓ is independent of v (and u). Thus we have $\gamma|_x = l_x \gamma_{[x]} l_y^{-1}$. □

We are now ready to compute the standard action of IMG$(z^2 + \mathrm{i})$ on $\{0,1\}^*$. The only critical point of this map is $z = 0$, which is preperiodic:

$$0 \xrightarrow{f} \mathrm{i} \xrightarrow{f} (\mathrm{i}-1) \overset{f}{\rightleftarrows} -\mathrm{i}.$$

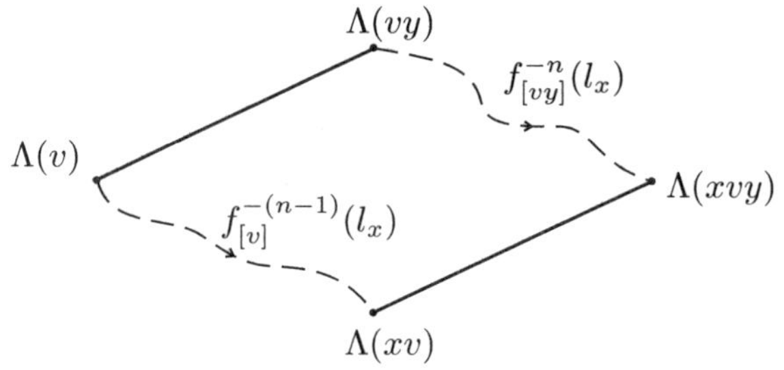

FIGURE 1. Isomorphism Λ between T and X^*

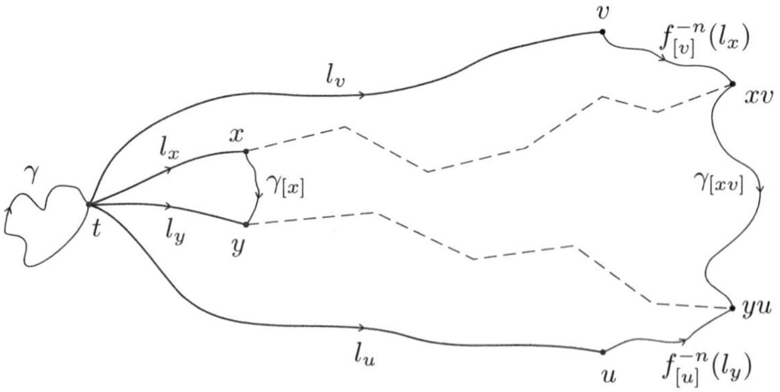

FIGURE 2. Self-similar action of Iterated Monodromy Group

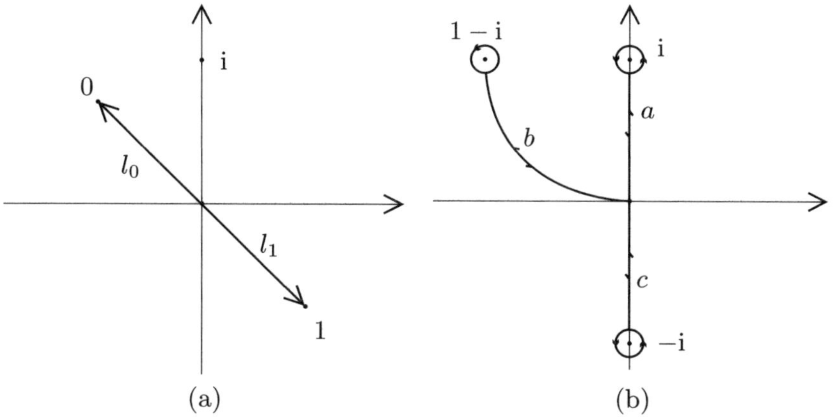

FIGURE 3. (a) Paths connecting $t = 0$ to its preimages (b) Generators of the fundamental group $\pi(M, t)$

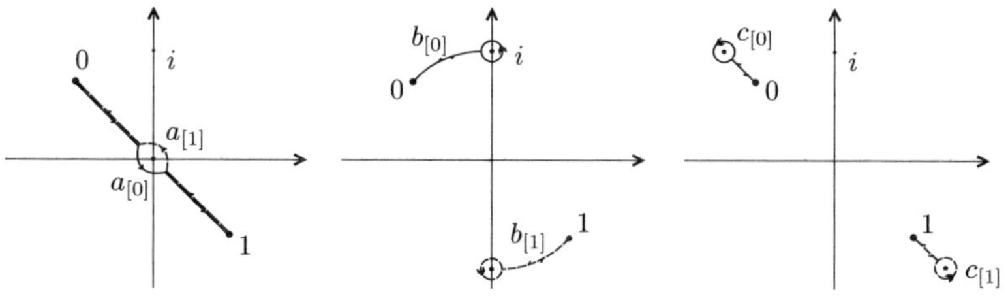

FIGURE 4. Preimages of the generating loops

and, hence, the postcritical set of f is $\{\mathrm{i}, \mathrm{i}-1, -\mathrm{i}\}$. Therefore the restriction of f on $M_1 = \mathbb{C} \setminus \{\mathrm{i}, \mathrm{i}-1, -\mathrm{i}, 0\}$ is a 2-fold covering of $M = \mathbb{C} \setminus \{\mathrm{i}, \mathrm{i}-1, -\mathrm{i}\}$.

Set $t = 0 \in \mathbb{C}$ as the base point. It has two preimages $e^{\mathrm{i}3\pi/4}$ and $e^{\mathrm{i}7\pi/4}$ which are identified with the letters 0 and 1, respectively (more precisely, we set $\Lambda(0) = e^{\mathrm{i}3\pi/4}$ and $\Lambda(1) = e^{\mathrm{i}7\pi/4}$). For the paths l_0 and l_1 connecting t to its preimages we choose the straight segments shown in Figure 3(a).

The fundamental group $\pi_1(M, t)$ is generated by the 3 loops a, b, c shown in Figure 3(b) going around i, $-$i and $\mathrm{i}-1$ respectively. Each of these loops has two preimages $a_{[i]}$, $b_{[i]}$ and $c_{[i]}$, $i = 0, 1$, shown in Figure 4.

According to formula (1.2) and Figures 3 and 4 the sections of the generators a, b, c at 0 and 1 satisfy:

$$a|_0 = l_0 a_{[0]} l_1^{-1} = 1, \qquad a|_1 = l_1 a_{[1]} l_0^{-1} = 1,$$
$$b|_0 = l_0 b_{[0]} l_0^{-1} = a, \qquad b|_1 = l_1 b_{[1]} l_1^{-1} = c,$$
$$c|_0 = l_0 c_{[0]} l_0^{-1} = b, \qquad c|_1 = l_1 c_{[1]} l_1^{-1} = 1,$$

where 1 denotes the trivial loop at t, which represents the identity element of $\mathrm{IMG}(z^2 + \mathrm{i})$.

Since a permutes the elements of $f^{-1}(t)$, while b and c do not, we obtain the following wreath recursion for the generators of $\mathrm{IMG}(z^2 + \mathrm{i})$

$$a = (1, 1)\sigma, \quad b = (a, c), \quad c = (b, 1), \tag{1.3}$$

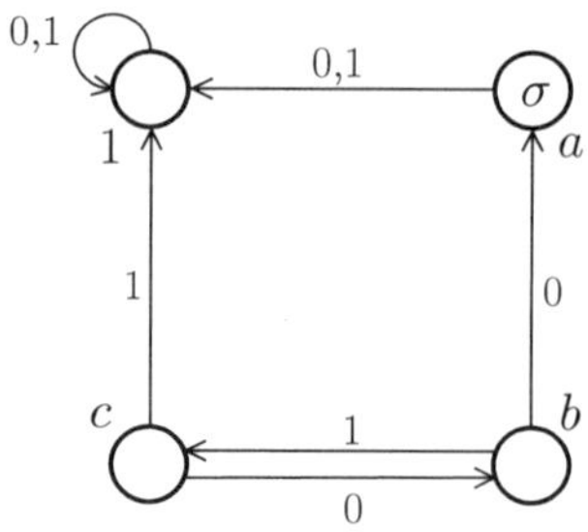

FIGURE 5. Automaton generating group $\mathrm{IMG}(z^2 + \mathrm{i})$

where σ is the nontrivial transposition in $\mathrm{Sym}(2)$.

These relations show that the set of all sections of the generators a, b, c of $\mathcal{G} = \mathrm{IMG}(z^2 + \mathrm{i})$ is $\{1, a, b, c\}$ and that the group $\mathcal{G}$ is generated by the states of the finite automaton shown in Figure 5.

We now say a few words about the relation between the dynamics of the map $z \mapsto z^2 + \mathrm{i}$ and the combinatorial properties of the action of $\mathcal{G}$ on the tree T.

Recall that if a group G acts on a set Y then the *Schreier graph* of this action is an oriented graph, whose set of vertices is Y and there is an edge from $y \in Y$ to $z \in Y$ labeled by $g \in G$ if and only if $g(y) = z$. It is convenient sometimes to forget about the labels and/or the orientation of the edges.

Every group acting on a rooted tree acts on each level of the tree. The Schreier graphs of such actions are of particular interest, since in many situations (such as the one we are in) they can be used to find the spectrum of the Markov operator on the boundary of the tree.

Recent results of Nekrashevych [19] show that the Schreier graphs of $\mathrm{IMG}(f)$ on the levels of the tree converge to the Julia set of the map f. Therefore the structure of the Julia set of f provides understanding of the structure of the Schreier graphs of $\mathrm{IMG}(f)$ (and vice versa). In our case the Julia set of $z^2 + \mathrm{i}$ is the dendrite shown in the top half of Figure 6. The bottom half of this figure displays the Schreier graph of $\mathcal{G}$ on level 8. The set of vertices of this graph is just $f^{-8}(0)$ and the vertices are connected according to the action of $\mathcal{G}$ (no loops are drawn though).

2. Branch groups

Another important class of subgroups of $\mathrm{Aut}\, X^*$ is the class of branch groups [3, 11]. Here we give basic definitions and prove that $IMG(z^2 + \mathrm{i})$ is a regular branch group. This shows that branch groups arise naturally in mathematics (not just as a way to construct groups with unusual properties).

Let G be a subgroup of $\mathrm{Aut}\, X^*$. Then for any vertex $v \in X^*$ one can define the subgroup of G consisting of all the elements in G fixing all words in X^* that do not have v as a prefix. This subgroup of G is called the *rigid stabilizer* of v and it is denoted by $\mathrm{rist}_G(v)$. Furthermore, the subgroup

$$\mathrm{rist}_G(n) = \left\langle \bigcup_{v \in X^n} \mathrm{rist}_G(v) \right\rangle$$

generated by the union of the rigid stabilizers of vertices at level n, is called the *rigid stabilizer of the nth level.* Since elements of rigid stabilizers of different vertices on

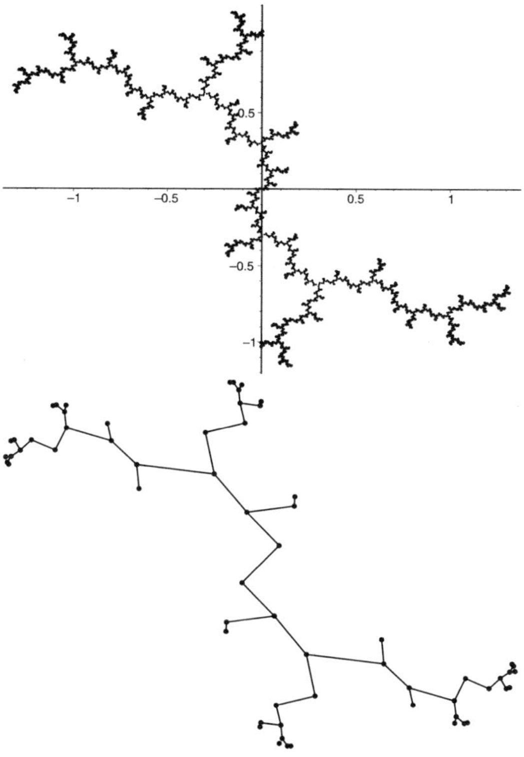

FIGURE 6. Schreier graph of $\mathcal{G}$ and Julia set of $z^2 + \mathrm{i}$

the same level commute we have

$$\mathrm{rist}_G(n) = \prod_{v \in X^n} \mathrm{rist}_G(v).$$

Note that, if G acts transitively on the levels of the tree, then all rigid stabilizers of the vertices on a fixed level are conjugate and, hence, isomorphic.

Definition 2.1. A group G of tree automorphisms of X^* that acts transitively on the levels of X^* is called a *branch group* if all rigid level stabilizers $\mathrm{rist}_G(n)$, $n \geq 0$, have finite index in G. If all rigid stabilizers are nontrivial then G is called a *weakly branch* group.

It is often easier to prove that a given group belongs to a more narrow class of regular (weakly) branch groups. Consider a self-similar group G and its normal subgroup $\mathrm{St}_G(1)$ consisting of all elements in G that stabilize the first level of X^*. There is a natural embedding

$$\Psi\colon \mathrm{St}_G(1) \hookrightarrow G \times G \times \cdots \times G$$

given by

$$g \stackrel{\Psi}{\mapsto} (g|_0, g|_1, \ldots, g|_{d-1}).$$

Definition 2.2. Let $K, K_0, \ldots, K_{d-1}$ be subgroups of a self-similar group G acting on X^*. We say that K *geometrically contains* $K_0 \times \cdots \times K_{d-1}$ and write

$$K_0 \times \cdots \times K_{d-1} \preceq K$$

if $K_0 \times \cdots \times K_{d-1} \leq \Psi(\mathrm{St}_G(1) \cap K)$.

Definition 2.3. A group G of tree automorphisms of X^* that acts transitively on the levels of the tree X^* is called a *regular weakly branch group* over its normal subgroup K if

$$K \times \cdots \times K \preceq K.$$

If, in addition, the index of K in G is finite then G is called a *regular branch group* over K.

It can be shown that if G is a regular (weakly) branch group than it is a (weakly) branch group.

Definition 2.4. A self-similar group G is called *self-replicating* if, for every vertex u in X^*, the map $\varphi_u : G_u \to G$ given by $\varphi_u(g) = g_u$ is onto (where G_u is the stabilizer of the vertex u in G).

Note that $\mathcal{G} = \mathrm{IMG}(z^2 + \mathrm{i})$ is self-replicating. This is clear from the equalities

$$b = (a, c), \quad c = (b, 1), \quad aba = (c, a), \quad aca = (1, b).$$

Consider the normal subgroup N of $\mathcal{G}$ defined by

$$N = \langle [a, b], [b, c] \rangle^{\mathcal{G}}.$$

By definition, $[g, h] = g^{-1}h^{-1}gh$ and $\langle \cdot \rangle^{\mathcal{G}}$ denotes normal closure in $\mathcal{G}$.

Theorem 2.5. *The group $\mathcal{G}$ is a regular branch group over N.*

Proof. First we observe that N has finite index in $\mathcal{G}$. Direct computation shows that $a^2 = b^2 = c^2 = (ac)^4 = (ab)^8 = (bc)^8 = 1$, so $\mathcal{G}/N$ is a homomorphic image of

$$\langle a, b, c \mid a^2 = b^2 = c^2 = (ac)^4 = [a, b] = [b, c] = 1 \rangle \cong \mathrm{C}_2 \times \mathrm{D}_4,$$

where C_2 is the cyclic group of order 2 and D_4 is the dihedral group of order 8.

Further, we have

$$[b, c] = ([a, b], 1) \quad [c, b^a] = ([b, c], 1)$$

Since $[c, b^a] = cb[b, a]cb[b, a] = [b, a]^{bc}[c, b][b, a] \in N$ we have that $([a, b], 1)$ and $([b, c], 1)$ are elements in N. The fractalness of $\mathcal{G}$ enables us to conjugate the sections in $([a, b], 1)$ and $([b, c], 1)$ by arbitrary elements in $\mathcal{G}$ without leaving N. Thus we get the inclusion $N \times 1 \preceq N$. The transitivity of $\mathcal{G}$ on the first level then implies

$$N \times N \preceq N.$$

The level transitivity of $\mathcal{G}$ can be obtained almost for free. The fact that N is nontrivial along with the fact that $N \times N \preceq N$ implies that N is infinite, and hence so is $\mathcal{G}$. But one can prove that a self-similar group acting on a binary rooted tree is infinite if and only if it acts transitively on all levels. □

3. L-presentation

The goal of this section is to prove the following result.

Theorem 3.1. *The group $\mathcal{G}$ has the following L-presentation*

$$\mathcal{G} \cong \langle a, b, c \mid \phi^n(a^2), \phi^n((ac)^4), \phi^n([c, ab]^2), \phi^n([c, bab]^2), \phi^n([c, ababa]^2), \phi^n([c, ababab]^2), \phi^n([c, bababab]^2), n \geq 0 \rangle, \tag{3.1}$$

where ϕ is the substitution defined on words in the free monoid over the alphabet $\{a, b, c\}$ by

$$\phi\colon \begin{cases} a \to b, \\ b \to c, \\ c \to aba. \end{cases}$$

In order to prove Theorem 3.1 we introduce some notation and prove a few intermediate results.

The group

$$\Gamma = \langle a, b, c \mid a^2, b^2, c^2, (ac)^4 \rangle$$

covers $\mathcal{G}$ (the relators of Γ are relators of $\mathcal{G}$). The action of $\mathcal{G}$ on the binary tree induces an action of the covering group Γ on the same tree, which is not faithful. Let Ω be the kernel of this action. Then, obviously, a set of generators of Ω as a normal subgroup in Γ, together with the relators in Γ constitutes a presentation for $\mathcal{G}$.

The embedding $\mathcal{G} \hookrightarrow \mathcal{G} \wr \mathrm{Sym}(2)$ induces a homomorphism

$$\Psi\colon \Gamma \to \Gamma \wr \mathrm{Sym}(2)$$

defined by

$$\Psi\colon \begin{cases} a \mapsto (1, 1)\sigma, \\ b \mapsto (a, c), \\ c \mapsto (b, 1). \end{cases}$$

Indeed, the relators of Γ are mapped to the trivial element $(1, 1)$ of $\Gamma \wr \mathrm{Sym}(2)$:

$$\Psi(a^2) = (1, 1)\sigma(1, 1)\sigma = (1, 1), \qquad \Psi(b^2) = (a, c)^2 = (a^2, c^2) = (1, 1),$$
$$\Psi(c^2) = (b, 1)^2 = (b^2, 1) = (1, 1), \qquad \Psi((ac)^4) = ((1, b)\sigma)^4 = (b^2, b^2) = (1, 1).$$

The homomorphism Ψ induces homomorphisms $\Psi_n\colon \Gamma \to \Gamma \wr (\wr_{i=1}^n \mathrm{Sym}(2))$ (here $\wr_{i=1}^n \mathrm{Sym}(2)$ denotes the iterated permutational wreath product) defined recursively by $\Psi_1 = \Psi$ and

$$\Psi_n\colon \Gamma \xrightarrow{\Psi_{n-1}} \Gamma \wr \left(\wr_{i=1}^{n-1} \mathrm{Sym}(2) \right) \xrightarrow{\Psi} (\Gamma \wr \mathrm{Sym}(2)) \wr \left(\wr_{i=1}^{n-1} \mathrm{Sym}(2) \right) = \Gamma \wr \left(\wr_{i=1}^{n} \mathrm{Sym}(2) \right).$$

If, for $g \in \Gamma$, we have $\Psi_n(g) = (g|_u, u \in X^n)\sigma_n$, with $g|_u \in \Gamma$ and $\sigma_n \in \wr_{i=1}^n \mathrm{Sym}(2)$ we call $g|_u$ the *section* of g at u.

For every $g \in \Gamma$ denote by $l(g)$ the length of the shortest word in a, b, c representing g in Γ. The following lemma shows that Γ possesses the so called *contraction* property.

Lemma 3.2. *For every $g \in \Gamma$ and $u \in X^2$*

$$l(g|_u) \leq \frac{l(g)+1}{2}. \tag{3.2}$$

PROOF. Observe that, because of the self-similarity, all generators satisfy inequality (3.2). Indeed,

$$\Psi_2(a) = (1,1,1,1)(02)(13), \quad \Psi_2(b) = (1,1,b,1)(01), \quad \Psi_2(c) = (a,c,1,1),$$

where, for ease of notation, the vertices on the second level are renamed by using the identifications $00 \leftrightarrow 0$, $01 \leftrightarrow 1$, $10 \leftrightarrow 2$, and $11 \leftrightarrow 3$.

All pairwise products of generators also satisfy inequality (3.2):

$$\begin{aligned} \Psi_2(a^2) = 1, \quad \Psi_2(b^2) &= 1, \quad \Psi_2(c^2) = 1, \\ \Psi_2(ab) &= \big(\Psi_1(c), \Psi_1(a)\big)\sigma = (b,1,1,1)(0213), \\ \Psi_2(ba) &= \big(\Psi_1(a), \Psi_1(c)\big)\sigma = (1,1,b,1)(0312), \\ \Psi_2(ac) &= \big(\Psi_1(1), \Psi_1(b)\big)\sigma = (1,1,a,c)(02)(13), \\ \Psi_2(ca) &= \big(\Psi_1(b), \Psi_1(1)\big)\sigma = (a,c,1,1)(02)(13), \\ \Psi_2(bc) &= \big(\Psi_1(ab), \Psi_1(c)\big) = (c,a,b,1)(01), \\ \Psi_2(cb) &= \big(\Psi_1(ba), \Psi_1(c)\big) = (a,c,b,1)(01). \end{aligned} \tag{3.3}$$

Any word w in a, b, c of length n can be split into a product of at most $(n+1)/2$ products of pairs of generators (if the length of w is odd one can pair the last letter in w with 1). Therefore the sections of w on the vertices of the second level are products of at most $(n+1)/2$ letters. Thus the inequality (3.2) holds for w as well. □

Define an increasing sequence of subgroups of Γ by

$$\Omega_n = \ker \Psi_n.$$

Lemma 3.3. *The kernel Ω of the canonical epimorphism $\Gamma \to \mathcal{G}$ satisfies*

$$\Omega = \bigcup_{n \geq 1} \Omega_n$$

PROOF. Let h be a word in a, b, c of length at most $2^n + 1$ representing the trivial element in $\mathcal{G}$. Then, since for any words $u, v \in X^*$

$$h|_{uv} = h|_u|_v, \tag{3.4}$$

by Lemma 3.2 we obtain that all sections of h have length at most 1 on the $2(n+1)$th level. Therefore they must be trivial, because h acts trivially on the tree. Hence, $h \in \Omega_{2(n+1)}$. □

The last lemma reduces the problem of finding generators for Ω to finding generators for Ω_n. We start from $\Omega_1 = \ker \Psi$ and, based on it, derive generators for Ω_n.

Let $H = \mathrm{St}_\Gamma(1)$ be the stabilizer of the first level of the tree in Γ.

Lemma 3.4. *The group H has the following presentation*

$$H = \langle \beta, \delta, \gamma, \rho \mid \beta^2 = \delta^2 = \gamma^2 = \rho^2 = (\rho\delta)^2 = 1 \rangle,$$

where $\beta = b$, $\delta = c$, $\gamma = aba$, $\rho = aca$.

PROOF. The index of H in Γ is 2 and the coset representatives are $\{1, a\}$. The Reidemeister–Schreier procedure gives the above presentation. □

Obviously, each Ω_n is a subgroup of H. Therefore one can restrict Ψ to H. Since H stabilizes the first level one can think of Ψ as a homomorphism $H \to \Gamma\times\Gamma$. This map is given by

$$\Psi\colon \begin{cases} \beta = b \to (a,c), \\ \gamma = aba \to (c,a), \\ \delta = c \to (b,1), \\ \rho = aca \to (1,b), \end{cases}$$

which mimics the corresponding embedding of the generators b, aba, c, aca of $\mathrm{St}_G(1)$ into $G \times G$.

Define the following words in Γ:

$$U_1 = (ba)^8, \qquad U_2 = [c, ab]^2, \qquad U_3 = [c, bab]^2,$$
$$U_4 = [c, ababa]^2, \qquad U_5 = [c, ababab]^2, \qquad U_6 = [c, bababab]^2.$$

Lemma 3.5. $\Omega_1 = \langle U_1, U_2, U_3, U_4, U_5, U_6\rangle^\Gamma$.

PROOF. In order to find a generating set for $\Omega_1 = \ker\Psi$ we first describe $\Psi(H)$. Since $\Psi(\delta) = (b,1)$ and $\Psi(\rho) = (1,b)$, we get

$$B \times B \trianglelefteq \Psi(H),$$

where $B = \langle b\rangle^\Gamma$. Furthermore, $\Psi(H)/(B\times B) \cong \langle (a,c),(c,a)\rangle \cong \mathrm{D}_4$. Therefore

$$\Psi(H) \cong (B\times B) \rtimes \mathrm{D}_4.$$

Now we provide a presentation for B.
Define

$$\xi_1 = b, \qquad \xi_2 = b^a, \qquad \xi_3 = b^c, \qquad \xi_4 = b^{ca},$$
$$\xi_5 = b^{ac}, \qquad \xi_6 = b^{aca}, \qquad \xi_7 = b^{cac}, \qquad \xi_8 = b^{acac}.$$

Since $\Gamma = C_2 * D_4$ where the cyclic group C_2 of order 2 is generated by b and the dihedral group D_4 of order 8 is generated by a and c, it is clear that B is generated by all conjugates of b by the elements in $D_4 = \langle a, c\rangle$. Thus $\{\ \xi_i \mid i = 1,\dots,8\ \}$ is a generating set for B. Moreover, it is clear that

$$B = \langle \xi_i, i = 1,\dots,8 \mid \xi_i^2 = 1, i = 1,\dots,8\rangle,$$

i.e., B is a free product of 8 copies of the cyclic group of order 2 (indeed, none of the b's in an expression of the form $\xi_{i_1}\xi_{i_2}\cdots\xi_{i_m}$ can be canceled in Γ when $i_j \neq i_{j+1}$ for $j = 1,\dots,m-1$).

Therefore $B\times B$ is generated by 16 elements, namely, $\tilde{\xi}_i = (\xi_i, 1)$ and $\hat{\xi}_i = (1, \xi_i)$ and is presented by

$$B \times B = \langle \tilde{\xi}_i, \hat{\xi}_j, i,j = 1,\dots,8 \mid \tilde{\xi}_i^2, \hat{\xi}_j^2, [\tilde{\xi}_i, \hat{\xi}_j], i,j = 1,\dots,8\rangle.$$

Now we compute the action of D_4 generated by $x = (a, c)$ and $y = (c, a)$ on $B \times B$.

$$
\begin{aligned}
\tilde{\xi}_1^x &= (b,1)^{(a,c)} = (aba,1) = \tilde{\xi}_2,\\
\tilde{\xi}_2^x &= (aba,1)^{(a,c)} = (b,1) = \tilde{\xi}_1,\\
\tilde{\xi}_3^x &= (cbc,1)^{(a,c)} = (acbca,1) = \tilde{\xi}_4,\\
\tilde{\xi}_4^x &= (acbca,1)^{(a,c)} = (cbc,1) = \tilde{\xi}_3,\\
\tilde{\xi}_5^x &= (cabac,1)^{(a,c)} = (acabaca,1) = \tilde{\xi}_7,\\
\tilde{\xi}_6^x &= (cacbcac,1)^{(a,c)} = (acacbcaca,1) = \tilde{\xi}_8,\\
\tilde{\xi}_7^x &= (acabaca,1)^{(a,c)} = (cabac,1) = \tilde{\xi}_5,\\
\tilde{\xi}_8^x &= (acacbcaca,1)^{(a,c)} = (cacbcac,1) = \tilde{\xi}_6,\\
\tilde{\xi}_1^y &= (b,1)^{(c,a)} = (cbc,1) = \tilde{\xi}_3,\\
\tilde{\xi}_2^y &= (aba,1)^{(c,a)} = (cabac,1) = \tilde{\xi}_5,\\
\tilde{\xi}_3^y &= (cbc,1)^{(c,a)} = (b,1) = \tilde{\xi}_1,\\
\tilde{\xi}_4^y &= (acbca,1)^{(c,a)} = (cacbcac,1) = \tilde{\xi}_6,\\
\tilde{\xi}_5^y &= (cabac,1)^{(c,a)} = (aba,1) = \tilde{\xi}_2,\\
\tilde{\xi}_6^y &= (cacbcac,1)^{(c,a)} = (acbca,1) = \tilde{\xi}_4,\\
\tilde{\xi}_7^y &= (acabaca,1)^{(c,a)} = (cacabacac,1) = (acacbcaca,1) = \tilde{\xi}_8,\\
\tilde{\xi}_8^y &= (acacbcaca,1)^{(c,a)} = (cacacbcacac,1) = (acabaca,1) = \tilde{\xi}_7.
\end{aligned}
\tag{3.5}
$$

The action on $\hat{\xi}_i$'s can be determined from the action on $\tilde{\xi}_i$. Namely, if $\tilde{\xi}_i^x = \tilde{\xi}_p$ and $\tilde{\xi}_i^y = \tilde{\xi}_q$, then

$$\hat{\xi}_i^x = \hat{\xi}_q \quad \text{and} \quad \hat{\xi}_i^y = \hat{\xi}_p. \tag{3.6}$$

Now we can write down a presentation for $\Psi(H)$.

$$
\begin{aligned}
\Psi(H) = \langle \tilde{\xi}_i, \hat{\xi}_j, i, j = 1, \dots, 8, x, y \mid \tilde{\xi}_i^2 = \hat{\xi}_j^2 = [\tilde{\xi}_i, \hat{\xi}_j] = 1, i, j = 1, \dots, 8,\\
x^2 = y^2 = (xy)^4 = 1, \text{relations (3.5) and (3.6)}\rangle.
\end{aligned}
$$

Note that relations (3.5) and (3.6) show that $\Psi(H) = \langle \tilde{\xi}_1, \hat{\xi}_1, x, y\rangle$. The kernel of Ψ is generated by the preimages of the relators of $\Psi(H)$. We have

$$\Psi(c) = \tilde{\xi}_1, \quad \Psi(aca) = \hat{\xi}_1, \quad \Psi(b) = x, \quad \Psi(aba) = y.$$

We can now start writing down the generators of $\ker\Psi$ as a normal subgroup in H.

$$x^2 \to b^2 = 1 \quad y^2 \to (aba)^2 = 1 \quad (xy)^4 \to (baba)^4 = (ba)^8 = U_1$$

We use the fact that $(ba)^8 \in \ker\Psi$ to simplify the further calculations.

Each of $\tilde{\xi}_i$'s and $\hat{\xi}_i$'s has the form $\tilde{\xi}_1^{z_i}$ and $\hat{\xi}_1^{w_i}$, respectively, for some $z_i, w_i \in \langle x, y\rangle$. The lifts of all elements from $\langle x, y\rangle$ constitute the normal closure $\langle b\rangle^{\mathrm{D}_8}$ of b in $\mathrm{D}_8 = \langle a, b\rangle$. Therefore the lifts of $\tilde{\xi}_i$'s will have form

$$\tilde{\xi}_i \to c^{z},$$

where z runs over all elements of $\langle b\rangle^{\mathrm{D}_8}$. In the same way the lifts of $\hat{\xi}_i$'s look like

$$\hat{\xi}_i \to (aca)^z = c^{az} = c^w,$$

where w runs over the complement of $\langle b\rangle^{\mathrm{D}_8}$ in D_8, which is just the coset $a\langle b\rangle^{\mathrm{D}_8}$.

Hence we get the remaining lifts of relators

$$\tilde{\xi}_i^2 \to (c^z)^2 = z^{-1}czz^{-1}cz = 1,$$
$$\hat{\xi}_i^2 \to (c^w)^2 = w^{-1}cww^{-1}cw = 1,$$
$$[\tilde{\xi}_i, \hat{\xi}_j], i, j = 1, \dots, 8 \to [c^z, c^w], z \in \langle b\rangle^{\mathrm{D}_8}, w \in a\langle b\rangle^{\mathrm{D}_8}$$

Now we would like to simplify the generators we obtained. First of all, since $c^2 = 1$ we immediately get

$$[c^z, c^w] = [c, c^{wz^{-1}}]^z$$

so we can get rid of z (because $wz^{-1} \in a\langle b\rangle^{\mathrm{D}_8}$). Furthermore,

$$[c, c^w] = cw^{-1}cw \cdot cw^{-1}cw = [c, w]^2$$

We can discard 3 more generators since

$$[c, ba]^2 = \left([c, ab]^{-2}\right)^c, \quad [c, bababa]^2 = \left([c, ababab]^{-2}\right)^c \quad [c, a]^2 = (ca)^4 = 1.$$

Thus we get 5 more generators for $\ker \Psi$:

$$[c, ab]^2 = U_2, \quad [c, bab]^2 = U_3, \quad [c, ababa]^2 = U_4,$$
$$[c, ababab]^2 = U_5, \quad [c, bababab]^2 = U_6.$$

These generators, together with $U_1 = (ba)^8$ generate $\ker \Psi$ as a normal subgroup in Γ. □

Lemma 3.6. $\Omega_n = \langle \phi^i(U_j), i = 0, \dots, n-1, j = 1, \dots, 6\rangle^\Gamma$

PROOF. We will use induction on n. For $n = 1$ the statement holds by Lemma 3.5. Assume it is true for some fixed n.

By the definition of Ω_{n+1} we have $\Psi(\Omega_{n+1}) \leq \Omega_n \times \Omega_n$. We will show that $\Psi(\Omega_{n+1}) \geq \Omega_n \times \Omega_n$. Observe that

$$\varphi_1\Big(\Psi\big(\phi^i(U_j)\big)\Big) = 1 \tag{3.7}$$

in Γ (recall that, for an element $h = (h|_0, h|_1)$ in H, $\varphi_1(h) = h|_1$. Indeed, since $\varphi_1\left(\Psi(\phi(\Gamma))\right) \leq \langle a, c\rangle = \mathrm{D}_4$ it's sufficient to check only that all U_i's are trivial in D_4. But this is true since all these words are squares of commutators and $[\mathrm{D}_4, \mathrm{D}_4] \cong \mathbb{Z}/2\mathbb{Z}$.

Equation (3.7) for $i = n+1$, together with the inductive assumption yields $\Omega_n \times 1 \leq \Psi(\Omega_{n+1})$. Since Ω_{n+1} is normal in Γ conjugation by a yields $1 \times \Omega_n \leq \Psi(\Omega_{n+1})$. Therefore

$$\Psi(\Omega_{n+1}) = \Omega_n \times \Omega_n.$$

Equation (3.7) also implies that

$$\Psi(\phi^{n+1}(U_j)) = (\phi^n(U_j), 1), \quad \Psi(\phi^{n+1}(U_j)^a) = (1, \phi^n(U_j)),$$

i.e.,

$$\Psi(\langle \phi^i(U_j), i = 1, \dots, n, j = 1, \dots, 6\rangle^\Gamma) = \Omega_n \times \Omega_n.$$

Therefore

$$\begin{aligned}\Omega_{n+1} &= \ker\Psi \cdot \langle \phi^i(U_j), i=1,\dots,n, j=1,\dots,6\rangle^\Gamma \\ &= \langle \phi^i(U_j), i=0,\dots,n, j=1,\dots,6\rangle^\Gamma.\end{aligned}$$ □

Lemmas 3.6 and 3.3 prove Theorem 3.1. Since $\phi((ac)^4) = (ba)^8 = U_1$, $\phi(a^2) = b^2$, $\phi(b^2) = c^2$ and $\phi(c^2) = (aba)^2 = 1$ the presentation in (3.1) is slightly simplified.

Corollary 3.7. *The group $\mathcal{G}$ embeds into an amenable finitely presented group of exponential growth*

$$\tilde{\mathcal{G}} = \langle a,b,c,s \mid a^2, (ac)^4, [c,ab]^2, [c,bab]^2, [c,ababa]^2, [c,ababab]^2, [c,bababab]^2, \\ a^s = b, b^s = c, c^s = aba\rangle,$$

which is an ascending HNN-extension of $\mathcal{G}$.

PROOF. By Theorem 3.1 and the fact that $\varphi_0(\Psi(\phi(u))) = u$ the substitution ϕ induces an injective endomorphism of $\mathcal{G}$. Thus the HNN-extension construction can be applied. Since $\tilde{\mathcal{G}}$ is an extension of the amenable group $\mathcal{G}$ (see Section 4) by the amenable group $\mathbb{Z}$ generated by s, $\tilde{\mathcal{G}}$ is amenable. The growth of $\tilde{\mathcal{G}}$ is exponential because it contains a free semigroup of rank 2 (it follows from the HNN-extension construction that, for example, s and sa generate such a semigroup). □

4. Self-affine measures and amenability

Although the amenability of $\mathcal{G}$ follows from the intermediate growth of this group, which was established in [7], we present here a different approach based on the tools developed in [6, 18]. More precisely, we construct a particular self-affine measure on $\mathcal{G}$, which proves the vanishing of the asymptotic entropy and, hence, amenability.

Let G be a self-similar group acting spherically transitively on a d-ary tree. Consider a nondegenerate probability measure μ on G (the support of μ generates G). Then for any $x \in X$ one can define a new probability measure $\mu|_x$ on G, which is called the *restriction* of μ on x. The details of the definition and proofs of relevant statements are given in [18] and here we only give the basic idea.

We consider a right random walk $g_n = h_1 h_2 \dots h_n$ on G determined by μ, i.e., $\{h_n\}$ is a sequence of independent variables identically distributed according to the measure μ. We consider $\mathcal{G}$ as embedded in $\mathcal{G} \wr \mathrm{Sym}(2)$ and keep track of the xth coordinate of the image of g_n in $G \wr \mathrm{Sym}(d)$. Recall that h_n is an automorphism of X^*. For $x \in X$, $h_n(x)$ denotes the action of h_n on x. Since

$$g_{n+1}|_x = g_n|_x \cdot h_{n+1}|_{g_n(x)}$$

the probability distribution of $g_{n+1}|_x$ is completely determined by $g_n|_x$ and $g_n(x)$. Therefore the induced random walk

$$(g_n|_x, g_n(x)) \tag{4.1}$$

on $G \times X$ is again a Markov chain. The last random walk is called a *random walk with internal degrees of freedom.* Since X is finite and G acts transitively on X, the subset $G \times \{x\} \subset G \times X$ is recurrent with respect to (4.1). Therefore one can consider the trace of the random walk (4.1) on $G \times \{x\}$, which is also a random walk. Finally, we define the measure $\mu|_x$ as the transition law for the last random walk on $G \times \{x\}$ considered as a copy of G.

There is a convenient way to compute $\mu|_x$ using the properties of the random walk (4.1). The random walk with internal degrees of freedom is governed by the matrix

$$M = (\mu_{xy})_{x,y\in X}$$

whose entries μ_{xy} are subprobability measures on G such that $\mu_{xy}(h)$ is a transition probability of getting to the state (gh, y) from the state (g, x).

With slight abuse of notation, we denote by g the δ-measure concentrated at g. Then the matrix M can be expressed as

$$M = \sum_{g\in \operatorname{supp}\mu} \mu(g)M^g,$$

where

$$M^g_{xy} = \begin{cases} g|_x, & y = g(x), \\ 0, & y \neq g(x), \end{cases}$$

The following theorem is proved in [18].

Theorem 4.1. *The measure $\mu|_x, x \in X$ can be expressed in terms of the matrix M as*

$$\mu|_x = \mu_{xx} + M_{x\bar{x}}(1 - M_{\bar{x}\bar{x}})^{-1}M_{\bar{x}x}, \tag{4.2}$$

where $M_{x\bar{x}}$ (resp., $M_{\bar{x}x}$) denotes the xth row (column) of M from which the entry μ_{xx} is removed, and $M_{\bar{x}\bar{x}}$ is the matrix obtained from M by removing the xth row and the xth column.

One can define $\mu|_w$ for any $w = x_1x_2\cdots x_n \in X^*$ by

$$\mu_w = (\cdots(\mu|_{x_1})|_{x_2}\cdots)|_{x_n}.$$

Definition 4.2. The nondegenerate probability measure μ on a self-similar group G is called *self-affine* (*self-similar* in [18]) if there is a word $w \in X^*$ such that

$$\mu|_w = \alpha e + (1-\alpha)\mu,$$

where $0 < \alpha < 1$ and e is the identity element in G.

For simplicity, we write α instead of αe.

Theorem 4.3 ([18]). *If a self-similar group G carries a self-affine nondegenerate measure μ with finite entropy, then it is amenable.*

In this section we construct such a measure on $\mathcal{G}$. Since this measure should be nondegenerate and have finite entropy the most natural place to look for it is the space Q of positive convex linear combinations of δ-measures concentrated on the generators a, b, and c, i.e., measure μ of the form

$$\mu = xa + yb + zc, \quad x + y + z = 1,\ x, y, z > 0.$$

Suppose we want this measure to be self-affine with respect to $x \in X$. By definition this means $\mu|_x = \alpha + (1-\alpha)\mu$ or, equivalently,

$$\mu = \frac{\mu|_x - \alpha}{1-\alpha}.$$

Since $\mu(e) = 0$ we get $\alpha = \mu|_x(e)$. Thus the measure μ is a fixed point of the transformation

$$\Phi\colon \mu \mapsto \frac{\mu|_x - \mu|_x(e)}{1-\mu|_x(e)}, \tag{4.3}$$

which is defined in [18] and used to prove amenability of a family of groups generalizing Basilica group $\mathrm{IMG}(z^2-1)$.

Let's compute $\mu|_0$ and the corresponding transformation Φ in the case of $\mathcal{G}$. The support of μ is $\{a,b,c\}$ and the corresponding matrices M^g are given by

$$M^a = \begin{pmatrix} 0 & e \\ e & 0 \end{pmatrix}, \quad M^b = \begin{pmatrix} a & 0 \\ 0 & c \end{pmatrix}, \quad M^c = \begin{pmatrix} b & 0 \\ 0 & e \end{pmatrix}. \tag{4.4}$$

Hence,

$$M = xM^a + yM^b + zM^c = \begin{pmatrix} ya+zb & x \\ x & yc+z \end{pmatrix}.$$

By Theorem 4.1

$$\mu|_0 = (ya+zb) + x^2(1-yc-z)^{-1}.$$

Since c has order 2 in $\mathcal{G}$ it is easy see that in the group algebra $\mathbb{R}\mathcal{G}$

$$(1-yc-z)^{-1} = \frac{y}{z^2-2z+1-y^2}\cdot c - \frac{z-1}{z^2-2z+1-y^2}.$$

Therefore

$$\mu|_0 = y\cdot a + z\cdot b + \frac{yx^2}{z^2-2z+1-y^2}\cdot c - \frac{(z-1)x^2}{z^2-2z+1-y^2}$$

and the transformation Φ takes the form

$$\begin{aligned}\Phi(xa+yb+zc) &= \frac{y\cdot a + z\cdot b + yx^2/(z^2-2z+1-y^2)\cdot c}{1+(z-1)x^2/(z^2-2z+1-y^2)} \\ &= \frac{y(z^2-2z+1-y^2)}{z^2-2z+1-y^2+x^2z-x^2}\cdot a \\ &\quad + \frac{z(z^2-2z+1-y^2)}{z^2-2z+1-y^2+x^2z-x^2}\cdot b \\ &\quad + \frac{yx^2}{z^2-2z+1-y^2+x^2z-x^2}\cdot c.\end{aligned}$$

Now we are interested in a fixed point of the rational map $F\colon \mathbb{R}^3 \to \mathbb{R}^3$ induced by Φ, which maps (x,y,z) to the coefficients of $\Phi(xa+yb+zc)$. Moreover, we are searching for such a fixed point only in the invariant simplex $x+y+z=1$, $x,y,z>0$. Fortunately, there is such a fixed point. If $\zeta \approx 0.4786202932$ is the unique real root of the polynomial $Z^3-6Z^2+11Z-4$, then the point

$$(\zeta, \zeta^2-4\zeta+2, -1+3\zeta-\zeta^2)$$

is fixed under F, which produces a self-affine nondegenerate probabilistic measure on $\mathcal{G}$ with finite support, proving amenability of $\mathcal{G}$. This point is unique in the simplex of nondegenerate measures. Indeed, from the equation $F_1(x,y,1-x-y) = x$, where F_1 is the first coordinate of F, we get

$$y = \tfrac{1}{4}(-x^2+x\pm x\sqrt{x^2-10x+9}). \tag{4.5}$$

Substitution in $F_2(x,y,1-x-y) = y$ yields

$$(-x^4+12x^3-39x^2+40x-12) \pm \sqrt{x^2-10x+9}(x^3-7x^2+12x-4) = 0. \tag{4.6}$$

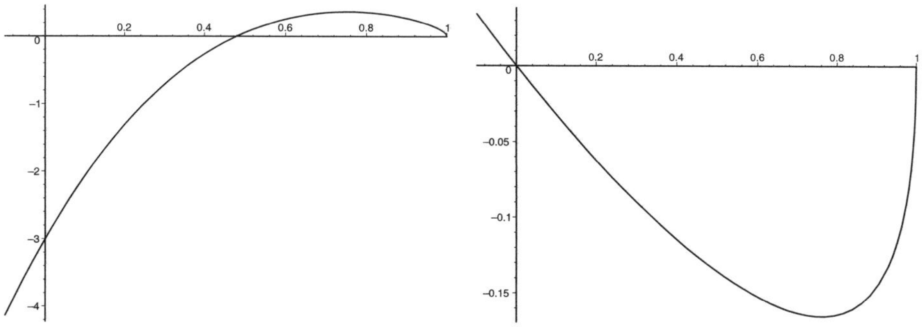

(a) (b)

FIGURE 7. Uniqueness of a self-affine measure

Moving the second summand to the righthand side and squaring both sides produces the equation $x(x-1)(x^3-6x^2+11x-4)=0$, whose unique real root on the interval $(0,1)$ is ζ. The graphs of the two functions in (4.6) are shown in Figure 7: (a) for "plus" and (b) for "minus." The solution comes from (a), so in (4.5) "plus" should be used. It is a routine to check that indeed $y=\zeta^2-4\zeta+2$.

5. Spectral properties and Schur complement

Let H be a Hilbert space and M be an operator on H. Let $H=H_0\oplus H_1$ and there are operators $A\in B(H_0)$, $D\in B(H_1)$, $B\colon H_1\to H_0$ and $C\colon H_0\to H_1$, such that the matrix of M in the basis consisting of the bases of H_0 and H_1 takes the form:

$$M=\begin{pmatrix} A & B\\ C & D\end{pmatrix}.$$

The following fact is of folklore type.

Proposition 5.1. *Let D be invertible. The operator M is invertible if and only if $S_1(M)=A-BD^{-1}C$ is invertible.*

The matrix $S_1(M)$ is called *the first Schur complement* of M.

PROOF. Indeed, the matrix

$$L=\begin{pmatrix} I_0 & 0\\ -D^{-1}C & D^{-1}\end{pmatrix}$$

is invertible. Therefore M is invertible if and only if

$$ML=\begin{pmatrix} A-BD^{-1}C & BD^{-1}\\ 0 & I_1\end{pmatrix}$$

is invertible, which is equivalent to the nonsingularity of $S_1(M)$. □

In our case the action of $\mathcal{G}$ on the boundary X^ω (the set of infinite sequences over X) of the tree X^* induces a unitary representation $\pi_g(f)(x)=f(g^{-1}x)$ of $\mathcal{G}$ in $\mathcal{H}=B\big(L_2(X^\omega)\big)$. The Markov operator $M=\frac{1}{3}(\pi_a+\pi_b+\pi_c)$ corresponding to this unitary representation plays an important role (we do not include inverse elements because all generators are of order 2). The usual method to find the spectrum of M for a self-similar group G is to approximate M with finite dimensional operators

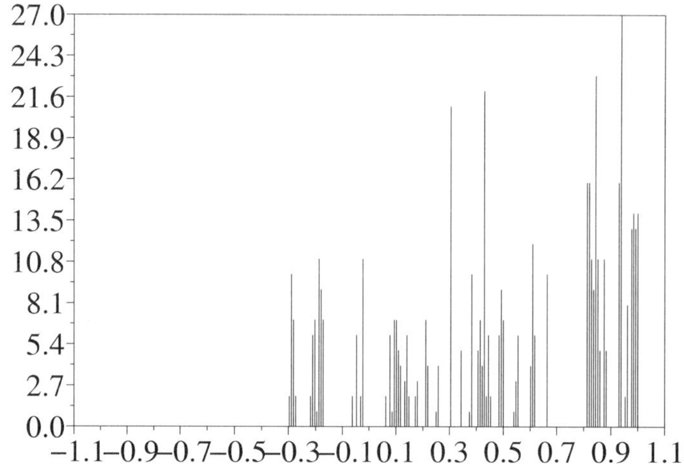

FIGURE 8. Histogram of the spectrum on the 9th level

arising from the action of $\mathcal{G}$ on the levels of the tree X^*. For more on this see [1]. For simplicity we write g for π_g.

Let $\mathcal{H}_n$ be the subspace of $\mathcal{H}$ spanned by the 2^n characteristic functions f_v, $v \in X^n$, of the cylindrical sets, corresponding to the 2^n vertices of the nth level. Then $\mathcal{H}_n$ is invariant under the action of $\mathcal{G}$ and $\mathcal{H}_n \subset \mathcal{H}_{n+1}$. Also $\mathcal{H}_n$ can be naturally identified with $L_2(X^n)$. By $\pi_g^{(n)}$ (or, with a slight abuse of notation, by g_n) we denote the restriction of π_g on $\mathcal{H}_n$. Then, for $n \geq 0$,

$$M_n = \tfrac{1}{3}(a_n + b_n + c_n)$$

are finite-dimensional operators whose spectra converge to the spectrum of M in the sense

$$\mathrm{sp}(M) = \overline{\bigcup_{n\geq 0} \mathrm{sp}(M_n)}.$$

Moreover, if P is the stabilizer of an infinite word from X^ω, then one can consider the Markov operator $M_{G/P}$ on the Schreier graph of G with respect to P. The following fact is observed in [1] and can be applied in the case of $\mathcal{G}$.

Theorem 5.2. *If G is amenable then*

$$\mathrm{sp}(M_{G/P}) = \mathrm{sp}(M).$$

Common practice for finding the spectrum of M, initiated in [1], is to consider a pencil of operators

$$\widetilde{M}(y, z, \lambda) = a + yb + zc - \lambda$$

and find the set $\mathrm{sp}(y, z, \lambda)$ of points (y, z, λ) such that $\widetilde{M}(y, z, \lambda)$ is not invertible. Then the spectrum of M is just the intersection of $\mathrm{sp}(y, z, \lambda)$ with the line $y = z = 1$, shrunk by a factor of $\frac{1}{3}$. We take 1 as the coefficient at a to simplify the computation. Otherwise one can divide it out (we restrict our attention to the case when x, y, z are nonzero).

Let us consider the corresponding pencil $\tilde{M}_n(y, z, \lambda) = a_n + yb_n + zc_n - \lambda$ and find its matrix in the basis $\{f_v : v \in X^n\}$. The orthogonal subspaces $H_n^{(i)} = \mathrm{span}(f_{iv}, v \in X^{n-1})$, $i = 0, 1$ span H_n and are naturally isomorphic to H_{n-1}. The self-similar structure of $\mathcal{G}$ gives the following operator recursion (which coincides with the recursion (4.4))

$$a_n = \begin{pmatrix} 0 & I_{n-1} \\ I_{n-1} & 0 \end{pmatrix}, \quad b_n = \begin{pmatrix} a_{n-1} & 0 \\ 0 & c_{n-1} \end{pmatrix}, \quad c_n = \begin{pmatrix} b_{n-1} & 0 \\ 0 & I_{n-1} \end{pmatrix}, \tag{5.1}$$

for $n > 0$, where I_{n-1} denotes the identity matrix of size 2^{n-1}. The matrices a_0, b_0 and c_0 are equal to the 1×1 matrix $[1]$. For any constant r, we write r instead of rI_n. Thus we have,

$$\widetilde{M}_n(y, z, \lambda) = a_n + yb_n + zc_n - \lambda = \begin{pmatrix} ya_{n-1} + zb_{n-1} - \lambda & 1 \\ 1 & yc_{n-1} + z - \lambda \end{pmatrix}.$$

By Proposition 5.1 in case $yc_{n-1} + z - \lambda$ is invertible the operator $\widetilde{M}_n(y, z, \lambda)$ is invertible if and only if $S_1(\widetilde{M}_n(y, z, \lambda))$ is invertible. The inverse of $yc_{n-1} + z - \lambda$ in $\mathbb{R}\mathcal{G}$ is

$$\frac{y}{-y^2 + z^2 - 2z\lambda + \lambda^2} \cdot c_{n-1} + \frac{-z + \lambda}{-y^2 + z^2 - 2z\lambda + \lambda^2}.$$

Hence, $yc_{n-1} + z - \lambda$ is not invertible if and only if $-y^2 + z^2 - 2z\lambda + \lambda^2 = (z - \lambda - y) \times (z - \lambda + y) = 0$. Denote the union of these 2 planes by Z_1. Note, that $\widetilde{M}_n(y, z, \lambda)$ is not necessary singular at each point of Z_1.

The first Schur complement of $\widetilde{M}$ is

$$\begin{aligned} &S_1(\widetilde{M}_n(y, z, \lambda)) \\ &\quad = ya_{n-1} + zb_{n-1} - \lambda - (yc_{n-1} + z - \lambda)^{-1} \\ &\quad = y \cdot a_{n-1} + z \cdot b_{n-1} + \frac{y}{-y^2 + z^2 - 2z\lambda + \lambda^2} \cdot c_{n-1} + \frac{-z + \lambda}{-y^2 + z^2 - 2z\lambda + \lambda^2} - \lambda. \end{aligned}$$

If $y = 0$ we get $S_1(\widetilde{M}_n(0, z, \lambda)) = zb_{n-1} + \frac{1}{\lambda - z} - \lambda$ is not invertible if and only if

$$\begin{aligned} \det &\begin{pmatrix} 1/(\lambda - z) - \lambda & z \\ z & 1/(\lambda - z) - \lambda \end{pmatrix} \\ &\qquad = \frac{1}{(\lambda - z)^2}(1 - (\lambda - z)(\lambda + z))(1 - \lambda + z)(1 + \lambda - z) = 0. \end{aligned}$$

Denote corresponding union of a hyperbola and two lines in $\mathbb{R}^3$ by Z_2. Note that $Z_2 \cap Z_1 = \varnothing$.

If $y \neq 0$ then

$$\begin{aligned} \frac{1}{y} S_1(\widetilde{M}_n(y, z, \lambda)) &= a_{n-1} + \frac{z}{y} \cdot b_{n-1} + \frac{1}{-y^2 + z^2 - 2z\lambda + \lambda^2} \cdot c_{n-1} \\ &\qquad - \frac{-\lambda y^2 + \lambda z^2 - 2z\lambda^2 + \lambda^3 + z - \lambda}{y(-y^2 + z^2 - 2z\lambda + \lambda^2)} \\ &= \widetilde{M}_{n-1}(F(y, z, \lambda)), \end{aligned}$$

where $F \colon \mathbb{R}^3 \to \mathbb{R}^3$ is the rational map defined by

$$F \colon (y, z, \lambda) \to \left(\frac{z}{y}, \frac{1}{-y^2 + z^2 - 2z\lambda + \lambda^2}, \frac{-\lambda y^2 + \lambda z^2 - 2z\lambda^2 + \lambda^3 + z - \lambda}{y(-y^2 + z^2 - 2z\lambda + \lambda^2)} \right).$$

Therefore the set $\mathrm{sp}_n(y, z, \lambda)$ of points (y, z, λ) where $\widetilde{M}_n(y, z, \lambda)$ is not invertible in this case ($y \neq 0$) is a preimage under F of the corresponding set

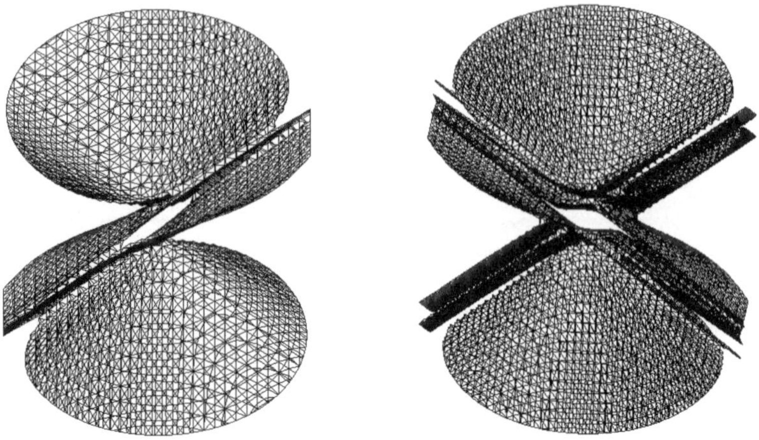

FIGURE 9. Part of the spectrum of $\widetilde{M}(y,z,\lambda)$

$\mathrm{sp}_{n-1}(y,z,\lambda)$. To summarize,

$$Z_2 \cup F^{-1}(\mathrm{sp}_{n-1}(y,z,\lambda)) \subset \mathrm{sp}_n(y,z,\lambda) \subset Z_1 \cup Z_2 \cup F^{-1}(\mathrm{sp}_{n-1}(y,z,\lambda)). \tag{5.2}$$

Since $\widetilde{M}_0(y,z,\lambda) = (1+y+z-\lambda)$ we have

$$\mathrm{sp}_0(y,z,\lambda) = \{(y,z,\lambda) : 1+y+z-\lambda = 0\}. \tag{5.3}$$

Denote this plane by P.

Equations (5.2) and (5.3) show that

$$F^{-n}(P) \cup \bigcup_{i=0}^{n-1} F^{-i}(Z_2) \subset \mathrm{sp}_n(y,z,\lambda) \subset F^{-n}(P) \cup \bigcup_{i=0}^{n-1} F^{-i}(Z_1 \cup Z_2). \tag{5.4}$$

Since Z_2 consists of points with $y=0$ every point (y,z,λ) from $F^{-1}(Z_2)$ must satisfy $z=0$. But the preimages of all such points are empty. Hence, $F^{-2}(Z_2) = \varnothing$ and $\bigcup_{i=0}^{n-1} F^{-i}(Z_2) = Z_2 \cup F^{-1}(Z_2)$. Denote the last subset by Z_3.

One can easily check that $P \subset F^{-1}(P)$ and, hence, $F^{-n}(P) = \bigcup_{i=0}^{n} F^{-i}(P)$. Therefore equation (5.4) transforms to

$$Z_3 \cup \bigcup_{i=0}^{n} F^{-i}(P) \subset \mathrm{sp}_n(y,z,\lambda) \subset F^{-n}(P) \cup Z_3 \cup \bigcup_{i=0}^{n-1} F^{-i}(P \cup Z_1).$$

Thus, the spectrum of the operator $\widetilde{M}$ on $L_2(X^\omega)$ satisfies

$$\overline{Z_3 \cup \bigcup_{i=0}^{\infty} F^{-i}(P)} \subset \mathrm{sp}(\tilde{M}(y,z,\lambda)) = \overline{\bigcup_{i=0}^{\infty} \mathrm{sp}_n(y,z,\lambda)} \subset \overline{Z_3 \cup \bigcup_{i=0}^{\infty} F^{-i}(P \cup Z_1)}.$$

Note, that the sets $A = \overline{Z_3 \cup \bigcup_{i=0}^{\infty} F^{-i}(P)}$ and $B = \overline{Z_3 \cup \bigcup_{i=0}^{\infty} F^{-i}(P \cup Z_1)}$ are almost invariant with respect to F, in the sense that

$$F^{-1}(A) \cup Z_2 = A, \quad F^{-1}(B) \cup Z_1 \cup Z_2 = B,$$

which is an analog of [17, Theorem 4.1] for the Basilica group.

The preimages of the plane P under F^4 and F^5 are shown in Figure 9.

Note that there are points in the spectrum of $\tilde{M}(y,z,\lambda)$ which do not belong to any preimage of the plane P. In particular, the point $(-\frac{1}{2},0,\frac{1}{2})$ belongs to Z_1 so it is not in the domain of F, but

$$\det \tilde{M}_n(-\tfrac{1}{2},0,\tfrac{1}{2}) = \det((a_{n-1}+1)(c_{n-1}+1)-4) = 0$$

since 4 is an eigenvalue of $(a_{n-1}+1)(c_{n-1}+1)$. However, this point could be in the closure of the union of all preimages of P.

On the other hand we can formulate a conjecture that the spectrum of $M = \frac{1}{3}(a+b+c)$ is the intersection of the line $y=z=1$ with $A = \overline{Z_3 \cup \bigcup_{i=0}^{\infty} F^{-i}(P)}$, shrunk by a factor of $\frac{1}{3}$. This conjecture survives at least up to the 6th level.

Note also that the map F is conjugate to a simpler map

$$G\colon (y,z,\lambda) \to \left(\frac{z}{y}, \frac{\lambda}{y}(-2+y\lambda), \frac{1}{\lambda}(-y+y\lambda^2-\lambda)\right)$$

by the conjugator map

$$(y,z,\lambda) \to \left(\frac{1}{y}, \frac{1}{z}, y+z-\lambda\right).$$

The histogram for the spectral density of the operator M_n acting on 9th level is shown in Figure 8.

Further steps are required to identify the spectrum of the pencil $\widetilde{M}(y,z,\lambda)$ and of M more precisely. This is related to the problem of finding invariant subsets of the rational map F. Perhaps the spectrum of M is just the intersection of the "strange attractor" of F with the line $y=z=1$, shrunk by factor of $\frac{1}{3}$. In any case here we have one more example when the spectral problem is related to the dynamics of a multidimensional rational map. There is a hope that the methods developed for this type of transformations (see, for instance [21]) could help to handle this case.

References

1. L. Bartholdi and R. I. Grigorchuk, *On the spectrum of Hecke type operators related to some fractal groups*, Tr. Mat. Inst. Steklova **231** (2000), 5–45 (Russian); English transl., Proc. Steklov Inst. Math. **2000**, no. 4(231), 1–41.
2. L. Bartholdi, R. I. Grigorchuk, and V. Nekrashevych, *From fractal groups to fractal sets*, Fractals in Graz 2001, Trends Math., Birkhäuser, Basel, 2003.
3. L. Bartholdi, R. I. Grigorchuk, and Z. Šunić, *Branch groups*, Handbook of Algebra, Vol. 3, North-Holland, Amsterdam, 2003.
4. L. Bartholdi and V. Nekrashevych, *Thurston equivalence of topological polynomials*, Acta Math. **197** (2006), no. 1, 1–51.
5. ———, *Iterated monodromy groups of quadratic polynomials*. I, available at `arXiv:math.GR/0611177`.
6. L. Bartholdi and B. Virág, *Amenability via random walks*, Duke Math. J. **130** (2005), no. 1, 39–56, available at `arXiv:math.GR/0305262`.
7. K.-U. Bux and R. Pérez, *On the growth of iterated monodromy groups*, Topological and Asymptotic Aspects of Group Theory, Contemp. Math., vol. 394, Amer. Math. Soc., Providence, RI, 2006, pp. 61–76, available at `arXiv:math.GR/0405456`.
8. R. I. Grigorchuk, *On Burnside's problem on periodic groups*, Funktsional. Anal. i Prilozhen. **14** (1980), no. 1, 53–54 (Russian); English transl., Functional Anal. Appl. **14** (1980), no. 1, 41–43.
9. ———, *Degrees of growth of finitely generated groups and the theory of invariant means*, Izv. Akad. Nauk SSSR Ser. Mat. **48** (1984), no. 5, 939–985 (Russian); English transl., Math. USSR-Izv. **25** (1985), no. 2, 259–300.

10. ———, *An example of a finitely presented amenable group that does not belong to the class EG*, Mat. Sb. **189** (1998), no. 1, 79–100 (Russian); English transl., Sb. Math. **189** (1998), no. 1-2, 75–95.
11. ———, *Just infinite branch groups*, New Horizons in Pro-p Groups, Progr. Math., vol. 184, Birkhäuser, Boston, MA, 2000, pp. 121–179.
12. R. I. Grigorchuk, V. Nekrashevych, and V. I. Sushchanskiĭ, *Automata, dynamical systems, and groups*, Tr. Mat. Inst. Steklova **231** (2000), 134–214 (Russian); English transl., Proc. Steklov Inst. Math. **2000**, no. 4(231), 128–203.
13. R. I. Grigorchuk and V. Nekrashevych, *Schur complement, operator algebras and self-similar groups*, available at `arXiv:math.GR/0612421`.
14. R. I. Grigorchuk and Z. Šunik, *Asymptotic aspects of Schreier graphs and Hanoi Towers groups*, C. R. Math. Acad. Sci. Paris **342** (2006), no. 8, 545–550, available at `arXiv:math.GR/0601592`.
15. R. I. Grigorchuk and A. Żuk, *On the asymptotic spectrum of random walks on infinite families of graphs*, Random Walks and Discrete Potential Theory (Cortona, 1997), Sympos. Math., vol. 39, Cambridge Univ. Press, Cambridge, 1999, pp. 188–204.
16. ———, *On a torsion-free weakly branch group defined by a three state automaton*, Internat. J. Algebra Comput. **12** (2002), no. 1-2, 223–246.
17. ———, *Spectral properties of a torsion-free weakly branch group defined by a three state automaton*, Computational and Statistical Group Theory (Las Vegas, NV/Hoboken, NJ, 2001), Contemp. Math., vol. 298, Amer. Math. Soc., Providence, RI, 2002, pp. 57–82.
18. V. A. Kaimanovich, *"Münchhausen trick" and amenability of self-similar groups*, Internat. J. Algebra Comput. **15** (2005), no. 5-6, 907–937.
19. V. Nekrashevych, *Self-similar groups*, Math. Surveys Monogr., vol. 117, Amer. Math. Soc., Providence, RI, 2005.
20. ———, *A minimal Cantor set in the space of* 3*-generated groups*, to appear in Geom. Dedicata.
21. N. Sibony, *Dynamique des applications rationnelles de* $\mathbf{P}^k$, Dynamique et géométrie complexes (Lyon, 1997), Panor. Synthèses, vol. 8, Soc. Math. France, Paris, 1999, pp. 97–185.

Department of Mathematics, Texas A&M University, College Station, TX 77843-3368, USA

E-mail address, R.Grigorchuk: `grigorch@math.tamu.edu`

E-mail address, D.Savchuk: `savchuk@math.tamu.edu`

E-mail address, Z. Šunić: `sunic@math.tamu.edu`

Centre de Recherches Mathématiques
CRM Proceedings and Lecture Notes
Volume **42**, 2007

On Scattering of Solitons for Wave Equation Coupled to a Particle

Valery Imaykin, Alexander Komech, and Boris Vainberg

ABSTRACT. We establish a long time soliton asymptotics for a nonlinear system of wave equation coupled to a charged particle. The coupled system has a six dimensional manifold of soliton solutions. We show that in the large time approximation, any solution, with an initial state close to the solitary manifold, is a sum of a soliton and a dispersive wave which is a solution of the free wave equation. It is assumed that the charge density satisfies the Wiener condition which is a version of the Fermi Golden Rule and that the total charge of the particle equals zero. The proof is based on a development of the general strategy introduced in the papers of Soffer and Weinstein, Buslaev and Perel′man, and others: symplectic projection in Hilbert space onto the solitary manifold, modulation equations for the parameters of the projection, and decay of the transversal component.

1. Introduction

Our paper concerns an old mathematical problem of nonlinear field-particle interaction. A charged particle radiates a field which acts on the particle, etc. This self-action is probably responsible for some crucial features of the process: asymptotically uniform motion and stability of the particle, increment of the particle's mass, etc. (see [23]). The problem has many different appearances: for a classical particle coupled to a scalar or Maxwell field, for coupled Maxwell – Schrödinger or Maxwell – Dirac equations, for the corresponding second-quantized equations, etc.

One of the main goals of a mathematical investigation of this problem is studying soliton type long time asymptotics and asymptotic stability of soliton solutions to the equations.

First results in this direction have been discovered for KdV equation and other *complete integrable equations*. For KdV equation, any solution with sufficiently

---

2000 *Mathematics Subject Classification.* Primary 35L70, 35P25; Secondary 35Q51, 81Uxx, 37K40.

The two first authors are supported partially by Austrian Science Foundation (FWF) Project (P19138-N13), by research grants of DFG (436 RUS 113/615/0-1(R)) and RFBR (01-01-04002).

The second author is supported partially by Max-Planck Institute of Mathematics in the Sciences (Leipzig), and Wolfgang Pauli Institute of Vienna University.

The third author is supported partially by the NSF grant DMS-0405927.

This is the final form of the paper.

smooth and rapidly decaying initial data converges to a finite sum of solitons moving to the right, and a dispersive wave moving to the left. A complete survey and proofs can be found in [5].

For nonintegrable equations the long time convergence of the solution to a soliton part and dispersive wave was obtained first by Soffer and Weinstein in the context of U(1)-invariant Schrödinger equation, [18–20]. The extension to translation invariant equations was obtained by Buslaev and Perel′man [1, 2] for 1D Schrödinger equation, and by Miller, Pego and Weinstein for 1D modified KdV and RLW equations, [15–17]. A fundamental role in all these results play the techniques introduced originally in the Weinstein paper [27].

In [1,2] the long time convergence is obtained for translation invariant and U(1)-invariant nonlinear Schrödinger equation: for any finite-energy solution $\psi(x,t)$ with initial data close to a soliton $\psi_{v_0}(x - v_0 t - a_0)e^{i\omega_0 t}$, the following asymptotics hold:

$$\psi(x,t) = \psi_{v_\pm}(x - v_\pm t - a_\pm)e^{i\omega_\pm t} + W_0(t)\psi_\pm + r_\pm(x,t), \quad t \to \pm\infty. \tag{1.1}$$

Here the first term of the right-hand side is a soliton with parameters $v_\pm$, $a_\pm$, $\omega_\pm$ close to v_0, a_0, ω_0, the function $W_0(t)\psi_\pm$ is a dispersive wave which is a solution to the free Schrödinger equation, and the remainder $r_\pm(x,t)$ converges to zero in the global L^2-norm. Recently Cuccagna extended the results to nD Schrödinger equations with $n \geq 3$, [3, 4].

We consider a scalar real-valued wave field $\psi(x)$ in $\mathbb{R}^3$, coupled to a relativistic particle with position q and momentum p, governed by

$$\begin{aligned} &\dot\psi(x,t) = \pi(x,t), \qquad &&\dot\pi(x,t) = \Delta\psi(x,t) - \rho\big(x - q(t)\big), \qquad x \in \mathbb{R}^3, \\ &\dot q(t) = \frac{p(t)}{\sqrt{1+p^2}}, &&\dot p(t) = \int \psi(x,t)\nabla\rho\big(x - q(t)\big)\,dx. \end{aligned} \tag{1.2}$$

This is a Hamilton system with the Hamilton functional

$$\mathcal{H}(\psi,\pi,q,p) = \frac{1}{2}\int \big(|\pi(x)|^2 + |\nabla\psi(x)|^2\big)\,dx + \int \psi(x)\rho(x-q)\,dx + \sqrt{1+p^2}. \tag{1.3}$$

The first two equations for the fields are equivalent to the wave equation with the source $\rho(x - q)$. We have set the mechanical mass of the particle and the speed of wave propagation equal to one. The case of the point particle corresponds to $\rho(x) = \delta(x)$ and then the interaction term in the Hamiltonian is simply $\psi(q)$. However, in this case the Hamiltonian is unbounded from below which leads to the ill-posedness of the problem, that is also known as ultraviolet divergence. Therefore we smooth the coupling by the function $\rho(x)$ following the "extended electron" strategy proposed by M. Abraham for the Maxwell field. In analogy to the Maxwell-Lorentz equations we call ρ the "charge distribution." Finally, the form of the last two equations in (1.2) is determined by the choice of the relativistic kinetic energy $\sqrt{1+p^2}$ in (1.3).

Let us write the system (1.2) as

$$\dot Y(t) = F\big(Y(t)\big), \quad t \in \mathbb{R}, \tag{1.4}$$

where $Y(t) := \big(\psi(x,t), \pi(x,t), q(t), p(t)\big)$. Below we always deal with column vectors but often write them as row vectors. The system (1.2) is translation-invariant and admits soliton solutions

$$Y_{a,v}(t) = (\psi_v(x - vt - a), \pi_v(x - vt - a), vt + a, p_v), \quad p_v = \frac{v}{\sqrt{1-v^2}} \tag{1.5}$$

for all $a, v \in \mathbb{R}^3$ with $|v| < 1$ (see (2.7) below). The states $S_{a,v} := Y_{a,v}(0)$ form the solitary manifold

$$\mathcal{S} := \{S_{a,v} : a, v \in \mathbb{R}^3, |v| < 1\}. \tag{1.6}$$

Our main result is the soliton-type asymptotics of type (1.1),

$$(\psi(x,t), \pi(x,t)) \sim (\psi_{v_\pm}(x - v_\pm t - a_\pm), \pi_{v_\pm}(x - v_\pm t - a_\pm)) + W^0(t)\mathbf{\Psi}_\pm, \quad t \to \pm\infty, \tag{1.7}$$

for solutions to (1.2) with initial data close to the solitary manifold $\mathcal{S}$. Here $W^0(t)$ is the dynamical group of the free wave equation, $\mathbf{\Psi}_\pm$ are the corresponding *asymptotic scattering states*, and the remainder converges to zero *in the global energy norm*, i.e., in the norm of the Sobolev space $\dot{H}^1(\mathbb{R}^3) \oplus L^2(\mathbb{R}^3)$, see Section 2. For the particle trajectory we prove that

$$\dot{q}(t) \to v_\pm, \quad q(t) \sim v_\pm t + a_\pm, \qquad t \to \pm\infty. \tag{1.8}$$

The results are established under the following conditions on the charge distributions: ρ is a real valued function of the Sobolev class $H^2(\mathbb{R}^3)$, compactly supported, and spherically symmetric, i.e.,

$$\rho, \nabla\rho, \nabla\nabla\rho \in L^2(\mathbb{R}^3), \quad \rho(x) = 0 \text{ for } |x| \geq R_\rho, \quad \rho(x) = \rho_1(|x|). \tag{1.9}$$

We require that all "nonzero modes" of the wave field are coupled to the particle. This is formalized by the Wiener condition

$$\hat{\rho}(k) = (2\pi)^{-3/2} \int e^{ikx} \rho(x)\, dx \neq 0 \quad \text{for all } k \in \mathbb{R}^3 \setminus \{0\}. \tag{1.10}$$

It is an analogue of the Fermi Golden Rule: the coupling term $\rho(x - q)$ is not orthogonal to the eigenfunctions e^{ikx} of the continuous spectrum of the linear part of the equation (cf. [21, 22]). As we will see, the Wiener condition (1.10) is very essential for our asymptotic analysis. Generic examples of the coupling function ρ satisfying (1.9) and (1.10) are given in [13]. Moreover, we will assume that $\hat{\rho}(0) = 0$, i.e.,

$$\int \rho(x)\, dx = 0, \tag{1.11}$$

that means that the total charge of the particle is zero (neutrality condition).

The problem under investigation was studied earlier for the Klein–Gordon equation [11] without the neutrality condition.

Remarks 1.1. (i) The asymptotic stability of the solitary manifold $\mathcal{S}$ is caused by the *radiation of energy to infinity* which appears as the *local energy decay*. Similarly, the decay for the linearized autonomous dynamics also is caused by the radiation. Analytically, the decay is related to the existence of the *continuous spectrum* of the linearized dynamics. The latter is a characteristic feature of infinite dimensional Hamilton systems.

(ii) The asymptotics (2.14) can be interpreted as the collision of the incident soliton, with a trajectory $v_- t + a_-$, with an incident wave $W^0(t)\mathbf{\Psi}_-$, which results in an outgoing soliton with a new trajectory $v_+ t + a_+$, and a new outgoing wave $W^0(t)\mathbf{\Psi}_+$. The collision process can be represented by the diagram of Figure 1. It suggests to introduce the (nonlinear) *scattering operator*

$$\mathbf{S}\colon (v_-, a_-, \mathbf{\Psi}_-) \mapsto (v_+, a_+, \mathbf{\Psi}_+). \tag{1.12}$$

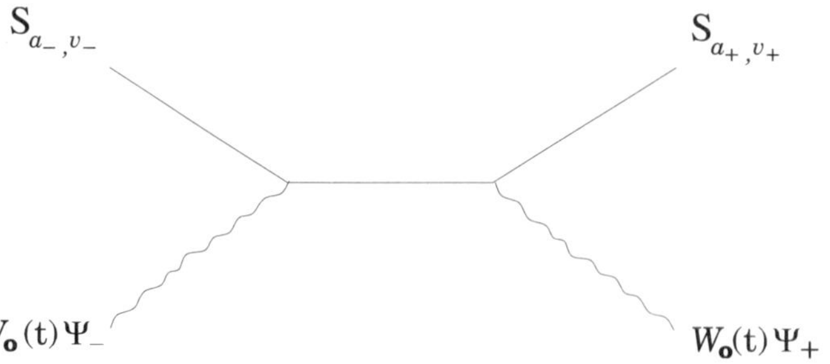

FIGURE 1. Wave-particle scattering

However, a correct definition of the operator is an open problem as well as the question on its *asymptotic completeness* (i.e., on its range).

Our paper is organized as follows. In Section 2, we formulate the main result. In Section 3, we introduce the symplectic projection onto the solitary manifold. The linearized equation is defined and studied in Sections 4 – 5. In Section 6, we split the dynamics in two components: along the solitary manifold, and in transversal directions, and we justify the estimate concerning the tangential component. Section 7 concerns the time decay of the linearized dynamics. The time decay of the transversal component is established in Sections 8 – 11 under an assumption on the time decay of the linearized dynamics. In Section 12, we prove the main result.

2. Main results

2.1. Existence of dynamics. To formulate our results precisely, we need some definitions. We introduce a suitable phase space for the Cauchy problem corresponding to (1.2) and (1.3). Let L^2 be the real Hilbert space $L^2(\mathbb{R}^3)$ with norm $|\cdot|$, and let $\dot{H}^1$ be the completion of the real space $C_0^\infty(\mathbb{R}^3)$ with the norm $|\nabla\psi(x)|$. Equivalently, using Sobolev's embedding theorem, $\dot{H}^1 = \{\psi(x) \in L^6(\mathbb{R}^3) : |\nabla\psi(x)| \in L^2\}$. Let us introduce the weighted Sobolev spaces L^2_α and H^1_α with the norms

$$|\psi|_\alpha = |(1+|x|)^\alpha\psi|, \quad \|\psi\|_\alpha = |(1+|x|)^\alpha\nabla\psi| + |(1+|x|)^\alpha\psi|.$$

Definition 2.1. (i) $\mathcal{E}$ is the Hilbert space $\dot{H}^1 \oplus L^2 \oplus \mathbb{R}^3 \oplus \mathbb{R}^3$ with the finite norm

$$\|Y\|_\mathcal{E} = |\nabla\psi| + |\pi| + |q| + |p| \quad \text{for } Y = (\psi, \pi, q, p).$$

(ii) $\mathcal{E}_\alpha$ is the space $H^1_\alpha \oplus L^2_\alpha \oplus \mathbb{R}^3 \oplus \mathbb{R}^3$ with the norm

$$\|Y\|_\alpha = \|Y\|_{\mathcal{E}_\alpha} = \|\psi\|_\alpha + |\pi|_\alpha + |q| + |p|. \tag{2.1}$$

(iii) $\mathcal{F}$ is the space $\dot{H}^1 \oplus L^2$ of fields $F = (\psi, \pi)$ with the finite norm

$$\|F\|_\mathcal{F} = |\nabla\psi| + |\pi|.$$

(iv) $\mathcal{F}_\alpha$ is the space $H^1_\alpha \oplus L^2_\alpha$ with the norm

$$\|F\|_\alpha = \|F\|_{\mathcal{F}_\alpha} = \|\psi\|_\alpha + |\pi|_\alpha. \tag{2.2}$$

Note that $\dot{H}^1$ is not contained in L^2 and for instance $|\psi_v| = \infty$, see below. However, $\mathcal{E}$ is the space of finite energy states (i.e., $\mathcal{H}(Y) < \infty$ for $Y \in \mathcal{E}$) due to the following estimates which are valid for arbitrary smooth $\psi(x)$ vanishing at infinity

$$
-\frac{1}{8\pi}\iint \mathrm{d}^3x\,\mathrm{d}^3y\,\frac{\rho(x)\rho(y)}{|x-y|}=\frac{1}{2}(\rho,\Delta^{-1}\rho)\le\frac{1}{2}|\nabla\psi|^2+\big(\psi(x),\rho(x-q)\big)
\le|\nabla\psi|^2-\frac{1}{2}(\rho,\Delta^{-1}\rho). \tag{2.3}
$$

The Hamiltonian functional $\mathcal{H}$ is continuous on the space $\mathcal{E}$, and the lower bound in (2.3) implies that the energy (1.3) is bounded from below. Note that the latter is not true if ρ is delta-function.

We consider the Cauchy problem for the Hamilton system (1.2) which we write as

$$
\dot{Y}(t)=F\big(Y(t)\big),\quad t\in\mathbb{R}:\qquad Y(0)=Y_0. \tag{2.4}
$$

Here $Y(t)=\big(\psi(t),\pi(t),q(t),p(t)\big)$, $Y_0=(\psi_0,\pi_0,q_0,p_0)$, and all derivatives are understood in the sense of distributions.

Proposition 2.2 ([13]). *Let (1.9) hold. Then*

(i) *For every $Y_0\in\mathcal{E}$, the Cauchy problem (2.4) has a unique solution $Y(t)\in C(\mathbb{R},\mathcal{E})$.*

(ii) *For every $t\in\mathbb{R}$, the map $U(t)\colon Y_0\mapsto Y(t)$ is continuous on $\mathcal{E}$.*

(iii) *The energy is conserved, i.e.,*

$$
\mathcal{H}\big(Y(t)\big)=\mathcal{H}(Y_0),\quad t\in\mathbb{R}, \tag{2.5}
$$

and the velocity is bounded:

$$
|\dot{q}(t)|\le\bar{v}<1,\quad t\in\mathbb{R}, \tag{2.6}
$$

with some $\bar{v}$ which depends on Y_0.

2.2. Solitary manifold and main result. Let us compute the solitons (1.5). The substitution to (1.2) gives the following stationary equations,

$$
\begin{aligned}
-v\cdot\nabla\psi_v(y)&=\pi_v(y), & -v\cdot\nabla\pi_v(y)&=\Delta\psi_v(y)-\rho(y)\\
v&=\frac{p_v}{\sqrt{1+p_v^2}}, & 0&=-\int\nabla\psi_v(y)\rho(y)\,dy
\end{aligned} \tag{2.7}
$$

Then the first two equations imply

$$
(-\Delta+(v\cdot\nabla)^2)\psi_v(y)=-\rho(y),\quad y\in\mathbb{R}^3. \tag{2.8}
$$

For $|v|<1$ the equation (2.8) defines a unique function $\psi_v\in\dot{H}^1(\mathbb{R}^3)$. If v is given and $|v|<1$, then p_v can be found from the third equation of (2.7). Further, functions ρ and ψ_v are even due to (1.9). Thus, $\nabla\psi_v$ is odd and the last equation of (2.7) holds. Hence, the soliton solution (1.5) exists and is defined uniquely for any couple (a,v) with $|v|<1$.

The function ψ_v can be computed by the Fourier transform. The soliton is given by the formulas

$$
\begin{aligned}
\psi_v(x)&=-\frac{1}{4\pi}\int\frac{\rho(y)\,\mathrm{d}^3y}{|\gamma(y-x)_{\|}+(y-x)_{\perp}|},\\
\pi_v(x)&=-v\cdot\nabla\psi_v(x),\quad p_v=\gamma v.
\end{aligned} \tag{2.9}
$$

Here we set $\gamma=1/\sqrt{1-v^2}$ and $x=x_{\|}+x_{\perp}$, where $x_{\|}\parallel v$ and $x_{\perp}\perp v$ for $x\in\mathbb{R}^3$.

From the neutrality condition (1.11) it follows that $\pi_v \in L^2_\alpha$ for $\alpha < \frac{3}{2}$, and $\psi_v \in H^1_\alpha$ for $\alpha < \frac{1}{2}$. Thus, for the soliton solutions $Y_{a,v}$ we have

$$Y_{a,v} \in \mathcal{E}_\alpha, \quad \alpha < \tfrac{1}{2}. \tag{2.10}$$

Definition 2.3. A soliton state is $S(\sigma) := (\psi_v(x-b), \pi_v(x-b), b, p_v)$, where $\sigma := (b, v)$ with $b \in \mathbb{R}^3$ and $|v| < 1$.

Obviously, the soliton solution admits the representation $S\big(\sigma(t)\big)$, where

$$\sigma(t) = \big(b(t), v(t)\big) = (vt + a, v). \tag{2.11}$$

Definition 2.4. A solitary manifold is the set $\mathcal{S} := \{S(\sigma) : b \in \mathbb{R}^3, |v| < 1\}$.

The main result of our paper is the following theorem.

Theorem 2.5. *Let the Wiener condition* (1.10) *and the neutrality condition* (1.11) *hold. Let* $\beta = 1+\delta$, $0 < \delta < \frac{1}{2}$ *and* $Y(t)$ *be the solution to the Cauchy problem* (2.4) *with the initial state* Y_0 *which is sufficiently close to the solitary manifold*:

$$Y_0 = S(\sigma_0) + Z_0, \quad d_\beta := \|Z_0\|_\beta \ll 1. \tag{2.12}$$

Then the asymptotics hold for $t \to \pm\infty$,

$$\dot q(t) = v_\pm + \mathcal{O}(|t|^{-1-\delta}), \quad q(t) = v_\pm t + a_\pm + \mathcal{O}(|t|^{-2\delta}); \tag{2.13}$$

(2.14)

$$\big(\psi(x,t), \pi(x,t)\big) = \big(\psi_{v_\pm}(x - v_\pm t - a_\pm), \pi_{v_\pm}(x - v_\pm t - a_\pm)\big) + W^0(t)\boldsymbol{\Psi}_\pm + r_\pm(x,t)$$

with

$$\|r_\pm(t)\|_{\mathcal{F}} = \mathcal{O}(|t|^{-\delta}). \tag{2.15}$$

It suffices to prove the asymptotics (2.14), (2.13) for $t \to +\infty$ since the system (1.2) is time reversible.

3. Symplectic projection

3.1. Symplectic structure and Hamilton form. The system (1.2) reads as the Hamilton system

$$\dot Y = J\mathcal{D}\mathcal{H}(Y), \quad J := \begin{pmatrix} 0 & 1 & 0 & 0 \\ -1 & 0 & 0 & 0 \\ 0 & 0 & 0 & 1 \\ 0 & 0 & -1 & 0 \end{pmatrix}, \quad Y = (\psi, \pi, q, p) \in \mathcal{E}, \tag{3.1}$$

where $\mathcal{DH}$ is the Fréchet derivative of the Hamilton functional (1.3). Let us identify the tangent space to $\mathcal{E}$, at every point, with $\mathcal{E}$. Consider the bilinear form Ω defined on $\mathcal{E}$ by $\Omega = \int \mathrm{d}\psi(x) \wedge \mathrm{d}\pi(x)\,\mathrm{d}x + \mathrm{d}q \wedge \mathrm{d}p$, i.e.,

$$\Omega(Y_1, Y_2) = \langle Y_1, JY_2\rangle, \quad Y_1, Y_2 \in \mathcal{E}, \tag{3.2}$$

where

$$\langle Y_1, Y_2\rangle := \langle \psi_1, \psi_2\rangle + \langle \pi_1, \pi_2\rangle + q_1 q_2 + p_1 p_2$$

and $\langle \psi_1, \psi_2\rangle = \int \psi_1(x)\psi_2(x)\,dx$, etc., if the integrals converge.

Definition 3.1. (i) $Y_1 \nmid Y_2$ means that Y_1 is symplectic orthogonal to Y_2, i.e., $\Omega(Y_1, Y_2) = 0$.

(ii) A projection operator $\boldsymbol{P}\colon \mathcal{E} \to \mathcal{E}$ is called symplectic orthogonal if $Y_1 \nmid Y_2$ for $Y_1 \in \operatorname{Ker} \boldsymbol{P}$ and $Y_2 \in \operatorname{Im} \boldsymbol{P}$.

3.2. Symplectic projection onto solitary manifold. Let us consider the tangent space $\mathcal{T}_{S(\sigma)}\mathcal{S}$ to the manifold $\mathcal{S}$ at a point $S(\sigma)$. The vectors $\tau_j := \partial_{\sigma_j} S(\sigma)$, where $\partial_{\sigma_j} := \partial_{b_j}$ and $\partial_{\sigma_{j+3}} := \partial_{v_j}$ with $j = 1, 2, 3$, form a basis in $\mathcal{T}_\sigma\mathcal{S}$. In detail,

$$\begin{aligned} \tau_j = \tau_j(v) &:= \partial_{b_j} S(\sigma) = (-\partial_j \psi_v(y), -\partial_j \pi_v(y), e_j, 0) \\ \tau_{j+3} = \tau_{j+3}(v) &:= \partial_{v_j} S(\sigma) = (\partial_{v_j} \psi_v(y), \partial_{v_j} \pi_v(y), 0, \partial_{v_j} p_v) \end{aligned} \quad j = 1, 2, 3, \tag{3.3}$$

where $y := x - b$ is the "moving frame coordinate," $e_1 = (1, 0, 0)$, etc. Let us stress that the functions τ_j will be considered always as the functions of y, not of x. The formulas (2.9) and the conditions (1.9), (1.11) imply that for $|v| < 1$

$$\tau_j(v) \in \mathcal{E}_\alpha, \qquad \alpha < \tfrac{3}{2}, \quad j = 1, 2, 3; \qquad \alpha < \tfrac{1}{2}, \quad j = 4, 5, 6. \tag{3.4}$$

Lemma 3.2. *The matrix with the elements $\Omega\big(\tau_l(v), \tau_j(v)\big)$ is nondegenerate for $|v| < 1$.*

The proof is made by a straightforward computation, see [11, proof of Lemma 3.2].

Now we show that in a small neighborhood of the soliton manifold $\mathcal{S}$ a "symplectic orthogonal projection" onto $\mathcal{S}$ is well-defined. Let us introduce the translations $T_a \colon (\psi(\cdot), \pi(\cdot), q, p) \mapsto (\psi(\cdot - a), \pi(\cdot - a), q + a, p)$, $a \in \mathbb{R}^3$. Note that the manifold $\mathcal{S}$ is invariant with respect to the translations.

Definition 3.3. Let us denote by $v(Y) := p/\sqrt{1 + p^2}$ where $p \in \mathbb{R}^3$ is the last component of the vector Y.

Lemma 3.4. *Let (1.9) hold, $\alpha > -\frac{1}{2}$ and $\bar{v} < 1$. Then*

(i) *there exists a neighborhood $\mathcal{O}_\alpha(\mathcal{S})$ of $\mathcal{S}$ in $\mathcal{E}_\alpha$ and a map $\mathbf{\Pi} \colon \mathcal{O}_\alpha(\mathcal{S}) \to \mathcal{S}$ such that $\mathbf{\Pi}$ is uniformly continuous on $\mathcal{O}_\alpha(\mathcal{S}) \cap \{Y \in \mathcal{E}_\alpha : v(Y) \leq \bar{v}\}$ in the metric of $\mathcal{E}_\alpha$,*

$$\mathbf{\Pi} Y = Y \quad \text{for } Y \in \mathcal{S}, \qquad \text{and} \qquad Y - S \nmid \mathcal{T}_S\mathcal{S}, \quad \text{where } S = \mathbf{\Pi} Y. \tag{3.5}$$

(ii) *$\mathcal{O}_\alpha(\mathcal{S})$ is invariant with respect to the translations T_a, and*

$$\mathbf{\Pi} T_a Y = T_a \mathbf{\Pi} Y, \quad \text{for } Y \in \mathcal{O}_\alpha(\mathcal{S}) \text{ and } a \in \mathbb{R}^3. \tag{3.6}$$

(iii) *For any $\bar{v} < 1$ there exists a $\tilde{v} < 1$ s.t. $|v(\mathbf{\Pi} Y)| < \tilde{v}$ when $|v(Y)| < \bar{v}$.*

(iv) *For any $\tilde{v} < 1$ there exists an $r_\alpha(\tilde{v}) > 0$ s.t. $S(\sigma) + Z \in \mathcal{O}_\alpha(\mathcal{S})$ if $\big|v\big(S(\sigma)\big)\big| < \tilde{v}$ and $\|Z\|_\alpha < r_\alpha(\tilde{v})$.*

For proof see [11, Lemma 3.4].

We will call $\mathbf{\Pi}$ the symplectic orthogonal projection onto $\mathcal{S}$.

Corollary 3.5. *The condition (2.12) implies that $Y_0 = S + Z_0$ where $S = S(\sigma_0) = \mathbf{\Pi} Y_0$, and*

$$\|Z_0\|_\beta \ll 1. \tag{3.7}$$

4. Linearization on the solitary manifold

Let us consider a solution to the system (1.2), and split it as the sum

$$Y(t) = S\big(\sigma(t)\big) + Z(t), \tag{4.1}$$

where $\sigma(t) = \big(b(t), v(t)\big) \in \mathbb{R}^3 \times \{|v| < 1\}$ is an arbitrary smooth function of $t \in \mathbb{R}$. In detail, denote $Y = (\psi, \pi, q, p)$ and $Z = (\Psi, \Pi, Q, P)$. Then (4.1) means that

$$\begin{aligned} \psi(x,t) &= \psi_{v(t)}\big(x - b(t)\big) + \Psi(x - b(t), t), \qquad & q(t) &= b(t) + Q(t) \\ \pi(x,t) &= \pi_{v(t)}\big(x - b(t)\big) + \Pi(x - b(t), t), \qquad & p(t) &= p_{v(t)} + P(t) \end{aligned} \tag{4.2}$$

Let us substitute (4.2) to (1.2), and linearize the equations in Z. Later we will choose $S\big(\sigma(t)\big) = \mathbf{\Pi} Y(t)$, i.e., $Z(t)$ is symplectic orthogonal to $\mathcal{T}_{S(\sigma(t))}\mathcal{S}$. However, this orthogonality condition is not needed for the formal process of linearization. The orthogonality condition will be important in Section 6, where we derive "modulation equations" for the parameters $\sigma(t)$.

Let us proceed to linearization. Setting $y = x - b(t)$ which is the "moving frame coordinate," we obtain from (4.2) and (1.2) that

$$\begin{aligned} \dot\psi &= \dot v \cdot \nabla_v \psi_{v(t)}(y) - \dot b \cdot \nabla \psi_{v(t)}(y) + \dot\Psi(y,t) - \dot b \cdot \nabla\Psi(y,t) \\ &= \pi_{v(t)}(y) + \Pi(y,t) \\ \dot\pi &= \dot v \cdot \nabla_v \pi_{v(t)}(y) - \dot b \cdot \nabla \pi_{v(t)}(y) + \dot\Pi(y,t) - \dot b \cdot \nabla\Pi(y,t) \\ &= \Delta\psi_{v(t)}(y) + \Delta\Psi(y,t) - \rho(y - Q) \\ \dot q &= \dot b + \dot Q = \frac{p_v + P}{\sqrt{1 + (p_v + P)^2}} \\ \dot p &= \dot v \cdot \nabla_v p_{v(t)} + \dot P = -\big\langle \nabla\big(\psi_{v(t)}(y) + \Psi(y,t)\big), \rho(y - Q)\big\rangle \end{aligned} \tag{4.3}$$

The equations are linear in Ψ and Π, hence it remains to extract linear terms in Q and P. First note that $\rho(y - Q) = \rho(y) - Q \cdot \nabla\rho(y) - N_2(Q)$, where $-N_2(Q) = \rho(y - Q) - \rho(y) + Q \cdot \nabla\rho(y)$; for $N_2(Q)$ the bound holds,

$$|N_2(Q)|_\beta \le C_\beta(\overline{Q}) Q^2, \tag{4.4}$$

uniformly in $|Q| \le \overline{Q}$ for any fixed $\overline{Q}$. Here β is the constant defined in Theorem 2.5. Second, the Taylor expansion gives

$$\frac{p_v + P}{\sqrt{1 + (p_v + P)^2}} = v + \nu\big(P - v(v \cdot P)\big) + N_3(v, P),$$

where $\nu = \nu_v := (1 + p_v^2)^{-1/2} = \sqrt{1 - v^2}$, and

$$|N_3(v, P)| \le C(\tilde v) P^2 \tag{4.5}$$

uniformly in $|v| \le \tilde v < 1$. Using the equations (2.7), we obtain from (4.3) the following equations for the components of the vector $Z(t)$:

$$\begin{aligned} \dot\Psi(y,t) &= \Pi(y,t) + \dot b \cdot \nabla\Psi(y,t) + (\dot b - v) \cdot \nabla\psi_v(y) - \dot v \cdot \nabla_v \psi_v(y) \\ \dot\Pi(y,t) &= \Delta\Psi(y,t) + \dot b \cdot \nabla\Pi(y,t) + Q \cdot \nabla\rho(y) + (\dot b - v) \cdot \nabla\pi_v(y) \\ &\qquad - \dot v \cdot \nabla_v \pi_v(y) + N_2 \\ \dot Q(t) &= \nu_v (E - v \otimes v) P + (v - \dot b) + N_3 \\ \dot P(t) &= \langle \Psi(y,t), \nabla\rho(y)\rangle + \langle \nabla\psi_v(y), Q \cdot \nabla\rho(y)\rangle - \dot v \cdot \nabla_v p_v + N_4(v, Z) \end{aligned} \tag{4.6}$$

where $N_4(v, Z) = \langle \nabla\psi_v, N_2(Q)\rangle + \langle \nabla\Psi, Q \cdot \nabla\rho\rangle + \langle \nabla\Psi, N_2(Q)\rangle$. Clearly, $N_4(v, Z)$ satisfies the following estimate

$$|N_4(v, Z)| \le C_\beta(\rho, \tilde v, \overline{Q}) \Big[Q^2 + \|\Psi\|_{-\beta} |Q|\Big], \tag{4.7}$$

uniformly in $|v| \le \tilde{v}$ and $|Q| \le \overline{Q}$ for any fixed $\tilde{v} < 1$. We can write the equations (4.6) as

$$\dot{Z}(t) = A(t)Z(t) + T(t) + N(t), \quad t \in \mathbb{R}. \tag{4.8}$$

Here the operator $A(t)$ depends on $\sigma(t) = \big(b(t), v(t)\big)$. We will use the parameters $v = v(t)$ and $w := \dot{b}(t)$. Then $A(t)$ can be written in the form

$$A(t)\begin{pmatrix}\Psi\\ \Pi\\ Q\\ P\end{pmatrix} = A_{v,w}\begin{pmatrix}\Psi\\ \Pi\\ Q\\ P\end{pmatrix} := \begin{pmatrix} w\cdot\nabla & 1 & 0 & 0\\ \Delta & w\cdot\nabla & \nabla\rho\cdot & 0\\ 0 & 0 & 0 & B_v\\ \langle\cdot,\nabla\rho\rangle & 0 & \langle\nabla\psi_v,\cdot\nabla\rho\rangle & 0\end{pmatrix}\begin{pmatrix}\Psi\\ \Pi\\ Q\\ P\end{pmatrix}, \tag{4.9}$$

where $B_v = \nu_v(E - v\otimes v)$. Furthermore, $T(t)$ and $N(t)$ in (4.8) stand for
(4.10)

$$T(t) = T_{v,w} = \begin{pmatrix}(w-v)\cdot\nabla\psi_v - \dot{v}\cdot\nabla_v\psi_v\\ (w-v)\cdot\nabla\pi_v - \dot{v}\cdot\nabla_v\pi_v\\ v-w\\ -\dot{v}\cdot\nabla_v p_v\end{pmatrix}, \quad N(t) = N(\sigma, Z) = \begin{pmatrix}0\\ N_2(Z)\\ N_3(v,Z)\\ N_4(v,Z)\end{pmatrix},$$

where $v = v(t)$, $w = w(t)$, $\sigma = \sigma(t) = \big(b(t), v(t)\big)$, and $Z = Z(t)$. The estimates (4.4), (4.5) and (4.7) imply that

$$\|N(\sigma, Z)\|_\beta \le C(\tilde{v}, \overline{Q})\|Z\|^2_{-\beta} \tag{4.11}$$

uniformly in σ and $\|Z\|_{-\beta} \le r_{-\beta}(\tilde{v})$ for any fixed $\tilde{v} < 1$ and $|v| < \tilde{v}$.

Remarks 4.1. (i) The term $A(t)Z(t)$ in the right-hand side of the equation (4.8) is linear in $Z(t)$, and $N(t)$ is a *high order term* in $Z(t)$. On the other hand, $T(t)$ is a zero order term which does not vanish at $Z(t) = 0$ since $S\big(\sigma(t)\big)$ generally is not a soliton solution if (2.11) does not hold (though $S\big(\sigma(t)\big)$ belongs to the solitary manifold).

(ii) Formulas (3.3) and (4.10) imply:

$$T(t) = -\sum_{l=1}^{3}[(w-v)_l\tau_l + \dot{v}_l\tau_{l+3}] \tag{4.12}$$

and hence $T(t) \in \mathcal{T}_{S(\sigma(t))}\mathcal{S}$, $t \in \mathbb{R}$. This fact suggests an unstable character of the nonlinear dynamics *along the solitary manifold.*

5. The linearized equation

Here we collect some Hamiltonian and spectral properties of the operator (4.9). The proofs of the statements of the section are similar to those of [11, Section 5]. First, let us consider the linear equation

$$\dot{X}(t) = A_{v,w}X(t), \quad t \in \mathbb{R} \tag{5.1}$$

with an arbitrary fixed v such that $|v| < 1$, and $w \in \mathbb{R}^3$.

Lemma 5.1. (i) *For any v, $|v| < 1$ and $w \in \mathbb{R}^3$ the equation (5.1) formally can be written as the Hamilton system (cf. (3.1)),*

$$\dot{X}(t) = JD\mathcal{H}_{v,w}\big(X(t)\big), \quad t \in \mathbb{R}, \tag{5.2}$$

where $D\mathcal{H}_{v,w}$ is the Fréchet derivative of the Hamilton functional

$$\mathcal{H}_{v,w}(X) = \frac{1}{2}\int [|\Pi|^2 + |\nabla\Psi|^2]\,dy + \int \Pi w\cdot\nabla\Psi\,dy + \int \rho(y)Q\cdot\nabla\Psi\,dy \\ + \tfrac{1}{2}P\cdot B_v P - \tfrac{1}{2}\langle Q\cdot\nabla\psi_v(y), Q\cdot\nabla\rho(y)\rangle, \\ X = (\Psi,\Pi,Q,P)\in\mathcal{E}. \tag{5.3}$$

(ii) *Energy conservation law holds for the solutions $X(t)\in C^1(\mathbb{R},\mathcal{E})$,*

$$\mathcal{H}_{v,w}\big(X(t)\big) = \text{const}, \quad t\in\mathbb{R}. \tag{5.4}$$

(iii) *The skew-symmetry relation holds,*

$$\Omega(A_{v,w}X_1, X_2) = -\Omega(X_1, A_{v,w}X_2), \quad X_1, X_2\in\mathcal{E}. \tag{5.5}$$

Lemma 5.2. *The operator $A_{v,w}$ acts on the tangent vectors $\tau_j(v)$ to the solitary manifold as follows,*

$$\begin{aligned} A_{v,w}[\tau_j(v)] &= (w-v)\cdot\nabla\tau_j(v), \\ A_{v,w}[\tau_{j+3}(v)] &= (w-v)\cdot\nabla\tau_{j+3}(v) + \tau_j(v), \end{aligned} \qquad j = 1,2,3. \tag{5.6}$$

We will apply Lemma 5.1 mainly to the operator $A_{v,v}$ corresponding to $w=v$. In that case the linearized equation has the following additional essential features.

Lemma 5.3. *Let us assume that $w=v$ and $|v|<1$. Then*
(i) *The tangent vectors $\tau_j(v)$ with $j=1,2,3$ are eigenvectors, and $\tau_{j+3}(v)$ are root vectors of the operator $A_{v,v}$, corresponding to zero eigenvalue, i.e.,*

$$A_{v,v}[\tau_j(v)] = 0, \quad A_{v,v}[\tau_{j+3}(v)] = \tau_j(v), \qquad j=1,2,3. \tag{5.7}$$

(ii) *The Hamilton function* (5.3) *is positive definite,*

$$\mathcal{H}_{v,v}(X) \geq 0. \tag{5.8}$$

Remark 5.4. Lemma 5.3(ii) together with energy conservation (5.4) imply the analyticity of the resolvent $(A_{v,v}-\lambda)^{-1}$ for $\operatorname{Re}\lambda > 0$.

Remark 5.5. For a soliton solution of the system (1.2) we have $\dot b = v$, $\dot v = 0$, and hence $T(t)\equiv 0$. Thus, the equation (5.1) is the linearization of the system (1.2) on a soliton solution. In fact, we do not linearize (1.2) on a soliton solution, but on a trajectory $S\big(\sigma(t)\big)$ with $\sigma(t)$ being nonlinear in t. We will show later that $T(t)$ is quadratic in $Z(t)$ if we choose $S\big(\sigma(t)\big)$ to be the symplectic orthogonal projection of $Y(t)$. Then (5.1) is again the linearization of (1.2).

6. Symplectic decomposition of the dynamics

Here we decompose the dynamics in two components: along the manifold $\mathcal{S}$ and in transversal directions. The equation (4.8) is obtained without any assumption on $\sigma(t)$ in (4.1). We are going to choose $S\big(\sigma(t)\big) := \mathbf{\Pi}\, Y(t)$, but then we need to know that

$$Y(t)\in\mathcal{O}_\alpha(\mathcal{S}), \quad t\in\mathbb{R}, \tag{6.1}$$

with some $\mathcal{O}_\alpha(\mathcal{S})$ defined in Lemma 3.5. It is true for $t=0$ by our main assumption (2.12) with sufficiently small $d_\beta > 0$. Then $S\big(\sigma(0)\big) = \mathbf{\Pi}\, Y(0)$ and $Z(0) = Y(0) - S\big(\sigma(0)\big)$ are well defined. We will prove below that (6.1) holds with $\alpha = -\beta$ if d_β

is sufficiently small. First, the a priori estimate (2.6) together with Lemma 3.4 iii) imply that $\mathbf{\Pi} Y(t) = S\big(\sigma(t)\big)$ with $\sigma(t) = \big(b(t), v(t)\big)$, and

$$|v(t)| \le \tilde{v} < 1, \quad t \in \mathbb{R} \tag{6.2}$$

if $Y(t) \in \mathcal{O}_{-\beta}(\mathcal{S})$. Denote by $r_{-\beta}(\tilde{v})$ the positive number from Lemma 3.4 iv) which corresponds to $\alpha = -\beta$. Then $S(\sigma) + Z \in \mathcal{O}_{-\beta}(\mathcal{S})$ if $\sigma = (b, v)$ with $|v| < \tilde{v}$ and $\|Z\|_{-\beta} < r_{-\beta}(\tilde{v})$. Note that (2.6) implies $\|Z(0)\|_{-\beta} < r_{-\beta}(\tilde{v})$ if d_β is sufficiently small. Therefore, $S\big(\sigma(t)\big) = \mathbf{\Pi} Y(t)$ and $Z(t) = Y(t) - S\big(\sigma(t)\big)$ are well defined for $t \ge 0$ so small that $\|Z(t)\|_{-\beta} < r_{-\beta}(\tilde{v})$. This is formalized by the following standard definition.

Definition 6.1. t_* is the "exit time,"

$$t_* = \sup\{t > 0 : \|Z(s)\|_{-\beta} < r_{-\beta}(\tilde{v}), 0 \le s \le t\}, \quad Z(s) = Y(s) - S\big(\sigma(s)\big). \tag{6.3}$$

One of our main goals is to prove that $t_* = \infty$ if d_β is sufficiently small. This would follow if we show that

$$\|Z(t)\|_{-\beta} < r_{-\beta}(\tilde{v})/2, \quad 0 \le t < t_*. \tag{6.4}$$

Note that

$$|Q(t)| \le \overline{Q} := r_{-\beta}(\tilde{v}), \quad 0 \le t < t_*. \tag{6.5}$$

Now $N(t)$ in (4.8) satisfies, by (4.11), the following estimate,

$$\|N(t)\|_{\beta} \le C_\beta(\tilde{v}) \|Z(t)\|^2_{-\beta}, \quad 0 \le t < t_*. \tag{6.6}$$

6.1. Longitudinal dynamics: Modulation equations. From now on we fix the decomposition $Y(t) = S\big(\sigma(t)\big) + Z(t)$ for $0 < t < t_*$ by setting $S\big(\sigma(t)\big) = \mathbf{\Pi} Y(t)$ which is equivalent to the symplectic orthogonality condition of type (3.5),

$$Z(t) \nmid \mathcal{T}_{S(\sigma(t))} \mathcal{S}, \quad 0 \le t < t_*. \tag{6.7}$$

This allows us to simplify drastically the asymptotic analysis of the dynamical equations (4.8) for the transversal component $Z(t)$. As the first step, we derive the longitudinal dynamics, i.e., the "modulation equations" for the parameters $\sigma(t)$. Let us derive a system of ordinary differential equations for the vector $\sigma(t)$. For this purpose, let us write (6.7) in the form

$$\Omega\big(Z(t), \tau_j(t)\big) = 0, \quad j = 1, \ldots, 6, 0 \le t < t_*, \tag{6.8}$$

where the vectors $\tau_j(t) = \tau_j\big(\sigma(t)\big)$ span the tangent space $\mathcal{T}_{S(\sigma(t))} \mathcal{S}$. Note that $\sigma(t) = \big(b(t), v(t)\big)$, where

$$|v(t)| \le \tilde{v} < 1, \quad 0 \le t < t_*, \tag{6.9}$$

by Lemma 3.4(iii). It would be convenient for us to use some other parameters (c, v) instead of $\sigma = (b, v)$, where $c(t) = b(t) - \int_0^t v(\tau)\, d\tau$ and

$$\dot{c}(t) = \dot{b}(t) - v(t) = w(t) - v(t), \quad 0 \le t < t_*. \tag{6.10}$$

We do not need an explicit form of the equations for (c, v) but the following statement.

Lemma 6.2 (cf. [11, Lemma 6.2]). *Let $Y(t)$ be a solution to the Cauchy problem* (2.4), *and* (4.1), (6.8) *hold. Then $\big(c(t), v(t)\big)$ satisfies the equation*

$$\begin{pmatrix} \dot c(t) \\ \dot v(t) \end{pmatrix} = \mathcal{N}\big(\sigma(t), Z(t)\big), \quad 0 \le t < t_*, \tag{6.11}$$

where

$$\mathcal{N}(\sigma, Z) = \mathcal{O}(\|Z\|^2_{-\beta}) \tag{6.12}$$

uniformly in $\sigma \in \{(b, v) : |v| \le \tilde v\}$.

6.2. Decay for the transversal dynamics. In Section 12 we will show that our main Theorem 2.5 can be derived from the following time decay of the transversal component $Z(t)$:

Proposition 6.3. *Let all conditions of Theorem 2.5 hold. Then $t_* = \infty$, and*

$$\|Z(t)\|_{-\beta} \le \frac{C(\rho, \overline{v}, d_\beta)}{(1 + |t|)^\beta}, \quad t \ge 0. \tag{6.13}$$

We will derive (6.13) in Sections 8 – 11 from our equation (4.8) for the transversal component $Z(t)$. This equation can be specified using Lemma 6.2. Indeed, the lemma implies that

$$\|T(t)\|_\beta \le C(\tilde v)\|Z(t)\|^2_{-\beta}, \quad 0 \le t < t_*, \tag{6.14}$$

by (4.10) since $w - v = \dot c$. Thus (4.8) becomes the equation

$$\dot Z(t) = A(t) Z(t) + \widetilde{N}(t), \quad 0 \le t < t_*, \tag{6.15}$$

where $A(t) = A_{v(t), w(t)}$, and $\widetilde{N}(t) := T(t) + N(t)$ satisfies the estimate

$$\|\widetilde{N}(t)\|_\beta \le C\|Z(t)\|^2_{-\beta}, \quad 0 \le t < t_*. \tag{6.16}$$

In all remaining part of our paper we will analyze mainly the *basic equation* (6.15) to establish the decay (6.13). We are going to derive the decay using the bound (6.16) and the orthogonality condition (6.7).

Let us comment on two main difficulties in proving (6.13). The difficulties are common for the problems studied in [2, 3]. First, the linear part of the equation is non-autonomous, hence we cannot apply directly known methods of scattering theory. Similarly to the approach of [2, 3], we reduce the problem to the analysis of the *frozen* linear equation,

$$\dot X(t) = A_1 X(t), \quad t \in \mathbb{R}, \tag{6.17}$$

where A_1 is the operator A_{v_1, v_1} defined in (4.9) with $v_1 = v(t_1)$ and a fixed $t_1 \in [0, t_*)$. Then we estimate the error by the method of majorants.

Second, even for the frozen equation (6.17), the decay of type (6.13) for all solutions does not hold without the orthogonality condition of type (6.7). Namely, by (5.7) the equation (6.17) admits the *secular solutions*

$$X(t) = \sum_1^3 C_j \tau_j(v_1) + \sum_1^3 D_j[\tau_j(v_1)t + \tau_{j+3}(v_1)] \tag{6.18}$$

which arise also by differentiation of the soliton (1.5) in the parameters a and v_1 in the moving coordinate $y = x - v_1 t$. Hence, we have to take into account the orthogonality condition (6.7) in order to avoid the secular solutions. For this purpose we will apply the corresponding symplectic orthogonal projection which kills the "runaway solutions" (6.18).

Remark 6.4. The solution (6.18) lies in the tangent space $\mathcal{T}_{S(\sigma_1)}\mathcal{S}$ with $\sigma_1 = (b_1, v_1)$ (for an arbitrary $b_1 \in \mathbb{R}$) that suggests an unstable character of the nonlinear dynamics *along the solitary manifold* (cf. Remark 4.1(iii)).

Definition 6.5. (i) Denote by $\mathbf{\Pi}_v$, $|v| < 1$, the symplectic orthogonal projection of $\mathcal{E}$ onto the tangent space $\mathcal{T}_{S(\sigma)}\mathcal{S}$, and $\boldsymbol{P}_v = \mathbf{I} - \mathbf{\Pi}_v$.

(ii) Denote by $\mathcal{Z}_v = \boldsymbol{P}_v\mathcal{E}$ the space symplectic orthogonal to $\mathcal{T}_{S(\sigma)}\mathcal{S}$ with $\sigma = (b, v)$ (for an arbitrary $b \in \mathbb{R}$).

Note that by the linearity,

$$\mathbf{\Pi}_v\, Z = \sum \mathbf{\Pi}_{jl}(v)\tau_j(v)\Omega(\tau_l(v), Z), \quad Z \in \mathcal{E}, \tag{6.19}$$

with some smooth coefficients $\mathbf{\Pi}_{jl}(v)$. Hence, the projector $\mathbf{\Pi}_v$, in the variable $y = x - b$, does not depend on b, and this explains the choice of the subindex in $\mathbf{\Pi}_v$ and $\boldsymbol{P}_v$.

Now we have the symplectic orthogonal decomposition

$$\mathcal{E}_\beta = \mathcal{T}_{S(\sigma)}\mathcal{S} + \mathcal{Z}_v, \quad \sigma = (b, v), \tag{6.20}$$

and the symplectic orthogonality (6.7) can be written in the following equivalent forms,

$$\mathbf{\Pi}_{v(t)}\, Z(t) = 0, \quad \boldsymbol{P}_{v(t)}Z(t) = Z(t), \qquad 0 \le t < t_*. \tag{6.21}$$

Remark 6.6. The tangent space $\mathcal{T}_{S(\sigma)}\mathcal{S}$ is invariant under the operator $A_{v,v}$ by Lemma 5.3(i), hence the space $\mathcal{Z}_v$ is also invariant by (5.5): $A_{v,v}Z \in \mathcal{Z}_v$ for *sufficiently smooth* $Z \in \mathcal{Z}_v$.

In the next section we prove the following proposition which is one of the main ingredients for proving (6.13). Let us consider the Cauchy problem for the equation (6.17) with $A = A_{v,v}$ for a fixed v, $|v| < 1$. Recall that the parameter $\beta = 1 + \delta$, $0 < \delta < \frac{1}{2}$ is also fixed.

Proposition 6.7. *Let the Wiener condition* (1.10) *and the neutrality condition* (1.11) *hold,* $|v_1| \le \tilde{v} < 1$, *and* $X_0 \in \mathcal{E}$. *Then*

(i) *The equation* (6.17), *with* $A_1 = A_{v_1,v_1}$, *admits the unique solution* $\mathrm{e}^{A_1 t}X_0 := X(t) \in C(\mathbb{R}, \mathcal{E})$ *with the initial condition* $X(0) = X_0$.

(ii) *For* $X_0 \in \mathcal{Z}_{v_1} \cap \mathcal{E}_\beta$, *the solution* $X(t)$ *has the following decay,*

$$\|\mathrm{e}^{A_1 t}X_0\|_{-\beta} \le \frac{C_\beta(\tilde{v})}{(1 + |t|)^\beta}\|X_0\|_\beta, \quad t \in \mathbb{R}. \tag{6.22}$$

Remark 6.8. The decay is provided by two fundamental facts which we will establish below:

(i) the null root space of the generator A_1 coincides with the $\mathcal{T}_{S(\sigma_1)}\mathcal{S}$ with $\sigma_1 = (b_1, v_1)$ (for an arbitrary $b_1 \in \mathbb{R}$), and

(ii) the spectrum of A_1 in the space $\mathcal{Z}_{v_1}$ is absolutely continuous.

7. Proof of Proposition 6.7

Since the proof repeat in main features that of [11], we sketch here only the main steps, the general strategy being the following one. The equation (6.17) is a system of four equations involving field components, Ψ and Π as well as vector components, Q and P. We apply Fourier–Laplace transform, express the field components in terms of the vector components from the first two equations and substitute to the third and the fourth equations. Then we obtain a closed system for the vector components alone and prove their decay. Finally, for the field components we come to a wave equation with a right hand side which has the established decay. This implies the corresponding decay for the field components.

So let us apply the Fourier–Laplace transform

$$\tilde{X}(\lambda) = \int_0^\infty \mathrm{e}^{-\lambda t} X(t)\, dt, \quad \mathrm{Re}\,\lambda > 0 \tag{7.1}$$

to (6.17). The integral converges in $\mathcal{E}$ by the energy conservation (5.4) for the linearized equation. The analyticity of $\tilde{X}(\lambda)$ and Paley–Wiener arguments should provide the existence of a $\mathcal{E}$-valued distribution $X(t)$, $t \in \mathbb{R}$, with a support in $[0, \infty)$. Formally,

$$X(t) = \frac{1}{2\pi} \int_{\mathbb{R}} \mathrm{e}^{\mathrm{i}\omega t} \tilde{X}(\mathrm{i}\omega + 0)\, d\omega, \quad t \in \mathbb{R}. \tag{7.2}$$

To prove the decay (6.22) we have to study the smoothness of $\tilde{X}(\mathrm{i}\omega + 0)$ at $\omega \in \mathbb{R}$. After Fourier–Laplace transform the equation (6.17) becomes

$$\lambda \tilde{X}(\lambda) = A_1 \tilde{X}(\lambda) + X_0, \quad \mathrm{Re}\,\lambda > 0.$$

In detail,

$$\left.\begin{aligned} \tilde{\Pi}(y) + v \cdot \nabla \tilde{\Psi}(y) - \lambda \tilde{\Psi}(y) &= -\tilde{\Psi}_0(y) \\ \Delta \tilde{\Psi}(y) - m^2 \tilde{\Psi}(y) + v \cdot \nabla \tilde{\Pi}(y) + \tilde{Q} \cdot \nabla \rho(y) - \lambda \tilde{\Pi}(y) &= -\Pi_0(y) \\ B_v \tilde{P} - \lambda \tilde{Q} &= -Q_0 \\ -\langle \nabla \tilde{\Psi}(y), \rho(y) \rangle + \langle \nabla \psi_v(y), \tilde{Q} \cdot \nabla \rho(y) \rangle - \lambda \tilde{P} &= -P_0 \end{aligned}\right\} \quad y \in \mathbb{R}^3. \tag{7.3}$$

We express the field $\tilde{\Psi}$ in terms of $\tilde{Q}$ from the first two equations (applying first Fourier transform), and substitute to the second two equations. Then we obtain a closed system of linear algebraic equations for the vector components $(\tilde{Q}(\lambda), \tilde{P}(\lambda))$. The system reads

$$M(\lambda) \begin{pmatrix} \tilde{Q} \\ \tilde{P} \end{pmatrix} = \begin{pmatrix} Q_0 \\ P_0' \end{pmatrix}, \quad \mathrm{Re}\,\lambda > 0, \tag{7.4}$$

where the matrix $M(\lambda)$ can be computed explicitly. In detail it is given in [11, formulas (13.18)–(13.20) with $m = 0$]. Further, $P_0' = P_0 + \Phi(\Psi_0, \Pi_0)$.

Lemma 7.1 ([11]). *The matrix $M(\lambda)$ is invertible for $\mathrm{Re}\,\lambda > 0$, the matrix function $M(\lambda)$ (respectively, $M^{-1}(\lambda)$) admits an analytic (respectively meromorphic) continuation to the entire complex plane $\mathbb{C}$.*

The vector components are given by the Fourier integral

$$\begin{pmatrix} Q(t) \\ P(t) \end{pmatrix} = \frac{1}{2\pi} \int \mathrm{e}^{\mathrm{i}\omega t} M^{-1}(\mathrm{i}\omega + 0) \begin{pmatrix} Q_0 \\ P_0' \end{pmatrix} d\omega \tag{7.5}$$

if it converges in the sense of distributions.

Proposition 7.2. *The matrix $M^{-1}(\mathrm{i}\omega)$ is smooth in $\omega \in \mathbb{R} \setminus 0$.*

The proof is based on the following

Lemma 7.3. *Let ρ satisfy the Wiener condition* (1.10)*, and $|v| < 1$. Then the limit matrix $M(\mathrm{i}\omega + 0)$ is invertible for $\omega \neq 0$, $\omega \in \mathbb{R}$.*

Remark 7.4. The proof of Lemma 7.3 is the unique point in the paper where the Wiener condition is indispensable.

The Proposition 7.2 alone is not sufficient for the proof of the convergence and decay of the integral. Namely, we need an additional information about a regularity of the matrix $M^{-1}(\mathrm{i}\omega)$ at its singular point $\omega = 0$ and some bounds at $|\omega| \to \infty$. We will analyze all the points separately.

I. First we study the asymptotic behavior of $M^{-1}(\lambda)$ at infinity. Let us recall that $M^{-1}(\lambda)$ was originally defined for $\mathrm{Re}\,\lambda > 0$, but it admits a meromorphic continuation to the entire complex plane $\mathbb{C}$ (see Lemma 7.1).

The following proposition is a very particular case of a general fundamental theorem about the bound for the truncated resolvent on the continuous spectrum. The bound plays a crucial role in the study of the long-time asymptotics of general linear hyperbolic PDEs, [26].

Proposition 7.5. *There exist a matrix R_0 and a matrix-function $R_1(\omega)$, such that*

$$M^{-1}(\mathrm{i}\omega) = \frac{R_0}{\omega} + R_1(\omega), \quad |\omega| \geq 1, \omega \in \mathbb{R},$$

where, for every $k = 0, 1, 2, \ldots$,

$$|\partial_\omega^k R_1(\omega)| \leq \frac{C_k}{|\omega|^2}, \quad |\omega| \geq 1, \omega \in \mathbb{R}. \tag{7.6}$$

II. Second, we consider the point $\omega = 0$ which is most singular. The point is an isolated pole of a finite degree by Lemma 7.1, hence the Laurent expansion holds,

$$M^{-1}(i\omega) = \sum_{k=0}^{n} L_k \omega^{-k-1} + h(\omega), \quad |\omega| < \varepsilon_0, \tag{7.7}$$

where L_k are 6×6 complex matrices, $\varepsilon_0 > 0$, and $h(\omega)$ is an analytic matrix-valued function for complex ω with $|\omega| < \varepsilon_0$.

The rest of the proof of Proposition 6.7 is similar to that of [11].

8. Frozen form of transversal dynamics

Now let us fix an arbitrary $t_1 \in [0, t_*)$, and rewrite the equation (6.15) in a "frozen form"

$$\dot{Z}(t) = A_1 Z(t) + (A(t) - A_1) Z(t) + \widetilde{N}(t), \quad 0 \leq t < t_*, \tag{8.1}$$

where $A_1 = A_{v(t_1), v(t_1)}$ and

$$A(t) - A_1 = \begin{pmatrix} [w(t) - v(t_1)] \cdot \nabla & 0 & 0 & 0 \\ 0 & [w(t) - v(t_1)] \cdot \nabla & 0 & 0 \\ 0 & 0 & 0 & B_{v(t)} - B_{v(t_1)} \\ 0 & 0 & \langle \nabla(\psi_{v(t)} - \psi_{v(t_1)}), \cdot \nabla\rho \rangle & 0 \end{pmatrix}.$$

The next trick is important since it allows us to kill the "bad terms" $[w(t)-v(t_1)]\cdot\nabla$ in the operator $A(t)-A_1$.

Definition 8.1. Let us change the variables $(y,t)\mapsto(y_1,t)=(y+d_1(t),t)$ where

$$d_1(t):=\int_{t_1}^t (w(s)-v(t_1))\,ds,\quad 0\le t\le t_1. \tag{8.2}$$

Next define

$$\begin{aligned} Z_1(t) &= \big(\Psi_1(y_1,t),\Pi_1(y_1,t),Q(t),P(t)\big):=\big(\Psi(y,t),\Pi(y,t),Q(t),P(t)\big)\\ &=\big(\Psi(y_1-d_1(t),t),\Pi(y_1-d_1(t),t),Q(t),P(t)\big). \end{aligned} \tag{8.3}$$

Then we obtain the final form of the "frozen equation" for the transversal dynamics

$$\dot Z_1(t)=A_1Z_1(t)+B_1(t)Z_1(t)+N_1(t),\quad 0\le t\le t_1, \tag{8.4}$$

where $N_1(t)=\widetilde N(t)$ expressed in terms of $y=y_1-d_1(t)$, and

$$B_1(t)=\begin{pmatrix} 0 & 0 & 0 & 0\\ 0 & 0 & 0 & 0\\ 0 & 0 & 0 & B_{v(t)}-B_{v(t_1)}\\ 0 & 0 & \langle\nabla(\psi_{v(t)}-\psi_{v(t_1)}),\nabla\rho\rangle & 0\end{pmatrix}.$$

At the end of this section, we will derive appropriate bounds for the "remainder terms" $B_1(t)Z_1(t)$ and $N_1(t)$ in (8.4). First, note that we have by Lemma 6.2,

$$|B_{v(t)}-B_{v(t_1)}|\le\Big|\int_{t_1}^t \dot v(s)\cdot\nabla_v B_{v(s)}\,ds\Big|\le C\int_t^{t_1}\|Z(s)\|^2_{-\beta}\,ds. \tag{8.5}$$

Similarly,

$$|\langle\nabla(\psi_{v(t)}-\psi_{v(t_1)}),\nabla\rho\rangle|\le C\int_t^{t_1}\|Z(s)\|^2_{-\beta}\,ds. \tag{8.6}$$

Let us recall the following well-known inequality: for any $\alpha\in\mathbb{R}$

$$(1+|y+x|)^\alpha\le(1+|y|)^\alpha(1+|x|)^{|\alpha|},\quad x,y\in\mathbb{R}^3. \tag{8.7}$$

Lemma 8.2. *For $(\Psi,\Pi,Q,P)\in\mathcal{E}_\alpha$ with any $\alpha\in\mathbb{R}$ the following estimate holds*:

$$\|(\Psi(y_1-d_1),\Pi(y_1-d_1),Q,P)\|_\alpha\le\|(\Psi,\Pi,Q,P)\|_\alpha(1+|d_1|)^{|\alpha|},\quad d_1\in\mathbb{R}^3. \tag{8.8}$$

PROOF. Let us check the estimate only for one component, say, for Π. One has by (8.7)

$$\begin{aligned} |\Pi(y_1-d_1,t)|^2_\alpha &= \int|\Pi(y_1-d_1,t)|^2(1+|y_1|)^{2\alpha}\,dy_1\\ &=\int|\Pi(y,t)|^2(1+|y+d_1|)^{2\alpha}\,dy\\ &\le\int|\Pi(y,t)|^2(1+|y|)^{2\alpha}(1+|d_1|)^{2|\alpha|}\,dy\le(1+|d_1|)^{2|\alpha|}|\Pi|^2_\alpha, \end{aligned}$$

and the lemma is proved. □

Corollary 8.3. *The following bound holds*

$$\|N_1(t)\|_\beta\le\|Z_1(t)\|^2_{-\beta}(1+|d_1(t)|)^{3\beta},\quad 0\le t\le t_1. \tag{8.9}$$

Indeed, applying the previous lemma twice, once for β and once for $-\beta$, we obtain from (6.16) that

$$\|N_1(t)\|_\beta \le (1+|d_1(t)|)^\beta \big\|\widetilde{N}(t,Z(t))\big\|_\beta \le (1+|d_1(t)|)^\beta \|Z\|^2_{-\beta}$$
$$\le (1+|d_1(t)|)^{3\beta}\|Z_1(t)\|^2_{-\beta}.$$

Corollary 8.4. *The following bound holds*

$$(8.10)\quad \|B_1(t)Z_1(t)\|_\beta \le C\|Z_1(t)\|_{-\beta}\int_t^{t_1}(1+|d_1(\tau)|)^{2\beta}\|Z_1(\tau)\|^2_{-\beta}\,\mathrm{d}\tau,\quad 0\le t\le t_1.$$

For the proof we apply Lemma 8.2 to (8.5) and (8.6) and use the fact that $B_1(t)Z_1(t)$ depends only on the finite-dimensional components of $Z_1(t)$.

9. Integral inequality

The equation (8.4) can be written in the integral form:

$$(9.1)\qquad Z_1(t)=\mathrm{e}^{A_1 t}Z_1(0)+\int_0^t \mathrm{e}^{A_1(t-s)}[B_1Z_1(s)+N_1(s)]\,\mathrm{d}s,\quad 0\le t\le t_1.$$

We apply the symplectic orthogonal projection $\boldsymbol{P}_1:=\boldsymbol{P}_{v(t_1)}$ to both sides, and get

$$\boldsymbol{P}_1Z_1(t)=\mathrm{e}^{A_1 t}\boldsymbol{P}_1Z_1(0)+\int_0^t \mathrm{e}^{A_1(t-s)}\boldsymbol{P}_1[B_1Z_1(s)+N_1(s)]\,\mathrm{d}s.$$

We have used here that $\boldsymbol{P}_1$ commutes with the group $\mathrm{e}^{A_1 t}$ since the space $\mathcal{Z}_1:=\boldsymbol{P}_1\mathcal{E}$ is invariant with respect to $\mathrm{e}^{A_1 t}$ by Proposition 6.7(ii) (cf. also Remark 6.6). Applying (6.22) we obtain that

$$(9.2)\quad \|\boldsymbol{P}_1Z_1(t)\|_{-\beta}$$
$$\le \frac{C}{(1+t)^\beta}\|\boldsymbol{P}_1Z_1(0)\|_\beta + C\int_0^t \frac{1}{(1+|t-s|)^\beta}\|\boldsymbol{P}_1[B_1Z_1(s)+N_1(s)]\|_\beta\,\mathrm{d}s.$$

The operator $\boldsymbol{P}_1=\mathbf{I}-\boldsymbol{\Pi}_1$ is continuous in $\mathcal{E}_\beta$ by (6.19). Hence, from (9.2) and (8.9), (8.10), we obtain that

$$(9.3)\quad \|\boldsymbol{P}_1Z_1(t)\|_{-\beta}\le \frac{C}{(1+t)^\beta}\|Z_1(0)\|_\beta$$
$$+C(\bar d_1(t_1))\int_0^t \frac{1}{(1+|t-s|)^\beta}\Big[\|Z_1(s)\|_{-\beta}\int_s^{t_1}\|Z_1(\tau)\|^2_{-\beta}\,\mathrm{d}\tau+\|Z_1(s)\|^2_{-\beta}\Big]\,\mathrm{d}s,$$
$$0\le t\le t_1,$$

where $\bar d_1(s):=\sup_{0\le t\le s}|d_1(t)|$, $0\le s\le t_1$. Since $\|Z_1(t)\|_{-\beta}\le C(\bar d_1)\|Z(t)\|_{-\beta}$ by Lemma 8.2, we can rewrite (9.3) as

$$(9.4)\quad \|\boldsymbol{P}_1Z_1(t)\|_{-\beta}\le \frac{C(\bar d_1)}{(1+t)^\beta}\|Z(0)\|_\beta$$
$$+C(\bar d_1)\int_0^t \frac{1}{(1+|t-s|)^\beta}\Big[\|Z(s)\|_{-\beta}\int_s^{t_1}\|Z(\tau)\|^2_{-\beta}\,\mathrm{d}\tau+\|Z(s)\|^2_{-\beta}\Big]\,\mathrm{d}s,$$
$$0\le t\le t_1,$$

We can replace here the constants $C(\bar d_1)$ by C if $\bar d_1$ is bounded for $t_1\ge 0$. In order to do this replacement, we reduce the exit time

Definition 9.1. t'_* is the exit time

$$t'_* = \sup\{t \in [0, t_*) : \bar{d}_1(s) \le 1, 0 \le s \le t\}. \tag{9.5}$$

Now (9.4) implies that for $t_1 < t'_*$

$$\|\boldsymbol{P}_1 Z_1(t)\|_{-\beta} \le \frac{C}{(1+t)^\beta}\|Z(0)\|_\beta + C\int_0^t \frac{1}{(1+|t-s|)^\beta}\left[\|Z(s)\|_{-\beta}\int_s^{t_1}\|Z(\tau)\|^2_{-\beta}\,\mathrm{d}\tau + \|Z(s)\|^2_{-\beta}\right]\mathrm{d}s, \quad 0 \le t \le t_1, \tag{9.6}$$

10. Symplectic orthogonality

Finally, we are going to change $\boldsymbol{P}_1 Z_1(t)$ by $Z(t)$ in the left-hand side of (9.6). We will prove that it is possible using again that $d_\beta \ll 1$ in (2.12). For the justification we reduce the exit time once more. First, we introduce the "majorant"

$$m(t) := \sup_{s\in[0,t]} (1+s)^\beta \|Z(s)\|_{-\beta}, \quad t \in [0, t_*). \tag{10.1}$$

Let us denote by ε a fixed positive number which we will specify below.

Definition 10.1. t''_* is the exit time

$$t''_* = \sup\{t \in [0, t'_*) : m(s) \le \varepsilon, 0 \le s \le t\}. \tag{10.2}$$

The following important bound (10.3) allows us to change the norm of $\boldsymbol{P}_1 Z_1(t)$ in the left-hand side of (9.6) by the norm of $Z(t)$.

Lemma 10.2. *For sufficiently small $\varepsilon > 0$, we have for $t_1 < t''_*$*

$$\|Z(t)\|_{-\beta} \le C\|\boldsymbol{P}_1 Z_1(t)\|_{-\beta}, \quad 0 \le t \le t_1, \tag{10.3}$$

where C depends only on ρ and $\bar{v}$.

PROOF. The proof is based on the symplectic orthogonality (6.21), i.e.,

$$\boldsymbol{\Pi}_{v(t)}\, Z(t) = 0, \quad t \in [0, t_1], \tag{10.4}$$

and on the fact that all the spaces $\mathcal{Z}(t) := \boldsymbol{P}_{v(t)}\mathcal{E}$ are almost parallel for all t (see Figure 2).

Namely, we first note that $\|Z(t)\|_{-\beta} \le C\|Z_1(t)\|_{-\beta}$ by Lemma 8.2, since $|d_1(t)| \le 1$ for $t \le t_1 < t''_* < t'_*$. Therefore, it suffices to prove that

$$\|Z_1(t)\|_{-\beta} \le 2\|\boldsymbol{P}_1 Z_1(t)\|_{-\beta}, \quad 0 \le t \le t_1. \tag{10.5}$$

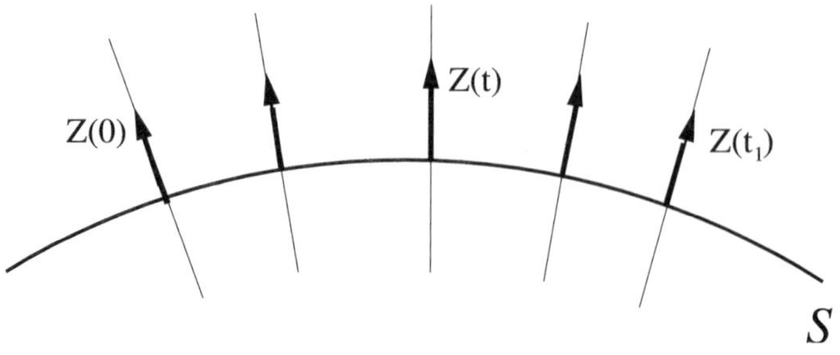

FIGURE 2. Symplectic orthogonality

This estimate will follow from

$$\|\boldsymbol{\Pi}_{v(t_1)}\, Z_1(t)\|_{-\beta} \le \tfrac{1}{2}\|Z_1(t)\|_{-\beta}, \quad 0 \le t \le t_1, \tag{10.6}$$

since $\boldsymbol{P}_1 Z_1(t) = Z_1(t) - \boldsymbol{\Pi}_{v(t_1)}\, Z_1(t)$. To prove (10.6), we write (10.4) as

$$\boldsymbol{\Pi}_{v(t),1}\, Z_1(t) = 0, \quad t \in [0, t_1], \tag{10.7}$$

where $\boldsymbol{\Pi}_{v(t),1}\, Z_1(t)$ is $\boldsymbol{\Pi}_{v(t)}\, Z(t)$ expressed in terms of the variable $y_1 = y + d_1(t)$. Hence, (10.6) follows from (10.7) if the difference $\boldsymbol{\Pi}_{v(t_1)} - \boldsymbol{\Pi}_{v(t),1}$ is small uniformly in t, i.e.,

$$\|\boldsymbol{\Pi}_{v(t_1)} - \boldsymbol{\Pi}_{v(t),1}\| < \tfrac{1}{2}, \quad 0 \le t \le t_1. \tag{10.8}$$

It remains to justify (10.8) for a sufficiently small $\varepsilon > 0$.

In order to prove the bound (10.8), we will need the formula (6.19) and the following relation which follows from (6.19):

$$\boldsymbol{\Pi}_{v(t),1}\, Z_1(t) = \sum \boldsymbol{\Pi}_{jl}\big(v(t)\big)\tau_{j,1}\big(v(t)\big)\Omega\big(\tau_{l,1}\big(v(t)\big), Z_1(t)\big), \tag{10.9}$$

where $\tau_{j,1}\big(v(t)\big)$ are the vectors $\tau_j\big(v(t)\big)$ expressed in the variables y_1. In detail (cf. (3.3)),

$$\begin{aligned} \tau_{j,1}(v) &:= \big(-\partial_j \psi_v(y_1 - d_1(t)), -\partial_j \pi_v(y_1 - d_1(t)), e_j, 0\big), \\ \tau_{j+3,1}(v) &:= \big(\partial_{v_j}\psi_v(y_1 - d_1(t)), \partial_{v_j}\pi_v(y_1 - d_1(t)), 0, \partial_{v_j} p_v\big), \end{aligned} \quad j = 1, 2, 3, \tag{10.10}$$

where $v = v(t)$. Since $|d_1(t)| \le 1$ and $\nabla \tau_j$ are smooth and sufficiently fast decaying at infinity functions, Lemma 8.2 implies that

$$\|\tau_{j,1}\big(v(t)\big) - \tau_j\big(v(t)\big)\|_\beta \le C|d_1(t)|^\beta, \quad 0 \le t \le t_1 \tag{10.11}$$

for all $j = 1, 2, \ldots, 6$. Furthermore,

$$\tau_j\big(v(t)\big) - \tau_j\big(v(t_1)\big) = \int_t^{t_1} \dot v(s) \cdot \nabla_v \tau_j\big(v(s)\big)\, \mathrm{d}s,$$

and therefore

$$\|\tau_j\big(v(t)\big) - \tau_j\big(v(t_1)\big)\|_\beta \le C \int_t^{t_1} |\dot v(s)|\, \mathrm{d}s, \quad 0 \le t \le t_1. \tag{10.12}$$

Similarly,

$$\begin{aligned} |\boldsymbol{\Pi}_{jl}\big(v(t)\big) - \boldsymbol{\Pi}_{jl}\big(v(t_1)\big)| &= \left| \int_t^{t_1} \dot v(s) \cdot \nabla_v \boldsymbol{\Pi}_{jl}\big(v(s)\big)\, \mathrm{d}s \right| \\ &\le C \int_t^{t_1} |\dot v(s)|\, \mathrm{d}s, \qquad 0 \le t \le t_1, \end{aligned} \tag{10.13}$$

since $\big|\nabla_v \boldsymbol{\Pi}_{jl}\big(v(s)\big)\big|$ is uniformly bounded by (6.9). Hence, the bounds (10.8) will follow from (6.19), (10.9) and (10.11)–(10.13) if we establish that $|d_1(t)|$ and the integral in the right-hand side of (10.12) can be made as small as we please by choosing $\varepsilon > 0$ sufficiently small.

To estimate $d_1(t)$, we note that

$$w(s) - v(t_1) = w(s) - v(s) + v(s) - v(t_1) = \dot c(s) + \int_s^{t_1} \dot v(\tau)\, \mathrm{d}\tau \tag{10.14}$$

by (6.10). Hence, (8.2), Lemma 6.2 and the definition (10.1) imply that

$$(10.15)\quad |d_1(t)| = \left|\int_{t_1}^{t} (w(s) - v(t_1))\,\mathrm{d}s\right| \le \int_t^{t_1} \Big(|\dot c(s)| + \int_s^{t_1} |\dot v(\tau)|\,\mathrm{d}\tau\Big)\mathrm{d}s$$
$$\le Cm^2(t_1)\int_t^{t_1}\Big(\frac{1}{(1+s)^{2\beta}} + \int_s^{t_1}\frac{\mathrm{d}\tau}{(1+\tau)^{2\beta}}\Big)\,\mathrm{d}s$$
$$\le Cm^2(t_1) \le C\varepsilon^2, \qquad 0 \le t \le t_1$$

since $t_1 < t_*''$. Similarly,

$$(10.16)\qquad \int_t^{t_1} |\dot v(s)|\,\mathrm{d}s \le Cm^2(t_1)\int_t^{t_1}\frac{\mathrm{d}s}{(1+s)^{2\beta}} \le C\varepsilon^2, \quad 0 \le t \le t_1.$$

The proof is completed. □

11. Decay of transversal component

Here we prove Proposition 6.3.

STEP (i). We fix $\varepsilon > 0$ and $t_*'' = t_*''(\varepsilon)$ for which Lemma 10.2 holds. Then the bound of type (9.6) holds with $\|\boldsymbol{P}_1 Z_1(t)\|_{-\beta}$ in the left-hand side replaced by $\|Z(t)\|_{-\beta}$:

$$(11.1)\quad \|Z(t)\|_{-\beta} \le \frac{C}{(1+t)^\beta}\|Z(0)\|_\beta$$
$$+ C\int_0^t \frac{1}{(1+|t-s|)^\beta}\Big[\|Z(s)\|_{-\beta}\int_s^{t_1}\|Z(\tau)\|^2_{-\beta}\,\mathrm{d}\tau + \|Z(s)\|^2_{-\beta}\Big]\,\mathrm{d}s, \qquad 0 \le t \le t_1$$

for $t_1 < t_*'$. This implies an integral inequality for the majorant

$$m(t) := \sup_{s\in[0,t]} (1+s)^\beta \|Z(s)\|_{-\beta}\,.$$

Namely, multiplying both sides of (11.1) by $(1+t)^\beta$, and taking the supremum in $t \in [0,t_1]$, we get

$$m(t_1) \le C\|Z(0)\|_\beta$$
$$+ C\sup_{t\in[0,t_1]}\int_0^t \frac{(1+t)^\beta}{(1+|t-s|)^\beta}\Big[\frac{m(s)}{(1+s)^\beta}\int_s^{t_1}\frac{m^2(\tau)\,\mathrm{d}\tau}{(1+\tau)^{2\beta}} + \frac{m^2(s)}{(1+s)^{2\beta}}\Big]\,\mathrm{d}s$$

for $t_1 \le t_*''$. Taking into account that $m(t)$ is a monotone increasing function, we get

$$(11.2)\qquad m(t_1) \le C\|Z(0)\|_\beta + C[m^3(t_1) + m^2(t_1)]I(t_1), \quad t_1 \le t_*''.$$

where

$$I(t_1) = \sup_{t\in[0,t_1]}\int_0^t \frac{(1+t)^\beta}{(1+|t-s|)^\beta}\Big[\frac{1}{(1+s)^\beta}\int_s^{t_1}\frac{\mathrm{d}\tau}{(1+\tau)^{2\beta}} + \frac{1}{(1+s)^{2\beta}}\Big]\,\mathrm{d}s \le \bar I < \infty, \qquad t_1 \ge 0.$$

Therefore, (11.2) becomes

$$(11.3)\qquad m(t_1) \le C\|Z(0)\|_\beta + C\bar I[m^3(t_1) + m^2(t_1)], \quad t_1 < t_*''.$$

This inequality implies that $m(t_1)$ is bounded for $t_1 < t''_*$, and moreover,

$$m(t_1) \le C_1 \|Z(0)\|_\beta, \quad t_1 < t''_*, \tag{11.4}$$

since $m(0) = \|Z(0)\|_\beta$ is sufficiently small by (3.7).

STEP (ii). The constant C_1 in the estimate (11.4) does not depend on t_*, t'_* and t''_* by Lemma 10.2. We choose d_β in (2.12) so small that $\|Z(0)\|_\beta < \varepsilon/(2C_1)$. It is possible due to (3.7). Then the estimate (11.4) implies that $t''_* = t'_*$ and therefore (11.4) holds for all $t_1 < t'_*$. Then the bound (10.15) holds for all $t < t'_*$. We choose ε so small that the right-hand side in (10.15) does not exceed one. Then $t'_* = t_*$. Therefore, (11.4) holds for all $t_1 < t_*$, hence (6.4) also holds if $\|Z(0)\|_\beta$ is sufficiently small. Finally, this implies that $t_* = \infty$, hence also $t''_* = t'_* = \infty$ and (11.4) holds for all $t_1 > 0$ if d_β is small enough. □

12. Soliton asymptotics

Here we prove our main Theorem 2.5 under the assumption that the decay (6.13) holds. First we will prove the asymptotics (2.13) for the vector components, and afterward the asymptotics (2.14) for the fields.

Asymptotics for the vector components. From (4.3) we have $\dot q = \dot b + \dot Q$, and from (6.15), (6.16), (4.9) it follows that $\dot Q = B_{v(t)}P + \mathcal{O}(\|Z\|^2_{-\beta})$. Thus,

$$\dot q = \dot b + \dot Q = v(t) + \dot c(t) + B_{v(t)}P(t) + \mathcal{O}(\|Z\|^2_{-\beta}). \tag{12.1}$$

Recall that $\beta = 1+\delta$, $0 < \delta < \frac{1}{2}$. The equation (6.11) and the estimates (6.12), (6.13) imply that

$$|\dot c(t)| + |\dot v(t)| \le \frac{C_1(\rho, \bar v, d_\beta)}{(1+t)^{2+2\delta}}, \quad t \ge 0. \tag{12.2}$$

Therefore, $c(t) = c_+ + \mathcal{O}(t^{-(1+2\delta)})$ and $v(t) = v_+ + \mathcal{O}(t^{-(1+2\delta)})$, $t \to \infty$. Since $|P| \le \|Z\|_{-\beta}$, the estimate (6.13), and (12.2), (12.1) imply that

$$\dot q(t) = v_+ + \mathcal{O}(t^{-\beta}). \tag{12.3}$$

Similarly,

$$b(t) = c(t) + \int_0^t v(s)\, ds = v_+ t + a_+ + \mathcal{O}(t^{-2\delta}), \tag{12.4}$$

hence the second part of (2.13) follows:

$$q(t) = b(t) + Q(t) = v_+ t + a_+ + \mathcal{O}(t^{-2\delta}), \tag{12.5}$$

since $Q(t) = \mathcal{O}(t^{-\beta})$ by (6.13).

Asymptotics for the fields. We apply the approach developed in [9], see also [6, 8, 10, 12]. For the field part of the solution, $F(t) = \big(\psi(x,t), \pi(x,t)\big)$ let us define the accompanying soliton field as $F_{\mathrm{v}(t)}(t) = \big(\psi_{\mathrm{v}(t)}(x - q(t)), \pi_{\mathrm{v}(t)}(x - q(t))\big)$, where we define now $\mathrm{v}(t) = \dot q(t)$, cf. (12.1). Then for the difference $Z(t) = F(t) - F_{\mathrm{v}(t)}(t)$ we obtain easily the equation [12, eq. (2.5)],

$$\dot Z(t) = AZ(t) - \dot{\mathrm{v}} \cdot \nabla_{\mathrm{v}} F_{\mathrm{v}(t)}(t), \quad A(\psi, \pi) = (\pi, \Delta\psi).$$

Then

$$Z(t) = W^0(t)Z(0) - \int_0^t W^0(t-s)[\dot{\mathrm{v}}(s) \cdot \nabla_{\mathrm{v}} F_{\mathrm{v}(s)}(s)]\, ds. \tag{12.6}$$

To obtain the asymptotics (2.14) it suffices to prove that $Z(t) = W^0(t)\Psi_+ + r_+(t)$ with some $\Psi_+ \in \mathcal{F}$ and $\|r_+(t)\|_{\mathcal{F}} = \mathcal{O}(t^{-\delta})$. This is equivalent to

$$W^0(-t)Z(t) = \mathbf{\Psi}_+ + r'_+(t), \tag{12.7}$$

where $\|r'_+(t)\|_{\mathcal{F}} = \mathcal{O}(t^{-\delta})$ since $W^0(t)$ is a unitary group in the Sobolev space $\mathcal{F}$ by the energy conservation for the free wave equation. Finally, (12.7) holds since (12.6) implies that

$$W^0(-t)Z(t) = Z(0) + \int_0^t W^0(-s)R(s)\,\mathrm{d}s, \quad R(s) = \dot{\mathrm{v}}(s)\cdot\nabla_{\mathrm{v}}F_{\mathrm{v}(s)}(s), \tag{12.8}$$

where the integral in the right-hand side of (12.8) converges in the Hilbert space $\mathcal{F}$ with the rate $\mathcal{O}(t^{-\delta})$. The latter holds since $\|W^0(-s)R(s)\|_{\mathcal{F}} = \mathcal{O}(s^{-\beta})$ by the unitarity of $W^0(-s)$ and the decay rate $\|R(s)\|_{\mathcal{F}} = \mathcal{O}(s^{-\beta})$ which follows from the asymptotics for the vector components. More precisely, differentiating the modulation equations (6.11) we obtain an estimate for $\dot{\mathrm{v}}(t) = \ddot{q}(t)$ providing the mentioned decay rate of $R(s)$. □

Acknowledgments. The authors thank V. Buslaev for detailed and numerous lectures on his results and fruitful discussions.

References

1. V. S. Buslaev and G. S. Perel′man, *On nonlinear scattering of states which are close to a soliton*, Astérisque **210** (1992), 49–63.
2. ———, *Scattering for the nonlinear Schrödinger equation: states close to a soliton*, Algebra i Analiz **4** (1992), no. 6, 63–102 (Russian); English transl., St. Petersburg Math. J. **4** (1993), no. 6, 1111–1142.
3. S. Cuccagna, *Stabilization of solutions to nonlinear Schrödinger equations*, Comm. Pure Appl. Math. **54** (2001), no. 9, 1110–1145.
4. ———, *On asymptotic stability of ground states of NLS*, Rev. Math. Phys. **15** (2003), no. 8, 877–903.
5. W. W. Eckhaus and A. van Harten, *The inverse scattering transformation and the theory of solitons: An introduction*, North-Holland Math. Stud., vol. 50, North-Holland, Amsterdam, 1981.
6. V. Imaikin, A. Komech, and P. A. Markowich, *Scattering of solitons of the Klein–Gordon equation coupled to a classical particle*, J. Math. Phys. **44** (2003), no. 3, 1202–1217.
7. V. Imaikin, A. Komech, and N. Mauser, *Soliton-type asymptotics for the coupled Maxwell–Lorentz equations*, Ann. Inst. H. Poincaré Phys. Théor. **5** (2004), 1117–1135.
8. V. Imaikin, A. Komech, and H. Spohn, *Soliton-like asymptotics and scattering for a particle coupled to Maxwell field*, Russ. J. Math. Phys. **9** (2002), no. 4, 428–436.
9. ———, *Scattering theory for a particle coupled to a scalar field*, Discrete Contin. Dyn. Syst. **10** (2004), no. 1-2, 387–396.
10. ———, *Rotating charge coupled to the Maxwell field: scattering theory and adiabatic limit*, Monatsh. Math. **142** (2004), no. 1-2, 143–156.
11. V. Imaikin, A. Komech, and B. Vainberg, *On scattering of solution for the Klein–Gordon equation coupled to a particle*, Comm. Math. Phys. **268**, no. 2, 321–367.
12. A. Komech, M. Kunze, and H. Spohn, *Effective dynamics for a mechanical particle coupled to a wave field*, Comm. Math. Phys. **203** (1999), no. 1, 1–19.
13. A. Komech, H. Spohn, and M. Kunze, *Long-time asymptotics for a classical particle interacting with a scalar wave field*, Comm. Partial Differential Equations **22** (1997), no. 1-2, 307–335.
14. A. Komech and H. Spohn, *Soliton-like asymptotics for a classical particle interacting with a scalar wave field*, Nonlin. Analysis **33** (1998), no. 1, 13–24.
15. J. Miller and M. Weinstein, *Asymptotic stability of solitary waves for the regularized long-wave equation*, Comm. Pure Appl. Math. **49** (1996), no. 4, 399–441.

16. R. L. Pego and M. I. Weinstein, *On asymptotic stability of solitary waves*, Phys. Lett. A **162** (1992), no. 3, 263 – 268.
17. ———, *Asymptotic stability of solitary waves*, Comm. Math. Phys. **164** (1994), no. 2, 305 – 349.
18. A. Soffer and M. I. Weinstein, *Multichannel nonlinear scattering for nonintegrable systems*, Integrable Systems and Applications (île d'Oléron, 1988) (M. Balabane, Lochak P., and C. Sulem, eds.), Lecture Notes in Physics, vol. 342, Springer, Berlin, 1989, pp. 312 – 327.
19. ———, *Multichannel nonlinear scattering in nonintegrable systems*, Comm. Math. Phys. **133** (1990), no. 1, 119 – 146.
20. ———, *Multichannel nonlinear scattering and stability.* II: *The case of anisotropic potential and data*, J. Differential Equations **98** (1992), no. 2, 376 – 390.
21. ———, *Resonances, radiation damping and instability in Hamiltonian nonlinear wave equations*, Invent. Math. **136** (1999), no. 1, 9 – 74.
22. ———, *Selection of the ground state for nonlinear Schrödinger equations*, Rev. Math. Phys. **16** (2004), no. 8, 977 – 1071.
23. H. Spohn, *Dynamics of charged particles and their radiation field*, Cambridge Univ. Press, Cambridge, 2004.
24. B. Vainberg, *Behavior of the solution of the Cauchy problem for a hyperbolic equation as* $t \to \infty$, Mat. Sb. (N.S.) **78 (120)** (1969), 542 – 578 (Russian); English transl., Math. USSR-Sb. **7** (1969), no. 4, 533 – 568.
25. ———, *The short-wave asymptotic behavior of solutions of stationary problems and the asymptotic behavior as* $t \to \infty$ *of solutions of nonstationary problems*, Uspehi Mat. Nauk **30** (1975), no. 2(182), 3 – 55 (Russian); English transl., Russian Math. Surveys **30** (1975), no. 2, 1 – 58.
26. ———, *Asymptotic methods in equations of mathematical physics*, Gordon & Breach Publishers, New York, 1989.
27. M. I. Weinstein, *Modulational stability of ground states of nonlinear Schrödinger equations*, SIAM J. Math. Anal. **16** (1985), no. 3, 472 – 491.

ZENTRUM MATHEMATIK, TECHNISCHEN UNIVERSITÄT MÜNCHEN, 85748 GARCHING, GERMANY
Current address: Fakultät für Mathematik, Universität Wien, Nordbergstraße 15, 1090 Wien, Austria
E-mail address: `imaikin@ma.tum.de`

FAKULTÄT FÜR MATHEMATIK, UNIVERSITÄT WIEN, NORDBERGSTRASSE 15, 1090 WIEN, AUSTRIA
Current address: Department of Mechanics and Mathematics, Moscow State University, Moscow 119992, GSP-2 Russia
E-mail address: `alexander.komech@univie.ac.at`

DEPARTMENT OF MATHEMATICS AND STATISTICS, UNIVERSITY OF NORTH CAROLINA — CHARLOTTE, 376 FRETWELL BLDG., 9201 UNIVERSITY CITY BLVD., CHARLOTTE, NC 28223-0001, USA
E-mail address: `brvainbe@email.uncc.edu`

Centre de Recherches Mathématiques
CRM Proceedings and Lecture Notes
Volume **42**, 2007

Purely Absolutely Continuous Spectrum for some Random Jacobi Matrices

Uri Kaluzhny and Yoram Last

Abstract. We consider random Jacobi matrices of the form

$$(J_\omega u)(n) = a_n(\omega)u(n+1) + b_n(\omega)u(n) + a_{n-1}(\omega)u(n-1)$$

on $\ell^2(\mathbb{N})$, where $a_n(\omega) = \tilde{a}_n + \alpha_n(\omega)$, $b_n(\omega) = \tilde{b}_n + \beta_n(\omega)$, $\{\tilde{a}_n\}$ and $\{\tilde{b}_n\}$ are sequences of bounded variation obeying $\tilde{a}_n \to 1$ and $\tilde{b}_n \to 0$, and $\{\alpha_n(\omega)\}$ and $\{\beta_n(\omega)\}$ are sequences of independent random variables on a probability space $(\Omega, \mathrm{d}P(\omega))$ obeying

$$\sum_{n=1}^{\infty} \int_\Omega \bigl(\alpha_n^2(\omega) + \beta_n^2(\omega)\bigr)\, \mathrm{d}P(\omega) < \infty$$

and $\int_\Omega \alpha_n(\omega)\, \mathrm{d}P(\omega) = \int_\Omega \beta_n(\omega)\, \mathrm{d}P(\omega) = 0$ for each n. We further assume that there exists $C_0 > 0$ such that $1/C_0 < a_n(\omega) < C_0$ and $-C_0 < b_n(\omega) < C_0$ for every n and P a.e. ω. We prove that, for P a.e. ω, J_ω has purely absolutely continuous spectrum on $(-2, 2)$.

1. Introduction

In this paper we study self-adjoint Jacobi matrices of the form

$$J(\{a_n\}_{n=1}^{\infty}, \{b_n\}_{n=1}^{\infty}) = \begin{pmatrix} b_1 & a_1 & 0 & \cdots \\ a_1 & b_2 & a_2 & \cdots \\ 0 & a_2 & b_3 & \ddots \\ \vdots & \vdots & \ddots & \ddots \end{pmatrix}, \tag{1.1}$$

where $a_j > 0$ and $b_j \in \mathbb{R}$. We consider only such matrices whose entries are bounded, so they define bounded self-adjoint operators on $\ell^2(\mathbb{N})$. We say that a sequence $\{c_n\}_{n=1}^{\infty} \subset \mathbb{R}$ is *of bounded variation* if $\sum_{n=1}^{\infty} |c_{n+1} - c_n| < \infty$. We denote by J_0 the free Laplacian on $\ell^2(\mathbb{N})$, which is the Jacobi matrix of the form (1.1) with $a_n = 1$ and $b_n = 0$. Our main result in this paper is the following:

2000 *Mathematics Subject Classification.* Primary: 81Q10; Secondary: 47B15, 82B44.

This research was supported in part by The Israel Science Foundation (Grant No. 188/02) and in part by Grant No. 2002068 from the United States – Israel Binational Science Foundation (BSF), Jerusalem, Israel.

This is the final form of the paper.

Theorem 1.1. *Let* $(\Omega, \mathrm{d}P(\omega))$ *be a probability space and let*

$$J_\omega = J(\{a_n(\omega)\}_{n=1}^\infty, \{b_n(\omega)\}_{n=1}^\infty),$$

where $a_n(\omega) = \tilde{a}_n + \alpha_n(\omega)$ *and* $b_n(\omega) = \tilde{b}_n + \beta_n(\omega)$. *Assume that*

(i) $\{\tilde{a}_n\}_{n=1}^\infty$ *and* $\{\tilde{b}_n\}_{n=1}^\infty$ *are real-valued sequences of bounded variation obeying* $\lim_{n\to\infty} \tilde{a}_n = 1$ *and* $\lim_{n\to\infty} \tilde{b}_n = 0$.

(ii) $\{\alpha_n(\omega)\}_{n=1}^\infty$ *and* $\{\beta_n(\omega)\}_{n=1}^\infty$ *are sequences of real-valued independent random variables on* $(\Omega, \mathrm{d}P(\omega))$ *obeying*

$$\int_\Omega \alpha_n(\omega)\, \mathrm{d}P(\omega) = \int_\Omega \beta_n(\omega)\, \mathrm{d}P(\omega) = 0$$

for each n *and*

$$\sum_{n=1}^\infty \int_\Omega (\alpha_n^2(\omega) + \beta_n^2(\omega))\, \mathrm{d}P(\omega) < \infty.$$

(iii) *There exists a constant* $C_0 > 0$ *such that* $1/C_0 < a_n(\omega) < C_0$ *and* $-C_0 < b_n(\omega) < C_0$ *for every* n *and* P *a.e.* ω.

Then, for P *a.e.* ω, J_ω *has purely absolutely continuous spectrum on* $(-2, 2)$ *with essential support* $(-2, 2)$.

Remarks. (1) Saying that the essential support of the absolutely continuous spectrum of J_ω is $(-2, 2)$ means that the absolutely continuous part of the spectral measure of J_ω is supported on $(-2, 2)$ and gives positive weight to any subset of $(-2, 2)$ that has positive Lebesgue measure.

(2) Our proof actually shows something a bit stronger (see Remark (3) to Theorem 2.1 below), namely, that for P a.e. fixed ω, the purity of the absolutely continuous spectrum in $(-2, 2)$ will be stable under changing any finite number of entries in J_ω.

Theorem 1.1 is essentially an extension of a result of Kiselev–Last–Simon [7, Theorem 8.1], who obtained this theorem for the special case $\tilde{a}_n = 1$, $\tilde{b}_n = 0$, $\alpha_n(\omega) = 0$ (also see [4, 8, 14] for related earlier results). Part of the extension is in considering the nontrivial off-diagonal part, rather than just $a_n(\omega) = 1$ for all n. While this extension is fairly straight forward, it does add some technical complexity to the problem (partly due to the fact that one-step transfer matrices depend on pairs of neighboring a_n's and are thus not independent of each other). The more important extension in Theorem 1.1 is in adding the decaying perturbation of bounded variation $J(\{\tilde{a}_n - 1\}_{n=1}^\infty, \{\tilde{b}_n\}_{n=1}^\infty)$.

We note that the Jacobi matrix $J(\{\tilde{a}_n\}_{n=1}^\infty, \{\tilde{b}_n\}_{n=1}^\infty)$, where $\{\tilde{a}_n\}_{n=1}^\infty$ and $\{\tilde{b}_n\}_{n=1}^\infty$ are sequences of bounded variation obeying $\lim_{n\to\infty} \tilde{a}_n = 1$ and $\lim_{n\to\infty} \tilde{b}_n = 0$, is well-known to have purely absolutely continuous spectrum on $(-2, 2)$ with essential support $(-2, 2)$. That is, adding the decaying perturbation of bounded variation $J(\{\tilde{a}_n - 1\}_{n=1}^\infty, \{\tilde{b}_n\}_{n=1}^\infty)$ to the free Laplacian J_0 doesn't change its absolutely continuous spectrum. This fact is the discrete version of Weidmann's theorem [17] (see, e.g., [16] for a proof). One may thus be tempted to think that such a perturbation of bounded variation may never be important for absolutely continuous spectrum. However, one of us have recently constructed [11] an example of a Jacobi matrix $J(\{a_n\}_{n=1}^\infty, \{b_n\}_{n=1}^\infty)$ with $a_n = 1$, $b_n = \tilde{b}_n + \hat{b}_n$, so that $\lim_{n\to\infty} \tilde{b}_n = \lim_{n\to\infty} \hat{b}_n = 0$, $\{\tilde{b}_n\}_{n=1}^\infty$ is of bounded variation, $J(\{a_n\}_{n=1}^\infty, \{\hat{b}_n\}_{n=1}^\infty)$ (like

$J(\{a_n\}_{n=1}^{\infty}, \{\tilde{b}_n\}_{n=1}^{\infty}))$ has purely absolutely continuous spectrum on $(-2, 2)$ with essential support $(-2, 2)$, but $J(\{a_n\}_{n=1}^{\infty}, \{b_n\}_{n=1}^{\infty})$ has empty absolutely continuous spectrum. In particular, adding a decaying perturbation of bounded variation to a Jacobi matrix can fully "destroy" its absolutely continuous spectrum (even though such a perturbation does not change the absolutely continuous spectrum when added to the free Laplacian J_0).

Theorem 1.1 is connected with a recent result of Breuer – Last [1] which says, roughly speaking, that absolutely continuous spectrum of Jacobi matrices which is associated with bounded generalized eigenfunctions is stable under square-summable random perturbations of the form $J(\{\alpha_n(\omega)\}_{n=1}^{\infty}, \{\beta_n(\omega)\}_{n=1}^{\infty})$. In particular, their result imply that, with probability one, J_ω of Theorem 1.1 has absolutely continuous spectrum with essential support $(-2, 2)$, but it doesn't exclude by itself the possibility of embedded singular spectrum in $(-2, 2)$. If the probability distributions of the random variables $\{\alpha_n(\omega)\}_{n=1}^{\infty}$ and $\{\beta_n(\omega)\}_{n=1}^{\infty}$ happen to be absolutely continuous with respect to the Lebesgue measure (in fact, it suffices for only some of the distributions to be absolutely continuous, for example, absolute continuity of the distribution of $\beta_1(\omega)$ or of the distributions of any consecutive pair $\beta_n(\omega)$, $\beta_{n+1}(\omega)$ would suffice), then one can use spectral averaging (see, e.g., [15, Theorem I.8]) to conclude almost sure purely absolutely continuous spectrum in $(-2, 2)$, namely, to recover Theorem 1.1 from the general result of [1]. The main ingredient of Theorem 1.1 which doesn't follow from the result of [1] is the purity of the absolutely continuous spectrum in $(-2, 2)$ even in cases where all of the probability distributions of the random variables $\{\alpha_n(\omega)\}_{n=1}^{\infty}$ and $\{\beta_n(\omega)\}_{n=1}^{\infty}$ are singular with respect to the Lebesgue measure. This ingredient is connected with a key technical difference between the analysis of the current paper and [1]. Here we develop, following the original approach of [7], estimates which are uniform in energy over subintervals of $(-2, 2)$ and which yield bounds on certain integrated (over energies) quantities. Such an approach cannot possibly work in the more general context considered by [1], where the obtained estimates are per individual energy and uniformity over energy ranges cannot hold. The current paper and [1] can thus be viewed as complementary to each other.

We note that much of the original Kiselev – Last – Simon result [7, Theorem 8.1] discussed above can be seen as a special case of the celebrated deterministic result of Deift – Killip [3], which has been considerably strengthened by Killip – Simon [6]. The Killip – Simon theorem says, among other things, that any Jacobi matrix J for which $J - J_0$ is Hilbert – Schmidt has absolutely continuous spectrum with essential support $(-2, 2)$ (this doesn't exclude singular spectrum embedded in $(-2, 2)$). Thus, by spectral averaging, one can recover [7, Theorem 8.1] for the case of absolutely continuous probability distributions. While various variants and extensions of the Killip – Simon result exist (see, e.g., [9, 10, 13, 18] and references therein), it seems that the sum-rule techniques underlying all these deterministic results cannot handle a general decaying perturbation of bounded variation, which may have an arbitrarily slow decay rate.

Stanislav Molchanov has been among the most important contributors to the theory of random Schrödinger operators ever since his seminal paper with Goldsheid and Pastur [5], which gave the first proof of Anderson localization for such an operator. It is a pleasure to dedicate this paper to him on the occasion of his 65th birthday.

2. Proof of Theorem 1.1

Our proof of Theorem 1.1 relies on controlling the asymptotic growth of the norms of the 2×2 transfer matrices associated with the problem. These are defined by

$$\mathcal{T}_{n,m}(E,\omega)=\mathcal{T}_n(E,\omega)\mathcal{T}_{n-1}(E,\omega)\dots\mathcal{T}_m(E,\omega),$$

where

$$\mathcal{T}_j(E,\omega)=\begin{pmatrix}(E-b_j(\omega))/a_j(\omega) & -a_{j-1}(\omega)/a_j(\omega)\\ 1 & 0\end{pmatrix}.$$

We use the following result of [12]:

Theorem 2.1. *Suppose there is some $m\in\mathbb{N}$ so that*

$$\liminf_{n\to\infty}\int_a^b\|\mathcal{T}_{n,m}(E,\omega)\|^p\,\mathrm{d}E<\infty$$

for some $p>2$. Then (a,b) is in the essential support of the absolutely continuous spectrum of J_ω and the spectrum of J_ω is purely absolutely continuous on (a,b).

Remarks. (1) This is essentially [12, Theorem 1.3]. While [12] only discusses Jacobi matrices with $a_n=1$, the result easily extends to our more general context.

(2) As noted in [12], this theorem is an extension of an idea of Carmona [2].

(3) While the fact that (a,b) is in the essential support of the absolutely continuous spectrum isn't explicitly stated in [12, Theorem 1.3], this easily follows from spectral averaging and the fact that the $\liminf_{n\to\infty}\int_a^b\|\mathcal{T}_{n,m}(E,\omega)\|^p\,\mathrm{d}E<\infty$ condition is invariant to changing any finite number of entries in the Jacobi matrix.

To prove Theorem 1.1 we fix an $\varepsilon>0$ and use Theorem 2.1 on the interval $\Delta=(-2+\varepsilon,2-\varepsilon)$. Since ε is arbitrary, Theorem 1.1 would follow.

One of the technical difficulties we need to overcome is the fact that each of the matrices $\mathcal{T}_n(E,\omega)$ depends on both $a_n(\omega)$ and $a_{n-1}(\omega)$. Thus, $\mathcal{T}_n(E,\omega)$'s for different n's are not independent. To overcome this, we introduce the matrices

$$\mathcal{A}_n(\omega)=\begin{pmatrix}1 & 0\\ 0 & a_{n-1}(\omega)\end{pmatrix}$$

and define

$$\mathcal{P}_n(E,\omega)=\mathcal{A}_{n+1}(\omega)\mathcal{T}_n(E,\omega)\mathcal{A}_n^{-1}(\omega)=\begin{pmatrix}(E-b_n(\omega))/a_n(\omega) & -1/a_n(\omega)\\ a_n(\omega) & 0\end{pmatrix}.$$

Now,

$$\mathcal{P}_{n,m}(E,\omega)\equiv\mathcal{P}_n\mathcal{P}_{n-1}\cdots\mathcal{P}_m=\mathcal{A}_{n+1}(\omega)\mathcal{T}_{n,m}(E,\omega)\mathcal{A}_m^{-1}(\omega),$$

and since $\|\mathcal{A}_n\|\le C_0$, $\|\mathcal{A}_n^{-1}\|\le C_0$, it suffice to prove that for a.e. $\omega\in\Omega$,

$$\liminf_{n\to\infty}\int_\Delta\|\mathcal{P}_{n,m}(E,\omega)\|^4\,\mathrm{d}E<\infty.$$

We choose an m so that for every $n\ge m$, $\sqrt{(1-\varepsilon/3)/(1+\varepsilon/3)}<\tilde{a}_n$ and $|\tilde{b}_n|<\varepsilon/3$. Then for every $E\in\Delta$, $|E-\tilde{b}_n|<2(1-\varepsilon/3)$, hence $|(E-\tilde{b}_n)/\tilde{a}_n|<2\sqrt{1-\varepsilon^2/9}$, and we can define $k_n\in(0,\pi)$ by

$$2\cos k_n=(E-\tilde{b}_n)/\tilde{a}_n.$$

Since $|\cos k_n| < \sqrt{1-\varepsilon^2/9}$, we have for every $n \geq m$ and $E \in \Delta$,

$$\sin k_n > \varepsilon/3. \tag{2.1}$$

We want to separate the deterministic and the random parts of $\mathcal{P}_n$. Since

$$\begin{aligned}\frac{E-\tilde{b}_n}{\tilde{a}_n} - \frac{E-b_n(\omega)}{a_n(\omega)} &= \frac{(E-\tilde{b}_n)(a_n(\omega)-\tilde{a}_n)+\tilde{a}_n(b_n(\omega)-\tilde{b}_n)}{\tilde{a}_n a_n(\omega)}\\ &= \frac{\alpha_n(\omega)}{a_n(\omega)}2\cos k_n + \frac{\beta_n(\omega)}{a_n(\omega)},\end{aligned}$$

we have

$$\frac{E-b_n(\omega)}{a_n(\omega)} = \frac{\tilde{a}_n}{a_n(\omega)}\left(2\cos k_n - \frac{\beta_n(\omega)}{\tilde{a}_n}\right)$$

and thus

$$\begin{aligned}\mathcal{P}_n(E,\omega) &= \begin{pmatrix} (E-b_n(\omega))/a_n(\omega) & -1/a_n(\omega) \\ a_n(\omega) & 0 \end{pmatrix}\\ &= \frac{\tilde{a}_n}{a_n(\omega)}\begin{pmatrix} 2\cos k_n - \beta_n(\omega)/\tilde{a}_n & -1/\tilde{a}_n \\ a_n^2(\omega)/\tilde{a}_n & 0 \end{pmatrix}\\ &= \frac{\tilde{a}_n}{a_n(\omega)}\left[\begin{pmatrix} 2\cos k_n & -1/\tilde{a}_n \\ \tilde{a}_n & 0 \end{pmatrix} + \frac{1}{\tilde{a}_n}\begin{pmatrix} -\beta_n(\omega) & 0 \\ a_n^2(\omega)-\tilde{a}_n^2 & 0 \end{pmatrix}\right].\end{aligned}$$

To control the growth of transfer matrices, [7] uses the EFGP transform, which is connected with the following equality:

$$\begin{pmatrix} 0 & \sin k \\ 1 & -\cos k \end{pmatrix}\begin{pmatrix} 2\cos k & -1 \\ 1 & 0 \end{pmatrix} = \begin{pmatrix} \cos k & \sin k \\ -\sin k & \cos k \end{pmatrix}\begin{pmatrix} 0 & \sin k \\ 1 & -\cos k \end{pmatrix}.$$

We use the following modification:

$$\begin{pmatrix} 0 & \sin k \\ \tilde{a} & -\cos k \end{pmatrix}\begin{pmatrix} 2\cos k & -1/\tilde{a} \\ \tilde{a} & 0 \end{pmatrix} = \begin{pmatrix} \cos k & \sin k \\ -\sin k & \cos k \end{pmatrix}\begin{pmatrix} 0 & \sin k \\ \tilde{a} & -\cos k \end{pmatrix}.$$

Denote

$$\mathcal{C}_n(E) = \begin{pmatrix} 0 & \sin k_n \\ \tilde{a}_n & -\cos k_n \end{pmatrix}, \qquad \text{so } \mathcal{C}_n(E)^{-1} = \begin{pmatrix} \cot k_n/\tilde{a}_n & 1/\tilde{a}_n \\ 1/\sin k_n & 0 \end{pmatrix},$$

$$\mathcal{B}_n(E) = \begin{pmatrix} \cos k_n & \sin k_n \\ -\sin k_n & \cos k_n \end{pmatrix}, \qquad \mathcal{D}_n(E) = \begin{pmatrix} 2\cos k_n & -1/\tilde{a}_n \\ \tilde{a}_n & 0 \end{pmatrix}.$$

Based on the previous computations, we have

$$\mathcal{P}_n(E,\omega) = \mathcal{D}_n(E) + \mathcal{Q}_n(E,\omega), \tag{2.2}$$

where

$$\mathcal{D}_n(E) = \mathcal{C}_n(E)^{-1}\mathcal{B}_n(E)\mathcal{C}_n(E) \tag{2.3}$$

and

$$\mathcal{Q}_n(E,\omega) = \frac{1}{a_n(\omega)}\left[-\alpha_n(\omega)\mathcal{D}_n(E) + \begin{pmatrix} -\beta_n(\omega) & 0 \\ a_n^2(\omega)-\tilde{a}_n^2 & 0 \end{pmatrix}\right]. \tag{2.4}$$

We define $\mathcal{D}_{i,j}(E) \equiv \mathcal{D}_i(E)\mathcal{D}_{i-1}(E)\cdots\mathcal{D}_j(E)$ for $i \geq j$. These matrices define the deterministic part of $\mathcal{P}_{n,m}(E,\omega)$ and are uniformly bounded in the following sense:

Lemma 2.2. *There exists a constant C_Δ, such that for every $E \in \Delta$ and $n \geq m$, $\|\mathcal{D}_{n,m}(E)\| \leq C_\Delta$ and $\|\mathcal{D}_{n,m}(E)^{-1}\| \leq C_\Delta$.*

PROOF. Since the determinant of $\mathcal{D}_{n,m}(E)$ is 1, we have $\|\mathcal{D}_{n,m}(E)\| = \|\mathcal{D}_{n,m}(E)^{-1}\|$ for every E. It is thus sufficient to consider $\|\mathcal{D}_{n,m}(E)\|$. From (2.3), we have $\mathcal{D}_{n,m} = \mathcal{C}_n^{-1}\mathcal{B}_n(\mathcal{C}_n\mathcal{C}_{n-1}^{-1})\mathcal{B}_{n-1}\cdots\mathcal{C}_m$, and thus, since $\|\mathcal{B}_i\| = 1$, we get

$$\|\mathcal{D}_{n,m}\| \leq \|\mathcal{C}_n^{-1}\|\Big(\prod_{j=m+1}^{n}\|\mathcal{C}_j\mathcal{C}_{j-1}^{-1}\|\Big)\|\mathcal{C}_m\|.$$

Since

$$\|\mathcal{C}_j\mathcal{C}_{j-1}^{-1}\| = \|(\mathcal{C}_{j-1}+\mathcal{C}_j-\mathcal{C}_{j-1})\mathcal{C}_{j-1}^{-1}\| \leq 1 + \|\mathcal{C}_j-\mathcal{C}_{j-1}\|\|\mathcal{C}_{j-1}^{-1}\|,$$
$$|\cos k_j - \cos k_{j-1}| = \tfrac{1}{2}\big|(E-\tilde{b}_j)/\tilde{a}_j - (E-\tilde{b}_{j-1})/\tilde{a}_{j-1}\big|,$$

and

$$|\sin k_j - \sin k_{j-1}| = \frac{|\cos k_j + \cos k_{j-1}|}{|\sin k_j + \sin k_{j-1}|}|\cos k_j - \cos k_{j-1}|,$$

we see that for some constant C, uniformly on Δ,

$$\|\mathcal{C}_j\mathcal{C}_{j-1}^{-1}\| \leq 1 + C(|\tilde{a}_j-\tilde{a}_{j-1}| + |\tilde{b}_j-\tilde{b}_{j-1}|).$$

Using the fact that $1+x \leq \mathrm{e}^x$ for $x \geq 0$, we thus conclude that

$$\|\mathcal{D}_{n,m}(E)\| \leq \|\mathcal{C}_n^{-1}(E)\|\|\mathcal{C}_m(E)\|\exp\Big(C\sum_{n=1}^{\infty}(|\tilde{a}_{n+1}-\tilde{a}_n| + |\tilde{b}_{n+1}-\tilde{b}_n|)\Big),$$

which is uniformly bounded on Δ. This proves the lemma. □

PROOF OF THEOREM 1.1. If $\{\boldsymbol{e}_1, \boldsymbol{e}_2\}$ is an orthonormal basis in $\mathbb{R}^2$, then for any 2×2 matrix $\mathcal{M}$,

$$\|\mathcal{M}\| \leq \|\mathcal{M}\boldsymbol{e}_1\| + \|\mathcal{M}\boldsymbol{e}_2\|.$$

Thus, if we show that there exists a constant C, such that for every $\boldsymbol{e} \in \mathbb{R}^2$ with $\|\boldsymbol{e}\| = 1$ and every $n \geq m$,

$$\int_\Omega\int_\Delta \|\mathcal{P}_{n,m}(E,\omega)\boldsymbol{e}\|^4\,\mathrm{d}E\,\mathrm{d}P(\omega) \leq C, \tag{2.5}$$

then it would follow that

$$\int_\Omega\int_\Delta \|\mathcal{P}_{n,m}(E,\omega)\|^4\,\mathrm{d}E\,\mathrm{d}P(\omega) < 16C,$$

and, by Fatou's lemma, we will have

$$\int_\Omega \liminf_{n\to\infty}\int_\Delta \|\mathcal{P}_{n,m}(E,\omega)\|^4\,\mathrm{d}E\,\mathrm{d}P(\omega) < 16C.$$

This would mean that, for P a.e. ω,

$$\liminf_{n\to\infty}\int_\Delta \|\mathcal{P}_{n,m}(E,\omega)\|^4\mathrm{d}E < \infty,$$

and, as discussed above, this would prove the theorem.

To prove the estimate (2.5), define $\boldsymbol{v}_l(E,\omega) = \mathcal{D}_{n,l}(E)\mathcal{P}_{l-1,m}(E,\omega)\boldsymbol{e}$. Using equation (2.2), we get

$$\boldsymbol{v}_{l+1} = \mathcal{D}_{n,l+1}(\mathcal{D}_l + \mathcal{Q}_l)\mathcal{P}_{l-1,m}\boldsymbol{e} = (\mathcal{I} + \mathcal{D}_{n,l+1}\mathcal{Q}_l\mathcal{D}_{n,l}^{-1})\boldsymbol{v}_l = (\mathcal{I} + \mathcal{R}_l)\boldsymbol{v}_l,$$

where

$$\begin{aligned}\mathcal{R}_l &\equiv \mathcal{D}_{n,l+1}\mathcal{Q}_l\mathcal{D}_{n,l}^{-1}\\ &= \frac{-\alpha_l}{a_l}\mathcal{I} + \frac{-\beta_l}{a_l}\mathcal{D}_{n,l+1}\begin{pmatrix}1 & 0\\ 0 & 0\end{pmatrix}\mathcal{D}_{n,l}^{-1} + \frac{a_l^2-\tilde{a}_l^2}{a_l}\mathcal{D}_{n,l+1}\begin{pmatrix}0 & 0\\ 1 & 0\end{pmatrix}\mathcal{D}_{n,l}^{-1}.\end{aligned}$$

We want to estimate the average of

$$\begin{aligned}\|\boldsymbol{v}_{l+1}\|^4 &= \langle(\mathcal{I}+\mathcal{R}_l)\boldsymbol{v}_l, (\mathcal{I}+\mathcal{R}_l)\boldsymbol{v}_l\rangle^2\\ &= (\|\boldsymbol{v}_l\|^2 + 2\langle\boldsymbol{v}_l, \mathcal{R}_l\boldsymbol{v}_l\rangle + \|\mathcal{R}_l\boldsymbol{v}_l\|^2)^2\\ &= \|\boldsymbol{v}_l\|^4 + \|\mathcal{R}_l\boldsymbol{v}_l\|^4 + 2\|\boldsymbol{v}_l\|^2\|\mathcal{R}_l\boldsymbol{v}_l\|^2 + 4\|\boldsymbol{v}_l\|^2\langle\boldsymbol{v}_l, \mathcal{R}_l\boldsymbol{v}_l\rangle\\ &\qquad + 4\|\mathcal{R}_l\boldsymbol{v}_l\|^2\langle\boldsymbol{v}_l, \mathcal{R}_l\boldsymbol{v}_l\rangle + 4\langle\boldsymbol{v}_l, \mathcal{R}_l\boldsymbol{v}_l\rangle^2.\end{aligned}$$

Since $\boldsymbol{v}_l(\omega)$ is independent of $\alpha_l(\omega)$ and $\beta_l(\omega)$, we have

$$\begin{aligned}\int_\Omega \|\boldsymbol{v}_l\|^2\langle\boldsymbol{v}_l, \mathcal{R}_l\boldsymbol{v}_l\rangle\,\mathrm{d}P(\omega) &= \int_\Omega \frac{-\alpha_l}{a_l}\,\mathrm{d}P(\omega)\int_\Omega \|\boldsymbol{v}_l\|^4\,\mathrm{d}P(\omega)\\ &+ \int_\Omega \frac{-\beta_l}{a_l}\,\mathrm{d}P(\omega)\int_\Omega \|\boldsymbol{v}_l\|^2\left\langle\boldsymbol{v}_l, \mathcal{D}_{n,l+1}\begin{pmatrix}1 & 0\\ 0 & 0\end{pmatrix}\mathcal{D}_{n,l}^{-1}\boldsymbol{v}_l\right\rangle\mathrm{d}P(\omega)\\ &+ \int_\Omega \frac{a_l^2-\tilde{a}_l^2}{a_l}\,\mathrm{d}P(\omega)\int_\Omega \|\boldsymbol{v}_l\|^2\left\langle\boldsymbol{v}_l, \mathcal{D}_{n,l+1}\begin{pmatrix}0 & 0\\ 1 & 0\end{pmatrix}\mathcal{D}_{n,l}^{-1}\boldsymbol{v}_l\right\rangle\mathrm{d}P(\omega).\end{aligned}$$

Since $1/C_0 < a_l(\omega) < C_0$, we see that

$$\begin{aligned}\frac{\alpha_l(\omega)}{a_l(\omega)} &= \frac{\alpha_l(\omega)}{\tilde{a}_l} + O\big(\alpha_l^2(\omega)\big),\\ \frac{\beta_l(\omega)}{a_l(\omega)} &= \frac{\beta_l(\omega)}{\tilde{a}_l} + O\big(\beta_l(\omega)\alpha_l(\omega)\big),\end{aligned}$$

and

$$\frac{a_l^2(\omega)-\tilde{a}_l^2}{a_l(\omega)} = 2\alpha_l(\omega) + O\big(\alpha_l^2(\omega)\big).$$

Thus, since $\alpha_l(\omega)$ and $\beta_l(\omega)$ are of zero mean, Lemma 2.2 implies that for some constant C, uniformly on Δ and in l,

$$\left|\int_\Omega \|\boldsymbol{v}_l\|^2\langle\boldsymbol{v}_l, \mathcal{R}_l\boldsymbol{v}_l\rangle\,\mathrm{d}P(\omega)\right| \le C\int_\Omega(\alpha_l^2+\beta_l^2)\,\mathrm{d}P(\omega)\int_\Omega\|\boldsymbol{v}_l\|^4\,\mathrm{d}P(\omega).$$

In a similar way, one can see that for a sufficiently large constant C,

$$\begin{aligned}\int_\Omega \langle\boldsymbol{v}_l, \mathcal{R}_l\boldsymbol{v}_l\rangle^2\,\mathrm{d}P(\omega) &\le C\int_\Omega(\alpha_l^2+\beta_l^2)\,\mathrm{d}P(\omega)\int_\Omega\|\boldsymbol{v}_l\|^4\,\mathrm{d}P(\omega),\\ \int_\Omega \|\boldsymbol{v}_l\|^2\|\mathcal{R}_l\boldsymbol{v}_l\|^2\,\mathrm{d}P(\omega) &\le C\int_\Omega(\alpha_l^2+\beta_l^2)\,\mathrm{d}P(\omega)\int_\Omega\|\boldsymbol{v}_l\|^4\,\mathrm{d}P(\omega),\\ \int_\Omega \|\mathcal{R}_l\boldsymbol{v}_l\|^4\,\mathrm{d}P(\omega) &\le C\int_\Omega(\alpha_l^2+\beta_l^2)\,\mathrm{d}P(\omega)\int_\Omega\|\boldsymbol{v}_l\|^4\,\mathrm{d}P(\omega)\end{aligned}$$

and

$$\left|\int_\Omega \|\mathcal{R}_l\boldsymbol{v}_l\|^2\langle\boldsymbol{v}_l, \mathcal{R}_l\boldsymbol{v}_l\rangle\,\mathrm{d}P(\omega)\right| \le C\int_\Omega(\alpha_l^2+\beta_l^2)\,\mathrm{d}P(\omega)\int_\Omega\|\boldsymbol{v}_l\|^4\,\mathrm{d}P(\omega).$$

Thus, we obtain that for some constant C, uniformly on Δ and in l,

$$\int_\Omega \|\boldsymbol{v}_{l+1}\|^4\,\mathrm{d}P(\omega) \le \left(1 + C\int_\Omega(\alpha_l^2+\beta_l^2)\,\mathrm{d}P(\omega)\right)\int_\Omega\|\boldsymbol{v}_l\|^4\,\mathrm{d}P(\omega).$$

Since

$$\sum_{n=1}^{\infty} \int_{\Omega} \left(\alpha_n^2(\omega) + \beta_n^2(\omega)\right) \mathrm{d}P(\omega) < \infty,$$

the inequality $1 + x \leq \mathrm{e}^x$ for $x > 0$ implies that

$$\prod_{l=m}^{n} \left(1 + C \int_{\Omega} \left(\alpha_l^2(\omega) + \beta_l^2(\omega)\right) \mathrm{d}P(\omega)\right)$$

is uniformly (in m and n) bounded from above by some constant $\widetilde{C}$, and so, uniformly on Δ and in n,

$$\int_{\Omega} \|\boldsymbol{v}_{n+1}(E,\omega)\|^4 \,\mathrm{d}P(\omega) \leq \widetilde{C} \int_{\Omega} \|\boldsymbol{v}_m(E,\omega)\|^4 \,\mathrm{d}P(\omega),$$

which implies

$$\int_{\Delta} \int_{\Omega} \|\boldsymbol{v}_{n+1}(E,\omega)\|^4 \,\mathrm{d}P(\omega)\,\mathrm{d}E \leq \widetilde{C} \int_{\Delta} \int_{\Omega} \|\boldsymbol{v}_m(E,\omega)\|^4 \,\mathrm{d}P(\omega)\,\mathrm{d}E.$$

Thus, by Fubini's theorem, we see that

$$\int_{\Omega} \int_{\Delta} \|\boldsymbol{v}_{n+1}(E,\omega)\|^4 \,\mathrm{d}E\,\mathrm{d}P(\omega) \leq \widetilde{C} \int_{\Omega} \int_{\Delta} \|\boldsymbol{v}_m(E,\omega)\|^4 \,\mathrm{d}E\,\mathrm{d}P(\omega).$$

Since $\boldsymbol{v}_{n+1}(E,\omega) = \mathcal{P}_{n,m}(E,\omega)\boldsymbol{e}$ and $\boldsymbol{v}_m(E,\omega) = \mathcal{D}_{n,m}(E)\boldsymbol{e}$, Lemma 2.2 implies that (2.5) holds. From this Theorem 1.1 follows. □

References

1. J. Breuer and Y. Last, *Stability of spectral types for Jacobi matrices under decaying random perturbations*, J. Funct. Anal., to appear.
2. R Carmona, *Exponential localization in one-dimensional disordered systems*, Duke Math. J. **49** (1982), no. 1, 191–213.
3. P. Deift and R. Killip, *On the absolutely continuous spectrum of one-dimensional Schrödinger operators with square summable potentials*, Comm. Math. Phys. **203** (1999), no. 2, 341–347.
4. F. Delyon, B. Simon, and B. Souillard, *From power pure point to continuous spectrum in disordered systems*, Ann. Inst. H. Poincaré Phys. Théor. **42** (1985), no. 3, 283–309.
5. I. Goldsheid, S. Molchanov, and L. Pastur, *A random homogeneous Schrdinger operator has a pure point spectrum*, Funkcional. Anal. i Priložen. **11** (1977), no. 1, 1–10 (Russian); English transl., Functional Anal. Appl. **11** (1977), no. 1, 1–8.
6. R. Killip and B. Simon, *Sum rules for Jacobi matrices and their applications to spectral theory*, Ann. of Math. (2) **158** (2003), no. 3, 253–321.
7. A. Kiselev, Y. Last, and B. Simon, *Modified Prüfer and EFGP transforms and the spectral analysis of one-dimensional Schrödinger operators*, Comm. Math. Phys. **194** (1998), no. 1, 1–45.
8. S. Kotani and N. Ushiroya, *One-dimensional Schrödinger operators with random decaying potentials*, Comm. Math. Phys. **115** (1988), no. 2, 247–266.
9. S. Kupin, *Spectral properties of Jacobi matrices and sum rules of special form*, J. Funct. Anal. **227** (2005), no. 1, 1–29.
10. A. Laptev, S. Naboko, and O. Safronov, *On new relations between spectral properties of Jacobi matrices and their coefficients*, Comm. Math. Phys. **241** (2003), no. 1, 91–110.
11. Y. Last, *Destruction of absolutely continuous spectrum by perturbation potentials of bounded variation*, Comm. Math. Phys., to appear.
12. Y. Last and B. Simon, *Eigenfunctions, transfer matrices, and absolutely continuous spectrum of one-dimensional Schrödinger operators*, Invent. Math. **135** (1999), no. 2, 329–367.
13. S. Molchanov, M. Novitskii, and B. Vainberg, *First KdV integrals and absolutely continuous spectrum for 1-D Schrödinger operator*, Comm. Math. Phys. **216** (2001), no. 1, 195–213.
14. B. Simon, *Some Jacobi matrices with decaying potential and dense point spectrum*, Comm. Math. Phys. **87** (1982/83), no. 2, 253–258.

15. ———, *Spectral analysis and rank one perturbations and applications*, Mathematical Quantum Theory (Vancouver, BC, 1993) (J. Feldman, R. Froese, and L. Rosen, eds.), CRM Proc. Lecture Notes, vol. 8, Amer. Math. Soc., Providence, RI, 1995, pp. 109–149; reprinted in ———, *Trace ideals and their applications*, 2nd ed., Math. Surveys Monogr., vol. 120, Amer. Math. Soc., Providence, RI, 2005.
16. ———, *Bounded eigenfunctions and absolutely continuous spectra for one-dimensional Schrödinger operators*, Proc. Amer. Math. Soc. **124** (1996), no. 11, 3361–3369.
17. J Weidmann, *Zur Spektraltheorie von Sturm-Liouville-Operatoren*, Math. Z. **98** (1967), 268–302.
18. A. Zlatoš, *Sum rules for Jacobi matrices and divergent Lieb–Thirring sums*, J. Funct. Anal. **225** (2005), no. 2, 371–382.

Institute of Mathematics, The Hebrew University, 91904 Jerusalem, Israel
E-mail address, U. Kaluzhny: kaluzhny@math.huji.ac.il
E-mail address, Y. Last: ylast@math.huji.ac.il

Centre de Recherches Mathématiques
CRM Proceedings and Lecture Notes
Volume **42**, 2007

The Parabolic Anderson Model and Its Universality Classes

Wolfgang König

ABSTRACT. We discuss the parabolic Anderson model, the Cauchy problem for the heat equation with random potential on $\mathbb{Z}^d$. After a general introduction following [13], we give heuristics for the long-time behavior of the model for general i.i.d. potentials. Following [17], we explain that exactly *four* qualitatively different types of intermittent behavior can occur. These four universality classes depend on the upper tail of the potential distribution: (1) tails at ∞ that are thicker than the double-exponential tails, (2) double-exponential tails at ∞ studied by Gärtner and Molchanov, (3) a new class called *almost bounded potentials*, and (4) potentials bounded from above studied by Biskup and König. The new class (3), which contains both unbounded and bounded potentials, is studied in both the annealed and the quenched setting. We show that intermittency occurs on unboundedly increasing islands whose diameter is slowly varying in time. The characteristic variational formulas describing the optimal profiles of the potential and of the solution are solved explicitly by parabolas, respectively, Gaussian densities.

1. Introduction and heuristics

1.1. The parabolic Anderson model. We consider the continuous solution $u\colon [0,\infty) \times \mathbb{Z}^d \to [0,\infty)$ to the Cauchy problem for the heat equation with random coefficients and localized initial datum,

$$\frac{\partial}{\partial t} u(t,z) = \Delta^{\mathrm{d}} u(t,z) + \xi(z) u(t,z) \quad \text{for } (t,z) \in (0,\infty) \times \mathbb{Z}^d, \tag{1.1}$$

$$u(0,z) = \delta_0(z), \qquad \text{for } z \in \mathbb{Z}^d. \tag{1.2}$$

2000 *Mathematics Subject Classification.* Primary 60H25, 82C44; Secondary 60F10, 35B40.
Key words and phrases. Parabolic Anderson problem, intermittency, random environment, quenched asymptotics, heat equation with random potential, universality.
Partially supported by the German Science Foundation, Schwerpunkt project SPP 1033, and by a Heisenberg grant awarded by the German Science Foundation, realized in 2003/04.
This note is an extended summary of a talk given at the Centre de Recherches Mathématiques, Montreal, on the Conference on Probability and Mathematical Physics in honor of Stanislav Molchanov's 65th birthday. It is a summary of [17] and [13]. Let me thank the organizers of that conference for the invitation and let me express my gratitude to my coauthors, J. Gärtner, R. van der Hofstad and P. Mörters, for the kind permission to use the material here.
This is the final form of the paper.

Here $\xi = (\xi(z) \colon z \in \mathbb{Z}^d)$ is an i.i.d. random potential with values in $[-\infty, \infty)$, and $\Delta^{\rm d}$ is the discrete Laplacian

$$\Delta^{\rm d} f(z) = \sum_{y \sim z} [f(y) - f(z)], \quad \text{for } z \in \mathbb{Z}^d,\ f \colon \mathbb{Z}^d \to \mathbb{R}.$$

The parabolic problem (1.1) is called the *parabolic Anderson model* (PAM). The operator $\Delta^{\rm d}+\xi$ appearing on the right is called the *Anderson Hamiltonian*; its spectral properties are well-studied in mathematical physics. Equation (1.1) describes a random mass transport through a random field of sinks and sources, corresponding to lattice points z with $\xi(z) < 0$, respectively, > 0. It is a linearized model for chemical kinetics [15], is equivalent to Burger's equation in hydrodynamics [6], and describes magnetic phenomena [19]. We refer the reader to [6, 15, 18] for more background.

The long-time behavior of the parabolic Anderson problem is well-studied in the mathematics and mathematical physics literature because it is the prime example of a model exhibiting an *intermittency effect*. This means, loosely speaking, that most of the total mass of the solution,

$$U(t) = \sum_{z \in \mathbb{Z}^{\rm d}} u(t, z), \quad \text{for } t > 0, \tag{1.3}$$

is concentrated on a small number of remote islands, called the *intermittent islands.* A manifestation of intermittency in terms of the moments of $U(t)$ is as follows. For $0 < p < q$, the main contribution to the qth moment of $U(t)$ comes from islands that contribute only negligibly to the $p^{\rm th}$ moments. Therefore, intermittency can be defined by the requirement,

$$\limsup_{t \to \infty} \frac{\langle U(t)^p \rangle^{1/p}}{\langle U(t)^q \rangle^{1/q}} = 0, \quad \text{for } 0 < p < q, \tag{1.4}$$

where $\langle \cdot \rangle$ denotes expectation with respect to ξ. Whenever ξ is truly random, the parabolic Anderson model is intermittent in this sense, see [15, Theorem 3.2] and Section 1.2 below.

The large-time behavior of the solution u to the PAM (1.1) is determined by the spectral properties of the Anderson Hamiltonian

$$\mathcal{H} = \Delta^{\rm d} + \xi. \tag{1.5}$$

Namely, since (under natural assumptions on ξ) the upper part of the spectrum of $\mathcal{H}$ in $\ell^2(\mathbb{Z}^d)$ is a pure point spectrum [3, 10], the solution u admits the spectral representation

$$u(t, \cdot) = \sum_i {\rm e}^{\lambda_i t} v_i(0) v_i(\cdot) \tag{1.6}$$

with respect to the random eigenvalues λ_i and the corresponding exponentially localized random eigenfunctions v_i. (For simplicity we ignore the possible occurrence of a continuous central part of the spectrum.) As t increases unboundedly, only summands with larger and larger eigenvalues will contribute to (1.6), and the corresponding eigenfunctions are expected to be localized more and more far from each other. Hence, for large t, the solution $u(t, \cdot)$ indeed looks like a weighted superposition of high peaks concentrated on distant islands.

The solution u to (1.1) describes a random particle flow in $\mathbb{Z}^d$ in the presence of random sources (lattice sites x with $\xi(x) > 0$) and random sinks (sites x with

$\xi(x) < 0)$.[1] Two competing effects are present: the diffusion mechanism governed by the Laplacian, and the local growth governed by the potential. The diffusion tends to make the random field $u(t, \cdot)$ flat, whereas the random potential ξ has a tendency to make it irregular.

The solution u to (1.1) also admits a branching particle dynamics interpretation. Imagine that initially, at time $t = 0$, there is a single particle at the origin, and all other sites are vacant. This particle moves according to a continuous-time symmetric random walk with generator Δ^{d}. When present at site x, the particle is split into two particles with rate $\xi_+(x)$ and is killed with rate $\xi_-(x)$, where $\xi_+ = (\xi_+(x))_{x\in\mathbb{Z}^d}$ and $\xi_- = \left(\xi_-(x)\right)_{x\in\mathbb{Z}^d}$ are independent random i.i.d. fields ($\xi_-(x)$ may attain the value ∞). Every particle continues from its birth site in the same way as the parent particle, and their movements are independent. Put $\xi(x) = \xi_+(x) - \xi_-(x)$. Then, given ξ_- and ξ_+, the expected number of particles present at the site x at time t is equal to $u(t, x)$. Here the expectation is taken over the particle motion and over the splitting resp. killing mechanism, but not over the random medium (ξ_-, ξ_+).

A very useful standard tool for the probabilistic investigation of (1.1) is the well-known *Feynman – Kac formula* for the solution u, which (after time-reversal) reads

$$(1.7) \qquad u(t, x) = \mathbb{E}_0\left[\exp\left\{\int_0^t \xi(X(s))\,\mathrm{d}s\right\}\delta_x(X(t))\right], \qquad (t, x) \in [0, \infty) \times \mathbb{Z}^d,$$

where $(X(s))_{s\in[0,\infty)}$ is continuous-time random walk on $\mathbb{Z}^d$ with generator Δ^{d} starting at $x \in \mathbb{Z}^d$ under $\mathbb{E}_x$.

Our main interest concerns the large-time behavior of the random field $u(t, \cdot)$. In particular, we consider the total mass, i.e., the random variable

$$(1.8) \qquad U(t) = \sum_{x\in\mathbb{Z}^d} u(t, x) = \mathbb{E}_0\left[\exp\left\{\int_0^t \xi(X(s))\,\mathrm{d}s\right\}\right], \qquad t > 0.$$

Note that $U(t)$ coincides with the value $\hat{u}(t, 0)$ of the solution $\hat{u}$ to the parabolic equation (1.1) with initial datum $\hat{u}(0, \cdot) = \mathbb{1}$ instead of $u(0, \cdot) = \delta_0$. One should have in mind that, because of this, our considerations below also concern the large-time asymptotics of $\hat{u}$.

We ask the following questions:

(1) What is the asymptotic behavior of $U(t)$ as $t \to \infty$?
(2) Where does the main mass of $u(t, \cdot)$ stem from? What are the regions that contribute most to $U(t)$? What are these regions determined by? How many of them are there and how far away are they from each other?
(3) What do the typical shapes of the potential $\xi(\cdot)$ and of the solution $u(t, \cdot)$ look like in these regions?

We call the regions that contribute the overwhelming part to the total mass $U(t)$ *relevant islands* or *relevant regions*. The notion of *intermittency* states that there does exist a small number of relevant islands which are far away from each other and carry asymptotically almost all the total mass $U(t)$ of $u(t, \cdot)$.

This effect may also be studied from the point of view of *typical paths* $X(s)$, $s \in [0, t]$, giving the main contribution to the expectation in the Feynman – Kac

[1] Sites x with $\xi(x) = -\infty$ may be allowed and interpreted as ('hard') traps or obstacles, sites with $\xi(x) \in (-\infty, 0)$ are sometimes called "soft" traps.

formula (1.8). On the one hand, the random walker X should move quickly and as far as possible through the potential landscape to reach a region of exceptionally high potential and then stay there up to time t. This will make the integral in the exponent on the right of (1.8) large. On the other hand, the probability to reach such a distant potential peak up to t may be rather small. Hence, the first order contribution to $U(t)$ comes from paths that find a good compromise between the high potential values and the far distance. This contribution is given by the height of the peak. The second order contribution to $U(t)$ is determined by the precise manner in which the optimal walker moves within the potential peak, and this depends on the geometric properties of the potential in that peak.

It is part of our study to understand the effect of intermittency for the parabolic Anderson model in great detail. We distinguish between the so-called *quenched* setting, where we consider $u(t,\cdot)$ almost surely with respect to the medium ξ, and the *annealed* one, where we average with respect to ξ. It is clear that the quantitative details of the answers to the above questions strongly depend on the distribution of the field ξ (more precisely, on the upper tail of the distribution of the random variable $\xi(0)$), and that different phenomena occur in the quenched and the annealed settings.

It will turn out that there is a universal picture present in the asymptotics of the parabolic Anderson model. Inside the relevant islands, after appropriate vertical shifting and spatial rescaling, the potential ξ will turn out to asymptotically approximate a universal, nonrandom shape, V, which is determined by a characteristic variational problem. The absolute height of the potential peaks and the diameter of the relevant islands are asymptotically determined by the upper tails of the random variable $\xi(0)$, while the number of the islands and their locations are random. Furthermore, after multiplication with an appropriate factor and rescaling, also the solution $u(t,\cdot)$ approaches a universal shape on these islands, namely the principal eigenfunction of the Hamiltonian $\Delta^{\mathrm{d}} + V$ with V the above universal potential shape. Remarkably, there are only four universal classes of potential shapes for the PAM in (1.1).

1.2. Intermittency. Let $\hat{u}$ denote the solution to the equation in (1.1) with initial datum $\hat{u}(0,\cdot) = \mathbb{1}$ instead of $u(0,\cdot) = \delta_0$, with an i.i.d. potential $\xi = (\xi(x))_{x\in\mathbb{Z}^d}$. Assume that all positive exponential moments of $\xi(0)$ are finite. Let $\mathrm{Prob}(\cdot)$ and $\langle\cdot\rangle$ denote probability and expectation w.r.t. ξ.

A first, rough, mathematical approach to intermittency consists in a comparison of the growth of subsequent moments of the ergodic field $\hat{u}(t,\cdot)$ as $t\to\infty$. Define

$$\Lambda_p(t) = \log\langle \hat{u}(t,0)^p\rangle, \qquad p\in\mathbb{N},$$

and write $f \ll g$ if $\lim_{t\to\infty}[g(t) - f(t)] = \infty$.

Definition 1.1. For $p\in\mathbb{N}\setminus\{1\}$, the homogeneous ergodic field $\hat{u}(t,\cdot)$ is called *p-intermittent* as $t\to\infty$, if

$$\frac{\Lambda_{p-1}}{p-1} \ll \frac{\Lambda_p}{p}. \tag{1.9}$$

Note that, by Hölder's inequality, always $\Lambda_{p-1}/(p-1) \le \Lambda_p/p$. If the finite moment Lyapunov exponents

$$\lambda_p = \lim_{t\to\infty}\frac{1}{t}\Lambda_p(t), \quad p\in\mathbb{N},$$

exist, then the strict inequality $\lambda_{p-1}/(p-1) < \lambda_p/p$ implies p-intermittency.

To explain the meaning of Definition 1.1, assume (1.9) for some $p \in \mathbb{N} \setminus \{1\}$ and choose a level function ℓ_p such that $\Lambda_{p-1}/(p-1) \ll \ell_p \ll \Lambda_p/p$. Then, by Chebyshev's inequality,

$$\operatorname{Prob}\big(\hat{u}(t,0) > e^{\ell_p(t)}\big) \le e^{-(p-1)\ell_p(t)}\langle \hat{u}(t,0)^{p-1}\rangle = \exp\{\Lambda_{p-1}(t) - (p-1)\ell_p(t)\},$$

and the expression on the right converges to zero as $t \to \infty$. In other words, the density of the homogeneous point process

$$\Gamma(t) = \big\{x \in \mathbb{Z}^d \colon \hat{u}(t,x) > e^{\ell_p(t)}\big\}$$

vanishes asymptotically as $t \to \infty$. On the other hand,

$$\big\langle \hat{u}(t,0)^p \mathbb{1}\big\{\hat{u}(t,0) \le e^{\ell_p(t)}\big\}\big\rangle \le e^{p\ell_p(t)} = e^{p\ell_p(t) - \Lambda_p(t)}\langle \hat{u}(t,0)^p\rangle = o(\langle \hat{u}(t,0)^p\rangle)$$

and, consequently,

$$\langle \hat{u}(t,0)^p\rangle \sim \big\langle \hat{u}(t,0)^p \mathbb{1}\big\{\hat{u}(t,0) > e^{\ell_p(t)}\big\}\big\rangle$$

as $t \to \infty$. Hence, by Birkhoff's ergodic theorem, for large t and large centered boxes B in $\mathbb{Z}^d$,

$$|B|^{-1} \sum_{x \in B} \hat{u}(t,x)^p \approx |B|^{-1} \sum_{x \in B \cap \Gamma(t)} \hat{u}(t,x)^p.$$

This means that the pth moment $\langle \hat{u}(t,0)^p\rangle$ is "generated" by the high peaks of $\hat{u}(t,\cdot)$ on the "thin" set $\Gamma(t)$ and therefore indicates the presence of intermittency in the above verbal sense. Unfortunately, this approach does not reflect the geometric structure of the set $\Gamma(t)$. This set might consist of islands or, e.g., have a net-like structure.

Theorem 1.2. *If $\xi = (\xi(x))_{x \in \mathbb{Z}^d}$ is a nondeterministic field of i.i.d. random variables with $\langle e^{t\xi(0)}\rangle < \infty$ for all $t > 0$, then the solution $\hat{u}(t,\cdot)$ is p-intermittent for all $p \in \mathbb{N} \setminus \{1\}$.*

This is part of [15, Theorem 3.2], where, for general homogeneous ergodic potentials ξ, necessary and sufficient conditions for p-intermittency of $\hat{u}(t,\cdot)$ have been given in spectral terms of the Hamiltonian (1.5).

1.3. Annealed second order asymptotics. Let us discuss, on a heuristic level, what the asymptotics of the moments of $U(t)$ are determined by, and how they can be described. For simplicity we restrict ourselves to the first moment.

The basic observation is that, as a consequence of the spectral representation (1.6),

$$U(t) \approx e^{t\lambda_t(\xi)} \tag{1.10}$$

(in the sense of logarithmic equivalence), where $\lambda_t(\varphi)$ denotes the principal (i.e., largest) eigenvalue of the operator $\Delta^{\mathrm{d}} + \varphi$ with zero boundary condition in the "macrobox" $B_t = [-t,t]^d \cap \mathbb{Z}^d$. Hence, we have to understand the large-time behavior of the exponential moments of the principal eigenvalue of the Anderson Hamiltonian $\mathcal{H}$ in a large, time-dependent box.

It turns out that the main contribution to $\langle e^{t\lambda_t(\xi)}\rangle$ comes from realizations of the potential ξ having high peaks on distant islands of some radius of order $\alpha(t)$ that is much smaller than t. But this implies that $\lambda_t(\xi)$ is close to the principal eigenvalue of $\mathcal{H}$ on one of these islands. Therefore, since the number of subboxes of

B_t of radius of order $\alpha(t)$ grows only polynomial in t and ξ is spatially homogeneous, we may expect that

$$\big\langle \mathrm{e}^{t\lambda_t(\xi)}\big\rangle \approx \big\langle \mathrm{e}^{t\lambda_{R\alpha(t)}(\xi)}\big\rangle$$

for R large as $t \to \infty$.

The choice of the scale function $\alpha(t)$ depends on asymptotic "stiffness" properties of the potential, more precisely of its tails at its essential supremum, and is determined by a large deviation principle, see (1.17) below. In Section 2 we shall see examples of potentials such that $\alpha(t)$ tends to 0, to ∞, or stays bounded and bounded away from zero as $t \to \infty$. In the present heuristics, we shall assume that $\alpha(t) \to \infty$, which implies the necessity of a spatial rescaling. In particular, after rescaling, the main quantities and objects will be described in terms of the continuous counterparts of the discrete objects we started with, i.e., instead of the discrete Laplacian, the continuous Laplace operator appears etc. The following heuristics can also be read in the case where $\alpha(t) \equiv 1$ by keeping the discrete versions for the limiting objects.

The optimal behavior of the field ξ in the "microbox" $B_{R\alpha(t)}$ is to approximate a certain (deterministic) shape φ after appropriate spatial scaling and vertical shifting. It easily follows from the Feynman – Kac formula (1.8) that

$$\mathrm{e}^{H(t)-2dt} \le \langle U(t)\rangle \le \mathrm{e}^{H(t)},$$

where

$$H(t) = \log\big\langle \mathrm{e}^{t\xi(0)}\big\rangle, \quad t > 0, \tag{1.11}$$

denotes the *cumulant generating function* of $\xi(0)$ (often called *logarithmic moment generating function*). Hence, the peaks of $\xi(\cdot)$ mainly contributing to $\langle U(t)\rangle$ have height of order $H(t)/t$. Together with Brownian scaling this leads to the ansatz

$$\bar\xi_t(\cdot) = \alpha(t)^2\Big[\xi\big(\lfloor \cdot\,\alpha(t)\rfloor\big) - \frac{H(t)}{t}\Big], \tag{1.12}$$

for the spatially rescaled and vertically shifted potential in the cube $Q_R = (-R,R)^d$. Now the idea is that the main contribution to $\langle U(t)\rangle$ comes from fields that are shaped in such a way that $\bar\xi_t \approx \varphi$ in Q_R, for some $\varphi\colon Q_R \to \mathbb{R}$, which has to be chosen optimally. Observe that

$$\bar\xi_t \approx \varphi \text{ in } Q_R \iff \xi(\cdot) \approx \frac{H(t)}{t} + \frac{1}{\alpha(t)^2}\varphi\Big(\frac{\cdot}{\alpha(t)}\Big) \text{ in } B_{R\alpha(t)}. \tag{1.13}$$

Let us calculate the contribution to $\langle U(t)\rangle$ coming from such fields. Using (1.10), we obtain

$$\begin{aligned}(1.14)\quad &\langle U(t)\mathbb{1}\{\bar\xi_t \approx \varphi \text{ in } Q_R\}\rangle\\ &\qquad\approx \mathrm{e}^{H(t)}\exp\Big\{t\lambda_{R\alpha(t)}\Big(\frac{1}{\alpha(t)^2}\varphi\Big(\frac{\cdot}{\alpha(t)}\Big)\Big)\Big\}\operatorname{Prob}(\bar\xi_t \approx \varphi \text{ in } Q_R).\end{aligned}$$

The asymptotic scaling properties of the discrete Laplacian, Δ^{d}, imply that

$$\lambda_{R\alpha(t)}\Big(\frac{1}{\alpha(t)^2}\varphi\Big(\frac{\cdot}{\alpha(t)}\Big)\Big) \approx \frac{1}{\alpha(t)^2}\lambda^{\mathrm{c}}_R(\varphi), \tag{1.15}$$

where $\lambda_R^{\rm c}(\varphi)$ denotes the principal eigenvalue of $\Delta + \varphi$ in the cube Q_R with zero boundary condition, and Δ is the usual Laplacian. This leads to

$$(1.16)\qquad \langle U(t)\mathbb{1}\{\bar{\xi}_t \approx \varphi \text{ in } Q_R\}\rangle \approx {\rm e}^{H(t)} \exp\Big\{\frac{t}{\alpha(t)^2}\lambda_R^{\rm c}(\varphi)\Big\} \operatorname{Prob}(\bar{\xi}_t \approx \varphi \text{ in } Q_R).$$

In order to achieve a balance between the second and the third factor on the right, it is necessary that the logarithmic decay rate of the considered probability is $t/\alpha(t)^2$. One expects to have a large deviation principle for the shifted, rescaled field, which reads

$$(1.17)\qquad \operatorname{Prob}\big(\bar{\xi}_t \approx \varphi \text{ in } Q_R\big) \approx \exp\Big\{-\frac{t}{\alpha(t)^2} I_R(\varphi)\Big\},$$

where the scale $\alpha(t)$ has to be determined in such a way that the rate function I_R is nondegenerate. Now substitute (1.17) into (1.16). Then the Laplace method tells us that the exponential asymptotics of $\langle U(t)\rangle$ is equal to the one of $\big\langle U(t)\,\mathbb{1}\{\bar{\xi}_t \approx \varphi \text{ in } Q_R\}$ with optimal φ. Hence, optimizing on φ and remembering that R is large, we arrive at

$$(1.18)\qquad \langle U(t)\rangle \approx {\rm e}^{H(t)} \exp\Big\{-\frac{t}{\alpha(t)^2}\chi\Big\},$$

where the constant χ is given in terms of the characteristic variational problem

$$(1.19)\qquad \chi = \lim_{R\to\infty} \inf_{\varphi\colon Q_R\to\mathbb{R}} [I_R(\varphi) - \lambda_R^{\rm c}(\varphi)].$$

The first term on the right of (1.18) is determined by the absolute height of the typical realizations of the potential and the second contains information about the shape of the potential close to its maximum in spectral terms of the Anderson Hamiltonian $\mathcal{H}$ in this region. More precisely, those realizations of ξ with $\bar{\xi}_t \approx \varphi_*$ in Q_R for large R and φ_* a minimizer in the variational formula in (1.19) contribute most to $\langle U(t)\rangle$. In particular, the geometry of the relevant potential peaks is hidden via χ in the second asymptotic term of $\langle U(t)\rangle$.

1.4. Quenched second order asymptotics. Here we explain, again on a heuristic level, the almost sure asymptotics of $U(t)$ as $t \to \infty$. Because of (1.10), it suffices to study the asymptotics of the principal eigenvalue $\lambda_t(\xi)$.

Like for the annealed asymptotics, the main contribution to $\lambda_t(\xi)$ comes from islands whose radius is of a certain deterministic, time-depending order, which we denote $\widetilde{\alpha}(t)$. As $t \to \infty$, the scale function $\widetilde{\alpha}(t)$ tends to zero, one, or ∞, respectively, if the scale function $\alpha(t)$ for the moments tends to these respective values (see also (1.22) below). However, $\widetilde{\alpha}(t)$ is roughly of logarithmic order in $\alpha(t)$ if $\alpha(t) \to \infty$, hence it is *much* smaller than $\alpha(t)$.

The relevant islands ("microboxes") have radius $R\widetilde{\alpha}(t)$, where R is chosen large. Let $z \in B_t$ denote the (certainly random) center of one of these islands $\widetilde{B} = z + B_{R\widetilde{\alpha}(t)}$ meeting the two requirements (1) the potential ξ is very large in $\widetilde{B}$ and (2) ξ has an optimal shape within $\widetilde{B}$. This is further explained as follows. Let $h_t = \max_{B_t} \xi$ be the maximal potential value in the large box B_t. (Then h_t is a priori random, but well approximated by deterministic asymptotics, which can be deduced from asymptotics of $H(t)$.) Then $\xi - h_t$ is roughly of finite order within the relevant "microbox" $\widetilde{B}$. Furthermore, $\xi - h_t$ should approximate a fixed deterministic shape in $\widetilde{B}$. Hence, we consider the shifted and rescaled field in the box $\widetilde{B}$,

$$\bar{\xi}_t(\cdot) = \widetilde{\alpha}(t)^2 \big[\xi\big(z + \cdot\, \widetilde{\alpha}(t)\big) - h_t\big], \quad \text{in } Q_R = (-R, R)^d. \tag{1.20}$$

Note that

$$\bar{\xi}_t \approx \varphi \text{ in } Q_R \iff \xi(z+\cdot) \approx h_t + \frac{1}{\widetilde{\alpha}(t)^2}\varphi\Big(\frac{\cdot}{\widetilde{\alpha}(t)}\Big) \text{ in } \widetilde{B} - z. \tag{1.21}$$

A crucial Borel – Cantelli argument shows that, for a given shape φ, with probability one, for any t sufficiently large, there does exist at least one box $\widetilde{B}$ having radius $R\widetilde{\alpha}(t)$ such that the event $\{\bar{\xi}_t \approx \varphi \text{ in } Q_R\}$ occurs if $I_R(\varphi) < 1$, where I_R is the rate function of the large deviation principle in (1.17). If $I_R(\varphi) > 1$, then this happens with probability 0. For the Borel – Cantelli argument to work, one needs the scale function $\widetilde{\alpha}(t)$ to be defined in terms of the annealed scale function $\alpha(t)$ in the following way:

$$\frac{\widetilde{\alpha}(t)}{\alpha\big(\widetilde{\alpha}(t)\big)^2} = d \log t, \tag{1.22}$$

i.e., $\widetilde{\alpha}(t)$ is the inverse of the map $t \mapsto t/\alpha(t)^2$, evaluated at $d \log t$. Note that the growth of $\widetilde{\alpha}(t)$ is roughly of logarithmic order of the growth of $\alpha(t)$, i.e., if the annealed relevant islands grow unboundedly, then the quenched relevant islands also grow unboundedly, but with much smaller velocity.

Hence, with probability one, for all large t, there is at least one box $\widetilde{B}$ in which the potential looks like the function on the right of (1.21). The contribution to $\lambda_t(\xi)$ coming from one of the boxes $\widetilde{B}$ is equal to the associated principal eigenvalue

$$\lambda_{\widetilde{B}-z}\Big(h_t + \frac{1}{\widetilde{\alpha}(t)^2}\varphi\Big(\frac{\cdot}{\widetilde{\alpha}(t)}\Big)\Big) \approx h_t + \frac{1}{\widetilde{\alpha}(t)^2}\lambda_R^{\mathrm{c}}(\varphi), \tag{1.23}$$

where we recall that $\lambda_R^{\mathrm{c}}(\varphi)$ is the principal Dirichlet eigenvalue of the operator $\Delta + \varphi$ in the "continuous" cube Q_R. Obviously, $\lambda_t(\xi)$ is asymptotically not smaller than the expression on the right of (1.23). In terms of the Feynman – Kac formula in (1.7), this lower estimate is obtained by inserting the indicator on the event that the random path moves quickly to the box $\widetilde{B}$ and stays all the time until t in that box.

It is an important technical issue to show that, asymptotically as $t \to \infty$, $\lambda_t(\xi)$ is also estimated from *above* by the right-hand side of (1.23), if φ is optimally chosen, i.e., if $\lambda_R^{\mathrm{c}}(\varphi)$ is optimized over all admissible φ and on R. This implies that the almost sure asymptotics of $U(t)$ are given as

$$\frac{1}{t}\log U(t) \approx \lambda_t(\xi) \approx h_t - \frac{1}{\widetilde{\alpha}(t)^2}\widetilde{\chi}, \quad t \to \infty, \tag{1.24}$$

where $\widetilde{\chi}$ is given in terms of the characteristic variational problem

$$\widetilde{\chi} = \lim_{R\to\infty} \inf_{\varphi\colon Q_R \to \mathbb{R}, I_R(\varphi)<1} [-\lambda_R^{\mathrm{c}}(\varphi)]. \tag{1.25}$$

This ends the heuristic derivation of the almost sure asymptotics of $U(t)$. Like in the annealed case, there are two terms, which describe the absolute height of the potential in the 'macrobox' B_t, and the shape of the potential in the relevant 'microbox' $\widetilde{B}$, more precisely the spectral properties of $\Delta^{\mathrm{d}} + \xi$ in that microbox. The interpretation is that, for R large and φ_* an approximate minimizer in (1.25), the main contribution to $U(t)$ comes from a small box $\widetilde{B}$ in B_t, with radius $R\widetilde{\alpha}(t)$, in which the shifted and rescaled potential $\overline{\xi}_t$ looks like φ_*. The condition $I_R(\varphi_*) < 1$ guarantees the existence of such a box, and $\lambda_R(\varphi_*)$ quantifies the contribution from that box.

Let us remark that the variational formulas in (1.25) and (1.19) are in close connection to each other. In particular, it can be shown that the minimizers of (1.25) are rescaled versions of the minimizers of (1.19). This means that, up to rescaling, the optimal potential shapes in the annealed and in the quenched setting are identical.

2. Examples of potentials

2.1. Double-exponential distributions. Consider a distribution which lies in the vicinity of the *double-exponential distribution*, i.e.,

$$\text{Prob}(\xi(0) > r) \approx \exp\{-\mathrm{e}^{r/\rho}\}, \quad r \to \infty, \tag{2.1}$$

with $\rho \in (0, \infty)$ a parameter. It turns out [16] that this class of potentials constitutes a critical class in the sense that the radius of the relevant islands stays finite as $t \to \infty$. This is related to the characteristic property of the double-exponential distribution that

$$\text{Prob}(\xi(x) > h) \approx \text{Prob}(\xi(y) > h - \rho \log 2, \xi(z) > h - \rho \log 2),$$

meaning that single-site potential peaks of height $h \gg 1$ occur with the same frequency as two-site potential peaks with height of the same order. Hence, no spatial rescaling is necessary, and we put $\alpha(t) = 1$.

For the boundary cases $\rho = \infty$ and $\rho = 0$ ("beyond" and "on this side of" the double-exponential distribution, respectively), [16] argued that the boundary cases $\alpha(t) \downarrow 0$ and $\alpha(t) \to \infty$ occur. In other words, the fields beyond the double-exponential (which includes, e.g., Gaussian fields) are simple in the sense that the main contribution comes from islands consisting of single lattice sites. Unbounded fields that are in the vicinity of the case $\rho = 0$ are called "almost bounded" in [16].

2.2. Survival probabilities. The case when the field ξ assumes the values $-\infty$ and 0 only has a nice interpretation in terms of survival probabilities and is therefore of particular importance. The fundamental papers [8, 9] by Donsker and Varadhan on the Wiener sausage contain apparently the first substantial annealed results on the asymptotics for the parabolic Anderson model. In the nineties, the thorough and deep work by Sznitman (see his monograph [20]), pushed the rigorous understanding of the quenched situation much further.

2.2.1. *Brownian motion in a Poisson field of traps.* We consider the continuous case, i.e., the version of (1.1) with $\mathbb{Z}^d$ replaced by $\mathbb{R}^d$ and the lattice Laplacian replaced by the usual Laplace operator. The field ξ is given as follows. Let $(x_i)_{i \in I}$ be the points of a homogeneous Poisson point process in $\mathbb{R}^d$, and consider the union $\mathcal{O}$ of the balls $B_a(x_i)$ of radius a around the Poisson points x_i. We define a random potential by putting

$$\xi(x) = \begin{cases} 0 & \text{if } x \notin \mathcal{O}, \\ -\infty & \text{if } x \in \mathcal{O}. \end{cases} \tag{2.2}$$

(There are more general versions of this type of potentials, but for simplicity we keep with that.) The set $\mathcal{O}$ receives the meaning of the set of "hard traps" or "obstacles." Let $T_{\mathcal{O}} = \inf\{t > 0 \colon X(t) \in \mathcal{O}\}$ denote the entrance time into $\mathcal{O}$ for a Brownian motion $(X(t))_{t\in[0,\infty)}$. Then we have the Feynman–Kac representation

$$u(t,x) = \mathbb{P}_0(T_{\mathcal{O}} > t, X(t) \in \mathrm{d}x)/\mathrm{d}x,$$

i.e., $u(t,x)$ is equal to the sub-probability density of $X(t)$ on survival in the Poisson field of traps by time t for Brownian motion starting from the origin. The total mass $U(t) = \mathbb{P}_0(T_{\mathcal{O}} > t)$ is the survival probability by time t. It is easily seen that the first moment of $U(t)$ coincides with a negative exponential moment of the volume of the Wiener sausage $\bigcup_{s\in[0,t]} B_a(X(s))$.

Donsker and Varadhan analyzed the leading asymptotics of $\langle U(t)\rangle$ by using their large deviation principle for Brownian occupation time measures. The relevant islands have radius of order $\alpha(t) = t^{1/(d+2)}$. To handle the quenched asymptotics of $U(t)$, Sznitman developed a coarse-graining scheme for Dirichlet eigenvalues on random subsets of $\mathbb{R}^d$, the so-called method of enlargement of obstacles (MEO). The MEO replaces the eigenvalues in certain complicated subsets of $\mathbb{R}^d$ by those in coarse-grained subsets belonging to a *discrete* class of much smaller combinatorial complexity such that control is kept on the relevant properties of the eigenvalue.

Qualitatively, the considered model falls into the class of bounded from above fields introduced in Section 2.3 with $\gamma = 0$.

2.2.2. *Simple random walk among Bernoulli traps.* This is the discrete version of Brownian motion among Poisson traps. Consider the i.i.d. field $\xi = (\xi(x))_{x\in\mathbb{Z}^d}$ where $\xi(x)$ takes the values 0 or $-\infty$ only. Again, $u(t,x)$ is the survival probability of continuous-time random walk paths from 0 to x among the set of traps $\mathcal{O} = \{y \in \mathbb{Z}^d \colon \xi(y) = -\infty\}$.

In their paper [9], Donsker and Varadhan also investigated the discrete case and described the logarithmic asymptotics of $\langle U(t)\rangle$ by proving and exploiting a large deviation principle for occupation times of random walks. Later Bolthausen [5] carried out a deeper analysis of $\langle U(t)\rangle$ in the two-dimensional case using refined large deviation arguments. Antal [1, 2] developed a discrete variant of the MEO and demonstrated its value by proving limit theorems for the survival probability $U(t)$ and its moments.

2.3. General fields bounded from above. In [4], a large class of potentials with $\operatorname{ess\,sup} \xi(0) < \infty$ is considered. Assume for simplicity that $\operatorname{ess\,sup} \xi(0) = 0$ and that the tail of $\xi(0)$ at 0 is given by

$$\text{Prob}(\xi(0) > -x) \approx \exp\{-Dx^{-\gamma/(1-\gamma)}\}, \quad x \downarrow 0, \tag{2.3}$$

with $D > 0$ and $\gamma \in [0,1)$ two parameters. The case $\gamma = 0$ contains simple random walk among Bernoulli traps as a particular case. The cumulant generating function is roughly $H(t) \approx -\text{const}t^\gamma$, and the annealed scale function is $\alpha(t) \approx t^\nu$ where $\nu = (1-\gamma)/(2+d-d\gamma)$. The power ν ranges from 0 to $1/(d+2)$ as the parameter γ ranges from 1 to 0.

It turns out in [4] that the rate function I_R (see (1.17)) is given by

$$I_R(\varphi) = \text{const} \int_{Q_R} |\varphi(x)|^{-\gamma/(1-\gamma)}\,\mathrm{d}x, \tag{2.4}$$

where in the case $\gamma = 0$ we interpret the integral as the Lebesgue measure of the support of φ. The characteristic variational formula for the annealed field shapes in (1.19) has been analyzed in great detail in the case $\gamma = 0$. In particular it was shown that the minimizer is unique and has compact support, and it was characterized in terms of Bessel functions. However, in the general case $\gamma \in (0,1)$, an analysis of (1.19) has not yet been carried out.

2.4. Gaussian fields and Poisson shot noise. Two important particular cases in the *continuous* version of the parabolic Anderson model are considered in [12,14] (see also [7] for first rough results). The continuous version of (1.1) replaces $\mathbb{Z}^d$ by $\mathbb{R}^d$ and the discrete lattice Laplacian by the usual Laplace operator. Unlike in the discrete case, where any distribution on $\mathbb{R}$ may be used for the definition of an i.i.d. potential, in the continuous case it is not easy to find examples of fields that can be expressed in easily manageable terms. Since a certain degree of regularity of the potential is required, the condition of spatial independence must be dropped.

In [12,14], two types of fields are considered: a Gaussian field ξ whose covariance function B has a parabolic shape around zero with $B(0) = \sigma^2 > 0$, and a so-called Poisson shot-noise field, which is defined as the superposition of copies of parabolic-shaped *positive* clouds around the points of a homogeneous Poisson point process in $\mathbb{R}^d$ (in contrast to the trap case of Section 2.2). A certain (mild) assumption on the decay of the covariance function (respectively of the cloud) at infinity ensures sufficient independence between regions that are far apart.

Both fields easily develop very high peaks on small islands (the Poisson shot-noise field is large where many Poisson points are close together). The annealed scale function is $\alpha(t) = t^{-1/4}$ for the Gaussian field and $\alpha(t) = t^{d/8}e^{-\sigma^2 t/4}$ for the Poisson field [12].

2.5. Almost bounded potentials. The potential class of this section, introduced and analyzed in [17], lies in the union of the boundary cases $\rho = 0$ of the double-exponential distribution (the "almost bounded" fields of Section 2.1) and $\gamma = 1$ for the general bounded from above fields, see Section 2.3. Examples of distributions that fall into this class look somewhat odd, but it turns out that the optimal potential shape and the optimal shape of the solution are perfectly parabolic, respectively Gaussian, which makes this class rather appealing. In particular, the appearing variational formulas can be easily solved explicitly and uniquely, see Section 3 below.

We now explain what kind of upper tail behavior is covered by the almost bounded case, arguing separately for the bounded and unbounded case, denoted by B and U, respectively. Suppose the distribution of the field $\xi(0)$ satisfies

$$\log \operatorname{Prob}\{\xi(0) > r\} \sim -\mathrm{e}^{f(r)}, \quad \text{as} \begin{cases} r \uparrow \infty & \text{in case U,} \\ r \uparrow 0 = \operatorname{ess\,sup} \xi(0), & \text{in case B.} \end{cases} \tag{2.5}$$

Here f is a positive, strictly increasing smooth function satisfying $f'(r) \uparrow \infty$ as $r \uparrow \infty$ in case U and $f'(r)r \uparrow \infty$ as $r \uparrow 0$ in case B. Note that typical representatives of the class of double-exponential-like distributions are $f(r) \approx cr$ as $r \uparrow \infty$, violating the condition in case U; and typical representatives of the class of bounded from above potentials as in Section 2.3 are $f(r) \approx -\gamma/(1-\gamma)\log|r|$ as $r \uparrow 0$, violating the condition in case B. This means that the asymptotics close to the essential supremum of this distribution is different from the previous classes. This class was introduced and analyzed in [17].

3. The universality classes

Now we consider the PAM in (1.1) for *arbitrary* potentials, with the aim of identifying *all universality classes* of intermittent behavior that can arise for different potential distributions. Our standing assumption is that the potentials $(\xi(z) : z \in \mathbb{Z}^d)$ are independent and identically distributed and that all positive exponential moments of $\xi(0)$ are finite, which is necessary and sufficient for the finiteness of the pth moments of $U(t)$ at all times.

We are going to explain that, under some mild regularity assumptions on the tails of $\xi(0)$ at its essential supremum, there are only four universality classes of asymptotic behaviors of the parabolic Anderson model: the double-exponential distribution with $\rho \in (0,\infty)$ respectively $\rho = \infty$ (see Section 2.1), general fields bounded from above (see Section 2.3), and the class that was described in Section 2.5.

Our basic assumption on the logarithmic moment generating function H in (1.11) is the following.

Assumption H. There are a function $\widehat{H}\colon (0,\infty) \to \mathbb{R}$ and a continuous auxiliary function $\kappa\colon (0,\infty) \to (0,\infty)$ such that

$$\lim_{t\uparrow\infty} \frac{H(ty) - yH(t)}{\kappa(t)} = \widehat{H}(y) \neq 0 \quad \text{for } y \neq 1. \tag{3.1}$$

The function $\widehat{H}$ extracts the asymptotic scaling properties of the cumulant generating function H. In the language of the theory of regular functions, the assumption is that the logarithmic moment generating function H is in the de Haan class. This does not leave many possibilities for $\widehat{H}$:

Proposition 3.1. *Suppose that Assumption* H *holds.*

(i) *There is a* $\gamma \geq 0$ *such that* $\lim_{t\uparrow\infty} \kappa(yt)/\kappa(t) = y^\gamma$ *for any* $y > 0$, *i.e.,* κ *is regularly varying of index* γ. *In particular,* $\kappa(t) = t^{\gamma+o(1)}$ *as* $t \to \infty$.

(ii) *There exists a parameter* $\rho > 0$ *such that, for every* $y > 0$,
 (a) $\widehat{H}(y) = \rho(y - y^\gamma)/(1-\gamma)$ *if* $\gamma \neq 1$,
 (b) $\widehat{H}(y) = \rho y \log y$ *if* $\gamma = 1$.

Informally, the function $\widehat{H}(y)$ is equal to $\rho y \log y$ in the double-exponential case with $\rho \in (0,\infty)$ in Section 2.1, and in the case of almost bounded potentials of Section 2.5, and it is equal to $\rho(y - y^\gamma)/(1-\gamma)$ in the case of general potentials bounded from above (see Section 2.3).

Our second regularity assumption is a mild supposition on the auxiliary function κ. This assumption is necessary only in the case $\gamma = 1$ (which will turn out to be the critical case).

Assumption K. The limit $\kappa_* = \lim_{t\to\infty} \kappa(t)/t \in [0,\infty]$ exists.

We now introduce a scale function $\alpha\colon (0,\infty) \to (0,\infty)$, by

$$\frac{\kappa\big(t\alpha(t)^{-d}\big)}{t\alpha(t)^{-d}} = \frac{1}{\alpha(t)^2}. \tag{3.2}$$

The function $\alpha(t)$ turns out to be the annealed scale function for the radius of the relevant islands in the parabolic Anderson model. We can easily say something about the asymptotics of $\alpha(t)$:

Lemma 3.2. *Suppose that Assumptions H) and (K) hold. If $\gamma \le 1$ and $\kappa_* < \infty$, then there exists a unique solution $\alpha\colon (0,\infty) \to (0,\infty)$ to (3.2), and it satisfies $\lim_{t\to\infty} t\alpha(t)^{-d} = \infty$. Moreover,*

(i) *If $\gamma = 1$ and $0 < \kappa_* < \infty$, then $\lim_{t\to\infty} \alpha(t) = 1/\sqrt{\kappa_*} \in (0,\infty)$.*
(ii) *If $\gamma = 1$ and $\kappa_* = 0$, then $\alpha(t) = t^{\nu+o(1)}$ as $t \to \infty$, where $\nu = (1-\gamma)/(d+2-d\gamma)$.*

Now we identify the four universality classes, ordered from thick to thin upper tails of $\xi(0)$. For the moment, we focus on the large time behavior of the pth moment $\langle U(t)^p\rangle$ for any $p > 0$. In all cases there is a scale function $\alpha\colon (0,\infty) \to (0,\infty)$ and a number $\chi \in \mathbb{R}$ such that

$$\text{(3.3)} \qquad \frac{1}{pt}\log\langle U(t)^p\rangle = \frac{H\big(pt\,\alpha(pt)^{-d}\big)}{pt\,\alpha(pt)^{-d}} - \frac{1}{\alpha(pt)^2}\big(\chi + o(1)\big), \qquad \text{as } t\uparrow\infty.$$

The scale function α describes how fast the expected total mass, which at time $t=0$ is localized at the origin, spreads, in the sense that

$$\text{(3.4)} \qquad \lim_{R\uparrow\infty}\liminf_{t\uparrow\infty} \frac{\alpha(t)^2}{t}\log\frac{\langle\sum_{z\in\mathbb{Z}^d} u(t,z)\mathbb{1}\{|z|\le R\alpha(t)\}\rangle}{\langle\sum_{z\in\mathbb{Z}^d} u(t,z)\rangle} = 0.$$

Moreover, in the three classes where the mass does not concentrate asymptotically in a single point, there exists $R > 0$ such that

$$\text{(3.5)} \qquad \liminf_{t\uparrow\infty} \frac{\alpha(t)^2}{t}\log\frac{\langle\sum_{z\in\mathbb{Z}^d} u(t,z)\mathbb{1}\{|z|\le R\,\alpha(t)\}\rangle}{\langle\sum_{z\in\mathbb{Z}^d} u(t,z)\rangle} < 0.$$

(1) $\boxed{\gamma > 1, \text{ or } \gamma = 1 \text{ and } \kappa^* = \infty.}$

This case is included in [16] as the upper boundary case $\rho = \infty$ in their notation, see Section 2.1. Here $\chi = 2d$, the scale function $\alpha(t) = 1$ is constant, and the first term on the right hand side in (3.3) dominates the sum, which diverges to infinity. The asymptotics in (3.4) can be strengthened to

$$\lim_{t\uparrow\infty}\frac{1}{t}\log\frac{\langle u(t,0)\rangle}{\langle\sum_{z\in\mathbb{Z}^d} u(t,z)\rangle} = 0,$$

i.e., the expected total mass remains essentially in the origin and the intermittent islands are single sites, a phenomenon of *complete localization.* We call this the *single-peak case.* □

(2) $\boxed{\gamma = 1 \text{ and } \kappa^* \in (0,\infty).}$

This case, the *double-exponential case*, is the main objective of [16], see Section 2.1. The prime example is the double exponential distribution with parameter $\rho \in (0,\infty)$,

$$\text{Prob}\{\xi(0) > r\} = \exp\{-\mathrm{e}^{r/\rho}\},$$

which implies $H(t) = \rho t\log(\rho t) - \rho t + o(t)$. Here $\alpha(t) \to 1/\sqrt{\kappa}^* \in (0,\infty)$, so that the size of the intermittent islands is constant in time. The first term on the right-hand side in (3.3) dominates the sum, which goes to infinity. Moreover,

$$\text{(3.6)} \qquad \chi = \min_{\substack{g\colon \mathbb{Z}^d\to\mathbb{R}\\ \sum g^2 = 1}} \Big\{\frac{1}{2}\sum_{\substack{x,y\in\mathbb{Z}^d\\ x\sim y}} \big(g(x)-g(y)\big)^2 - \rho\sum_{x\in\mathbb{Z}^d} g^2(x)\log g^2(x)\Big\},$$

where we write $x \sim y$ if x and y are neighbors. This variational problem is difficult to Analise. It has a solution, which is unique for sufficiently large values of ρ, and heuristically this minimizer represents the shape of the solution. As noted in [11], for any family of minimizers g_ρ, as $\rho \uparrow \infty$, g_ρ converges to δ_0, which links to the single-peak case. Furthermore, as $\rho \downarrow 0$, the minimizers g_ρ are asymptotically given by

$$g_\rho^2(\lfloor x/\sqrt{\rho}\rfloor) = (1+o(1))\mathrm{e}^{-|x|^2}\pi^{-d/2},$$

uniformly on compacts and in $L^1(\mathbb{R}^d)$. Consequently,

$$\chi = \rho\, d\Big(1 - \frac{1}{2}\log\frac{\rho}{\pi} + o(1)\Big) \quad \text{as } \rho \downarrow 0. \qquad \square$$

(3) $\boxed{\gamma = 1 \text{ and } \kappa^* = 0.}$

Potentials in this class are called *almost bounded* in [16] and may be seen as the degenerate case for $\rho = 0$ in their notation. This class contains both bounded and unbounded potentials, and is analyzed for the first time in [17], see Section 2.5. The scale function $\alpha(t)$ and hence the diameter of the intermittent islands goes to infinity and is slowly varying, in particular it is slower than any power of t. The first term on the right hand side in (3.3) dominates the sum, which may go to infinity or zero. Moreover,

$$\chi = \min_{\substack{g\in H^1(\mathbb{R}^d)\\ \|g\|_2=1}} \Big\{\int_{\mathbb{R}^d} |\nabla g(x)|^2\,\mathrm{d}x - \rho\int g^2(x)\log g^2(x)\,\mathrm{d}x\Big\}. \tag{3.7}$$

This variational formula is obviously the continuous variant of (3.6), and it is much easier to solve. There is a unique minimizer, given by

$$g_\rho(x) = \Big(\frac{\rho}{\pi}\Big)^{d/4} \exp\Big(-\frac{\rho}{2}|x|^2\Big),$$

representing the rescaled shape of the solution on an intermittent island. In particular, $\chi = \rho d(1 - \frac{1}{2}\log(\rho/\pi))$, which is the asymptotics of (3.6) as $\rho \downarrow 0$. Hence, on the level of variational problems, (3) is the boundary case of (2) for $\rho \downarrow 0$. $\square$

(4) $\boxed{\gamma < 1.}$

This is the case of *potentials bounded from above*, which is treated in [4], see Section 2.3. Indeed, in [4], it is assumed that there exists a nondecreasing function $\alpha(t)$ and a nonpositive function $\widetilde{H}\colon (0,\infty) \to (-\infty, 0]$ such that

$$\lim_{t\uparrow\infty} \frac{\alpha(t)^{d+2}}{t} H\Big(\frac{t}{\alpha(t)^d}\, y\Big) = \widetilde{H}(y),$$

uniformly on compact sets in $(0,\infty)$. It is not too difficult to see that this assumption holds if Assumption H holds for the index $\gamma < 1$, for α defined by (1.22) and

$$\widetilde{H}(y) = \frac{\rho}{\gamma - 1}\, y^\gamma.$$

Here $\alpha(t) \to \infty$ as $t \mapsto \alpha(t)$ is regularly varying with index $(1-\gamma)/(d+2-d\gamma)$. The potential ξ is necessarily bounded from above. In this case, the two terms on the right hand side in (3.3) are of the *same order*, and (3.3) converges to zero. Moreover

$$\chi = \inf_{\substack{g\in H^1(\mathbb{R}^d)\\ \|g\|_2=1}} \Big\{\int_{\mathbb{R}^d} |\nabla g(x)|^2\,\mathrm{d}x - \rho\int_{\mathbb{R}^d} \frac{g^{2\gamma}(x) - g^2(x)}{\gamma - 1}\,\mathrm{d}x\Big\}. \tag{3.8}$$

In the lower boundary case where $\gamma = 0$, the functional $\int g^{2\gamma}$ must be replaced by the Lebesgue measure of supp(g). In this case the formula is well-known and well-understood. In particular, the minimizer exists, is unique up to spatial shifts, and has compact support. To the best of our knowledge, for $\gamma \in (0, 1)$, the formula in (3.8) has not been analyzed explicitly, unless in $d = 1$. One can also show [17] that (3.8) converges to (3.7), as would follow from interchanging the limit $\gamma \uparrow 1$ with the infimum on g. This means that, on the level of variational formulas, (3) is the boundary case of (4) for $\gamma \uparrow 1$. □

The variational problems in (3.6), (3.7), and (3.8) encode the asymptotic shape of the rescaled and normalized solution $u(t, \cdot)$ in the centered ball with radius of order $\alpha(t)$. Informally, the main contribution to $\langle U(t) \rangle$ comes from the events that

$$\frac{u(t, \lceil \cdot \alpha(t) \rceil)}{\| u(t, \lceil \cdot \alpha(t) \rceil) \|_2} \approx g,$$

where g is a minimizer in the definition of χ. To the best of our knowledge this heuristics has not been made rigorous in any nontrivial case so far. Note that in case (1), formally, (3.6) holds with $\rho = \infty$ and hence the optimal g is $\mathbb{1}_0$.

References

1. P. Antal, *Trapping problems for the simple random walk*, Thesis, ETH, Zürich, 1994.
2. ———, *Enlargement of obstacles for the simple random walk*, Ann. Probab. **23** (1995), no. 1, 1061–1101.
3. M. Aizenman and S. Molchanov, *Localization at large disorder and extreme energies: an elementary derivation*, Comm. Math. Phys. **157** (1993), no. 2, 245–278.
4. M. Biskup and W. König, *Long-time tails in the parabolic Anderson model with bounded potential*, Ann. Probab. **29** (2001), no. 2, 636–682.
5. E. Bolthausen, *Localization of a two-dimensional random walk with an attractive path interaction*, Ann. Probab. **22** (1994), 875–918.
6. R. A. Carmona and S. A. Molchanov, *Parabolic Anderson problem and intermittency*, Mem. Amer. Math. Soc. **108** (1994), no. 518.
7. ———, *Stationary parabolic Anderson model and intermittency*, Probab. Theory Related Fields **102** (1995), 433–453.
8. M. Donsker and S. R. S. Varadhan, *Asymptotics for the Wiener sausage*, Comm. Pure Appl. Math. **28** (1975), 525–565.
9. ———, *On the number of distinct sites visited by a random walk*, Comm. Pure Appl. Math. **32** (1979), 721–747.
10. J. Fröhlich, F. Martinelli, E. Scoppola, and T. Spencer, *Constructive proof of localization in the Anderson tight binding model*, Comm. Math. Phys. **101** (1985), no. 1, 21–46.
11. J. Gärtner and F. den Hollander, *Correlation structure of intermittency in the parabolic Anderson model*, Probab. Theory Related Fields **114** (1999), 1–54.
12. J. Gärtner and W. König, *Moment asymptotics for the continuous parabolic Anderson model*, Ann. Appl. Probab. **10** (2000), no. 3, 192–217.
13. ———, *The parabolic Anderson model*, Interacting Stochastic Systems (J.-D. Deuschel and A. Greven, eds.), Springer, 2005, 153–179.
14. J. Gärtner, W. König, and S. Molchanov, *Almost sure asymptotics for the continuous parabolic Anderson model*, Probab. Theory Related Fields **118** (2000), no. 4, 547–573.
15. J. Gärtner and S. A. Molchanov, *Parabolic problems for the Anderson model.* I: *Intermittency andrelated topics*, Comm. Math. Phys. **132** (1990), no. 3, 613–655.
16. ———, *Parabolic problems for the Anderson model.* II: *Second-order asymptotics and structure of high peaks*, Probab. Theory Related Fields **111** (1998), no. 1, 17–55.
17. R. van der Hofstad, W. König, and P. Mörters, *The universality classes in the parabolic Anderson model*, Comm. Math. Phys., to appear.
18. S. Molchanov, *Lectures on random media*, Lectures on Probability Theory (Saint-Flour, 1992) (P. Bernard, ed.), Lectures Notes in Math., vol. 1581, Springer, New York, 1994, pp. 242–411.

19. S. Molchanov and A. Ruzmaikin, *Lyapunov exponents and distributions of magnetic fields in dynamo models*, The Dynkin Festschrift: Markov Processes and their Applications (M. Freidlin, ed.), Birkhäuser, 1994, pp. 287 – 306.
20. A.-S. Sznitman, *Brownian motion obstacles and random media*, Springer Monogr. Math., Springer, Berlin, 1998.

Mathematisches Institut, Universität Leipzig, Augustusplatz 10/11, 04109 Leipzig, Germany

E-mail address: koenig@math.uni-leipzig.de

Centre de Recherches Mathématiques
CRM Proceedings and Lecture Notes
Volume **42**, 2007

An Inverse Problem for Gibbs Fields

Leonid Koralov

ABSTRACT. According to classical results of Minlos and Ruelle, given a potential of pair interaction that is regular and stable, and a small value of activity, one can define the corresponding Gibbs field (a measure on the space of configurations of points in $\mathbb{R}^d$ or $\mathbb{Z}^d$).

In this paper we consider a problem that may be viewed as a converse to the result of Minlos and Ruelle. Namely, we show that for a sufficiently small constant $\bar{\rho}_1$ and a sufficiently small function $\bar{\rho}_2(x)$, $x \in \mathbb{Z}^d$, there are a pair potential and a value of activity, such that $\bar{\rho}_1$ is the density and $\bar{\rho}_2$ is the pair correlation function of the corresponding Gibbs field.

1. Introduction

Let us consider a translation invariant measure μ on the space of particle configurations on the lattice $\mathbb{Z}^d$. For a given configuration each site can be occupied by one particle or be empty. An m-point correlation function $\rho_m(x_1, \ldots, x_m)$ is the probability of finding m different particles at positions $x_1, \ldots, x_m \in \mathbb{Z}^d$. The following natural question has been extensively discussed in physical and mathematical literature: given $\rho_1(x_1) \equiv \bar{\rho}_1$ and $\rho_2(x_1, x_2) = \bar{\rho}_2(x_1 - x_2)$, does there exist a measure μ, for which these are the first correlation function (density) and the pair correlation function, respectively?

There are well-know conditions on $\bar{\rho}_1$ and $\bar{\rho}_2$ that are easy to verify and are necessary (but not sufficient) for the existence of the translation invariant measure. Let $\xi(x)$ be a stationary random field with values 0 and 1. Its distribution can be viewed as a measure on the space of particle configurations. Let $\bar{\rho}_1$ and $\bar{\rho}_2(x)$ be its first two correlation functions. Then the mean of the field ξ is $E\xi(x) \equiv \bar{\rho}_1$, and its covariance is given by

$$C(x) = E(\xi(x) - \bar{\rho}_1)(\xi(0) - \bar{\rho}_1) = \begin{cases} \bar{\rho}_1 - \bar{\rho}_1^2 & \text{if } x = 0 \\ \bar{\rho}_2(x) - \bar{\rho}_1^2 & \text{otherwise.} \end{cases}$$

2000 *Mathematics Subject Classification.* 60G55, 60G60.

Key words and phrases. Gibbs field, Gibbs measure, cluster functions, pair potential, correlation functions, Ursell functions.

The work of the author is partially supported by NSF Research Grant.

This is the final form of the paper.

It is well-known that positive-definiteness of $C(x)$ is necessary and sufficient for the existence of some random field with such a covariance. For example, if $C(x)$ is positive-definite, one can always take ξ to be a Gaussian field. However, the requirement that ξ only takes values 0 and 1 imposes additional requirements on the correlation functions. In particular, there is a simple condition on the variance of the number of particles in a given region, which follows from the fact that the number of particles (sites where $\xi(x) = 1$) is integer. The original statement of this condition can be found the the paper by Yamada [12], and a more general version in the paper of Costin and Lebowitz [3].

In the series of papers [5–7] Lenard provided a set of relations on the functions ρ_m which are necessary and sufficient for the existence of a measure with such correlation functions. However, given ρ_1 and ρ_2, it is not clear how to check if there are some ρ_3, ρ_4,... for which these relations hold.

There are several recent papers which demonstrate the existence of particular types of point processes (measures on the space of particle configurations), which correspond to given ρ_1 and ρ_2 under certain conditions on ρ_1 and ρ_2. In particular, one dimensional point processes of renewal type are considered by Costin and Lebowitz in [3], while determinantal processes are considered by Soshnikov in [10]. In [1] Ambartzumian and Sukiasian prove the existence of a point process corresponding to a sufficiently small density and correlation function. Recently Costin and Lebowitz [3], and Caglioti et al. [2] suggested generalizations of their results. In [11] Stillinger and Torquato consider fields over a space with finitely many points. Besides, for the lattice model, they discuss possible existence of a pair potential for a given density and correlation function using cluster expansion without addressing the issue of convergence.

In this paper we show that if ρ_1 and ρ_2 are small (in a certain sense), there exists a measure on the space of configurations for which ρ_1 is the density and ρ_2 is the pair correlation function. Moreover, this measure is the Gibbs measure corresponding to some pair potential and some value of activity. In a sense, this is the converse of the classical statement that a given potential of pair interaction and a sufficiently small value of activity determine a translation invariant Gibbs measure on the space of particle configurations in $\mathbb{Z}^d$ (or $\mathbb{R}^d$) and the sequence of infinite volume correlation functions.

2. Gibbs random fields

Let us remind the reader of the construction of the Gibbs random field corresponding to a given potential of pair interaction and a given value of activity. Let $\Phi(x)$, $x \in \mathbb{Z}^d$ be a real-valued function (potential of pair interaction), and let $U(x_1, \ldots, x_n) = \sum_{1 \le i < j \le n} \Phi(x_i - x_j)$ be the total potential energy of the configuration $(x_1, \ldots, x_n)$. We assume that $\Phi(x) = \Phi(-x) \ge c_0 > -\infty$ for all x and that $\Phi(0) = +\infty$. The full list of assumptions on $\Phi(x)$ will be given below.

Let Λ be a finite subset of $\mathbb{Z}^d$. The grand canonical ensemble is defined by a measure on $\bigcup_{n=0}^{\infty} \Lambda^n$, whose restriction on Λ^n is equal to

$$\nu(x_1, \ldots, x_n) = \frac{z^n}{n!} e^{-U(x_1, \ldots, x_n)}.$$

The parameter $z > 0$ is called the activity. The inverse temperature, which is the factor usually present in front of the function U, is set to be equal to one (or, equivalently, incorporated into the function U). The total mass of the measure is the grand partition function

$$\Xi(\Lambda, z, \Phi) = \sum_{n=0}^{\infty} \frac{z^n}{n!} \sum_{(x_1,\dots,x_n)\in\Lambda^n} e^{-U(x_1,\dots,x_n)}.$$

The m-point correlation function is defined as the probability of finding m different particles at positions $x_1, \dots, x_m \in \Lambda$,

$$\rho_m^\Lambda(x_1, \dots, x_m) = \Xi(\Lambda, z, \Phi)^{-1} \sum_{n=0}^{\infty} \frac{z^{m+n}}{n!} \sum_{(y_1,\dots,y_n)\in\Lambda^n} e^{-U(x_1,\dots,x_m,y_1,\dots,y_n)}.$$

The corresponding measure on the space of all configurations of particles on the set Λ (Gibbs measure) will be denoted by μ^Λ. Given another set $\Lambda_0 \subseteq \Lambda$, we can consider the measure $\mu^\Lambda_{\Lambda_0}$ obtained as a restriction of the measure μ^Λ to the set of particle configurations on Λ_0.

Given a potential of pair interaction $\Phi(x)$, we define $g(x) = e^{-\Phi(x)} - 1$, $x \in \mathbb{Z}^d$. We shall make the following standard assumptions:

(2.1) $$g(x) \geq -a > -1 \quad \text{for } x \neq 0.$$

(2.2) $$g(0) = -1; \qquad g(x) = g(-x) \quad \text{for all } x; \qquad \sum_{x\neq 0} |g(x)| \leq c < \infty.$$

Clearly, any function $g(x)$ which satisfies (2.1)–(2.2) defines a potential of pair interaction via

$$\Phi(x) = -\ln(g(x) + 1).$$

It is well-known ([8, 9]) that when $\Lambda \to \mathbb{Z}^d$ in a suitable manner (for example, $\Lambda = [-k, k]^d$ and $k \to \infty$) the following two limits exist for sufficiently small z:

(a) There is a probability measure $\mu^{\mathbb{Z}^d}$ on the space of all configurations on $\mathbb{Z}^d$, such that

(2.3) $$\mu^\Lambda_{\Lambda_0} \to \mu^{\mathbb{Z}^d}_{\Lambda_0} \quad \text{as } \Lambda \to \mathbb{Z}^d$$

for any finite set $\Lambda_0 \subset \mathbb{Z}^d$.

(b) All the correlation functions converge to the infinite volume correlation functions. Namely,

(2.4) $$\rho_m^\Lambda(x_1, \dots, x_m) \to \rho_m(x_1, \dots, x_m) \quad \text{as } \Lambda \to \mathbb{Z}^d.$$

The infinite volume correlation functions are the probabilities with respect to the measure $\mu^{\mathbb{Z}^d}$ of finding m different particles at positions $x_1, \dots, x_m \in \mathbb{Z}^d$. To make these statements precise we formulate them as a lemma.

Lemma 2.1 ([8, 9])**.** *Assuming that* (2.1) *and* (2.2) *hold, there is a positive* $\bar{z} = \bar{z}(a, c)$*, such that* (2.3) *and* (2.4) *hold for all* $0 < z \leq \bar{z}$ *when* $\Lambda = [-k, k]^d$ *and* $k \to \infty$.

Thus, a pair potential defines a sequence of infinite volume correlation functions for sufficiently small values of activity. Note that $\rho_m(x_1, \dots, x_m) = 0$ if $x_i = x_j$ for $i \neq j$, since two distinct particles can not occupy the same position. Also note that all the correlation functions are translation invariant,

$$\rho_m(x_1, \dots, x_m) = \rho_m(0, x_2 - x_1, \dots, x_m - x_1).$$

Thus, ρ_1 is a constant, ρ_2 can be considered as a function of one variable, etc. Let $\bar{\rho}_m$ be the function of $m-1$ variables, such that

$$\rho_m(x_1,\ldots,x_m) = \bar{\rho}_m(x_2-x_1,\ldots,x_m-x_1). \tag{2.5}$$

3. Formulation of the result

The main result of this paper is the following theorem.

Theorem 3.1. *Let $0<r<1$ be a constant. Given any sufficiently small constant $\bar{\rho}_1$ and any function $\bar{\rho}_2(x)$, such that $\bar{\rho}_2(0)=0$ and $\sum_{x\neq 0}|\bar{\rho}_2(x)-\bar{\rho}_1^2|\leq r\bar{\rho}_1^2$, there are a potential $\Phi(x)$, which satisfies (2.1) – (2.2), and a value of activity z, such that $\bar{\rho}_1$ and $\bar{\rho}_2(x)$ are the first and the second correlation functions respectively for the system defined by (z,Φ).*

Remark. As will be seen from the proof of the theorem, the pair potential and the activity corresponding to given $\bar{\rho}_1$ and $\bar{\rho}_2(x)$ are unique, if we restrict consideration to sufficiently small values of Φ and z. The method of the proof allows one to explore the properties of the pair potential based on the properties of the correlation function.

The main idea of the proof is the following. Assuming that a pair potential and a value of the activity exist, we express the correlation functions (or, rather, the cluster functions, which are closely related to the correlation functions) in terms of the pair potential and the activity. This relationship can be viewed as an equation for unknown Φ and z. Then we use the contracting mapping principle to demonstrate that this equation has a solution. New estimates on the Ursell functions (see below) are needed to prove that the right hand side of the equation on Φ and z is indeed a contraction.

4. Cluster functions and Ursell functions

In this section we state a useful expression for cluster functions in terms of the pair potential. The cluster functions are closely related to the correlation functions. The reader is referred to [9, Chapter 4] for a more detailed exposition. We also formulate a lemma on the growth of Ursell functions. Its proof can be found in [4].

Let A be the complex vector space of sequences ψ,

$$\psi = \big(\psi_m(x_1,\ldots,x_m)\big)_{m\geq 0}$$

such that, for each $m\geq 1$, ψ_m is a bounded function on $\mathbb{Z}^{md}$, and ψ_0 is a complex number. It will be convenient to represent a finite sequence $(x_1,\ldots,x_m)$ by a single letter $X=(x_1,\ldots,x_m)$. We shall write

$$\psi(X) = \psi_m(x_1,\ldots,x_m).$$

Let now $\psi^1,\psi^2\in A$. We define

$$\psi^1 * \psi^2(X) = \sum_{Y\subseteq X}\psi^1(Y)\psi^2(X\backslash Y),$$

where the summation is over all subsequences Y of X and $X\backslash Y$ is the subsequence of X obtained by striking out the elements of Y in X.

Let A_+ be the subspace of A formed by the elements ψ such that $\psi_0=0$. Let $\mathbf{1}$ be the unit element of A ($\mathbf{1}_0=1$, $\mathbf{1}_m\equiv 0$ for $m\geq 1$).

We define the mapping Γ of A_+ onto $\mathbf{1}+A_+$:

$$\Gamma\varphi = \mathbf{1} + \varphi + \frac{\varphi * \varphi}{2!} + \frac{\varphi * \varphi * \varphi}{3!} + \cdots .$$

The mapping Γ has an inverse Γ^{-1} on $\mathbf{1}+A_+$:

$$\Gamma^{-1}(\mathbf{1}+\varphi') = \varphi' - \frac{\varphi' * \varphi'}{2} + \frac{\varphi' * \varphi' * \varphi'}{3} - \cdots .$$

It is easy to see that $\Gamma\varphi(X)$ is the sum of the products $\varphi(X_1)\cdots\varphi(X_r)$ corresponding to all the partitions of X into subsequences $X_1,\dots,X_r$. If $\varphi \in A_+$ and $\psi = \Gamma\varphi$, the first few components of ψ are

$$\psi_0 = 1; \quad \psi_1(x_1) = \varphi_1(x_1); \quad \psi_2(x_1,x_2) = \varphi_2(x_1,x_2) + \varphi_1(x_1)\varphi_1(x_2).$$

Let Φ be a pair correlation function which satisfies (2.1) – (2.2), and let $z \le \bar{z}(a,c)$. Note that the sequence of correlation functions $\rho = (\rho_m)_{m\ge 0}$ (with $\rho_0 = 1$) is an element of $\mathbf{1}+A_+$.

Definition 4.1. *The cluster functions $\omega_m(x_1,\dots,x_m)$, $m \ge 1$ are defined by*

$$\omega = \Gamma^{-1}\rho.$$

Thus,

$$\omega_1(x_1) = \rho_1(x_1); \quad \omega_2(x_1,x_2) = \rho_2(x_1,x_2) - \rho_1(x_1)\rho_1(x_2),$$

or, equivalently,

$$\overline{\omega}_1 = \bar{\rho}_1; \quad \overline{\omega}_2(x) = \bar{\rho}_2(x) - \bar{\rho}_1^2$$

where $\overline{\omega}_m$ are defined as in (2.5).

Let $\psi \in \mathbf{1}+A_+$ be defined by

$$\psi_0 = 1; \quad \psi_m(x_1,\dots,x_m) = e^{-U(x_1,\dots,x_m)}.$$

Define also

$$\varphi = \Gamma^{-1}\psi.$$

Definition 4.2. *The functions ψ_m and φ_m are called Boltzmann factors and Ursell functions, respectively.*

Lemma 4.3 ([9]). *The cluster functions can be expressed in terms of the Ursell functions as follows*

$$\omega_m(x_1,\dots,x_m) = z^m \sum_{n=0}^{\infty} \frac{z^n}{n!} \sum_{y_1,\dots,y_n \in \mathbb{Z}^d} \varphi_{m+n}(x_1,\dots,x_m,y_1,\dots,y_n).$$

We next state a lemma which provides a bound on the growth of Ursell functions in terms of the function $g(x) = \exp\bigl(-\Phi(x)\bigr) - 1$.

Lemma 4.4 ([4]). *Suppose that functions $g_1(x)$ and $g_2(x)$ satisfy (2.2) with $c < 1$. Let $\varphi^k = \bigl(\varphi^k_m(x_1,\dots,x_m)\bigr)_{m\ge 0}$, $k=1,2$ be the corresponding Ursell functions. Then there exist constants q_1 and q_2 such that*

$$\sum_{y_1,\dots,y_n \in \mathbb{Z}^d} |\varphi^k_{1+n}(0,y_1,\dots,y_n)| \le n! q_1^{n+1}, \quad k = 1,2,$$

$$\sum_{y_1,\dots,y_n \in \mathbb{Z}^d} |\varphi^1_{1+n}(0,y_1,\dots,y_n) - \varphi^2_{1+n}(0,y_1,\dots,y_n)| \le n! q_2^{n+1} \sum_{x\ne 0} |g_1(x) - g_2(x)|.$$

5. Proof of the main result

Let us recast the main theorem in terms of the cluster functions and examine a system of equations, which relates the first two cluster functions with the pair potential and the activity. Theorem 3.1 can clearly be reformulated as follows

Proposition 5.1. *Let $0 < r < 1$ be a constant. Given any sufficiently small constant $\overline{\omega}_1$ and any function $\overline{\omega}_2(x)$, such that $\overline{\omega}_2(0) = -\overline{\omega}_1^2$ and $\sum_{x \neq 0} |\overline{\omega}_2(x)| \leq r\overline{\omega}_1^2$, there are a potential $\Phi(x)$, which satisfies (2.1) – (2.2), and a value of activity z, such that $\overline{\omega}_1$ and $\overline{\omega}_2(x)$ are the first and the second cluster functions respectively for the system defined by (z, Φ).*

Consider the power expansions for ω_1 and ω_2, which are provided by Lemma 4.3. Let us single out the first term in both expansions. Note the translation invariance of the functions ω_m and φ_m and the fact that $\varphi(x_1, x_2) = g(x_1 - x_2)$.

$$\overline{\omega}_1 = z + z^2 \sum_{n=1}^{\infty} \frac{z^{n-1}}{n!} \sum_{y_1,\ldots,y_n \in \mathbb{Z}^d} \varphi_{1+n}(0, y_1, \ldots, y_n), \tag{5.1}$$

$$\overline{\omega}_2(x) = z^2 g(x) + z^3 \sum_{n=1}^{\infty} \frac{z^{n-1}}{n!} \sum_{y_1,\ldots,y_n \in \mathbb{Z}^d} \varphi_{2+n}(0, x, y_1, \ldots, y_n). \tag{5.2}$$

Let

$$A(z, g) = \sum_{n=1}^{\infty} \frac{z^{n-1}}{n!} \sum_{y_1,\ldots,y_n \in \mathbb{Z}^d} \varphi_{1+n}(0, y_1, \ldots, y_n),$$

$$B(z, g)(x) = \sum_{n=1}^{\infty} \frac{z^{n-1}}{n!} \sum_{y_1,\ldots,y_n \in \mathbb{Z}^d} \varphi_{2+n}(0, x, y_1, \ldots, y_n).$$

Thus equations (5.1) and (5.2) can be rewritten as follows

$$z = \overline{\omega}_1 - z^2 A(z, g)\ , \tag{5.3}$$

$$g = \frac{\overline{\omega}_2}{z^2} - zB(z, g). \tag{5.4}$$

Instead of looking at (5.3) – (5.4) as a formula defining $\overline{\omega}_1$ and $\overline{\omega}_2$ by a given pair potential and the activity, we can instead consider the functions $\overline{\omega}_1$ and $\overline{\omega}_2$ fixed, and g and z unknown. Thus, Proposition 5.1 follows from the following.

Proposition 5.2. *If $\overline{\omega}_1$ and $\overline{\omega}_2$ satisfy the assumptions of Proposition 5.1, then the system (5.3) – (5.4) has a solution (z, g), such that the function g satisfies (2.1) – (2.2) and $z \leq \overline{z}(a, c)$.*

Now we concentrate on the proof of Proposition 5.2. We shall need the following notations. Let $\mathcal{G}$ be the space of functions g, which satisfy (2.2) with some $c < \infty$. Let $\|g\| = \sum_{x \neq 0} |g(x)|$. This is not a norm, since $\mathcal{G}$ is not a linear space, however $d(g_1, g_2) = \|g_1 - g_2\|$ is a metric on the space $\mathcal{G}$. Let $\mathcal{G}_c$ be the set of elements of $\mathcal{G}$ for which $\|g\| \leq c$. Note that if $c < 1$ then all elements of $\mathcal{G}_c$ satisfy (2.1) with $a = c$.

We also define $I_{z_0}^{a_1,a_2} = [a_1 z_0, a_2 z_0]$. Let $D = I_{z_0}^{a_1,a_2} \times \mathcal{G}_c$. Note that if $c < 1$ then $(z, g) \in D$ implies that $z \leq \overline{z}(c, c)$ if z_0 is sufficiently small. Thus, the infinite volume correlation functions and cluster functions are correctly defined for $(z, g) \in D$ if z_0 is sufficiently small.

Let us define an operator Q on the space of pairs $(z,g) \in D$ by $Q(z,g) = (z',g')$, where

$$z' = \overline{\omega}_1 - z^2 A(z,g), \tag{5.5}$$

$$g'(x) = \frac{\overline{\omega}_2(x)}{z^2} - zB(z,g)(x) \quad \text{for } x \neq 0; \qquad g'(0) = -1. \tag{5.6}$$

We shall prove the following lemma.

Lemma 5.3. *Let $0 < r < 1$ be a constant. There exist positive constants $a_1 < 1$, $a_2 > 1$, and $c < 1$ such that the equation $(z,g) = Q(z,g)$ has a solution $(z,g) \in D$ for all sufficiently small z_0 if $\overline{\omega}_1 = z_0$, $\overline{\omega}_2(0) = -z_0^2$, and $\sum_{x\neq 0} |\overline{\omega}_2(x)| \leq r z_0^2$.*

Before we prove this lemma, let us verify that it implies Proposition 5.2. Let $0 < r < 1$ be fixed and let $\overline{\omega}_1$ be sufficiently small for the statement of Lemma 5.3 to be valid. Let $\overline{\omega}_2$ be such that $\overline{\omega}_2(0) = -\overline{\omega}_1^2$ and $\sum_{x\neq 0} |\overline{\omega}_2(x)| \leq r\overline{\omega}_1^2$. Let (z,g) be the solution of $(z,g) = Q(z,g)$, whose existence is guaranteed by Lemma 5.3. Let $\overline{\omega}_1'$ and $\overline{\omega}_2'$ be the first two cluster functions corresponding to the pair (z,g). Note that $\overline{\omega}_1$ and $\overline{\omega}_1'$ satisfy the same equation

$$z = \overline{\omega}_1 - z^2 A(z,g); \quad z = \overline{\omega}_1' - z^2 A(z,g).$$

Therefore, $\overline{\omega}_1 = \overline{\omega}_1'$. The functions $\overline{\omega}_2$ and $\overline{\omega}_2'$ also satisfy the same equation

$$g(x) = \frac{\overline{\omega}_2(x)}{z^2} - zB(z,g)(x); \quad g(x) = \frac{\overline{\omega}_2'(x)}{z^2} - zB(z,g)(x); \qquad \text{for } x \neq 0.$$

Thus, $\overline{\omega}_2(x) = \overline{\omega}_2'(x)$ for $x \neq 0$. The fact that $\overline{\omega}_2(0) = \overline{\omega}_2'(0)$ follows from

$$\overline{\omega}_2(0) = -\overline{\omega}_1^2 = -\overline{\omega}_1'^2 = \overline{\omega}_2'(0).$$

Thus it remains to prove Lemma 5.3. The proof will be based on the fact that for small z_0 the operator $Q\colon D \to D$ is a contraction in an appropriate metric. Define

$$d_{z_0}(z_1,z_2) = \frac{h|z_1 - z_2|}{z_0}.$$

The value of the constant h will be specified later. Now the metric on D is given by

$$\rho\big((z_1,g_1),(z_2,g_2)\big) = d_{z_0}(z_1,z_2) + d(g_1,g_2).$$

Lemma 5.3 clearly follows from the contracting mapping principle and the following lemma.

Lemma 5.4. *Let $0 < r < 1$ be a constant. There exist positive constants $a_1 < 1$, $a_2 > 1$, and $c < 1$ such that for all sufficiently small z_0 the operator Q acts from the domain D into itself and is uniformly contracting in the metric ρ for some value of $h > 0$, provided that $\overline{\omega}_1 = z_0$, $\overline{\omega}_2(0) = -z_0^2$, and $\sum_{x\neq 0} |\overline{\omega}_2(x)| \leq r z_0^2$.*

PROOF. Take $c = (r+2)/3$, $a_1 = \sqrt{2r3(r+1)}$, $a_2 = 2$. We shall need certain estimates on the values of $A(z,g)$ and $B(z,g)$ for $(z,g) \in D$. Namely, there exist universal constants $u_1, \dots, u_6$, such that for sufficiently small z_0 we have

$$\sup_{(z,g)\in D} |A(z,g)| \le u_1, \tag{5.7}$$

$$\sup_{(z,g)\in D} \sum_{x\neq 0} |B(z,g)(x)| \le u_2, \tag{5.8}$$

$$\sup_{(z_1,g),(z_2,g)\in D} |A(z_1,g) - A(z_2,g)| \le u_3|z_1 - z_2|, \tag{5.9}$$

$$\sup_{(z,g_1),(z,g_2)\in D} |A(z,g_1) - A(z,g_2)| \le u_4 d(g_1,g_2), \tag{5.10}$$

$$\sup_{(z_1,g),(z_2,g)\in D} \sum_{x\neq 0} |B(z_1,g)(x) - B(z_2,g)(x)| \le u_5|z_1 - z_2|, \tag{5.11}$$

$$\sup_{(z,g_1),(z,g_2)\in D} \sum_{x\neq 0} |B(z,g_1)(x) - B(z,g_2)(x)| \le u_6 d(g_1,g_2). \tag{5.12}$$

These estimates immediately follow from Lemma 4.4 above. The fact that $QD \subseteq D$ is guaranteed by the inequalities

$$z_0 + (a_2 z_0)^2 u_1 \le a_2 z_0 \ , \tag{5.13}$$

$$z_0 - (a_2 z_0)^2 u_1 \ge a_1 z_0 \ , \tag{5.14}$$

$$\frac{r z_0^2}{(a_1 z_0)^2} + a_2 z_0 u_2 \le c. \tag{5.15}$$

It is clear that (5.13) – (5.15) hold for sufficiently small z_0. Let us now demonstrate that for some h and for all sufficiently small z_0 we have

$$\rho\big(Q(z_1,g_1), Q(z_2,g_2)\big) \le \frac{1}{2}\rho\big((z_1,g_1),(z_2,g_2)\big) \quad \text{if } (z_1,g_1),(z_2,g_2) \in D. \tag{5.16}$$

First, taking (5.7), (5.9), and (5.10) into account, we note that

$$\begin{aligned}
&d_{z_0}\big(z_1^2 A(z_1,g_1), z_2^2 A(z_2,g_2)\big)\\
&\quad\le d_{z_0}\big(z_1^2 A(z_1,g_1), z_2^2 A(z_1,g_1)\big) + d_{z_0}\big(z_2^2 A(z_1,g_1), z_2^2 A(z_2,g_1)\big)\\
&\qquad\qquad + d_{z_0}\big(z_2^2 A(z_2,g_1), z_2^2 A(z_2,g_2)\big)\\
&\quad\le \frac{u_1 h |z_1^2 - z_2^2|}{z_0} + \frac{u_3 h (a_2 z_0)^2 |z_1 - z_2|}{z_0} + \frac{u_4 h (a_2 z_0)^2 d(g_1,g_2)}{z_0}.
\end{aligned}$$

If h is fixed, the right hand side of this inequality can be estimated from above, for all sufficiently small z_0, by

$$\tfrac{1}{6}\big(d_{z_0}(z_1,z_2) + d(g_1,g_2)\big).$$

Similarly,

$$\begin{aligned}
&\sum_{x\neq 0} |z_1 B(z_1,g_1)(x) - z_2 B(z_2,g_2)(x)|\\
&\quad\le \sum_{x\neq 0} |z_1 B(z_1,g_1)(x) - z_2 B(z_1,g_1)(x)| + \sum_{x\neq 0} |z_2 B(z_1,g_1)(x) - z_2 B(z_2,g_1)(x)|\\
&\qquad\qquad + \sum_{x\neq 0} |z_2 B(z_2,g_1)(x) - z_2 B(z_2,g_2)(x)|\\
&\quad\le u_2|z_1 - z_2| + u_5 a_2 z_0 |z_1 - z_2| + u_6 a_2 z_0 d(g_1,g_2).
\end{aligned}$$

Again, if h is fixed, the right-hand side of this inequality can be estimated from above, for all sufficiently small z_0, by

$$\tfrac{1}{6}\big(d_{z_0}(z_1, z_2) + d(g_1, g_2)\big).$$

Finally,

$$\sum_{x\neq 0}\left|\frac{\overline{\omega}_2(x)}{z_1^2} - \frac{\overline{\omega}_2(x)}{z_2^2}\right| \le r z_0^2\left|\frac{1}{z_1^2} - \frac{1}{z_2^2}\right| \le \frac{2a_2|z_1 - z_2|}{a_1^4 z_0}.$$

We can now take $h = 12a_2/a_1^4$, which implies that the right-hand side of the last inequality can be estimated from above by $\frac{1}{6}d_{z_0}(z_1, z_2)$. We have thus demonstrated the validity of (5.16), which means that the operator Q is uniformly contracting. This completes the proof of the lemma. □

References

1. R. V. Ambartzumian and H. S. Sukiasian, *Inclusion-exclusion and point processes*, Acta Appl. Math. **22** (1991), no. 1, 15 – 31.
2. E. Caglioti, T. Kuna, J. L. Lebowitz, and E. R. Speer, *Point processes with specified low order correlations*, Markov Process. Related Fields **12** (2006), no. 2, 257 – 272.
3. O. Costin and J. L. Lebowitz, *On the construction of particle distributions with specified single and pair densities*, J. Phys. Chem. B **108** (2004), no. 51, 19614 – 19618, available at `arXiv:cond-mat0405519`.
4. L. Koralov, *Existence of pair potential corresponding to specified density and pair correlation*, Lett. Math. Phys. **71** (2005), no. 2, 135 – 148.
5. A. Lenard, *Correlation functions and the uniqueness of the state in classical statistical mechanics*, Comm. Math. Phys. **30** (1973), 35-44.
6. ———, *States of classical statistical mechanical systems of infinitely many particles*. I, Arch. Rational Mech. Anal. **59** (1975), no. 3, 219–239.
7. ———, *States of classical statistical mechanical systems of infinitely many particles*. II: *Characterization of correlation measures*, Arch. Rational Mech. Anal. **59** (1975), no. 3, 241 – 256.
8. R. A. Minlos, *Gibbs' limit distribution*, Funkcional. Anal. i Priložen. **1** (1967), no. 2, 60 – 73 (Russian).
9. D. Ruelle, *Statistical mechanics*: *Rigorours results*, W. A. Benjamin, Inc., New York – Amsterdam, 1969.
10. A. Soshnikov, *Determinantal random point fields*, Uspekhi Mat. Nauk **55** (2000), no. 5(335), 107 – 160 (Russian); English transl., Russian Math. Surveys **55** (2000), no. 5, 923 – 975.
11. F. H. Stillinger and S. Torquato, *Pair correlation function realizability*: *Lattice model implications*, J. Phys. Chem. B **108** (2004), no. 51, 19589 – 19594.
12. M. Yamada, *Geometrical study of the pair distribution function in the many-body problem*, Progr. Theoret. Phys. **25** (1961), 579 – 594.

Department of Mathematics, Princeton University, Princeton, NJ 08544, USA

Current address: Department of Mathematics, University of Maryland, College Park, MD 20742-4015, USA

E-mail address: `koralov@math.umd.edu`

Centre de Recherches Mathématiques
CRM Proceedings and Lecture Notes
Volume **42**, 2007

Hierarchical Anderson Model

Evgenij Kritchevski

Happy birthday Stanislas Molchanov!

1. Introduction

In this article, we will review the spectral localization problem in the hierarchical Anderson model. We will present the original result of Molchanov (see Theorem 1.3 below) on localization at arbitrary disorder in any spectral dimension, when the random perturbations are i.i.d. with a Cauchy distribution, as well as generalization to the case of mixed Cauchy distributions. We will also prove the new result (see Theorem 1.6 below), that in spectral dimension $d \leq 4$, localization holds for very general distributions of the perturbation. This is an improvement of the main theorem of [5]. Our notation will be consistent with [5] and this article should be a complementary reading to [5] and [8]. For additional information about the hierarchical structures and the hierarchical Anderson model we refer the reader to [1, 2, 4–8].

We first recall the definition of the model and its basic properties. Let X be an infinite countable set. Throughout the paper, δ_x will denote the Kronecker delta function at $x \in X$. A partition $\mathcal{P}$ of X is a collection of its disjoint subsets whose union is equal to X. Let $\mathbf{n} = (n_r)_{r\geq 0}$ be a sequence of positive integers and $\mathbf{P} = (\mathcal{P}_r)_{r\geq 0}$ a sequence of partitions of X. The elements of $\mathcal{P}_r$ are called "clusters" of rank r. We say that $(X, \mathbf{P}, \mathbf{n})$ is a hierarchical structure if the following hold:

(1) $n_0 = 1$ and every $Q \in \mathcal{P}_0$ has exactly one element.
(2) For $r \geq 1$, every $Q \in \mathcal{P}_r$ is a disjoint union of n_r clusters in $\mathcal{P}_{r-1}$.
(3) Given $x, y \in X$, there is a cluster Q of some rank containing both x and y.

Let us state some immediate consequences of this definition. Every cluster of rank $r \geq 0$ has size $N_r := \prod_{s=0}^{r} n_s$. Given $x \in X$ and $r \geq 0$, there is a unique cluster of rank r containing x. We denote this cluster by $Q_r(x)$. The map

$$d(x, y) := \min\{r : y \in Q_r(x)\},$$

is a metric on X and $Q_r(x) = \{y : d(x, y) \leq r\}$. Note that $Q_r(x) = Q_r(y)$ whenever $d(x, y) \leq r$. Given an integer $n \geq 2$, a hierarchical structure is called *homogeneous*

2000 *Mathematics Subject Classification.* 47B80, 47A55, 93A13.
This work was supported by FQRNT and ISM grants.
This is the final form of the paper.

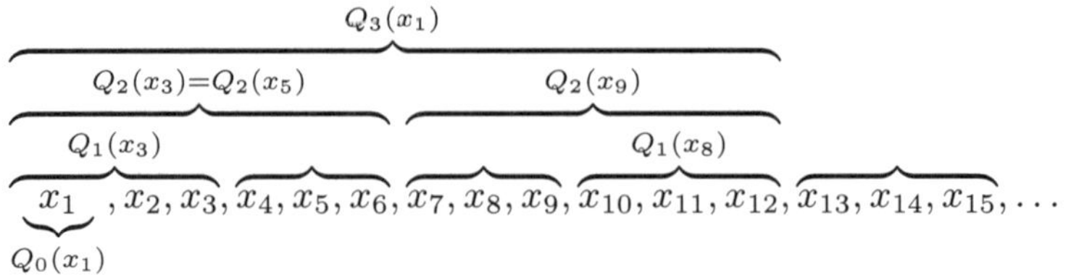

here $d(x_1, x_1) = 0$, $d(x_1, x_3) = 1$, $d(x_1, x_6) = 2$, $d(x_1, x_8) = 3$, etc.

FIGURE 1. A hierarchical structure

of degree n if $n_r = n$ for all $r \geq 1$. Figure 1 illustrates the first three levels of hierarchy in a hierarchical structure with $n_1 = 3$, $n_2 = 2$, $n_3 = 3$.

The free Laplacian on the hierarchical structure $(X, \mathbf{P}, \mathbf{n})$ is defined as follows. For each $r \geq 0$, let $E_r \colon l^2(X) \to l^2(X)$ be the averaging operator

$$(E_r\psi)(x) := \frac{1}{N_r} \sum_{d(x,y)\leq r} \psi(y).$$

Let $\mathbf{p} = (p_r)_{r\geq 1}$ be a sequence of positive numbers such that $\sum_{r=1}^{\infty} p_r = 1$. In the sequel we set $p_0 := 0$ and

$$\lambda_r := \sum_{s=0}^{r} p_s, \qquad r = 0, 1, \dots, \infty.$$

The hierarchical Laplacian Δ on $l^2(X)$ is defined by

$$\Delta := \sum_{r=0}^{\infty} p_r E_r.$$

Clearly, Δ is a bounded self-adjoint operator and $0 \leq \Delta \leq 1$.

A hierarchical model is a hierarchical structure $(X, \mathbf{P}, \mathbf{n})$ together with the hierarchical Laplacian Δ. The basic spectral theorem for Δ is the following.

Theorem 1.1. (1) *The spectrum of* Δ *is given by*

$$\sigma(\Delta) = \{\lambda_r : r = 0, \dots, \infty\}.$$

Each λ_r, $r < \infty$, *is an eigenvalue of* Δ *of infinite multiplicity. The point* $\lambda_\infty = 1$ *is not an eigenvalue.*

(2) $E_r - E_{r+1}$ *is the orthogonal projection onto the eigenspace of* λ_r *and*

$$\Delta = \sum_{r=0}^{\infty} \lambda_r (E_r - E_{r+1}).$$

(3) *For every* $x \in X$, *the spectral measure for* δ_x *and* Δ *is given by*

$$\mu = \sum_{r=0}^{\infty} \left(\frac{1}{N_r} - \frac{1}{N_{r+1}} \right) \delta(\lambda_r),$$

where $\delta(\lambda_r)$ *stands for the Dirac unit mass at* λ_r. *Note that* μ *does not depend on* x.

A proof can be found in [5] or [8].

We now define the hierarchical Anderson model associated to $(X, \mathbf{P}, \mathbf{n})$ and the hierarchical Laplacian Δ. Consider the probability space $(\Omega, \mathcal{F}, \mathbb{P})$ where $\Omega := \mathbb{R}^X$, $\mathcal{F}$ is the usual Borel σ-algebra in Ω, and $\mathbb{P}$ is a given probability measure on $(\Omega, \mathcal{F})$. For $\omega \in \Omega$, we set

$$V_\omega := \sum_{x \in X} \omega(x) \langle \delta_x \mid \cdot \rangle \delta_x.$$

V_ω is a self-adjoint (possibly unbounded) multiplication operator on $l^2(X)$. Let

$$H_\omega := \Delta + V_\omega, \qquad \omega \in \Omega.$$

The family of self-adjoint operators $\{H_\omega\}_{\omega \in \Omega}$ indexed by the events of the probability space $(\Omega, \mathcal{F}, \mathbb{P})$ is called the *hierarchical Anderson model.* The main problem is to understand the spectral properties of H_ω. The first natural question is to describe the spectrum $\sigma(H_\omega)$ of H_ω as a set. Note that, unlike the spectrum of the standard discrete Laplacian on $\mathbb{Z}^d$, the spectrum $\sigma(\Delta)$ of the hierarchical Laplacian is a disconnected set.

Lemma 1.2. *Suppose that the random variables $\{\omega(x) : x \in X\}$ are i.i.d. with distribution ν, i.e., $\mathbb{P} = \bigotimes_{x \in X} \nu$. Let S be the support of ν and $\mathrm{ch}(S)$ the convex hull of S. Then, for $\mathbb{P}$-a.e. $\omega \in \Omega$,*

$$\sigma(\Delta) + S \subset \sigma(H_\omega) \subset \big(\sigma(\Delta) + \mathrm{ch}(S)\big) \cap ([0,1] + S).$$

In particular, if S is connected, then $\sigma(H_\omega) = \sigma(\Delta) + S$, for $\mathbb{P}$-a.e. $\omega \in \Omega$.

The proof of Lemma 1.2 (given in the appendix) is a straight forward adaptation of the proof of the corresponding well-known result for the standard Anderson model on $\mathbb{Z}^d$, where the spectrum is almost surely equal to $[0, 4d] + \mathrm{supp}(\nu)$ (cf., for example, [3, Proposition V.3.5]).

We denote by $\sigma_{\mathrm{ac}}(H_\omega)$ the absolutely continuous part of the spectrum of H_ω and by $\sigma_{\mathrm{cont}}(H_\omega)$ the continuous part. In [8], Molchanov proved the following localization result:

Theorem 1.3. *Suppose that*

$$\sum_{r=1}^{\infty} p_r r^{1+\varepsilon} < \infty, \tag{1.1}$$

for some $\varepsilon > 0$. If the random variables $\{\omega(x) : x \in X\}$ are i.i.d. with a Cauchy distribution,

$$k_{a,h}(e) de := \frac{1}{\pi} \frac{h}{(e-a)^2 + h^2} de,$$

for some parameters $a \in \mathbb{R}$, $h > 0$, then $\sigma_{\mathrm{cont}}(H_\omega) = \varnothing$ for $\mathbb{P}$-a.e. ω.

A proof will be given in Section 3. A generalization of Molchanov's theorem will be stated and proved in Section 4.

Molchanov's theorem is remarkable for several reasons. First of all, it is a result valid for any disorder. Given a coupling constant $c > 0$, the random operator $\Delta + cV_\omega$ can be rewritten as $\Delta + \sum_{x \in X} c\omega(x) \langle \delta_x \mid \cdot \rangle \delta_x$. If $\omega(x)$ has a Cauchy distribution $k_{a,h}$, then $c\omega(x)$ has a Cauchy distribution $k_{ca,ch}$. Hence, under the hypotheses of the theorem, $\Delta + cV_\omega$ has pure point spectrum for $\mathbb{P}$-a.e. ω. The next observation is that the condition (1.1) is very weak, it particular it does not at all depend on n_r. To appreciate the generality of (1.1), one has to look at the density of states and at the spectral dimension of Δ. Let $x_0 \in X$ be given and consider the increasing sequence of clusters $Q_r(x_0)$, $r \geq 0$. Let P_r be the orthogonal projection onto the N_r-dimensional subspace

$$l^2\big(Q_r(x_0)\big) := \{\psi \in l^2(X) : \psi(x) = 0 \text{ for } x \notin Q_r(x_0)\}.$$

Let $e_1^{(r)} \leq e_2^{(r)} \leq \cdots \leq e_{N_r}^{(r)}$ be the eigenvalues of the restricted Laplacian $P_r \Delta P_r$ acting on $l^2\big(Q_r(x_0)\big)$ and

$$\nu_r := \frac{1}{N_r} \sum_{s=1}^{r} \delta(e_s^{(r)}),$$

the corresponding counting measure.

Proposition 1.4. *The weak-* limit* $\lim_{r\to\infty} \nu_r$ *exists and is equal to* μ.

A detailed proof can be found in [5]. In the original paper [8], an analogous result is proved, but with a different definition of ν_r. The content of the proposition is that the spectral measure μ can be naturally interpreted as the integrated density of states of the operator Δ.

If

$$\lim_{t\downarrow 0} \frac{\log \mu([1-t, 1])}{\log t} = d/2,$$

then the number d is called the *spectral dimension* of Δ. This definition is motivated by analogy with the edge asymptotics of the density of states of the standard discrete Laplacian on $\mathbb{Z}^d$, for which the spectral and spatial dimensions coincide.

The relation $\sum_{y\in X} \langle \delta_x \mid \Delta\delta_y \rangle = 1$ yields that Δ generates a random walk on X. We recall that the random walk on $\mathbb{Z}^d$ generated by the standard discrete Laplacian is recurrent if $d = 1, 2$ and transient if $d > 2$. The corresponding result for the hierarchical Laplacian is:

Proposition 1.5. *Consider a homogeneous hierarchical structure of degree* $n \geq 2$. *Suppose that there exist constants* $C_1 > 0$, $C_2 > 0$ *and* $\rho > 1$ *such that*

$$C_1 \rho^{-r} \leq p_r \leq C_2 \rho^{-r},$$

for r *big enough. Then:*

(1) *The spectral dimension of this model is*

$$d(n, \rho) = 2\frac{\log n}{\log \rho}.$$

Hence $0 < d(n, \rho) \leq 2$ *iff* $n \leq \rho$.

(2) *The random walk generated by* Δ *is recurrent if* $0 < d(n, \rho) \leq 2$ *and transient if* $d(n, \rho) > 2$.

See [5, 8] for a proof. We can now see that Theorem 1.3 holds in arbitrary spectral dimension. Indeed, let $n \geq 2$ be given and let d be a positive real number. Then there is a unique $\rho > 1$ with $d = d(n, \rho)$. Set $p_r = C\rho^{-r}$, where C is a normalization constant making $\sum_{r=1}^{\infty} p_r = 1$. Then p_r satisfies (1.1). Therefore, we can construct homogeneous hierarchical Anderson models of arbitrary spectral dimension for which Theorem 1.3 holds.

Since Cauchy random variables play a very special role in the theory of random Schrödinger operators, it is natural to ask whether one can extend Theorem 1.3 to distributions other than Cauchy. A partial answer to this question is that one can prove localization for very general distributions of $\omega(x)$, at any disorder, provided that one imposes stronger decay conditions on p_r than (1.1). We will prove the following new result:

Theorem 1.6. *Assume that there exists a sequence $u_r > 0$ with $\sum_{r=1}^{\infty} u_r^{-1} < \infty$ and*

$$\sum_{r=1}^{\infty} p_r u_r \sqrt{N_r} < \infty. \tag{1.2}$$

Then:

(1) *For all $\omega \in \Omega$, $\sigma_{\mathrm{ac}}(H_\omega) = \varnothing$.*
(2) *If $\mathbb{P}$ is conditionally a.c. then $\sigma_{\mathrm{cont}}(H_\omega) = \varnothing$ for $\mathbb{P}$-a.e. ω.*

A proof will be given in Section 3.

Remark 1.7. The condition (1.2) is more demanding than (1.1). The required decay of p_r imposes an upper bound on the spectral dimension of Δ. Theorem 1.6 and Proposition 1.5 allow us to construct hierarchical models with spectral dimension $d \leq 4$ that exhibit localization at arbitrary disorder. If $(X, \mathbf{P}, \mathbf{n})$ is a homogeneous hierarchical structure of degree $n \geq 2$ and $p_r = C\rho^{-r}$ with $\rho > \sqrt{n}$, then the hypothesis (1.2) is fulfilled for $u_r = r^{1+\varepsilon}$. Given $0 < \mathrm{d} < 4$ one can adjust $\rho > \sqrt{n}$ to make $d(n, \rho) = d$. If $p_r = Cr^{-2-\varepsilon}n^{-r/2}$, then the model has spectral dimension $\mathrm{d} = 4$ and (1.2) is verified for $u_r = r^{1+\varepsilon/2}$. One can also construct trivial models with $d = 0$ by taking p_r to decrease faster than ρ^{-r} for any ρ. We emphasize that homogeneity of the hierarchical structure is not required for Theorem 1.6.

Remark 1.8. Theorem 1.6 is an improvement of the main localization theorem of [5]. In [5], instead of (1.2), it is assumed that $\sum_{r=1}^{\infty} p_r N_{r-1} u_{r-1} u_r < \infty$, which imposes the spectral dimension to be ≤ 2.

2. Localization criterion

In this section we will formulate and prove a sufficient condition for H_ω to have $\sigma_{\mathrm{ac}}(H_\omega) = \varnothing$ for all $\omega \in \Omega$, and a sufficient condition for H_ω to have $\sigma_{\mathrm{cont}}(H_\omega) = \varnothing$ for $\mathbb{P}$-a.e. ω. Both Theorems 1.3 and 1.6 will follow from this localization criterion and will be proved in the next section.

Consider the truncated hierarchical Laplacian

$$\Delta_r := \sum_{s=0}^{r} p_s E_s, \quad r \geq 0,$$

and set

$$H_{\omega,r} := V_\omega + \Delta_r, \quad r \geq 0.$$

Fix $\omega \in \Omega$. Note that for any $Q_r \in \mathcal{P}_r$, the subspace $l^2(Q_r)$ is invariant for $H_{\omega,r}$. This is the main reason for working with the truncated hierarchical Laplacian. Let $\sigma(\omega, Q_r)$ be the set of the eigenvalues of the restricted operator $H_{\omega,r} \restriction l^2(Q_r)$ and $\sigma_\omega := \bigcup \sigma(\omega, Q_r)$ where the union is over all clusters of all ranks. Clearly, σ_ω is a countable subset of $\mathbb{R}$, and hence of zero Lebesgue measure. For $z \in \mathbb{C}\backslash\sigma_\omega$, $r \geq 0$, and $x, y \in X$, we set

$$G_{\omega,r}(x, y; z) := \langle \delta_x \mid (H_{\omega,r} - z)^{-1}\delta_y \rangle.$$

For $z \in \mathbb{C}\backslash\sigma_\omega$, $r \geq 0$ and $t \in X$, let $g_{\omega,r}(t;z)$ be the average of $G_{\omega,r}(\cdot,t;z)$ over the cluster $Q_r(t)$, i.e.,

$$g_{\omega,r}(t;z) := \frac{1}{N_r} \sum_{d(t',t)\leq r} G_{\omega,r}(t',t;z).$$

Since the joint spectral measure for δ_t, $\delta_{t'}$ and $H_{\omega,r}$ is real, $G_{\omega,r}(t',t;z) = G_{\omega,r}(t,t';z)$ and

$$g_{\omega,r}(t;z) = \frac{1}{N_r} \sum_{d(t',t)\leq r} G_{\omega,r}(t,t';z) = \frac{1}{N_r}\langle \delta_t \mid (H_{\omega,r} - z)^{-1}\mathbf{1}_{Q_r(t)}\rangle. \tag{2.1}$$

For $Q_r \in \mathcal{P}_r$, let

$$\gamma_\omega(Q_r;z) := \frac{1}{N_r}\langle \mathbf{1}_{Q_r} \mid (H_{\omega,r} - z)^{-1}\mathbf{1}_{Q_r}\rangle.$$

We denote by m the Lebesgue measure on $\mathbb{R}$. For $x \in X$ and $\omega \in \Omega$, we denote by μ_x^ω the spectral measure for $\Delta + V_\omega$ and δ_x, by $\mu_{x,\mathrm{cont}}^\omega$ the continuous part of μ_x^ω and by $\mu_{x,\mathrm{ac}}^\omega$ the a.c. part. The main spectral localization criterion for the hierarchical Anderson model is the following

Theorem 2.1. *Assume that there exists a sequence $u_r > 0$ with $\sum_{r=1}^\infty u_r^{-1} < \infty$ and*

$$\sum_{r=1}^\infty p_r u_r < \infty. \tag{2.2}$$

Let $x \in X$ and a Borel set $B \subset \mathbb{R}$ be given. Then:

(1) *If for a fixed $\omega \in \Omega$,*

$$\sum_{s=1}^\infty p_s |\gamma_\omega(Q_s(x);e)| < \infty, \tag{2.3}$$

for m-a.e. $e \in B$, then $\mu_{x,\mathrm{ac}}^\omega(B) = 0$.

(2) *If $\mathbb{P}$ has a conditional density along the xth coordinate, and (2.3) holds for $\mathbb{P}\otimes m$-a.e. $(\omega,e) \in \Omega \times B$, then $\mu_{x,\mathrm{cont}}^\omega(B) = 0$ for $\mathbb{P}$-a.e. $\omega \in \Omega$.*

(3) *Assume that for $\mathbb{P}\otimes m$-a.e. $(\omega,e) \in \Omega \times B$, there is a finite constant $C_{\omega,e,x}$ such that*

$$|\gamma_\omega(Q_r(x);e)| \leq C_{\omega,e,x} u_r, \tag{2.4}$$

for all $r \geq 0$. If $\mathbb{P}$ has a conditional density along the xth coordinate, then $\mu_{x,\mathrm{cont}}^\omega(B) = 0$ for $\mathbb{P}$-a.e. $\omega \in \Omega$.

To facilitate the exposition, we will first prove three simple propositions (Propositions 2.2, 2.3, and 2.4). Then we will review the Simon – Wolff localization criterion (Theorem 2.5) and give the proof of Theorem 2.1.

Proposition 2.2. *Let $\omega \in \Omega$, $x, y \in X$, $z \in \mathbb{C}\backslash\sigma_\omega$ and $r \geq 0$ be given. Then*

$$G_{\omega,r}(x,y;z) = G_{\omega,0}(x,y;z) - \sum_{s=d(x,y)}^{r} p_s N_{s-1} g_{\omega,s-1}(x;z) g_{\omega,s}(y;z), \tag{2.5}$$

and

$$g_{\omega,r}(x;z) = \frac{1}{N_r} G_{\omega,0}(x,x;z) \prod_{s=1}^{r} (1 - p_s\gamma_\omega(Q_s(x);z)). \tag{2.6}$$

PROOF. The formula (2.5) holds for $r = 0$ since $p_0 = 0$. For $s \geq 1$, the resolvent identity yields

$$(H_{\omega,s} - z)^{-1}\delta_y - (H_{\omega,s-1} - z)^{-1}\delta_y = -(H_{\omega,s-1} - z)^{-1}p_s E_s(H_{\omega,s} - z)^{-1}\delta_y.$$

Observe that $E_s(H_{\omega,s} - z)^{-1}\delta_y = g_{\omega,s}(y;z)\mathbf{1}_{Q_s(y)}$. Taking $\langle \delta_x \mid \cdot \rangle$ in the above equation yields

$$G_{\omega,s}(x,y;z) - G_{\omega,s-1}(x,y;z) = -p_s g_{\omega,s}(y;z)\langle \delta_x \mid (H_{\omega,s-1} - z)^{-1}\mathbf{1}_{Q_s(y)}\rangle. \tag{2.7}$$

Note that by (2.1),

$$\langle \delta_x \mid (H_{\omega,s-1} - z)^{-1}\mathbf{1}_{Q_s(y)}\rangle = \begin{cases} N_{s-1}g_{\omega,s-1}(x;z), & \text{if } d(x,y) \leq s, \\ 0, & \text{if } d(x,y) > s. \end{cases}$$

The formula (2.5) follows after adding (2.7) for $s = 1, 2, \ldots, r$.

The proof of (2.6) is similar. The resolvent identity yields

$$\langle \delta_x \mid (H_{\omega,r} - z)^{-1}\mathbf{1}_{Q_r(x)}\rangle = \big(1 - p_r\gamma_\omega(Q_r(x);z)\big)\langle \delta_x \mid (H_{\omega,r-1} - z)^{-1}\mathbf{1}_{Q_{r-1}(x)}\rangle,$$

and (2.6) follows after iterating the above formula. □

Proposition 2.3. *Let $u_r > 0$ be a sequence with $\sum_{r=1}^{\infty} u_r^{-1} < \infty$. Let $\omega \in \Omega$ and $x \in X$ be fixed. Then for m-a.e. $e \in \mathbb{R}$, there exists a finite constant C_e such that*

$$\Big(\sum_{d(x,y)\leq s} |g_{\omega,s}(y;e)|^2\Big)^{1/2} \leq C_e u_s, \tag{2.8}$$

for all $s \geq 0$.

PROOF. Since $l^2(Q_r(x))$ is an N_r-dimensional invariant subspace for $H_{\omega,r}$ and since $\|\mathbf{1}_{Q_r(x)}\|_2 = \sqrt{N_r}$, we have by Lemma A.1(see appendix) that for $M_r > 0$,

$$m(\{e \in \mathbb{R} \setminus \sigma_\omega : \|(H_{\omega,r} - e)^{-1}\mathbf{1}_{Q_r(x)}\|_2^2 \geq M_r\}) \leq \frac{4N_r}{\sqrt{M_r}}.$$

Let $M_r := (u_r N_r)^2$. Then $\sum_{r=1}^{\infty} N_r M_r^{-1/2} < \infty$. By the Borel–Cantelli lemma, for m-a.e. $e \in \mathbb{R} \setminus \sigma_\omega$, there exists a finite constant C_e such that

$$\|(H_{\omega,r} - e)^{-1}\mathbf{1}_{Q_r(x)}\|_2 < C_e\sqrt{M_r}, \tag{2.9}$$

for all $r \geq 0$. Observe that

$$\begin{aligned}\Big(\sum_{d(x,y)\leq s} |g_{\omega,s}(y;e)|^2\Big)^{1/2} &= \Big(\sum_{d(x,y)\leq s} \Big|\frac{1}{N_s}\langle \delta_y \mid (H_{\omega,s} - e)^{-1}\mathbf{1}_{Q_s(y)}\rangle\Big|^2\Big)^{1/2} \\ &= \frac{1}{N_s}\Big(\sum_{d(x,y)\leq s} |\langle \delta_y \mid (H_{\omega,s} - e)^{-1}\mathbf{1}_{Q_s(x)}\rangle|^2\Big)^{1/2} \\ &= \frac{1}{N_s}\|(H_{\omega,s} - e)^{-1}\mathbf{1}_{Q_s(x)}\|_2.\end{aligned}$$

Inequality (2.9) yields (2.8). □

The key step in proving Theorem 2.1 is:

Proposition 2.4. *Assume that there exists a sequence* $u_r > 0$ *with* $\sum_{r=1}^{\infty} u_r^{-1} < \infty$ *and* $\sum_{r=1}^{\infty} p_r u_r < \infty$. *Let* $B \subset \mathbb{R}$ *be a Borel set and let* $\omega \in \Omega$, $x \in X$ *be fixed. Suppose that*

$$\sum_{s=1}^{\infty} p_s |\gamma_\omega(Q_s(x); e)| < \infty, \tag{2.10}$$

for m*-a.e.* $e \in B$. *Then, for* m*-a.e.* $e \in B$,

$$\sup_{r \geq 0} \sum_{y \in X} |G_{\omega,r}(x,y;e)|^2 < \infty. \tag{2.11}$$

PROOF. For all $e \in \mathbb{C} \backslash \sigma_\omega$, the representation formula (2.5) and Cauchy–Schwarz inequality yield

$$\Big(\sum_{y \in X} |G_{\omega,r}(x,y;e)|^2\Big)^{1/2} \\ \leq |G_{\omega,0}(x,x;e)| + \sum_{s=1}^{r} p_s N_{s-1} |g_{\omega,s-1}(x;e)| \Big(\sum_{d(x,y) \leq s} |g_{\omega,s}(y;e)|^2\Big)^{1/2}.$$

Hence, by Proposition 2.3, for m-a.e. $e \in \mathbb{R}$,

$$\Big(\sum_{y \in X} |G_{\omega,r}(x,y;e)|^2\Big)^{1/2} \leq |G_{\omega,0}(x,x;e)| + C_e \sum_{s=1}^{r} p_s u_s N_{s-1} |g_{\omega,s-1}(x;e)|. \tag{2.12}$$

Hypothesis (2.10) implies that for m-a.e. $e \in B$, the product $\prod_{s=1}^{\infty}(1 - p_s \gamma_\omega(Q_s(x); e))$ converges. For such e, it follows from (2.6) that there is a finite constant C'_e such that

$$N_s |g_{\omega,s}(x;e)| \leq C'_e, \tag{2.13}$$

for all $s \geq 0$.

Combination of (2.12) with (2.13) yields the estimate

$$\Big(\sum_{y \in X} |G_{\omega,r}(x,y;e)|^2\Big)^{1/2} \leq |G_{\omega,0}(x,x;e)| + C_e C'_e \sum_{s=1}^{r} p_s u_s,$$

and the result follows. □

Before proving Theorem 2.1, let us recall the Simon–Wolff localization criterion. Define the function $G_{\omega,x} \colon \mathbb{R} \to [0, +\infty]$ by

$$G_{\omega,x}(e) := \int_{\mathbb{R}} \frac{\mathrm{d}\mu_x^\omega(\lambda)}{(e-\lambda)^2} = \lim_{\epsilon \downarrow 0} \|(\Delta + V_\omega - e - i\epsilon)^{-1} \delta_x\|^2.$$

By the Theorem of de la Vallée Poussin,

$$\mathrm{d}\mu_{x,\mathrm{ac}}^\omega(e) = \pi^{-1} \Big(\lim_{\epsilon \downarrow 0} \varepsilon \|(\Delta + V_\omega - e - i\epsilon)^{-1} \delta_x\|^2\Big)\, \mathrm{d}e.$$

Hence, if for a fixed $\omega \in \Omega$ we have that $G_{\omega,x}(e) < \infty$ for m-a.e. $e \in \mathbb{R}$, then $\mu_{x,\mathrm{ac}}^\omega = 0$.

The Simon–Wolff localization criterion is summarized in:

Theorem 2.5. *Assume that* $\mathbb{P}$ *has a conditional density along the* x*th coordinate. Let* $B \subset \mathbb{R}$ *be a Borel set such that* $G_{\omega,x}(e) < \infty$ *for* $\mathbb{P} \otimes m$*-a.e.* $(\omega, e) \in \Omega \times B$. *Then* $\mu_{x,\mathrm{cont}}^\omega(B) = 0$ *for* $\mathbb{P}$*-a.e.* $\omega \in \Omega$.

Theorem 2.5 is a well-known consequence of the rank-1 Simon – Wolff theorem [9] and the conditional Fubini's theorem.

Now we can prove the main localization criterion.

PROOF OF THEOREM 2.1. Fix $\omega \in \Omega$ and fix $e \in \mathbb{R}\backslash\sigma_\omega$ for which (2.2) holds. Proposition 2.4 yields that then we also have the bound (2.11). By monotone convergence

$$\int_{\mathbb{R}} \frac{\mathrm{d}\mu_x^\omega(\lambda)}{(e-\lambda)^2} = \lim_{\varepsilon\downarrow 0}\int_{\mathbb{R}} \frac{\mathrm{d}\mu_x^\omega(\lambda)}{(e-\lambda)^2+\varepsilon^2} = \sup_{\varepsilon>0}\int_{\mathbb{R}} \frac{\mathrm{d}\mu_x^\omega(\lambda)}{(e-\lambda)^2+\varepsilon^2}.$$

Since for any $z \in \mathbb{C}\backslash\mathbb{R}$,

$$\lim_{r\to\infty} \|(H_{\omega,r}-z)^{-1}-(H_\omega-z)^{-1}\| = 0,$$

we have that the weak-* limit $\lim_{r\to\infty}\mu_{x,r}^\omega$ equals μ_x^ω, where $\mu_{x,r}^\omega$ is the spectral measure for $H_{\omega,r}$ and δ_x. Therefore

$$\begin{aligned}\int_{\mathbb{R}} \frac{d\mu_x^\omega(\lambda)}{(e-\lambda)^2} &= \sup_{\varepsilon>0}\lim_{r\to\infty}\int_{\mathbb{R}} \frac{\mathrm{d}\mu_x^{\omega,r}(\lambda)}{(e-\lambda)^2+\varepsilon^2} \le \sup_{\varepsilon>0,r\ge 1}\int_{\mathbb{R}} \frac{\mathrm{d}\mu_x^{\omega,r}(\lambda)}{(e-\lambda)^2+\varepsilon^2}\\ &= \sup_{r\ge 1}\int_{\mathbb{R}} \frac{\mathrm{d}\mu_x^{\omega,r}(\lambda)}{(e-\lambda)^2} = \sup_{r\ge 1}\|(H_{\omega,r}-e)^{-1}\delta_x\|^2\\ &= \sup_{r\ge 1}\sum_{y\in X}|G_{\omega,r}(x,y;e)|^2 < \infty.\end{aligned}$$

In the final equality we used the fact that $\{\delta_y : y \in X\}$ is an orthonormal basis for $l^2(X)$. Since $m(\sigma_\omega)=0$ and since the bound (2.11) holds for m-a.e. $e \in \mathbb{R}\setminus\sigma_\omega$, we have that for every fixed $\omega\in\Omega$, $G_{\omega,x}(e)<\infty$ for m-a.e. $e\in\mathbb{R}$. This proves part (1). Part (2) follows from the fact that $G_{\omega,x}(e)<\infty$ for $\mathbb{P}\otimes m$-a.e. $(\omega,e)\in\Omega\times\mathbb{R}$ and the Simon – Wolff criterion. Part (3) is an immediate consequence of part (2) and assumption (2.2). □

3. Proofs of localization theorems

In this section we show how to apply the localization criterion (Theorem 2.1) to prove Theorems 1.6 and 1.3.

PROOF OF THEOREM 1.6. Since $\gamma_\omega(Q_s(x);e)=\sum_{d(x,y)\le s} g_{\omega,s}(y;e)$, Cauchy – Schwarz inequality and Proposition 2.3 yield that for m-a.e. $e\in\mathbb{R}$,

$$|\gamma_\omega(Q_s(x);e)| \le C_e u_s\sqrt{N_s}. \tag{3.1}$$

Hence $\sum_{s\ge 0} p_s|\gamma_\omega(Q_s(x);e)|<\infty$ and the result follows from parts (1) and (2) of Theorem 2.1. □

The proof of Molchanov's Theorem 1.3 requires more work. Let us introduce some notation. It is convenient to form the complex number $v=a+ih\in\mathbb{C}_+$ and to denote by k_v the Cauchy distribution $k_{a,h}$. If ν is a Borel probability measure on the Riemann sphere $\mathbb{S}:=\mathbb{C}\cup\{\infty\}$, and $\tau\colon\mathbb{S}\to\mathbb{S}$ is a Borel measurable map, then $\tau\mu$ will denote the induced Borel probability measure on $\mathbb{S}$:

$$(\tau\nu)(B) := \nu\big(\tau^{-1}(B)\big),$$

for Borel sets $B \subset \mathbb{S}$. If $n \geq 2$ is a integer, and ν is a Borel probability measure on $\mathbb{S}$, we let

$$A_n\nu := \tau(\underbrace{\nu \star \ldots \star \nu}_{n \text{ times}}),$$

where $\tau(z) = z/n$, and $\star$ is the usual convolution of measures. Hence, if $Y_1, \ldots Y_n$ are i.i.d. random variables on $\mathbb{S}$ with distribution ν, then $A_n\nu$ is the distribution of $(Y_1 + \cdots + Y_n)/n$. The following proposition, whose proof is left an an easy exercise for the reader, summarizes some of the well-known classical facts about Cauchy distributions.

Proposition 3.1. *Cauchy distributions have the following properties:*

(1) *If $v_1, v_1 \in \mathbb{C}_+$, then*

$$k_{v_1} \star k_{v_2} = k_{v_1+v_2}.$$

(2) *If $\tau(z) = -\bar{z}$ or if $\tau(z) = (az+b)/(cz+d)$, where $a, b, c, d \in \mathbb{R}$, $ad-bc > 0$, then*

$$\tau k_v = k_{\tau v},$$

for all $v \in \mathbb{C}_+$.

(3) *For all $v \in \mathbb{C}_+$ and integers $n \geq 2$,*

$$A_n k_v = k_v.$$

We will use fractional linear transformations of the special form

$$\tau_p(z) = \frac{z}{1+pz},$$

where $p \geq 0$. Note that $\tau_{p+p'} = \tau_p \circ \tau_{p'}$. The transformations τ_p are important in the hierarchical Anderson model because of the following recursive relation, proved by Molchanov in [8].

Proposition 3.2. *Let $\omega \in \Omega$ and $z \in \mathbb{C} \setminus \sigma_\omega$ be given. Then for all $x \in X$,*

$$\gamma_\omega(Q_0(x); z) = \frac{1}{\omega(x) - z}. \tag{3.2}$$

Moreover, for all $r \geq 1$ and $Q_r \in \mathcal{P}_r$, we have

$$\gamma_\omega(Q_r; z) = \tau_{p_r}\Bigg(\frac{1}{n_r} \sum_{Q^j_{r-1} \subset Q_r} \gamma_\omega(Q^j_{r-1}; z)\Bigg), \tag{3.3}$$

where the sum contains n_r terms, one for each rank $r-1$ subcluster Q^j_{r-1} of Q_r.

PROOF. The formula (3.2) is clear since $G_{\omega,0}(x, y; z) := (\omega(x) - z)^1 \langle \delta_x \mid \delta_y \rangle$. For $r \geq 1$, the resolvent identity yields

$$(H_{\omega,r} - z)^{-1}\mathbf{1}_{Q_r} - (H_{\omega,r-1} - z)^{-1}\mathbf{1}_{Q_r} = -p_r\gamma_\omega(Q_r; z)(H_{\omega,r-1} - z)^{-1}\mathbf{1}_{Q_r}.$$

Taking $\langle \mathbf{1}_{Q_r} \mid \cdot \rangle$ in the above equation yields

$$N_r\gamma_\omega(Q_r; z) - \sum_{Q^j_{r-1} \subset Q_r} N_{r-1}\gamma_\omega(Q^j_{r-1}; z) = -p_r\gamma_\omega(Q_r; z) \sum_{Q^j_{r-1} \subset Q_r} N_{r-1}\gamma_\omega(Q^j_{r-1}; z).$$

The formula (3.3) follows after dividing by N_r in the above equation and then solving for $\gamma_\omega(Q_r; z)$. ☐

PROOF OF THEOREM 1.3. Since for every fixed ω, $m(\sigma_\omega) = 0$, we have that for $\mathbb{P}\otimes m$-a.e. $(\omega, e) \in \Omega \times \mathbb{R}$, $e \notin \sigma_w$. Hence, there is a Borel measurable set $E \subset \mathbb{R}$ with the properties

(1) $m(\mathbb{R}\backslash E) = 0$, and
(2) for every fixed $e \in E$, there is a set $\Omega_e \in \mathcal{F}$ with $\mathbb{P}(\Omega_e) = 1$ and such that $e \notin \sigma_w$ for all $\omega \in \Omega_e$.

Let us fix $e \in E$. Then for all $\omega \in \Omega_e$, and hence for $\mathbb{P}$-a.e. $\omega \in \Omega$, all the representation formulas in Propositions 2.2 and 3.2 are valid. It follows from (3.2) and from Proposition 3.1 that $\{\gamma_\omega(Q_0(x); e) : x \in X\}$ are i.i.d. Cauchy random variables with distribution k_{v_0}, where $v_0(e) = -(e - a + ih)^{-1}$. Moreover, (3.3) and Proposition 3.1 yield that for $r \geq 1$, $\{\gamma_\omega(Q_r; e) : Q_r \in P_r\}$ are i.i.d. Cauchy random variables with distribution k_{v_r}, where $v_r(e) = \tau_{\lambda_r} v_0(e)$. Since $\lambda_r \to \lambda_\infty = 1$ as $r \to \infty$, the closure of the orbit

$$V(e) := \bigcup_{r\geq 0}\{v_r(e)\},$$

is equal to $V(e) \cup \{\tau_1 v_0(e)\}$, a compact set in $\mathbb{C}_+$. Since $\sup_{r\geq 0}|v_r(e)| < \infty$, there is a constant $K(e) < \infty$, such that

$$\mathbb{P}(|\gamma_\omega(Q_r; e)| > u) \leq \frac{K(e)}{u},$$

for all real $u > 0$, integer $r \geq 0$ and $Q_r \in P_r$. Let us now fix $x \in X$ and take $u_r = r^{1+\varepsilon}$. By the Borel-Cantelli lemma, for $\mathbb{P}$-a.e. $\omega \in \Omega_e$, there exists a finite constant $L_\omega(e)$ such that

$$|\gamma_\omega(Q_r(x); e)| \leq L_\omega(e) u_r, \tag{3.4}$$

for all $r \geq 0$. Part (3) of Theorem 2.1 yields that $\mu^\omega_{x,\text{cont}} = 0$ for $\mathbb{P}$-a.e. $\omega \in \Omega$. Since $x \in X$ is arbitrary, the result follows. □

Note that, in the proof given above, (3.4) is a significant improvement of (3.1) since the factor $\sqrt{N_r}$ is no longer present.

4. A generalization of Molchanov's theorem

In this section we generalize Theorem 1.3 to convex combinations of Cauchy distributions. Let $\mathcal{M}$ denote the set of Borel probability measures on $\mathbb{C}_+$. Given $\nu \in \mathcal{M}$ and a Borel set $B \subset \mathbb{R}$, we set

$$k_\nu(B) := \int k_v(B) d\nu(v).$$

Then k_ν is a Borel probability measure on $\mathbb{R}$ and for every bounded Borel function $f \colon \mathbb{R} \to \mathbb{C}$,

$$\int_{\mathbb{R}} f \, \mathrm{d}k_\nu = \int \left(\int_{\mathbb{R}} f \, \mathrm{d}k_v \right) \mathrm{d}\nu(v).$$

We call k_ν a *mixed Cauchy distribution.* Note that the usual Cauchy distribution k_{a+ih} is a special case of the above general defintion: $k_{a+ih} = k_\nu$ where $\nu = \delta(a+ih)$. Let S be the set of mixed Cauchy distributions k_ν, such that the distance from $\mathbb{R}$ to the support of ν is strictly positive.

Theorem 4.1. *Assume that there exists a sequence $u_r > 0$ with $\sum_{r=1}^{\infty} u_r^{-1} < \infty$ and*

$$\sum_{r=1}^{\infty} p_r u_r < \infty.$$

If the random variables $\{\omega(x) : x \in X\}$ are i.i.d. with a mixed Cauchy distribution $k_\nu \in S$, then $\sigma_{\mathrm{cont}}(H_\omega) = \varnothing$ for $\mathbb{P}$-a.e. ω.

Remark 4.2. Theorem 4.1 is valid at arbitrary disorder. Indeed, if $\omega(x)$ has a distribution $k_\nu \in S$, and $c > 0$, then $c\omega$ has again a distribution of the form $k_{\nu'} \in S$. Precisely, $\nu' = \tau\nu$, where $\tau(z) = cz$.

Remark 4.3. If f is any compactly supported continuous probability density on $\mathbb{R}$, and $\varepsilon > 0$, then there is a mixed Cauchy distribution $k_\nu \in S$ s.t.

$$\sup_{e\in\mathbb{R}} |f(e) - \mathrm{d}k_\nu/\mathrm{d}e| < \varepsilon.$$

To see this, recall that the harmonic function $F(v) = \int_{\mathbb{R}} f(e)\,\mathrm{d}k_v(e)$, $v \in \mathbb{C}_+$, converges uniformly to f as $\Im v \downarrow 0$, and note that the integral can be approximated by finite Riemann sums.

The proof of Theorem 4.1 parallels that of Molchanov's Theorem 1.3. First, we need a generalization of Proposition 3.1 to the case of mixed Cauchy distributions.

Proposition 4.4. *Mixed Cauchy distributions have the following properties:*

(1) *If $\nu_1, \nu_1 \in \mathcal{M}$, then*

$$k_{\nu_1} \star k_{\nu_2} = k_{\nu_1 \star \nu_2}.$$

(2) *If $\tau(z) = -\bar{z}$ or if $\tau(z) = (az+b)/(cz+d)$, where $a, b, c, d \in \mathbb{R}$, $ad - bc > 0$, then*

$$\tau k_\nu = k_{\tau\nu},$$

for all $\nu \in \mathcal{M}$.

(3) *For all $\nu \in \mathcal{M}$ and integers $n \geq 2$,*

$$A_n k_\nu = k_{A_n \nu}.$$

The proof is a straightforward consequence of the definitions together with Proposition 3.1 and Fubini's theorem.

Proof of Theorem 1.3. Let E be the set of full Lebesgue measure as in the beginning of the proof of Theorem 1.3, and let $e \in E$ be fixed. It follows from (3.2) and from Proposition 4.4 that $\{\gamma_\omega(Q_0(x); e) : x \in X\}$ are i.i.d. mixed Cauchy random variables with distribution k_{ν_0}, where $\nu_0 = \tau_0\nu$, $\tau_0(z) = -(e - \bar{z})^{-1}$. Since the distance from the support of ν to $\mathbb{R}$ is strictly positive, there is a closed disk $D_0 \subset \mathbb{C}_+$ such that $\mathrm{supp}(\nu_0) \subset D_0$. The relation (3.3) and Proposition 4.4 yield that for $r \geq 1$, $\{\gamma_\omega(Q_r; e) : Q_r \in P_r\}$ are i.i.d. mixed Cauchy random variables with distribution k_{ν_r}, where

$$\nu_r = \tau_{p_r} A_{n_r} \nu_{r-1}.$$

For $r \geq 0$, let $D_r := \tau_{\lambda_r} D_0$. Since τ_{λ_r} is a fractional linear transformation taking $\mathbb{C}_+$ onto itself, D_r is again a closed disk in $\mathbb{C}_+$. We claim that for all $r \geq 0$,

$$\mathrm{supp}(\nu_r) \subset D_r. \tag{4.1}$$

The proof is by induction. By construction of D_0, (4.1) holds for $r = 0$. Let $r \geq 1$ and suppose (4.1) holds for $r-1$. Since $\text{supp}(\nu_{r-1}) \subset D_{r-1}$, and D_{r-1} is convex, we have that $\text{supp}(A_{n_r}\nu_{r-1}) \subset D_{r-1}$. Hence $\text{support}(\tau_{p_r}A_{n_r}\nu_{r-1}) \subset \tau_{p_r}D_{r-1} = D_r$, and (4.1) holds for r, proving the claim.

Let D be the closure of $\bigcup_{r\geq 0} D_r$. D is a compact subset of $\mathbb{C}_+$ and for all $r \geq 0$, $\text{supp}(\nu_r) \subset D$. It follows that there is a constant $K(e) < \infty$, such that

$$\mathbb{P}(|\gamma_\omega(Q_r;e)| > u) \leq \frac{K(e)}{u},$$

for all real $u > 0$, integer $r \geq 0$ and $Q_r \in P_r$. The rest of the proof is the same as in Theorem 1.3, with the more general sequence u_r instead of $r^{1+\varepsilon}$. □

Appendix A

PROOF OF LEMMA 1.2. Let $e \in S$. Given a cluster Q_r or rank r and given $\varepsilon > 0$, consider the event

$$B(Q_r, \varepsilon) := \{\omega : \max_{x\in Q_r} |\omega(x) - e| < \varepsilon\}.$$

Since $\nu\big((e-\varepsilon, e+\varepsilon)\big) > 0$ and the random variables $\{\omega(x) : x \in Q_r\}$ are independent, we have that

$$\mathbb{P}\big(B(Q_r,\varepsilon)\big) = \Big(\nu\big((e-\varepsilon, e+\varepsilon)\big)\Big)^{N_r} > 0.$$

According to the hierarchical structure, X is a countable union of disjoint clusters of rank r: $X = \bigcup_{j=1}^{\infty} Q_{r,j}$. Let $B_{r,\varepsilon} := \limsup_j B(Q_{r,j},\varepsilon)$. Since the events $\{B(Q_{r,j},\varepsilon), j = 1,\dots,\infty\}$ are independent and $\sum_{j=1}^{\infty} \mathbb{P}\big(B(Q_{r,j},\varepsilon)\big) = \infty$, the second Borel–Cantelli Lemma yields that $\mathbb{P}(B_{r,\varepsilon}) = 1$. Letting $B := \bigcap_{r\geq 0} B_{r,1/(r+1)}$, we also have $\mathbb{P}(B) = 1$. Hence, for $\omega \in B$, there is a sequence $(Q_{r,j_r} : r \geq 0)$ of clusters of increasing rank with the property that

$$\max_{x\in Q_{r,j_r}} |\omega(x) - e| < \frac{1}{r+1}, \tag{A.1}$$

for all $r \geq 0$. Let $\lambda \in \sigma(\Delta)$. Then $\lambda = \lambda_s$ for some $s \in \{1,\dots,\infty\}$. For $r \geq 0$, we let $q_r = \min(r,s)$. It is easy to see that for each $r \geq 0$, there exists a unit eigenvector ψ_r of Δ corresponding to the eigenvalue λ_{q_r}, and such that $\psi_r(x) = 0$ for $x \notin Q_{r,j_r}$. Indeed, take a unit vector ψ_r constant on every rank q_r subcluster of Q_{r,j_r}, zero outside Q_{r,j_r} and such that $E_{q_r+1}\psi_r = 0$. Then

$$\lim_{r\to\infty} \|\Delta\psi_r - \lambda_s\psi_r\| = 0,$$

and, by (A.1),

$$\lim_{r\to\infty} \|H_\omega\psi_r - (\lambda_s + e)\psi_r\| = 0,$$

for $\omega \in B$. Hence, ψ_r is a Weil sequence for the operator H_ω and $\lambda_s + e \in \sigma(H_\omega)$. We have shown that for $\mathbb{P}$-a.e. $\omega \in \Omega$, $\sigma(\Delta) + S \subset \sigma(H_\omega)$. The inclusion $\sigma(H_\omega) \subset \big(\sigma(\Delta) + \text{ch}(S)\big) \cap ([0,1] + S)$ follows from general facts about the spectrum of a sum of two self-adjoint operators. □

Lemma A.1. *Let A be a Hermitian $N \times N$ matrix and $v \in \mathbb{C}^N$. Then for all $M > 0$*

$$m(\{e : \|(A-e)^{-1}v\|_2^2 \geq M\}) \leq 4\sqrt{\frac{N}{M}}\|v\|_2.$$

PROOF (Following [8]). Let $\lambda_1, \dots, \lambda_N$ be the eigenvalues of A and $\psi_1, \dots, \psi_N$ the corresponding orthonormal basis of eigenvectors. In this basis, we have $v = \sum_{i=1}^N v_i\psi_i$, $\|v\|_2^2 = \sum_{i=1}^N |v_i|^2$ and $(A-e)^{-1}v = \sum_{i=1}^N (\lambda_i - e)^{-1} v_i \psi_i$. For each $i = 1, \dots, N$, consider the open interval $\Delta_i = (\lambda_i - v_i/\sqrt{M}, \lambda_i + v_i/\sqrt{M})$. The total length of these N intervals is

$$m\Big(\bigcup_{i=1}^N \Delta_i\Big) \leq \sum_{i=1}^N m(\Delta_i) = \frac{2}{\sqrt{M}} \sum_{i=1}^N |v_i| \leq \frac{2}{\sqrt{M}} \sqrt{N} \|v\|_2.$$

Chebyshev inequality yields

$$\begin{aligned} m\Big(\Big\{e \in \mathbb{R} \setminus \bigcup_{i=1}^N \Delta_i : \|(A-e)^{-1}v\|_2^2 \geq M\Big\}\Big) \\ &\leq \frac{1}{M} \int_{\mathbb{R}\setminus\bigcup_{i=1}^N \Delta_i} \|(A-e)^{-1}v\|_2^2 \mathrm{d}e = \frac{1}{M} \int_{\mathbb{R}\setminus\bigcup_{i=1}^N \Delta_i} \sum_{i=1}^N \Big|\frac{v_i}{\lambda_i - e}\Big|^2 \mathrm{d}e \\ &\leq \frac{1}{M} \sum_{i=1}^N \int_{\mathbb{R}\setminus\Delta_i} \left|\frac{v_i}{\lambda_i - e}\right|^2 \mathrm{d}e = \frac{1}{M} \sum_{i=1}^N 2 \int_{|v_i|/\sqrt{M}}^{\infty} \frac{|v_i|^2}{t^2} \mathrm{d}t \\ &= \frac{2}{M} \sqrt{M} \sum_{i=1}^N |v_i| \leq \frac{2}{\sqrt{M}} \sqrt{N} \|v\|_2, \end{aligned}$$

and the result follows. □

Acknowledgments. We are grateful to Vojkan Jakšić for suggesting this research project and for giving us constructive feedback. We would also like to thank Stanislav Molchanov for very helpful discussions and enthusiastic encouragements.

References

1. P. M. Bleher and Ja. G. Sinai, *Investigation of the critical point in models of the type of Dyson's hierarchical models*, Comm. Math. Phys. **33** (1973), no. 1, 23–42.
2. A. Bovier, *The density of states in the Anderson model at weak disorder: a renormalization group analysis of the hierarchical model*, J. Statist. Phys. **59** (1990), no. 3-4, 745–779.
3. R. Carmona and J. Lacroix, *Spectral theory of random Schrödinger operators*, Probab. Appl., Birkhäuser, Boston, MA, 1990.
4. F. J. Dyson, *Existence of a phase-transition in a one-dimensional Ising ferromagnet*, Comm. Math. Phys. **12** (1969), 91–107.
5. E. Kritchevski, *Spectral localization in the hierarchical Anderson model*, Proc. Amer. Math. Soc., to appear.
6. ———, Ph.D. thesis, McGill University, in preparation.
7. S. Molchanov, *Lectures on random media*, Lectures on Probability Theory (Saint-Flour, 1992) (P. Bernard, ed.), Lectures Notes in Math., vol. 1581, Springer, New York, 1994, pp. 242–411.
8. ———, *Hierarchical random matrices and operators. Application to Anderson model*, Multidimensional Statistical Analysis and Theory of Random Matrices (Bowling Green, OH, 1996), VSP, Utrecht, 1996, pp. 179–194.
9. B. Simon and T. Wolff, *Singular continuum spectrum under rank one perturbations and localization for random Hamiltonians*, Comm. Pure Appl. Math. **39** (1986), no. 1, 75–90.

DEPARTMENT OF MATHEMATICS AND STATISTICS, MCGILL UNIVERSITY, 805 SHERBROOKE STREET WEST, MONTREAL, QC H3A 2K6, CANADA

E-mail address: ekritc@math.mcgill.ca

Centre de Recherches Mathématiques
CRM Proceedings and Lecture Notes
Volume **42**, 2007

Integral Representations of Solutions of Periodic Elliptic Equations

Peter Kuchment

ABSTRACT. The paper discusses relations between the structure of the complex Fermi surface below the spectrum of a second order periodic elliptic equation and integral representations of certain classes of its solutions. These integral representations are analogs of those previously obtained by S. Agmon, S. Helgason, and other authors for solutions of the Helmholtz equation (i.e., for generalized eigenfunctions of Laplace operator). In a previous joint work with Y. Pinchover we described all solutions that can be represented as integrals of positive Bloch solutions over the imaginary Fermi surface, with a hyperfunction as a "measure." Here we characterize the class of solutions such that the corresponding hyperfunction is a distribution on the Fermi surface.

1. Introduction

This paper is devoted to integral representations of solutions of second order elliptic periodic differential equations. These representations are analogs of those for solutions of the Helmholtz equation in $\mathbb{R}^n$

$$-\Delta u - k^2 u = 0 \quad \text{in } \mathbb{R}^n, \tag{1.1}$$

where $k \in \mathbb{C}^* := \mathbb{C} \setminus \{0\}$. Such representations have been obtained by S. Helgason [15,16], M. Hashizume et al. [14], M. Morimoto [30], and S. Agmon [2,3]. In these results, solutions were expanded into exponential ones

$$e_\xi(x) := \exp(\mathrm{i}\xi \cdot x). \tag{1.2}$$

Here

$$\xi \in S := \{\xi \in \mathbb{C}^n \mid \xi = k\omega, \omega \in S^{n-1} \subset \mathbb{R}^n\}, \tag{1.3}$$

2000 *Mathematics Subject Classification.* Primary 35B05, 35C15, 58J15; Secondary 35J15, 35P05, 58J50.

Key words and phrases. Elliptic operator, spectrum, Floquet theory, integral representation, periodic operator.

The work of the author was partially supported by the NSF Grant DMS 0406022 and by Grant No. 1999208 from the United States–Israel Binational Science Foundation (BSF). The author expresses his gratitude to the NSF and BSF for this support. The content of this paper does not necessarily reflect the position or the policy of the federal government of the USA, and no official endorsement should be inferred.

This is the final form of the paper.

and $\xi \cdot x = \sum_{j=1}^{n} \xi_j x_j$. These expansions can be written as

$$u(x) = \int_S e_\xi(x)\, \mathrm{d}\phi(\xi) = \langle \phi(\xi), e_\xi(x) \rangle, \tag{1.4}$$

where $\phi(\xi)$ is a functional on the sphere S. In particular, it was understood what classes of solutions correspond to different classes of functionals (e.g., hyperfunctions, distributions, measures) [2, 3].

Such representations are related to the L. Ehrenpreis' *fundamental principle* [11,32] for constant coefficient operators, which in the particular case of (1.1) claims that any solution of (1.1) can be represented as an integral with respect to the parameter ξ of the exponential solutions

$$e_\xi(x) := \exp(\mathrm{i}\xi \cdot x) \quad \xi \in \Sigma. \tag{1.5}$$

Here

$$\Sigma := \{\xi \in \mathbb{C}^n \mid \xi^2 = k^2\},$$

is the characteristic variety of the operator in the left-hand side of (1.1) (see the details and more precise formulation in [11,32]). The representation (1.5) is highly non-unique, due to existence of functionals orthogonal to analytic functions on Σ. On the other hand, Σ is an analytic subset of $\mathbb{C}^n$, uniquely determined for $k \neq 0$ by its spherical subset S. Thus, one can expect the possibility of a unique representation like the one in (1.4). It is crucial here that Σ is irreducible and that S is sufficiently massive, so S determines Σ uniquely (otherwise it would not be possible to obtain the representation of all solutions using only $\xi \in S$). Moreover, S is a rather simple analytic manifold. This enables one to obtain rather explicit descriptions of the needed spaces of test functions and functionals.

In this paper, we consider the case of second order periodic elliptic equations (see the exact description of the class of equations in the next section). For such (and more general) periodic equations, an analog of the "fundamental principle" was obtained in [19, 33] for solutions with some growth restrictions. Here, instead of exponential solutions one needs to use the so called *Floquet – Bloch* solutions. The analog of the characteristic manifold Σ is the Fermi surface F (see [4, 19, 35] and Definitions 2.1 and 2.2 below for these notions). However, finding an analog of (1.4) for the periodic case is far from being straightforward. In particular, one wonders what should be the analog of the sphere S. It is natural to expect that when zero belongs to the spectrum of the operator, one might try to use the *real part* of the Fermi surface, while if zero does not belong to the spectrum, the *imaginary part* might be appropriate. Due to the complicated structure of the Fermi surface, this idea is not easy to implement. As it was shown in the joint work [21] with Y. Pinchover, for second order elliptic equations with *positive* generalized principal eigenvalue Λ_0 (see (2.9)), an appropriate variety is provided by the analysis of the cone of positive solutions done by S. Agmon and by V. Lin and Y. Pinchover [1, 19, 26]. No results of this type are known so far above Λ_0. Another difficulty is in proving irreducibility of the Fermi surface F, which happens to be a very hard problem (it also arises in many other spectral considerations [6, 13, 18, 20, 23–25]). Fortunately, as the reader can see from this paper and from [21], by appropriately restricting the growth of the solutions, we manage to work near a single irreducible component of F and hence avoid proving the irreducibility of F. Consequently, we prove a representation theorem (Theorem 3.1) that characterizes classes of the solutions that have integral expansion analogous to (1.4) into *positive* Bloch solutions with hyperfunctions or distributions as "measures." The hyperfunction case was investigated in [21] and is presented here without a proof. The distribution result that we could not obtain in [21] is new and is proved in the present paper. In order to prove it, additional analytic techniques need to be involved, in particular results on peak sets in A^∞ functional algebras in several complex variables [8, 9]. We are grateful to A. Tumanov for pointing us to the relevant literature.

The proofs of the results of this paper are based upon the techniques of the Floquet theory developed in [19] (the reader can find all necessary preliminary information in the next section). The methods that were used to prove the "fundamental principle" [11, 33] provide a crucial technical approach. In particular, solutions of the equation are treated in the dual sense, i.e., as functionals on appropriate function spaces that are orthogonal to the range of the dual operator.

The outline of the paper is as follows. The next section introduces necessary notations and preliminary results from the Floquet theory and the theory of positive solutions of periodic elliptic equations. It mostly (but not entirely) repeats the corresponding sections from [21, 22] and is included for the reader's convenience. Section 3 contains the proof of the integral representation (Theorem 3.1) that describes the sets of solutions allowing integral representations with distributional and hyperfunction "measures." The last section contains acknowledgments.

2. Notations and preliminary results

Due to the nature of this section, most of it repeats some parts of [19, 21, 22]. We regret the necessity of doing this, but otherwise reading the rest of the paper would probably become impossible without constant referring to [21].

In this paper we consider second order elliptic operators on $\mathbb{R}^n$ with *real* periodic coefficients of the form

$$P(x, \partial) = -\sum_{i,j=1}^{n} a_{ij}(x)\partial_i\partial_j + \sum_{i=1}^{n} b_i(x)\partial_i + c(x). \tag{2.1}$$

It is assumed that the uniform ellipticity condition

$$\sum_{i,j=1}^{n} a_{ij}(x)\zeta_i\zeta_j \geq a\sum_{i=1}^{n} \zeta_i^2$$

is satisfied for all $x, \zeta \in \mathbb{R}^n$, where a is a positive constant. In the notation $P(x, D)$ we used the standard convention $D = -\mathrm{i}\partial/\partial x$.

We will assume sufficient smoothness of the coefficients, namely that $a_{ij} \in C^2(\mathbb{R}^n)$, $b_i \in C^1(\mathbb{R}^n)$ and $c \in C(\mathbb{R}^n)$. In fact, it is sufficient to assume that both the operator P and its dual (the formal adjoint) P^* have Hölder continuous coefficients.[1] Here the duality is provided by the bilinear (rather than the sesquilinear) form

$$\langle g, f \rangle := \int_{\mathbb{R}^n} f(x)g(x)\,\mathrm{d}x.$$

So, the dual operator P^* has similar properties to the ones of P.

The coefficients of P are assumed to be periodic with respect to a lattice Γ in $\mathbb{R}^n$. In what follows, the particular choice of the lattice is irrelevant and can always

[1] See [21, Section 6]. We only need that both operators P and P^* define Fredholm mappings between the Sobolev space $H^2(\mathbb{T}^n)$ and $L_2(\mathbb{T}^n)$, where $\mathbb{T}^n = \mathbb{R}^n/\Gamma$.

be reduced by change of variables to the case $\Gamma = \mathbb{Z}^n$, which we will assume from now on. We will always use the word "periodic" in the meaning of "Γ-periodic."

We now introduce some standard notions and results from Floquet theory of periodic PDEs [4, 10, 19, 21, 22, 35].

We denote by $K = [0,1]^n$ the standard fundamental domain (the *Wigner–Seitz cell*) of the lattice $\Gamma = \mathbb{Z}^n$, and by $B = [-\pi, \pi]^n$ the *first Brillouin zone*, which is a fundamental domain of the reciprocal (dual) lattice $\Gamma^* = (2\pi\mathbb{Z})^n$. We naturally identify Γ-periodic functions with functions on $\mathbb{T}^n = \mathbb{R}^n/\Gamma$.

Definition 2.1. A function $u(x)$ representable as a finite sum of the form

$$u(x) = e^{ik\cdot x}\Big(\sum_{j=(j_1,\ldots,j_n)\in\mathbb{Z}^n_+} x^j p_j(x)\Big) \tag{2.2}$$

with nonzero Γ-periodic functions $p_j(x)$ is called a *Floquet function with a quasi-momentum* $k \in \mathbb{C}^n$. Here $x^j = x_1^{j_1} x_2^{j_2} \cdots x_n^{j_n}$. The maximum value of $|j| = \sum_{l=1}^n j_l$ in the representation (2.2) is said to be the *order* of the Floquet function. Floquet functions of zero order are called *Bloch functions*.

The set introduced below plays in the periodic case the role of the characteristic variety Σ for constant coefficient operators.

Definition 2.2. The (complex) *Fermi surface* F_P of the operator P (at the zero energy level) consists of all vectors $k \in \mathbb{C}^n$ (called *quasimomenta*) such that the equation $Pu = 0$ has a nonzero Bloch solution $u(x) = e^{ik\cdot x}p(x)$, where $p(x)$ is a Γ-periodic function.

Introducing a spectral parameter λ, one arrives at the notion of the *Bloch variety*:

Definition 2.3. The (complex) *Bloch variety* B_P of the operator P consists of all pairs $(k, \lambda) \in \mathbb{C}^{n+1}$ such that the equation $Pu = \lambda u$ has a nonzero Bloch solution $u(x) = e^{ik\cdot x}p(x)$ with the quasimomentum k.

The Bloch variety B_P can be treated as the graph of a multivalued function $\lambda(k)$ (so called *dispersion relation*) that assigns to any quasimomentum k the spectrum of the operator $P(x, D+k)$ on the torus $\mathbb{T}^n$. Since for operators of the type (2.1), these spectra are known to be discrete (as in particular the discussion below will show), we can single out continuous branches λ_j of $\lambda(k)$. These branches are usually called the *band functions* (see [4, 19, 35]). The Fermi surfaces now become the level sets of the dispersion relation.

Lemma 2.4 ([19, Theorems 3.1.5, 3.1.7, and 4.4.2]). (1) *The Fermi and Bloch varieties are the sets of all zeros of entire functions of a finite order in* $\mathbb{C}^n$ *and* $\mathbb{C}^{n+1}$, *respectively.*

(2) *A quasimomentum* k *belongs to* F_{P^*} *if and only if* $-k \in F_P$. *Analogously,* $(k,\lambda) \in B_{P^*}$ *if and only if* $(-k, \lambda) \in B_P$. *In other words, the dispersion relations* $\lambda(k)$ *and* $\lambda^*(k)$ *for the operators* P *and* P^* *are related as follows:*

$$\lambda^*(k) = \lambda(-k). \tag{2.3}$$

The Fermi surface F_P is periodic with respect to the reciprocal lattice $\Gamma^* = (2\pi\mathbb{Z})^n$. It is often convenient to factor out this periodicity by considering the (analytic) exponential mapping $\rho\colon \mathbb{C}^n \to (\mathbb{C}^*)^n$, where

$$z = \rho(k) = \rho(k_1, \ldots, k_n) = (\exp ik_1, \ldots, \exp ik_n).$$

This mapping can be identified with the quotient map $\mathbb{C}^n \to \mathbb{C}^n/\Gamma^*$. We also introduce the complex torus

$$T := \rho(\mathbb{R}^n) = \{z \in \mathbb{C}^n \mid |z_j| = 1,\, j = 1, 2, \dots, n\}. \tag{2.4}$$

Definition 2.5. We call the image $\Phi_P := \rho(F_P)$ under the mapping ρ of the Fermi surface F_P the *Floquet surface* of the operator P.

In the Floquet theory for PDEs, this Floquet surface is the set of all Floquet multipliers of Floquet – Bloch solutions of the equation $Pu = 0$.

The following analog $\mathcal{U}$ of the Fourier transform (see [19, Section 2.2; 35]), which we will call the *Floquet transform*,[2] is the main tool in the Floquet theory:

$$f(x) \to \mathcal{U}f(z,x) := \sum_{\gamma\in\Gamma} f(x-\gamma) z^{\gamma} \quad z \in (\mathbb{C}^*)^n. \tag{2.5}$$

It is often convenient to use for the Floquet transform $\mathcal{U}$ the quasimomentum coordinate k instead of the multiplier $z = \rho(k)$.

For a point $z \in (\mathbb{C}^*)^n$, we denote by $E_{m,z}$ the closed subspace of the Sobolev space $H^m(K)$ formed by the restrictions of functions $v \in H^m_{\text{loc}}(\mathbb{R}^n)$ that satisfy the Floquet condition $v(x+\gamma) = z^{\gamma} v(x)$ for any $\gamma \in \Gamma$. One can show [19, Theorem 2.2.1] that

$$\mathcal{E}_m := \bigcup_{z\in(\mathbb{C}^*)^n} E_{m,z} \tag{2.6}$$

forms a holomorphic subbundle of the trivial bundle $(\mathbb{C}^*)^n \times H^m(K)$. As any infinite dimensional analytic Hilbert bundle over a Stein domain, it is trivializable (see [19, Chapter 1] and Lemma 2.6 below). One can notice that for $m = 0$ the bundle $\mathcal{E}_0$ coincides with the whole $(\mathbb{C}^*)^n \times L^2(K)$.

We collect now several statements from in [19, Theorems 1.3.2, 1.3.3, 1.5.23, and 2.2.2; 35, Theorem XIII.97]

Lemma 2.6. (1) *As any infinite-dimensional analytic Hilbert bundle over a Stein domain, the bundle $\mathcal{E}_m$ is analytically trivial.*

(2) *For any nonnegative integer m, the operator*

$$\mathcal{U}\colon H^m(\mathbb{R}^n) \to L^2(T, \mathcal{E}_m)$$

is an isometric isomorphism, where $L^2(T, \mathcal{E}_m)$ denotes the space of square integrable sections over the complex torus T of the bundle $\mathcal{E}_m$, equipped with the natural topology of a Hilbert space.

(3) *Let the space*

$$\Theta^m := \left\{ f \in H^m_{\text{loc}}(R^n) \,\middle|\, \sup_{\gamma\in\Gamma}\{\|f\|_{H^m(K+\gamma)} e^{b|\gamma|}\} < \infty, \forall b > 0 \right\}$$

be equipped with the natural Fréchet topology. Then

$$\mathcal{U}\colon \Theta^m \to \Gamma\big((\mathbb{C}^*)^n, \mathcal{E}_m\big)$$

is a topological isomorphism, where $\Gamma\big((\mathbb{C}^*)^n, \mathcal{E}_m\big)$ is the space of all analytic sections over $(\mathbb{C}^*)^n$ of the bundle $\mathcal{E}_m$, equipped with the topology of uniform convergence on compacta.

---

[2]It is sometimes also called the *Gel′fand transform*, due to Gel′fand's work [12].

(4) *Under the transform $\mathcal{U}$, the operator*

$$P\colon H^2(\mathbb{R}^n) \to L^2(\mathbb{R}^n)$$

becomes the operator of multiplication by a holomorphic Fredholm morphism $P(z)$ between the fiber bundles $\mathcal{E}_2$ and $\mathcal{E}_0$. Here $P(z)$ acts on the fiber of $\mathcal{E}_m$ over the point $z \in T$ as the restriction to this fiber of the operator P acting between $H^2(K)$ and $L^2(K)$.

Let us now mention another common way of looking at $P(z)$. If $z = \exp ik$, then commuting with the exponent $\exp ik\cdot x$ one reduce the bundle $\mathcal{E}_m$ to the trivial one with the fiber $H^m(\mathbb{T}^n)$, where as before $\mathbb{T}^n = \mathbb{R}^n/\Gamma$. On the other hand, the operator $P(z)$ takes the form $P(x, D+k)$ acting between Sobolev spaces on the torus $\mathbb{T}^n$. In other words, the options are either dealing with the restriction of a fixed operator to an analytically "rotating" subspace, or with a polynomial family of operators between fixed spaces.

We will need to see how the structure of the Floquet solutions (see Definition 2.1), and in general, the structure of functions of Floquet type (2.2) reacts to the Floquet transform. For instance, in the constant coefficient case, where the role of the Floquet solutions is played by the exponential polynomials

$$e^{ik\cdot x} \sum_{|j|\le N} p_j x^j,$$

such functions are Fourier transformed into distributions supported at the point $(-k)$. The next statement shows that under the Floquet transform, each Floquet type function of the form (2.2) corresponds, in a similar way, to a (vector valued) distribution supported at the quasimomentum $(-k)$.

Every Floquet type function u (see (2.2)), being of exponential growth, determines a (continuous linear) functional on the space Θ^0. If it satisfies the equation $Pu = 0$ for a periodic elliptic operator of order m, then as such a functional it is orthogonal to the range of the dual operator $P^*\colon \Theta^m \to \Theta^0$. According to Lemma 2.6, after the Floquet transform any such functional becomes a functional on $\Gamma\big((\mathbb{C}^*)^n, \mathcal{E}_0\big)$, which is orthogonal to the range of the Fredholm morphism $P^*(z)\colon \mathcal{E}_m \to \mathcal{E}_0$ generated by the dual operator $P^*\colon \Theta^m \to \Theta^0$. The following auxiliary result describes all such functionals.

Lemma 2.7 ([21, Lemma 8]). *A continuous linear functional u on Θ^0 is generated by a function of the Floquet form (2.2) with a quasimomentum k if and only if after the Floquet transform it corresponds to a functional on $\Gamma\big((\mathbb{C}^*)^n, \mathcal{E}_0\big)$ which is a distribution ϕ that is supported at the point $\nu = \exp(-ik)$, i.e., has the form*

$$\langle \phi, f\rangle = \sum_{|j|\le N} \left\langle q_j, \left.\frac{\partial^{|j|} f}{\partial z^j}\right|_\nu \right\rangle \quad f \in \Gamma\big((\mathbb{C}^*)^n, \mathcal{E}_0\big),$$

where $q_j \in L^2(K)$. The orders N of the Floquet function (2.2) and of the corresponding distribution ϕ are the same.

Everything discussed so far applies to essentially any elliptic periodic scalar or matrix operators of any order, not necessarily to the ones of the form (2.1) (see [19, 21, 22]). However, there is a special construction that applies only to operators (2.1) and which will play a crucial role in our considerations. Its properties were studied in detail in [1, 26, 34].

Consider the function $\Lambda(\xi)\colon \mathbb{R}^n \to \mathbb{R}$ defined by the condition that the equation

$$Pu = \Lambda(\xi)u$$

has a *positive* Bloch solution of the form

$$u_\xi(x) = \mathrm{e}^{\xi\cdot x} p_\xi(x), \tag{2.7}$$

where $p_\xi(x)$ is Γ-periodic.

Lemma 2.8 ([21, Lemma 12]). (1) *The value $\Lambda(\xi)$ is uniquely determined for any $\xi \in \mathbb{R}^n$.*

(2) *The function $\Lambda(\xi)$ is bounded from above, strictly concave, analytic, and has a nonzero gradient at all points except at its maximum point.*

(3) *Consider the operator*

$$P(\xi) = \mathrm{e}^{-\xi\cdot x} P \mathrm{e}^{\xi\cdot x} = P(x, D - \mathrm{i}\xi)$$

on the torus $\mathbb{T}^n$. Then $\Lambda(\xi)$ is the principal eigenvalue of $P(\xi)$ with a positive eigenfunction p_ξ. Moreover, $\Lambda(\xi)$ is algebraically simple.

(4) *The Hessian of $\Lambda(\xi)$ is nondegenerate at all points.*

One should note that since the function $\Lambda(\xi)$ is analytic, it is actually defined in a neighborhood of $\mathbb{R}^n$ in $\mathbb{C}^n$. This remark will be used in what follows.

Let us denote

$$\Lambda_0 := \max_{\xi\in\mathbb{R}^n} \Lambda(\xi). \tag{2.8}$$

It follows from [1,26] that an alternative definition of Λ_0 can be

$$\Lambda_0 = \sup\{\lambda \in \mathbb{R} \mid \exists u > 0 \text{ such that } (P - \lambda)u = 0 \text{ in } \mathbb{R}^n\}, \tag{2.9}$$

and that in the self-adjoint case Λ_0 coincides with the bottom of the spectrum of the operator P. The common name for Λ_0 is *the generalized principal eigenvalue* of the operator P in $\mathbb{R}^n$.

In our main result, we will need to assume that Λ_0 is strictly positive. In the self-adjoint case such an assumption has a clear spectral interpretation: the bottom of the spectrum is strictly positive. In the next lemma, we provide some known conditions for the nonnegativity or positivity of Λ_0 for not necessarily self-adjoint operators of the form (2.1).

Lemma 2.9 ([21, Lemma 13]). *Consider an operator P of the form* (2.1)

(1) *$\Lambda_0 \geq 0$ if and only if the operator P admits a positive (super)solution. This condition is satisfied in particular when $c(x) \geq 0$.*

(2) *$\Lambda_0 \geq 0$ if and only if the operator P admits a positive solution of the form* (2.7).

(3) *$\Lambda_0 = 0$ if and only if the equation $Pu = 0$ admits exactly one normalized positive solution in $\mathbb{R}^n$.*

(4) *If $c(x) = 0$, then $\Lambda_0 = 0$ if and only if $\int_{\mathbb{T}^n} b(x)\psi(x)\,\mathrm{d}x = 0$, where ψ is the principal eigenfunction of P^* on $\mathbb{T}^n$ (with principal eigenvalue zero). In particular, divergence form operators satisfy this condition.*

(5) *Let $\xi \in \mathbb{R}^n$, and assume that $u_\xi(x) = \mathrm{e}^{\xi\cdot x}p_\xi(x)$ and $u^*_{-\xi}$ are positive Bloch solutions of the equations $Pu = 0$ and $P^*u = 0$, respectively. Denote by ψ the*

periodic function $u_\xi u^*_{-\xi}$. *Consider the function*

$$\tilde{b}_i(x) := b_i(x) - 2\sum_{j=1}^{n} a_{ij}(x)\{\xi_j + [p_\xi(x)]^{-1}\partial_j p_\xi(x)\},$$

and denote

$$\gamma = (\gamma_1, \dots, \gamma_n) := \left(\int_{\mathbb{T}^n} \tilde{b}_1(x)\psi(x)\,\mathrm{d}x, \dots, \int_{\mathbb{T}^n} \tilde{b}_n(x)\psi(x)\,\mathrm{d}x\right).$$

Then $\Lambda_0 = 0$ *if and only if* $\gamma = 0$.

Let us discuss also some additional properties that will play an important role in the sequel. Assume that $\Lambda_0 > 0$. Then Lemma 2.8 implies that the zero level set

$$\Xi := \{\xi \in \mathbb{R}^n \mid \Lambda(\xi) = 0\} \tag{2.10}$$

is a strictly convex compact analytic surface in $\mathbb{R}^n$ of dimension $n-1$. The manifold Ξ consists of all $\xi \in \mathbb{R}^n$ such that the equation $Pu = 0$ admits a positive Bloch solution $u_\xi(x) = \mathrm{e}^{\xi\cdot x}p_\xi(x)$. Moreover, the set of all such positive Bloch solutions is the set of all *minimal positive solutions* of the equation $Pu = 0$ in $\mathbb{R}^n$ [1, 26].[3] We denote by G the convex hull of Ξ, and by $\mathring{G}$ its interior ($\mathring{G}$ is nonempty if and only if $\Lambda_0 > 0$).

Lemma 2.10 ([21, Lemma 14]). *Suppose that* $\Lambda_0 > 0$. *There exists a neighborhood* W *of* G *in* $\mathbb{C}^n$ *and an analytic function*

$$W \ni \xi \mapsto p_\xi(\cdot) \in H^2(\mathbb{T}^n)$$

such that for any $\xi \in W$ *the function of* x

$$u_\xi(x) = \exp(\xi \cdot x)p_\xi(x)$$

is a nonzero Bloch solution of the equation $Pu = \Lambda(\xi)u$ *with a quasimomentum* $-\mathrm{i}\xi$. *Moreover, one can choose the function* p *in such a way that it is positive for all* $\xi \in \Xi$.

Comparing Ξ with the Fermi surface F_P, one sees that

$$-\mathrm{i}\Xi \subset F_P.$$

The next result specifies further the relation between these two varieties:

Lemma 2.11 ([21, Lemma 15]). *Let* $\Lambda_0 \geq 0$. *Then*

(1) *The intersection of the complex Fermi surface* F_P *with the tube*

$$\mathcal{T} := \{k \in \mathbb{C}^n \mid \operatorname{Im} k = (\operatorname{Im} k_1, \dots, \operatorname{Im} k_n) \in -G\} \tag{2.11}$$

coincides with the union of the surface $-\mathrm{i}\Xi$ *with its translations by the vectors of the reciprocal lattice* Γ^*, *i.e., consists of vectors* $k = -\mathrm{i}\xi + \gamma$ *where* $\xi \in \Xi$ *and* $\gamma \in \Gamma^*$. *Moreover, up to a multiplicative constant, any nonzero Bloch solution with a quasimomentum in the above intersection is a positive Bloch solution.*

---

[3] It is also established that a function u is a positive solution of the equation $Pu = 0$ in $\mathbb{R}^n$ if and only if there exists a positive finite measure μ on Ξ such that

$$u(x) = \int_\Xi u_\xi(x)\,\mathrm{d}\mu(\xi).$$

(2) *If* $\Lambda_0 > 0$, *then the intersection of* F_P *with a sufficiently small neighborhood of* $-\mathrm{i}\Xi$ *is a (smooth) analytic manifold that coincides with the set of zeros of the function* $\Lambda(\mathrm{i}k)$.

Analogously to the Floquet surface $\Phi = \Phi_P$, we define the surface

$$\Psi := \rho(-\mathrm{i}\Xi) = \{z \mid z = (\exp \xi_1, \ldots, \exp \xi_n), \xi \in \Xi\}, \tag{2.12}$$

and the tubular domain

$$V := \rho(\mathcal{T}), \tag{2.13}$$

where $\mathcal{T}$ was defined in (2.11). The results of Lemmas 2.10 and 2.11 can be restated in terms of these new objects:

Lemma 2.12. *Let* $\Lambda_0 \geq 0$. *Then*

(1) $\Phi \cap V = \Psi$.

If $\Lambda_0 > 0$, *then*

(2) *The intersection of* Φ *with a sufficiently small neighborhood of* Ψ *is a (smooth) connected analytic manifold.*

(3) *The intersections of* Φ *with neighborhoods of the tube* V *form a basis of neighborhoods of* Ψ *in* Φ.

(4) *For a sufficiently small neighborhood* Φ_ε *of* Ψ *in* Φ *there exists an analytic function* $p : \Phi_\varepsilon \to H^2(\mathbb{T}^n)$ *such that for any* $z \in \Phi_\varepsilon$ *the function of* x

$$u_z(x) = z^x p(z, x)$$

is a nonzero Bloch solution of the equation $Pu = 0$.

3. Representation of solutions by hyperfunctions and distributions

The main result of this paper (Theorem 3.1 below) is analogous to the results of [2, 3] that characterize the classes of solutions of the Helmholtz equation that can be represented by means of distributions or hyperfunctions on S (see also the introduction to our paper). In order to state it, we need to introduce a new object. Let us denote by $h(\omega)$, $\omega \in S^{n-1}$ the indicator function of the convex domain G introduced in the previous section. Namely,

$$h(\omega) := \sup_{\xi \in G}(\omega \cdot \xi), \tag{3.1}$$

where $\omega \cdot \xi = \sum_{j=1}^{n} \omega_j \xi_j$ is the inner product in $\mathbb{R}^n$. The next main theorem will be stated in terms of this function.

Theorem 3.1. *Suppose that* $\Lambda_0 > 0$.

(1) *Let* u *be a solution of the equation* $Pu = 0$ *in* $\mathbb{R}^n$ *satisfying for some* N *the estimate*

$$|u(x)| \leq C(1 + |x|)^N \mathrm{e}^{h(x/|x|)|x|}. \tag{3.2}$$

Then u *can be represented as*

$$u(x) = \langle \mu(\xi), u_\xi(x) \rangle, \tag{3.3}$$

where u_ξ *is the analytic positive Bloch solution corresponding to* $\xi \in \Xi$ *(see Lemma 2.10), and* $\mu(\xi)$ *is a distribution on* Ξ. *The converse statement is also true: for any distribution* μ *on* Ξ, *the function* $u(x)$ *in* (3.3) *is a solution of the equation* $Pu = 0$ *in* $\mathbb{R}^n$ *which satisfies for some* N *the growth condition* (3.2).

(2) *Let u be a solution of the equation $Pu = 0$ in $\mathbb{R}^n$ satisfying for any $\varepsilon > 0$ the estimate*

$$|u(x)| \leq C_\varepsilon \exp[(h(x/|x|) + \varepsilon)|x|], \tag{3.4}$$

where C_ε is a constant depending only on ε and u. Then u can be represented as in (3.3) with $\mu(\xi)$ being a hyperfunction (analytic functional) on Ξ. The converse statement is also true: for any hyperfunction μ on Ξ, the function $u(x)$ in (3.3) is a solution of the equation $Pu = 0$ in $\mathbb{R}^n$ which satisfies the growth condition (3.4)

PROOF. The second statement of the theorem is proved in our paper [21] with Y. Pinchover. So, we concentrate now on the proof of the first one. The proof consists of three major parts: defining appropriate function spaces and interpreting the corresponding class of solutions as functionals; proving Paley – Wiener type theorems for this class of spaces (Lemma 3.3 below); constructing a specific exact sequence of topological spaces. The last step, i.e., constructing and proving exactness of a sequence (Lemma 3.5) is usually the most technical one.

Let us make first of all the following remark:

Remark 3.2. Using a standard elliptic argument (Schauder type estimate) and periodicity of the equation, it is standard to show that a solution satisfies for some N the pointwise growth condition (3.2) if and only if it satisfies for some (different) N the following L_2 growth condition:

$$u(x)(1 + |x|)^{-N} \mathrm{e}^{-h(x/|x|)|x|} \in L^2(\mathbb{R}^n). \tag{3.5}$$

Let us now return to the proof of the theorem. Assume first that a function u has the representation (3.3) with a distribution μ. Then it is obvious that it is a solution of the equation $Pu = 0$. We only need to establish the estimate (3.2). Due to compactness of Ξ, the distribution μ can be represented as a finite sum of terms of the form $D^k \mu_k(\xi)$, where D^k is a constant coefficient homogeneous linear differential operator of order k with respect to ξ and μ_k is a measure on Ξ. So, it is sufficient to establish (3.2) for such a term only. In other words, we need an estimate of the function $v(x) = \langle \mu_k(\xi), D_\xi^k u_\xi(x) \rangle$. According to Lemma 2.7, $D_\xi^k u_\xi(x)$ is an analytically depending on $\xi \in \Xi$ Floquet solution of $Pu = 0$ of order k. This means that it satisfies an estimate of the type (3.2) with $N = k$. Then the estimate for $v(x)$ follows, since μ_k is a finite measure. Hence $u(x)$, being the sum of a finitely many such terms, also satisfies (3.2).

Suppose now that u satisfies (3.2). We need to prove that u can be represented as in (3.3). In order to do so, we need first to interpret this class of solutions in dual terms.

Consider the following Fréchet spaces of test functions:

$$W_m := \{\phi \in H^m_{\text{loc}}(\mathbb{R}^n) \mid \langle \phi \rangle_{m,N} < \infty \ \forall N > 0\},$$

where

$$\langle \phi \rangle_{m,N} := \sup_{\gamma \in \Gamma} \{\|\phi\|_{H^m(K+\gamma)} (1 + |\gamma|)^N \mathrm{e}^{h(\gamma/|\gamma|)|\gamma|}\}.$$

The operator P^* clearly maps continuously W_2 into W_0. It is also clear that due to (3.2), the linear functional

$$\langle u, \phi \rangle := \int_{\mathbb{R}^n} u(x)\phi(x)\, \mathrm{d}x$$

is continuous on the space W_0. Since $Pu = 0$, Schauder elliptic estimates together with the periodicity of the operator show that estimates similar to (3.2) hold also for the derivatives of u. One observes by a simple argument that u is a continuous functional on W_0, which annihilates the range of the dual operator $P^*: W_2 \to W_0$. Now we can apply Floquet theory arguments analogous to the ones used in [19, Section 3.2] or in [21] to obtain (3.3). However, some technical details needed in the cases considered in [19, 21] and in this paper are significantly different, so we provide the details of this derivation.

First of all, we need to obtain a Paley-Wiener type theorem for the Floquet transform in the spaces W_m. Let us denote by V^* the tube that consists of all points $z \in (\mathbb{C}^*)^n$ such that $z^{-1} = (z_1^{-1}, \dots, z_n^{-1}) \in V$, where the tube V is defined in (2.13). We introduce the space $A^\infty(V^*)$ of holomorphic functions on the tube V^* that are infinitely differentiable up to its boundary ∂V^*. Analogously, if $\mathcal{E}$ is a holomorphic Banach bundle in a neighborhood of V^*, we denote by $A^\infty(V^*, \mathcal{E})$ the space of sections of $\mathcal{E}$ over the (closed) tube V^* that are holomorphic in the interior and infinitely differentiable up to the boundary of V^*. This space is equipped with the natural Fréchet space topology. The following statement is a Paley–Wiener type theorem for the transform $\mathcal{U}$ in the spaces W_m.

Lemma 3.3. (1) *The operator*

$$\mathcal{U}\colon W_m \to A^\infty(V^*, \mathcal{E}_m).$$

is a topological isomorphism.

(2) *Under the transform $\mathcal{U}$, the operator*

$$P^*\colon W_2 \to W_0$$

becomes the operator $\mathcal{P}(z)$ of multiplication by a holomorphic Fredholm morphism between the fiber bundles $\mathcal{E}_2$ and $\mathcal{E}_0$:

$$A^\infty(V^*, \mathcal{E}_2) \xrightarrow{\mathcal{P}(z)} A^\infty(V^*, \mathcal{E}_0).$$

Here $\mathcal{P}(z)$ acts on each fiber of $\mathcal{E}_2$ as the restriction to this fiber of the operator P^* acting between $H^2(K)$ and $L^2(K)$.

Before proving this lemma, we first obtain the following auxiliary statement:

Lemma 3.4. *Let H be a complex Hilbert space and $W(H)$ be the Fréchet space of sequences $f = \{f_\gamma\}$, $f_\gamma \in H$, $\gamma \in \Gamma$ such that the semi-norm*

$$\phi_N(f) := \sup_{\gamma \in \Gamma}\{\|f_\gamma\|_H (1 + |\gamma|)^N e^{h(\gamma/|\gamma|)|\gamma|}\}$$

is finite for any N. Here, as before, h is the indicator function (3.1).

Then a sequence $f = \{f_\gamma\}$ belongs to $W(H)$ if and only if the function

$$\hat{f}(z) := \sum_{\gamma \in \Gamma} f_{-\gamma} z^\gamma \tag{3.6}$$

belongs to $A^\infty(V^*, H)$. The mapping $f \mapsto \hat{f}$ is an isomorphism of the space $W(H)$ onto $A^\infty(V^*, H)$.

Proof. Let $f \in W(H)$. We will show that the series (3.6) converges uniformly on V^* as a series of H-valued functions on V^*. This will imply that $\hat{f}$ is analytic in V^* and continuous up to the boundary. Then we will check that the same holds for the derivatives of the series, which will imply that $\hat{f} \in A^\infty(V^*, H)$.

Taking into account that any $z \in V^*$ can be represented as $z = \mathrm{e}^{-\mathrm{i}k}$ with $\operatorname{Im} k \in G$, and thus $\operatorname{Im} k \cdot \gamma \le h(\gamma/|\gamma|)|\gamma|$, we can estimate

$$\begin{aligned}(3.7)\quad \|\widehat{f}(z)\| \le \sum_{\gamma\in\Gamma} \|f_{-\gamma}\| \mathrm{e}^{-\operatorname{Im} k\cdot\gamma} &= \sum_{\gamma\in\Gamma} \|f_\gamma\| \mathrm{e}^{\operatorname{Im} k\cdot\gamma} \\ &\le \sum_{\gamma\in\Gamma} (1+|\gamma|)^{-n-1} \|f_\gamma\| (1+|\gamma|)^{n+1} \mathrm{e}^{h(\gamma/|\gamma|)|\gamma|} \\ &\le \left[\sum_{\gamma\in\Gamma} (1+|\gamma|)^{-n-1}\right] \phi_{n+1}(f).\end{aligned}$$

Since the series $\sum_{\gamma\in\Gamma}(1+|\gamma|)^{-n-1}$ converges, this implies the analyticity in V^* and continuity up to the boundary of $\widehat{f}(z)$. Multiple differentiation with respect to k amounts to multiplying the coefficients of (3.6) by a polynomial with respect to γ factor. Due to the definition of the space $W(H)$, one can get an estimate from above similar to (3.7), but with the seminorm $\phi_{n+d+1}(f)$ instead of $\phi_{n+1}(f)$, where d is the order of differentiation. Thus, in fact the function is infinitely smooth up to the boundary. These estimates also prove that the mapping $f \in W(H) \mapsto \widehat{f} \in A^\infty(V^*, H)$ is continuous.

Let us prove the surjectivity of this mapping. Assume that $s(z) \in A^\infty(V^*, H)$. Let $z = \exp \mathrm{i}k$, then s as a function of k is periodic with respect to the reciprocal lattice Γ^*. Expanding it into the Fourier series, we get

$$(3.8)\qquad s(z) = \sum_{\gamma\in\Gamma} s_{-\gamma} z^\gamma = \sum_{\gamma\in\Gamma} s_\gamma z^{-\gamma},$$

where $s_{-\gamma} \in H$. We need to show now that $\{s_\gamma\} \in W(H)$. For this purpose, we use the standard formulas for the Fourier coefficients:

$$s_\gamma = \frac{1}{(2\pi)^n} \int_B s(\mathrm{e}^{\mathrm{i}(\beta - \mathrm{i}\alpha)}) \mathrm{e}^{\mathrm{i}(\beta-\mathrm{i}\alpha)\cdot\gamma}\, \mathrm{d}\beta, \quad \forall\, \alpha \in G,$$

where B is the first Brillouin zone, and we write $z = \exp \mathrm{i}k = \exp[\mathrm{i}(\beta - \mathrm{i}\alpha)]$, $\alpha \in G$.

Integrating by parts l times with respect to β, where $l = (l_1, \dots, l_n)$ is a multi-index, we obtain analogously

$$(3.9)\qquad s_\gamma = \frac{(-\mathrm{i}\gamma)^{-l}}{(2\pi)^n} \int_B \frac{\partial^l s}{\partial \beta^l} (\mathrm{e}^{\mathrm{i}(\beta-\mathrm{i}\alpha)}) \mathrm{e}^{\mathrm{i}(\beta-\mathrm{i}\alpha)\cdot\gamma}\, \mathrm{d}\beta \quad \forall \alpha \in G.$$

Now straightforward norm estimate in (3.9) gives

$$(3.10)\qquad \|s_\gamma\|_H \le C \max_{z\in V^*} \left\| \frac{\partial^l s}{\partial \beta^l}(z) \right\|_H \gamma^{-l} \mathrm{e}^{-\alpha\cdot\gamma}$$

for any multi-index l and any $\alpha \in G$. Optimizing with respect to $\alpha \in G$, we get

$$(3.11)\qquad \|s_\gamma\|_H \le C_N (1+|\gamma|)^{-N} \mathrm{e}^{-h(\gamma/|\gamma|)|\gamma|}$$

for any N. This means that $f := \{s_\gamma\}$ belongs to $W(H)$ and by its construction $\hat{f} = s(z)$. This proves Lemma 3.4. □

Let us now complete the proof of Lemma 3.3.

We start proving the first claim of the lemma. Let a function $F(x)$ belong to W_m. Consider a sequence $f = \{f_\gamma\}$ of elements of $H^m(K)$ defined as follows:

$$f_\gamma(x) = F(x+\gamma) \quad x \in K, \gamma \in \Gamma.$$

Then clearly the condition $F \in W_m$ is equivalent to two conditions: the first one that $f \in W(H^m(K))$, and second that $F \in H^m_{\mathrm{loc}}(\mathbb{R}^n)$, i.e., that the functions f_γ defined on shifted copies of the fundamental domain K, fit smoothly across the boundaries.

Analogously, the requirement that a section ϕ belongs to $A^\infty(V^*, \mathcal{E}_m)$ consists of two conditions. The first one that $\phi \in A^\infty(V^*, H^m(K))$ and the second that it is a section of the subbundle $\mathcal{E}_m \subset V^* \times H^m(K)$.

We can notice now that the Floquet transform on W_m is the restriction of the transform $f \mapsto \widehat{f}$ of Lemma 3.4 from the larger space $W(H^m(K))$. Thus, Lemma 3.4 claims that this transform is an isomorphism of $W(H^m(K))$ onto $A^\infty(V^*, H^m(K))$. On the other hand, the second conditions: the fitting of f_γ across the boundaries and being a section of the subbundle $\mathcal{E}$, are intertwined by the Floquet transform, according to the first statement of Lemma 2.6. This proves the first claim of the lemma.

Now, the second claim of Lemma 3.3 follows from the third one of Lemma 2.6. Lemma 3.3 is proved. □

Let us now return to the proof of Theorem 3.1. We remind the reader that we have a solution u with the estimate (3.2), for which we need to prove the representation (3.3). Let us apply the Floquet transform $\mathcal{U}$. Then the image $\mathcal{U}u$ of the solution u under the Floquet transform is a continuous linear functional on $A^\infty(V^*, \mathcal{E}_0)$, which is in the cokernel of the operator

$$A^\infty(V^*, \mathcal{E}_2) \xrightarrow{\mathcal{P}(z)} A^\infty(V^*, \mathcal{E}_0).$$

This, indeed is a one-to-one correspondence between solutions of the required class and such functionals. Thus, we need to describe all such functionals. Let $u_z(\cdot) = z^x p(z, \cdot)$ be the Bloch solution of the equation $Pu = 0$ introduced in Lemma 2.12. We will also employ the space $C^\infty(\Psi)$ with the standard topology, where the smooth variety Ψ is introduced in (2.12). Consider the mapping

$$t\colon A^\infty(V^*, \mathcal{E}_0) \to C^\infty(\Psi)$$

that for a section $f(z, x) \in A^\infty(V^*, \mathcal{E}_0)$ of the bundle $\mathcal{E}_0$ produces

$$t_f(z) = \langle f(z^{-1}, \cdot), u_z(\cdot)\rangle = \int_{\mathbb{T}^n} f(z^{-1}, x) u_z(x)\, dx.$$

Here $z^{-1} = (z_1^{-1}, \ldots, z_n^{-1})$.

As we will see soon, the following lemma will finish the proof of the theorem:

Lemma 3.5. *The mapping t is a topological homomorphism and the following sequence is exact:*

$$A^\infty(V^*, \mathcal{E}_2) \xrightarrow{\mathcal{P}(z)} A^\infty(V^*, \mathcal{E}_0) \xrightarrow{t} C^\infty(\Psi) \to 0. \tag{3.12}$$

Proof of the lemma. Continuity of $\mathcal{P}(z)$ is already established. Continuity of t is obvious. The complex property of the sequence (3.12) (i.e., that $t\mathcal{P}(z) = 0$) follows from the construction of t. Thus, the only thing that requires proof is exactness in the second and third terms. The topological homomorphism property will follow then from exactness and the open mapping theorem. So, we only need to prove that: i) any section $\phi \in A^\infty(V^*, \mathcal{E}_0)$ such that $t\phi = 0$ belongs to the range of $\mathcal{P}(z)$ and ii) any function $f \in C^\infty(\Psi)$ is in the range of t.

Let us start with the first of these tasks. So, let $\phi \in A^\infty(V^*, \mathcal{E}_0)$ be such that $t\phi = 0$. Consider the inverse $\mathcal{P}^{-1}(z)$ to the morphism $\mathcal{P}(z)$. It is defined (and hence holomorphic) in a neighborhood V_ϵ^* of the tube V^*, except for an analytic submanifold, whose intersection with V^* is Ψ (see Lemma 2.12). Let us consider the function $f = \mathcal{P}^{-1}(z)\phi(z)$. The only thing now to prove is that this function does not have any singularities along Ψ. This is a local question, so let us return in a neighborhood of a point of Ψ to the quasi-momenta coordinates k and consider the structure of the inverse $\mathcal{P}^{-1}(z)$. As it was shown in the proof of [21, Lemma 21], the inverse has the form $B(k)/\Lambda(k)$, where $B(k)$ is an analytic operator-valued function. This means that $f(k) = (B(k)\phi(k))/\Lambda(k)$. The condition $t\phi = 0$ guarantees that the numerator $g(k) = B(k)\phi(k) \in A^\infty(V^*, H)$ vanishes on Ψ, where H is a Hilbert space. Our goal is to prove that this is sufficient for its smooth divisibility (on ∂V^*) by Λ. We recall here that Λ is analytic in a vicinity of ∂V^* and has simple zeros along Ψ (Lemmas 2.8 and 2.11). We notice that it is sufficient to prove this for scalar functions, i.e., for $H = \mathbb{C}$. This can be justified in many different ways. For instance, the statement is local, and locally, due to the Fredholm nature of the morphism $\mathcal{P}(z)$, one can project the problem onto a finite dimensional subspace, using a lemma by M. Atiyah [5] (see also [19, Lemma 1.2.11 and Theorem 1.3.9; 36, Lemma 2.1]), which will reduce it to a finite-dimensional, and thus also to scalar case. So, we will assume in this part of the proof that $g \in A^\infty$ is a scalar function. According to a result of [17, 27] (see also [29, 31] and [29, Chapter VI, Theorem 1.1′]), it is sufficient to check the divisibility at each point of Ψ on the level of formal Taylor series. So, let us pick a point k of Ψ and introduce coordinates $x \in \mathbb{R}^{n-1}$ in the tangent space $T_k(\Psi) \in i\mathbb{R}^n$. The complexification $T_k^c(\Psi)$ of this tangent space is a part of the tangent space to the boundary of the tube. Let us chose coordinates $y \in \mathbb{R}^{n-1}$ in $T_k^c(\Psi) \cap \mathbb{R}^n$ that correspond to the coordinates x in $T_k(\Psi)$. An extra coordinate t in $T_k(\Psi) \cap \mathbb{R}^n$ is required to obtain the whole tangent space $T_k(\partial V^*)$. Let us denote by $\hat{g}(x, y, t)$ and $\hat{\Lambda}(x, y, t)$ the formal Taylor series of g and Λ at the point k. Then we know that $\hat{g}(x, 0, 0) = 0$ and $\hat{\Lambda}(x, 0, 0) = 0$ (formal power series versions of vanishing of functions g and Λ on Ψ). Recall that $\hat{g}(x, y, t)$ is the series for a CR-function g on the boundary (since g is the boundary value of an analytic function). This means that $\hat{g}(x, y, t)$ satisfies Cauchy–Riemann conditions with respect to the variable $z = x + iy \in \mathbb{C}^{n-1}$. Then uniqueness of analytic continuation[4] claims that $\hat{g}(x, 0, 0) = 0$ for all x implies $\hat{g}(x, y, 0) = 0$ for all (x, y). The same is true for $\hat{\Lambda}$, due to analyticity of Λ. Now, in coordinates $z = x + iy, t$ we are dealing with the formal series $\hat{g}(z, t)$ and $\hat{\Lambda}(z, t)$, both of which vanish at $t = 0$ and such that $\hat{\Lambda}$ has zero of first order at $t = 0$. Then, vanishing of $\hat{g}(z, 0)$ guarantees divisibility in formal series of $\hat{g}$ by $\hat{\Lambda}$. As it was explained above, this implies smooth divisibility of g by Λ and thus finishes the proof of exactness in the second term of the sequence (3.12).

Let us now prove the exactness in the third term of the sequence. First of all, we notice that the vector-function u_z, defined on Ψ only, can be extended to an analytic vector-function (which we will denote the same way) on V_ϵ^* for some small epsilon. Indeed, as it is shown in [21], V_ϵ^* is a Stein manifold. Then, according to the Corollary 1 from the Bishop's Theorem 3.3 in [36] (see the original theorem in [7]),

---

[4]The uniqueness of analytic continuation in this power series setting is straightforward to derive algebraically directly from the Cauchy–Riemann conditions for power series.

the restriction mapping to an analytic subset of a Stein variety is surjective. Thus, the required extension of u_z exists. Let also $v(z)$ be a holomorphic family such that $tv(z)|_\Psi = 1$ (it is not hard to prove the existence of such a family). Consider a function $\phi(z) \in C^\infty(\Psi)$. Notice that the domain V^* is strictly pseudo-convex and the complexifications of the tangent spaces to the submanifold $\Psi \subset \partial V^*$ are parts of the tangent spaces to ∂V^*. Thus, Ψ and ∂V^* satisfy the conditions of [9] needed for Ψ to be an A^∞ interpolation variety, and hence the restriction mapping $A^\infty(V^*) \mapsto C^\infty(\Psi)$ is surjective. Hence, there exists a function $\psi \in A^\infty(V^*)$ such that $\psi|_\Psi = \phi$. Now taking $f = \psi(z)v(z) \in A^\infty(V^*, \mathcal{E}_0)$ guarantees that $tf = \phi$. This finishes the proof of the lemma. □

It is easy now to finish the proof of the theorem. Indeed, after the Floquet transform solution u becomes a continuous linear functional on $A^\infty(V^*, \mathcal{E}_0)$ that annihilates the range of the operator of multiplication by $\mathcal{P}(z)$. Lemma 3.5 implies that such a functional can be pushed down to the space $C^\infty(\Psi)$. Any such functional is a distribution μ. Hence, the action $\langle u, \phi\rangle$ of the functional u on a function $\phi \in W_0$ can be obtained as

$$\langle u, \phi\rangle = \langle \mu(z), t(z)(\mathcal{U}\phi)\rangle.$$

Applying now the explicit formulas for the transforms $\mathcal{U}$ and t, one arrives to the representation (3.3). Indeed,

$$\begin{aligned} t_{(\mathcal{U}\phi)}(z) &= \int_K \mathcal{U}\phi(z^{-1}, x) u_z(x)\,\mathrm{d}x \\ &= \sum_{\gamma\in\Gamma} \int_{K-\gamma} \phi(x) z^{-\gamma} u_z(x+\gamma)\,\mathrm{d}x = \int_{\mathbb{R}^n} \phi(x) u_z(x)\,\mathrm{d}x. \end{aligned} \tag{3.13}$$

In this calculation we used the property of the Bloch solutions

$$u_z(x+\gamma) = z^\gamma u_z(x).$$

Therefore,

$$\langle u, \phi\rangle = \big\langle \langle \mu(z), u_z\rangle, \phi\big\rangle,$$

which concludes the proof of the theorem. □

Acknowledgments. The author expresses his gratitude to Y. Pinchover, the co-author of the previous papers [21, 22], with whom this manuscript has been discussed on numerous occasions and who has made many suggestions that have improved the text, and to A. Tumanov for helpful information on pick sets results.

References

1. S. Agmon, *On positive solutions of elliptic equations with periodic coefficients in* $\mathbb{R}^n$*, spectral results and extensions to elliptic operators on Riemannian manifolds*, Differential Equations (Birmingham, AL, 1983) (I. W. Knowles and R. T. Lewis, eds.), North-Holland Math. Stud., vol. 92, North-Holland, Amsterdam, 1984, pp. 7–17.
2. ———, *A representation theorem for solutions of the Helmholtz equation and resolvent estimates for the Laplacian*, Analysis, et cetera (P. H. Rabinowitz and E. Zehnder, eds.), Academic Press, Boston, MA, 1990, pp. 39–76.
3. ———, *Representation theorems for solutions of the Helmholtz equation on* $\mathbb{R}^n$, Differential Operators and Spectral Theory (V. Buslaev, M. Solomyak, and D. Yafaev, eds.), Amer. Math. Soc. Transl. Ser. 2, vol. 189, Amer. Math. Soc., Providence, RI, 1999, pp. 27–43.
4. N. W. Ashcroft and N. D. Mermin, *Solid state physics*, Holt, Rinehart and Winston, New York–London, 1976.
5. M. F. Atiyah, *K-theory*, W. A. Benjamin, New York–Amsterdam, 1967.

6. D. Bättig, D. Knörrer, and E. Trubowitz, *A directional compactification of the complex Fermi surface*, Compositio Math. **79** (1991), no. 2, 205 – 229.
7. E. Bishop, *Analytic functions with values in a Fréchet space*, Pacific J. Math. **12** (1962), 1177 – 1192.
8. J. Chaumat and A.-M. Chollet, *Ensembles pics pour $A^\infty(D)$*, Ann. Inst. Fourier (Grenoble) **29** (1979), no. 3, 171 – 200.
9. ______, *Caractérisation et propriétés des ensembles localement pics de $A^\infty(D)$*, Duke Math. J. **47** (1980), no. 4, 763 – 787.
10. M. S. P. Eastham, *The spectral theory of periodic differential equations*, Scottish Acad. Press Ltd., Edinburgh – London, 1973.
11. L. Ehrenpreis, *Fourier analysis in several complex variables*, Pure Appl. Math., vol. 17, Wiley, New York, 1970.
12. I. M. Gel′fand, *Expansion in eigenfunctions of an equation with periodic coefficients*, Dokl. Akad. Nauk SSSR **73** (1950), 1117 – 1120 (Russian).
13. D. Gieseker, H. Knörrer, and E. Trubowitz, *The geometry of algebraic Fermi curves*, Perspect. Math., vol. 14, Academic Press, Boston, MA, 1992.
14. M. Hashizume, A. Kowata, K. Minemura, and Okamoto K., *An integral representation of an eigenfunction of the Laplacian on the Euclidean space*, Hiroshima Math. J. **2** (1972), 535 – 545.
15. S. Helgason, *Eigenspaces of the Laplacian; integral representations and irreducibility*, J. Functional Analysis **17** (1974), 328 – 353.
16. ______, *Groups and geometric analysis: Integral geometry, invariant differential operators, and spherical functions*, Pure Appl. Math., vol. 113, Academic Press, Orlando, FL, 1984.
17. L. Hörmander, *On the division of distributions by polynomials*, Ark. Mat. **3** (1958), 555 – 568.
18. H. Knörrer and E. Trubowitz, *A directional compactification of the complex Bloch variety*, Comment. Math. Helv. **65** (1990), no. 1, 114 – 149.
19. P Kuchment, *Floquet theory for partial differential equations*, Oper. Theory Adv. Appl., vol. 60, Birkhäuser, Basel, 1993.
20. ______, *On some spectral problems of mathematical physics*, Partial Differential Equations and Inverse Problems (C. Conca, R. Manasevich, G. Uhlmann, and M. S. Vogelius, eds.), Contemp. Math., vol. 362, Amer. Math. Soc., Providence, RI, 2004, pp. 241 – 276.
21. P. Kuchment and Y. Pinchover, *Integral representations and Liouville theorems for solutions of periodic elliptic equations*, J. Funct. Anal. **181** (2001), no. 2, 402 – 446.
22. ______, *Liouville theorems and spectral edge behavior on abelian coverings of compact manifolds*, Trans. Amer. Math. Soc., to appear.
23. P. Kuchment and B. Vainberg, *On embedded eigenvalues of perturbed periodic Schrödinger operators*, Spectral and Scattering Theory (Newark, DE, 1997), Plenum, New York, 1998, pp. 67 – 75.
24. ______, *On absence of embedded eigenvalues for Schrödinger operators withperturbed periodic potentials*, Comm. Partial Differential Equations **25** (2000), no. 9-10, 1809 – 1826.
25. ______, *On the structure of eigenfunctions corresponding to embedded eigenvalues of locally perturbed periodic graph operators*, Comm. Math. Phys. **268** (2006), no. 3, 673 – 686.
26. V. Ya. Lin and Y. Pinchover, *Manifolds with group actions and elliptic operators*, Mem. Amer. Math. Soc. **540** (1994).
27. S. Łojasiewicz, *Sur le problème de la division*, Studia Math. **8** (1959), 87 – 136.
28. B. Malgrange, *Division des distributions*, exposés 21 – 25, Séminaire Schwartz, Vol. 4, 1959 – 1960.
29. ______, *Ideals of differentiable functions*, Tata Inst. Fund. Res. Studies in Math, vol. 3, Tata Inst. Fund. Res., Bombay; Oxford Univ. Press, London, 1967.
30. M. Morimoto, *Analytic functionals on the sphere and their Fourier – Borel transformations*, Complex Analysis (Warsaw, 1979), Banach Center Publ., vol. 11, PWN, Warsaw, 1983, pp. 223 – 250.
31. V. P. Palamodov, *Structure of polynomial ideals and their quotients in spaces of infinitely differentiable functions*, Dokl. Akad. Nauk SSSR **141** (1961), 1302 – 1305 (Russian).
32. V. Palamodov, *Linear differential equations with constant coefficients*, Grundlehren Math. Wiss., vol. 168, Springer, New York, 1970.
33. ______, *Harmonic synthesis of solutions of elliptic equations with periodic coefficients*, Ann. Inst. Fourier (Grenoble) **43** (1993), no. 3, 751 – 768.

34. R. G. Pinsky, *Second order elliptic operators with periodic coefficients: criticality theory, perturbations, and positive harmonic functions*, J. Funct. Anal. **129** (1995), no. 1, 80 – 107.
35. M. Reed and S. Simon, *Methods of modern mathematical physics.* IV: *Analysis of operators*, Academic Press, New York – London, 1978.
36. M. G. Zaidenberg, S. G. Krein, P. Kuchment, and A. A. Pankov, *Banach bundles and linear operators*, Uspehi Mat. Nauk **30**, no. 5(185), 101 – 157 (Russian); English transl., Russian Math. Surveys **30** (1975), no. 5, 115 – 175.

DEPARTMENT OF MATHEMATICS, TEXAS A&M UNIVERSITY, COLLEGE STATION, TX 77843-3368, USA

E-mail address: kuchment@math.tamu.edu

Centre de Recherches Mathématiques
CRM Proceedings and Lecture Notes
Volume **42**, 2007

Inverse Spectral Problems for Schrödinger Operators with Energy Depending Potentials

A. Laptev, R. Shterenberg, and V. Sukhanov

ABSTRACT. We study an inverse problem for a class of Schrödinger operators with energy depending potentials. In particular, we show that introduction of the discrete spectrum generically does not lead to singularities of the corresponding soliton solutions. In our last chapter we derive some new trace formulas which could be considered as generalization of a standard trace formulas for Schrödinger operators.

1. Introduction

In this paper we consider the inverse problem for the operator

$$-\psi''_{xx}(x,k) + \big(2ku(x) + v(x)\big)\psi(x,k) = k^2\psi(x,k), \quad x \in \mathbb{R}. \tag{1.1}$$

In what follows we assume that the potential functions u and v are real-valued, smooth and exponentially decay at infinity together with all their derivatives

$$\left|\frac{\mathrm{d}^j}{\mathrm{d}x^j}u(x)\right|, \quad \left|\frac{\mathrm{d}^j}{\mathrm{d}x^j}\,u(x)\right| \le C_j\exp(-\varepsilon|x|) \tag{1.2}$$

for $j \in \mathbb{N}$ and some $\varepsilon > 0$. In (1.1) k is a spectral parameter and if for some $k \in \mathbb{C}$ there is a L^2 solution of this equation, then we say that k is an eigenvalue.

Such operators are sometimes called Schrödinger operators with energy depending potentials and the scattering problem for them has been considered in the papers by Jaulent [3], Jaulent and Jean [4], Kaup [5] and also by Sattinger and Szmigielski [6, 7]. In [3, 4, 6] the authors used the Gel′fand – Levitan – Marchenko approach for the inverse problem on the line given in [2].

In fact, the inverse problem has been completely solved in [6] for the regular case (when the eigenvalues are absent) by using a so-called "vanishing lemma." If the data of the inverse problem have bound state, then the situation with solvability

---

2000 *Mathematics Subject Classification.* Primary 35P15; Secondary 35L15, 47A75, 35J10.

A. Laptev thanks a partial support by the SPECT ESF European program.

The work of R. Shterenberg was supported in part by NSF grant DMS-0314129.

V. Sukhanov and R. Shterenberg are grateful to the Royal Institute of Technology in Stockholm for its hospitality, for financial support from Gustafsson's foundation and from the Royal Swedish Academy of Science.

A support by the RFFI grant 05-01-01076 is also greatly acknowledged.

This is the final form of the paper.

becomes significantly more complicated. The main problem is that method for constructing the solution at the first step automatically gives a number of singularities. As it has been found in [6], for one-soliton solution these singularities generically cancel in the final formulas and for a two-soliton or more solutions problem, the cancellation of singularities has been stated as a conjecture.

The main aim of this paper is to clarify the situation where there is a finite number of eigenvalues. As it has been done in [7] we will use the Riemann–Hilbert as a method of solving inverse problem. The matrix solution of this problem has singularities in x if the data of the inverse problem contain the discrete spectrum. However we prove (see Theorem 4.1) that these singularities generically cancel at the second step of our construction.

In Section 4 we consider a class of reflectionless potentials and prove that the solutions u and v are generically smooth functions. We also consider a so-called "dressing up" transform (addition of one eigenvalue to the data of the inverse problem) and also prove that such a transform generically does not lead to singularities of the potential functions u and v.

In the end of the paper we derive some new trace formulas involving the potential functions u and v. If $u \equiv 0$ then these formulas coincide with the corresponding trace formulas for Schrödinger operators.

2. Properties of scattering data

In this section we recall some standard fact, cf. [7]. We begin with defining the Jost solution of the equation (1.1). Let us consider its two solutions satisfying the properties

$$f(x,k) = \mathrm{e}^{\mathrm{i}kx}\big(1+o(1)\big), \qquad x \to +\infty, \tag{2.1}$$

$$g(x,k) = \mathrm{e}^{-\mathrm{i}kx}\big(1+o(1)\big), \quad x \to -\infty. \tag{2.2}$$

We also introduce

$$\alpha(x) = \exp\left\{\mathrm{i}\int_x^\infty u(s)\,\mathrm{d}s\right\}, \quad \alpha_0 = \lim_{x\to-\infty}\alpha(x). \tag{2.3}$$

Since u is a real function $\alpha_0^{-1} = \overline{\alpha_0}$.

For a real-valued k we have two pairs of linear independent solutions:

$$\{f(x,k), \overline{f(x,k)}\}, \quad \{g(x,k), \overline{g(x,k)}\}.$$

In particular, the functions f and $\bar f$ could be written as linear combinations of g and $\bar g$

$$f(x,k) = a(k)\overline{g(x,k)} + b(k)g(x,k), \tag{2.4}$$

$$\overline{f(x,k)} = \overline{a(k)}g(x,k) + \overline{b(k)g(x,k)}. \tag{2.5}$$

Let us define the Wronskian

$$W[\varphi,\psi](x,k) = \varphi'_x(x,k)\psi(x,k) - \varphi(x,k)\psi'_x(x,k). \tag{2.6}$$

Proposition 2.1. *The following fundamental identity holds true*

$$|a(k)|^2 - |b(k)|^2 = 1. \tag{2.7}$$

PROOF. The Wronskian of the functions f and $\bar f$ is constant with respect to x. Comparing the asymptotic behavior of $W[f,\bar f]$ at $+\infty$ and $-\infty$ we complete the proof. □

In follows from (1.1) that the function f satisfies the following Volterra equation

$$f(x,k) = \mathrm{e}^{\mathrm{i}kx} - \int_x^{\infty} \frac{\sin\big(k(x-y)\big)}{k}\big(ku(x)+v(x)\big)f(y,k)\,\mathrm{d}y. \tag{2.8}$$

Therefore f can be written as a convergent series obtained by standard iterations and this immediately implies that f is an analytic function with respect to $k \in \mathbb{C}_+$ in the upper half plane and can be continuously extended on the real line. In particular, this also implies the analyticity of the scattering coefficients $a(k)$ in $\mathbb{C}_+$ and therefore $\overline{a(\bar{k})}$ in $\mathbb{C}_-$. Similarly we find that the solution g is an analytic function in $\mathbb{C}_+$.

It is well known that if $u \equiv 0$ then $f(x,k) \to \exp(\mathrm{i}kx)$ as $k \to \infty$ ($\operatorname{Im} k > 0$). If $u \not\equiv 0$ then we obtain the following.

Proposition 2.2. *For* $\operatorname{Im} k > 0$ *functions* $f(x,k)$ *and* $g(x,k)$ *satisfy following asymptotic relations*

$$f(x,k) = \mathrm{e}^{\mathrm{i}kx}\exp\Big\{\mathrm{i}\int_x^{\infty} u(s)\,\mathrm{d}s\Big\}\big(1+o(1)\big), \qquad k \to \infty; \tag{2.9}$$

$$g(x,k) = \mathrm{e}^{-\mathrm{i}kx}\exp\Big\{\mathrm{i}\int_{-\infty}^{x} u(s)\,\mathrm{d}s\Big\}\big(1+o(1)\big), \qquad k \to \infty. \tag{2.10}$$

Similarly we find asymptotic properties of the functions $a(k)$ and $b(k)$.

Proposition 2.3. *The scattering coefficients* a *and* b *satisfy the asymptotic behavior as* $k \to \infty$

$$\lim_{k\to\infty} a(k) = \alpha_0^{-1}, \quad \operatorname{Im} k \geq 0,$$
$$\lim_{k\to\infty} b(k) = 0, \qquad \operatorname{Im} k = 0.$$

From the asymptotical behavior of f at $\pm\infty$ we immediately find

Proposition 2.4. *Zeros of the scattering coefficient* a *in* $\mathbb{C}_+$ *coincide with the eigenvalues of the operator* (1.1).

Remark 2.5. Due to the assumption (1.2) the scattering coefficient $a(k)$ is also an analytical function for $\operatorname{Im} k > -\varepsilon$, $k \neq 0$, where $\varepsilon > 0$ is introduced in (1.2). Point $k = 0$ can be a pole for function $a(k)$ or it is regular point. Therefore it can only have a finite number of zeros in $\mathbb{C}_+$.

3. Riemann – Hilbert problem

Here we describe the inverse problem by using Riemann – Hilbert approach, see also [7]. Let us assume that the scattering coefficient $a(k)$ has N simple zeros $\{\varkappa_j\}_{j=1}^N$, $\varkappa_j \in \mathbb{C}_+$. Denote by $\Psi(x,k) = \binom{\psi_1(x,k)}{\psi_2(x,k)}$ the vector defined by

$$\Psi(x,k) = \begin{pmatrix} f(x,k) \\ g(x,k)/a(k) \end{pmatrix}, \quad \text{for } k \in \mathbb{C}_+,$$

$$\Psi(x,k) = \begin{pmatrix} \overline{g(x,\bar{k})/a(\bar{k})} \\ \overline{f(x,\bar{k})} \end{pmatrix}, \quad \text{for } k \in \mathbb{C}_-.$$

By using the analytical properties of the functions f and g and also the properties of the scattering coefficients a and b described in the previous section, we find that the vector Ψ is meromorphic in $\mathbb{C}_\pm$ and can be continuously extended to the boundary.

Let $\Psi^+(x,k)$ and $\Psi^-(x,k)$ be the limits of the function $\Psi(x,k)$ as k approaches the real axis from upper and lower complex half plane respectively. One finds that the this vector satisfies following relation

$$\Psi^+(x,k) = G(k)\Psi^-(x,k), \tag{3.1}$$

$$G(k) = \begin{pmatrix} 1 & -\overline{r(k)} \\ r(k) & 1-|r(k)|^2 \end{pmatrix}, \tag{3.2}$$

where

$$r(k) = -\frac{\overline{b(k)}}{a(k)}, \quad k \in \mathbb{R}. \tag{3.3}$$

It follows from Proposition 2.1 that if $k \in \mathbb{R}\setminus\{0\}$, then $|r(k)| < 1$. Moreover, if we denote by $\sigma_3 = \left(\begin{smallmatrix} 1 & 0 \\ 0 & -1 \end{smallmatrix}\right)$, then using (2.9) and (2.10) we find

$$\Psi(x,k) = \exp\{\mathrm{i}\sigma_3 kx\} \begin{pmatrix} \alpha(x) \\ \big(\alpha(x)\big)^{-1} \end{pmatrix} (1+o(1)), \quad \text{as } k \to \infty. \tag{3.4}$$

For the discrete spectrum we obtain

$$\operatorname{res}_{k=\varkappa_j} \psi_2(x,k) = C_j\psi_1(x,\varkappa_j), \quad \operatorname{res}_{k=\overline{\varkappa_j}} \psi_1(x,k) = \overline{C_j}\psi_2(x,\overline{\varkappa_j}), \quad C_j \neq 0, j=1,\dots,N. \tag{3.5}$$

Therefore meromorthic vector function $\Psi(x,k)$ is the solution of the Riemann–Hilbert problem: (3.1), (3.4), (3.5). In order to exclude the direct appearance of the function $\alpha(x)$ from it we consider its matrix version which is slightly different. Namely, let us introduce a meromorthic in $\mathbb{C}_+$ and $\mathbb{C}_-$ matrix

$$\Xi = \begin{pmatrix} \xi_{11}(x,k) & \xi_{12}(x,k) \\ \xi_{21}(x,k) & \xi_{22}(x,k) \end{pmatrix}, \tag{3.6}$$

such that

$$\Xi^+ = G(k)\Xi^-, \quad k \in \mathbb{R}, \tag{3.7}$$

$$\Xi = \exp\{\mathrm{i}\sigma_3 kx\} \begin{pmatrix} 1 & 0 \\ 0 & 1 \end{pmatrix} (1+o(1)), \quad k \to \infty.$$

$$\begin{aligned} &\operatorname{res}_{k=\varkappa_j} \xi_{21}(x,k) = C_j\xi_{11}(x,\varkappa_j), && \operatorname{res}_{k=\varkappa_j} \xi_{22}(x,k) = C_j\xi_{12}(x,\varkappa_j), \\ &\operatorname{res}_{k=\overline{\varkappa_j}} \xi_{11}(x,k) = \overline{C_j}\xi_{21}(x,\overline{\varkappa_j}), && \operatorname{res}_{k=\overline{\varkappa_j}} \xi_{12}(x,k) = \overline{C_j}\xi_{22}(x,\overline{\varkappa_j}), \\ & && C_j \neq 0, j=1,\dots,N. \end{aligned} \tag{3.8}$$

As before we denote by $\Xi^\pm(x,k)$, $k \in \mathbb{R}$, the limits in k of $\Xi(x,k)$ from the upper and the lower complex half plane.

The asymptotic formulas for the functions $\xi_{js}(x,k)$, $j,s = 1,2$, as $k \to \infty$, are given by

$$\xi_{js}(x,k) = \exp\{(-1)^{j-1}\mathrm{i}kx\}\left(\delta_{js} + \frac{\varphi_{js}(x)}{k} + O(k^{-2})\right).$$

The vector Ψ and the matrix Ξ are related as follows

$$\Psi(x,k) = \begin{pmatrix} \psi_1(x,k) \\ \psi_2(x,k) \end{pmatrix} = \alpha(x)\begin{pmatrix} \xi_{11}(x,k) \\ \xi_{21}(x,k) \end{pmatrix} + \big(\alpha(x)\big)^{-1}\begin{pmatrix} \xi_{12}(x,k) \\ \xi_{22}(x,k) \end{pmatrix}. \tag{3.9}$$

Let us now substitute the functions $\psi_1(x,k)$, $\psi_2(x,k)$ into the equation (1.1). Since they are solutions of this equation, then comparing the terms of order k^0, we obtain

$$\begin{aligned}(3.10)\qquad \frac{\big(\alpha(x)\big)''}{\alpha(x)} + 2\mathrm{i}\big(\varphi_{11}(x)\big)' + 2\mathrm{i}\big((\alpha(x))^{-2}\big)'\varphi_{12}(x) \\ + 2\mathrm{i}(\alpha(x))^{-2}(\varphi_{12}(x))' = v(x),\end{aligned}$$

$$\begin{aligned}(3.11)\qquad \frac{\Big(\big(\alpha(x)\big)^{-1}\Big)''}{\big(\alpha(x)\big)^{-1}} - 2\mathrm{i}\big(\varphi_{22}(x)\big)' - 2i\big((\alpha(x))^{2}\big)'\varphi_{21}(x) \\ - 2\mathrm{i}\big(\alpha(x)\big)^2\big(\varphi_{21}(x)\big)' = v(x).\end{aligned}$$

Note that due to the symmetry of the Riemann – Hilbert problem, the equation (3.11) is complex conjugated to the equation (3.10). Note also that the function

$$(3.12)\qquad m(x,k) = \det \Xi(x,k) = \xi_{11}(x,k)\xi_{22}(x,k) - \xi_{12}(x,k)\xi_{21}(x,k)$$

is analytic in $\mathbb{C}_\pm$ and solves the trivial Riemann – Hilbert problem

$$m^+(x,k) = m^-(x,k), \quad k \in \mathbb{R}, \qquad m(x,k) \to 1, \quad k \to \infty.$$

Thus $m(x,k) \equiv 1$ and therefore comparing the terms of order k^{-1} and in view of (3.12) we derive

$$(3.13)\qquad \varphi_{11}(x,k) + \varphi_{22}(x,k) = 0.$$

Using (3.10), (3.11) and (3.13) we obtain that the function $t(x) := \big(\alpha(x)\big)^2$ satisfies the Riccati equation

$$(3.14)\qquad (t(x))' + 2\mathrm{i}\varphi_{21}(x)(t(x))^2 + 2\mathrm{i}\varphi_{12}(x) = 0, \qquad t(x) \to 1, \quad \text{as } x \to +\infty.$$

Solving this equation leads to finding the potential u

$$u(x) = \frac{\mathrm{i}}{2}\,\frac{t'(x)}{t(x)}.$$

The latter formula is not very explicit and relies on the solution of the equation (3.14). Therefore we use a different approach (see [6]). Obviously $f(x,0) = \overline{f(x,0)}$ and thus

$$\alpha(x)\psi_{11}^+(x,0) + \big(\alpha(x)\big)^{-1}\psi_{12}^+(x,0) = \alpha(x)\psi_{21}^-(x,0) + \big(\alpha(x)\big)^{-1}\psi_{22}^-(x,0).$$

Thus

$$(3.15)\qquad t(x) = \big(\alpha(x)\big)^2 = \exp\Big\{-2\mathrm{i}\int_\infty^x u(s)\,\mathrm{d}s\Big\} = \frac{\xi_{22}^-(x,0) - \xi_{12}^+(x,0)}{\xi_{11}^+(x,0) - \xi_{21}^-(x,0)}.$$

Remark 3.1. We emphasize that $|t(x)| = 1$ and thus *potential u is regular.*

Remark 3.2. Assume now that the inverse problem does not contain eigenvalues. If reflection coefficient $r(k)$ satisfies the conditions which are usually imposed for the Schrödinger operator then the similar arguments as in the Schrödinger operator case show that the Riemann – Hilbert problem can be solved and thus potentials u and v can be uniquely reconstructed. We do not go into details. The complete proof via Gel′fand – Levitan equations can be found in [6]. The close result in terms of Riemann – Hilbert problem has been obtained in [7].

4. Class of reflectionless potentials

In [6] a particular example of one-soliton solution was constructed. It was pointed out that the determinant of the system which appears when solving Riemann–Hilbert problem (3.7)–(3.8) (or the corresponding Gel′fand–Levitan problem) has zeros if the eigenvalues are included.

It has been proved in [6] that if there is only one soliton solution then this singularity cancels after substituting it into (3.9).

In the present section we show that this cancellation is generic. Note also that, in general, one cannot expect the global solvability. The difficulties appearing here are similar to the difficulties when considering the Schrödinger operator with complex potential. We prove that generically we obtain smooth reflectionless potentials but if we choose some "wrong" coefficients, then potential functions u and v can be singular. Already the case of one-soliton potential shows that, in general, the set of "wrong" coefficients although "small" but its structure is sufficiently complicated.

Assume that $r(k) = 0$ and the set of simple eigenvalues $\{\varkappa_j\}$, $\mathrm{Im}(\varkappa_j) > 0$, $j = 1, \dots, N$, is given. Since the potentials u and v are real functions we obtain

$$\xi_{11} = \mathrm{e}^{ikx}\left(1 + \sum_{j=1}^{N} \frac{h_j(x)}{k - \overline{\varkappa_j}}\right), \qquad \xi_{12} = \mathrm{e}^{ikx} \sum_{j=1}^{N} \frac{w_j(x)}{k - \overline{\varkappa_j}}, \tag{4.1}$$

$$\xi_{21} = \mathrm{e}^{-ikx} \sum_{j=1}^{N} \frac{\overline{w_j}(x)}{k - \varkappa_j}, \qquad \xi_{22} = \mathrm{e}^{-ikx}\left(1 + \sum_{j=1}^{N} \frac{\overline{h_j}(x)}{k - \varkappa_j}\right). \tag{4.2}$$

We define $h := (h_1, \dots, h_N)^t$, $w := (w_1, \dots, w_N)^t$. Let $B = B(x)$ be the block $(2N \times 2N)$-matrix such that

$$\begin{aligned} &B_{11} = \overline{B_{22}} = A, && B_{12} = \overline{B_{21}} = J; \\ &A_{js} = -\frac{1}{\varkappa_j - \overline{\varkappa_s}}, && J = \mathrm{diag}\{C_j^{-1} e^{-2i\varkappa_j x}\}. \end{aligned} \tag{4.3}$$

It follows from (3.8) that

$$B\begin{pmatrix} h \\ \overline{w} \end{pmatrix} = \begin{pmatrix} 1 \\ 0 \end{pmatrix}. \tag{4.4}$$

By using the properties (4.3), we find that the determinant $d(x) := \det B(x)$ is real-valued. Zeros of d define points of singularity of vector-valued functions h and w. We are going to show that these singularities generically cancel.

Let us define

$$p = d\left(\sum_j \frac{\overline{w_j}}{\varkappa_j} - \sum_j \frac{h_j}{\overline{\varkappa_j}}\right). \tag{4.5}$$

Note that functions $d\,h_j$ and $d\,w_j$, $j = 1, \dots, N$, do not have singularities. It follows from (3.15), (4.1)–(4.2) that

$$\alpha^2 = \frac{d + \bar{p}}{d + p}. \tag{4.6}$$

The function α is regular and thus the potential u is regular. We are going to prove the regularity of $\psi_1(x,k)$, $\psi_2(x,k)$. It follows from (3.9) that it is sufficient to prove regularity of the vector-valued function $s := \alpha^2 h + w$.

Theorem 4.1. *Let $d(x_0) = 0$. Assume that*

(1) $d'(x_0) \neq 0$.
(2) $\dim \ker B(x_0) = 1$.
(3) $p(x_0) \neq 0$.

Then $(sd)(x_0) = 0$ and thus s and, correspondingly, ψ_1 and ψ_2 are regular functions at x_0.

PROOF. From (4.4) and the equality $d(x_0) = 0$ we obtain that vector $\left(\frac{dh}{d\overline{w}}\right)(x_0)$ belongs to the kernel of $B(x_0)$. It follows from the structure of B (see (4.3)) that the vector $\left(\frac{dw}{d\bar{h}}\right)(x_0)$ also belongs to the kernel of $B(x_0)$ (recall that d is real-valued). Due to the condition (2) there exists a constant σ such that

$$\left(\frac{dh}{d\overline{w}}\right)(x_0) = \sigma \left(\frac{dw}{d\bar{h}}\right)(x_0). \tag{4.7}$$

Note that if $\left(\frac{dh}{d\overline{w}}\right)(x_0) = 0$ then theorem is proved. Otherwise, it follows from (4.7) that $(dh)(x_0) = \sigma(dw)(x_0) = |\sigma|^2(dh)(x_0)$ and thus,

$$|\sigma| = 1. \tag{4.8}$$

Now, the direct calculations show that (4.5) – (4.8), the equality $d(x_0) = 0$ and the condition (3) imply

$$(ds)(x_0) = 0.$$

Indeed,

$$\begin{aligned}(ds)(x_0) &= \alpha^2(x_0)(dh)(x_0) + (dw)(x_0) = \left(\frac{\overline{p(x_0)}}{p(x_0)}\sigma + 1\right)(dw)(x_0)\\ &= \left(\left(\frac{\sum_j (dw_j)(x_0)/\overline{\varkappa_j} - \overline{\sigma}\sum_j (d\overline{w_j})(x_0)/\varkappa_j}{\sum_j (d\overline{w_j})(x_0)/\varkappa_j - \sigma\sum_j (dw_j)(x_0)/\overline{\varkappa_j}}\right)\sigma + 1\right)(dw)(x_0) = 0.\end{aligned}$$

Now it follows from (1) that s is a regular function at x_0. □

5. "Dressing up" process

Now we proceed to the general case. Let u and v be regular potentials corresponding to some scattering data with eigenvalues $\varkappa_j$, $j = 1, \dots, N$ and coefficient of reflection $r(k)$. Let matrix $\{\xi_{js}\}$, $j, s = 1, 2$, solve the Riemann – Hilbert problem (3.7) – (3.8). We would like to consider the possibility to add one more eigenvalue. So, let $\varkappa_0 \in \mathbb{C}_+$, $\varkappa_0 \neq \varkappa_j$, $j = 1, \dots, N$. We are going to construct new potentials $\widetilde{u}$, $\widetilde{v}$ by extended initial data. Let $\{\widetilde{\xi}_{js}\}$, $j, s = 1, 2$, solve the Riemann – Hilbert problem (3.7) – (3.8) with additional condition

$$\begin{aligned} &\operatorname{res}_{k=\varkappa_0} \widetilde{\xi}_{21}(x,k) = C_0\widetilde{\xi}_{11}(x,\varkappa_0), &&\operatorname{res}_{k=\varkappa_0} \widetilde{\xi}_{22}(x,k) = C_0\widetilde{\xi}_{12}(x,\varkappa_0),\\ &\operatorname{res}_{k=\overline{\varkappa_0}} \widetilde{\xi}_{11}(x,k) = \overline{C_0}\widetilde{\xi}_{21}(x,\overline{\varkappa_0}), &&\operatorname{res}_{k=\overline{\varkappa_0}} \widetilde{\xi}_{12}(x,k) = \overline{C_0}\widetilde{\xi}_{22}(x,\overline{\varkappa_0}),\\ & && C_0 \neq 0. \end{aligned} \tag{5.1}$$

The solution can be written in the following form.

$$\begin{pmatrix}\widetilde{\xi}_{11}\\ \widetilde{\xi}_{21}\end{pmatrix}=\begin{pmatrix}\xi_{11}\\ \xi_{21}\end{pmatrix}\left(1+\frac{h_1}{k-\varkappa_0}+\frac{h_2}{k-\overline{\varkappa_0}}\right)+\begin{pmatrix}\xi_{12}\\ \xi_{22}\end{pmatrix}\left(\frac{w_1}{k-\varkappa_0}+\frac{w_2}{k-\overline{\varkappa_0}}\right), \tag{5.2}$$

$$\begin{pmatrix}\widetilde{\xi}_{12}\\ \widetilde{\xi}_{22}\end{pmatrix}=\begin{pmatrix}\xi_{11}\\ \xi_{21}\end{pmatrix}\left(\frac{\overline{w_2}}{k-\varkappa_0}+\frac{\overline{w_1}}{k-\overline{\varkappa_0}}\right)+\begin{pmatrix}\xi_{12}\\ \xi_{22}\end{pmatrix}\left(1+\frac{\overline{h_2}}{k-\varkappa_0}+\frac{\overline{h_1}}{k-\overline{\varkappa_0}}\right). \tag{5.3}$$

Since $\widetilde{\xi}_{11}, \widetilde{\xi}_{12}$ are analytic in $\mathbb{C}_+$ and $\widetilde{\xi}_{22}, \widetilde{\xi}_{21}$ are analytic in $\mathbb{C}_-$ we get two additional equations

$$\xi_{11}(\varkappa_0)h_1+\xi_{12}(\varkappa_0)w_1=0,\quad \xi_{21}(\overline{\varkappa_0})h_2+\xi_{22}(\overline{\varkappa_0})w_2=0. \tag{5.4}$$

Similarly to (3.15) we have

$$\widetilde{\alpha}^2:=\exp\left\{-2\mathrm{i}\int_\infty^x \widetilde{u}\,\mathrm{d}s\right\}=\frac{\widetilde{\xi}_{22}^-(0)-\widetilde{\xi}_{12}^+(0)}{\widetilde{\xi}_{11}^+(0)-\widetilde{\xi}_{21}^-(0)}. \tag{5.5}$$

The potential $\widetilde{u}$ is always regular. Note that $\widetilde{\alpha}^2=\overline{\widetilde{\alpha}^{-2}}$. Recall also that

$$\xi_{11}\xi_{22}-\xi_{12}\xi_{21}=1, \tag{5.6}$$

and due to the fact that both potentials u and v are real

$$\xi_{11}(k)=\overline{\xi_{22}(\bar{k})},\quad \xi_{12}(k)=\overline{\xi_{21}(\bar{k})}. \tag{5.7}$$

Using (5.4) we express h_1 and w_2 via w_1 and h_2 correspondingly. Substituting it into (5.2), (5.3) and taking into account (5.6), (5.7) we arrive after simple calculations at the equalities

$$\begin{aligned}\begin{pmatrix}\widetilde{\xi}_{11}\\ \widetilde{\xi}_{21}\end{pmatrix}&=\begin{pmatrix}\xi_{11}\\ \xi_{21}\end{pmatrix}+\frac{\chi_1}{k-\varkappa_0}\left(\xi_{11}(\varkappa_0)\begin{pmatrix}\xi_{12}\\ \xi_{22}\end{pmatrix}-\xi_{12}(\varkappa_0)\begin{pmatrix}\xi_{11}\\ \xi_{21}\end{pmatrix}\right)\\&\quad+\frac{\chi_2}{k-\overline{\varkappa_0}}\left(\overline{\xi_{11}(\varkappa_0)}\begin{pmatrix}\xi_{11}\\ \xi_{21}\end{pmatrix}-\overline{\xi_{12}(\varkappa_0)}\begin{pmatrix}\xi_{12}\\ \xi_{22}\end{pmatrix}\right),\end{aligned} \tag{5.8}$$

$$\begin{aligned}\begin{pmatrix}\widetilde{\xi}_{12}\\ \widetilde{\xi}_{22}\end{pmatrix}&=\begin{pmatrix}\xi_{12}\\ \xi_{22}\end{pmatrix}+\frac{\overline{\chi_2}}{k-\varkappa_0}\left(\xi_{11}(\varkappa_0)\begin{pmatrix}\xi_{12}\\ \xi_{22}\end{pmatrix}-\xi_{12}(\varkappa_0)\begin{pmatrix}\xi_{11}\\ \xi_{21}\end{pmatrix}\right)\\&\quad+\frac{\overline{\chi_1}}{k-\overline{\varkappa_0}}\left(\overline{\xi_{11}(\varkappa_0)}\begin{pmatrix}\xi_{11}\\ \xi_{21}\end{pmatrix}-\overline{\xi_{12}(\varkappa_0)}\begin{pmatrix}\xi_{12}\\ \xi_{22}\end{pmatrix}\right).\end{aligned} \tag{5.9}$$

Here, $\chi_1=w_1/\xi_{11}(\varkappa_0)$, $\chi_2=h_2/\overline{\xi_{11}(\varkappa_0)}$. Substituting it into (5.1) we obtain the system

$$\widetilde{B}\begin{pmatrix}\chi_1\\ \chi_2\end{pmatrix}=\begin{pmatrix}C_0\xi_{11}(\varkappa_0)\\ C_0\xi_{12}(\varkappa_0)\end{pmatrix}, \tag{5.10}$$

where $\widetilde{B}$ is 2×2-matrix with the following entries

$$\begin{aligned}\widetilde{B}_{11}&=\overline{\widetilde{B}_{22}}=1-C_0\left(\frac{\xi_{12}(k)\xi_{11}(\varkappa_0)-\xi_{12}(\varkappa_0)\xi_{11}(k)}{k-\varkappa_0}\right)\Bigg|_{k=\varkappa_0},\\ \widetilde{B}_{12}&=\overline{\widetilde{B}_{21}}=C_0\frac{|\xi_{11}(\varkappa_0)|^2-|\xi_{12}(\varkappa_0)|^2}{\overline{\varkappa_0}-\varkappa_0}.\end{aligned} \tag{5.11}$$

Denote $\tilde{d} := \det \widetilde{B}$. Obviously, $\tilde{d}$ is real-valued. In general, $\tilde{d}$ must have zeros and thus χ_1, χ_2 are singular. But the theorem below shows that generically the vector-valued function (the solutions of the equation)

$$(5.12) \qquad \widetilde{\Psi}(x,k) = \begin{pmatrix} \widetilde{\psi}_1(x,k) \\ \widetilde{\psi}_2(x,k) \end{pmatrix} := \widetilde{\alpha}(x) \begin{pmatrix} \widetilde{\xi}_{11}(x,k) \\ \widetilde{\xi}_{21}(x,k) \end{pmatrix} + (\widetilde{\alpha}(x))^{-1} \begin{pmatrix} \widetilde{\xi}_{12}(x,k) \\ \widetilde{\xi}_{22}(x,k) \end{pmatrix}$$

will be regular and so will be the potential $\widetilde{v}$.

Define

$$(5.13) \qquad \widetilde{s}(x) := \tilde{d}(x)(\widetilde{\alpha}^2(x)\chi_1(x) + \overline{\chi}_2(x)),$$

and

$$(5.14) \qquad \widetilde{p}(x) := \tilde{d}(x)(\widetilde{\xi}^+_{11}(x,0) - \widetilde{\xi}^-_{21}(x,0)).$$

Theorem 5.1. *Let $\tilde{d}(x_0) = 0$. Assume that*

(1) $\tilde{d}'(x_0) \neq 0$.
(2) $\dim\ker \widetilde{B}(x_0) = 1$.
(3) $\widetilde{p}(x_0) \neq 0$.

Then $\widetilde{\xi}$ and thus $\widetilde{v}$ is regular at point x_0.

PROOF. It follows from (5.8), (5.9), and (5.12) that to prove regularity of $\widetilde{\Psi}$ at point x_0 it is sufficient to show that the function $\widetilde{\alpha}^2(x)\chi_1(x) + \overline{\chi}_2(x)$ is regular at x_0. Due to the condition (1) of the theorem the result will follow if we prove the equality $\widetilde{s}(x_0) = 0$. The same arguments as in the proof of Theorem 4.1 show that conditions (1), (2) together with the equality $\tilde{d}(x_0) = 0$ imply the identity

$$(5.15) \qquad (\tilde{d}\overline{\chi_2})(x_0) = \widetilde{\sigma}(\tilde{d}\chi_1)(x_0)$$

for some constant $\widetilde{\sigma}$ such that

$$(5.16) \qquad |\widetilde{\sigma}| = 1.$$

Substituting (5.8), (5.9), (5.15) into (5.5) and taking into account (5.16) and condition (3) of the theorem we obtain that

$$(5.17) \qquad \widetilde{\alpha}^2(x_0) = \frac{\Big(\tilde{d}(x)(\widetilde{\xi}^-_{22}(x,0) - \widetilde{\xi}^+_{12}(x,0))\Big)(x_0)}{\Big(\tilde{d}(x)(\widetilde{\xi}^+_{11}(x,0) - \widetilde{\xi}^-_{21}(x,0))\Big)(x_0)} = -\widetilde{\sigma}.$$

From here and (5.13), (5.15) we obtain the result. □

6. Trace formulas

Let us rewrite the equation (1.1)

$$-f''_{xx}(x,k) + (2ku(x) + v(x))f(x,k) = k^2 f(x,k)$$

in the terms of function $h(x)$

$$f = \exp\Big(\mathrm{i}kx - \mathrm{i}\int_\infty^x u(t)\,\mathrm{d}t + \int_\infty^x h(t)\,\mathrm{d}t\Big).$$

Then

$$(6.1) \qquad u^2 - h^2 - 2\mathrm{i}kh + 2\mathrm{i}uh + v + \mathrm{i}u' - h' = 0.$$

Due to asymptotic properties of the function $f(x,k)$ function $h(x,k)$ has the following asymptotic expansion as $k\to\infty$

$$h(x)=\sum_{j=1}^{\infty}\frac{h_j}{k^j}.$$

Substituting it to the equation (6.1) we have

$$u^2+v-2ih_1+iu'=0,$$

$$-\sum_{p=1}^{l-1}h_ph_{l-p}-2ih_{l+1}+2iuh_l-h_l'=0,\quad l>0.$$

In particular, we obtain

$$\begin{aligned}
h_1&=\frac{1}{2i}(u^2+v+iu'),\\
h_2&=\frac{1}{2i}(u^3+vu)+uu'+\frac{1}{4}(v'+iu''),\\
h_3&=\frac{1}{2i}(2ih_2-h_1^2-h_2')\\
&=\frac{1}{2i}\left(u^4+vu^2+\frac{1}{4}(u^2+v)^2-h_2'\right)+\frac{1}{4}(vu)'+\frac{5}{12}(u^3)'-\frac{1}{8i}(2uu''+(u')^2).
\end{aligned}$$

On the other hand, due to analytical properties of the function $a(k)$ it follows that

$$a(k)=\prod_{1}^{n}\frac{k-k_l}{k-\overline{k_l}}\exp\left(i\int_{-\infty}^{\infty}u(x)\,dx+\frac{1}{2\pi i}\int_{-\infty}^{\infty}\frac{\ln|a(s)|^2}{s-k}\,ds\right).$$

Let us consider asymptotic expansion of $a(k)$ as $k\to\infty$

$$\ln a(k)=i\int_{-\infty}^{\infty}u(x)\,dx+\sum_{1}^{\infty}\frac{a_j}{k^j}.$$

A simple calculation shows that

$$a_j=\frac{-2i}{j}\sum_{l=1}^{n}\operatorname{Im}(k_l^j)-\frac{1}{2\pi i}\int_{-\infty}^{\infty}\ln|a(s)|^2s^{j-1}\,ds. \tag{6.2}$$

Note that for $\operatorname{Im}k>0$

$$\int_{-\infty}^{\infty}h(x)\,dx=i\int_{-\infty}^{\infty}u(x)\,dx-\ln a(k). \tag{6.3}$$

From (6.2) and (6.3) it follows that

$$\int_{-\infty}^{\infty}h_j(x)\,dx=\frac{2i}{j}\sum_{l=1}^{n}\operatorname{Im}(k_l^j)+\frac{1}{2\pi i}\int_{-\infty}^{\infty}\ln|a(s)|^2s^{j-1}\,ds.$$

These relations are trace formulas for equation (1.1). In particular, first three trace formulas have the following form

$$\int_{-\infty}^{\infty} \frac{1}{2\mathrm{i}}(u^2+v)\,\mathrm{d}x = 2\mathrm{i}\sum_{l=1}^{n} \mathrm{Im}(k_l) + \frac{1}{2\pi\mathrm{i}}\int_{-\infty}^{\infty} \ln|a(s)|^2\,\mathrm{d}s,$$

$$\int_{-\infty}^{\infty} \frac{1}{2\mathrm{i}}(u^3+vu)\,\mathrm{d}x = \mathrm{i}\sum_{l=1}^{n} \mathrm{Im}(k_l^2) + \frac{1}{2\pi\mathrm{i}}\int_{-\infty}^{\infty} \ln|a(s)|^2 s\,\mathrm{d}s,$$

$$\int_{-\infty}^{\infty} \frac{1}{2\mathrm{i}}\Big(u^4+vu^2+\frac{1}{4}\big((u^2+v)^2+(u')^2\big)\Big)\,\mathrm{d}x$$

$$= \frac{2}{3}\mathrm{i}\sum_{l=1}^{n} \mathrm{Im}(k_l^3) + \frac{1}{2\pi\mathrm{i}}\int_{-\infty}^{\infty} \ln|a(s)|^2 s^2\,\mathrm{d}s.$$

References

1. P. Deift and E. Trubowitz, *Inverse scattering on the line*, Comm. Pure Appl. Math. **32** (1979), no. 2, 121–25.
2. L. D. Faddeev, *The inverse problem in the quantum theory of scattering*, Uspehi Mat. Nauk **14** (1959), no. 4(88), 57–119 (Russian).
3. M. Jaulent, *On an inverse scattering problem with an energy dependent potential*, Ann. Inst. H. Poincaré Sect. A **17** (1972), 363–378.
4. M. Jaulent and C. Jean, *The inverse problem for the one-dimensional Schrödinger operator with an energy dependent potential*. I, Ann. Inst. H. Poincaré Sect. A **25** (1976), no. 2, 105–118; II, 119–137.
5. D. Kaup, *A higher-order water-wave equation and the method of solving it*, Progr. Theoret. Phys. **54** (1975), no. 2, 396–408.
6. D. Sattinger and J. Sznigielski, *Energy dependent scattering theory*, Differential Integral Equations **8** (1995), no. 5, 945–959.
7. ———, *A Riemann–Hilbert problem for an energy dependent Schrödinger operator*, Inverse Problems **12** (1996), no. 6, 1003–1025.

Department of Mathematics, Royal Institute of Technology, 10044 Stockholm, Sweden

E-mail address: laptev@math.kth.se

Department of Mathematics, University of Wisconsin, Madison, 480 Lincoln Drive, Madison, WI 53706, USA

E-mail address: shterenb@math.wisc.edu

Department of Mathematical Physics, Institute of Physics, St Petersburg University, Ulyanov str. 1, St Petersburg Petrodworetz, 198904 Russia

E-mail address: vvsukhanov@mail.ru

Centre de Recherches Mathématiques
CRM Proceedings and Lecture Notes
Volume **42**, 2007

Theory of Point Processes and Some Basic Notions in Energy Level Statistics

Nariyuki Minami

ABSTRACT. Although it is not explicitly stated in physics literature, energy level statistics is based upon the hypothesis that the spectrum of a quantum Hamiltonian, after suitably normalized (unfolded), and observed in thermodynamic or semiclassical limit, looks like a typical realization of a stationary point process. In this survey article, we present a purely phenomenological view of energy level statistics based on the theory of point processes.

1. Introduction

Energy level statistics is a way to observe the discrete spectra of quantum Hamiltonians, in which one is interested in statistical fluctuation of energy levels (eigenvalues) looked at as a whole, rather than the location of individual levels nor the asymptotic or "average" distribution of levels. It originates in the study by nuclear physicists in early 60s ([33]), in which the excitation spectra of heavy nuclei were modeled by the spectra of random matrices. This led to an extensive study by Dyson, Mehta et al. of random matrices. Their theory was also applied to fluctuation of spectra of small metallic particles ([10]). Later, Berry and Tabor ([3]) proposed to apply energy level statistics to semi-classical Hamiltonians in order to distinguish between regular and irregular spectra, a classification of discrete spectra of quantum Hamiltonians made by Percival ([32]) according to the property of the underlying classical Hamiltonian dynamics describing the system being integrable or non-integrable. They argued that the spectrum of a quantum Hamiltonian which are obtained through quantization of a classical completely integrable Hamiltonian looks, after suitably normalized, like a typical realization of the Poisson point process. (This assertion is now called the "Berry – Tabor conjecture" [17].) They also conjectured that if a classical Hamiltonian dynamical system has chaotic behavior, then the spectrum of its quantization has fluctuation property similar to random matrices. This conjecture was later confirmed numerically ([5]), but it is yet to be clarified mathematically. Fluctuation of spectra was investigated

2000 *Mathematics Subject Classification.* Primary :81Q50; Secondary 82B44, 60G55, 60H25.
Key words and phrases. level statistics, stationary point process, Palm measure, Palm – Khinchin formula, unfolding, level clustering, quantum chaos.
This is the final form of the paper.

also for one-dimensional random Schrödinger operators ([28, 29, 34, 35]), and for multidimensional Anderson tight binding models ([23]). In connection to Anderson transition, it is believed that the spectrum of random Schrödinger operator, considered on a large finite domain, looks like Poisson point process in the energy region where Anderson localization takes place, while in the energy region where delocalization of wave functions is expected, it looks like the spectrum of a random matrix.

In order to give a meaning to the vague notion of "fluctuation of levels," suppose we are given a (random) Hamiltonian H with discrete spectrum

$$\cdots < E_n < E_{n+1} < \cdots < E_{n+m} < \cdots .$$

Among various statistics of the spectrum, one is often interested in spacings between consecutive levels, $S_i = E_{i+1} - E_i$, and wishes to know their statistical characteristics. (see e.g. §1.4 of Mehta [20].) But the following questions immediately arise:

(1) The "density" of levels is in general different in different energy regions. For example, if $\{E_i\}$ are eigenvalues of the Dirichlet Laplacian $-\Delta_D$ on a bounded domain $\Omega \subset \mathbb{R}^d$, the cumulative distribution function

$$N(E) := \sum_i 1_{\{E_i \le E\}}$$

has the asymptotic behavior $N(E) \sim \overline{N}(E) := CE^{d/2}$ $(E \to \infty)$ with a positive constant C. Hence for $d > 2$, levels are distributed more densely for higher energy region, while for $d = 1$, they are distributed more sparsely. Thus in general, S_i's cannot have the same statistical nature.

(2) When one is observing the spectrum in the high energy region, the exact numbering of levels is practically meaningless.

(3) Even if the density of levels is uniform over the entire energy axis, it is not obvious if S_i's have the same probability distribution so that one can discuss the "typical" level spacing.

In physics literature, the first obstacle is overcome by "unfolding" of the spectrum. In the case of Dirichlet Laplacian for example, this amounts to considering the sequence $\{\lambda_i\}_i$, where $\lambda_i := \overline{N}(E_i)$, instead of $\{E_i\}$. By doing so, we obtain a sequence which has asymptotically uniform distribution. In fact, as $L \to \infty$,

$$\sum_i 1_{\{\lambda_i \le L\}} = \sum_i 1_{\{\overline{N}(E_i) \le L\}} = N(\overline{N}^{-1}(L)) \sim \overline{N}(\overline{N}^{-1}(L)) = L.$$

In §5 of this article, we shall discuss the notion of unfolding in a slightly generalized setting.

In order that level statistics makes sense, the unfolded levels $\{\lambda_i\}_i$ are expected to be distributed homogeneously over $\mathbb{R}^1$. Namely, the statistical nature of levels should not depend on the choice of the origin: for any choice of a real number λ, the sequence

$$\ldots, \lambda_n - \lambda, \lambda_{n+1} - \lambda, \ldots, \lambda_{n+m} - \lambda, \ldots$$

as a whole should look like

$$\ldots, \lambda_n, \lambda_{n+1}, \ldots, \lambda_{n+m}, \ldots .$$

Thus we should look at the levels $\{\lambda_i\}$ not one by one, but as a (random) configuration on $\mathbb{R}^1$. In particular, the exact numbering of levels is not essential, and we are thus led to formulate the basic assumption of energy level statistics, which is not clearly stated in physics literature:

After suitably unfolded, the discrete spectrum of a quantum Hamiltonian looks like a typical realization of a stationary point process on $\mathbb{R}^1$.

The purpose of the present article is to present a phenomenological theory of basic notions in statistics of energy spectrum, viewed as a realization of a fictitious stationary point process on $\mathbb{R}$. Much of materials are taken from author's notes ([25–27]) published in preliminary forms, with additional new comments and improvements.

The outline of the article is as follows: In §2, we summarize briefly some basic facts about stationary point processes on the real line, especially the Palm–Khinchin formula. Relations among various statistics such as $E(k, L)$, $F(k, L)$ or $P(S)$ (see, e.g., [2, 20]) are derived as direct corollaries of Palm-Khinchin formula. In §3, we assume that our point process is ergodic. Then as a result of ergodic theorem, various statistics taken along the energy axis coincide with their counterparts taken on the ensemble (namely the probability space at the basis). It will be stressed that even if the unfolded spectrum $\{\lambda_i\}$ is realized as a sample of ergodic point process, the probability distribution of spacings $s_i := \lambda_{i+1} - \lambda_i$ can differ from i to i, and that *the* spacing should be observed not under the probability measure $\mathbb{P}$ given at the beginning, but under the Palm distribution $\widehat{\mathbb{P}}$ which is naturally associated to $\mathbb{P}$ when $\{\lambda_i\}$ has unit mean density. In the next section §4, we consider a deterministic family of sequences of real numbers, which we consider samples of a fictitious stationary point process. This time, there is no ensemble average, but one can consider various statistics taken along $\mathbb{R}$ and deduce relations among them. In particular, a deterministic analogue of Palm-Khinchin formula is obtained. The level statistics discussed in this section is what is usually done in quantum chaology. In §5, we shall discuss two notions of unfolding. While the first one is rather well defined and applied in physics literature (see, e.g., [14]), the second idea, which is essentially due to Berry and Tabor ([3]), is not commonly used. Unfortunately, the present author is unable to clarify the general relationship of these two notions. For some examples, these two notions of unfolding give the same results, but there are also some cases where they lead to different mathematical problems. In the final section §6, we give sufficient conditions for repulsion and clustering of levels in a wide sense.

2. Stationary point processes on the real line; statistics over the ensemble

2.1. Generalities and Palm – Khinchin formula. Let $M_p(\mathbb{R})$ be the totality of integer valued Radon measures on $\mathbb{R}$, and let $\mathcal{M}_p(\mathbb{R})$ be the σ-field on $M_p(\mathbb{R})$generated by mappings

$$M_p(\mathbb{R}) \ni N \mapsto N(I) \in \mathbb{Z}_+$$

for all bounded intervals $I \subset \mathbb{R}$. $\mathcal{M}_p(\mathbb{R})$ is equal to the Borel σ-field generated by the topology of vague convergence of Radon measures in $M_p(\mathbb{R})$ (see [7, 8, 31]). This topology is known to be metrizable.

Definition 2.1. A random variable N_ω defined on a probability space $(\Omega, \mathcal{F}, \mathbb{P})$ and taking values in $\big(M_p(\mathbb{R}), \mathcal{M}_p(\mathbb{R})\big)$ is called a point process on $\mathbb{R}$. It is said to be *stationary*, if its probability law induced on the space $\big(M_p(\mathbb{R}), \mathcal{M}_p(\mathbb{R})\big)$ is invariant under the translation $N(\cdot) \mapsto N(\cdot + x)$ for all $x \in \mathbb{R}$.

To each sample N_ω of a point process, we can associate sequences of random variables $\{x_n(\omega)\}_{n=-\infty}^{\infty}$ and $\{m_n(\omega)\}_{n=-\infty}^{\infty}$ so that

$$N_\omega(\mathrm{d}x) = \sum_{n\in\mathbb{Z}} m_n(\omega)\delta_{x_n(\omega)}(\mathrm{d}x),$$

$\delta_a(\mathrm{d}x)$ being the unit mass concentrated at a. Here the random variables $m_n(\omega) \geq 1$ are integer valued, and $x_n(\omega)$'s satisfy

$$\text{(2.1)} \qquad \cdots < x_{-1}(\omega) < x_0(\omega) \leq 0 < x_1(\omega) < x_2(\omega) < \cdots .$$

$m_n(\omega)$ is called the multiplicity of the point $x_n(\omega)$. The point process N_ω is said to be *simple* if $m_n(\omega) = 1$ almost surely for all $n \in \mathbb{Z}$. In the following, we shall assume that $m := \mathbb{E}\big[N(0,1]\big] < \infty$, and call it the mean density of the point process N_ω. (For notational simplicity, we have written $N(0,1]$ instead of $N\big((0,1]\big)$ or $N_\omega\big((0,1]\big)$.)

Now let us summarize some of the fundamental properties of stationary point processes which are derived directly from the definition. For detailed proofs, the reader is referred to the book by Daley and Vere-Jones [7, 8].

Proposition 2.2. *For any $t \in \mathbb{R}$ and any $x \geq 0$, one has $\mathbb{E}\big[N(t,t+x]\big] = mx$.*

Indeed, by the stationarity we have $\mathbb{E}\big[N(t,t+x]\big] = \mathbb{E}\big[N(0,x]\big]$, which we shall denote by $M(x)$. Then for any $x, y \geq 0$,

$$M(x+y) = \mathbb{E}\big[N(0,x+y]\big] = \mathbb{E}\big[N(0,x]\big] + \mathbb{E}\big[N(x,x+y]\big] = M(x) + M(y).$$

Under the assumption $m = M(1) < \infty$, $M(x)$ is right continuous in x, hence we must have $M(x) = mx$ by a well known argument.

Proposition 2.3. *The limit*

$$\lambda := \lim_{h\searrow 0} \frac{1}{h}\mathbb{P}(N(0,h] > 0)$$

exists and satisfies $\lambda \leq m$. λ is called the intensity *of the stationary point process N.*

For the proofs of limit theorems of this kind, the following lemma is repeatedly used:

Lemma 2.4. *For any function $g(x)$ defined on $[0,\infty)$ which satisfies the conditions*

(1) $\lim_{x\searrow 0} g(x) = g(0) = 0$;
(2) (sub-additivity) $g(x+y) \leq g(x) + g(y)$ $(x, y \geq 0)$,

the limit

$$\alpha := \lim_{x\searrow 0} \frac{1}{x} g(x) \leq \infty$$

exists and satisfies

$$\alpha = \sup_{x>0} \frac{1}{x} g(x).$$

For example, Proposition 2.3 is easily proved if you consider the function

$$\phi(x) := \mathbb{P}(N(0,x] > 0),$$

which obviously satisfies the conditions of the lemma and $\phi(x) \le mx$. Hence the desired limit λ exists and we have $\lambda \le m$.

Proposition 2.5. *If the stationary point process N_ω is simple, then $\lambda = m$.*

From a given stationary point process

$$N_\omega(\mathrm{d}x) = \sum_{n\in\mathbb{Z}} m_n(\omega)\delta_{x_n(\omega)}(\mathrm{d}x),$$

we obtain a new point process

$$N^*_\omega(\mathrm{d}x) := \sum_{n\in\mathbb{Z}} \delta_{x_n(\omega)}(\mathrm{d}x),$$

which is stationary and simple. Obviously N and N^* have the same intensity λ, and by Proposition 2.5, λ is equal to the mean density of N^*.

Proposition 2.6. *Under the assumption $m < \infty$, a stationary point process N is simple if and only if it satisfies the condition*

$$\mathbb{P}(N(0,h] \ge 2) = o(h) \quad (h \searrow 0).$$

As an important example of a point process, let us introduce the stationary Poisson point process:

Definition 2.7. A point process N is called the stationary Poisson point process with parameter $c > 0$, if it satisfies the following two conditions:

(1) For any finite family $A_1, \dots, A_n$ of disjoint Borel subsets of $\mathbb{R}$, $N(A_1), \dots, N(A_n)$ are independent random variables;
(2) If A is a bounded Borel subset of $\mathbb{R}$ with Lebesgue measure $|A|$, then $N(A)$ obeys the Poisson distribution with mean $c|A|$, namely

$$\mathbb{P}(N(A) = k) = e^{-c|A|}\frac{(c|A|)^k}{k!}, \quad k = 0, 1, 2, \dots.$$

It is easy to see that a stationary Poisson point process N is "stationary" in the sense of Definition 2.1. Moreover since

$$\mathbb{P}(N(0,h] \ge 2) = 1 - \mathrm{e}^{-ch} - ch\cdot \mathrm{e}^{-ch} = \mathcal{O}(h^2)$$

as $h \searrow 0$, we see that N is simple by Proposition 2.6, which has been not quite obvious from Definition 2.7. Of course, we have $\lambda = m = c$.

Proposition 2.8. *For $k = 1, 2, \dots$, the limits*

$$\pi_k := \lim_{h\searrow 0} \mathbb{P}(N(0,h] = k \mid N(0,h] > 0)$$

exist, and satisfy $\sum_{k\ge 1}\pi_k = 1$.

Intuitively, π_k is the conditional probability that a point has multiplicity k, given that the origin is occupied by that point. When the stationary point process N is simple, Propositions 2.3 and 2.6 show that $\pi_k = 0$ for $k \ge 2$.

Proposition 2.9. *For each $x > 0$ and $k = 0, 1, 2, \dots$, the limit*

$$Q_k(x) := \lim_{h \searrow 0} \mathbb{P}(N(0, x] \le k \mid N(-h, 0] > 0)$$

exists, and is right-continuous and non-decreasing as a function of x.

Now we can state the celebrated Palm – Khinchin formula:

Theorem 2.10 (the Palm – Khinchin formula). *If N is a simple stationary point process (so that $\lambda = m$) such that*

$$\mathbb{P}(N(-\infty, 0] = N(0, \infty) = \infty) = 1, \tag{2.2}$$

then one has

$$\mathbb{P}(N(0, x] \le k) = \lambda \int_x^\infty q_k(u)\, du, \quad k = 0, 1, 2, \dots, \tag{2.3}$$

where we have set

$$q_k(x) := Q_k(x) - Q_{k-1}(x) = \lim_{h \searrow 0} \mathbb{P}(N(0, x] = k \mid N(-h, 0] > 0),$$

with the convention $Q_{-1}(x) \equiv 0$. Moreover for $k \ge 1$,

$$R_k(x) := 1 - \sum_{j=0}^{k-1} q_j(x) = 1 - Q_{k-1}(x) \tag{2.4}$$

is a probability distribution function on $(0, \infty)$ with mean

$$\int_0^\infty x \, d_x R_k(x) = \frac{k}{\lambda}. \tag{2.5}$$

Note that if we let $x = 0$ in (2.3), we get

$$\lambda \int_0^\infty q_k(u)\, du = 1, \tag{2.6}$$

and hence

$$\mathbb{P}(N(0, x] > k) = \lambda \int_0^x q_k(u)\, du. \tag{2.7}$$

We omit the proof of this theorem as well as of Propositions 2.5 – 2.9, leaving them to [7, 8]. We only note that the fact $R_k(\infty) = 1$ is a consequence of the condition (2.2). In fact, we can write intuitively

$$R_k(x) = 1 - \mathbb{P}(N(0, x] \le k - 1 \mid N(\{0\}) > 0) = \mathbb{P}(N(0, x] \ge k \mid N(\{0\}) > 0).$$

If we let $x \to \infty$, we should have

$$R_k(\infty) = \mathbb{P}(N(0, \infty) \ge k \mid N(\{0\}) > 0).$$

According to (2.2), this probability should be equal to 1 for all k.

2.2. Factorial moment measures of a point process. Given a (not necessarily stationary) point process $N_\omega(dx)$ on $\mathbb{R}$, we define its kth factorial $N_\omega^{[k]}$, which is a random measure on $\mathbb{R}^k$, by letting

$$N_\omega^{[k]}(B) := (N_\omega \times \cdots \times N_\omega)(B \cap \mathbb{R}_k^*), \tag{2.8}$$

where B runs through $\mathcal{B}(\mathbb{R}^k)$, the totality of k-dimensional Borel sets, and we have set

$$\mathbb{R}_k^* := \{(x_1, \ldots, x_k) \in \mathbb{R}^k \mid x_i \neq x_j \ (i \neq j)\}.$$

When the point process N_ω is simple and is written in the form

$$N_\omega(dx) = \sum_{n \in \mathbb{Z}} \delta_{x_n(\omega)}(dx), \tag{2.9}$$

with $\{x_n(\omega)\}$ satisfying (2.1), then we have, for any intervals $I_1, \ldots, I_k$,

$$N_\omega^{[k]}(I_1 \times \cdots \times I_k) = \sharp\big\{\big(x_{i_1}(\omega), \ldots, x_{i_k}(\omega)\big) \bigm| x_{i_j}(\omega) \text{ are distinct and } x_{i_j}(\omega) \in I_j \ (j = 1, \ldots, k)\big\}. \tag{2.10}$$

Also we define the kth factorial moment measure $m^{[k]}(dx_1, \ldots, dx_k)$ as the expectation of $N^{[k]}$, namely

$$m^{[k]}(B) := \mathbb{E}[N^{[k]}(B)], \quad B \in \mathcal{B}(\mathbb{R}^k). \tag{2.11}$$

Suppose $B \in \mathcal{B}(\mathbb{R}^k)$ is of the form

$$B = B_1^{k_1} \times B_2^{k_2} \times \cdots \times B_\ell^{k_\ell}, \tag{2.12}$$

where $B_1, \ldots, B_\ell \in \mathcal{B}(\mathbb{R})$ are disjoint and $k_1 + \cdots + k_\ell = k$. Then it is easy to see that

$$N_\omega^{[k]}(B) = N_\omega(B_1)^{[k_1]} N_\omega(B_2)^{[k_2]} \cdots N_\omega(B_\ell)^{[k_\ell]}, \tag{2.13}$$

where for integers $n \geq 0$ and $j \geq 1$, we let

$$n^{[j]} = \begin{cases} n(n-1)\cdots(n-j+1) & \text{for } n \geq 1 \\ 0 & \text{for } n = 0. \end{cases}$$

As an important example, consider the case where N_ω is the Poisson point process with parameter c. When $B \in \mathcal{B}(\mathbb{R}^k)$ is of the form (2.12), then by (2.13) and by the definition of the Poisson point process, one has

$$\mathbb{E}[N^{[k]}(B_1^{k_1} \times \cdots \times B_\ell^{k_\ell})] = \mathbb{E}[N(B_1)^{[k_1]}] \cdots \mathbb{E}[N(B_1)^{[k_\ell]}] = \prod_{j=1}^{\ell} (c|B_j|)^{k_j}. \tag{2.14}$$

This can be rephrased as follows: when the one-dimensional Borel sets $B_1, B_2, \ldots, B_k$ are such that B_i and B_j $(i \neq j)$ are either disjoint or identical, then

$$\mathbb{E}[N^{[k]}(B_1 \times \cdots \times B_k)] = c^k \prod_{j=1}^{k} |B_j|. \tag{2.15}$$

This can be used to prove the following

Proposition 2.11. *If N is the stationary Poisson point process with parameter c, then for any sequence $A_1, A_2, \ldots, A_k \in \mathcal{B}(\mathbb{R})$ of k Borel sets, one has*

$$m^{[k]}(A_1 \times \cdots \times A_k) = \mathbb{E}[N^{[k]}(A_1 \times \cdots \times A_k)] = c^k \prod_{j=1}^{k} |A_j|.$$

PROOF. Let $\{B_1, \ldots, B_m\}$ be the disjoint partition of the set $\bigcup_{i=1}^n A_i$ generated by $\{A_1, \ldots, A_n\}$, so that we can write

$$A_i = \bigcup_{p \in J_i} B_p$$

with a $J_i \subset \{1, 2, \ldots, m\}$. Then we obtain

$$\begin{aligned}\mathbb{E}[N^{[k]}(A_1 \times \cdots \times A_k)] &= \mathbb{E}\left[N^{[k]}\left(\bigcup_{(p_1,\ldots,p_k)\in J_1\times\cdots\times J_k} (B_{p_1} \times \cdots \times B_{p_k})\right)\right] \\ &= \sum_{(p_1,\ldots,p_k)\in J_1\times\cdots\times J_k} \mathbb{E}[N^{[k]}(B_{p_1} \times \cdots \times B_{p_k})] \\ &= \sum_{(p_1,\ldots,p_k)\in J_1\times\cdots\times J_k} c^k |B_{p_1}| \cdot |B_{p_2}| \cdots |B_{p_k}| = c^k \prod_{j=1}^{k} |A_j|\end{aligned}$$

as desired. □

2.3. Applications to level statistics. As we already noted, the unfolded spectrum of a quantum Hamiltonian is often implicitly assumed to be a realization $\{x_i\}$ of some stationary point process. In the context of quantum chaos, this identification is often something which might be called an intentional confusion. For example in [2], the authors define $E(k, S)$, where $k = 0, 1, 2, \ldots,$ and $S > 0$, to be "the probability to find k (unfolded) levels in a randomly chosen interval of length L." They even write it as $E(k, S, \hat{x})$ in order to indicate that this interval is of the form $(x, x+S]$ with x randomly chosen from $(\hat{x}, c\hat{x}]$ $(c > 1)$ according to the uniform distribution. This is actually a right way to consider the statistics of levels, because the quantal spectrum under consideration does not have the true stochasticity. (We shall return to the discussion of this situation in §4.) But immediately, they omit $\hat{x}$ "to keep the notation simple," and continue their discussion as if the spectrum is a truly random sequence of levels. In this subsection, we shall assume *explicitly* that the unfolded quantal spectrum under consideration is actually represented by a simple stationary point process N_ω defined on a probability space $(\Omega, \mathcal{F}, \mathbb{P})$. Thus we consider $E(k, S)$ to be the probability to find k levels in a *fixed* interval $(x, x+S]$ of length S. By the stationarity, this probability does not depend on the choice of x, so we define

$$E(k, S) := \mathbb{P}(N(0, S] = k). \tag{2.16}$$

Following [2] (see also [13]), we also consider $F(k, S)$, "the probability to count k levels in an interval of length S which starts at an arbitrary quantal level x_i," and $P(k, S)$, "the probability to obtain exactly k levels in an interval of length S beginning at an arbitrary quantal level x_i and ending at the quantal level x_{i+k+1}." $F(k, S)$ should be interpreted as the conditional probability to count k levels in the interval $(0, S]$, given that the origin 0 is occupied by a level. Since the event

$\{N(\{0\}) = 1\}$ has probability zero, we are led to define this conditional probability as

$$F(k,S) := \lim_{h \searrow 0} \mathbb{P}(N(0,S] = k \mid N(-h,0] > 0), \tag{2.17}$$

or $F(k,S) := q_k(S)$ in the notation of Theorem 2.10. To give a meaning to $P(k,S)$, we interpret $P(k,S)\,\mathrm{d}S$ as the conditional probability, given that the origin is occupied by a level, say x_i, to find x_{i+k+1} between S and $S + \mathrm{d}S$. This leads us to consider

$$\mathbb{P}(k,S)\,\mathrm{d}S = \lim_{h \searrow 0} \mathbb{P}(N(0,S] \leq k \mid N(-h,0] > 0) \\ - \lim_{h \searrow 0} \mathbb{P}(N(0,S+dS] \leq k \mid N(-h,0] > 0).$$

In view of this, we define

$$P(k,S)\,\mathrm{d}S := -\mathrm{d}Q_k(S). \tag{2.18}$$

When $k = 0$, we shall write $E(S)$, $F(S)$, and $P(S)$ respectively for $E(0,S)$, $F(0,S)$, and $P(0,S)$. $E(S)$ is often called the "gap probability." $P(S)$, when it exists, is the probability density of the spacing to the level which lies next to the one occupying the origin, because $F(S)$ is the probability that this spacing is greater than S.

We now apply the Palm–Khinchin formula, to obtain

$$\begin{aligned} E(k,S) &= \mathbb{P}(N(0,S] \leq k) - \mathbb{P}(N(0,S] \leq k-1) \\ &= \lambda \int_S^\infty \{q_k(x) - q_{k-1}(x)\}\,\mathrm{d}x \\ &= \lambda \int_S^\infty \{F(k,x) - F(k-1,x)\}\,\mathrm{d}x, \end{aligned}$$

for $k \geq 1$, and

$$E(S) = E(0,S) = \lambda \int_S^\infty F(0,x)\,\mathrm{d}x$$

for $k = 0$, where λ is the intensity of the point process N. Summing over k, we get

$$\lambda \int_S^\infty F(k,x)\,\mathrm{d}x = \sum_{j=0}^{k} E(j,S),$$

or

$$F(k,S) = -\frac{1}{\lambda}\frac{\mathrm{d}^+}{\mathrm{d}S} \sum_{j=0}^{k} E(j,S), \tag{2.19}$$

which is formula (13) of [2]. Recall that $F(k,S)$ is right-continuous in S, so that differentiation from the right is possible. If both sides of (2.19) are again differentiable in S, we can write

$$\begin{aligned} P(k,S) &= -\frac{\mathrm{d}}{\mathrm{d}S}\sum_{i=0}^{k} F(i,S) = \frac{1}{\lambda}\frac{\mathrm{d}^2}{\mathrm{d}S^2}\sum_{i=0}^{k}\sum_{j=0}^{i} E(j,S) \\ &= \frac{1}{\lambda}\frac{\mathrm{d}^2}{\mathrm{d}S^2}\sum_{i=0}^{k}(k-i+1)E(i,S). \end{aligned} \tag{2.20}$$

In particular, letting $k = 0$,

$$P(S) = \frac{1}{\lambda} \frac{\mathrm{d}^2}{\mathrm{d}S^2} E(S). \tag{2.21}$$

The following relations are also valid:

$$\sum_{k=0}^{\infty} E(k, S) = 1; \tag{2.22}$$

$$\sum_{k=0}^{\infty} F(k, S) = 1; \tag{2.23}$$

$$\int_0^{\infty} P(k, S)\,\mathrm{d}S = 1\ ; \tag{2.24}$$

$$\int_0^{\infty} F(k, S)\,\mathrm{d}S = \frac{1}{\lambda}; \tag{2.25}$$

$$\int_0^{\infty} SP(k, S)\,\mathrm{d}S = \frac{k+1}{\lambda}. \tag{2.26}$$

(2.22) and (2.24) are obvious from the definitions of $E(k, S)$ and $P(k, S)\,\mathrm{d}S$. (2.25) and (2.26) follow from (2.6) and (2.5) respectively. (2.23) can be verified in the following way. Since $F(k, S) = Q_k(S) - Q_{k-1}(S)$ and $Q_{-1}(S) = 0$, we have

$$\sum_{k=0}^{\infty} F(k, S) = \lim_{k\to\infty} Q_k(S) = \sup_{k\geq 0} Q_k(S).$$

On the other hand, by the definition of $Q_k(S)$ (Proposition 2.9) and $\mathbb{P}(N(-h, 0] > 0) \sim \lambda h$, we can write

$$Q_k(S) = \lim_{h\searrow 0} \frac{1}{\lambda h} \mathbb{P}(N(0, S] \leq k, N(-h, 0] > 0).$$

Now it is not difficult to show that $\mathbb{P}(N(0, S] \leq k, N(-h, 0] > 0)$, as a function of $h > 0$, satisfies the condition of Lemma 2.4. (The proof of Proposition 2.9 depends on this fact.) Hence

$$Q_k(S) = \sup_{h>0} \frac{1}{\lambda h} \mathbb{P}(N(0, S] \leq k, N(-h, 0] > 0).$$

Interchanging $\sup_k$ and $\sup_{h>0}$, we see

$$\begin{aligned}\sum_{k=0}^{\infty} F(k, S) &= \sup_{h>0} \sup_{k} \frac{1}{\lambda h} \mathbb{P}(N(0, S] \leq k, N(-h, 0] > 0) \\ &= \sup_{h>0} \frac{1}{\lambda h} \mathbb{P}(N(-h, 0] > 0) = 1,\end{aligned}$$

by Proposition 2.3.

Let us now look for a representation of the *number variance* $\Sigma^2(S)$ in terms of $F(k, S)$ or $P(k, S)$. $\Sigma^2(S)$ is defined as the variance of the random variable $N(0, S]$, i.e.,

$$\Sigma^2(S) := \mathbb{E}\left[N(0, S]^2\right] - (\lambda S)^2. \tag{2.27}$$

We can compute, using Theorem 2.10 and $\lambda S = \mathbb{E}[N(0,S]]$,

$$\begin{aligned}\mathbb{E}[N(0,S]^2] - \lambda S &= \sum_{k=2}^{\infty} k(k-1)\mathbb{P}(N(0,S] = k) \\ &= \sum_{k=2}^{\infty} k(k-1)\{\mathbb{P}(N(0,S] > k-1) - \mathbb{P}(N(0,S] > k)\} \\ &= 2\sum_{k=1}^{\infty} k\mathbb{P}(N(0,S] > k) \\ &= 2\lambda \int_0^S \sum_{k=1}^{\infty} kq_k(u)\,\mathrm{d}u.\end{aligned}$$

Thus we obtain

$$\Sigma^2(S) = \lambda S - (\lambda S)^2 + 2\lambda \int_0^S \left(\sum_{k=1}^{\infty} kF(k,u)\right) \mathrm{d}u,$$

or

$$\Sigma^2(S) = \lambda S + 2\lambda \int_0^S \left(\sum_{k=1}^{\infty} kF(k,u) - \lambda u\right) \mathrm{d}u.$$

Formally, this can be rewritten as

$$\Sigma^2(S) = \lambda S + 2\lambda \int_0^S (S-u)\left(\sum_{k=0}^{\infty} P(k,u) - \lambda\right) \mathrm{d}u. \tag{2.28}$$

Here

$$= \rho_2(u) := \sum_{k=0}^{\infty} P(k,u) \tag{2.29}$$

and

$$Y_2(S) := \rho_2(u) - \lambda, \tag{2.30}$$

which are in general singular, are often called *two level correlation* and *two level cluster function* respectively.

Example 2.12. Let N be the stationary Poisson point process with parameter $c > 0$. Then by the independence (see Definition 2.7),

$$F(k,S) = q_k(S) = \mathbb{P}(N(0,S] = k) = E(k,S) = \mathrm{e}^{-cS}\frac{(cS)^k}{k!}; \tag{2.31}$$

$$P(k,S) = -\frac{\mathrm{d}}{\mathrm{d}S}\sum_{i=0}^{k} F(i,S) = c\mathrm{e}^{-cS}\frac{(cS)^k}{k!}, \tag{2.32}$$

and hence

$$\rho_2(u) = c; \quad Y_2(u) = 0; \quad \Sigma_2(S) = cS. \tag{2.33}$$

Example 2.13. For each ω in the probability space $(\Omega, \mathcal{B}[0,1), \mathbb{P})$, where $\Omega = [0,1)$, $\mathcal{B}[0,1)$ is the Borel σ-field on $[0,1)$, and $\mathbb{P}$ is the Lebesgue measure, define

$$N_\omega(\mathrm{d}x) := \sum_{n=-\infty}^{\infty} \delta_{(n+\omega)/\lambda}(\mathrm{d}x), \tag{2.34}$$

where $\lambda > 0$. The resulting point process is obviously simple and stationary. It is just the random translation of the lattice $\lambda^{-1}\mathbb{Z}$.

For any bounded measurable function $f(x)$ with compact support, one has

$$\mathbb{E}\Big[\int_{-\infty}^{\infty} f(x) N_\omega(\mathrm{d}x)\Big] = \int_0^1 \sum_{n=-\infty}^{\infty} f\big(\lambda^{-1}(n+\omega)\big)\,\mathrm{d}\omega = \lambda \int_{-\infty}^{\infty} f(x)\,\mathrm{d}x.$$

In particular letting $f(x) = 1_{(0,1]}(x)$,

$$\mathbb{E}\big[N(0,1]\big] = \lambda.$$

Now fix $k \geq 0$ and $x > 0$. Then for sufficiently small $h > 0$, $N_\omega(-h,0] > 0$ implies $N_\omega(0,x] \leq k$ when $x < (k+1)/\lambda$, and $N_\omega(0,x] > k$ when $x \geq (k+1)/\lambda$. On the other hand, $\mathbb{P}(N(-h,0] > 0) = \lambda h$ for small $h > 0$. Hence

$$Q_k(x) = \lim_{h \searrow 0} \frac{1}{\lambda h} \mathbb{P}(N(0,x] \leq k, N(-h,0] > 0) = 1_{[0,(k+1)/\lambda)}(x). \tag{2.35}$$

Thus we get

$$F(k,S) = q_k(S) = 1_{[k/\lambda,(k+1)/\lambda)}(S); \quad P(k,S) = \delta\Big(S - \frac{k+1}{\lambda}\Big). \tag{2.36}$$

By direct computation, one can verify

$$\Sigma^2(S) = \{\lambda S\}(1 - \{\lambda S\}), \tag{2.37}$$

where $\{\lambda S\}$ is the fractional part of λS.

2.4. Superposition of independent point processes. In several situations, we can consider the spectrum of a quantum system as a superposition of spectra of independent or very weakly interacting subsystems (see, e.g., [4, 14, 15], and also [23, 28]). In our language, this amounts to consider a point process of the form

$$N(\mathrm{d}x) = N_1(\mathrm{d}x) + N_2(\mathrm{d}x) + \cdots + N_p(\mathrm{d}x), \tag{2.38}$$

where $N_1(\mathrm{d}x), N_2(\mathrm{d}x), \ldots, N_p(\mathrm{d}x)$ are stationary point processes which are statistically independent from each other. Let m_i, m, λ_i, λ be respectively the mean density and the intensity N_i and N respectively. Also let $\Sigma_i^2(S)$, $\Sigma^2(S)$, $Y_2^{(i)}(u)$, $Y_2(u)$ be the corresponding number variances and two level cluster functions, and let $E_i(S)$, $E(S)$, $F_i(S)$, $F(S)$ be defined similarly.

From (2.38), we obviously have

$$m = m_1 + m_2 + \cdots + m_p \tag{2.39}$$

and

$$\Sigma^2(S) = \Sigma_i^2(S) + \Sigma_2^2(S) + \cdots + \Sigma_p^2(S) \tag{2.40}$$

from the independence. For small $h > 0$, we have from (2.38), the independence, and Proposition 2.3,

$$\begin{aligned}
\mathbb{P}(N(0,h] > 0) &= 1 - \mathbb{P}(N(0,h] = 0) = 1 - \prod_{i=1}^{p} \mathbb{P}(N_i(0,h] = 0) \\
&= 1 - \prod_{i=1}^{p} (1 - \lambda_i h + o(h)) = \Big(\sum_{i=1}^{p} \lambda_i\Big) h + o(h),
\end{aligned}$$

whence

$$\lambda = \sum_{i=1}^{p} \lambda_i. \tag{2.41}$$

Now suppose $N_1, \dots, N_p$ are all simple, so that $\mathbb{P}(N_i(0,h] \geq 2) = o(h)$ as $h \searrow 0$ for $i = 1, \dots, p$. Since

$$\{N(0,h] \geq 2\} \subset \bigcup_{1 \leq i \leq p} \{N_i(0,h] \geq 2\} \cup \bigcup_{1 \leq i < j \leq p} \{N_i(0,h] \geq 1, N_j(0,h] \geq 1\},$$

we see that

$$\begin{aligned} \mathbb{P}(N(0,h] \geq 2) &\leq \sum_{i=1}^{p} \mathbb{P}(N_i(0,h] \geq 2) + \sum_{i<j} \mathbb{P}(N_i(0,h] \geq 1)\mathbb{P}(N_j(0,h] \geq 1) \\ &= o(h) + \mathcal{O}(h^2) = o(h), \end{aligned}$$

so that by Proposition 2.6, the superposition N is also simple. In this case, if we compare (2.40) with (2.28) and (2.29), we obtain (as in [15])

$$\lambda Y_2(S) = \sum_{i=1}^{p} \lambda_i Y_2^{(i)}(u). \tag{2.42}$$

If we apply (2.19) to

$$E(S) = \mathbb{P}(N(0,S] = 0) = \prod_{i=1}^{p} \mathbb{P}(N_i(0,S] = 0) = \prod_{i=1}^{p} E_i(S), \tag{2.43}$$

we get

$$F(S) = \sum_{i=1}^{p} \frac{\lambda_i}{\lambda} F_i(S) \prod_{j \neq i} E_i(S). \tag{2.44}$$

(Compare [4, 14].)

Remark 2.14. It is natural to consider the case where the system is decomposed into infinitely many non-interacting small subsystems, so that the spectrum of the system is a superposition of a large number of subspectra. To model this situation, suppose for each $n = 1, 2, \dots$, we are given mutually independent point processes $N_1^{(n)}(dx), \dots, N_{p_n}^{(n)}(dx)$ such that $p_n \to \infty$ and that

$$\lim_{n \to \infty} \max_{1 \leq i \leq p_n} \mathbb{P}(N_i^{(n)}(I) \geq 1) = 0 \tag{2.45}$$

holds for each bounded interval I. Such a triangular array of point processes is called *infinitesimal.* The class of all point processes which are obtained as a weak limit of the superposition

$$N^{(n)}(dx) = N_1^{(n)}(dx) + \cdots + N_{p_n}^{(n)}(dx) \tag{2.46}$$

from an infinitesimal triangular array coincides with the class of all infinitely divisible point processes (see [18]), among which the Poisson point process is only a special example. That is, independent superposition of a large number of sparse point processes does not necessarily imply the limiting Poissonian nature, as is often naively supposed. In physics literature, the possibility of obtaining non-Poissonian limits was pointed out in a recent work of Makino and Tasaki ([14]).

It is known ([18]) that when each constituent $N_j^{(n)}(dx)$ is a simple stationary point process, the superposition (2.46) converges weakly to a Poisson point process as $n \to \infty$, if and only if one has

$$\lim_{n\to\infty} \sum_{j=1}^{p_n} \lambda_j^{(n)}\{1 - Q_0^{(n,j)}(u)\} = 0 \tag{2.47}$$

for any $u > 0$, where $\lambda_j^{(n)}$ is the intensity of $N_j^{(n)}(dx)$ and

$$Q_0^{(n,j)}(u) := \lim_{h\searrow 0} \mathbb{P}(N_j^{(n)}(0,u] = 0 \mid N_j^{(n)}(-h,0] > 0).$$

The condition (2.47) is exactly the same as "Case 1" which is formulated in [14].

3. Ergodic point processes; statistics along the sequence

3.1. Ergodic point processes on $\mathbb{R}$. In this section, we strengthen the assumption on the point process N_ω, and suppose that our probability space $(\Omega, \mathcal{F}, \mathbb{P})$ is equipped with a measure preserving *ergodic* flow $\{\theta_t; t \in \mathbb{R}\}$ such that

$$N_{\theta_t\omega}(A) = N_\omega(t + A) \tag{3.1}$$

holds for all $t \in \mathbb{R}$ and any Borel set $A \in \mathcal{B}(\mathbb{R})$. Here $t + A = \{t + x; x \in A\}$ is the translation of the set A. A point process satisfying the condition (3.1) is called a $\{\theta_t\}$-ergodic point process on the probability space $(\Omega, \mathcal{F}, \mathbb{P}; \{\theta_t\})$. As before, we shall assume $m := \mathbb{E}\big[N(0,1]\big] < \infty$.

Example 3.1. Suppose N is a stationary Poisson point process with parameter $c > 0$, and let P_N be its law induced on the measurable space $\big(M_p(\mathbb{R}), \mathcal{M}_p(\mathbb{R})\big)$. Let us then forget the original probability space on which $N = N_\omega$ was given, and take $(M_p(\mathbb{R}), \mathcal{M}_p(\mathbb{R}), P_N)$ as our new probability space, the mapping $M_p(\mathbb{R}) \ni \omega \mapsto N_\omega$ being the identity, and the transformation $\theta_t\omega$ being the spatial translation of $\omega \in M_p(\mathbb{R})$. Then the flow $\{\theta_t\}$ is ergodic, and the new point process satisfies (3.1) and the conditions of Definition 2.1 with the same parameter $c > 0$. Thus a stationary Poisson point process can always be supposed to be ergodic.

Example 3.2. Define the transformations $\{\theta_t; t \in \mathbb{R}\}$ on the unit circle $\Omega = [0,1)$ by $\theta_t\omega = \{t + \omega\}$. It is well known that $\{\theta_t; t \in \mathbb{R}\}$ is an ergodic flow on the probability space $(\Omega, \mathcal{B}[0,1), \mathbb{P})$ of Example 2.13, and it is easy to see that the point process defined by (2.34) is a $\{\theta_t\}$-ergodic point process.

Let N_ω be an ergodic point process, which we consider as representing an unfolded quantal spectrum. The individual ergodic theorem then tells us that various statistics taken along the spectral sequence has non-random limits with probability one. In the following proposition, we formulate some of such limiting relations, for later comparison with level statistics in §4:

Proposition 3.3. (i) *With probability one, we have*

$$\lim_{L\to\infty} \frac{1}{L} N_\omega(0,L] = m. \tag{3.2}$$

(ii) *For any $n \geq 1$, any bounded intervals $I_1, \ldots, I_n$, and any integers $k_1, \ldots, k_n \geq 0$, define*

$$F_\omega := \{t \geq 0; N_\omega(t + I_j) = k_j, j = 1, \ldots, n\}.$$

Then we have with probability one

$$\lim_{L\to\infty}\frac{1}{L}\int_0^L 1_{F_\omega}(t)\,\mathrm{d}t = \mathbb{P}(N(I_j)=k_j, j=1,\dots,n). \tag{3.3}$$

In particular, for any $k=0,1,\dots,$ and $c>0$, the limit

$$\pi_k(c) := \lim_{L\to\infty}\frac{1}{L}\int_0^L 1_{\{N_\omega(t,t+c]=k\}}\,\mathrm{d}t \tag{3.4}$$

exists almost surely, and is equal to $\mathbb{P}(N(0,c]=k)$.

(iii) *For any $k\geq 1$ and any bounded intervals $I_1,\dots,I_n$, we have*

$$\begin{aligned}\lim_{L\to\infty}\frac{1}{L}\int_0^L N_\omega^{[k]}\big((t+I_1)\times\cdots\times(t+I_k)\big)\,\mathrm{d}t \\ = \lim_{L\to\infty}\frac{1}{L}\int_0^L N_{\theta_t\omega}^{[k]}(I_1\times\cdots\times I_k)\,\mathrm{d}t = m^{[k]}(I_1\times\cdots\times I_k).\end{aligned} \tag{3.5}$$

In particular, for any $k=0,1,\dots,$ and $c>0$, the limit

$$\mu_k(c) := \lim_{L\to\infty}\frac{1}{L}\int_0^L (N_\omega(t,t+c])^{[k]}\,\mathrm{d}t \tag{3.6}$$

exists almost surely, and is equal to $\mathbb{E}\big[(N_\omega(0,c])^{[k]}\big]$, the kth factorial moment of the random variable $N_\omega(0,c]$.

3.2. The Palm measure. To each ergodic point process N_ω defined on a probability space $(\Omega,\mathcal{F},\mathbb{P};\{\theta_t\})$, there corresponds a measure $\widehat{\mathbb{P}}$ on $(\Omega,\mathcal{F})$ called the Palm measure of the point process N_ω. The Palm measure clarifies the basic structure of a given ergodic point process. Although $\widehat{\mathbb{P}}$ and $\mathbb{P}$ are mutually singular, it will turn out that the level spacing should be observed under $\widehat{\mathbb{P}}$, but not under $\mathbb{P}$.

The basic materials of this section are taken from [31], to which the reader is referred for detailed proofs.

Theorem 3.4 (Existence and definition of the Palm measure). *If N_ω is a $\{\theta_t\}$-ergodic point process defined on the probability space $(\Omega,\mathcal{F},\mathbb{P};\{\theta_t\})$, then there is a unique measure $\widehat{\mathbb{P}}$ on $(\Omega,\mathcal{F})$ with the following property: for any jointly measurable function $f(\omega,s)\geq 0$ on $\Omega\times\mathbb{R}$,*

$$\int_\Omega \mathbb{P}(\mathrm{d}\omega)\int_{\mathbb{R}} N_\omega(\mathrm{d}s)f(\theta_s\omega,s) = \int_\Omega \widehat{\mathbb{P}}(\mathrm{d}\omega)\int_{\mathbb{R}}\mathrm{d}s\, f(\omega,s). \tag{3.7}$$

If we let $f(\omega,s)=g(\theta_{-s}\omega,s)$ with $g(\omega,s)\geq 0$ jointly measurable, the defining condition (3.7) of the Palm measure is equivalent to

$$\int_\Omega \mathbb{P}(\mathrm{d}\omega)\int_{\mathbb{R}} N_\omega(\mathrm{d}s)g(\omega,s) = \int_\Omega \widehat{\mathbb{P}}(\mathrm{d}\omega)\int_{\mathbb{R}}\mathrm{d}s\, g(\theta_{-s}\omega,s). \tag{3.8}$$

In particular, taking $g(\omega,s)=1_A(s)$, with $A\in\mathcal{B}(\mathbb{R})$, we get

$$\mathbb{E}[N(A)] = |A|\widehat{\mathbb{P}}(\Omega).$$

Letting $A=(0,1]$, we obtain

Proposition 3.5.

$$\widehat{\mathbb{P}}(\Omega) = \mathbb{E}\big[N(0,1]\big] = m.$$

Hence $\widehat{\mathbb{P}}$ is a finite measure if $m < \infty$, but is a probability measure only if $m = 1$.

Proposition 3.6. *The Palm measure* $\widehat{\mathbb{P}}$ *is concentrated on the set*

$$\widehat{\Omega} := \{\omega \in \Omega \mid N_\omega(\{0\}) > 0\}.$$

To see this, take a strictly positive function u with $\int_{\mathbb{R}} u(s)\,\mathrm{d}s = 1$. Then for $k = 0, 1, \ldots,$

$$\begin{aligned}\widehat{\mathbb{P}}(N(\{0\}) = k) &= \int_\Omega \widehat{\mathbb{P}}(\mathrm{d}\omega) \int_{\mathbb{R}} \mathrm{d}s\, 1_{\{N_\omega(\{0\})=k\}} u(s) \\ &= \int_\Omega \mathbb{P}(\mathrm{d}\omega) \int_{\mathbb{R}} N_\omega(\mathrm{d}s) 1_{\{N_{\theta_s\omega}(\{0\})=k\}} u(s) \\ &= \int_\Omega \mathbb{P}(\mathrm{d}\omega) \int_{\mathbb{R}} N_\omega(\mathrm{d}s) 1_{\{N_\omega(\{s\})=k\}} u(s).\end{aligned}$$

This shows $\widehat{\mathbb{P}}(N(\{0\}) = 0) = 0$, and when N is simple, also $\widehat{\mathbb{P}}(N(\{0\}) \geq 2) = 0$.

For non-negative continuous functions $\varphi(x)$ on $\mathbb{R}$, we set

$$\mathcal{L}_N(\varphi) := \mathbb{E}\big[\exp(-N(\varphi))\big], \tag{3.9}$$

where

$$N_\omega(\varphi) := \int_{\mathbb{R}} \varphi(x) N_\omega(\mathrm{d}x),$$

and call it the Laplace functional of the (not necessarily stationary) point process N. The probability law of a point process is uniquely determined by its Laplace functional.

By definition, a sequence $\{N^{(n)}\}$ of point processes converges weakly to a point process N if their law induced on the metric space $M_p(\mathbb{R})$ converge weakly to that of N. It is well known that this is equivalent to the convergence of $\mathcal{L}_{N^{(n)}}(\varphi)$ to $\mathcal{L}_N(\varphi)$ for all φ in the class $C_0^+(\mathbb{R})$, the totality of non-negative continuous functions with a compact support. The following proposition then tells us that the conditional law $\mathbb{P}(N \in \bullet \mid N(-h, 0] > 0)$ of a simple ergodic point process N converges weakly to the probability law $\widehat{\mathbb{P}}(N \in \bullet)/\widehat{\mathbb{P}}(\Omega)$ as $h \searrow 0$.

Proposition 3.7. *Let a point process* N_ω *be* $\{\theta_t\}$*-ergodic and simple, so that* $\widehat{\mathbb{P}}(\Omega) = m = \lambda < \infty$. *Then for any* $\varphi \in C_0^+(\mathbb{R})$, *one has*

$$\frac{1}{\widehat{P}(\Omega)} \int_\Omega \exp(-N_\omega(\varphi))\widehat{\mathbb{P}}(\mathrm{d}\omega) = \lim_{h \searrow 0} \int_\Omega \exp(-N_\omega(\varphi))\mathbb{P}(\mathrm{d}\omega \mid N(-h, 0] > 0). \tag{3.10}$$

PROOF. If we set $H(\omega) := \exp[-N_\omega(\varphi)]$, then $H(\theta_t\omega)$ is continuous in t. Hence letting $g(\omega, s) = H(\omega) 1_{(-h,0]}(s)$ in (3.8), we get

$$\frac{1}{\lambda h} \int_\Omega \mathbb{P}(\mathrm{d}\omega) H(\omega) N_\omega(-h, 0] = \frac{1}{\lambda h} \int_\Omega \widehat{\mathbb{P}}(\mathrm{d}\omega) \int_{-h}^0 \mathrm{d}s\, H(\theta_{-s}\omega).$$

Since N_ω is simple, we have $m = \lambda$, $\mathbb{P}(N(-h, 0] > 0) \sim \lambda h = \mathbb{E}\big[N(-h, 0]\big]$ as $h \searrow 0$. Hence $N_\omega(-h, 0]$ in the above equality can be replaced by $1_{\{N_\omega(-h,0]>0\}}$ in the limit $h \searrow 0$, and the conclusion follows. $\square$

Remark 3.8. If we let $f(\omega, s) = 1_{(-\infty,0]}(s)1_{\{N_\omega(0,x-s]=k\}}$ in (3.7), we obtain the equality

$$\mathbb{P}(N(0,x] \le k) = \int_x^\infty \widehat{\mathbb{P}}(N(0,s] = k)\,\mathrm{d}s. \tag{3.11}$$

Since the integrand on the right-hand side of (3.11) is easily seen to be right continuous, we can conclude from Palm – Khinchin formula (Theorem 2.10) that

$$\widehat{\mathbb{P}}(N(0,x] = k) = \lambda q_k(x) := \lambda \lim_{h \searrow 0} \mathbb{P}(N(0,x] = k \mid N(-h,0] > 0) \tag{3.12}$$

holds for $k = 0, 1, 2, \dots$, and $x \ge 0$. (Note that $q_k(x)$ is also right continuous by Proposition 2.9.) This is not a direct consequence of the weak convergence of $\mathbb{P}(N \in \bullet \mid N(-h,0] > 0)$ to $\widehat{\mathbb{P}}(N \in \bullet)/\lambda$. In fact, weak convergence of a sequence of point process is in general equivalent to the convergence

$$\mathbb{P}(N^{(n)}(I_j) = k_j, j = 1, 2, \dots, p) \longrightarrow \mathbb{P}(N(I_j) = k_j, j = 1, 2, \dots, p)$$

for all $p \ge 1$, all bounded intervals $I_1, \dots, I_p$ such that $\mathbb{P}(N(\partial I_j) > 0) = 0$, and for all integers $k_1, \dots, k_p$. But $\widehat{\mathbb{P}}$ always has a mass at the end point 0 of $(0, x]$.

Example 3.9 (Palm measure of the stationary Poisson point process). Let N_ω be a stationary Poisson point process with parameter $c > 0$ which is $\{\theta_t\}$-ergodic. (See Example 3.1.) Then Proposition 3.7 can be applied to obtain its Palm measure in the following way. If we define for $h > 0$,

$$N_\omega^{1,h}(\cdot) := N_\omega(\cdot \cap (-h, 0]); \quad N_\omega^{2,h}(\cdot) := N_\omega(\cdot \setminus (-h, 0]),$$

then $N_\omega^{1,h}$ and $N_\omega^{2,h}$ are independent. Hence for each $\varphi \in C_0^+(\mathbb{R})$,

$$\begin{aligned}
&\int_\Omega \exp[-N_\omega(\varphi)]\mathbb{P}(\mathrm{d}\omega \mid N(-h,0] > 0) \\
&\quad = \int_\Omega \exp[-N_\omega^{1,h}(\varphi)] \exp[-N_\omega^{2,h}(\varphi)]\mathbb{P}(\mathrm{d}\omega \mid N(-h,0] > 0) \\
&\quad = \mathbb{E}\big[\exp[-N_\omega^{2,h}(\varphi)]\big]\mathbb{E}\big[1_{\{N(-h,0]>0\}} \exp[-N^{1,h}(\varphi)]\big]/\mathbb{P}(N(-h,0] > 0).
\end{aligned}$$

It is easy to see that this converges to

$$\frac{1}{c}\mathrm{e}^{-\varphi(0)}\widehat{\mathbb{E}}\big[\exp[-N(\varphi)]\big].$$

as $h \searrow 0$. Hence, under the probability measure $(1/c)\widehat{\mathbb{P}}$, N has the distribution of the stationary Poisson point process with an extra non-random point placed at the origin.

Example 3.10. For the point process treated in Examples 2.13 and 3.2, it is easy to see that $\widehat{\mathbb{P}}(\mathrm{d}\omega) = \lambda\delta_0(d\omega)$. Under $(1/\lambda)\widehat{\mathbb{P}}$, N is the non-random lattice $\lambda\mathbb{Z}$.

In the remaining part of this section, we assume that our ergodic point process N_ω is simple and has the representation (2.9), with $x_n(\omega)$'s satisfying (2.1). Its Palm measure is concentrated on the set

$$\widehat{\Omega} := \{\omega \in \Omega \mid x_0(\omega) = 0\}.$$

In terms of $x_0(\omega)$ and $x_1(\omega)$, the defining condition (3.7) can be rephrased in the following way:

Proposition 3.11. *For any jointly measurable function $g(\omega, s) \geq 0$ on $\Omega \times [0, \infty)$,*

$$\int_\Omega \mathbb{P}(\mathrm{d}\omega) g\big(\theta_{x_0(\omega)}\omega, -x_0(\omega)\big) = \int_{\widehat{\Omega}} \widehat{\mathbb{P}}(\mathrm{d}\omega) \int_0^{x_1(\omega)} \mathrm{d}s\, g(\omega, s). \tag{3.13}$$

In particular, if we let $g(\omega, s) = f(\omega)$, we get

$$\int_\Omega \mathbb{P}(\mathrm{d}\omega) f(\theta_{x_0(\omega)}\omega) = \int_{\widehat{\Omega}} \widehat{\mathbb{P}}(\mathrm{d}\omega) x_1(\omega) f(\omega), \tag{3.14}$$

and letting $g(\omega, s) = f(\theta_s\omega)$ and noting $g\big(\theta_{x_0(\omega)}\omega, -x_0(\omega)\big) = f(\omega)$, we obtain

$$\int_\Omega \mathbb{P}(\mathrm{d}\omega) f(\omega) = \int_{\widehat{\Omega}} \widehat{\mathbb{P}}(\mathrm{d}\omega) \int_0^{x_1(\omega)} \mathrm{d}s\, f(\theta_s\omega). \tag{3.15}$$

PROOF. The function $a_0(\omega, s) = 1_{\{x_0(\omega) \leq s < x_1(\omega)\}}$ satisfies the relation

$$a_0(\theta_s\omega, -s) = a_0(\omega, s).$$

If we take

$$f(\omega, s) = a_0(\omega, -s) g(\omega, -s)$$

in (3.7), we get

$$\int_\Omega \mathbb{P}(\mathrm{d}\omega) \int_\mathbb{R} N_\omega(\mathrm{d}s)\, a_0(\theta_s\omega, -s) g(\theta_s\omega, -s) = \int_{\widehat{\Omega}} \widehat{\mathbb{P}}(\mathrm{d}\omega) \int_\mathbb{R} \mathrm{d}s\, a_0(\omega, -s) g(\omega, -s).$$

But

$$\begin{aligned}\int_\mathbb{R} N_\omega(\mathrm{d}s)\, a_0(\theta_s\omega, -s) g(\theta_s\omega, -s) &= \int_\mathbb{R} N_\omega(\mathrm{d}s)\, a_0(\omega, s) g(\theta_s\omega, -s) \\ &= g\big(\theta_{x_0(\omega)}\omega, -x_0(\omega)\big),\end{aligned}$$

and since $x_0(\omega) = 0$ for $\omega \in \widehat{\Omega}$ on which $\widehat{\mathbb{P}}$ is concentrated, we have

$$\int_\mathbb{R} \mathrm{d}s\, a_0(\omega, -s) g(\omega, -s) = \int_0^{x_0(\omega)} \mathrm{d}s\, g(\omega, s).$$

Hence we obtain (3.13). □

As corollaries, we obtain the following formulas for the probability distributions of x_0 and x_1 ([31]):

Proposition 3.12. *If we define the measure $\widehat{F}(\mathrm{d}v)$ on $(0, \infty)$ by $\widehat{F}(\mathrm{d}v) = \widehat{\mathbb{P}}(x_1 \in \mathrm{d}v)$, then*

(0) $$\int_0^\infty v\widehat{F}(\mathrm{d}v) = 1;$$

(i) $$\mathbb{P}(-x_0 \in \mathrm{d}s, x_1 - x_0 \in \mathrm{d}v) = 1_{[0,v)}(\mathrm{d}s)\widehat{F}(\mathrm{d}v);$$

(ii) $$\mathbb{P}(x_1 - x_0 \in \mathrm{d}v) = v\widehat{F}(\mathrm{d}v);$$

(iii) $$\mathbb{P}(-x_0 \in \mathrm{d}s) = \mathbb{P}(x_1 \in \mathrm{d}s) = p(s)\,\mathrm{d}s, \quad \textit{where } p(s) = \widehat{F}\big((s, \infty)\big).$$

In particular, the distributions of x_0 and x_1 are absolutely continuous.

PROOF. Applying (3.13) to the function of the form $g(\omega,s) = h(x_1(\omega) - x_0(\omega), s)$, with $h(v,s)$ an arbitrary non-negative Borel function on $\mathbb{R}_+^2$, we obtain, after noting $g(\theta_{x_0(\omega)}\omega, s) = g(\omega,s)$ and that $x_0(\omega) = 0$ for $\omega \in \widehat{\Omega}$,

$$\mathbb{E}\big[h\big(x_1(\omega) - x_0(\omega), -x_0(\omega)\big)\big] = \int_0^\infty \mathrm{d}s \int_{\widehat{\Omega}} \widehat{\mathbb{P}}(\mathrm{d}\omega) 1_{[0,x_1(\omega))}(s) h(x_1(\omega), s), \tag{3.16}$$

which proves the assertion (i). When $h \equiv 1$, (3.16) reduces to the assertion (0), and when $h(v,s) = \psi(v)$, with $\psi \geq 0$ Borel, then we have from (3.16)

$$\mathbb{E}\big[\psi\big(x_1(\omega) - x_0(\omega)\big)\big] = \int_{\widehat{\Omega}} \widehat{\mathbb{P}}(\mathrm{d}\omega)\, x_1(\omega)\psi\big(x_1(\omega)\big),$$

proving (ii). On the other hand if h is of the form $h(v,s) = \varphi(s)$, we get

$$\mathbb{E}\big[\varphi\big(-x_0(\omega)\big)\big] = \int_0^\infty \mathrm{d}s\, \varphi(s) \widehat{\mathbb{P}}(x_1 > s),$$

or

$$\mathbb{P}(-x_0 \in \mathrm{d}s) = \widehat{F}\big((s,\infty)\big)\, \mathrm{d}s.$$

Finally, let $h(v,s) = 1_{(u,\infty)}(v-s)$, where $u > 0$ is arbitrarily fixed. Then we have

$$g\big(\theta_{x_0(\omega)}\omega, -x_0(\omega)\big) = 1_{(u,\infty)}\big(x_1(\omega)\big)$$

for $\omega \in \Omega$, and

$$g(\omega,s) = 1_{(u+s,\infty)}\big(x_1(\omega)\big)$$

for $\omega \in \widehat{\Omega}$ and $s \geq 0$. Thus (3.16) reduces to

$$\mathbb{P}(x_1 > u) = \int_u^\infty \widehat{\mathbb{P}}(x_1 > s)\, \mathrm{d}s,$$

which completes the proof of (iii). □

Now we state a theorem which describes the fundamental structure of ergodic simple point processes. For its proof, we refer the reader to [31] or [12].

Theorem 3.13. *Let $N_\omega = \sum_{n\in\mathbb{Z}} \delta_{x_n(\omega)}$ be a simple, $\{\theta_t\}$-ergodic point process. Then for each $n \in \mathbb{Z}$, the mapping θ_{x_n}, defined by $\theta_{x_n(\omega)}\omega$, maps Ω onto $\widehat{\Omega}$. If we denote by $\hat{\theta}_n$ the restriction of θ_{x_n} to $\widehat{\Omega}$, then $\{\hat{\theta}_n \mid n \in \mathbb{Z}\}$ is an ergodic group of measure preserving transformation on $(\widehat{\Omega}, (1/\lambda)\widehat{\mathbb{P}})$.*

Let $\tau_n(\omega) := x_n(\omega) - x_{n-1}(\omega)$ $(n \in \mathbb{Z})$. Then obviously $\tau_n = \tau \circ \hat{\theta}_1^{n-1}$ holds on $\widehat{\Omega}$, for all $n \in \mathbb{Z}$. By the theorem above, $\{\tau_n\}$, which is the sequence of spacings between points $\{x_n(\omega)\}$, is stationary and ergodic under the probability measure $(1/\lambda)\widehat{\mathbb{P}}$.

Example 3.14. When the random variables $\{\tau_n\}$ are independent, the point process N_ω is called a *renewal process*. The stationary Poisson point process (Example 3.1) and the randomly shifted periodic lattice (Examples 2.13 and 3.2) are both renewal processes. For the case of the Poisson point process, the distribution of τ_n under $(1/\lambda)\widehat{\mathbb{P}}$ is the exponential distribution $\lambda e^{-\lambda t}\, dt$.

Remark 3.15. If N_ω is a renewal process, the distributions of τ_n for $n \neq 1$ under $\mathbb{P}$ and $(1/\lambda)\widehat{\mathbb{P}}$ coincide. But according to (ii) of Proposition 3.12(ii), $\tau_1 = x_1 - x_0$ and τ_n $(n \neq 1)$ have different distribution in general. When $\{x_n\}$ is Poissonian, this phenomena is called the "waiting time paradox" ([7,8]). In fact, the only renewal process such that τ_1 and τ_n $(n \neq 1)$ are identically distributed under $\mathbb{P}$ is the randomly shifted lattice $\lambda^{-1}\mathbb{Z}$ discussed in Examples 2.13, 3.2 and 3.10. This is seen by solving the equation

$$\frac{1}{\lambda}\widehat{F}(\mathrm{d}v) = v\hat{F}(\mathrm{d}v).$$

3.3. The level spacing distribution. Let $N_\omega = \sum_{n\in\mathbb{Z}}\delta_{x_n(\omega)}$ be a simple, ergodic point process. In the previous section, we saw that the spacings $\tau_n(\omega) := x_n(\omega) - x_{n-1}(\omega)$ of points are identically distributed under the probability measure $(1/\lambda)\widehat{\mathbb{P}}$. But under the original probability measure $\mathbb{P}$, the distribution of τ_n can be different for different n. To see this, first note that $\tau_n(\omega) = \tau_n(\theta_{x_0}\omega)$. By (3.14) of Proposition 3.11, for any $t \geq 0$,

$$P(\tau_n(\omega) > t) = P(\tau_n \circ \theta_{x_0} > t) = \int_{\widehat{\Omega}} \widehat{\mathbb{P}}(\mathrm{d}\omega)x_1(\omega)1_{\{\tau_n(\omega)>t\}}. \tag{3.17}$$

If N is not a renewal process, x_1 and τ_n $(n \neq 1)$ are correlated under $\widehat{\mathbb{P}}$. Hence the right hand side of (3.17) can be different for different n.

This poses a question: Suppose we are given a sample

$$\cdots < x_{-1}(\omega) < x_0(\omega) \leq 0 < x_1(\omega) < x_2(\omega) < \cdots$$

of an ergodic point process, which we regard as an unfolded quantal spectrum. If you wish to determine the level spacing distribution experimentally, then what you naturally do would be to collect a large set of data $\tau_k(\omega) = x_k(\omega) - x_{k-1}(\omega)$, $1 \leq k \leq n$, of level spacings, and then draw a histogram of their frequency, expecting that it will approximately represent some smooth curve when the data size n is large enough and the histogram is fine enough. Mathematically speaking, this amounts to considering the empirical distribution

$$F^n_\omega(\mathrm{d}t) := \frac{1}{n}\sum_{k=1}^{n}\delta_{\tau_k(\omega)}(\mathrm{d}t), \tag{3.18}$$

and expecting this to converge, as $n \to \infty$, to some fixed distribution $F(\mathrm{d}t)$ almost surely. But if the random variables τ_n are observed under the probability measure $\mathbb{P}$, they are not identically distributed. Hence mixing these to form empirical distributions seems meaningless. On the other hand, if the data $\{\tau_n\}$ are observed under $(1/\lambda)\widehat{\mathbb{P}}$, they form a stationary ergodic sequence. In this case, the ergodic theorem works well, and $F^n_\omega(\mathrm{d}t)$ converges to $(1/\lambda)\widehat{\mathbb{P}}(\tau_1 \in \mathrm{d}t)$ as $n \to \infty$, $\widehat{\mathbb{P}}$-almost surely. But $\widehat{\mathbb{P}}$ and $\mathbb{P}$ are mutually singular. Which of them should be considered as the real law governing $\{\tau_n\}$?

Fortunately, this puzzle is easily solved:

Proposition 3.16. *The empirical distribution $F^n_\omega(\mathrm{d}t)$ defined by (3.18) converges weakly, as $n \to \infty$, to $(1/\lambda)\widehat{\mathbb{P}}(\tau_1 \in \mathrm{d}t)$ for $\mathbb{P}$-almost all $\omega \in \Omega$.*

PROOF. For each $t \geq 0$, define

$$A_t = \left\{\omega \in \Omega \;\middle|\; \lim_{n\to\infty}\frac{1}{n}\sum_{k=1}^{n}1_{\{\tau_k(\omega)>t\}} = \frac{1}{\lambda}\widehat{\mathbb{P}}(\tau_1 > t)\right\}.$$

As was already pointed out, $\tau_k = \tau_1 \circ \hat{\theta}_1^{k-1}$ holds on $(\widehat{\Omega}, (1/\lambda)\widehat{\mathbb{P}})$, $\hat{\theta}_1$ being an ergodic automorphism. Hence by the individual ergodic theorem,

$$\lim_{n\to\infty} \frac{1}{n} \sum_{k=1}^{n} 1_{\{\tau_k(\omega)>t\}} = \lim_{n\to\infty} \frac{1}{n} \sum_{k=1}^{n} 1_{\{\tau_1>t\}} \circ \hat{\theta}_1^{k-1}(\omega) = \frac{1}{\lambda}\widehat{\mathbb{P}}(\tau_1 > t)$$

holds for $(1/\lambda)\widehat{\mathbb{P}}$-almost all $\omega \in \widehat{\Omega}$, namely

$$\frac{1}{\lambda}\widehat{\mathbb{P}}(\widehat{\Omega} \cap A_t) = 1.$$

It remains to show $\mathbb{P}(A_t) = 1$. But since $\omega \in A$ if and only if $\theta_{x_0}\omega \in A$, we have from Proposition 3.11,

$$\begin{aligned}\mathbb{P}(A) &= \int_{\Omega} \mathbb{P}(\mathrm{d}\omega) 1_A(\omega) = \int_{\Omega} \mathbb{P}(\mathrm{d}\omega) 1_A(\theta_{x_0}\omega) \\ &= \int_{\widehat{\Omega}} \widehat{\mathbb{P}}(\mathrm{d}\omega) x_1(\omega) 1_A(\omega) = \int_{\widehat{\Omega}} \widehat{\mathbb{P}}(\mathrm{d}\omega) x_1(\omega) = 1,\end{aligned}$$

the last equality following from Proposition 3.12(0). □

Remark 3.17. By the Palm-Khinchin formula, we have

$$\mathbb{P}(N(0,x] = 0) = \int_x^\infty \widehat{\mathbb{P}}(N(0,s] = 0)\,\mathrm{d}s = \int_x^\infty \widehat{\mathbb{P}}(x_1 > t)\,\mathrm{d}t.$$

Thus if we put $\pi_0(x) := \mathbb{P}(N(0,x] = 0)$, and $\rho(t) := \widehat{\mathbb{P}}(x_1 > t)$, we get

$$\pi_0(x) = \int_x^\infty \rho(t)\,\mathrm{d}t = 1 - \int_0^x \rho(t)\,\mathrm{d}t, \tag{3.19}$$

or

$$\rho(x) = -\frac{\mathrm{d}^+}{\mathrm{d}x}\pi_0(x). \tag{3.20}$$

The consideration in the proof of Proposition 3.16 can be easily generalized to yield the following assertion:

Proposition 3.18. *For any non-negative measurable function f on $(\Omega, \mathcal{F}, \mathbb{P})$, one has*

$$\lim_{L\to\infty} \frac{1}{L} \int_{(0,L]} N_\omega(\mathrm{d}t) f(\theta_t\omega) = \int_{\widehat{\Omega}} \widehat{\mathbb{P}}(\mathrm{d}\omega) f(\omega) \tag{3.21}$$

for $\mathbb{P}$-almost all $\omega \in \Omega$.

PROOF. If we let

$$A = \left\{\omega \in \Omega \;\middle|\; \lim_{n\to\infty} \frac{1}{n} \sum_{k=1}^{n} f(\theta_{x_k}\omega) = \frac{1}{\lambda} \int_{\widehat{\Omega}} \widehat{\mathbb{P}}(\mathrm{d}\omega) f(\omega)\right\},$$

then by the ergodic theorem applied to $(\widehat{\Omega}, \widehat{\mathbb{P}}, \hat{\theta}_1)$, we have $\widehat{\mathbb{P}}(\widehat{\Omega} \cap A) = \lambda$. But since $\omega \in A$ is equivalent to $\theta_{x_0}\omega \in A$, the argument in Proposition 3.16 gives $\mathbb{P}(A) = 1$. On the other hand, we have

$$\lim_{L\to\infty} \frac{1}{L} N_\omega(0, L] = \lambda$$

$\mathbb{P}$-almost surely. Hence for $\mathbb{P}$-almost all $\omega \in \Omega$, we have $N_\omega(0, L] \to \infty$ and

$$\begin{aligned}\lim_{L\to\infty} \frac{1}{L} \int_{(0,L]} N_\omega(\mathrm{d}x) f(\theta_t\omega) &= \lim_{L\to\infty} \frac{1}{L} \sum_j 1_{(0,L]}\big(x_j(\omega)\big) f(\theta_{x_j}\omega) \\ &= \lim_{L\to\infty} \frac{N_\omega(0,L]}{L} \cdot \lim_{n\to\infty} \frac{1}{n} \sum_{j=1}^n f(\theta_{x_j}\omega) \\ &= \lambda \cdot \frac{1}{\lambda} \int_{\widehat{\Omega}} \widehat{\mathbb{P}}(\mathrm{d}\omega) f(\omega) = \int_{\widehat{\Omega}} \widehat{\mathbb{P}}(\mathrm{d}\omega) f(\omega). \qquad \square\end{aligned}$$

Let us apply this proposition to the limiting k-level correlation, for later comparison with level statistics in §§4 and 6.

Proposition 3.19. *For realizations $\{x_n(\omega)\}$ of a simple ergodic point process, define the k-level correlation*

$$\begin{aligned}(3.22)\quad & R_L^\omega(I_2, \dots, I_k) \\ &:= \frac{1}{L} \sharp\big\{ (x_{i_1}(\omega), \dots, x_{i_k}(\omega)) \mid x_{i_1}(\omega) \le L,\ i_1, \dots, i_k \text{ are distinct} \\ &\qquad\qquad \text{and } x_{i_j}(\omega) - x_{i_1}(\omega) \in I_j,\ j = 2, \dots, k \big\},\end{aligned}$$

where $L > 0$ and $I_2, \dots, I_k$ are any bounded intervals. Then

$$(3.23)\qquad \lim_{L\to\infty} R_L^\omega(I_2, \dots, I_k) = \int_{\widehat{\Omega}} \widehat{\mathbb{P}}(\mathrm{d}\omega) N_\omega^{[k-1]}(I_2^* \times \cdots \times I_k^*) =: R_k(I_2, \dots, I_k)$$

holds for $\mathbb{P}$-almost all $\omega \in \Omega$, where we have set $I_j^* := I_j \setminus \{0\}$.

PROOF. Noting (2.10), we can write

$$\begin{aligned}& LR_L^\omega(I_2, \dots, I_k) \\ &= \sum_i 1_{(0,L]}(x_i(\omega)) \sharp\big\{ (x_{i_2}(\omega), \dots, x_{i_k}(\omega)) \mid i_2, \dots, i_k \text{ are distinct}, \\ &\qquad\qquad x_{i_j}(\omega) \ne x_{i_1}(\omega),\ x_{i_j}(\omega) \in I_j + x_{i_j}(\omega), j = 2, \dots, k \big\} \\ &= \int N_\omega(\mathrm{d}s) 1_{(0,L]}(s) N_\omega^{[k-1]}\big((I_2 + s) \setminus \{s\}\big) \times \cdots \times \big((I_k + s) \setminus \{s\}\big) \\ &= \int N_\omega(\mathrm{d}s) 1_{(0,L]}(s) N_{\theta_s\omega}^{[k-1]}(I_2^* \times \cdots \times I_k^*),\end{aligned}$$

from which the conclusion follows by Proposition 3.18. $\square$

Remark 3.20. In case $I_2 = \cdots = I_k = (0, c]$, let us write $R_k(I_2, \dots, I_k) = R_k(c)$. By the Palm-Khinchin formula, we can compute

$$\begin{aligned}\frac{1}{k!} \mathbb{E}\big[(N(0,L])^{[k]}\big] &= \sum_{n\ge k} \binom{n-1}{k-1} \mathbb{P}(N(0,c] \ge n) \\ &= \sum_{n\ge k} \binom{n-1}{k-1} \int_0^c \widehat{\mathbb{P}}(N(0,s] = n-1)\,\mathrm{d}s \\ &= \frac{1}{(k-1)!} \int_0^c \int_{\widehat{\Omega}} \widehat{\mathbb{P}}(\mathrm{d}\omega) (N_\omega(0,s])^{[k-1]},\end{aligned}$$

namely

$$\mu_k(c) = k \int_0^c R_k(s)\,\mathrm{d}s. \tag{3.24}$$

See [6] for related formulas.

Example 3.21. If N_ω is the stationary Poisson point process with parameter $\lambda > 0$, then by Example 3.9 and Proposition 2.11,

$$R_k(I_2, \ldots, I_k) = \lambda \int_\Omega \mathbb{P}(\mathrm{d}\omega) N_\omega^{[k-1]}(I_2 \times \cdots \times I_k) = \lambda^k \prod_{j=2}^{k} |I_j|. \tag{3.25}$$

In particular, $R_k(c) = \lambda^k c^{k-1}$.

Remark 3.22. If $\{x_j(\omega)\}$ is a renewal process, then $R_2(c) = \lambda^2 c$ implies that N is Poissonian. Indeed, if we set

$$F(c) = \frac{1}{\lambda}\widehat{P}(N(0,c] > 0) = \frac{1}{\lambda}\widehat{P}(x_{n+1} - x_n \le c),$$

which does not depend on n, then

$$\begin{aligned}\frac{1}{\lambda}R_2(c) &= \frac{1}{\lambda}\int_{\widehat{\Omega}} \widehat{\mathbb{P}}(\mathrm{d}\omega) N_\omega(0,c] = \frac{1}{\lambda}\sum_{n=1}^{\infty} \widehat{\mathbb{P}}(N(0,c] \ge n)\\ &= \frac{1}{\lambda}\sum_{n=1}^{\infty} \widehat{\mathbb{P}}\Big(\sum_{j=0}^{n-1}(x_{j+1} - x_j) \le c\Big) = \sum_{n=1}^{\infty}(F^{n*})(c),\end{aligned}$$

where F^{n*} is the n-fold convolution of F. Hence if $\mathcal{L}(\xi) = \int_0^\infty \mathrm{e}^{-\xi c}\mathrm{d}F(c)$ is the Laplace transform of $\mathrm{d}F(c)$, then from $R_2(c) = \lambda^2 c$, we get

$$\sum_{n=1}^{\infty}\big(\mathcal{L}(\xi)\big)^n = \frac{\mathcal{L}(\xi)}{1 - \mathcal{L}(\xi)} = \frac{1}{\lambda}\int_0^\infty \mathrm{e}^{-\xi c}\mathrm{d}R_2(c) = \frac{\lambda}{\xi},$$

namely $\mathcal{L}(\xi) = \lambda/(\lambda + \xi)$. Hence $\mathrm{d}F(c) = \lambda e^{-\lambda c}\mathrm{d}c$. But the Poisson point process is the only renewal process for which $x_{n+1} - x_n$ obeys the exponential distribution.

4. Statistics for asymptotically uniformly distributed sequences

Suppose we are given a finite or infinite sequence

$$0 \le x_1(L) < x_2(L) < \cdots < x_n(L) < \cdots$$

depending on a parameter L, which we shall let tend to infinity. We think of this sequence $\{x_j(L)\}_j$ as the unfolded spectrum of a quantum Hamiltonian depending on a parameter, or more directly a realization of some fictitious simple ergodic point process with mean density (or intensity) λ, if it is asymptotically uniformly distributed in the following sense:

Condition 1. There is a positive constant $A \le \infty$ and $\lambda < \infty$ such that for any $0 \le \alpha < A$,

$$\sharp\{j \ge 1 \mid x_j(L) \le \alpha L\} \sim \lambda\alpha L \quad (L \to \infty). \tag{4.1}$$

In particular, when the sequence $\{x_j(L)\}_j$ does not depend on L and satisfies

$$\sharp\{j \geq 1 \mid x_j \leq L\} \sim \lambda L \quad (L \to \infty), \tag{4.2}$$

then it satisfies Condition 1.

For each $c > 0$ and for each interval $I = (a, b]$, we consider the following quantity (statistics):

$$N_L(I) := \sharp\{j \geq 1 \mid x_j(L) \in I\}; \tag{4.3}$$

$$\pi_k(L; I) := \frac{1}{L}\int_0^L 1_{\{N_L(t+I)=k\}}\,dt \qquad (k \geq 0); \tag{4.4}$$

$$\mu_k(L; I) := \frac{1}{L}\int_0^L N_L(t+I)^{[k]}\,dt = \sum_{j\geq k} j^{[k]}\pi_j(L; I) \qquad (k \geq 1); \tag{4.5}$$

$$\rho(L; c) := \frac{1}{L}\sharp\{j \geq 1 \mid x_j(L) \leq L, x_{j+1}(L) - x_j(L) > c\} \quad (c > 0); \tag{4.6}$$

$$R_k(L; c) := \frac{1}{L}\sharp\{(x_{i_1}(L), \ldots, x_{i_k}(L)) \mid i_1, \ldots, i_k \text{ are distinct}, \tag{4.7}$$
$$x_{i_1}(L) \leq L, x_{i_j}(L) - x_{i_1}(L) \in (0, c]\} \qquad (c > 0, k \geq 2).$$

As an immediate consequence of Condition 1, we obtain the following

Proposition 4.1. *For any* $c > 0$,

$$\mu_1(c) := \lim_{L\to\infty} \frac{1}{L}\int_0^L N_L(t, t+c]\,dt = \lim_{L\to\infty} \mu_1(L; (0, c]) = \lambda c. \tag{4.8}$$

PROOF. By the definition,

$$\mu_1(L; (0, c]) = \frac{1}{L}\sum_j \int_0^L 1_{(t,t+c]}(x_j(L))\,dt = \frac{1}{L}\sum_j \int_0^L 1_{[x_j(L)-c, x_j(L))}(t)\,dt. \tag{4.9}$$

But when $L > c > 0$, one has

$$\int_0^L 1_{[x_j(L)-c, x_j(L))}(t)\,dt = \begin{cases} c, & \text{if } c \leq x_j(L) \leq L \\ 0, & \text{if } x_j(L) > L + c, \end{cases}$$

while

$$\int_0^L 1_{[x_j(L)-c, x_j(L))}(t)\,dt \leq c, \quad \text{if } x_j(L) < c \text{ or } L < x_j(L) \leq L + c.$$

Hence for each fixed $c > 0$, it is obvious from Condition 1 that

$$\mu_1(L; (0, c]) \sim c\frac{1}{L}\sharp\{j \geq 1 \mid x_j(L) \leq L\} \quad \text{as } L \to \infty,$$

so that

$$\mu_1(c) = \lim_{L\to\infty} \mu_1(L; (0, c]) = \lambda c,$$

which was to be proved. □

Except for this trivial case, Condition 1 does not imply that any of the statistics (4.4) – (4.7) converges to some limit as $L \to \infty$. But when some of these convergences occur, we can prove equalities between them, which may be called deterministic analogues of the Palm – Khinchin formula, as the following two propositions show. (Compare Remarks 3.17 and 3.20.)

Proposition 4.2. *Let $k \geq 2$ and assume that*

$$\sum_{x_{j_1}(L)<\cdots<x_{j_k}(L)<c} \big(c - x_{j_k}(L)\big) = o(L) \quad (L \to \infty) \tag{4.10}$$

holds for any $c > 0$.

(i) *If the limit*

$$\mu_k(c) := \lim_{L\to\infty} \mu_k(L;(0,c]) \tag{4.11}$$

exists for each $c > 0$ and is differentiable with respect to c, then

$$R_k(c) := \lim_{L\to\infty} R_k(L;c) \tag{4.12}$$

also exists and one has

$$R_k(c) = \frac{1}{k}\frac{\mathrm{d}}{\mathrm{d}c}\mu_k(c). \tag{4.13}$$

(ii) *If on the other hand, $R_k(c)$ exists for each $c > 0$, then $\mu_k(c)$ also exists and*

$$\mu_k(c) = k\int_0^c R_k(c')\mathrm{d}c'. \tag{4.14}$$

Remark 4.3. The condition (4.10) is obviously satisfied when the sequence $\{x_j(L)\}_j$ does not depend on L.

Proof of Propsition 4.2. Since

$$\begin{aligned}\frac{1}{k!}(N_L(t,t+c])^{[k]} &= \sum_{x_{j_1}(L)<\cdots<x_{j_k}(L)} 1_{(t,t+c]}\big(x_{j_1}(L)\big)\cdots 1_{(t,t+c]}\big(x_{j_k}(L)\big)\\ &= \sum_{x_{j_1}(L)<\cdots<x_{j_k}(L)} 1_{[x_{j_1}(L)-c,x_{j_1}(L))}(t)\cdots 1_{[x_{j_k}(L)-c,x_{j_k}(L))}(t),\end{aligned}$$

we can compute

$$\begin{aligned}&\frac{L}{k!}\mu_k(L;(0,c])\\ &\quad= \sum_{x_{j_1}(L)<\cdots<x_{j_k}(L)} \left|[0,L]\cap\bigcap_{p=1}^k [x_{j_p}(L)-c, x_{j_p}(L))\right|\\ &\quad= \sum_{x_{j_1}(L)<\cdots<x_{j_k}(L)} \big\{(L\wedge x_{j_1}(L)) - (x_{j_k}(L)-c)_+\big\}_+\\ &\quad= \sum_{\substack{x_{j_1}(L)<\cdots<x_{j_k}(L)\\ x_{j_1}(L)\leq L, x_{j_k}(L)\geq c}} \{c - x_{j_k}(L) + x_{j_1}(L)\}_+ + \sum_{L<x_{j_1}(L)<\cdots<x_{j_k}(L)} \{c - x_{j_k}(L) + L\}_+\\ &\qquad\qquad + \sum_{x_{j_1}(L)<\cdots<x_{j_k}(L)<c} x_{j_1}(L)\end{aligned}$$

$$
\begin{aligned}
&= \sum_{x_{j_1}(L)<\cdots<x_{j_k}(L)\le L} \{c - x_{j_k}(L) + x_{j_1}(L)\}_+ \\
&\quad + \Bigg[\sum_{\substack{x_{j_1}(L)<\cdots<x_{j_k}(L) \\ x_{j_1}(L)\le L<x_{j_k}(L)}} \{c - x_{j_k}(L) + x_{j_1}(L)\}_+ + \sum_{L<x_{j_1}(L)<\cdots<x_{j_k}(L)} \{c - x_{j_k}(L) + L\}_+ \Bigg] \\
&\quad + \sum_{x_{j_1}(L)<\cdots<x_{j_k}(L)<c} [x_{j_1}(L) - \{c - x_{j_k}(L) + x_{j_1}(L)\}_+].
\end{aligned}
$$

The third term equals

$$
- \sum_{x_{j_1}(L)<\cdots<x_{j_k}(L)<c} (c - x_{j_k}(L)),
$$

and is $o(L)$ as $L \to \infty$ by the assumption. On the other hand, the second term, which is enclosed in the big bracket and which we denote by $M(L)$, can be estimated in two ways. First of all, we have

$$
\begin{aligned}
M(L) &\le \sum_{\substack{x_{j_1}(L)<\cdots<x_{j_k}(L); \\ x_{j_1}(L)\le L<x_{j_k}(L) \text{ or} \\ L<x_{j_1}(L)<\cdots<x_{j_k}(L)\le L+c}} \left(c - x_{j_k}(L) + x_{j_1}(L)\right)_+ \\
&= \sum_{\substack{x_{j_1}(L)<\cdots<x_{j_k}(L); \\ x_{j_1}(L)\le L<x_{j_k}(L) \text{ or} \\ L<x_{j_1}(L)<\cdots<x_{j_k}(L)\le L+c}} \int_{-\infty}^{\infty} 1_{(t,t+c]}\left(x_{j_1}(L)\right) \cdots 1_{(t,t+c]}\left(x_{j_k}(L)\right) dt.
\end{aligned}
$$

Suppose $x_{j_1}(L) \le L < x_{j_k}(L)$. Then we have $1_{(t,t+c]}\left(x_{j_1}(L)\right) = 0$ for $t > L + c$ and $1_{(t,t+c]}\left(x_{j_k}(L)\right) = 0$ for $t < L - c$. If, on the other hand, $L < x_{j_1}(L) < \cdots < x_{j_k}(L) \le L+c$, then we have $1_{(t,t+c]}\left(x_{j_k}(L)\right) = 0$ for $t > L+c$ and $1_{(t,t+c]}\left(x_{j_1}(L)\right) = 0$ for $t < L - c$. Hence

$$
\begin{aligned}
M(L) &\le \sum_{x_{j_1}(L)<\cdots<x_{j_k}(L)} \int_{L-c}^{L+c} 1_{(t,t+c]}\left(x_{j_1}(L)\right) \cdots 1_{(t,t+c]}\left(x_{j_k}(L)\right) dt \\
&= \frac{1}{k!} \int_{L-c}^{L+c} (N_L(t, t+c])^{[k]}\, dt,
\end{aligned}
$$

which is of $o(L)$ $(L \to \infty)$ when $\mu_k(c)$ exists.

Secondly we also have the inequality

$$
\begin{aligned}
M(L) &\le c \sum_{\substack{x_{j_1}(L)<\cdots<x_{j_k}(L); \\ x_{j_1}(L)\le L<x_{j_k}(L) \text{ or} \\ L<x_{j_1}(L)<\cdots<x_{j_k}(L)\le L+c}} 1_{\{x_{j_k}(L)-x_{j_1}(L)\le c\}} \\
&\le c \sum_{x_{j_1}(L)<\cdots<x_{j_k}(L)\le L+c} 1_{\{x_{j_k}(L)-x_{j_1}(L)\le c\}} - c \sum_{x_{j_1}(L)<\cdots<x_{j_k}(L)\le L} 1_{\{x_{j_k}(L)-x_{j_1}(L)\le c\}}.
\end{aligned}
$$

Obviously, the right-hand side is of $o(L)$ as $L \to \infty$ when $R_k(c)$ exists. Therefore, under either of the conditions of (i) and (ii), one has

$$
\frac{1}{k!} L\mu_k(L; (0, c]) = \sum_{x_{j_1}(L)<\cdots<x_{j_k}(L)\le L} \{c - x_{j_k}(L) + x_{j_1}(L)\}_+ + o(L). \tag{4.15}
$$

Let us now prove the assertion (i). From the above remark and

$$\{c+\delta - x_{j_k}(L) + x_{j_1}(L)\}_+ - \{c - x_{j_k}(L) + x_{j_1}(L)\}_+ \geq \delta 1_{\{x_{j_k}(L)-x_{j_1}(L)\leq c\}},$$

we see

$$\frac{1}{k!}\big(\mu_k(c+\delta) - \mu_k(c)\big) \geq \delta \limsup_{L\to\infty} \frac{1}{L} \sum_{x_{j_1}(L)<\cdots<x_{j_k}(L)\leq L} 1_{\{x_{j_k}(L)-x_{j_1}(L)\leq c\}}.$$

Similarly, from the inequality

$$\{c - x_{j_k}(L) + x_{j_1}(L)\}_+ - \{c - \delta - x_{j_k}(L) + x_{j_1}(L)\}_+ \leq \delta 1_{\{x_{j_k}(L)-x_{j_1}(L)\leq c\}},$$

it follows that

$$\frac{1}{k!}\big(\mu_k(c) - \mu_k(c-\delta)\big) \leq \delta \liminf_{L\to\infty} \frac{1}{L} \sum_{x_{j_1}(L)<\cdots<x_{j_k}(L)\leq L} 1_{\{x_{j_k}(L)-x_{j_1}(L)\leq c\}}.$$

Under the assumption of the differentiability of $\mu_k(c)$, we can let $\delta \searrow 0$, to conclude

$$\lim_{L\to\infty} \frac{1}{L} \sum_{x_{j_1}(L)<\cdots<x_{j_k}(L)\leq L} 1_{\{x_{j_k}(L)-x_{j_1}(L)\leq c\}} = \frac{1}{k!}\frac{\mathrm{d}}{\mathrm{d}c}\mu_k(c).$$

But the limit on the left-hand side is easily seen to coincide with

$$\frac{1}{(k-1)!}\lim_{L\to\infty} R_k(L;c)$$

and hence

$$R_k(c) = \frac{1}{k}\frac{\mathrm{d}}{\mathrm{d}c}\mu_k(c),$$

as was to be proved.

Let us turn to the proof of (ii). Let $\ell \in \mathbb{N}$, and write $\delta = c/\ell$ and $\Delta = x_{j_k}(L) - x_{j_1}(L)$. Then we have

$$\delta \sum_{m=1}^{\ell} 1_{\{c\geq \Delta+m\delta\}} \leq (c-\Delta)_+ \leq \delta \sum_{m=0}^{\ell} 1_{\{c\geq \Delta+m\delta\}},$$

and hence

$$\begin{aligned}\frac{c}{\ell}\sum_{m=1}^{\ell}\frac{1}{L} \sum_{x_{j_1}(L)<\cdots<x_{j_k}(L)\leq L} & 1_{\{x_{j_k}(L)-x_{j_1}(L)\leq c-mc/\ell\}} \\ &\leq \frac{1}{L}\sum_{x_{j_1}(L)<\cdots<x_{j_k}(L)\leq L} \{c - x_{j_k}(L) + x_{j_1}(L)\}_+ \\ &\leq \frac{c}{\ell}\sum_{m=0}^{\ell}\frac{1}{L}\sum_{x_{j_1}(L)<\cdots<x_{j_k}(L)\leq L} 1_{\{x_{j_k}(L)-x_{j_1}(L)\leq c-mc/\ell\}}.\end{aligned}$$

Letting $L \to \infty$ and noting (4.15), we obtain

$$\begin{aligned}(4.16)\qquad \frac{1}{(k-1)!}\frac{c}{\ell}\sum_{m=1}^{\ell} R_k\Big(c\Big(1-\frac{m}{\ell}\Big)\Big) &\le \frac{1}{k!}\liminf_{L\to\infty}\mu_k(L;(0,c]) \\ &\le \frac{1}{k!}\limsup_{L\to\infty}\mu_k(L;(0,c]) \\ &\le \frac{1}{(k-1)!}\frac{c}{\ell}\sum_{m=0}^{\ell} R_k\Big(c\Big(1-\frac{m}{\ell}\Big)\Big),\end{aligned}$$

for all $\ell \in \mathbf{N}$. Finally letting $\ell \to \infty$, we arrive at

$$\frac{1}{k!}\mu_k(c) = \frac{1}{k!}\lim_{L\to\infty}\mu_k(L;(0,c]) = c\int_0^1 R_k\big(c(1-x)\big)\,\mathrm{d}x = \int_0^c R_k(c')\,\mathrm{d}c',$$

or

$$\mu_k(c) = k\int_0^c R_k(c')\,\mathrm{d}c',$$

completing the proof. □

Proposition 4.4. *Suppose $\pi_0(c) := \lim_{L\to\infty}\pi_0(L;(0,c])$ exists for each $c > 0$ and is differentiable with respect to c. Then $\rho(c) = \lim_{L\to\infty}\rho(L;c)$ also exists, and is given by*

$$(4.17)\qquad \rho(c) = -\frac{\mathrm{d}}{\mathrm{d}c}\pi_0(c).$$

Conversely if $\rho(c)$ exists for each $c > 0$, then $\pi_0(c)$ also exists, and is given by

$$(4.18)\qquad \pi_0(c) = 1 - \int_0^c \rho(c')\,\mathrm{d}c'.$$

PROOF. If $x_j(L) \le L$ and $x_{j+1}(L) - x_j(L) > c$, then for any $\delta \in (0,c)$, we have the implication

$$t \in \big[x_j(L)-\delta, x_j(L)\big) \implies N_L(t;\delta) > 0,\, N_L(t+\delta; c-\delta) = 0,$$

the intervals $\big[x_j(L)-\delta, x_j(L)\big)$ corresponding to these js being disjoint. Hence if we set

$$n(c;L) = \sharp\{j \ge 1 \mid x_j(L) \le L, x_{j+1}(L) - x_j(L) > c\},$$

then it holds that

$$\begin{aligned}\delta n(L;c) &= \sum_{\substack{x_j(L)\le L\\ x_{j+1}(L)-x_j(L)>c}} \int_{-\delta}^{L} 1_{[x_j(L)-\delta, x_j(L))}(t)\,\mathrm{d}t \\ &\le \int_{-\delta}^{L} 1_{\{N_L(t,t+\delta]>0,\ N_L(t+\delta,t+c]=0\}}\,\mathrm{d}t \\ &= \int_{-\delta}^{L}\big[1_{\{N_L(t+\delta,t+c]=0\}} - 1_{\{N_L(t,t+c]=0\}}\big]\,\mathrm{d}t \\ &= \int_0^{L+\delta} 1_{\{N_L(t,t+c-\delta]=0\}}\,\mathrm{d}t - \int_{-\delta}^{L} 1_{\{N_L(t,t+c]=0\}}\,\mathrm{d}t \\ &\le \delta + \int_0^{L} 1_{\{N_L(t,t+c-\delta]=0\}}\,\mathrm{d}t - \int_0^{L} 1_{\{N_L(t,t+c]=0\}}\,\mathrm{d}t,\end{aligned}$$

namely

$$\delta\rho(L;c) \le \frac{\delta}{L} + \pi_0(L;(0,c-\delta]) - \pi_0(L;(0,c]).$$

Letting $L \to \infty$, we get

$$\delta \limsup_{L\to\infty} \rho(L;c) =\le \pi_0(c-\delta) - \pi_0(c).$$

Since $\delta > 0$ is arbitrary, this shows

$$\limsup_{L\to\infty} \rho(L;c) \le -\frac{\mathrm{d}}{\mathrm{d}c}\pi_0(c).$$

On the other hand, if $N_L(t,t+\delta] > 0$ and $N_L(t+\delta,t+c+\delta) = 0$, then there is an $x_j(L) \in (t,t+\delta]$ such that $x_{j+1}(L) - x_j(L) > c$. Hence the set

$$\{t \in [0, L-\delta] \mid N_L(t,t+\delta] > 0, N_L(t+\delta,t+c+\delta] = 0\}$$

is a union of finitely many disjoint intervals I_ℓ each of length no greater than δ, and for each I_ℓ, there corresponds an $x_j(L) \le L$ satisfying $x_{j+1}(L) - x_j(L) > c$. Hence the number of these intervals does not exceed $n(c;L)$, and so

$$\begin{aligned}
\delta n(c;L) &\ge \sum_\ell |I_\ell| \\
&= \int_0^{L-\delta} 1_{\{N_L(t,t+\delta]>0,\ N_L(t+\delta,t+c+\delta]=0\}}\,\mathrm{d}t \\
&= \int_0^{L-\delta} \left[1_{\{N_L(t+\delta,t+c+\delta]=0\}} - 1_{\{N_L(t,t+c+\delta]=0\}}\right]\mathrm{d}t \\
&= \int_\delta^{L} 1_{\{N_L(t,t+c]=0\}}\,\mathrm{d}t - \int_0^{L-\delta} 1_{\{N_L(t,t+c+\delta]=0\}}\,\mathrm{d}t \\
&\ge -\delta + L\pi_0(L;(0,c]) - L\pi_0(L;(0,c+\delta]).
\end{aligned}$$

Dividing both sides by L and letting $L \to \infty$, we get

$$\delta \liminf_{L\to\infty} \rho(L;c) \ge \pi_0(c) - \pi_0(c+\delta),$$

and since $\delta > 0$ is arbitrary, we obtain

$$\liminf_{L\to\infty} \rho(L;c) \ge -\frac{\mathrm{d}}{\mathrm{d}c}\pi_0(c),$$

which completes the proof of the first assertion.

Conversely suppose that the limit $\rho(c)$ exists for each $c > 0$. As we have already seen, the following two inequalities are valid for any $L > 0$ and $\delta \in (0,c)$:

$$\delta\rho(L;c) \le \frac{\delta}{L} + \pi_0(L;(0,c-\delta]) - \pi_0(L;(0,c]); \tag{4.19}$$

$$\delta\rho(L;c) \ge -\frac{\delta}{L} + \pi_0(L;(0,c]) - \pi_0(L;(0,c+\delta]). \tag{4.20}$$

Now let $0 < c_1 < c_2$ be given and let $\ell \in \mathbb{N}$ be large enough so that $\delta := (c_2 - c_1)/\ell < c_1$. If in (4.19) we take $c = c_1 + k\delta$, $k = 1, 2, \ldots, \ell$, then

$$\delta\rho(L;c_1+k\delta) \le \frac{\delta}{L} + \pi_0(L;(0,c_1+(k-1)\delta]) - \pi_0(L;(0,c_1+k\delta]).$$

Summing over k, one obtains

$$\delta \sum_{k=1}^{\ell} \rho(L; c_1 + k\delta) \leq \frac{c_2 - c_1}{L} + \pi_0(L; (0, c_1]) - \pi_0(L; (0, c_2]).$$

Letting first $L \to \infty$ and then $\ell \to \infty$, we arrive at

$$\limsup_{L\to\infty} \pi_0(L; (0, c_2]) + \int_{c_1}^{c_2} \rho(c')\, dc' \leq 1. \tag{4.21}$$

In the same way, taking $c = c_1 + k\delta$, $k = 0, 1, \ldots, \ell - 1$, in (4.20), and summing over k, we obtain

$$\liminf_{L\to\infty} \pi_0(L; (0, c_2]) + \int_{c_1}^{c_2} \rho(c')\, dc' \geq \liminf_{L\to\infty} \pi_0(L; (0, c_1]). \tag{4.22}$$

Denote $\underline{\pi}_0(c) := \liminf_{L\to\infty} \pi_0(L; (0, c])$. We shall now verify that $\underline{\pi}_0(c) \nearrow 1$ as $c \searrow 0$. Suppose, on the contrary, that

$$\alpha := \lim_{c\searrow 0} \underline{\pi}_0(c) < 1.$$

Then for any $\beta \in (\alpha, 1)$ and $c > 0$, we must have

$$\limsup_{L\to\infty} \frac{1}{L} \int_0^L 1_{\{N_L(t,t+c]>0\}}\, dt > 1 - \beta.$$

Now the set

$$\{t \in (0, L] \mid N_L(t, t + c] > 0\}$$

is a disjoint union of intervals $I_1, \ldots, I_m$, and by the definition of this set, we see

$$\sharp\{j \geq 1 \mid x_j(L) \leq L\} \geq \frac{1}{c} \sum_{j=1}^{m} |I_j|,$$

and hence

$$\lambda = \lim_{L\to\infty} \frac{1}{L} \sharp\{j \geq 1 \mid x_j(L) \leq L\} \geq \frac{1}{c} \limsup_{L\to\infty} \frac{1}{L} \sum_{j=1}^{m} |I_j| \geq \frac{1 - \beta}{c}.$$

Letting $c \searrow 0$, we arrive at a contradiction.

Having shown $\lim_{c\searrow 0} \underline{\pi}_0(c) = 1$, we can let $c_1 \searrow 0$ in (4.21) and (4.22), to obtain

$$\limsup_{L\to\infty} \pi_0(L; (0, c]) + \int_0^c \rho(c')\, dc' \leq 1 \leq \liminf_{L\to\infty} \pi_0(L; (0, c]) + \int_0^c \rho(c')\, dc', \tag{4.23}$$

which show the existence of $\pi_0(c)$ and the relation

$$\pi_0(c) = 1 - \int_0^c \rho(c')\, dc'. \qquad \square$$

Remark 4.5. Proposition 4.4 above shows the close relationship between the "two possible ways of describing the randomness of the spectrum," proposed by Sinai in the introduction of [40].

5. Two notions of "unfolding"

5.1. "Unfoldable" family of sequences. In general, energy levels of a quantum Hamiltonian have a different local density in different parts of the energy axis. Hence, as we noted in the introduction, in order to discuss the fluctuating properties of energy levels, it is necessary to "unfold" the quantal spectrum under consideration, so that the transformed sequence has uniform asymptotic distribution. To formulate this notion, let us first define a class of spectra which can be unfolded.

Suppose we are given a finite or infinite increasing sequence $\{E_j(h); j \geq 1\}$ of real numbers depending on a parameter $h > 0$. We shall say that this family is *unfoldable* in the limit $h \searrow 0$ if the following condition is satisfied:

Condition 2. There is an $\alpha > 0$ and a non-decreasing function $\nu(E) \geq 0$ on $\mathbb{R}$ with $\lim_{E \to -\infty} \nu(E) = 0$ such that for any $E \in \mathbb{R}$, one has

$$N_h(E) := \sharp\{j \geq 1; E_j(h) \leq E\} \sim \nu(E) h^{-\alpha} \tag{5.1}$$

as $h \searrow 0$.

Below we give some examples of quantum Hamiltonians whose spectrum is unfoldable in the sense defined above.

Example 5.1. Let $H(\hbar) := -\hbar^2 \Delta_D$, where Δ_D is the Laplacian on a connected, bounded domain $D \subset \mathbb{R}^d$ with smooth boundary, considered under the Dirichlet boundary condition. If $0 < e_1 < e_2 < \cdots$ are the eigenvalues of $H(1) = -\Delta_D$, then the spectrum $\{E_j(\hbar)\}_j$ of $H(\hbar)$ is given by $E_j(\hbar) = \hbar^2 e_j$. As is well known (Weyl's asymptotic), there is a universal constant $c_d > 0$, depending only on the dimension of the space, such that

$$N_1(E) := \sharp\{j \geq 1; e_j \leq E\} \sim c_d |D| E^{d/2}, \quad E \to \infty. \tag{5.2}$$

Hence for any fixed $E > 0$, we have

$$N_\hbar(E) := \sharp\{j \geq 1; E_j(\hbar) \leq E\} = N_1(E/\hbar^2) \sim (c_d |D| E^{d/2}) \hbar^{-d} \tag{5.3}$$

as $E \to \infty$. Thus $\{E_j(\hbar)\}_j$ is unfoldable in the (semiclassical) limit $\hbar \searrow 0$ with $\alpha = d$ and $\nu(E) = c_d |D| (E \vee 0)^{d/2}$.

Example 5.2 (Regular spectra ([3,32])). Suppose that the classical Hamiltonian system associated to $H(p,q)$ is completely integrable. Then one can transform the momentum-coordinate variable $(p,q) \in \mathbb{R}^d \times \mathbb{R}^d$ into the action – angle variable $(I, \varphi) = (I_1, \ldots, I_d; \varphi_1, \ldots, \varphi_d)$, and $H(p,q) = H(I)$ depends only on the action variable (see [1]). Following Percival ([32]), Berry and Tabor ([3]), we shall say that $H(\hbar)$, the quantization of $H(p,q)$, has regular spectrum if its energy levels are (approximately) given by quantizing the action variables $I_1, \ldots, I_d$ appearing in $H(I)$. Namely, we suppose that the energy levels of $H(\hbar)$ are

$$E_{\mathbf{n}}(\hbar) = H\left(\hbar(\mathbf{n} + \tfrac{1}{4}\boldsymbol{\alpha})\right); \quad \mathbf{n} = (n_1, \ldots, n_d) \in \mathbb{Z}_+^d, \tag{5.4}$$

where $\boldsymbol{\alpha} \in \mathbb{Z}_+^d$ is called the Maslov index. (In the following discussion, we put $\boldsymbol{\alpha} = \mathbf{0}$ as in [3].) This procedure is called the EBK quantization after the names of Einstein, Brillouin and Keller, and it gives the exact spectrum in such concrete examples as the harmonic oscillators or free motion on flat tori ([37,41]).

Suppose that $H(I)$, as a function on $\mathbb{R}_+^d$ satisfies the following conditions:

(1) $H(r\mathbf{x})$ is increasing in $r \geq 0$ for any $\mathbf{x} \in \mathbb{R}_+^d \setminus \{\mathbf{0}\}$;
(2) $H(\mathbf{0}) = 0$;

(3) $\{\mathbf{x}; H(\mathbf{x}) \leq E\}$ is compact for any $E > 0$.

Then $\{E_{\mathbf{n}}(\hbar); \mathbf{n} \in \mathbb{Z}_+^d\}$ is unfoldable in the limit $\hbar \searrow 0$ with $\alpha = d$ and

$$\nu(E) = \int \cdots \int_{\mathbb{R}_+^d} 1_{\{H(I) \leq E\}} \, \mathrm{d}I_1 \cdots \mathrm{d}I_d. \tag{5.5}$$

Example 5.3 (Ergodic Schrödinger operators). Let $\{V_\omega(x); x \in \mathbb{R}^d \text{ or } \mathbb{Z}^d\}$ be an ergodic random field defined on some probability space $(\Omega, \mathcal{F}, \mathbb{P})$. Consider the random Schrödinger operator $H_\omega := -\Delta + V_\omega(x)$ and its restriction H_ω^L to the cube $[-L/2, L/2]^d$ under Dirichlet boundary condition. Here Δ denotes the continuum or discrete Laplacian. Let $\{E_j^\omega(L)\}_j$ be the spectrum of H_ω^L. Then under mild conditions on the random potential $\{V_\omega(x)\}_x$, there exists a non-decreasing function $\mathcal{N}(E) \geq 0$ (the integrated density of states) such that $\mathcal{N}(-\infty) = 0$ and that for $\mathbb{P}$-almost all $\omega \in \Omega$, one has

$$\lim_{L \to \infty} \frac{1}{|\Lambda|} \sharp\{j \geq 1; E_j^\omega(L) \leq E\} = \mathcal{N}(E) \tag{5.6}$$

for all $E \in \mathbb{R}$. (See [42] for a review of recent results.) This shows that for $\mathbb{P}$-almost all $\omega \in \Omega$, the spectrum $\{E_j^\omega(L)\}_j$ is unfoldable in the (thermodynamic) limit $h := 1/L \searrow 0$, with $\alpha = d$ and $\nu(E) = \mathcal{N}(E)$.

5.2. The unfolding of the first kind. In general, the purpose of unfolding is to transform a given family of unfoldable sequences of levels into another family of sequences $\{x_j(L)\}_j$ which is asymptotically uniformly distributed so as to satisfy Condition 1. In this subsection, we assume that the function $\nu(E)$ appearing in Condition 2 is continuous in E.

For each $t \in \big(0, \nu(+\infty)\big)$, define

$$\nu^{-1}(t) := \inf\{E; \nu(E) > t\}. \tag{5.7}$$

Then we have $\nu\big(\nu^{-1}(t)\big) = t$ and the equivalence

$$\nu(E) \leq t \iff E \leq \nu^{-1}(t).$$

If we let $e_j(h) := \nu\big(E_j(h)\big)$, then for each $t > 0$,

$$\begin{aligned} \sharp\{j \geq 1; e_j(h) \leq t\} &= \sharp\{j \geq 1; E_j(h) \leq \nu^{-1}(t)\} \\ &= N_h\big(\nu^{-1}(t)\big) \sim \nu\big(\nu^{-1}(t)\big) h^{-\alpha} = t h^{-\alpha} \end{aligned}$$

as $h \searrow 0$ by Condition 2. Hence if we let $L = h^{-\alpha}$, and $x_j(L) = L e_j(L^{-1/\alpha})$, then we have for any $a > 0$,

$$\sharp\{j \geq 1; x_j(L) \leq aL\} = \sharp\{j \geq 1; e_j(L^{-1/\alpha}) \leq a\} \sim a(L^{-1/\alpha})^{-\alpha} = aL$$

as $L \to \infty$, so that the parametrized sequence $\{x_j(L)\}_j$ satisfies Condition 1 with $\lambda = 1$.

5.3. The unfolding of the second kind. The unfolding formulated above is what is usually practiced in quantum chaos (see, e.g., [14]), and looks like a quite natural idea in that it leads one to directly observe energy levels, although under a non-linear transformation $e_j = \nu(E_j)$. But with regard to this, we notice a seemingly strange remark made by Berry and Tabor in their celebrated paper [3]. They argue that the energy E itself "*is an unsuitable parameter to describe the levels, for two reasons. The first is that the* mean level density $\bar{n}(E)$ *depends on energy.... The second reason is that the statistics of the levels might themselves* *depend on the region of energy being studied.*" As a solution to this problem, Berry and Tabor propose to observe the quantum system at a fixed energy value E, and to "quantize the Planck's constant $\hbar$." They apply this procedure only to regular spectra discussed in Example 5.2, but in a general setting, it would be formulated in the following way: In addition to Condition 2, let us assume that $E_j(h)$ is a non-decreasing function of $h > 0$ for each j. Let us fix E such that $E_0 < E < E_1$, where

$$E_0 := \inf\{E; \nu(E) > 0\} \geq -\infty \tag{5.8}$$

and

$$E_1 := \sup\{E; \nu(E) < \nu(\infty)\} \leq \infty. \tag{5.9}$$

Now define

$$h_j(E) := \inf\{h > 0; E_j(h) > E\} \tag{5.10}$$

for each $j \geq 1$, where we let $h_j(E) = 0$ if $E_j(h) > E$ for all $h > 0$, and $h_j(E) = +\infty$ if $E_j(h) \leq E$ for all $h > 0$. Actually, the last two possibilities are ruled out by the following argument: Since $\nu(E) > 0$, Condition 2 implies in particular that

$$\lim_{h \searrow 0} \sharp\{j \geq 1; E_j(h) \leq E\} = \infty. \tag{5.11}$$

If $h_j(E) = 0$ for some i, then $E_j(h) > E$ for all $j \geq i$ and $h > 0$. Hence

$$\sharp\{j \geq 1; E_j(h) \leq E\} \leq i - 1,$$

contradicting (5.11), and so $h_j(E) > 0$ must hold for all $j \geq 1$. On the other hand, since $E < E_1$, there is an $E' > E$ such that $\nu(E) < \nu(E') < \nu(\infty)$. Again by Condition 2, we have

$$\lim_{h \searrow 0} \sharp\{j \geq 1; E < E_j(h) \leq E'\} = \infty. \tag{5.12}$$

This shows that for only finitely many j's, one can have

$$\sup_{h>0} E_j(h) \leq E.$$

Thus there is a $j_0 \geq 1$ such that $h_j(E) < \infty$ for all $j \geq j_0$. We now define

$$x_j = x_j(E) := \nu(E) h_j(E)^{-\alpha} \tag{5.13}$$

for $j \geq j_0$. In the sequel we assume $j_0 = 1$. By the definition, we have the following equivalence and implication for each $L > 0$:

$$\begin{aligned} x_j < L \iff h_j(E) > (\nu(E)/L)^{1/\alpha} &\implies E_j\big((\nu(E)/L)^{1/\alpha}\big) \leq E \\ &\implies h_j(E) \geq (\nu(E)/L)^{1/\alpha} \iff x_j \leq L. \end{aligned}$$

This shows

$$\sharp\{j \geq 1; x_j < L\} \leq \sharp\{j \geq 1; E_j\big((\nu(E)/L)^{1/\alpha}\big) \leq E\} \leq \sharp\{j \geq 1; x_j \leq L\}.$$

But by Condition 2, we have

$$\lim_{L \to \infty} \frac{1}{L} \sharp \Big\{j \geq 1; E_j\big((\nu(E)/L)^{1/\alpha}\big) \leq E\Big\} = \lim_{L \to \infty} \frac{1}{L} \nu(E) \Bigg\{\bigg(\frac{\nu(E)}{L}\bigg)^{1/\alpha}\Bigg\}^{-\alpha} = 1,$$

so that

$$\limsup_{L \to \infty} \frac{1}{L} \sharp\{j \geq 1; x_j < L\} \leq 1 \leq \liminf_{L \to \infty} \sharp\{j \geq 1; x_j \leq L\},$$

namely

$$\sharp\{j \geq 1; x_j \leq L\} \sim L, \quad L \to \infty.$$

Since $\{x_j\}$ itself does not depend on L, we obtain a sequence satisfying Condition 1 with $\lambda = 1$.

5.4. Examples of unfolding. Unfortunately, the present author is unable to clarify the general relationship between unfoldings of first and second kinds. For the time being, let us content ourselves by discussing several examples of quantal spectra to which both kinds of unfoldings can be applied.

5.4.1. *The quantum billiard.* For the Laplacian examined in Example 5.1, we have $E_j(\hbar) = \hbar^2 e_j$, $\alpha = d$, and $\nu(E) = c_d|D|E^{d/2}$. Since $\nu(E)$ is continuous and $E_j(\hbar)$ is increasing in $\hbar$, both kinds of unfoldings can be applied. To perform the unfolding of the first kind, we put as in §5.2,

$$e_j(\hbar) := \nu(E_j(\hbar)) = c_d|D|\hbar^d e_j^{d/2}; \tag{5.14}$$

and

$$x_j(L) := Le_j(L^{-1/d}) = c_d|D|e_j^{d/2}. \tag{5.15}$$

The resulting sequence $\{x_j(L)\}$ does not depend on the parameter L.

To perform the unfolding of the second kind, let us fix an $E > 0$. Then we put

$$\hbar_j(E) := \inf\{\hbar; \hbar^2 e_j = E\} = (E/e_j)^{1/2} \tag{5.16}$$

and

$$x_j(E) := \nu(E)\hbar_j(E)^{-d} = c_d|D|E^{d/2}(E/e_j)^{-d/2} = c_d|D|e_j^{d/2}. \tag{5.17}$$

Thus both kinds of unfoldings yield the same sequence.

5.4.2. *A one-dimensional Schrödinger operator with δ-potentials in the semi-classical limit.* Consider the one-dimensional Schrödinger operator

$$H_v(\hbar) = -\hbar^2 \frac{\mathrm{d}^2}{\mathrm{d}x^2} + v\sum_{s=1}^{n} \delta(x - x_s), \quad 0 \leq x \leq 1 \tag{5.18}$$

under Dirichlet boundary conditions at $x = 0, 1$. Here $v > 0$, and

$$0 \equiv x_0 < x_1 < \cdots < x_n < x_{n+1} \equiv 1.$$

Let $\{E_j(\hbar)\}_j$ be the spectrum of $H_v(\hbar)$. Then we have for each $E > 0$

$$N_\hbar(E) := \sharp\{j \geq 1; E_j(\hbar) \leq E\} \sim \frac{1}{\pi}\sqrt{E}\hbar^{-1}, \quad \hbar \searrow 0. \tag{5.19}$$

Thus Condition 2 is satisfied with $\alpha = 1$ and $\nu(E) = \frac{1}{\pi}\sqrt{E}$, which is continuous. Moreover, we can verify that each energy level $E_j(\hbar)$ is non-decreasing in $\hbar > 0$ by the min-max principle. Hence both kinds of unfoldings are applicable to our spectrum $\{E_j(\hbar)\}$: For each $L > 0$, the unfolded levels of the first kind are given by

$$x_j(L) := L\nu(E_j(L^{-1})) = \frac{L}{\pi}\sqrt{E_j(L^{-1})}, \tag{5.20}$$

while for each fixed $E > 0$, the unfolded levels of the second kind are given by

$$x_j(E) := \nu(E)\hbar_j(E)^{-1} = \frac{\sqrt{E}}{\pi}\hbar_j(E)^{-1}, \tag{5.21}$$

where $\hbar_j(E)$ is defined by the equation $E_j(\hbar) = E$.

It was shown in [22] that E is an energy level of $H_v(\hbar)$, if and only if it is a zero of the function

$$\Gamma_\hbar(E) := \sum_{p=0}^{n}\left(\frac{\hbar\sqrt{E}}{v}\right)^{n-p} \sum_{1\le j_1<\cdots<j_p\le n} \prod_{s=1}^{p+1} \sin\left\{(x_{j_s}-x_{j_{s-1}})\frac{\sqrt{E}}{\hbar}\right\}. \tag{5.22}$$

This shows on one hand that for a fixed $L>0$, $\xi>0$ is an unfolded level of the first kind if and only if it is a zero of the function $G_L(\xi) := \Gamma_{L^{-1}}\big((\pi\xi/L)^2\big)$, while on the other hand $\zeta>0$ is an unfolded level of the second kind if and only if it is a zero of the function $\gamma_E(\zeta) := \Gamma_{\sqrt{E}/\pi\zeta}(E)$. If we define

$$f(z) := \prod_{s=1}^{n+1} \sin\{\pi(x_s-x_{s-1})z\}, \tag{5.23}$$

then we can write

$$G_L(\xi) = f(\xi) + g_L(\xi) \quad \text{and} \quad \gamma_E(\zeta) = f(\zeta) + \delta_E(\zeta), \tag{5.24}$$

where $g_L(\xi)$ is small for large L and for $\xi = \mathcal{O}(L)$, while $\delta_E(\zeta)$ is small for large values of ζ. Hence by the argument in [22] using Rouché's theorem, the statistics for the zeros of the functions $G_L(\xi)$ and $\gamma_E(\zeta)$ are both reduced to the statistics for the zeros of the function $f(z)$, which are

$$\left\{\frac{m}{x_s-x_{s-1}}; m\ge 1, s=0,1,\dots,n\right\}. \tag{5.25}$$

Thus two kinds of unfoldings applied to the spectrum $\{E_j(\hbar)\}$ lead to the same level statistics.

To proceed, consider as in §4 the counting function

$$N(t,t+c] := \sum_{s=0}^{n}\sum_{m=1}^{\infty} 1_{(t,t+c]}\left(\frac{m}{x_s-x_{s-1}}\right), \tag{5.26}$$

and the probability

$$\pi_k(L;(0,c]) := \frac{1}{L}\int_0^L 1_{\{N(t,t+c]=k\}}\, dt \tag{5.27}$$

for $k=0,1,\dots$, $c>0$, and $L>0$. Then we have

$$M_c \le N(t,t+c] \le M_c+n+1, \tag{5.28}$$

where

$$M_c := \sum_{s=1}^{n+1}[c(x_s-x_{s-1})].$$

Moreover if x_s-x_{s-1}, $s=1,2,\dots,n+1$, are rationally independent, one can show ([22]) that

$$\begin{aligned}\pi_k(c) &:= \lim_{L\to\infty}\pi_k(L;(0,c]) \\ &= \sum_{A\in\mathcal{F}(k-M_c)}\left(\prod_{s\in A}\{c(x_s-x_{s-1})\}\right)\left(\prod_{s\in A^c}(1-\{c(x_s-x_{s-1})\})\right)\end{aligned} \tag{5.29}$$

for each $k = M_c, M_c + 1, \ldots, M_c + n + 1$. Here $\{a\}$ is the fractional part of a, and $\mathcal{F}(j)$ is the totality of subsets of $\{0, 1, \ldots, n\}$ having cardinality j. In particular, if $c > 0$ is small, then $M_c = 0$ and

$$\pi_0(c) = \prod_{s=1}^{n+1} \big(1 - c(x_s - x_{s-1})\big). \tag{5.30}$$

We can now apply Proposition 4.4, to obtain

$$\rho(c) = -\frac{d}{dc}\pi_0(c) = \left(\sum_{s=1}^{n+1} \frac{x_s - x_{s-1}}{1 - c(x_s - x_{s-1})}\right) \prod_{s=1}^{n} \big(1 - c(x_s - x_{s-1})\big). \tag{5.31}$$

For large n and for typical configuration $\{x_s\}$ of the δ-potentials, $\pi_k(c)$, $k = 0, 1, \ldots$, and $\rho(c)$ are different from, but very close to being Poissonian. In fact, if $X_1, X_2, \ldots$ are independent random variables which are uniformly distributed on $(0, 1)$, and if $0 < x_1^{(n)} < \cdots < x_n^{(n)} < 1$ is a rearrangement of $X_1, \ldots, X_n$ according to their magnitude, then one can show ([22]) that with probability one,

$$\lim_{n\to\infty} \pi_k\big(c; x_1^{(n)}, \ldots, x_n^{(n)}\big) = e^{-c}\frac{c^k}{k!}, \quad k = 0, 1, \ldots, \tag{5.32}$$

and

$$\lim_{n\to\infty} \rho\big(c; x_1^{(n)}, \ldots, x_n^{(n)}\big) = e^{-c}. \tag{5.33}$$

5.4.3. *Regular spectra.* Under the assumptions on the classical Hamiltonian of Example 5.2, $\nu(E)$ is a continuous function of E, and $E_{\mathbf{n}}(\hbar) = H(\hbar\mathbf{n})$ is increasing in $\hbar$ for each $\mathbf{n}$. Hence both kinds of unfolding are applicable to our regular spectrum $\{E_{\mathbf{n}}(\hbar); \mathbf{n} \in \mathbb{Z}_+^d\}$. To apply the unfolding of the *second* kind, we first note that for each $\mathbf{x} \in \mathbb{R}_+^d \setminus \{\mathbf{0}\}$ and $E > 0$, the equation $H(\hbar\mathbf{x}) = E$ can be solved for $\hbar$. If we denote this solution by $\hbar_E(\mathbf{x})$, then the unfolded energy level is given by $x_{\mathbf{n}}(E) = U_E(\mathbf{n})$, where we have defined

$$U_E(\mathbf{x}) = \nu(E)\hbar_E(\mathbf{x})^{-d}, \quad \mathbf{x} \in \mathbb{R}_+^d. \tag{5.34}$$

Since $\beta\hbar_E(\beta\mathbf{x}) = \hbar_E(\mathbf{x})$ for any $\beta > 0$, we see that $U_E(\mathbf{x})$ is a homogeneous function of degree d, namely

$$U_E(\beta\mathbf{x}) = \beta^d U_E(\mathbf{x}), \tag{5.35}$$

for any $\beta > 0$. (See [3].)

To perform the level statistics of §4 for this sequence $\{U_E(\mathbf{n})\}_{\mathbf{n}}$, which is asymptotically uniformly distributed, we need to consider for $I = (a, b]$

$$N_L(t + I) = \sum_{\mathbf{n}\in\mathbb{Z}_+^d} 1_{(t+a,t+b]}\big(U_E(\mathbf{n})\big). \tag{5.36}$$

If we define

$$\theta(\mathbf{x}) = \frac{\mathbf{x}}{|\mathbf{x}|}, \quad \mathbf{x} \in \mathbb{R}_+^d \setminus \{\mathbf{0}\} \tag{5.37}$$

and

$$\Pi_I(t) := \{\mathbf{x} \in \mathbb{R}_+^d \mid (t+a)^{1/d} U_E(\theta(\mathbf{x}))^{-1/d} < |\mathbf{x}| \le (t+b)^{1/d} U_E(\theta(\mathbf{x}))^{-1/d}\}, \tag{5.38}$$

then

$$N_L(t + I) = \sum_{\mathbf{n}\in\mathbb{Z}_+^d} 1_{\Pi_I(t)}(\mathbf{n}), \tag{5.39}$$

which is the number of lattice points contained in the d-dimensional domain $\Pi_I(t)$. It is then easy to see $|\Pi_I(t)| = b - a = |I|$, so that as t gets large, the domain $\Pi_I(t)$ expands in the space $\mathbb{R}_+^d$, but at the same time gets thinner and thinner to keep its volume constant. Hence provided the boundary of $\Pi_I(t)$ is not too regular, e.g., flat or spherical, then each point in $\mathbb{Z}_+^d$ would randomly belong to $\Pi_I(t)$ if t were chosen at random from a long interval $[0, L]$, so that the total number of lattice points in $\Pi_I(t)$, namely $N_L(t + I)$, would obey Poisson distribution with mean $|I|$. Namely we expect that

$$\pi_k(I) := \lim_{L\to\infty} \pi_k(L; I) = \mathrm{e}^{-|I|}\frac{|I|^k}{k!}, \quad k = 0, 1, 2, \ldots \tag{5.40}$$

holds. This is conceivable if we recall how Poisson's law of small numbers is proved in elementary probability theory. Actually, it is a difficult problem to rigorously justify this intuition. In fact, no example of classically integrable Hamiltonian is known for which (5.40) is verified.

Let us now apply the unfolding of the *first* kind to our regular spectrum. This gives

$$x_{\mathbf{n}}(L) := L \cdot \nu\big(E_{\mathbf{n}}(L^{-1/d})\big) = L \cdot (\nu \circ H)(L^{-1/d}\mathbf{n}), \quad L > 0, \mathbf{n} \in \mathbb{Z}_+^d,$$

as unfolded levels. To proceed further, we assume that our Hamiltonian $H(\mathbf{x})$ satisfies the following

Condition 3. For any $\mathbf{x}, \mathbf{z} \in \mathbb{R}_+^d$ and $\beta > 0$, $H(\beta\mathbf{z}) \le H(\beta\mathbf{x})$ implies $H(\mathbf{z}) \le H(\mathbf{x})$.

Under this condition, the function $G := \nu \circ H$ is homogeneous with degree d. In fact for any $\beta > 0$,

$$\begin{aligned} G(\beta\mathbf{x}) &= \int\cdots\int_{\mathbb{R}_+^d} 1_{\{H(I)\le H(\beta\mathbf{x})\}}\, \mathrm{d}I_1\cdots\mathrm{d}I_d \\ &= \beta^d \int\cdots\int_{\mathbb{R}_+^d} 1_{\{H(\beta\mathbf{z})\le H(\beta\mathbf{x})\}}\, \mathrm{d}z_1\cdots\mathrm{d}z_d = \beta^d G(\mathbf{x}). \end{aligned}$$

Thus we have $x_{\mathbf{n}}(L) = G(\mathbf{n})$, and hence replacing U_E by $G = \nu \circ H$, we are lead to the same lattice point problem as before.

If in particular, the Hamiltonian $H(\mathbf{x})$ itself is a homogeneous function with degree $\gamma > 0$, say, then Condition 3 is automatically satisfied and for each $E > 0$, we have $\nu(E) = c_H E^{d/\gamma}$ with

$$c_H = \int\cdots\int_{\mathbb{R}_+^d} 1_{\{H(\mathbf{x})\le 1\}}\, \mathrm{d}x_1\cdots\mathrm{d}x_d,$$

and $\hbar_E(\mathbf{x}) = E^{1/\gamma}H(\mathbf{x})^{-1/\gamma}$. Hence

$$U_E(\mathbf{x})\nu(E)\hbar_E(\mathbf{x})^{-d} = c_H H(\mathbf{x})^{d/\gamma} = \nu(H(\mathbf{x})) = G(\mathbf{x}).$$

Thus the unfolded levels $x_{\mathbf{n}}(E) = U_E(\mathbf{n})$ of the second kind does not depend on E, and coincide with the unfolded levels $x_{\mathbf{n}}(L) = G(\mathbf{n})$ of the first kind.

At this point, it is appropriate to discuss the limit theorem of Sinai [40] and Major [16] (see also [24]). They consider the case $d = 2$ in which the curve

$$r = f(\theta), \quad 0 \le \theta \le \pi/2$$

defining the boundary of the domain

$$\Pi_I(t) := \{\mathbf{x} \in \mathbb{R}^2_+ \mid \sqrt{t+a}f(\theta) < r \le \sqrt{t+b}f(\theta)\}$$

is very random. It was then proved by Major ([16]) that (5.40) actually holds for almost all realization of this random curve, and by the present author ([24]) that

$$\mu_k(I) := \lim_{L\to\infty} \mu_k(L; I) = |I|^k, \quad k = 1, 2, \dots, \tag{5.41}$$

also holds almost surely, which is a slightly stronger assertion than (5.40). Since the randomness assumed in these works is so strong that the typical realizations of the curve is non-smooth, the direct connection of these limit theorems to the question of level statistics is lost. But the analysis in [16, 24, 40] suggests us that it would be possible to prove that $\mu_k(I) = |I|^k$, $k = 1, 2, \dots, K$, should hold almost surely with some finite K if the curve has only a finite-dimensional randomness. This was actually done by VanderKam ([41]), the significance of which we shall discuss again in the next section.

5.4.4. *Schrödinger operator with white noise potential in the thermodynamic limit.* Let $\{B_\omega(t); t \ge 0\}$ be the standard Brownian motion defined on a probability space $(\Omega, \mathcal{F}, \mathbb{P})$, and consider for each $L > 0$ the Schrödinger operator

$$H^L_\omega := -\frac{\mathrm{d}^2}{\mathrm{d}t^2} + B'_\omega(t), \quad 0 \le t \le L \tag{5.42}$$

under Dirichlet boundary conditions at $x = 0, L$. Here $B'_\omega(t)$ is the formal derivative of the sample function $B_\omega(t)$. Although $B'_\omega(t)$ is singular, H^L_ω is well defined as a random self-adjoint operator acting in $L^2([0, L]; \mathrm{d}t)$, which has a purely discrete spectrum. Compared to previous definitions (e.g., [9, 21]) of the operators of the form

$$H = -\frac{\mathrm{d}^2}{\mathrm{d}t^2} + Q'(t) \tag{5.43}$$

with a non-smooth $Q(t)$, the recent method by Savchuk and Shkalikov ([36]) seems to be the most ingenious in that it yields Sturm's oscillation theorem through almost identical argument as in the classical theory ([30, 43]).

Now let

$$E^\omega_1(L) < E^\omega_2(L) < \cdots < E^\omega_j(L) < \cdots$$

be the eigenvalues of H^L_ω. As a special case of an ergodic Schrödinger operator (see Example 5.3), there exists a non-decreasing function $\mathcal{N}(E)$ such that for almost all $\omega \in \Omega$,

$$\sharp\{j \ge 1; E^\omega_j(L) \le E\} \sim \mathcal{N}(E)L, \quad L \to \infty$$

holds for every $E \in \mathbb{R}$. Actually, it is known ([9]) that

$$\mathcal{N}(E) = \left\{\sqrt{2\pi}\int_0^\infty \exp(-\tfrac{1}{6}u^3 - 2Eu)\frac{\mathrm{d}u}{\sqrt{u}}\right\}^{-1}. \tag{5.44}$$

Hence for almost all $\omega \in \Omega$, the spectrum of H^L_ω is unfoldable with $h = L^{-1}$, $\alpha = 1$, and $\nu(E) = \mathcal{N}(E)$, the latter being continuous in E. The first kind unfolding is carried through by letting

$$x^\omega_j(L) := L\mathcal{N}(E^\omega_j(L)). \tag{5.45}$$

If we consider the statistics $\pi_k(L; I)$ defined by (4.4) for this family of sequences $\{x_j^\omega(L)\}_j$, then in view of the results by Molchanov ([28]) and by McKean ([19]) (see the remark below), we would expect that

$$\lim_{L\to\infty} \pi_k(L;(0,c]) = \mathrm{e}^{-c}\frac{c^k}{k!}$$

holds almost surely for each $c > 0$, or at least that $\{x_j^\omega(L); j \geq 1\}$ converges in law to the Poisson point process on $(0, \infty)$ with mean density 1. But for the moment, this is only a conjecture.

Remark 5.4. McKean actually proved that as $L \to \infty$, the distribution of $x_1^\omega(L) = L\mathcal{N}(E_1^\omega(L))$ converges weakly to the exponential distribution with mean 1. See [11] for related topics.

Let $\psi_E^\omega(t)$ be the solution of the equation $H_\omega\psi = E\psi$ satisfying the initial condition $\psi(0) = 0$, $\psi'(0) = 1$. Then by the Sturm's oscillation theorem, $E = E_j^\omega(L)$ if and only if $t = L$ is the jth zero of the function $\psi_E^\omega(t)$. Moreover $E_j^\omega(L)$ is increasing in $h := L^{-1}$, and hence we can apply the unfolding of the second kind. For this purpose, fix an energy value $E \in \mathbb{R}$ and let $L_j^\omega(E)$ be defined by the equation $E_j^\omega(L) = E$. Then $L_j^\omega(E)$ has the following meaning. If the stochastic processes $\{r_E(t)\}$ and $\{\theta_E(t)\}$ are introduced through the Prüfer's transformation

$$\psi_E(t) = r_E(t)\sin\theta_E(t); \quad \psi_E'(t) = r_E(t)\cos\theta_E(t), \tag{5.46}$$

then $\{\theta_E(t)\}_t$ is a diffusion process on the circle $\mathbb{R}/\pi\mathbb{Z}$ ([9, 19]) with $\theta_E(0) = 0$ (mod π), and $L_j(E)$ ($j \geq 1$) are the jth hitting times to the origin 0 of this process. Namely letting $L_0(E) := 0$, we have

$$L_j(E) := \inf\{t > L_{j-1}(E); \theta_E(t) = 0 \pmod{\pi}\} \tag{5.47}$$

for $j \geq 1$. By the strong Markov property of $\{\theta_E(t)\}_t$, $\tau_j(E) := L_j(E) - L_{j-1}(E)$, $j = 1, 2, \ldots$, are independent and identically distributed random variables, and $L_j^\omega(E) = \sum_{k=1}^{j} \tau_k(E)$. Moreover one has $\mathbb{E}[\tau_1(E)] = \mathcal{N}(E)^{-1}$ ([9]). Hence the unfolded levels of the second kind, given by

$$x_j = x_j(E) := \mathcal{N}(E)L_j^\omega(E), \quad j = 1, 2, \ldots \tag{5.48}$$

constitute a renewal process on $(0, \infty)$ with mean density 1, so that the corresponding level statistics is rather trivial. For example, one has

$$\lim_{L\to\infty} \rho(L;c) = \lim_{L\to\infty} \frac{1}{L}\sharp\{j \geq 1; x_j \leq L, x_{j+1} - x_j > c\} = \mathbb{P}\left(\frac{\tau_1(E)}{\mathbb{E}[\tau_1(E)]} > c\right) \tag{5.49}$$

with probability one. In this case, the result of unfolding of the second kind does depend on the fixed energy value E, the distribution of $\tau_1(E)/\mathbb{E}[\tau_1(E)]$ being dependent on E. As we let $E \to -\infty$, this distribution tends to the exponential distribution $1_{(0,\infty)}(x)e^{-x}\,dx$, as was proved by McKean ([19]). Accordingly the renewal process $\{x_j(E)\}_j$ converges weakly to the Poisson point process on $(0, \infty)$ with mean density 1.

6. Clustering and repulsion of levels

As we noted in §5.4.3, no concrete example of regular spectrum is known for which Berry-Tabor conjecture is literally true in the sense that

$$\rho(c) := \lim_{L\to\infty} \rho(L;c) = \mathrm{e}^{-c} \quad c > 0 \tag{6.1}$$

holds for the unfolded spectrum. The assertion (6.1) on the level spacing distribution function $\rho(c)$ is what Berry and Tabor called the "level clustering." We have also seen in §5.4.2 that there is even an example of quantum Hamiltonian for which level statistics is carried out in full details, yet yielding a level spacing distribution function $\rho(c)$ which is close to, but different from e^{-c}. This situation suggests us that we should give some wider, or looser definition to the term "level clustering" than the strict assertion like (6.1). To formulate this problem, suppose we are given a sequence $\{x_j(L)\}_j$ depending on a parameter $L > 0$ which satisfies Condition 1 with $\lambda = 1$, and which we regard as an unfolded quantal spectrum. For this sequence $\{x_j(L)\}_j$, let us consider statistics defined by (4.4)–(4.7) with $I = (0, c]$. Except that we always have $\mu_1(c) := \lim_{L\to\infty} \mu_1(L; (0, c]) = c$, none of these is guaranteed to converge to some limit as $L \to \infty$. Let us then define

$$\underline{\pi}_k(c) := \liminf_{L\to\infty} \pi_k(L; (0, c]), \qquad \overline{\pi}_k(c) := \limsup_{L\to\infty} \pi_k(L; (0, c]); \tag{6.2}$$

$$\underline{\mu}_k(c) := \liminf_{L\to\infty} \mu_k(L; (0, c]), \qquad \overline{\mu}_k(c) := \limsup_{L\to\infty} \mu_k(L; (0, c]); \tag{6.3}$$

$$\underline{\rho}(c) := \liminf_{L\to\infty} \rho(L; c), \qquad \overline{\rho}(c) := \limsup_{L\to\infty} \rho(L; c)\ ; \tag{6.4}$$

and

$$\underline{R}_k(c) := \liminf_{L\to\infty} R_k(L; c), \qquad \overline{R}_k(c) := \limsup_{L\to\infty} R_k(L; c)\ ; \tag{6.5}$$

When the corresponding upper and lower limits coincide, we shall denote their common value by $\pi_k(c)$, $\mu_k(c)$, $\rho(c)$, or $R_k(c)$ as before.

Now one of the characteristic of the exponential distribution is that its density is strictly positive at the origin $c = 0$. Hence if (6.1) were true, we would have relatively many closely adjacent pairs of levels. This situation would be well described in the following way.

Definition 6.1. We shall say that the unfolded spectrum $\{x_j(L)\}_j$ exhibits level clustering [resp. repulsion] in the wide sense if the condition

$$\liminf_{c\searrow 0} \frac{1-\overline{\rho}(c)}{c} > 0 \quad \left[\text{resp. } \limsup_{c\searrow 0} \frac{1-\underline{\rho}(c)}{c} = 0\right] \tag{6.6}$$

holds.

We can give sufficient conditions for repulsion or clustering in the wide sense in terms of $\mu_k(L; (0, c])$, $k = 2, 3$, as follows.

Proposition 6.2. (i) *If $\overline{\mu}_2(c) = o(c^2)$ as $c \searrow 0$, then the spectrum exhibits level repulsion in the wide sense.*

(ii) *If for some $a > 0$, $\underline{\mu}_2(c) \geq ac^2$ for all $c > 0$ and if $\overline{\mu}_3(c) = o(c^2)$ as $c \searrow 0$, then the spectrum exhibits level clustering in the wide sense.*

PROOF. Applying [6, Lemma 2.2] to the probability distribution $\pi_k(L; (0, c])$ $(k \geq 0)$ on non-negative integers, we obtain the following equality which is valid for $n \geq 1$ and $k \geq 0$:

$$P_n(L; c) := \sum_{j\geq n} \pi_j(L; (0, c]) = S_{n,k}(c) + (-1)^{k+1} T, \tag{6.7}$$

where

$$S_{n,k}(c) = \sum_{j=n}^{n+k} (-1)^{j-n} \binom{j-1}{n-1} \frac{1}{j!} \mu_j(L; (0,c]), \tag{6.8}$$

and $T = T_{n,k} \geq 0$ is given by

$$T_{n,k} = \sum_{j \geq n+k+1} \binom{j-1}{n-1} \binom{j-n-1}{k} P_j(L; c), \tag{6.9}$$

though this explicit expression for T will not be used in the following argument. When $n = 1$, formula (6.7) reduces to

$$1 - \pi_0(L; (0,c]) = \sum_{j=1}^{k+1} (-1)^{j-1} \frac{1}{j!} \mu_j(L; (0,c]) + (-1)^{k+1} T, \tag{6.10}$$

and hence

$$\pi_0(L; (0,c]) \leq 1 - \mu_1(L; (0,c]) + \frac{1}{2!} \mu_2(L; (0,c]) + \cdots + (-1)^k \frac{1}{k!} \mu_k(L; (0,c]) \tag{6.11}$$

when k is even, and

$$\pi_0(L; (0,c]) \geq 1 - \mu_1(L; (0,c]) + \frac{1}{2!} \mu_2(L; (0,c]) + \cdots + (-1)^k \frac{1}{k!} \mu_k(L; (0,c]) \tag{6.12}$$

when k is odd. (These are special cases of [6, Theorem 2.3].)

Given $L > 0$, define N by $x_N(L) \leq L < x_{N+1}(L)$, namely by

$$N := n(L) = \sharp\{j \geq 1; x_j(L) \leq L\}, \tag{6.13}$$

and put $x_0(L) = 0$. Then we have

$$\begin{aligned} (6.14)\quad & L\pi_0(L; (0,c]) \\ &= \sum_{j=0}^{N-1} (x_{j+1}(L) - x_j(L) - c)_+ + \{(x_{N+1}(L) - x_N(L) - c)_+ \wedge (L - x_N(L))\} \\ &= \sum_{j=0}^{N-1} (x_{j+1}(L) - x_j(L) - c) + \sum_{j=0}^{N-1} \Big(c - (x_{j+1}(L) - x_j(L))\Big)_+ \\ &\qquad + \{(x_{N+1}(L) - x_N(L) - c)_+ \wedge (L - x_N(L))\} \\ &\leq x_N(L) - cN + c \cdot \sharp\{j \geq 0; x_{j+1}(L) \leq L, x_{j+1}(L) - x_j(L) \leq c\} \\ &\qquad + \big(L - x_N(L)\big) \\ &= L - c \cdot \sharp\{j \geq 0; x_{j+1}(L) \leq L, x_{j+1}(L) - x_j(L) > c\}. \end{aligned}$$

Suppose that the condition of (ii) is satisfied. Then dividing both sides of (6.14) by L and letting $L \to \infty$, we obtain from (6.12)

$$1 - c\overline{\rho}(c) \geq \underline{\pi}_0(c) \geq 1 - \mu_1(c) + \frac{1}{2} \underline{\mu}_2(c) - \frac{1}{3!} \overline{\mu}_3(c) \geq 1 - c + \frac{a}{2} c^2 + o(c^2), \tag{6.15}$$

or

$$c(1 - \overline{\rho}(c)) \geq \frac{ac^2}{2} + o(c^2).$$

Hence we get

$$\liminf_{c \searrow 0} \frac{1 - \overline{\rho}(c)}{c} \geq \frac{a}{2} > 0. \tag{6.16}$$

Next suppose that the condition of (i) is satisfied. For each $0 < \delta < 1$, we have

$$
\begin{aligned}
(6.17)\quad L\pi_0(L;(0,c]) \\
&\geq x_N(L) - cN + (L - x_N(L) - c) \\
&\quad + \sum_{j=0}^{N-1} 1_{(0,\delta c]}\big(x_{j+1}(L) - x_j(L)\big)\{c - \big(x_{j+1}(L) - x_j(L)\big)\} \\
&\geq (L-c) - cN \\
&\quad + (1-\delta)c[N - \sharp\{j \geq 0; x_{j+1}(L) \leq L, x_{j+1}(L) - x_j(L) > \delta c\}].
\end{aligned}
$$

Again dividing by L and noting $N/L \to 1$ as $L \to \infty$, we obtain from (6.11)

$$
(6.18)\quad 1 - c + (1-\delta)c[1 - \underline{\rho}(\delta c)] \leq \overline{\pi}_0(c) \leq 1 - \mu_1(c) + \frac{1}{2}\overline{\mu}_2(c) = 1 - c + o(c^2),
$$

and hence

$$
(6.19)\qquad \limsup_{c \searrow 0} \frac{1 - \underline{\rho}(c)}{c} = \limsup_{c \searrow 0} \frac{1 - \underline{\rho}(\delta c)}{\delta c} = 0,
$$

as desired. □

Corollary 6.3. (i) *If $\overline{R}_2(c) = o(c)$ as $c \searrow 0$, then the spectrum exhibits level repulsion in the wide sense.*

(ii) *If for some $a > 0$, $\underline{R}_2(c) \geq ac$ for all $c > 0$ and if $\overline{R}_3(c) = o(c)$ as $c \searrow 0$, then the spectrum exhibits level clustering in the wide sense.*

PROOF. By the inequality (4.16) in the proof of Proposition 4.2, we see that

$$
(6.20)\quad kc\frac{1}{\ell}\sum_{\ell=0}^{m} \underline{R}_k\Big(c\Big(1 - \frac{m}{\ell}\Big)\Big) \leq \underline{\mu}_k(c) \leq \overline{\mu}_k(c) \leq kc\frac{1}{\ell}\sum_{\ell=0}^{m} \overline{R}_k\Big(c\Big(1 - \frac{m}{\ell}\Big)\Big)
$$

holds for all integer $\ell > 1$. Letting $\ell \to \infty$, we get

$$
(6.21)\qquad k\int_0^c \underline{R}_k(c')\,\mathrm{d}c \leq \underline{\mu}_k(c) \leq \overline{\mu}_k(c) \leq k\int_0^c \overline{R}_k(c')\,\mathrm{d}c'.
$$

It is now obvious that the conditions of the corollary imply those of Proposition 6.2. □

If we impose conditions on $R_k(c)$, we get more precise estimates on $\overline{\rho}(c)$ and $\underline{\rho}(c)$.

Proposition 6.4. *If the limits $R_j(c)$ exist for $j = 2, \ldots, k$, $k \geq 2$, then we have*

$$
(6.22)\qquad \overline{\rho}(c) \leq 1 + \sum_{j=1}^{k}(-1)^j \frac{1}{j!} R_{j+1}(c)
$$

when k is even, and

$$
(6.23)\qquad \underline{\rho}(c) \geq 1 + \sum_{j=1}^{k}(-1)^j \frac{1}{j!} R_{j+1}(c)
$$

when k is odd.

PROOF. Let $n(L) := \sum_j 1_{(0,L]}(x_j(L))$ and define the probability distribution $\{\rho_k(L;c)\}_{k\geq 0}$ by

$$\rho_k(L;c) := \frac{1}{n(L)}\sum_j 1_{(0,L]}\big(x_j(L)\big)1_{\{N_L(x_j(L),x_j(L)+c]=k\}}. \tag{6.24}$$

Then

$$\rho_0(L;c) = \frac{n(L)}{L}\rho(L;c). \tag{6.25}$$

Note that $n(L) \sim L$, and hence the upper and lower limits of $\rho_0(L;c)$ as $L \to \infty$ coincide with $\overline{\rho}(c)$ and $\underline{\rho}(c)$ respectively. If we let

$$r_k(L;c) := \frac{L}{n(L)}R_k(L;c) \tag{6.26}$$

for $k \geq 2$, then it is not difficult to see

$$\begin{aligned} r_k(L;c) &= \frac{1}{n(L)}\sum_j 1_{(0,L]}\big(x_j(L)\big)\big(N_L(x_j(L),x_j(L)+c]\big)^{[k-1]} \\ &= \sum_{i\geq k-1} i^{[k-1]}\rho_i(L;c). \end{aligned} \tag{6.27}$$

Namely $r_k(L;c)$ is the $(k-1)$th factorial moment of the probability distribution $\{\rho_i(L;c)\}_{i\geq 0}$. Thus again by [6, Lemma 2.2], we can write

$$Q_n(L;c) := \sum_{j\geq n}\rho_i(L;c) = \sigma_{n,k}(c) + (-1)^{k+1}\tau, \tag{6.28}$$

where $\tau \geq 0$ and

$$\sigma_{n,k}(c) = \sum_{j=n}^{n+k}(-1)^{j-n}\binom{j-1}{n-1}\frac{1}{j!}r_{j+1}(L;c). \tag{6.29}$$

When $n = 1$, this reduces to

$$1-\rho_0(L;c) = \sum_{j=1}^{k}(-1)^{j-1}\frac{1}{j!}r_{j+1}(L;c) + (-1)^k\tau, \tag{6.30}$$

which is valid for $k \geq 1$. The assertion of the proposition immediately follows from this equality. $\square$

Example 6.5. In [41], VanderKam considered the level statistics for the Laplacian on a flat d-dimensional torus, which is the quantization of the uniform motion on that torus. The corresponding Hamiltonian, expressed in the action variables, is a positive definite quadratic form

$$P(x_1,\dots,x_d) = {}^t\mathbf{x}A\mathbf{x}, \tag{6.31}$$

where A is a $d\times d$ real symmetric, positive definite matrix, and the eigenvalues of the Laplacian are values taken by P on the lattice $\hbar\mathbb{Z}^d$. To eliminate the obvious degeneracy, $\mathbf{x}$ and $-\mathbf{x}$ are identified. Let us denote this equivalence relation by$\sim$. Then the spectrum $\{P(\hbar\mathbf{n});\mathbf{n}\in\mathbb{Z}^d/\sim\}$ is unfoldable with $\alpha = d$ and $\nu(E) = c_P E^{d/2}$, where c_P is the volume of the domain $\{\mathbf{x}\in\mathbb{R}^d; P(\mathbf{x})\leq 1\}/\sim$. Since $P(\mathbf{x})$

is a homogeneous function of degree 2, both kinds of unfoldings discussed in §5 give the same unfolded sequence $\{\xi_{\mathbf{n}}; \mathbf{n} \in \mathbb{Z}^d\}$, where

$$\xi_{\mathbf{n}} := c_P P(\mathbf{n})^{d/2}. \tag{6.32}$$

Then VanderKam proves that for almost all realization of the coefficient matrix A of $P(\mathbf{x})$, one has $R_k(c) = c^{k-1}$ for $k = 2, \ldots, [d/2]$, for the correlation functions of the unfolded spectrum (6.32). Hence if $d \geq 6$, Corollary 6.3 tells us that $\{\xi_{\mathbf{n}}\}$ exhibits level clustering in the wide sense. Moreover, by Proposition 6.2, we can assert that when d is large and $c > 0$ is small, $\overline{\rho}(c)$ and $\underline{\rho}(c)$ are very close to each other, and are approximated by e^{-c}.

Acknowledgments. The author is grateful to Professor V. Jakšić for giving him the opportunity to make this contribution, and to Dr. E. Giere for careful reading of the manuscript and useful comments.

References

1. V. I. Arnol'd, *Mathematical methods of classical mechanics*, Grad. Texts in Math., vol. 60, Springer, New York, 1978.
2. R. Aurich, A. Bäcker, and F. Steiner, *Mode fluctuations as fingerprints of chaotic and non-chaotic systems*, Internat. J. Modern Phys. B **11** (1997), no. 7, 805–849.
3. M. V. Berry and M. Tabor, *Level clustering in the regular spectrum*, Proc. Roy. Soc. London Ser. A **356** (1977), no. 1686, 375–394.
4. M. V. Berry and M. Robnik, *Semiclassical level spacings when regular and chaotic orbits coexist*, J. Phys. A **17** (1984), no. 12, 2413–2421.
5. O. Bohigas and M.-J. Giannoni, *Chaotic motion and random matrix theories*, Mathematical and Computational Methods in Nuclear Physics (Granada, 1983) (J. S. Dehesa, J. M. G. Gomez, and A. Polls, eds.), Lecture Notes in Phys., vol. 209, Springer, Berlin, 1984, pp. 1–99.
6. H. Cramér, M. R. Leadbetter, and R. J. Serfling, *On distribution function — moment relationships in a stationary point process*, Z. Wahrscheinlichkeitstheorie und Verw. Gebiete **18** (1971), 1–8.
7. D. J. Daley and D. Vere-Jones, *An introduction to the theory of point processes*, Springer Ser. Statist., Springer, New York, 1988.
8. ———, *An introduction to the theory of point processes.* I: *Elementary theory and methods*, 2nd ed., Probab. Appl. (N. Y.), Springer, New York, 2003.
9. M. Fukushima and S. Nakao, *On spectra of the Schrödinger operator with a white Gaussian noise potential*, Z. Wahrscheinlichkeitstheorie und Verw. Gebiete **37** (1977), no. 3, 267–274.
10. L. P. Gor'kov and G. M. Èliashberg, *Minute metallic particles in an electromagnetic field*, Soviet Physics JETP **21** (1965), no. 5, 940–947.
11. L. N. Grenkova, S. A. Molchanov, and Yu. N. Sudarev, *On the basic states of one-dimensional disordered structures*, Comm. Math. Phys. **90** (1983), no. 1, 101–123.
12. A. Hanen, *Processus ponctuels stationaires et flots spéciaux*, Ann. Inst. H. Poincaré Sect. B (N.S.) **7** (1971), 23–30.
13. H. Hasegawa, *Level statistics for quantum systems: Introduction to quantum chaos*, Butsuri-gaku Saizensen, vol. 28, Kyoritsu Shuppan, Tokyo, 1991 (Japanese).
14. H. Makino and S. Tasaki, *Level spacing statistics of classically integrable system: Investigation along the lines of Berry–Robnik approach*, Phys. Rev. E (3) **67** (2003), 066205.
15. ———, *Long-range spectral statistics of classically integrable systems: Investigation along the line of Berry–Robnik approach*, Progr. Theoret. Phys. **114** (2005), no. 5, 929–941.
16. P. Major, *Poisson law for the number of lattice points in a random strip with finite area*, Probab. Theory Related Fields **92** (1992), no. 4, 423–464.
17. J. Marklof, *The Berry–Tabor conjecture*, European Congress of Mathematics (Barcelona, 2000), Progr. Math., vol. 202, Birkhäuser, Basel, pp. 421–427.
18. K. Matthes, J. Kerstan, and J. Mecke, *Infinitely divisible point processes*, John Wiley & Sons, New York, 1978.

19. H. P. McKean, *A limit law for ground state of Hill's equation*, J. Statist. Phys. **74** (1994), no. 5-6, 1227 – 1232.
20. M. L. Mehta, *Random matrices*, 3rd ed., Pure Appl. Math., vol. 142, Elsevier, Amsterdam, 2004.
21. N. Minami, *Schrödinger operators with potential which is the derivative of a temporally homogeneous Lévy process*, Probability Theory and Mathematical Statistics (Kyoto, 1986) (S. Watanabe and Yu. V. Prokhorov, eds.), Lecture Notes in Math., vol. 1299, Springer, Berlin, 1988, pp. 298 – 304.
22. ———, *Level clustering in a finite system* (1994), no. 116, 359 – 368.
23. ———, *Local fluctuation of the spectrum of a multidimensional Anderson tight binding model*, Comm. Math. Phys. **177** (1996), no. 3, 709 – 725.
24. ———, *On the Poisson limit theorems of Sinai and Major*, Comm. Math. Phys. **213** (2000), no. 1, 203 – 247.
25. ———, *Level statistics for quantum Hamiltonians: Some preliminary ideas toward mathematical justification of the theory of Berry and Tabor*, Proceedings of the Second ISAAC Congress (Fukuoka, 1999), Int. Soc. Anal. Appl. Comput., vol. 7, Kluwer, Dordrecht, 2000, pp. 7 – 14.
26. ———, *On level clustering in regular spectra*, Sūrikaisekikenkyūsho Kōkyūroku (2000), no. 1156, 113 – 130.
27. ———, *Mathematical foundation of energy level statistics in terms of point process theory*, Bussei Kenkyuu **73** (2000), no. 6, 957 – 1011 (Japanese).
28. S. A. Molchanov, *The local structure of the spectrum of the one-dimensional Schrödinger operator*, Comm. Math. Phys. **78** (1980/81), no. 3, 429 – 446.
29. S. A. Molchanov and A. Ya. Reznikova, *Limit theorems for random partitionings*, Teor. Veroyatnost. i Primenen. **27** (1982), no. 2, 296 – 307 (Russian); English transl., Theory Probab. Appl. **27** (1982), no. 2, 310 – 323.
30. K. Nagai, *The integrated density of states of one-dimensional random Schrödinger operator with white noise potential and background*, Tsukuba J. Math. **30** (2006), no. 2, 383 – 400.
31. J. Neveu, *Processus ponctuels*, École d'Été de Probabilités de Saint-Flour (1976), Lecture Notes in Math., vol. 598, Springer, Berlin, 1976, pp. 249 – 445.
32. I. C. Percival, *Regular and irregular spectra*, J. Phys. B **6** (1973), no. 9, L229 – L232.
33. C. E. Porter (ed.), *Statistical theory of spectra: Fluctuations*, Academic Press, New York, 1965.
34. A. Ya. Reznikova, *The central limit theorem for the spectrum of random Jacobi matrices*, Theory Probab. Appl. **25** (1980), no. 3, 504 – 513.
35. ———, *The central limit theorem for the spectrum of the random one-dimensional Schrödinger operator*, J. Statist. Phys. **25** (1981), no. 2, 291 – 308.
36. A. M. Savchuk and A. A. Shkalikov, *Sturm – Liouville operators with singular potentials*, Mat. Zametki **66** (1999), no. 6, 897 – 912 (Russian); English transl., Math. Notes **66** (1999), no. 6, 741 – 753.
37. P. Sarnak, *Values at integers of binary quadratic forms*, Harmonic Analysis and Number Theory (Montréal, 1996) (S. W. Drury and M. R. Murty, eds.), CMS Conf. Proc., vol. 21, Amer. Math. Soc., Providence, RI, 1997, pp. 181 – 203.
38. Ya. G. Sinaĭ, *Mathematical problems in the theory of quantum chaos*, Chaos (Woods Hole, MA, 1989) (D. K. Campbell, ed.), American Inst. Phys., New York, 1990, pp. 395 – 414.
39. ———, *Mathematical problems in the theory of quantum chaos*, (Unknown Month 1989), Geometric Aspects of Functional Analysis (J. Lindenstrauss and Milman V. D., eds.), Lecture Notes in Math., vol. 1469, Springer, Berlin, 1991, pp. 41 – 59.
40. ———, *Poisson distribution in a geometric problem*, Dynamical Systems and Statistical Mechanics (Moscow, 1991) (A. B. Sossinsky, ed.), Adv. Soviet Math., vol. 3, Amer. Math. Soc., Providence, RI, 1991, pp. 199 – 214.
41. J. M. VanderKam, *Correlations of eigenvalues on multi-dimensional flat tori*, Comm. Math. Phys. **210** (2000), no. 1, 203 – 223.
42. I. Veselić, *Integrated density of states and Wegner estimates for random Schrödinger operators*, Spectral Theory of Schrödinger Operators (Mexico City, 2001) (R. del Río and C. Villegas-Blas, eds.), Contemp. Math., vol. 340, Amer. Math. Soc., Providence, RI, 2004,

pp. 97 – 183.

43. J. Weidmann, *Spectral theory of ordinary differential operators*, Lecture Notes in Math., vol. 1258, Springer, Berlin, 1987.

INSTITUTE OF MATHEMATICS, UNIVERSITY OF TSUKUBA, TSUKUBA-SHI, IBARAKI 305-8571, JAPAN

Current address: School of Medecine, Keio University, Hiyoshi, Kouhoku-ku, Yokohama, 223-8521, Japan

E-mail address: `minami@sakura.cc.tsukuba.ac.jp`

Centre de Recherches Mathématiques
CRM Proceedings and Lecture Notes
Volume **42**, 2007

On the Law of Addition of Random Matrices: Covariance and the Central Limit Theorem for Traces of Resolvent

L. Pastur and V. Vasilchuk

ABSTRACT. We consider the ensemble of $n \times n$ random matrices $H_n = A_n + U_n^\dagger B_n U_n$, where A_n and B_n are Hermitian (real symmetric), having the limiting Normalized Counting Measure of eigenvalues (NCM), and U_n is unitary (orthogonal), uniformly distributed over $U(n)$ $(O(n))$. We give first a more short and transparent proof of our result of [9] on the existence of the NCM of H_n. This is based on a new inequality, analogous to the Poincaré–Nash inequality for Gaussian random variables. We then find the leading term of the covariance and establish the Central Limit Theorem for traces of resolvent of H_n as $n \to \infty$.

1. Introduction

The paper deals with Hermitian (real symmetric) $n \times n$ random matrices

$$H_n = A_n + U_n^\dagger B_n U_n, \tag{1.1}$$

where A_n and B_n are Hermitian (real symmetric) matrices such that if $\{\lambda_l^{A_n}\}_{l=1}^n$ and $\{\lambda_l^{B_n}\}_{l=1}^n$ are eigenvalues of A_n and B_n and N_{A_n} and N_{B_n} are their Normalized Counting Measures (NCM), defined as

$$\begin{aligned} N_{A_n}(\Delta) &= \#\{\lambda_l^{A_n} \in \Delta, l = 1, \dots, n\}n^{-1}, \\ N_{B_n}(\Delta) &= \#\{\lambda_l^{B_n} \in \Delta, l = 1, \dots, n\}n^{-1} \end{aligned} \tag{1.2}$$

for any interval $\Delta \subset \mathbb{R}$, then we have the weak convergence

$$N_{A_n} \to N_A, \quad N_{B_n} \to N_B \tag{1.3}$$

of N_{A_n} and N_{B_n} to probability measures N_A and N_B. We assume further that U_n in (1.1) is the random unitary matrix, whose probability law is given by the normalized to unity Haar measure on the unitary group $\mathrm{U}(n)$ in the case of Hermitian A_n and B_n and U_n is the random orthogonal matrix whose probability law is given by corresponding Haar measure on $\mathrm{O}(n)$ in the case of real symmetric A_n and B_n.

2000 *Mathematics Subject Classification.* Primary 15A52; Secondary 60F05.
This is the final form of the paper.

We mention also that following the general ideas of random matrix theory one can define analogous matrices using quaternion real A_n and B_n and symplectic U_n (see, e.g., [7] for corresponding definitions).

We will confine ourselves to the technically simplest case of Hermitian A_n and B_n and unitary U_n in (1.1). Our results for symmetric and quaternion real matrices and the groups $\mathrm{O}(n)$ and $\mathrm{S}(n)$ have a similar form, although their proof is more involved technically.

Matrices of the above form are of interest by themselves (see, e.g., [9] for various motivations) and arise also in the free probability studies and related combinatorial problems (see review works [11, 12] and references therein).

Our goal in this paper is the analysis of the eigenvalue distribution of H_n of (1.1), given that of A_n and B_n. The simplest but important for practically any random matrix study is the problem of weak convergence of the Normalized Counting Measures of eigenvalues $\{\lambda_l^{H_n}\}_{l=1}^n$ of H_n

$$N_n(\Delta) = \#\{\lambda_l^{H_n} \in \Delta, l = 1, \dots, n\}n^{-1} \tag{1.4}$$

to a non-random measure as $n \to \infty$. Following general ideas of spectral theory, we study N_n via the resolvent

$$G_n(z) = (H_n - z)^{-1}, \quad \Im z \neq 0, \tag{1.5}$$

of H_n and its normalized trace

$$g_n(z) = n^{-1} \operatorname{Tr} G_n(z), \tag{1.6}$$

related to the Normalized Counting Measures of eigenvalues of H_n by spectral theorem

$$g_n(z) = \int \frac{N_n(d\lambda)}{\lambda - z}, \quad \Im z \neq 0. \tag{1.7}$$

Here and below integrals without limits denote integrals over $\mathbb{R}$. To study the asymptotic behavior of g_n we use an approach, based on certain differentiation formulas (matrix analogs of the integration by parts), leading to certain identities for the moments of g_n and to bounds for the variance of g_n, allowing one to convert the identities into functional equations, determining uniquely g_n, hence the limiting measure (see [4, 8] for various realizations of the approach). The approach proved to be rather efficient in studies of Normalized Counting Measures of eigenvalues of a wide variety of random matrix ensembles since the paper [6], where it was introduced. For its modern forms see [4, 8]. A version of the approach was also used in [9] to treat random matrices (1.1). Here we partly use the ideas of [9], namely we relay again on a differentiation formula (see Proposition 2.1) to derive identities for the moments of g_n. However, to obtain the bounds for the variance of g_n we use not the same differentiation formula, applied to more complex expressions containing the resolvent, but on a "unitary" analog of the well-known Poincaré–Nash inequality for Gaussian random variables [2, 5]. This makes the proofs of bounds more short and transparent and enlarges the domain of validity of the bounds. Moreover, this allows us to find the leading term of the covariance of $g_n(z_1)$ and $g_n(z_2)$ for $\Im z_{1,2} \neq 0$, and to prove the Central Limit Theorem for any finite collection $\{g_n(z_l)\}_{l=1}^p$.

An important property of a wide variety of random matrix ensembles, the ensemble (1.1) in particular, is that the variance of

$$g_n(z) = \frac{1}{n}\sum_{l=1}^{n}\frac{1}{\lambda_l^{H_n} - z}, \qquad \Im z \neq 0$$

is of the order $O(n^{-2})$, rather than $O(n^{-1})$, as in the case, where $\lambda_l^{H_n}$ are replaced by the i.i.d. random variables. This implies that the Central Limit Theorem is valid not for $\sqrt{n}g_n$ but for ng_n.

The paper is organized as follows. In Section 2 we present our basic tools and rederive a system of functional equations, obtained in [9] and determining uniquely the limiting Normalized Counting Measures of eigenvalues of H_n. In Section 3 we outline the derivation of a compact form of the leading term of the covariance of $g_n(z_{1,2})$ and in Section 4 we do the same for the Central Limit Theorem. We mention in conclusion that our presentation is a kind of announcement. The detailed proofs and their discussion will be publishes elsewhere [10].

2. Convergence of Normalized Counting Measures of eigenvalues

We denote $\mathrm{E}\{\dots\}$ the expectation with respect to the normalized to unity Haar measure of $U(n)$. Our proofs are based on the following two facts on this expectation. The first one is the differentiation formula, given by

Proposition 2.1. *Denote $\mathcal{H}_n$ the linear space of $n\times n$ Hermitian matrices. Let $\Phi\colon \mathcal{H}_n \to \mathbb{C}$ be a continuously differentiable function. Then the following relation holds for any element $X \in \mathcal{H}_n$:*

$$\mathrm{E}\{\Phi'(U^\dagger MU)\cdot[X, U^\dagger MU]\} = 0,$$

where

$$[M_1, M_2] = M_1M_2 - M_1M_2$$

is the commutator of $M_{1,2} \in \mathcal{H}_n$.

The proof of the proposition is given in [9]. The second important technical mean is a "unitary" analog of the Poincaré–Nash inequality.

Proposition 2.2. *Let $\Phi\colon \mathrm{U}(n) \to \mathbb{C}$ be a continuously differentiable function. Then*

$$\mathrm{Var}\{\Phi(U_n)\} := \mathrm{E}\{|\Phi(U_n)|^2\} - |\,\mathrm{E}\{\Phi(U_n)\}|^2 \leq \frac{1}{n}\sum_{j,k=1}^{n}\mathrm{E}\{|\Phi'(U_n)\cdot E^{(j,k)}U_n|^2\},$$

where $\{E^{(j,k)}\}_{j,k=1}^n$ is canonical basis in the space $\mathcal{M}_n$ of all $n\times n$ matrices: $E^{(j,k)} = \{E^{(j,k)}_{pq}\}_{p,q=1}^n$, $E^{(j,k)}_{pq} = \delta_{jp}\delta_{kq}$.

The proof of proposition will be given in [10].

We will use the resolvent identity for resolvents G_1 and G_2 of two Hermitian matrices M_1 and M_2:

$$G_2(z) - G_1(z) = G_1(z)(M_1 - M_2)G_2(z) = G_2(z)(M_1 - M_2)G_1(z), \tag{2.1}$$

and the formula for the derivative of the resolvent of a Hermitian matrix M:

$$G'\cdot X = -GXG, \tag{2.2}$$

valid for any $X \in \mathcal{H}_n$.

We will also need the notion of the Nevanlinna functions (see, e.g., [1]). Namely, an analytic in $\mathbb{C} \setminus \mathbb{R}$ function f is a Nevanlinna function if

$$\overline{f(z)} = f(\bar{z}), \quad \Im f(z)\Im z > 0, \quad \Im z \neq 0. \tag{2.3}$$

Any Nevanlinna function admits the representation

$$f(z) = az + b + \int \frac{1 + \mu z}{\mu - z} m(\mathrm{d}\mu), \tag{2.4}$$

where $a \geq 0$, $b \in \mathbb{R}$, m is a finite non-negative measure and we write here and below integrals without limits for the integrals over $\mathbb{R}$. The representation takes the form

$$f(z) = \int \frac{m(\mathrm{d}\mu)}{\mu - z}, \tag{2.5}$$

with a finite non-negative m if and only if $\sup_{y \geq 1} |yf(\mathrm{i}y)| < \infty$, and in this case

$$\lim_{y \to \infty} |yf(\mathrm{i}y)| = m(\mathbb{R}) < \infty.$$

We have

Theorem 2.3. *Consider the random matrices of the form* (1.1) *and assume* (1.3). *Then there exists a non-random probability measure N and for any continuous and bounded function $\varphi\colon \mathbb{R} \to \mathbb{C}$ we have with probability* 1:

$$\lim_{n \to \infty} \int \varphi(\lambda) N_{H_n}(\mathrm{d}\lambda) = \int \varphi(\lambda) N(\mathrm{d}\lambda). \tag{2.6}$$

Moreover, the Stieltjes transform

$$f(z) = \int \frac{N(d\lambda)}{\lambda - z}, \quad \Im z \neq 0$$

of N is a unique solution of the system

$$\begin{cases} f(z) = f_A\big(h_B(z)\big), \\ f(z) = f_B(h_A(z)), \\ (f(z))^{-1} = z - h_A(z) - h_B(z), \end{cases} \tag{2.7}$$

where

$$f_{A,B}(z) = \int \frac{N_{A,B}(\mathrm{d}\lambda)}{\lambda - z},$$

$f(z)$ is a Nevanlinna function and $h_{A,B}(z)$ are analytic in $\mathbb{C} \setminus \mathbb{R}$ functions verifying the conditions

$$f(z) = -z^{-1} + o(z^{-1}), \quad h_{A,B}(z) = z + o(z), \quad z \to \infty. \tag{2.8}$$

Remark 2.4. An analogous theorem was proved in [9] under the additional condition

$$\sup_n \int |\lambda| N_{A_n}(\mathrm{d}\lambda) < \infty. \tag{2.9}$$

for the convergence in probability. It was also proved ([9, Theorem 3.1]) that if

$$\sup_n \int |\lambda|^4 N_{A_n, B_n}(\mathrm{d}\lambda) < \infty \tag{2.10}$$

then N_{H_n} converges to N weakly with probability 1.

We will prove the theorem in two steps. On the first step we will prove the convergence with probability 1 of N_n under the condition (2.10). This is the analog of [9, Theorem 3.1]. On the second step we will remove condition (2.10) and prove the convergence in probability, using some truncation procedure and a compactness argument.

On the both steps we need the following

Lemma 2.5. *The system* (2.7) *for the triple* (f, h_A, h_B) *is uniquely soluble in the class of triples* (f, h_A, h_B), *where* f *is a Nevanlinna function, satisfying* (2.8), *and* $h_{A,B}$ *are analytic in* $\mathbb{C} \setminus \mathbb{R}$ *and satisfy* (2.8).

PROOF. Indeed, introduce

$$r(z) = -\frac{1}{f(z)}, \tag{2.11}$$

where we have in view of (2.8)

$$r(z) = z + o(z), \quad z \to \infty,$$

we can rewrite (2.7) as

$$\begin{cases} r(z) - r_A(h_B(z)) = 0, \\ r(z) - r_B(h_A(z)) = 0, \\ r(z) - h_A(z) - h_B(z) = -z, \end{cases} \tag{2.12}$$

where

$$r_{A,B}(z) = -\frac{1}{f_{A,B}(z)}, \quad r_{A,B}(z) = z + o(z), \quad z \to \infty. \tag{2.13}$$

This implies the following linear system for the derivatives $r'(z)$, $h'_{A,B}(z)$:

$$\begin{cases} r'(z) - r'_A(h_B(z))h'_B(z) = 0, \\ r'(z) - r'_B(h_A(z))h'_A(z) = 0, \\ r'(z) - h'_A(z) - h'_B(z) = -1. \end{cases} \tag{2.14}$$

In view of (2.8) and (2.13) the derivatives $r'_{A,B}(z)$ have the asymptotics

$$r'_{A,B}(z) = 1 + o(1), \quad z \to \infty.$$

This and condition (2.8) imply that the determinant $J(z)$ of (2.14) has the same asymptotics:

$$J(z) = r'_A(h_B(z)) + r'_B(h_A(z)) - r'_A(h_B(z))r'_B(h_A(z)) = 1 + o(1), \quad z \to \infty. \tag{2.15}$$

Now suppose that the system (2.7) has two different solutions, hence there are two solutions $(r^{(1)}, h^{(1)}_{A,B})$ and $(r^{(2)}, h^{(2)}_{A,B})$ of (2.12). Using the integral representation of the Nevanlinna functions $r_{A,B}(z)$ (see (2.4)) we can obtain the linear system (2.14) with zero r.h.s. for the differences $r^{(1)} - r^{(2)}$, $h^{(1)}_{A,B} - h^{(2)}_{A,B}$. Its easy that in view of condition (2.8) the asymptotic of the determinant of this system is the same as that of J. Thus, this system has only trivial solution for z in some neighborhood of infinity and in view of analyticity of solutions in $\mathbb{C} \setminus \mathbb{R}$ $(r^{(1)}, h^{(1)}_{A,B})$ and $(r^{(2)}, h^{(2)}_{A,B})$ coincide. □

Corollary 2.6. *The linear system* (2.14) *has the unique solution*

$$r'(z) = \frac{1}{\left(r'_A\left(h_B(z)\right)\right)^{-1} + \left(r'_B\left(h_A(z)\right)\right)^{-1} - 1}, \tag{2.16}$$
$$h'_{A,B}(z) = \frac{r'(z)}{r'_{B,A}\left(h_{A,B}(z)\right)}.$$

In what follows we will omit the subscript n in A_n, B_n and $G_n(z)$ and the argument z of the resolvent in the cases where there will be no confusion.

PROOF OF THEOREM 2.3. Denote $\{G_{jk}\}_{j,k=1}^n$ the matrix of G. Taking in Proposition 2.1 $\Phi = G_{ac}$ and using (2.2), we obtain

$$\mathrm{E}\{(G[X, U^\dagger BU]G)_{ac}\} = 0, \quad a, c = 1, \ldots, n.$$

Then take $X = E^{(a,b)}$, rewrite this relation in the form

$$\mathrm{E}\{G_{aa}(U^\dagger BUG)_{bc}\} = \mathrm{E}\{(GU^\dagger BU)_{aa}G_{bc}\},$$

and apply to the result the operation $n^{-1}\sum_{a=1}^n$. This yields the matrix equality

$$\mathrm{E}\{\delta_B G\} = \mathrm{E}\{gU^\dagger BUG\},$$

where

$$\delta_B = n^{-1}\operatorname{Tr} GU^\dagger BU. \tag{2.17}$$

Introduce the centralized quantities

$$g_n^\circ = g_n - \mathrm{E}\{g_n\}, \quad \delta_B^\circ = \delta_B - \mathrm{E}\{\delta_B\}, \tag{2.18}$$

and rewrite $U^\dagger BUG$ using (2.1):

$$\mathrm{E}\{\delta_B\}\,\mathrm{E}\{G\} + \mathrm{E}\{\delta_B^\circ G\} = \mathrm{E}\{g_n\}(I + z\,\mathrm{E}\{G\} - \mathrm{E}\{AG\}) + \mathrm{E}\{g_n^\circ U^\dagger BUG\},$$

or, after regrouping terms

$$f_n(z)(A - h_{B_n} I)\,\mathrm{E}\{G\} = f_n(z) I + (\mathrm{E}\{g_n^\circ U^\dagger BUG\} - \mathrm{E}\{\delta_B^\circ G\}), \tag{2.19}$$

where

$$f_n(z) = \mathrm{E}\{g_n(z)\}, \quad h_{B_n}(z) = z - \frac{\mathrm{E}\{\delta_B(z)\}}{f_n(z)}. \tag{2.20}$$

In view of relations;

$$f_n(z) = -\frac{1}{z} + o\left(\frac{1}{z}\right), \quad |g_n(z)| \le \frac{1}{|\Im z|}, \tag{2.21}$$
$$|\delta_B(z)| \le \frac{m_B^{(1)}}{|\Im z|}, \quad m_B^{(k)} = \sup_n \int |\lambda|^k N_{B,n}(d\lambda),$$

valid for any z in the domain

$$\Gamma_{\alpha,\beta} = \{z \in \mathbb{C} : |\Re z| \le \alpha|\Im z|, |\Im z| \ge \beta\}, \quad \beta \ge (\alpha+1)m_B^{(1)} \tag{2.22}$$

and $k = 0, \ldots, 4$, we have

$$|\Im h_{B_n}(z)| \ge \beta - (\alpha+1)m_B^{(1)}$$

and, hence, the matrix $A - h_B(z)I$ in (2.19) has the inverse

$$G_A\left(h_{B_n}(z)\right) = (A - h_{B_n}(z)I)^{-1} \tag{2.23}$$

uniformly in n bounded for $z \in \Gamma_{\alpha,\beta}$:

$$\|G_A(h_{B_n}(z))\| \le (\beta - (\alpha+1)m_B^{(1)})^{-1}. \tag{2.24}$$

Thus, applying to (2.19) the inverse G_A and then the operation $n^{-1}\operatorname{Tr}$, we obtain

$$f_n^2(z) = f_n(z)n^{-1}\operatorname{Tr} G_A(h_{B_n}(z)) + R = f_n(z)f_{A_n}(h_{B_n}(z)) + R, \tag{2.25}$$

where

$$f_{A_n}(z) = n^{-1}\operatorname{Tr}\, G_A(z), \tag{2.26}$$

$$R = \mathrm{E}\{g^\circ n^{-1}\operatorname{Tr}\, G_A U^\dagger BUG\} - \mathrm{E}\{\delta_B^\circ n^{-1}\operatorname{Tr}\, G_A G\}. \tag{2.27}$$

Now Schwartz inequality for the expectation $\mathrm{E}\{\dots\}$ and (2.24) yield for $z \in \Gamma_{\alpha,\beta}$

$$|R| \le \frac{m_B^{(1)}\operatorname{Var}^{1/2}\{g_n(z)\} + \operatorname{Var}^{1/2}\{\delta_B(z)\}}{\beta(\beta-(\alpha+1)m_B^{(1)})}. \tag{2.28}$$

Besides, we have by using Proposition 2.2:

$$\begin{aligned}\operatorname{Var}\{g_n(z)\} &\le \frac{1}{n^3}\sum_{j,k=1}^{n} \mathrm{E}\{|\operatorname{Tr} GU^\dagger(E^{(k,j)}B - BE^{(j,k)})UG|^2\}\\ &= \frac{1}{n^3}\sum_{j,k=1}^{n} \mathrm{E}\{((BUG^2U^\dagger)_{jk} - (UG^2U^\dagger B)_{kj})\overline{(BUG^2U^\dagger)_{jk}} - \overline{(UG^2U^\dagger B)_{kj}}\}\\ &\le \frac{4m_B^{(2)}}{n^2|\Im z|^4},\end{aligned} \tag{2.29}$$

and analogously

$$\operatorname{Var}\{\delta_B(z)\} \le \frac{8m_B^{(4)}}{n^2|\Im z|^4}. \tag{2.30}$$

Thus, we have uniformly in $z \in \Gamma_{\alpha,\beta}$

$$R = O(n^{-1}). \tag{2.31}$$

Now a standard compactness argument leads to the first equation of the system (2.7) in the limit $n \to \infty$. The second equation can be obtained by interchanging A and B. The third equation follows from the resolvent identity (2.1) for the pair $(H_n, 0)$. Furthermore, relations (2.21) imply the corresponding asymptotics (2.8) and hence the uniqueness of solution in view of Lemma 2.5. Besides, bound (2.29) for the variance $\operatorname{Var}\{g_n(z)\}$ and the Borel–Cantelli lemma imply the convergence with probability 1 $g_n(z) \to f(z)$ as $n \to \infty$. Thus, we completed the first step of the proof.

On the second step we introduce for any $T > 0$ the truncated matrices A_n^T and B_n^T replacing eigenvalues A_n and B_n lying in (T, ∞) by T and eigenvalues lying in $(-\infty, -T]$ by $-T$. Denoting $N_{A_n}^T$ and $N_{B_n}^T$ the NCM's of A_n^T and B_n^T, we have the inequalities

$$\|A_n^T\| \le T, \quad \|B_n^T\| \le T, \tag{2.32}$$

and the weak convergence

$$N^T_{A_n} \to N^T_A, \qquad N^T_{B_n} \to N^T_B, \qquad n \to \infty,$$
$$N^T_A \to N_A, \qquad N^T_B \to N_B, \qquad T \to \infty. \tag{2.33}$$

Besides, since

$$\operatorname{rank}\big(A_n + U_n^\dagger B_n U_n - (A^T_n + U_n^\dagger B^T_n U_n)\big) \le \operatorname{rank}(A_n - A^T_n) + \operatorname{rank}(B_n - B^T_n),$$

we have

$$|g_n(z) - g^T_n(z)| \le \frac{\pi}{|\Im z|}\Big(N_{A_n}(\mathbb{R}\setminus(-T,T)) + N_{B_n}(\mathbb{R}\setminus(-T,T))\Big) = o(1), \quad T \to \infty.$$

According to the first step for any $T < \infty$ we have the limit

$$\lim_{n\to\infty} E\{|g^T_n(z) - f^T(z)|\} = 0,$$

valid uniformly in $\Gamma_{\alpha,\beta}$ of (2.22), where f^T is the unique solution of (2.7) with $f^T_{A,B}$, the Stieltjes transforms of $N^T_{A,B}$, instead of $f_{A,B}$. Since by Lemma 2.5 the system (2.7) is uniquely soluble for all $T \le \infty$, a standard compactness argument yields for any sequence $T_k \to \infty$:

$$\lim_{T_k\to\infty} f^{T_k}(z) = f(z)$$

uniformly in $\Gamma_{\alpha,\beta}$. On other hand, in view of (2.33)

$$\lim_{T_k\to\infty} f^{T_k}_{A,B}(z) = f_{A,B}(z).$$

Moreover, the truncation construction implies for any $L > 0$:

$$N^T_{A,B}\big(\mathbb{R}\setminus(-L,L)\big) \le N_{A,B}\big(\mathbb{R}\setminus(-L,L)\big).$$

As a result, the functions $f^T_{A,B}(z)$ have uniform in T asymptotics

$$f^T_{A,B}(z) = -\frac{1}{z}(1 + o(1)), \quad z \to \infty.$$

Hence they are invertible in some neighborhood of infinity of $\Gamma_{\alpha,\beta}$ uniformly in $T \le \infty$ and if $z^T_{A,B}$ are their inverses, defined as

$$z^T_{A,B}\big(f^T_{A,B}(z)\big) = z,$$

then

$$\lim_{T_k\to\infty} z^{T_k}_{A,B}(w) = z_{A,B}(w), \quad z_{A,B}(f_{A,B}(z)) = z.$$

Thus, we can apply $z_{A,B}$ to the two first equation of the system (2.7) and obtain two convergent in some neighborhood of infinity sequences of analytic functions

$$\{h^{T_k}_{B,A}\}_{k=1}^\infty, \quad h^{T_k}_{B,A}(z) = z^{T_k}_{A,B}(f^{T_k}(z)), \quad \lim_{T_k\to\infty} h^{T_k}_{B,A}(z) = h_{B,A}(z),$$

where

$$h_{B,A}(z) = z_{A,B}(f(z))$$

On other hand, the inverses $z_{A,B}$ have the asymptotics

$$z_{A,B}(f) = -\frac{1}{f} + o\Big(\frac{1}{f}\Big), \quad f \to 0,$$

thus

$$h_{A,B}(f) = z_{B,A}(f) = -\frac{1}{f} + o\Big(\frac{1}{f}\Big), \quad f \to 0.$$

It follows then from the third equation of (2.7) that we can define the inverse $z(f)$ of $f(z)$ in some neighborhood of infinity with the asymptotics

$$z(f) = \frac{1}{f} + h_A(f) + h_B(f) = -\frac{1}{f} + o\Big(\frac{1}{f}\Big), \quad f \to 0.$$

This implies the asymptotic

$$f(z) = -\frac{1}{z} + o\Big(\frac{1}{z}\Big), \quad z \to \infty$$

which leads to (2.8). Hence, we can use Lemma 2.5. Now the bound

$$|g_n(z) - f(z)| \leq |g_n(z) - g^{T_k}(z)| + |g_n^{T_k}(z) - f^{T_k}(z)| + |f(z) - f^{T_k}(z)|.$$

and the above argument implies the convergence of $\{g_n\}$ to f with probability 1 on a compact set on $\mathbb{C} \setminus \mathbb{R}$, hence (2.6). □

Remark 2.7. (1) It was proved in the recent paper [3] that the functions $h_{A,B}$ in system (2.7) are not only analytic but are also Nevanlinna functions, obeying asymptotics (2.8)

$$h_{A,B}(z) = z + o(z), \quad z \to \infty.$$

Using this we can carry out all the procedures in the proofs of the above theorem not only in the region $\Gamma_{\alpha,\beta}$ but in the $\mathbb{C} \setminus \mathbb{R}$. Indeed, it is known that Nevanlinna functions obeying this asymptotic have the property

$$|\Im h_{A,B}(z)| \geq |\Im z|, \quad z \in \mathbb{C} \setminus \mathbb{R}.$$

Thus, once we have proved convergence to the limiting system (2.7), we have $h_{B_n}(z) \to h_B(z)$, $z \in \mathbb{C} \setminus \mathbb{R}$ and, hence, the matrix $G_A\big(h_{B_n}(z)\big)$ of (2.23) will admit the bound:

$$\|G_A(h_{B_n}(z))\| \leq C \tag{2.34}$$

uniformly in z in a compact set $K \subset \mathbb{C} \setminus \mathbb{R}$ and sufficiently large n with an n-independent C, finite for $\Im z \neq 0$. Since Proposition 2.2 provides bounds (2.29) and (2.30) for variances for any $z \in \mathbb{C} \setminus \mathbb{R}$, we can estimate all above remainders for any $z \in \mathbb{C} \setminus \mathbb{R}$.

(2) Writing for the remainder (2.27) the bound

$$\begin{aligned}|R| \leq \operatorname{Var}^{1/2}\{g_n(z)\} \operatorname{Var}^{1/2}\{n^{-1} \operatorname{Tr} G_A U^\dagger B U G\}\\ + \operatorname{Var}^{1/2}\{\delta_B(z)\} \operatorname{Var}^{1/2}\{n^{-1} \operatorname{Tr} G_A G\},\end{aligned}$$

and arguing as in obtaining (2.29) and (2.30) it can be shown that $R = O(n^{-2})$ instead (2.31).

3. Covariance of traces of resolvent

Theorem 3.1. *Consider random matrices* (1.1) *and assume* (1.3) *and* (2.10). *Then we have for* g_n *of* (1.5) – (1.7) *and* $z_{1,2} \in \mathbb{C} \setminus \mathbb{R}$

$$\operatorname{Cov}\{g_n(z_1), g_n(z_2)\} = \frac{1}{n^2} S(z_1, z_2) + r_n(z_1, z_2), \tag{3.1}$$

where

$$S_n(z_1, z_2) = \frac{\partial^2}{\partial z_1 \partial z_2} \log \frac{(h_{A_n}(z_1) - h_{A_n}(z_2))(h_{B_n}(z_1) - h_{B_n}(z_2))}{(z_1 - z_2)(r_n(z_1) - r_n(z_2))}, \tag{3.2}$$

where

$$r_n(z) = -\frac{1}{f_n(z)}, \tag{3.3}$$

and $r_n(z_1, z_2)$ *admits the bound*

$$|r_n(z_1, z_2)| \leq C/n^3,$$

where C *is independent of* n *and is finite if* $\min\{|\Im z_1|, |\Im z_2|\} > 0$.

PROOF. In what follows we will omit the subscript n in A_n, B_n and $G_n(z)$. Taking in the Proposition 2.1 $\Phi = g_n^\circ(z_1)G_{ac}(z_2)$ and using (2.2), we obtain

$$\mathrm{E}\{g_n^\circ(z_1)\big(G(z_2)[X, U^\dagger BU]G(z_2)\big)_{ac}\} + \mathrm{E}\{\big(n^{-1}\operatorname{Tr} G(z_1)[X, U^\dagger BU]G(z_1)\big)G_{ac}(z_2)\} = 0.$$

As in the previous section, we take $X = E^{(a,b)}$ and apply to the above relation the operation $n^{-1}\sum_{a=1}^n$. This yields the matrix relation

$$\mathrm{E}\{g_n^\circ(z_1)\delta_B(z_2)g_n(z_2)\} = \mathrm{E}\{g_n^\circ(z_1)g_n(z_2)U^\dagger BUG(z_2)\} + \frac{1}{n^2}\mathrm{E}\{[U^\dagger BU, G^2(z_1)]g_n(z_2)\}.$$

Using (2.18) and (2.1), and recalling notation (2.20), we obtain

$$\begin{aligned} f_n(z_2)(A - h_{B_n}(z_2)I)\,\mathrm{E}\{g_n^\circ(z_1)g_n(z_2)\} &= \mathrm{E}\{g_n^\circ(z_1)g_n^\circ(z_2)U^\dagger BUG(z_2)\} - \mathrm{E}\{g_n^\circ(z_1)\delta_B^\circ(z_2)G(z_2)\} \\ &\quad + n^{-2}\,\mathrm{E}\{[U^\dagger BU, G^2(z_1)]g_n(z_2)\}. \end{aligned}$$

Passing in this relation centralized quantities

$$U^\dagger BUG = (U^\dagger BUG)^\circ + \mathrm{E}\{U^\dagger BUG\}, \quad G = G^\circ + \mathrm{E}\{G\},$$

to the resolvent $G_A\big(h_{B_n}(z_2)\big)$ of (2.23) and applying to the result the operation $n^{-1}\operatorname{Tr}$, we obtain

$$\begin{aligned} \operatorname{Cov}\{g_n(z_1), g_n(z_2)\} &= \mathrm{E}\{g_n^\circ(z_1)g_n(z_2)\} \\ &= a_A(z_2)\,\mathrm{E}\{g_n^\circ(z_1)g_n(z_2)\} - b_A(z_2)\,\mathrm{E}\{g_n^\circ(z_1)\delta_B^\circ(z_2)\} \\ &\quad + n^{-2}\gamma_{AB}(z_1, z_2) + R_1, \end{aligned} \tag{3.4}$$

where

$$\begin{aligned} a_A(z_2) &= \frac{1}{f_n(z_2)}n^{-1}\operatorname{Tr} G_A\big(h_{B_n}(z_2)\big)\,\mathrm{E}\{U^\dagger BUG(z_2)\}, \\ b_A(z_2) &= \frac{1}{f_n(z_2)}n^{-1}\operatorname{Tr} G_A\big(h_{B_n}(z_2)\big)\,\mathrm{E}\{G(z_2)\}, \\ \gamma_{AB}(z_1, z_2) &= \frac{1}{f_n(z_2)}n^{-1}\operatorname{Tr} G_A\big(h_{B_n}(z_2)\big)\,\mathrm{E}\{[U^\dagger BU, G^2(z_1)]G(z_2)\}, \\ R_1 &= \frac{1}{f_n(z_2)}\big(\mathrm{E}\{g_n^\circ(z_1)g_n^\circ(z_2)\big(n^{-1}\operatorname{Tr} G_A\big(h_{B_n}(z_2)\big)U^\dagger BUG(z_2)\big)^\circ\} \\ &\quad - \mathrm{E}\{g_n^\circ(z_1)\delta_B^\circ(z_2)\big(n^{-1}\operatorname{Tr} G_A\big(h_{B_n}(z_2)\big)G(z_2)\big)^\circ\big)\}. \end{aligned} \tag{3.5}$$

Besides, using relations (2.1) and (2.19), we obtain

$$
\begin{aligned}
a_A(z_2) &= \frac{\mathrm{E}\{\delta_B(z_2)\}}{f_n^2(z_2)} n^{-1} \operatorname{Tr} G_A^2\big(h_{B_n}(z_2)\big) + R_2, \\
b_A(z_2) &= \frac{1}{f_n(z_2)} n^{-1} \operatorname{Tr} G_A^2\big(h_{B_n}(z_2)\big) + R_3, \\
\gamma_{AB}(z_1, z_2) &= \frac{1}{f_n(z_2)} n^{-1} \operatorname{Tr} G_A\big(h_{B_n}(z_2)\big) \mathrm{E}\{[G^2(z_1), A]G(z_2)\},
\end{aligned} \tag{3.6}
$$

where

$$
R_2 = n^{-1} \operatorname{Tr}\Big(\frac{\mathrm{E}\{\delta_B(z_2)\}}{f_n^3(z_2)} G_A^2\big(h_{B_n}(z_2)\big) - \frac{1}{f_n^2(z_2)} G_A\big(h_{B_n}(z_2)\big)\Big) \times \big(\mathrm{E}\{g_n^\circ(z_2) U^\dagger B U G(z_2) - \delta_B^\circ(z_2) G(z_2)\big)\},
$$

and

$$
R_3 = n^{-1} \operatorname{Tr} \frac{1}{f_n(z_2)^2} G_A^2(h_{B_n}(z_2)) \,\mathrm{E}\{g_n^\circ(z_2) U^\dagger B U G(z_2) - \delta_B^\circ(z_2) G(z_2)\}.
$$

On other hand, bounds (2.29) and (2.30) for variances and the bound (2.34) for the norm of $G_A(h_{B_n}(z_2))$ yield

$$
R_2 = O(n^{-1}), \quad R_3 = O(n^{-1})
$$

for $z_2 \in \mathbb{C} \setminus \mathbb{R}$, and we prove in Appendix that for $z_{1,2} \in \mathbb{C} \setminus \mathbb{R}$

$$
R_1 = O(n^{-3}).
$$

Here and below the symbol $O(n^{-p})$, $p > 0$ denote a quantity admitting a bound Cn^{-p}, where C is n-independent and finite if $\min\{|\Im z_1|, |\Im z_2|\} > 0$. This and (3.4) yield

$$
\begin{aligned}
\mathrm{E}\{g_n(z_1) g_n^\circ(z_2)\} = \alpha_A(z_2) \mathrm{E}\{g_n(z_1) g_n^\circ(z_2)\} - \beta_A(z_2) \mathrm{E}\{g_n^\circ(z_1) \delta_B^\circ(z_2)\} \\
+ n^{-2} \gamma_{AB}(z_1, z_2) + O(n^{-3}),
\end{aligned} \tag{3.7}
$$

where in view of (2.20), (3.6), and the relation $G'(z) = G^2(z)$, valid for the resolvent of any Hermitian matrix, we have

$$
\begin{aligned}
\alpha_A(z_2) &= \frac{\mathrm{E}\{\delta_B(z_2)\}}{f_n^2(z_2)} n^{-1} \operatorname{Tr} G_A'\big(h_{B_n}(z_2)\big) = \frac{\mathrm{E}\{\delta_B(z_2)\}}{f_n^2(z_2)} f_{A_n}'\big(h_{B_n}(z_2)\big), \\
\beta_A(z_2) &= \frac{1}{f_n(z_2)} n^{-1} \operatorname{Tr} G_A'\big(h_{B_n}(z_2)\big) = \frac{1}{f_n(z_2)} f_{A_n}'\big(h_{B_n}(z_2)\big), \\
\gamma_{AB}(z_1, z_2) &= \frac{1}{f_n(z_2)} n^{-1} \operatorname{Tr} G_A\big(h_{B_n}(z_2)\big) \frac{\partial}{\partial z_1} \mathrm{E}\{[G(z_1), A] G(z_2)\}.
\end{aligned}
$$

Furthermore, using (2.20), (2.25) and (3.3), we obtain

$$
\begin{aligned}
\alpha_A(z_2) &= \frac{h_{B_n}(z_2) - z}{r_n(z_2)} r_{A_n}'\big(h_{B_n}(z_2)\big) + O(n^{-1}), \quad n \to \infty, \\
\beta_A(z_2) &= -\frac{1}{r_n(z_2)} r_{A_n}'\big(h_{B_n}(z_2)\big) + O(n^{-1}), \quad n \to \infty,
\end{aligned}
$$

where

$$
r_{A_n, B_n}(z) = -\frac{1}{f_{A_n, B_n}(z)}, \quad r_{A_n, B_n}'(z) = \frac{f_{A_n, B_n}'(z)}{f_{A_n, B_n}^2(z)}. \tag{3.8}
$$

We show in Appendix that

$$(3.9)\quad \gamma_{AB}(z_1,z_2)=\frac{\partial}{\partial z_1}\Big(\frac{1}{z_1-z_2}-\frac{1}{h_{B_n}(z_1)-h_{B_n}(z_2)} - r'_{A_n}(h_{B_n}(z_2))\frac{h_{B_n}(z_1)-h_{B_n}(z_2)-z_1+z_2}{(z_1-z_2)\big(r_n(z_1)-r_n(z_2)\big)}\Big), + O(n^{-1}),\quad n\to\infty.$$

On other hand, interchanging A and B we obtain from (3.7)

$$(3.10)\quad \mathrm{E}\{g_n(z_1)g_n^\circ(z_2)\}=\alpha_B(z_2)\,\mathrm{E}\{g_n(z_1)g_n^\circ(z_2)\}-\beta_B(z_2)\,\mathrm{E}\{g_n^\circ(z_1)\delta_B^\circ(z_2)\} + n^{-2}\gamma_{BA}(z_1,z_2)+O(n^{-3}),$$

where

$$\alpha_B(z_2)=\frac{h_{A_n}(z_2)-z}{r_n(z_2)}r'_B\big(h_{A_n}(z_2)\big)+O(n^{-1}),\quad n\to\infty,$$

$$\beta_B(z_2)=-\frac{1}{r_n(z_2)}r'_B\big(h_{A_n}(z_2)\big)+O(n^{-1}),\qquad n\to\infty,$$

and γ_{BA} is given by (3.9) with exchanged A and B. This and the resolvent identity (2.1) for the pair $(H_n,0)$, implying

$$z_2\,\mathrm{E}\{g_n^\circ(z_1)g_n(z_2)\}=\mathrm{E}\{g_n^\circ(z_1)\delta_A^\circ(z_2)\}+\mathrm{E}\{g_n^\circ(z_1)\delta_B^\circ(z_2)\},$$

allow us to show that the triple lead to the linear system for the triple

$$(\mathrm{E}\{g_n^\circ(z_1)g_n(z_2)\},\mathrm{E}\{g_n^\circ(z_1)\delta_A^\circ(z_2)\},\mathrm{E}\{g_n^\circ(z_1)\delta_B^\circ(z_2)\})$$

satisfies the system

$$(3.11)\quad \begin{cases} \big(1-\alpha_A(z_2)\big)\,\mathrm{E}\{g_n^\circ(z_1)g_n(z_2)\}+\beta_A(z_2)\,\mathrm{E}\{g_n^\circ(z_1)\delta_B^\circ(z_2)\} \\ \qquad\qquad = n^{-2}\gamma_{AB}(z_1,z_2)+O(n^{-3}), \\ \big(1-\alpha_B(z_2)\big)\,\mathrm{E}\{g_n^\circ(z_1)g_n(z_2)\}+\beta_B(z_2)\,\mathrm{E}\{g_n^\circ(z_1)\delta_A^\circ(z_2)\} \\ \qquad\qquad = n^{-2}\gamma_{BA}(z_1,z_2)+O(n^{-3}), \\ z_2\,\mathrm{E}\{g_n^\circ(z_1)g_n(z_2)\}-\mathrm{E}\{g_n^\circ(z_1)\delta_A^\circ(z_2)\}-\mathrm{E}\{g_n^\circ(z_1)\delta_B^\circ(z_2)\}=0. \end{cases}$$

The determinant $D(z_2)$ of the system coincides asymptotically up to a factor with the determinant J (2.15):

$$D(z_2)=\frac{1}{r(z_2)}J(z_2)+o(1),\quad n\to\infty.$$

Thus, (3.11) is uniquely soluble for sufficiently large n and its solution is

$$\mathrm{E}\{g_n^\circ(z_1)g_n(z_2)\}=-\frac{1}{n^2}\frac{\gamma_{AB}(z_1,z_2)\beta_B(z_2)+\gamma_{B,A}(z_1,z_2)\beta_A(z_2)}{D(z_2)}+O(n^{-3})$$

$$=\frac{1}{n^2}\Big(\frac{\gamma_{AB}(z_1,z_2)}{h'_{B_n}(z_2)}+\frac{\gamma_{B,A}(z_1,z_2)}{h'_{A_n}(z_2)}\Big)+O(n^{-3}).$$

Finally, using (2.14) and (2.16) we obtain from the above and (3.9):

$$\mathrm{E}\{g_n^\circ(z_1)g_n(z_2)\}=-\frac{1}{n^2}\frac{\partial}{\partial z_1}\Big(\frac{h'_{A_n}(z_2)}{h_{A_n}(z_1)-h_{A_n}(z_2)}+\frac{h'_{B_n}(z_2)}{h_{B_n}(z_1)-h_{B_n}(z_2)} - \frac{1}{z_1-z_2}-\frac{r'_n(z_2)}{r_n(z_1)-r_n(z_2)}\Big)+O(n^{-3}).$$

It is easy to see that the asymptotic formula coincides with (3.1)–(3.2). □

4. Central limit theorem for traces of resolvent

Theorem 4.1. *Consider the ensemble* (1.1) *and assume* (1.3) *and* (2.10). *Then we have for* $g_n(z)$, $z \in \mathbb{C} \setminus \mathbb{R}$ *of* (1.7) *the convergence in distribution*

$$ng_n(z) - \mathrm{E}\{ng_n(z)\} \to \alpha(z) + \mathrm{i}\beta(z),$$

where $\alpha(z)$ *and* $\beta(z)$ *are Gaussian random variables with zero mean and covariances:*

$$\begin{aligned}
\mathrm{Var}\{\alpha(z)\} &= 2^{-1}\Re\big(S(z,z) + S(z,\bar z)\big),\\
\mathrm{Var}\{\beta(z)\} &= -2^{-1}\Re\big(S(z,z) - S(z,\bar z)\big),\\
\mathrm{Cov}\{\alpha(z),\beta(z)\} &= 2^{-1}\Im\big(S(z,z) - S(z,\bar z)\big),
\end{aligned}$$

where

$$S(z_1,z_2) = \frac{\partial^2}{\partial z_1 \partial z_2}\log\frac{(h_A(z_1) - h_A(z_2))(h_B(z_1) - h_B(z_2))}{(z_1 - z_2)(r(z_1) - r(z_2))}$$

is the limiting form of $S_n(z_1,z_2)$ *of* (3.2) *as* $n \to \infty$.

Proof. Consider $\Re ng_n^\circ(z)$ and its characteristic function

$$F_n(t) = \mathrm{E}\{\psi(t)\}, \quad \psi(t) = \exp\{\mathrm{i}tn\Re g_n^\circ(z)\}. \tag{4.1}$$

To prove convergence of $\Re ng_n^\circ(z)$ to the Gaussian random variable $\alpha(z)$ with zero mean and variance $\mu(z) = \mathrm{Var}\{\alpha(z)\}$, it suffices to obtain the asymptotic relation

$$\frac{\mathrm{d}}{\mathrm{d}t}F_n(t) = -\mu(z)tF_n(t) + o(1), \quad n \to \infty. \tag{4.2}$$

uniformly in t on a compact interval.

Differentiating (4.1) with respect to t, we obtain

$$\frac{\mathrm{d}}{\mathrm{d}t}F_n(t) = \frac{in}{2}(\mathrm{E}\{\psi(t)g_n^\circ(z_1)\} + \mathrm{E}\{\psi(t)g_n^\circ(z_2)\}),$$

where $z_1 = z$, $z_2 = \bar z$. To find $\mathrm{E}\{n\psi(t)g_n^\circ(z_1)\}$ we use Proposition 2.1 with $\Phi = \psi^\circ(t)G_{ac}(z_1)$ and repeat the procedure, similar to that in the proof of the preceding theorem. This leads to the linear system

$$\begin{cases}
(1-\alpha_A(z_1))\,\mathrm{E}\{n\psi^\circ(t)g_n(z_1)\} + \beta_A(z_1)\,\mathrm{E}\{n\psi^\circ(t)\delta_B^\circ(z_1)\} = C_1,\\
(1-\alpha_B(z_1))\,\mathrm{E}\{n\psi^\circ(t)g_n(z_1)\} + \beta_B(z_1)\,\mathrm{E}\{n\psi^\circ(t)\delta_A^\circ(z_1)\} = C_2,\\
z_1\,\mathrm{E}\{n\psi^\circ(t)g_n(z_1)\} - \mathrm{E}\{n\psi^\circ(t)\delta_A^\circ(z_1)\} - \mathrm{E}\{n\psi^\circ(t)\delta_B^\circ(z_1)\} = 0,
\end{cases}$$

where the functions $\alpha_{A,B}$, $\beta_{A,B}$, γ_{AB}, $\gamma_{B,A}$ are defined in the preceding section and

$$\begin{aligned}
C_1 &= \frac{\mathrm{i}tF_n(t)}{2}(\gamma_{AB}(z_1,z_1) + \gamma_{AB}(z_2,z_1)) + O(n^{-1}),\\
C_2 &= \frac{\mathrm{i}tF_n(t)}{2}(\gamma_{B,A}(z_1,z_1) + \gamma_{B,A}(z_2,z_1)) + O(n^{-1}).
\end{aligned}$$

Solving this system we obtain

$$\begin{aligned}
&\mathrm{E}\{n\psi^\circ(t)g_n(z_1)\}\\
&= \frac{itF_n(t)}{2}\left(\frac{\gamma_{AB}(z_1,z_1)}{h'_B(z_1)} + \frac{\gamma_{B,A}(z_1,z_1)}{h'_A(z_1)} + \frac{\gamma_{AB}(z_2,z_1)}{h'_B(z_1)} + \frac{\gamma_{B,A}(z_2,z_1)}{h'_A(z_1)}\right) + O(n^{-1})\\
&= \frac{itF_n(t)}{2}\big(S_n(z_1,z_1)+S_n(z_1,z_2)\big) + O(n^{-1})\\
&= \frac{itF_n(t)}{2}\big(S(z_1,z_1)+S(z_1,z_2)\big) + o(1).
\end{aligned}$$

This relation and analogous relation for $\mathrm{E}\{n\psi^\circ(t)g_n(z_2)\}$ lead to(4.2) with

$$\mu(z) = 4^{-1}\big(S(z_1,z_1)+S(z_2,z_2)+2S(z_1,z_2)\big).$$

Thus, we have proved the theorem for $\Re n g_n^\circ(z)$. Other assertions can be proved analogously. □

Appendix A

Lemma A.1. *Let R_1 and $\gamma_{AB}(z_1,z_2)$ be defined in* (3.5). *Then for $z_{1,2}\in\mathbb{C}\backslash\mathbb{R}$:*

$$\begin{aligned}
R_1 = \frac{1}{f_n(z_2)}\big(&\mathrm{E}\{g_n^\circ(z_1)g_n^\circ(z_2)\big(n^{-1}\operatorname{Tr}G_A\big(h_B(z_2)\big)U^\dagger BUG_n(z_2)\big)^\circ\}\\
&-\mathrm{E}\{g_n^\circ(z_1)\delta_B^\circ(z_2)\big(n^{-1}\operatorname{Tr}G_A\big(h_B(z_2)\big)G_n(z_2)\big)^\circ\}\big) = O(n^{-3}), \quad n\to\infty,
\end{aligned}$$

and $\gamma_{AB}(z_1,z_2)$ is given by (3.9).

Proof. We will prove first the bound

$$\mathrm{E}\{g^\circ(z_1)g^\circ(z_2)\big(n^{-1}\operatorname{Tr}G_A\big(h_B(z_2)\big)U^\dagger BUG(z_2)\big)^\circ\} = O(n^{-3}). \tag{A.1}$$

Note that

$$\begin{aligned}
&\big|\mathrm{E}\{g^\circ(z_1)g^\circ(z_2)\big(n^{-1}\operatorname{Tr}G_A\big(h_{B_n}(z_2)\big)U^\dagger BUG(z_2)\big)^\circ\}\big|\\
&\qquad= \big|\mathrm{Cov}\{g^\circ(z_1)g^\circ(z_2), n^{-1}\operatorname{Tr}G_A\big(h_{B_n}(z_2)\big)U^\dagger BUG(z_2\}\big|\\
&\qquad\le \mathrm{Var}^{1/2}\{g^\circ(z_1)g^\circ(z_2)\}\,\mathrm{Var}^{1/2}\{n^{-1}\operatorname{Tr}G_A\big(h_{B_n}(z_2)\big)U^\dagger BUG(z_2)\}.
\end{aligned}$$

Using Proposition 2.2 and the bound (2.34), we obtain $z_2\in\mathbb{C}\setminus\mathbb{R}$

$$\mathrm{Var}\left\{\frac{1}{n}\operatorname{Tr}G_A\big(h_{B_n}(z_2)\big)U^\dagger BUG(z_2)\right\} \le \frac{8C^2m_B^{(4)}}{n^2|\Im z_2|^4}, \quad z_2\in\mathbb{C}\setminus\mathbb{R}. \tag{A.2}$$

Likewise, using again Proposition 2.2 and (2.29), we obtain for $z_2\in\mathbb{C}\setminus\mathbb{R}$

$$\begin{aligned}
&\mathrm{Var}\{g^\circ(z_1)g^\circ(z_2)\}\\
&\le \frac{1}{n^3}\sum_{j,k=1}^n \mathrm{E}\{|g^\circ(z_1)\operatorname{Tr}G(z_2)U^\dagger(E^{(k,j)}B+BE^{(j,k)})UG(z_2)\\
&\qquad\qquad + g^\circ(z_2)\operatorname{Tr}G(z_1)U^\dagger(E^{(k,j)}B+BE^{(j,k)})UG(z_1)|^2\}\\
&\le \left(\sqrt{\frac{4m_B^{(2)}}{n^2|\Im z_2|^4}\mathrm{Var}\{g(z_1)\}} + \sqrt{\frac{4m_B^{(2)}}{n^2|\Im z_1|^4}\mathrm{Var}\{g(z_2)\}}\right)^2\\
&\le \frac{64(m_B^{(2)})^2}{n^4|\Im z_2|^4|\Im z_1|^4}.
\end{aligned}$$

This bound and (A.2) lead to (A.1). Analogous argument yields

$$\mathrm{E}\{g_n^\circ(z_1)\delta_B^\circ(z_2)\big(n^{-1}\operatorname{Tr} G_A\big(h_B(z_2)\big)G_n(z_2)\big)^\circ\} = O(n^{-3}).$$

Hence, we get the first assertion of the lemma.

To prove the second assertion we note that

$$\begin{aligned}(A.3)\quad &\gamma_{AB}(z_1, z_2)\\ &= \frac{1}{f_n(z_2)}\frac{1}{n}\frac{\partial}{\partial z_1}\operatorname{Tr}\big(G_A\big(h_B(z_2)\big)(\mathrm{E}\{G(z_1)AG(z_2)\} - \mathrm{E}\{AG(z_1)G(z_2)\})\big),\end{aligned}$$

where we used the resolvent identity to write

$$\mathrm{E}\{AG(z_1)G(z_2)\} = \frac{1}{z_1 - z_2}(A\,\mathrm{E}\{G(z_1)\} - A\,\mathrm{E}\{G(z_2)\}).$$

To find the term $\mathrm{E}\{G(z_1)AG(z_2)\}$ in (A.3) we use Proposition 2.1 with $\Phi = \big(G(z_1)AG(z_2)\big)_{ac}$ to write

$$\mathrm{E}\{\big(G(z_1)[X, U^\dagger BU]G(z_1)AG(z_2)\big)_{ac}\} - \mathrm{E}\{\big(G(z_1)AG(z_2)[X, U^\dagger BU]G(z_2)\big)_{ac}\} = 0.$$

Then, taking $X = E^{(a,b)}$ and applying to the result the operation $n^{-1}\sum_{a=1}^n$, we obtain the following matrix relation

$$\begin{aligned}(A.4)\quad &\mathrm{E}\{g(z_1)U^\dagger BUG(z_1)AG(z_2)\}\\ &- \mathrm{E}\{\delta_B(z_1)G(z_1)AG(z_2)\} + \mathrm{E}\{\big(n^{-1}\operatorname{Tr} G(z_1)AG(z_2)\big)U^\dagger BUG(z_2)\}\\ &\qquad - \mathrm{E}\{(n^{-1}\operatorname{Tr} G(z_1)AG(z_2)U^\dagger BU)G(z_2)\} = 0.\end{aligned}$$

Applying to the above formula the operation $n^{-1}\operatorname{Tr}$ on (A.4) and regrouping terms according to (2.18), we arrive at

$$\begin{aligned}&\mathrm{E}\Big\{\frac{1}{n}\operatorname{Tr} G(z_1)AG(z_2)U^\dagger BU\Big\}\\ &\qquad = \mathrm{E}\Big\{\frac{1}{n}\operatorname{Tr} G(z_1)AG(z_2)\Big\}\frac{\mathrm{E}\{\delta_B(z_1)\} - \mathrm{E}\{\delta_B(z_2)\}}{f(z_1) - f_n(z_2)} + R_4,\end{aligned}$$

where

$$\begin{aligned}R_4 = \frac{1}{f_n(z_1) - f_n(z_2)}\big(&\mathrm{E}\{\big(g^\circ(z_1) - g^\circ(z_2)\big)(n^{-1}\operatorname{Tr} G(z_1)AG(z_2)U^\dagger BU)\}\\ &- \mathrm{E}\{\big(\delta_B^\circ(z_1) - \delta_B^\circ(z_2)\big)\big(n^{-1}\operatorname{Tr} G(z_1)AG(z_2)\big)\}\big).\end{aligned}$$

Furthermore, regrouping again terms in (A.4) according to (2.18) and the resolvent identity and applying to the result the resolvent $G_A\big(h_B(z_1)\big)$ of (2.23), we obtain

$$\begin{aligned}\mathrm{E}\{G(z_1)AG(z_2)\} &= AG_A(h_B(z_1))\,\mathrm{E}\{G(z_2)\}\\ &+ \frac{1}{f_n(z_1)}G_A(h_B(z_1))\Big(\frac{\mathrm{E}\{\delta_B(z_2)\}}{f_n(z_2)} - \frac{\mathrm{E}\{\delta_B(z_1) - \delta_B(z_2)\}}{f_n(z_1) - f_n(z_2)}\Big)\\ &\qquad \times \mathrm{E}\{n^{-1}\operatorname{Tr} G(z_1)AG(z_2)\}\,\mathrm{E}\{G(z_2)\} + R_5,\end{aligned}$$

where

$$R_5 = \frac{1}{f_n(z_1)} G_A(h_B(z_1))\big(\mathrm{E}\{g^\circ(z_1)G(z_1)AG(z_2)U^\dagger BU\} \\ - \mathrm{E}\{\delta_B^\circ(z_1)(n^{-1}\operatorname{Tr} G(z_1)AG(z_2)U^\dagger BU)\} \\ - \mathrm{E}\{\big(n^{-1}\operatorname{Tr} G(z_1)AG(z_2)\big)^\circ U^\dagger BUG(z_2)\} \\ - \mathrm{E}\{(n^{-1}\operatorname{Tr} G(z_1)AG(z_2)U^\dagger BU)^\circ G(z_2)\}\big).$$

The resolvent identity implies

$$\mathrm{E}\Big\{\frac{1}{n}\operatorname{Tr} G(z_1)AG(z_2)\Big\} = \frac{\mathrm{E}\{\delta_A(z_1)\} - \mathrm{E}\{\delta_A(z_2)\}}{z_1 - z_2},$$

and (2.19) leads to the relation

$$\text{(A.5)} \quad \mathrm{E}\{G(z)\} = G_A\big(h_{B_n}(z)\big) \\ + \frac{1}{f_n(z)}\mathrm{E}\big\{g^\circ(z)G_A\big(h_{B_n}(z)\big)U^\dagger BUG(z) - \delta_B^\circ(z)G_A\big(h_{B_n}(z)\big)G(z)\big\}.$$

Thus, we have

$$\text{(A.6)} \quad \mathrm{E}\{G(z_1)AG(z_2)\} = AG_A(h_{B_n}(z_1))G_A(h_{B_n}(z_2)) \\ + \left(\frac{\mathrm{E}\{\delta_B(z_2)\}}{f_n(z_2)} - \frac{\mathrm{E}\{\delta_B(z_1)\}}{f_n(z_1)}\right)\frac{\mathrm{E}\{\delta_A(z_1) - \delta_A(z_2)\}}{(f_n(z_1) - f_n(z_2))(z_1 - z_2)} \\ \times G_A(h_{B_n}(z_1))G_A(h_{B_n}(z_2)) + R_6,$$

where

$$R_6 \\ = R_5 + \left(A + \left(\frac{\mathrm{E}\{\delta_B(z_2)\}}{f_n(z_2)} - \frac{\mathrm{E}\{\delta_B(z_1)\}}{f_n(z_1)}\right)\frac{\mathrm{E}\{\delta_A(z_1) - \delta_A(z_2)\}}{(f_n(z_1) - f_n(z_2))(z_1 - z_2)} I\right) G_A\big(h_{B_n}(z_1)\big) \\ \times \frac{1}{f_n(z_2)}\mathrm{E}\big\{g^\circ(z_2)G_A(h_{B_n}(z_2))U^\dagger BUG(z_2) - \delta_B^\circ(z_2)G_A\big(h_{B_n}(z_2)\big)G(z_2)\big\}$$

and

$$\text{(A.7)} \quad \mathrm{E}\{AG(z_1)G(z_2)\} = \left(1 - \left(\frac{\mathrm{E}\{\delta_B(z_2)\}}{\mathrm{E}\{g(z_2)\}} - \frac{\mathrm{E}\{\delta_B(z_1)\}}{\mathrm{E}\{g(z_1)\}}\right)\frac{1}{z_1 - z_2}\right) \\ \times AG_A(h_{B_n}(z_1))G_A(h_{B_n}(z_2)) + R_7,$$

where

$$R_7 = \frac{1}{z_1 - z_2} A \\ \times \left(\frac{1}{f_n(z_1)}\mathrm{E}\big\{g^\circ(z_1)G_A\big(h_{B_n}(z_1)\big)U^\dagger BUG(z_1) - \delta_B^\circ(z_1)G_A\big(h_{B_n}(z_1)\big)G(z_1)\big\} \right. \\ \left. - \frac{1}{f_n(z_2)}\mathrm{E}\big\{g^\circ(z_2)G_A\big(h_{B_n}(z_2)\big)U^\dagger BUG(z_2) - \delta_B^\circ(z_2)G_A\big(h_{B_n}(z_2)\big)G(z_2)\big\}\right).$$

Now (A.6), (A.7), an the resolvent identity yield

$$
\begin{aligned}
&\mathrm{E}\{G(z_1)AG(z_2) - AG(z_1)G(z_2)\} \\
&= \frac{1}{z_1 - z_2}\left(\frac{\mathrm{E}\{\delta_B(z_2)\}}{f_n(z_2)} - \frac{\mathrm{E}\{\delta_B(z_1)\}}{f_n(z_1)}\right)\left(\frac{\mathrm{E}\{\delta_A(z_1) - \delta_A(z_2)\}}{z_1 - z_2} I - A\right) \\
&\qquad \times G_A(h_{B_n}(z_1))G_A(h_{B_n}(z_2)) + R_6 - R_7 \\
&= \left(\frac{\mathrm{E}\{\delta_B(z_2)\}}{f_n(z_2)} - \frac{\mathrm{E}\{\delta_B(z_1)\}}{f_n(z_1)}\right)\frac{f_n(z_1)f_n(z_2)}{(f_n(z_1) - f_n(z_2))(z_1 - z_2)} \\
&\qquad \times \left(\frac{1}{f_n(z_1)}G_A(h_{B_n}(z_1)) - \frac{1}{f_n(z_2)}G_A(h_{B_n}(z_2))\right) + R_6 - R_7.
\end{aligned}
$$

This relation and (2.20) allows us to write (A.3)

$$
\begin{aligned}
\text{(A.8)}\quad &\gamma_{AB}(z_1, z_2) \\
&= \frac{\partial}{\partial z_1}\frac{1}{f_n(z_2)}\frac{h_{B_n}(z_1) - h_{B_n}(z_2) - (z_1 - z_2)}{(z_1 - z_2)}\left(\frac{1}{f_n(z_2)} - \frac{1}{f_n(z_1)}\right)^{-1} \\
&\times \left(\frac{1}{f_n(z_1)}\frac{1}{n}\operatorname{Tr} G_A\big(h_{B_n}(z_1)\big)G_A\big(h_{B_n}(z_2)\big) - \frac{1}{f_n(z_2)}\frac{1}{n}\operatorname{Tr} G_A^2\big(h_{B_n}(z_2)\big)\right) \\
&\qquad + \frac{\partial}{\partial z_1}R_8,
\end{aligned}
$$

where

$$
R_8 = \frac{1}{f_n(z_2)}\frac{1}{n}\operatorname{Tr} G_A\big(h_{B_n}(z_2)\big)(R_6 - R_7).
$$

Now (2.34), (2.29), and (2.30) yield for $z_{1,2} \in \mathbb{C}\setminus\mathbb{R}$:

$$
R_8 = O(n^{-1}), \quad n \to \infty,
$$

and the analyticity of R_8 in $z_{1,2}$ and the Cauchy theorem lead to an analogous bound for the derivative

$$
\frac{\partial}{\partial z_1}R_8 = O(n^{-1}), \quad n \to \infty.
$$

Now, to finish the proof of second assertion, we have to transform (A.8), using (3.3), (3.8) and the formula

$$
\frac{1}{n}\operatorname{Tr} G_A\big(h_{B_n}(z_1)\big)G_A\big(h_{B_n}(z_2)\big) = \frac{f_n(z_1) - f_n(z_2)}{h_{B_n}(z_1) - h_{B_n}(z_2)}. \qquad \square
$$

References

1. N. I. Akhiezer and I. M. Glazman, *Theory of linear operators in Hilbert space*, Dover, New York, 1993.
2. L. H. Y. Chen, *An inequality for the multivariate normal distribution*, J. Multivariate Anal. **12** (1982), no. 2, 306–315.
3. G. Chistyakov and F. Götze, *The arithmetic of distributions in free probability theory*, available at `arXiv:math.OA/0508245`.
4. A. Khorunzhy, B. Khoruzhenko, and L. Pastur, *On the* $1/n$ *corrections to the Green functions of random matrices with independent entries*, J. Phys. A **28** (1995), no. 1, L31–L35.
5. M. Ledoux, *The concentration of measure phenomenon*, Math. Surveys Monogr., vol. 89, Amer. Math. Soc., Providence, RI, 2001.
6. V. Marchenko and L. Pastur, *Distribution of eigenvalue in certain setss of random matrices*, Mat. Sb. (N.S.) **72(114)** (1967), 507–536 (Russian); English transl., Math. USSR Sbornik **1** (1967), 457–483.

7. M. L. Mehta, *Random matrices*, 2nd ed., Academic Press, Boston, 1991.
8. L. Pastur, *A simple approach to the global regime of the random matrix theory*, Mathematical Results in Statistical Mechanics (Marseille, 1998) (S. Miracle-Sole, J. Ruiz, and V. Zagrebnov, eds.), World Sci. Publ., Singapore, 1999, pp. 429–454.
9. L. Pastur and V. Vasilchuk, *On the law of addition of random matrices*, Comm. Math. Phys. **214** (2000), no. 2, 249–286.
10. ———, *Law of addition of random matrices: Asymptotics of linear eigenvalue statistics*, in preparation.
11. D. Voiculescu (ed.), *Free probability theory*, Fields Inst. Commun., vol. 12, Amer. Math. Soc., Providence, RI, 1997.
12. ———, *Lectures on free probability theory*, Lectures on Probability Theory and Statistics (Saint-Flour, 1998), Lecture Notes in Math., vol. 1738, Springer, Berlin, 2000, pp. 279–349.

B. Verkin Institute for Low Temperature Physics and Engineering, 47 Lenin Ave., Kharkiv, 61103 Ukraine and Department of Mathematics, University of Wales Swansea, Singleton Park, Swansea, SA2 8PP, Wales, UK

E-mail address: lpastur@ilt.kharkov.ua; L.Pastur@Swansea.ac.uk

B. Verkin Institute for Low Temperature Physics and Engineering, 47 Lenin Ave., Kharkiv, 61103 Ukraine

E-mail address: vasilchuk@ilt.kharkov.ua

Centre de Recherches Mathématiques
CRM Proceedings and Lecture Notes
Volume **42**, 2007

Green's Functions of Generalized Laplacians

Philippe Poulin

Abstract. The Green's function of a typical discrete Laplacian, Δ, in dimension d is given by

$$G(n-m,z)=\int_{\mathbb{T}^d}\frac{\mathrm{e}^{\mathrm{i}(n-m)\cdot x}}{\Phi(x)-z}\,\mathrm{d}x,$$

where $m,n\in\mathbb{Z}^d$, $z\in\mathbb{C}_+$, and $\Phi(x)$ is the symbol of Δ. Using the stationary phase method we study the decay of $G(n,e+\mathrm{i}0)$ when $|n|\to\infty$ for values of energy, e, inside the range of $\Phi(x)$, where $\Phi(x)$ is analytic. We focus on two specific examples: the standard Laplacian and the Molchanov – Vainberg Laplacian.

1. Introduction

The present text, which corresponds to the first part of the author's doctoral thesis and details his contribution to the *MolchanovFest*, is based on [7]; indeed, it presents more general results.

There has been recent interest in establishing decay properties of Green's functions of discrete Laplacians for values of energy inside their associated spectra [2, 6, 8]. Such a priori estimates are useful in several contexts, for instance in scattering theory of the Anderson model.[1] In concrete examples the problem consists of investigating the decay of $G(n,e+\mathrm{i}0)$ when $|n|\to\infty$, where

$$G(n,z)=\int_{\mathbb{T}^d}\frac{\mathrm{e}^{\mathrm{i}n\cdot x}}{\Phi(x)-z}\,\mathrm{d}x,$$

$n\in\mathbb{Z}^d$, $e\in\operatorname{ran}\Phi$, and Φ is analytic. Such a decay is calculated in [7] for the Molchanov – Vainberg Laplacian. The mentioned calculation takes its origin in [8] and works for strictly convex constant-energy surfaces,

$$\Gamma(e)=\{x\in\mathbb{T}^d;\Phi(x)=e\}.$$

We generalize this study to nonconvex surfaces. To this end, we take special care of a supplemental parameter when reviewing the stationary phase method. Our review is an adaptation of [9, 10] based on the observation in [5] that the method

2000 *Mathematics Subject Classification.* Primary 47B39; Secondary 34L40.
This research was supported in part by NSERC.
This is the final form of the paper.

[1]For a list of papers using such estimates, see the introduction in [7].

also gives a decay for Fourier transforms over nonconvex surfaces. In the sequel we prove in detail this last observation, which leads to the treatment of a large class of generalized Laplacians.

Results. Consider firstly the standard Laplacian. An elementary computation shows that except for $e \in \{-2d, -2d+4, \dots, 2d-4, 2d\} \cup \{0\}$, $\Gamma(e)$ is a regular surface without planar point. The stationary phase method then yields

$$G(n, e + \mathrm{i}0) = O(|n|^{-1/2})$$

in any dimension. Without pretending that this result is optimal, it improves a naive estimate based on Bessel functions, which is presented in the appendix.

Secondly, consider the Molchanov–Vainberg Laplacian, which is defined in Section 7. Then, except for $e \in \{-2^d, 0, 2^d\}$, $\Gamma(e)$ is a regular, strictly convex surface. The classical stationary phase method thus yields the optimal decay,

$$G(n, e + \mathrm{i}0) = O(|n|^{-(d-1)/2}),$$

valid in any dimension. This is why Molchanov and Vainberg introduced this Laplacian in the literature in their work about scattering theory of the Anderson model [6].

Acknowledgments. I would like to thank Vojkan Jakšić for the invitation to the *MolchanovFest*; Boris Vainberg for useful discussions when writing [7]; and Marco Merkli, who helped me to enhance the introduction and took interest in this text.

2. Terminology

An *oscillatory integral* is an integral of the form

$$\int_{\mathbb{R}^d} \mathrm{e}^{\mathrm{i}r\varphi(x)} f(x)\,\mathrm{d}x,$$

where $r \in \mathbb{R}$ and $\varphi(x)$ is real-valued. Its *amplitude* and *phase* are $f(x)$ and $\varphi(x)$, respectively. The phase is *stationary at* x_0 if $\nabla\varphi(x_0) = 0$. The stationary phase point, x_0, is *nondegenerate* if in addition $\det \mathrm{D}_x^2\varphi(x_0) \neq 0$.

Our plan is to establish (1) the rapid decay of an oscillatory integral whose phase is nowhere stationary, for a given compactly supported amplitude; (2) the polynomial decay of an oscillatory integral whose stationary phase points are not degenerate; (3) similar results for Cauchy principal values of oscillatory integrals. Given an analytic function, $\Phi(x)$, on $\mathbb{T}^d$, we then consider Fourier transforms of functions over level surfaces of the form

$$\Gamma(e) = \{x \in \mathbb{T}^d; \Phi(x) = e\}$$

and investigate their decay when the $\Gamma(e)$'s are regular, compact, and admit at every point at least $\kappa > 0$ nonvanishing principal curvatures. Then, we do a similar study for Cauchy principal values of such Fourier transforms. Finally, we deduce the decay of Green's functions of generalized Laplacians for energies inside their associated spectra.

We adopt the following conventions: most of our theorems establish the existence of a neighborhood on which a certain phenomenon occurs. For sake of simplicity (and without loss of generality), we consider only nonempty, bounded, open, cubic neighborhoods and we call them *cubes.*[2]

Given a real-valued phase, $\varphi(x,t)$, we define

$$I_f(r,t) = \int_{\mathbb{R}^d} \mathrm{e}^{\mathrm{i}r\varphi(x,t)} f(x,t)\,\mathrm{d}x, \tag{2.1}$$

where $r > 0$, $t \in \mathbb{R}^m$, and $f(x,t)$ is a complex-valued function – provided that this integral makes sense.

Smooth is used in this text for *infinitely differentiable.* The vector space of all (complex-valued) smooth functions on $\mathbb{R}^d$ is denoted by $C^\infty(\mathbb{R}^d)$. It contains two important subspaces: $C^\omega(\mathbb{R}^d)$, consisting of all analytic functions on $\mathbb{R}^d$, and $C_c^\infty(\mathbb{R}^d)$, consisting of all compactly supported smooth functions on $\mathbb{R}^d$.

The *transpose* of a linear transformation on $C_c^\infty(\mathbb{R}^d)$ is defined by duality with respect to the following bracket:

$$\big(f(x) \mid g(x)\big) = \int_{\mathbb{R}^d} f(x)g(x)\,\mathrm{d}x.$$

For instance, denoting by $\partial_{x^{(j)}}$ the differentiation with respect to $x^{(j)}$, integration by parts gives $\partial^{\mathrm{t}}_{x^{(j)}} = -\partial_{x^{(j)}}$. Moreover, denoting by $F(x)$ the multiplication by a smooth function of the same name, $F(x)^{\mathrm{t}} = F(x)$.

Our notations regarding asymptotic behavior are standard: for instance,

$$f(r) = O(r^{-\alpha}) \quad \text{when } r \to \infty$$

means the existence of a positive constant, C, such that $|f(r)| \leqslant Cr^{-\alpha}$ when r is sufficiently large. If such a constant (depending on α) exists for any $\alpha \geqslant 0$, we write

$$f(r) = O(r^{-\infty}) \quad \text{when } r \to \infty.$$

Finally,

$$f(r) \sim \sum_{j=0}^{\infty} a_j r^{-j} \quad \text{when } r \to \infty$$

means that for any $N \geqslant 0$

$$f(r) - \sum_{j=0}^{N} a_j r^{-j} = O(r^{-N-1}).$$

If the function f also depends on a parameter, t, we say that the previous estimates/asymptotics are *uniform* in t if the constants C may be chosen independently of t.

3. Oscillatory integral without stationary phase point

The following theorems are stated in the way they are used when studying Fourier transforms over surfaces. They establish the existence of neighborhoods on which a certain phenomenon occurs, given a fixed phase. The given phase, $\varphi(x,t)$, is supposed to be smooth in $(x,t) \in \mathbb{R}^d \times \mathbb{R}^m$.

---

[2]Recall that a *cubic neighborhood* in $\mathbb{R}^d$ is a subset of the form $I_1 \times \cdots \times I_d$, where each I_d is an interval.

Lemma. *Let $d = 1$. Suppose $\partial_x \varphi \neq 0$ at a given $(x_0, t_0) \in \mathbb{R} \times \mathbb{R}^m$. Then, there exists an arbitrarily small cube, $U \times B$, containing (x_0, t_0) such that the following holds: if the amplitude, $f(x,t)$, is smooth in the neighborhood of $\mathbb{R} \times \overline{B}$ and vanishes on $U^{\mathrm{c}} \times \overline{B}$, then*

$$|I_f(r,t)| = O(r^{-\infty})$$

uniformly in $t \in \overline{B}$.[3]

PROOF. By continuity, $\partial_x \varphi(x,t) \neq 0$ on a certain cube $U' \times B'$ containing (x_0, t_0). The operator $D = 1/(\partial_x \varphi(x,t)) \circ \partial_x$ is thus well defined on this cube, where it satisfies $D\mathrm{e}^{\mathrm{i}r\varphi(x,t)} = \mathrm{i}r\mathrm{e}^{\mathrm{i}r\varphi(x,t)}$ and $D^{\mathrm{t}} = -\partial_x \circ 1/(\partial_x \varphi(x,t))$. Let $U \times B \ni (x_0, t_0)$ be a cube whose closure is in $U' \times B'$. If $f(x,t)$ satisfies the asserted properties, then for any $N \geqslant 0$ and $t \in \overline{B}$

$$\begin{aligned} |I_f(r,t)| &= \frac{1}{r^N} \left| \int_U (D^N \mathrm{e}^{\mathrm{i}r\varphi(x,t)}) f(x,t) \, \mathrm{d}x \right| \\ &= \frac{1}{r^N} \left| \int_U \mathrm{e}^{\mathrm{i}r\varphi(x,t)} (D^{\mathrm{t}})^N f(x,t) \, \mathrm{d}x \right| \leqslant \frac{C_N}{r^N}, \end{aligned}$$

where the constant C_N does not depend on $t \in \overline{B}$. □

The multidimensional analogue follows:

Theorem 3.1. *Suppose $\nabla_x \varphi(x_0, t_0) \neq 0$ for a given $(x_0, t_0) \in \mathbb{R}^d \times \mathbb{R}^m$. Then, there exists an arbitrarily small cube, $U \times B$, containing (x_0, t_0) such that the following holds: if the amplitude, $f(x,t)$, is smooth in the neighborhood of $\mathbb{R}^d \times \overline{B}$ and vanishes on $U^{\mathrm{c}} \times \overline{B}$, then*

$$|I_f(r,t)| = O(r^{-\infty})$$

uniformly in $t \in \overline{B}$.

PROOF. By assumption, $\partial_{x^{(k)}} \varphi(x_0, t_0) \neq 0$ for a certain $1 \leqslant k \leqslant d$, say, for $k = 1$. Interpreting $(x^{(2)}, \ldots, x^{(d)}, t)$ as a parameter, there exists an arbitrarily small cube, $U_1 \times (U_2 \times \cdots \times U_d \times B)$, containing $(x_0^{(1)}; x_0^{(2)}, \ldots, x_0^{(d)}, t_0)$ on which the previous lemma applies. Hence, if $f(x,t)$ satisfies the asserted properties, then

$$\left| \int_{U_1} \mathrm{e}^{\mathrm{i}r\varphi(x,t)} f(x,t) \, \mathrm{d}x^{(1)} \right| \leqslant \frac{C_N}{r^N}$$

uniformly in $(x^{(2)}, \ldots, x^{(d)}, t) \in \overline{U_2} \times \cdots \times \overline{U_d} \times \overline{B}$. In particular,

$$\begin{aligned} |I_f(r,t)| &= \left| \int_{U_2 \times \cdots \times U_d} \int_{U_1} \mathrm{e}^{\mathrm{i}r\varphi(x,t)} f(x,t) \, \mathrm{d}x^{(1)} \, \mathrm{d}x^{(2)} \ldots \, \mathrm{d}x^{(d)} \right| \\ &\leqslant |U_2 \times \cdots \times U_d| \frac{C_N}{r^N} \end{aligned}$$

uniformly in $t \in \overline{B}$. □

Scholium. A similar result holds when considering oscillatory integrals over the torus,

$$\int_{\mathbb{T}^d} \mathrm{e}^{\mathrm{i}r\varphi(x,t)} f(x,t) \, \mathrm{d}x,$$

[3] U^{c} denotes the complementary of U with respect to a set determined by the context; here, $U^{\mathrm{c}} = \mathbb{R} \setminus U$. For the definition of $I_f(r,t)$, see (2.1).

without stationary phase point. One then assumes $f(\cdot,t)$ is periodic for every t (instead of being compactly supported). The same proof works, due to the absence of boundary term when integrating by parts a smooth periodic function.

Cauchy principal value. We now turn our attention to Cauchy principal values of oscillatory integrals without stationary phase point. To this end, we now consider a parameter $(e,t)\in\mathbb{R}\times\mathbb{R}^m$ and a phase, $\varphi(x;e,t)$, smooth in $(x;e,t)\in\mathbb{R}^d\times(\mathbb{R}\times\mathbb{R}^m)$. We thus let

$$I_f(r;e,t)=\int_{\mathbb{R}^d}\mathrm{e}^{\mathrm{i}r\varphi(x;e,t)}f(x;e,t)\,\mathrm{d}x$$

and study

$$\mathrm{p.\,v.}\int_{|\eta-e|<\delta}\frac{1}{\eta-e}I_f(r;\eta,t)\,\mathrm{d}\eta=\lim_{\varepsilon\downarrow 0}\int_{\varepsilon<|\eta-e|<\delta}\frac{1}{\eta-e}I_f(r;\eta,t)\,\mathrm{d}\eta$$

for a given $\delta>0$. Our results are based on the following elementary estimate:

Lemma. *Suppose $f(h)$ is a continuously differentiable function in the neighborhood of $[-\delta,\delta]$, where $\delta>0$ is given. Then, for any $\varepsilon\in\,]0,\delta[$*

$$\left|\int_{\varepsilon\leqslant|h|\leqslant\delta}\frac{f(h)}{h}\,\mathrm{d}h\right|\leqslant 2\delta\max_{|h|\leqslant\delta}|f'(h)|.$$

In particular,

$$\left|\mathrm{p.\,v.}\int_{-\delta}^{\delta}\frac{f(h)}{h}\,\mathrm{d}h\right|\leqslant 2\delta\max_{|h|\leqslant\delta}|f'(h)|.$$

The proof, based on the Mean Value Theorem, is left to the reader.

Theorem 3.2. *Suppose $\nabla_x\varphi(x_0,0,t_0)\neq 0$ for a given $(x_0,t_0)\in\mathbb{R}^d\times\mathbb{R}^m$. Then, there exists an arbitrarily small cube, $U\times B$, containing (x_0,t_0) and an arbitrarily small $\delta>0$ such that the following holds: if the amplitude, $f(x,h,t)$, is smooth in the neighborhood of $\mathbb{R}^d\times[-\delta,\delta]\times\overline{B}$ and vanishes on $U^{\mathrm{c}}\times[-\delta,\delta]\times\overline{B}$, then*

$$\left|\mathrm{p.\,v.}\int_{-\delta}^{\delta}\frac{I_f(r,h,t)}{h}\,\mathrm{d}h\right|=O(r^{-\infty})$$

uniformly in $t\in\overline{B}$.

PROOF. By the lemma it suffices to estimate $\partial_h I_f(r,h,t)$. In fact, the dominated convergence theorem implies

$$\partial_h I_f(r,h,t)=\int_{\mathbb{R}^d}\mathrm{e}^{\mathrm{i}r\varphi(x,h,t)}\big(\partial_h f(x,h,t)+\mathrm{i}rf(x,h,t)\partial_h\varphi(x,h,t)\big)\,\mathrm{d}x.$$

Hence, by Theorem 3.1 there exist an arbitrarily small cube, $U\times B$, containing (x_0,t_0) and an arbitrarily small $\delta>0$ such that for $f(x,h,t)$ as stipulated

$$|\partial_h I_f(r,h,t)|\leqslant\frac{C_N}{r^N}$$

uniformly in $(h,t)\in[-\delta,\delta]\times\overline{B}$. The result follows. □

Applying the above to the phase $\widetilde{\varphi}(x,h;e,t)=\varphi(x,h+e,t)$ and the amplitude $\tilde{f}(x,h;e,t)=f(x,h+e,t)$, one obtains:

Corollary. *Suppose* $\nabla_x\varphi(x_0, e_0, t_0) \neq 0$ *for a given* $(x_0, e_0, t_0) \in \mathbb{R}^d \times \mathbb{R} \times \mathbb{R}^m$. *Then, there exist an arbitrarily small cube,* $U \times B$, *containing* (x_0, t_0) *and an arbitrarily small* $\delta > 0$ *such that the following holds: if the amplitude,* $f(x, e, t)$, *is smooth in the neighborhood of* $\mathbb{R}^d \times [e_0 - \delta, e_0 + \delta] \times \overline{B}$ *and vanishes on* $U^c \times [e_0 - \delta, e_0 + \delta] \times \overline{B}$, *then*

$$\left| \text{p.v.} \int_{|\eta - e| < \delta} \frac{I_f(r, \eta, t)}{\eta - e} \, d\eta \right| = O(r^{-\infty})$$

uniformly in $(e, t) \in [e_0 - \delta, e_0 + \delta] \times \overline{B}$.

4. Oscillatory integral with nondegenerate stationary phase points

Quadratic phase. We now investigate the decay of oscillatory integrals admitting nondegenerate stationary phase points. Via the *Morse lemma*, which is proved below, our study reduces to oscillatory integrals whose phases are canonical, nondegenerate quadratic forms:

$$Q(x) = \sum_{j=1}^{s} (x^{(j)})^2 - \sum_{k=s+1}^{d} (x^{(k)})^2,$$

where $0 \leqslant s \leqslant d$.

Exceptionally, in the next three lemmas the variable $r \in \mathbb{R}$ is allowed to be negative. Moreover $d = 1$ (so x varies in $\mathbb{R}$).

Lemma. *Suppose* $|r| > 1$. *For all* $l \in \mathbb{N}$ *there exist constants* $c_j^{(l)} \in \mathbb{C}$ *independent of* $|r|$ *such that*

$$\int_{-\infty}^{\infty} e^{-(1-ir)x^2} x^l \, dx = |r|^{-(l+1)/2} \sum_{j=0}^{\infty} c_j^{(l)} |r|^{-j}.$$

PROOF. Let us denote by $(\cdot)^{1/2}$ the branch of the square root whose singular cut is the positive imaginary axis. Let $z = (1 - ir)^{1/2} x$. Then,

$$\int_{\mathbb{R}} e^{-(1-ir)x^2} x^l \, dx = \left(\int_{\gamma} e^{-z^2} z^l \, dz \right) [(1 - ir)^{1/2}]^{-l-1}, \tag{4.1}$$

where γ is the oriented path $(1 - ir)^{1/2}\mathbb{R}$. Observe that

$$(1 - ir)^{1/2} = \sigma_r |r|^{1/2} \left(\frac{1}{r} - i \right)^{1/2}, \tag{4.2}$$

where

$$\sigma_r = \begin{cases} 1 & \text{if } r > 1 \\ i & \text{if } r < -1 \end{cases}$$

(so σ_r does not depend on $|r|$). Moreover the Taylor expansion

$$[(w - i)^{1/2}]^{-l-1} = \sum_{j=0}^{\infty} a_j^{(l)} w^j \tag{4.3}$$

is valid for all $|w| < 1$, by choice of the branch of $z^{1/2}$. Thus, substituting $w = \operatorname{sgn}(r)/|r|$ in the above, the result follows from (4.1), (4.2), and (4.3). □

Lemma. *Suppose $f(x,t)$ is smooth in the neighborhood of $\mathbb{R}\times\overline{B}$ and vanishes on $U^{c}\times\overline{B}$, where $U\subset\mathbb{R}$ and $B\subset\mathbb{R}^{m}$ are cubes. For any $l\in\mathbb{N}$ there exists a $C_l>0$ independent of $|r|$ satisfying*

$$\left|\int_{-\infty}^{\infty}\mathrm{e}^{\mathrm{i}rx^2}x^l f(x,t)\,\mathrm{d}x\right|\leqslant C_l|r|^{-(l+1)/2}$$

uniformly in $t\in\overline{B}$.

PROOF. Let $\chi(x)$ be a compactly supported smooth function on $\mathbb{R}$ such that $0\leqslant\chi(x)\leqslant 1$ everywhere and

$$\chi(x)=\begin{cases}0 & \text{if } |x|\geqslant 2\\ 1 & \text{if } |x|\leqslant 1.\end{cases}$$

For an arbitrarily fixed $\varepsilon>0$, $\int_{\mathbb{R}}\mathrm{e}^{\mathrm{i}rx^2}x^l f(x,t)\,\mathrm{d}x=\mathrm{I}+\mathrm{II}$, where

$$\mathrm{I}=\int_{|x|<2\varepsilon}\mathrm{e}^{\mathrm{i}rx^2}x^l f(x,t)\chi\Big(\frac{x}{\varepsilon}\Big)\,\mathrm{d}x,$$

$$\mathrm{II}=\int_{|x|>\varepsilon}\mathrm{e}^{\mathrm{i}rx^2}x^l f(x,t)\Big(1-\chi\Big(\frac{x}{\varepsilon}\Big)\Big)\,\mathrm{d}x.$$

Concerning I, there exist constants (generically denoted by Const) independent of $t\in\overline{B}$, but depending on l, satisfying

$$|\mathrm{I}|\leqslant \mathrm{Const}\int_{-2\varepsilon}^{2\varepsilon}|x|^l\,\mathrm{d}x=\mathrm{Const}\,\varepsilon^{l+1}.$$

Concerning II, let $D=(1/x)\circ\partial_x$, which is well defined on the support of the integrand. Since $1/(2\mathrm{i}r)D$ fixes $\mathrm{e}^{\mathrm{i}rx^2}$,

$$\mathrm{II}=\Big(\frac{1}{2\mathrm{i}r}\Big)^N\int_{|x|>\varepsilon}\mathrm{e}^{\mathrm{i}rx^2}(D^{\mathrm{t}})^N x^l f(x,t)\Big(1-\chi\Big(\frac{x}{\varepsilon}\Big)\Big)\,\mathrm{d}x.$$

For $F(x,y,t)=f(x,t)\big(1-\chi(y)\big)$, notice that $(D^{\mathrm{t}})^N x^l F(x,x/\varepsilon,t)$ is of the form

$$x^{l-2N}F_0(x,x/\varepsilon,t)+x^{l-2N+1}\varepsilon^{-1}F_1(x,x/\varepsilon,t)+\cdots+x^{l-N}\varepsilon^{-N}F_N(x,x/\varepsilon,t),$$

where $F_0,\ldots,F_N$ are bounded on $\mathbb{R}\times\mathbb{R}\times\overline{B}$. Hence, for a fixed $N>l+1$ and for $|x|>\varepsilon$,

$$|(D^{\mathrm{t}})^N x^l F(x,x/\varepsilon,t)|\leq \mathrm{Const}|x|^{l-N}\varepsilon^{-N}.$$

Therefore,

$$|\mathrm{II}|\leq \mathrm{Const}|r|^{-N}\int_{|x|>\varepsilon}|x|^{l-N}\varepsilon^{-N}\,\mathrm{d}x=\mathrm{Const}|r|^{-N}\varepsilon^{l-2N+1}.$$

In total,

$$|\mathrm{I}+\mathrm{II}|\leqslant \mathrm{Const}\,(\varepsilon^{l+1}+|r|^{-N}\varepsilon^{l-2N+1})$$

for such an N. Choosing $\varepsilon=|r|^{-1/2}$ then completes the proof. □

Lemma. *Consider a function, $f(x,t)$, smooth in the neighborhood of $\mathbb{R}\times\overline{B}$ (where $B\subset\mathbb{R}^m$ is a cube) and vanishing when $|x|<\varepsilon$ (where $\varepsilon>0$). Suppose that for all $N\geqslant 0$ there exist a $D_N>0$ and an $\alpha_N>0$ such that*

$$|\partial_x^N f(x,t)|\leqslant D_N(1+|x|^{\alpha_N})$$

uniformly in $t \in \overline{B}$. *Then,*

$$\int_{-\infty}^{\infty} \mathrm{e}^{(\mathrm{i}r-1)x^2} f(x,t)\,\mathrm{d}x = O(r^{-\infty})$$

uniformly in $t \in \overline{B}$.

PROOF. The derivation $D = (1/x) \circ \partial_x$ is well defined on the support of the integrand. Notice that $1/(2\mathrm{i}r)D$ fixes $\mathrm{e}^{\mathrm{i}rx^2}$ and $D^{\mathrm{t}} = -\partial_x \circ (1/x)$. Then,

$$\left|\int_{\mathbb{R}} \mathrm{e}^{(\mathrm{i}r-1)x^2} f(x,t)\,\mathrm{d}x\right| = \left(\frac{1}{2|r|}\right)^N \left|\int_{|x|\geqslant\varepsilon} \mathrm{e}^{\mathrm{i}rx^2} (D^{\mathrm{t}})^N (\mathrm{e}^{-x^2} f(x,t))\,\mathrm{d}x\right|.$$

Our assumption on the derivatives of $f(x,t)$ makes the integral on the right side of the previous equation uniformly bounded in $t \in \overline{B}$, where N is arbitrarily fixed. The result follows. □

In dimension one we are interested in the phase x^2, so let

$$I_f(r,t) = \int_{-\infty}^{\infty} \mathrm{e}^{\mathrm{i}rx^2} f(x,t)\,\mathrm{d}x.$$

According to our convention $r > 0$. However, since we are also interested in the phase $-x^2$, we allow r to be negative. We are ready to compute the asymptotic expansion of $I_f(r,t)$ (resp. $I_f(-r,t)$) when $|r| \to \infty$:

Theorem 4.1. *Consider an amplitude,* $f(x,t)$, *smooth in the neighborhood of* $\mathbb{R} \times \overline{B}$ *and vanishing on* $U^{\mathrm{c}} \times \overline{B}$, *where* $U \subset \mathbb{R}$ *and* $B \subset \mathbb{R}^m$ *are cubes. There exist constants,* $a_j(t)$, *depending smoothly on* $t \in \overline{B}$ *such that*

$$\int_{-\infty}^{\infty} \mathrm{e}^{\mathrm{i}rx^2} f(x,t)\,\mathrm{d}x \sim |r|^{-1/2} \sum_{j=0}^{\infty} a_j(t)|r|^{-j/2}$$

uniformly in $t \in \overline{B}$ *when* $r \to \infty$. *The same result holds when* $r \to -\infty$, *with different constants* $a_j(t)$.

PROOF. Let $\chi(x)$ be a smooth, bounded, compactly supported function such that $\chi(x) = 1$ on an interval containing $\overline{U} \cup \{0\}$. Then, for $f(x,t)$ as stated

$$I_f(r,t) = \int_{\mathbb{R}} \mathrm{e}^{(\mathrm{i}r-1)x^2} (\mathrm{e}^{x^2} f(x,t))\chi(x)\,\mathrm{d}x.$$

By Taylor's theorem (in dimension 1), for any fixed N there exists a polynomial,

$$P_t(x) = \sum_{l=0}^{N} b_l(t)x^l,$$

and a smooth remainder, $R_t(x)$, both depending smoothly on $(x,t) \in \mathbb{R} \times \overline{B}$, such that

$$\mathrm{e}^{x^2} f(x,t) = P_t(x) + x^{N+1} R_t(x).$$

Thus, $I_f(r,t)$ decomposes into $\mathrm{I}+\mathrm{II}+\mathrm{III}$, where

$$\mathrm{I}=\sum_{l=0}^{N} b_l(t)\int_{\mathbb{R}} \mathrm{e}^{(\mathrm{i}r-1)x^2}x^l\,\mathrm{d}x,$$
$$\mathrm{II}=\int_{\mathbb{R}} \mathrm{e}^{(\mathrm{i}r-1)x^2}x^{N+1}R_t(x)\chi(x)\,\mathrm{d}x,$$
$$\mathrm{III}=\int_{\mathbb{R}} \mathrm{e}^{(\mathrm{i}r-1)x^2}P_t(x)(\chi(x)-1)\,\mathrm{d}x.$$

Concerning I the first lemma gives the existence of $c_j^{(l)}\in\mathbb{C}$ such that for any $|r|>1$

$$\begin{aligned}\mathrm{I}&=\sum_{l=0}^{N} b_l(t)|r|^{-(l+1)/2}\sum_{j=0}^{\infty} c_j^{(l)}|r|^{-j}\\&=|r|^{-1/2}\sum_{l=0}^{N} b_l(t)|r|^{-l/2}\sum_{j\leqslant(N-l)/2} c_j^{(l)}|r|^{-j}+\sum_{l=0}^{N} b_l(t)|r|^{-(l+1)/2}\sum_{j>(N-l)/2} c_j^{(l)}|r|^{-j}.\end{aligned}$$

Notice that

$$\sum_{l=0}^{N} b_l(t)|r|^{-(l+1)/2}\sum_{j>(N-l)/2} c_j^{(l)}|r|^{-j}=O(|r|^{-N/2-1})$$

uniformly in $t\in\overline{B}$ when $r\to\infty$ (resp. $r\to-\infty$). Consequently, we have found coefficients, $a_k(t)$, smooth in $t\in\overline{B}$, satisfying

$$\mathrm{I}=|r|^{-1/2}\sum_{k=0}^{N} a_k(t)|r|^{-k/2}+O(|r|^{-N/2-1})$$

uniformly in $t\in\overline{B}$.

Moreover, the last two lemmas give respectively

$$\mathrm{II}=O(|r|^{-N/2-1})\quad\text{and}\quad\mathrm{III}=O(|r|^{-\infty})$$

uniformly in $t\in\overline{B}$. In total,

$$I_f(r,t)=|r|^{-1/2}\sum_{k=0}^{N} a_k(t)|r|^{-k/2}+O(|r|^{-N/2-1})$$

uniformly in $t\in\overline{B}$ when $r\to\infty$ (resp. $r\to-\infty$), as desired. □

Our treatment of a parameter $t\in\mathbb{R}^m$ permits to deduce from the previous theorem its multidimensional analogue by induction. Indeed, let us consider the phase

$$Q(x)=\sum_{j=1}^{s}\left(x^{(j)}\right)^2-\sum_{k=s+1}^{d}\left(x^{(k)}\right)^2,$$

where $x\in\mathbb{R}^d$ and $0\leqslant s\leqslant d$. Returning to the convention that $r>0$, we let

$$I_f(r,t)=\int_{\mathbb{R}^d}\mathrm{e}^{\mathrm{i}rQ(x)}f(x,t)\,\mathrm{d}x.$$

Then,

Corollary. *Consider an amplitude, $f(x,t)$, smooth in the neighborhood of $\mathbb{R}^d \times \overline{B}$ and vanishing on $U^c \times \overline{B}$, where $U \subset \mathbb{R}^d$ and $B \subset \mathbb{R}^m$ are cubes. There exist constants $a_j(t)$ depending smoothly on $t \in \overline{B}$ such that*

$$\int_{\mathbb{R}^d} \mathrm{e}^{\mathrm{i}rQ(x)} f(x,t)\,\mathrm{d}x \sim r^{-d/2} \sum_{j=0}^{\infty} a_j(t) r^{-j/2}$$

uniformly in $t \in \overline{B}$ when $r \to \infty$.

PROOF. Suppose the result holds for a certain $d-1$. Let $x_1 = \big(x^{(1)}, \dots, x^{(d-1)}\big)$ and $Q_1(x_1)$ be defined by

$$Q(x) = Q_1(x_1) \pm (x^{(d)})^2.$$

Recall that U may be written $U_1 \times \cdots \times U_d$. Then, by the inductive hypothesis, for an arbitrary $N \geqslant 0$

$$\begin{aligned} I_f(r,t) &= \int_{U_d} \mathrm{e}^{\pm \mathrm{i}r(x^{(d)})^2} \int_{\mathbb{R}^{d-1}} \mathrm{e}^{\mathrm{i}rQ_1(x_1)} f(x_1; x^{(d)}, t)\,\mathrm{d}x_1\,\mathrm{d}x^{(d)} \\ &= \int_{U_d} \mathrm{e}^{\pm \mathrm{i}r(x^{(d)})^2} \Big(r^{-(d-1)/2} \sum_{k=0}^{N+1} b_k(x^{(d)}, t) r^{-k/2} + O\big(r^{-(d+N+1)/2}\big)\Big)\,\mathrm{d}x^{(d)} \end{aligned}$$

uniformly in $(x^{(d)}, t) \in \overline{U_d} \times \overline{B}$. The following estimates then hold when $r \to \infty$, uniformly in $t \in \overline{B}$:

$$\begin{aligned} I_f(r,t) &= \sum_{k=0}^{N+1} r^{-(d+k-1)/2} \int_{U_d} \mathrm{e}^{\pm \mathrm{i}r(x^{(d)})^2} b_k(x^{(d)}, t)\,\mathrm{d}x^{(d)} + O\big(r^{-(d+N+1)/2}\big) \\ &= \sum_{k=0}^{N+1} r^{-(d+k-1)/2} \Big(r^{-1/2} \sum_{l=0}^{N} c_{k,l}(t) r^{-l/2}\Big) + O\big(r^{-(d+N+1)/2}\big) \\ &= r^{-d/2} \sum_{j=0}^{N} a_j(t) r^{-j/2} + O\big(r^{-(d+N+1)/2}\big). \end{aligned}$$

Since N is arbitrary, this completes the proof. □

Morse lemma. The investigation of oscillatory integrals with nondegenerate stationary phase points reduces to the above case by means of *Morse's lemma*, which we now prove.

Let the anticipated phase, $\phi(h,t)$, be a smooth function in $(h,t) \in \mathbb{R}^d \times \mathbb{R}^m$ satisfying

$$\phi(0,t) = 0, \quad \nabla_h \phi(0,t) = 0, \quad \text{and} \quad \det \mathrm{D}_h^2 \phi(0,t) \neq 0$$

for *all* $t \in B$, where $B \subset \mathbb{R}^m$ is a given cube containing a given $t_0 \in \mathbb{R}^m$. Such a function can be expressed *like* a quadratic form, but with coefficients varying smoothly in (h,t):

Lemma. *Under the above circumstances there exist functions, $\phi_{jk}(h,t)$, smooth in $(h,t) \in \mathbb{R}^d \times \mathbb{R}^m$ and satisfying*

$$\phi(h,t) = \sum_{j=1}^{d} \sum_{k=1}^{d} \phi_{jk}(h,t) h^{(j)} h^{(k)},$$

where $\phi_{jk}(h,t) = \phi_{kj}(h,t)$.

PROOF. Using our hypotheses on $\phi(h,t)$, the fundamental theorem of calculus and integration by parts give for any $t \in B$

$$\phi(h,t) = \int_0^1 \partial_s\big(\phi(sh,t)\big)\,\mathrm{d}s = \int_0^1 (1-s)\partial_s^2\big(\phi(sh,t)\big)\,\mathrm{d}s.$$

Expanding $\partial_s^2\big(\phi(sh,t)\big)$ in the above gives the result. □

The next step consists of applying *Lagrange's algorithm*, which is better understood using matrices. Using the standard basis on $\mathbb{R}^d$, h is represented by a column (also denoted by h), while the "quadratic form" given by the previous lemma is represented by a $d \times d$ matrix, denoted by $\Phi(h,t)$. The (j,k)th element of $\Phi(h,t)$ is then given by the function $\phi_{jk}(h,t)$, so the previous lemma gives

$$\phi(h,t) = h^{\mathrm{t}}\Phi(h,t)h.$$

For any given $t \in B$ the rank of $\Phi(0,t)$ is read through the Hessian of $\phi(h,t)$ using the following, straightforward relation:

Lemma. *Under the above circumstances, for* any $t \in B$, $\mathrm{D}_h^2\phi(0,t) = 2\Phi(0,t)$.

In order to perform Lagrange's algorithm one uses the following elementary line/column operations:

- Given a scalar $c \neq 0$, to multiply the jth row and then the jth column by c, which is denoted by $\mathrm{CL}_j(c)$;
- Given a scalar $c \in \mathbb{C}$, to add c times the kth row to the jth row and then c times the kth column to the jth column, which is denoted by $\mathrm{CL}_{jk}(c)$;
- To interchange the jth row with the kth row and then the jth column with the kth column, which is denoted by CL_{jk}.

By the previous lemma, since $\det \mathrm{D}_h^2\phi(0,t) \neq 0$ and since $\Phi(0,t)$ consists of smooth elements, there exists a cube $V_0 \times B_0 \subset \mathbb{R}^d \times B$ containing $(0,t_0)$ such that $\Phi(h,t)$ is invertible for all $(h,t) \in V_0 \times B_0$.

Without loss of generality $\phi_{11}(h,t) \neq 0$ on a certain cube $V_1 \times B_1 \subseteq V_0 \times B_0$ containing $(0,t_0)$. Otherwise, $\phi_{11}(h,t)$ vanishes at $(0,t_0)$. However, considering the Laplace expansion of the above determinant along the first row, there exists an index $1 \leq k \leq d$ such that $\phi_{1k}(0,t_0) \neq 0$. If $\phi_{kk}(0,t_0) = 0$, one replaces $\Phi(h,t)$ with $\mathrm{CL}_{1k}(1)\Phi(h,t)$; if $\phi_{kk}(0,t_0) \neq 0$, one replaces $\Phi(h,t)$ with $\mathrm{CL}_{1k}\Phi(h,t)$. In both cases the resulting upper-left element does not vanish at $(0,t_0)$, and hence on a certain cube $V_1' \times B_1' \subseteq V_0 \times B_0$ containing $(0,t_0)$,—which then takes the place of $V_1 \times B_1$.

Then, one may reduce the upper left element to ± 1 (depending on the sign of $\phi_{11}(h,t)$, which does not change on $V_1 \times B_1$) by applying $\mathrm{CL}_1(|\phi_{11}(h,t)|^{-1/2})$ to (the possibly refreshed) $\Phi(h,t)$. Finally, this resulting constant on the upper left corner permits to cancel the rest of the first line and column, by applying $CL_{k1}(f(h,t))$ for $k = 2,\ldots,d$ successively—where $f(h,t)$ is equal to the element to cancel (up to the sign). All these operations are represented by matrices having smooth elements in $(h,t) \in V_1 \times B_1$. They transform $\Phi(h,t)$ in a block diagonal matrix, having ± 1 as its first block and a square $(d-1) \times (d-1)$ matrix as its second block.

Repeating this procedure for the second block, $\Phi(h,t)$ is transformed in a block diagonal matrix having ± 1 as its first two blocks and a square $(d-2)\times(d-2)$ matrix as its third block. All the operations used for this second step are represented by matrices having smooth elements in $(h,t) \in V_2 \times B_2$, where

$$(0,t_0) \in V_2 \times B_2 \subseteq V_1 \times B_1 \subseteq V_0 \times B_0.$$

So on and so forth one transforms $\Phi(h,t)$ into a diagonal matrix having elements ± 1 only. All the required operations are smooth (in the previous sense) for (h,t) varying in a cube $V_d \times B_d$ containing $(0,t_0)$. One then applies permutations CL_{jk}, so the resulting matrix becomes $\mathrm{diag}(1,\dots,1,-1,\dots,-1)$, where the element 1 is repeated, say, s times. Since s does not depend on $(h,t) \in V_d \times B_d$, the previous lemma and the Sylvester Inertia Theorem permit to recover s from the signature of $\mathrm{D}_h^2\phi(0,t_0)$, which is then $(s, d-s, 0)$.

In summary, we have proved:

Lemma. *Under the above circumstances there exist a cube, $V_d \times B_d$, containing $(0,t_0)$ and a nonsingular linear map, $Q(h,t)$, whose matrix elements are smooth in $(h,t) \in V_d \times B_d$ such that*

$$Q(h,t)^{\mathrm{t}}\Phi(h,t)Q(h,t) = \mathrm{diag}(1,\dots,1,-1,\dots,-1).$$

In the above the element 1 *is repeated s times, where $(s, d-s, 0)$ is the signature of $\mathrm{D}_h^2\phi(0,t_0)$.*

It is thus tempting to consider the *nonlinear* mapping $h \mapsto Q(h,t)^{-1}h$ defined on V_d as a potential change of variables given a fixed $t \in B_d$. Considering $t \in B_d$ as a parameter, $\{Q(h,t)^{-1}h\}_{t\in B_d}$ is indeed a family of smooth mappings depending smoothly on t in the strong sense that $(h,t) \mapsto Q(h,t)^{-1}h$ is jointly smooth on $V_d \times B_d$. Moreover, all these mappings map 0 to 0. They really consist of invertible changes of variables when restricting suitably the ranges of h and t, as shown below:

Lemma. *Under the above circumstances there exists a cube, $V' \times B'$, such that*

$$(0,t_0) \in V' \times B' \subset V_d \times B_d,$$

on which

$$(h,t) \mapsto (Q(h,t)^{-1}h, t)$$

is a smooth diffeomorphism. In addition, this diffeomorphism maps $V' \times B'$ onto an open bounded set, hence contained in a certain cube $\widetilde{V'} \times B'$.

Proof. Let $F(h,t) = (Q(h,t)^{-1}h, t)$ for $(h,t) \in V_d \times B_d$. A direct computation shows that

$$\det \mathrm{D}_{(h,t)}F(0,t_0) = \det Q(0,t_0)^{-1},$$

which is not zero. Thus, $F(h,t)$ is a local diffeomorphism in a neighborhood of $(0,t_0)$. Choosing a cube, $V' \times B' \ni (0,t_0)$, whose closure is included in this last neighborhood (and in $V_d \times B_d$) then yields the result. □

The previous diffeomorphism maps $V' \times B'$ onto a bounded open set $\widetilde{\mathcal{D}} \subseteq \widetilde{V'} \times B'$. We want to fix $t \in B'$, so let $\widetilde{\mathcal{D}_t} = \{h \in \mathbb{R}^d; (h,t) \in \widetilde{\mathcal{D}}\}$. Then, $\widetilde{\mathcal{D}_t} \subseteq \widetilde{V'}$ is an open set in $\mathbb{R}^d$. Let us consider the restriction

$$V' \to \widetilde{\mathcal{D}_t}, \quad h \mapsto Q(h,t)^{-1}h,$$

which is also a smooth diffeomorphism. It may be used as a smooth invertible change of variables by setting

$$\tilde{h} = Q(h,t)^{-1}h$$

for $h \in V'$. Notice that 0 is then mapped to 0. Let P_t be the inverse change of variables, so

$$h = P_t(\tilde{h})$$

for $\tilde{h} \in \widetilde{\mathcal{D}_t}$. Then, $\tilde{h} = P_t^{-1}(h) = Q(h,t)^{-1}h$, which implies $h = Q(h,t)\tilde{h}$ and hence

$$\begin{aligned}\phi(P_t(\tilde{h}),t) &= \phi(h,t) = \phi(Q(h,t)\tilde{h},t) \\ &= \tilde{h}^{\mathrm{t}}Q(h,t)^{\mathrm{t}}\Phi(h,t)Q(h,t)\tilde{h} = \sum_{j=1}^{s}(\tilde{h}^{(j)})^2 - \sum_{k=s+1}^{d}(\tilde{h}^{(k)})^2.\end{aligned}$$

We have thus proved the *Morse lemma* with special care of the parameter t:

Theorem 4.2. *Given a cube $B \subset \mathbb{R}^m$ containing a fixed t_0, suppose $\phi(h,t)$ is smooth in $(h,t) \in \mathbb{R}^d \times B$ and satisfies*

$$\phi(0,t) = 0, \quad \nabla_h\phi(0,t) = 0, \quad \textit{and} \quad \det \mathrm{D}_h^2\phi(0,t) \neq 0$$

for all $t \in B$. Then, there exists a cube $V' \times B' \subset \mathbb{R}^d \times B$ containing $(0,t_0)$ such that the following holds: for all $t \in B'$ there exists an invertible change of variables

$$\tilde{h} = P_t^{-1}(h)$$

on V', smooth and with smooth inverse, mapping 0 to 0, which satisfies

$$\phi(P_t(\tilde{h}),t) = \sum_{j=1}^{s}(\tilde{h}^{(j)})^2 - \sum_{k=s+1}^{d}(\tilde{h}^{(k)})^2.$$

The resulting family of changes of variables, $\{P_t(\tilde{h})\}_{t\in B'}$, depends diffeomorphically on $t \in B'$ in the following sense: setting $\widetilde{\mathcal{D}_t} = P_t^{-1}(V')$,

$$\widetilde{\mathcal{D}} = \bigcup_{t\in B'} \widetilde{\mathcal{D}_t} \times \{t\}$$

is an open set in $\mathbb{R}^d \times \mathbb{R}^m$ (contained in a cube $\widetilde{V}' \times B'$) on which

$$\widetilde{\mathcal{D}} \to V' \times B', \quad (\tilde{h},t) \mapsto (P_t(\tilde{h}),t)$$

is a diffeomorphism.

Continuation. The corollary of Theorem 4.1 joined with the Morse lemma finally yield:

Theorem 4.3. *Suppose $\varphi(x,t)$ is smooth in $(x,t) \in \mathbb{R}^d \times \mathbb{R}^m$ and satisfies*

$$\nabla_x\varphi(x_0,t_0) = 0 \quad \textit{and} \quad \det \mathrm{D}_x^2\varphi(x_0,t_0) \neq 0.$$

Then, there exists an arbitrarily small cube, $U \times B$, containing (x_0,t_0) such that the following holds: if $f(x,t)$ is smooth in the neighborhood of $\mathbb{R}^d \times \overline{B}$ and vanishes on $U^{\mathrm{c}} \times \overline{B}$, then

$$I_f(r,t) \sim \mathrm{e}^{\mathrm{i}r\theta(t)}r^{-d/2}\sum_{j=0}^{\infty}a_j(t)r^{-j/2},$$

where $\theta(t)$ is real-valued, the $a_j(t)$'s are complex-valued, and all these functions are smooth in $t \in \overline{B}$. Moreover, these estimates are uniform in $t \in \overline{B}$.

PROOF. By the Implicit Function Theorem there exists a smooth function

$$B_0 \subset \mathbb{R}^m \to \mathbb{R}^d, \quad t \mapsto \mathrm{x}(t)$$

defined on a cube $B_0 \ni t_0$ such that $(\nabla_x \varphi)(\mathrm{x}(t), t) = 0$ for all $t \in B_0$ and $\mathrm{x}(t_0) = x_0$. Without loss of generality we also suppose

$$\det \mathrm{D}_x^2 \varphi(\mathrm{x}(t), t) \neq 0$$

for all $t \in B_0$. Let

$$\phi(h,t) = \varphi(\mathrm{x}(t) + h, t) - \varphi(\mathrm{x}(t), t),$$

which is smooth in $(h,t) \in \mathbb{R}^d \times B_0$. Then, $\phi(h,t)$ satisfies the hypotheses of Morse's lemma. Hence, there exists a cube, $V' \times B'$, containing $(0, t_0)$ and whose closure is in $\mathbb{R}^d \times B_0$, and a family of diffeomorphisms

$$P_t^{-1} \colon V' \to \widetilde{\mathcal{D}_t} \subseteq \widetilde{V}' \quad (\text{where } t \in B')$$

such that, letting $\tilde{h} = P_t^{-1}(h)$, one obtains

$$\phi(P_t(\tilde{h}), t) = \sum_{j=1}^{s} (\tilde{h}^{(j)})^2 - \sum_{k=s+1}^{d} (\tilde{h}^{(k)})^2.$$

Notice that $0 = x_0 - \mathrm{x}(t_0) \in V'$. Consequently, there exists an arbitrarily small cube, $U \times B \ni (x_0, t_0)$, such that

$$(x - \mathrm{x}(t), t) \in V' \times B'$$

for all $(x,t) \in \overline{U} \times \overline{B}$. In other words, $U \times B$ is mapped onto a region whose closure is in $V' \times B'$ via the change of variables $(h,t) = (x - \mathrm{x}(t), t)$.

Let us consider an amplitude, $f(x,t)$, satisfying the asserted properties. Since for a fixed $t \in \overline{B}$ the integrand in

$$I_f(r,t) = \int_{\mathbb{R}^d} \mathrm{e}^{\mathrm{i}r\varphi(x,t)} f(x,t)\, \mathrm{d}x$$

is supported in U, it follows that the right-hand side in

$$I_f(r,t) = \int_{\mathbb{R}^d} \mathrm{e}^{\mathrm{i}r\varphi(h+\mathrm{x}(t),t)} f(h + \mathrm{x}(t), t)\, \mathrm{d}h = \mathrm{e}^{\mathrm{i}r\varphi(\mathrm{x}(t),t)} \int_{\mathbb{R}^d} \mathrm{e}^{\mathrm{i}r\phi(h,t)} f(h + \mathrm{x}(t), t)\, \mathrm{d}h$$

is supported in V'. Then, by Morse's lemma the change of variables $\tilde{h} = P_t^{-1}(h)$ is available, yielding

$$I_f(r,t) = \mathrm{e}^{\mathrm{i}r\varphi(\mathrm{x}(t),t)} \int_{\widetilde{\mathcal{D}_t}} \mathrm{e}^{\mathrm{i}rQ(\tilde{h})} f(P_t(\tilde{h}) + \mathrm{x}(t), t) J_t(\tilde{h})\, \mathrm{d}\tilde{h},$$

where $J_t(\tilde{h})$ is the Jacobian and

$$Q(\tilde{h}) = \sum_{j=1}^{s} (\tilde{h}^{(j)})^2 - \sum_{j=s+1}^{d} (\tilde{h}^{(j)})^2.$$

Since the amplitude in the above extends smoothly on $\mathbb{R}^d \times \overline{B}$ and vanishes for $\tilde{h} \notin \widetilde{V}'$, the corollary of Theorem 4.1 then completes the proof. □

Finally, the following result is an interesting application of our treatment of a parameter:

Theorem 4.4. *Suppose $\varphi(x,t)$ is smooth in $(x,t) \in \mathbb{R}^d \times \mathbb{R}^m$ and satisfies*

$$\nabla_x \varphi(x_0, t_0) = 0 \quad \textit{and} \quad \operatorname{rank} \mathrm{D}_x^2 \varphi(x_0, t_0) = \kappa,$$

where $\kappa \geqslant 1$. Then, there exists an arbitrarily small cube, $U \times B$, containing (x_0, t_0) such that the following holds: if $f(x,t)$ is smooth in a neighborhood of $\mathbb{R}^d \times \overline{B}$ and vanishes on $U^{\mathrm{c}} \times \overline{B}$, then

$$I_f(r,t) = O(r^{-\kappa/2})$$

uniformly in $t \in \overline{B}$.

Proof. Since the rank of $\mathrm{D}_x^2 \varphi(x_0, t_0)$ is equal to the order of its largest nonzero principal minor, there exist indices, $j_1, \dots, j_\kappa$, such that, letting

$$\xi = \left(x^{(j_1)}, \dots, x^{(j_\kappa)}\right),$$

$\det \mathrm{D}_\xi^2 \varphi(x_0, t_0) \neq 0$. After permuting the variables we write

$$x = (\xi, \chi), \quad \text{and} \quad x_0 = (\xi_0, \chi_0),$$

with the obvious definitions of χ, ξ_0, and χ_0. Interpreting (χ, t) as a parameter the previous theorem gives the result. □

Remark. Using the previous decomposition one cannot derive the complete asymptotics of the considered oscillatory integral, since the resulting coefficients would be oscillatory integrals themselves! Their decay is not known a priori.

Cauchy principal value. We now derive similar results for Cauchy principal values of oscillatory integrals with nondegenerate stationary phase points. To this end let us consider first

$$\mathrm{p.\,v.} \int_{-\infty}^{\infty} \mathrm{e}^{\mathrm{i}rh} \frac{f(h)}{h} \, dh = \lim_{\varepsilon \downarrow 0} \int_{|h| > \varepsilon} \mathrm{e}^{\mathrm{i}rh} \frac{f(h)}{h} \, dh,$$

where the amplitude, $f(h)$, is smooth in $h \in \mathbb{R}$ and compactly supported.

Notice that the lemma of Theorem 3.1 generalizes to a complex-valued phase provided that the path of integration remains in $\overline{\mathbb{C}_+}$, explicitly: *Given a smooth regular*[4] *path, $\gamma(t)$, lying in the closure of $\mathbb{C}_+$, and an amplitude, $f(z) = f(x,y)$, smooth in z and compactly supported along this path,*

$$\int_\gamma \mathrm{e}^{\mathrm{i}rz} f(z) \, \mathrm{d}z = O(r^{-\infty})$$

when $r \to \infty$.

Lemma. *Suppose $f(h)$ is smooth in $h \in \mathbb{R}$, compactly supported, and analytic at 0. Then,*

$$\mathrm{p.\,v.} \int_{-\infty}^{\infty} \mathrm{e}^{\mathrm{i}rh} \frac{f(h)}{h} \, dh = \pi \mathrm{i} f(0) + O(r^{-\infty})$$

when $r \to \infty$.

Proof. For $\varepsilon > 0$, let C_ε be a smooth regular path starting at -2ε, going through $[-2\varepsilon, -\varepsilon]$, then avoiding the origin, but staying in

$$\{z \in \mathbb{C}; |\mathrm{Re}(z)| < \varepsilon \text{ and } 0 < \mathrm{Im}\, z < \varepsilon\},$$

[4]A smooth path, $\gamma(t)$, is *regular* if $\gamma'(t) \neq 0$ for any t.

and finally going through $[\varepsilon, 2\varepsilon]$. Since

$$\mathrm{e}^{\mathrm{i}rz}\frac{f(z)}{z} = \frac{f(0)}{z} + \text{analytic}$$

in a punctured neighborhood of the origin,

$$\lim_{\varepsilon\downarrow 0}\int_{C_\varepsilon} \mathrm{e}^{\mathrm{i}rz}\frac{f(z)}{z}\,\mathrm{d}z = -\pi\mathrm{i}f(0).$$

Thus, letting $\gamma_\varepsilon = \,]-\infty, -2\varepsilon] * C_\varepsilon * [2\varepsilon, \infty[$ (where $*$ denotes the concatenation), which is smooth and regular,

$$\text{p. v.}\int_{\mathbb{R}} \mathrm{e}^{\mathrm{i}rh}\frac{f(h)}{h}\,\mathrm{d}h - \pi\mathrm{i}f(0) = \lim_{\varepsilon\downarrow 0}\int_{\gamma_\varepsilon} \mathrm{e}^{\mathrm{i}rz}\frac{f(z)}{z}\,\mathrm{d}z = \int_\gamma \mathrm{e}^{\mathrm{i}rz}\frac{f(z)}{z}\,\mathrm{d}z,$$

where $\gamma = \gamma_{\varepsilon_0}$ for a fixed, but small enough $\varepsilon_0 > 0$. The result follows from the statement preceding the lemma. □

The analyticity assumption may be removed in the following way – where a parameter $t \in \mathbb{R}^m$ is also introduced for later purpose:

Theorem 4.5. *Given $\delta > 0$ and a cube $B \subset \mathbb{R}^m$, consider a function, $f(h,t)$, smooth in a neighborhood of $\mathbb{R} \times \overline{B}$ and vanishing on $]-\delta, \delta[^{\mathrm{c}} \times \overline{B}$. Then,*

$$\text{p. v.}\int_{-\infty}^{\infty} \mathrm{e}^{\mathrm{i}rh}\frac{f(h,t)}{h}\,\mathrm{d}h = \pi\mathrm{i}f(0,t) + O(r^{-\infty})$$

uniformly in $t \in \overline{B}$, when $r \to \infty$.

PROOF. Let $\chi(x)$ be a smooth function in $x \in \mathbb{R}$, compactly supported, such that $0 \leqslant \chi(x) \leqslant 1$ on $\mathbb{R}$ and $\chi(x) \equiv 1$ in a neighborhood of 0. The considered principal value then decomposes into $\mathrm{I} + \mathrm{II} + \mathrm{III}$, where

$$\mathrm{I} = \int_{\mathbb{R}} \mathrm{e}^{\mathrm{i}rh}\frac{f(h,t)}{h}(1 - \chi(h))\,\mathrm{d}h,$$
$$\mathrm{II} = \int_{\mathbb{R}} \mathrm{e}^{\mathrm{i}rh}\frac{f(h,t) - f(0,t)}{h}\chi(h)\,\mathrm{d}h,$$

and

$$\mathrm{III} = f(0,t)\,\text{p. v.}\int_{\mathbb{R}} \mathrm{e}^{\mathrm{i}rh}\frac{\chi(h)}{h}\,\mathrm{d}h.$$

The lemma of Theorem 3.1 implies $\mathrm{I} = O(r^{-\infty})$ and $\mathrm{II} = O(r^{-\infty})$ uniformly in $t \in \overline{B}$.[5] Since $\chi(x)$ is analytic at 0 and $\chi(0) = 1$, the result follows from the lemma. □

We now turn our attention to a phase, $\varphi(h,t)$, smooth in $(h,t) \in \mathbb{R} \times \mathbb{R}^m$, such that $\partial_h\varphi(0,t_0) \neq 0$. Our study reduces to the previous theorem by a change of variables, regarding which the following elementary result is helpful: *Given a cube $B \subset \mathbb{R}^m$, suppose $f(h,t)$ is smooth in a neighborhood of $\{0\} \times \overline{B}$ and satisfies, for $t \in \overline{B}$, $f(0,t) = 0$ and $\partial_h f(0,t) \neq 0$. Then, $h\partial_h f(h,t)/f(h,t)$ admits a smooth extension in a neighborhood of $\{0\} \times \overline{B}$, which is equal to 1 when $h = 0$.*

[5] More precisely, the result follows from the proof of this lemma, by noticing that $U' \times B'$ may be chosen to be equal to $\mathbb{R} \times \mathbb{R}^m$ — since the derivative of the phase is $1 \neq 0$ everywhere.

Lemma. *Suppose $\partial_h\varphi(0,t_0)\neq 0$ for a certain $t_0\in\mathbb{R}^m$. There exist a $\delta>0$ and an arbitrarily small cube, B, containing t_0 such that the following holds: if $f(h,t)$ is smooth in a neighborhood of $\mathbb{R}\times\overline{B}$ and vanishes on $]-\delta,\delta[^{\mathrm{c}}\times\overline{B}$, then*

$$\mathrm{p.\,v.}\int_{\mathbb{R}}\mathrm{e}^{\mathrm{i}r\varphi(h,t)}\frac{f(h,t)}{h}\,\mathrm{d}h=\sigma\pi\mathrm{i}f(0,t)\mathrm{e}^{\mathrm{i}r\varphi(0,t)}+O(r^{-\infty})$$

uniformly in $t\in\overline{B}$, where $\sigma=\operatorname{sgn}\bigl(\partial_h\varphi(0,t_0)\bigr)$.

PROOF. By continuity there exists a cube, $]-\delta,\delta[\times B'\ni(0,t_0)$, on which $\partial_h\varphi(h,t)\neq 0$. Let $\varphi_t(h)=\varphi(h,t)-\varphi(0,t)$ and $F(h,t)=(\varphi_t(h),t)$, which is smooth in $(h,t)\in\mathbb{R}\times\mathbb{R}^m$. Then,

$$\det\mathrm{D}_{(h,t)}F(h,t)=\partial_h\varphi(h,t)\neq 0$$

on $]-\delta,\delta[\times B'$. Hence, $F(h,t)$ admits a smooth inverse, $G(\tilde h,t)=(\psi_t(\tilde h),t)$, defined on $F(]-\delta,\delta[\times B')$. Notice that $\varphi_t(0)=0=\psi_t(0)$ and

$$\psi_t'(0)=\frac{1}{\varphi_t'(0)}\neq 0.$$

In particular, $\sigma=\operatorname{sgn}\bigl(\psi_t'(0)\bigr)$. Moreover, by the statement preceding the lemma $\tilde h\psi_t'(\tilde h)/\psi_t(\tilde h)$ has a smooth extension to a neighborhood of $\{0\}\times\overline{B}$ (where $\overline{B}\subset B'$ is arbitrarily chosen), which is equal to 1 when $\tilde h=0$. For $f(h,t)$ of the stipulated form and $t\in\overline{B}$, the change of variables $\tilde h=\varphi_t(h)$ then gives

$$\begin{aligned}\mathrm{p.\,v.}\int_{\mathbb{R}}\mathrm{e}^{\mathrm{i}r\varphi(h,t)}\frac{f(h,t)}{h}\,\mathrm{d}h&=\mathrm{p.\,v.}\int_{\mathbb{R}}\mathrm{e}^{\mathrm{i}r\tilde h}\mathrm{e}^{\mathrm{i}r\varphi(0,t)}\frac{f(\psi_t(\tilde h),t)}{\psi_t(\tilde h)}|\psi_t'(\tilde h)|\,\mathrm{d}\tilde h\\&=\sigma\mathrm{e}^{\mathrm{i}r\varphi(0,t)}\,\mathrm{p.\,v.}\int_{\mathbb{R}}\mathrm{e}^{\mathrm{i}r\tilde h}\frac{g(\tilde h,t)}{\tilde h}\,\mathrm{d}\tilde h,\end{aligned}$$

where

$$g(\tilde h,t)=f(\psi_t(\tilde h),t)\frac{\tilde h\psi_t'(\tilde h)}{\psi_t(\tilde h)}.$$

Notice that $g(0,t)=f(0,t)$. Moreover, $g(\tilde h,t)$ vanishes when $t\in\overline{B}$ and $\tilde h\notin\,]-\tilde\delta,\tilde\delta[$ for a certain $\tilde\delta>0$. The previous theorem thus applies, which completes the proof. □

Let us now consider a phase, $\varphi(x,h,t)$, smooth in $(x,h,t)\in\mathbb{R}^d\times\mathbb{R}\times\mathbb{R}^m$, and

$$I_f(r,h,t)=\int_{\mathbb{R}^d}\mathrm{e}^{\mathrm{i}r\varphi(x,h,t)}f(x,h,t)\,\mathrm{d}x.$$

Theorem 4.6. *Suppose $\nabla_x\varphi(x_0,0,t_0)=0$, $\det\mathrm{D}_x^2\varphi(x_0,0,t_0)\neq 0$, and*

$$\partial_h\varphi(x_0,0,t_0)\neq 0.$$

Then, there exist a $\delta>0$ and an arbitrarily small cube, $U\times B$, containing (x_0,t_0) such that the following holds: if $f(x,h,t)$ is smooth in the neighborhood of $\mathbb{R}^d\times\mathbb{R}\times\overline{B}$ and vanishes on $(U\times\,]-\delta,\delta[)^{\mathrm{c}}\times\overline{B}$, then

$$\mathrm{p.\,v.}\int_{\mathbb{R}}\frac{1}{h}I_f(r,h,t)\,\mathrm{d}h\sim\mathrm{e}^{\mathrm{i}r\theta(t)}r^{-d/2}\sum_{j=0}^{\infty}a_j(t)r^{-j/2}$$

uniformly in $t\in\overline{B}$, where $\theta(t)$ and $a_j(t)$ are smooth in the neighborhood of $\overline{B}$.

Proof. Interpreting $(x,t) \in \mathbb{R}^d \times \mathbb{R}^m$ as a parameter, the previous lemma holds, since $\partial_h\varphi(x_0,0,t_0) \neq 0$. Moreover, interpreting $(h,t) \in \mathbb{R} \times \mathbb{R}^m$ as a parameter, Theorem 4.3 also holds, since $\nabla_x\varphi(x_0,0,t_0) = 0$ and $\det \mathrm{D}_x^2\varphi(x_0,0,t_0) \neq 0$. Consequently, there exist a $\delta > 0$ and an arbitrarily small cube, $U \times B$, containing (x_0,t_0) such that for $f(x,h,t)$ of the stipulated form

$$\text{p. v.} \int_{\mathbb{R}} \mathrm{e}^{\mathrm{i}r\varphi(x,h,t)} \frac{f(x,h,t)}{h}\, \mathrm{d}h = \sigma\pi\mathrm{i} f(x,0,t)\mathrm{e}^{\mathrm{i}r\varphi(x,0,t)} + O(r^{-\infty}) \tag{4.4}$$

uniformly in $(x,t) \in \overline{U} \times \overline{B}$ and

$$\int_{\mathbb{R}^d} \mathrm{e}^{\mathrm{i}r\varphi(x,0,t)} f(x,0,t)\, \mathrm{d}x \sim \mathrm{e}^{\mathrm{i}r\theta(t)} r^{-d/2} \sum_{j=0}^{\infty} a_j(t) r^{-j/2} \tag{4.5}$$

uniformly and smoothly in $t \in \overline{B}$, both when $r \to \infty$.

By Fubini's and the dominated convergence theorems

$$\text{p. v.} \int_{\mathbb{R}} \frac{1}{h} I_f(r,h,t)\, \mathrm{d}h = \int_{\mathbb{R}^d} \text{p. v.} \int_{\mathbb{R}} \mathrm{e}^{\mathrm{i}r\varphi(x,h,t)} \frac{f(x,h,t)}{h}\, \mathrm{d}h\, \mathrm{d}x,$$

where the dominator is given by the lemma of Theorem 3.2. By the equation (4.4), the above is equal to

$$\int_{\mathbb{R}^d} \sigma\pi\mathrm{i} f(x,0,t)\mathrm{e}^{\mathrm{i}r\varphi(x,0,t)}\, \mathrm{d}x + O(r^{-\infty})$$

uniformly in $t \in \overline{B}$. The equation (4.5) then yields the result. □

In the same way as we derived from Theorem 3.2 its corollary,

Corollary. *Suppose* $\nabla_x\varphi(x_0,e_0,t_0) = 0$, $\det \mathrm{D}_x^2\varphi(x_0,e_0,t_0) \neq 0$, *and*

$$\partial_e\varphi(x_0,e_0,t_0) \neq 0.$$

Then, there exist a $\delta > 0$ *and an arbitrarily small cube* $U \times B$ *containing* (x_0,t_0) *such that the following holds: if* $f(x,e,t)$ *is smooth in the neighborhood of* $\mathbb{R}^d \times \mathbb{R} \times \overline{B}$ *and vanishes on* $(U \times]e_0-\delta, e_0+\delta[)^{\mathrm{c}} \times \overline{B}$, *then*

$$\text{p. v.} \int_{\mathbb{R}} \frac{1}{\eta - e} I_f(r,\eta,t)\, \mathrm{d}\eta \sim \mathrm{e}^{\mathrm{i}r\theta(e,t)} r^{-d/2} \sum_{j=0}^{\infty} a_j(e,t) r^{-j/2}$$

uniformly in $(e,t) \in [e_0-\delta, e_0+\delta] \times \overline{B}$, *where* $\theta(e,t)$ *and* $a_j(e,t)$ *are smooth in the neighborhood of* $[e_0-\delta, e_0+\delta] \times \overline{B}$.

More generally, by the argument used in Theorem 4.4 (which we repeat!),

Theorem 4.7. *Suppose* $\nabla_x\varphi(x_0,e_0,t_0) = 0$, $\operatorname{rank} \mathrm{D}_x^2\varphi(x_0,e_0,t_0) = \kappa$, *and*

$$\partial_e\varphi(x_0,e_0,t_0) \neq 0.$$

Then, there exist a $\delta > 0$ *and an arbitrarily small cube,* $U \times B$, *containing* (x_0,t_0) *such that the following holds: if* $f(x,e,t)$ *is smooth in the neighborhood of* $\mathbb{R}^d \times \mathbb{R} \times \overline{B}$ *and vanishes on* $(U \times]e_0-\delta, e_0+\delta[)^{\mathrm{c}} \times \overline{B}$, *then*

$$\text{p. v.} \int_{\mathbb{R}} \frac{1}{\eta - e} I_f(r,\eta,t)\, \mathrm{d}\eta = O(r^{-\kappa/2})$$

uniformly in $(e,t) \in [e_0-\delta, e_0+\delta] \times \overline{B}$.

PROOF. There exist indices, $j_1, \dots, j_\kappa$, such that, for

$$\xi = \left(x^{(j_1)}, \dots, x^{(j_\kappa)}\right),$$

$\det \mathrm{D}^2_\xi(x_0, t_0) \neq 0$. After permuting the variables we write

$$x = (\xi, \chi) \quad \text{and} \quad x_0 = (\xi_0, \chi_0)$$

with the obvious definitions of χ, ξ_0, and χ_0. Interpreting (χ, t) as a parameter, the above corollary gives the existence of a $\delta > 0$ and an arbitrarily small cube, $U' \times (U'' \times B) \ni (\xi_0; \chi_0, t_0)$, such that, letting $U = U' \times U'' \ni x_0$, for $f(x,t)$ of the stipulated form

$$\text{p.v.} \int_{\mathbb{R}} \frac{1}{\eta - e} \int_{\mathbb{R}^\kappa} \mathrm{e}^{\mathrm{i}r\varphi(\xi;\chi,\eta,t)} f(\xi; \chi, \eta, t)\, \mathrm{d}\xi\, \mathrm{d}\eta = O(r^{-\kappa/2})$$

uniformly in $(\chi, e, t) \in \overline{U''} \times [e_0 - \delta, e_0 + \delta] \times \overline{B}$. Therefore,

$$\int_{\mathbb{R}^{d-\kappa}} \text{p.v.} \int_{\mathbb{R}} \frac{1}{\eta - e} \int_{\mathbb{R}^\kappa} \mathrm{e}^{\mathrm{i}r\varphi(\xi;\chi,\eta,t)} f(\xi; \chi, \eta, t)\, \mathrm{d}\xi\, \mathrm{d}\eta\, \mathrm{d}\chi = O(r^{-\kappa/2})$$

uniformly in $(e,t) \in [e_0 - \delta, e_0 + \delta] \times \overline{B}$. By Fubini's and the dominated convergence theorems this last integral is equal to

$$\int_{\mathbb{R}^d} \text{p.v.} \int_{\mathbb{R}} \frac{1}{\eta - e} \mathrm{e}^{\mathrm{i}r\varphi(x;\eta,t)} f(x; \eta, t)\, \mathrm{d}\eta\, \mathrm{d}x,$$

and hence (for the same reasons) to

$$\text{p.v.} \int_{\mathbb{R}} \int_{\mathbb{R}^d} \frac{1}{\eta - e} \mathrm{e}^{\mathrm{i}r\varphi(x;\eta,t)} f(x; \eta, t)\, \mathrm{d}x\, \mathrm{d}\eta,$$

where both times the dominator is given by the lemma of Theorem 3.2. The proof is thus complete. □

5. Fourier transforms over level surfaces of analytic functions

We now consider a real-valued function, $\Phi(x)$, analytic in $x \in \mathbb{R}^d$ and its level surface $\Gamma(e) = \{x \in \mathbb{R}^d; \Phi(x) = e\}$.

Given a function, $f(x)$, summable on $\Gamma(e)$, its Fourier transform over $\Gamma(e)$ is defined as

$$\mathcal{F}(\Gamma(e), f)(n) = \int_{\Gamma(e)} \mathrm{e}^{\mathrm{i}n\cdot x} f(x)\, \mathrm{dS}(x)$$

for $n \in \mathbb{Z}^d$, where $\mathrm{dS}(x)$ denotes the element of surface on $\Gamma(e)$. In this section we derive the decay of $\mathcal{F}(\Gamma(e), f)(n)$ when $|n| \to \infty$ and show its uniformity when e varies on an appropriate interval. We then derive an analogous result for the Cauchy principal value of such a Fourier transform.

Let $a' < b'$ and $\mathcal{S}' = \bigcup_{e \in]a',b'[} \Gamma(e)$ be given. We assume:

Assumption A.
- $\Phi(x)$ is real-valued and analytic in $\mathbb{R}^d$;
- $\nabla\Phi(x) \neq 0$ for all $x \in \mathcal{S}'$;
- $\Gamma(e)$ is compact for all $e \in]a', b'[$.

The second statement in Assumption A and the Implicit Function Theorem ensure that $\Gamma(e)$ is a *regular smooth surface* for any $e \in]a', b'[$. In particular, $\Gamma(e)$ may be covered by real-analytic local parametrizations, $U \xrightarrow{\sigma} \Gamma(e)$, where each

$U \subset \mathbb{R}^{d-1}$ is open, each $\sigma(u)$ is a smooth homeomorphism between U and $\sigma(U)$ in the topology of $\Gamma(e)$, and

$$\bigcup_{(U,\sigma)} \sigma(U) = \Gamma(e).$$

Moreover, if two of the previous local parametrizations, (U,σ) and (V,τ), have a nonempty overlap, $\mathcal{D} = \sigma(U) \cap \tau(V) \subseteq \Gamma(e)$, then the change of parametrizations $\tau^{-1} \circ \sigma(u)$ is a real-analytic diffeomorphism from $\sigma^{-1}(\mathcal{D})$ to $\tau^{-1}(\mathcal{D})$.

Given a local parametrization, (U,σ), the restriction of the Fourier transform to $\sigma(U)$ gives

$$\mathcal{F}(\sigma(U), f)(n) = \int_{\sigma(U)} \mathrm{e}^{\mathrm{i}n\cdot x} f(x)\,\mathrm{dS}(x) = \int_U \mathrm{e}^{\mathrm{i}n\cdot\sigma(u)} f(\sigma(u))J(u)\,\mathrm{d}u$$

for any $f(x)$ summable on $\sigma(U)$, where $J(u)\,\mathrm{d}u$ is the element of surface. Of course the decay of such an integral when $|n| \to \infty$ is studied by means of the stationary phase method.

Let $n = r\omega$ be the polar form of $n \in \mathbb{Z}^d$, so $r = |n|$ and $\omega \in \mathrm{S}^{d-1}$. The *phase* in the previous integral then becomes $\varphi(u,\omega) = \omega \cdot \sigma(u)$. Let us consider a point, $x_0 = \sigma(u_0)$, in $\sigma(U)$. Remarkably, the fact that u_0 is or is not a stationary phase point depends on *intrinsic properties of* $\Gamma(e)$ *only*; if u_0 is stationary, the rank of $\mathrm{D}_u^2\varphi(u_0,\omega)$ is also intrinsic. Indeed,

Theorem 5.1. *In the above circumstances u_0 is stationary if and only if ω is perpendicular to $\Gamma(e)$ at $x_0 = \sigma(u_0)$. Then, the rank of the Hessian of $\varphi(u,\omega)$ at u_0 is equal to the number of nonvanishing principal curvatures of $\Gamma(e)$ at x_0.*

Proof. Notice that u_0 is stationary iff

$$\nabla_u \omega \cdot \sigma(u_0) = \big(\omega \cdot \mathrm{d}_{u^{(1)}}\sigma(u_0), \ldots, \omega \cdot \mathrm{d}_{u^{(d-1)}}\sigma(u_0)\big) = 0.$$

Since the tangent plane of $\Gamma(e)$ at x_0 is generated by

$$\{\mathrm{d}_{u^{(j)}}\sigma(u_0);\, j = 1, \ldots, d-1\},$$

the first statement follows.

Suppose u_0 is a stationary phase point and consider any other local parametrization, (V,τ), of a neighborhood of $x_0 = \sigma(u_0)$. Let $\mathcal{D} = \sigma(U) \cap \tau(V)$ and $F(v) = \sigma^{-1} \circ \tau(v)$. Then, $F(v)$ is a smooth diffeomorphism from $\tau^{-1}(\mathcal{D})$ to $\sigma^{-1}(\mathcal{D})$ satisfying

$$\omega \cdot \tau(v) = \omega \cdot \sigma\big(F(v)\big). \tag{5.1}$$

Let $v_0 = \tau^{-1}(x_0)$, so $\big(\nabla_u(\omega\cdot\sigma)\big)\big(F(v_0)\big) = \nabla_u(\omega\cdot\sigma)(u_0) = 0$. Then, the chain rule applied to (5.1) gives

$$\mathrm{D}_v^2(\omega \cdot \tau)(v_0) = \mathrm{D}_u^2(\omega \cdot \sigma)(u_0)\big(\mathrm{D}_v F(v_0)\big)^2.$$

Since $F(v)$ is a diffeomorphism, $\mathrm{D}_v F(v_0)$ is invertible and hence

$$\operatorname{rank} \mathrm{D}_v^2(\omega \cdot \tau)(v_0) = \operatorname{rank} \mathrm{D}_u^2(\omega \cdot \sigma)(u_0),$$

which shows that this last rank is intrinsic.

Finally, since u_0 is a stationary phase point, ω is perpendicular to $\Gamma(e)$ at x_0, so

$$\omega = \frac{\pm\nabla_x\Phi(x_0)}{\|\nabla_x\Phi(x_0)\|}.$$

Moreover, there exists a $j \in \{1, \dots, d\}$ such that $\omega^{(j)} \neq 0$. Suppose without loss of generality $\omega^{(d)} \neq 0$, and hence $\partial_{x^{(d)}}\Phi(x_0) \neq 0$. Let $w = (x^{(1)}, \dots, x^{(d-1)})$ and $w_0 = (x_0^{(1)}, \dots, x_0^{(d-1)})$. By the Implicit Function Theorem there exists a function, $h(w)$, smooth in the neighborhood of w_0, such that $\Phi\big(w, h(w)\big) = e$ and $h(w_0) = x_0^{(d)}$. Hence, $\gamma(w) = \big(w, h(w)\big)$ gives a smooth local parametrization of a neighborhood of x_0 as a graph of a smooth function. Differential geometry then shows that rank $\mathrm{D}_w^2 h(w)$ is equal to the number of nonvanishing principal curvatures at $x_0 \in \Gamma(e)$. Since

$$\mathrm{D}_w^2(\omega \cdot \gamma)(w) = \omega^{(d)}\mathrm{D}_w^2 h(w),$$

the proof is complete. □

Joint system of parametrizations. Given a fixed $e_0 \in \,]a', b'[$, we now construct a system of parametrizations for $\Gamma(e_0)$ compatible with all $\Gamma(e)$'s for e varying in a small neighborhood of e_0 (so the derived estimate for Fourier transforms will be uniform in e). We make the following hypothesis:

Assumption B. For every $e \in \,]a', b'[$, $\Gamma(e)$ admits at least κ nonvanishing principal curvatures at any point, where $\kappa \geqslant 1$ is a fixed integer.

The plan is the following: starting from an arbitrary system of real-analytic parametrizations for $\Gamma(e_0)$, we will parametrize $\Gamma(e)$ using the local coordinates of $\Gamma(e_0)$, by lifting them orthogonally to $\Gamma(e_0)$ (for e very close to e_0).

Let $\{(\mathcal{U}_\beta, \gamma_\beta)\}_{\beta=1}^M$ be a system of real-analytic parametrizations covering $\Gamma(e_0)$. Since $\Gamma(e_0)$ is compact, we assume M is finite. Without loss of generality, we also assume $\gamma_\beta(u)$ is analytic in a neighborhood of $\overline{\mathcal{U}_\beta}$, so the expression $\gamma_\beta(u)$ for $u \in \partial\mathcal{U}_\beta$ makes sense.

Let $x_0 \in \Gamma(e_0)$, say, $x_0 = \gamma_\beta(u_0)$ for a given $u_0 \in \mathcal{U}_\beta$ and a given $1 \leqslant \beta \leqslant M$. Since $\nabla_x\Phi(x)$ is perpendicular to $\Gamma(e_0)$ at any $x \in \Gamma(e_0)$, we need to solve

$$\Phi\Big(\gamma_\beta(u) + \lambda\nabla_x\Phi\big(\gamma_\beta(u)\big)\Big) - e = 0 \tag{5.2}$$

in the neighborhood of $(u, e, \lambda) = (u_0, e_0, 0)$. The derivative at 0 with respect to λ of the left-hand side in (5.2) is $\|\nabla_x\Phi(\gamma_\beta(u))\|^2$, which is strictly positive. Hence, there exists an analytic function, $\lambda(u, e)$, defined on an arbitrarily small cube, $\mathcal{U}' \times \,]e_0 - \Delta', e_0 + \Delta'[\, \ni (u_0, e_0)$, satisfying

$$\gamma_\beta(u) + \lambda(u, e)\nabla_x\Phi\big(\gamma_\beta(u)\big) \in \Gamma(e), \quad \lambda(u_0, e_0) = 0. \tag{5.3}$$

Remarkably, $\lambda(u, e)$ depends on $\gamma_\beta(u)$ only, not on its local coordinates u:

Lemma. *Suppose $\gamma_{\beta'}(u_0') = x_0$ and let $u' = \gamma_{\beta'}^{-1} \circ \gamma_\beta(u)$, where u varies in*

$$\gamma_\beta^{-1}\big(\gamma_\beta(\mathcal{U}_\beta) \cap \gamma_{\beta'}(\mathcal{U}_{\beta'})\big).$$

Define $\lambda'(u', e)$ as above with respect to $\gamma_{\beta'}(u')$ — while $\lambda(u, e)$ was defined with respect to $\gamma_\beta(u)$. Then, $\lambda(u, e) = \lambda'(u', e)$ when e is close enough to e_0.

The proof, based on the uniqueness part of the Implicit Function Theorem,[6] is immediate. Since the considered surfaces are compact, *this last lemma ensures the existence of an analytic function, $\Lambda(x, e)$, defined on $\Gamma(e_0) \times \,]e_0 - \Delta', e_0 + \Delta'[$ (where $\Gamma(e_0)$ is endowed with its surface structure), such that*

$$x + \Lambda(x, e)\nabla_x\Phi(x) \in \Gamma(e) \quad \text{and} \quad \Lambda(x_0, e_0) = 0. \tag{5.4}$$

[6]Of course, this uniqueness holds *locally*.

Incidentally, the function $\lambda(u,e)$ has the following, immediate properties:

Lemma. *For every $u \in \mathcal{U}'$, $\lambda(u,e_0) = 0$. In particular, $\partial_{u^{(j)}}\lambda(u,e_0) = 0$ for any $j = 1,\ldots,d-1$.*

Lemma. $\partial_e\lambda(u_0,e_0) > 0$.

PROOF. In fact, since $\Phi(x_0 + \lambda(u_0,e)\nabla_x\Phi(x_0)) = e$, the chain rule gives

$$\partial_e\lambda(u_0,e_0) = \frac{1}{\|\nabla_x\Phi(x_0)\|^2}. \qquad \square$$

The equation (5.3) permits to define the following local parametrization of $\Gamma(e)$:

$$\sigma(u,e) = \gamma_\beta(u) + \lambda(u,e)\nabla_x\Phi\big(\gamma_\beta(u)\big),$$

where $u \in \mathcal{U}'$ and $|e - e_0| < \Delta'$. Notice that $\sigma(u,e)$ is real-analytic. Indeed,

Lemma. *$\sigma(u,e)$ is a real-analytic diffeomorphism in a neighborhood of (u_0,e_0).*

PROOF. Notice that the ambient space, $\mathbb{R}^d$, is generated by the tangent vectors of $\Gamma(e_0)$ at x_0 (namely, $\mathrm{d}_{u^{(j)}}\gamma_\beta(u_0)$ for $1 \leqslant j \leqslant d-1$) and $\nabla_x\Phi(x_0)$. The penultimate lemma and a direct computation show that the columns of the matrix $\mathrm{D}_{(u,e)}\sigma(u_0,e_0)$ in canonical basis are

$$\begin{array}{ll} \mathrm{d}_{u^{(j)}}\gamma_\beta(u_0) & \text{for } 1 \leqslant j \leqslant d-1 \\ \partial_e\lambda(u_0,e_0)\nabla_x\Phi(x_0) & \text{for } j = d, \end{array}$$

which are linearly independent by the last lemma. The Inverse Mapping Theorem then completes the proof. $\square$

We thus select an arbitrarily small cube, $\mathcal{U}\times]e_0-\Delta, e_0+\Delta[\ni (u_0,e_0)$, contained in $\mathcal{U}' \times]e_0 - \Delta', e_0 + \Delta'[$, such that $\sigma(u,e)$ is a real-analytic diffeomorphism in the neighborhood of $\overline{\mathcal{U}} \times [e_0 - \Delta, e_0 + \Delta]$.

Let $\varphi(u;e,\omega) = \omega\cdot\sigma(u,e)$ be the anticipated phase, where $u \in \mathcal{U}$, $|e - e_0| < \Delta$, and $\omega \in \mathrm{S}^{d-1}$. By Theorem 5.1, the associated stationary phase points and $\operatorname{rank}\mathrm{D}_u^2\varphi$ at these points are intrinsic properties of $\Gamma(e)$. Let $\omega_0 \in \mathrm{S}^{d-1}$ be arbitrarily fixed. It appears that if x_0 is a stationary phase point, then $\partial_e\varphi(u_0;e_0,\omega_0)$ is also "intrinsic" in the following sense:

Theorem 5.2. *Suppose $x_0 = \sigma(u_0,e_0)$ is a stationary phase point. Then,*

$$\partial_e\varphi(u_0;e_0,\omega_0) = \frac{\pm 1}{\|\nabla_x\Phi(x_0)\|}.$$

PROOF. By Theorem 5.1, since x_0 is stationary, $\omega_0 = \pm\nabla_x\Phi(x_0)/\|\nabla_x\Phi(x_0)\|$. In particular,

$$\partial_e\varphi(u_0;e_0,\omega_0) = \omega_0\cdot\partial_e\sigma(u_0,e_0) = \frac{\pm 1}{\|\nabla_x\Phi(x_0)\|}\nabla_x\Phi(x_0)\cdot\partial_e\sigma(u_0,e_0).$$

Since $\Phi\big(\sigma(u,e)\big) = e$, the chain rule gives $\nabla_x\Phi(x_0)\cdot\partial_e\sigma(u_0,e_0) = 1$, which completes the proof. $\square$

Consequently, $\partial_e\varphi(u_0; e_0, \omega_0) \neq 0$ if x_0 is a stationary phase point. Hence, whether $\nabla_u\varphi(u_0; e_0, \omega_0) \neq 0$ or

$$\nabla_u\varphi(u_0; e_0, \omega_0) = 0,$$
$$\operatorname{rank} \mathrm{D}^2_u\varphi(u_0; e_0, \omega_0) \geqslant \kappa,$$
$$\partial_e\varphi(u_0; e_0, \omega_0) \neq 0.$$

By Theorems 3.1, 4.4, the corollary of Theorem 3.2, and Theorem 4.7 there exists an arbitrarily small cube, $U \times B \subset \mathcal{U} \times \mathbb{R}^d$, containing (u_0, ω_0) and an arbitrarily small $0 < \delta < \Delta$ such that the following holds:

If $f(u; e, \omega)$ is smooth in the neighborhood of $\mathbb{R}^{d-1} \times \mathbb{R} \times \overline{B}$ and vanishes on $(U \times]e_0 - \delta, e_0 + \delta[)^{\mathrm{c}} \times \overline{B}$, then

$$I_f(r; e, \omega) = O(r^{-\kappa/2})$$

and

$$\mathrm{p.\,v.} \int_{\mathbb{R}} \frac{1}{\eta - e} I_f(r; \eta, \omega)\, \mathrm{d}\eta = O(r^{-\kappa/2}),$$

where both estimates are uniform in $(e, \omega) \in [e_0 - \delta, e_0 + \delta] \times \overline{B}$.

Of course, the previous cube and associated constructions depend on the fixed e_0, ω_0, and $x_0 \in \Gamma(e_0)$. Repeating this procedure for all $x \in \Gamma(e_0)$, where e_0 and ω_0 are still fixed, one obtains a system of real-analytic diffeomorphisms, $\{(U_x \times]e_0 - \delta_x, e_0 + \delta_x[, \sigma_x)\}_{x\in\Gamma(e_0)}$, and a family of cubes, $\{B_x\}_{x\in\Gamma(e_0)}$, satisfying the above properties, from which one extracts a finite subsystem,

$$\{(U_\alpha \times]e_0 - \delta_\alpha, e_0 + \delta_\alpha[, \sigma_\alpha)\}_{\alpha=1}^N,$$

which covers $\Gamma(e_0)$:

$$\Gamma(e_0) = \bigcup_{\alpha=1}^N \sigma_\alpha(U_\alpha, e_0).$$

We limit our considerations to an arbitrarily small cube, B, whose closure is inside $\bigcap_{\alpha=1}^N B_\alpha$, and to $\{(U_\alpha \times]e_0 - \delta', e_0 + \delta'[, \sigma_\alpha)\}_{\alpha=1}^N$, where $0 < \delta' < \min_{\alpha=1}^N \delta_\alpha$ is arbitrarily small.

By the equation (5.4) the local parametrizations

$$\sigma_\alpha(u, e) = \gamma_{\beta_\alpha}(u) + \lambda_\alpha(u, e)\nabla_x\Phi(\gamma_{\beta_\alpha}(u))$$

yield the following, arguably global parametrization of $\Gamma(e)$:

$$\Sigma(x, e) = x + \Lambda(x, e)\nabla_x\Phi(x), \tag{5.5}$$

where $x \in \Gamma(e_0)$ and $|e - e_0| < \delta'$. In particular, when x varies over the entire $\Gamma(e_0)$, $\Sigma(x, e)$ describes a whole "closed" surface included in $\Gamma(e)$ (which should therefore corresponds to $\Gamma(e)$, as long as $\Gamma(e_0)$ and $\Gamma(e)$ have the same number of connected components). In other words, one expects

$$\Gamma(e) = \bigcup_{\alpha=1}^N \sigma_\alpha(U_\alpha, e). \tag{5.6}$$

However, it is easier to shorten δ' in order to *prove* the previous relation.

This may be done in the following way: suppose by contradiction there does not exist a δ such that the relation (5.6) holds for all $e \in]e_0 - \delta, e_0 + \delta[$, where $0 < \delta \leqslant \delta'$. Then there exist a sequence, $\{e_n\}$, converging to e_0, and points, x_n, on $\Gamma(e_n)$, such that $x_n \notin \bigcup_{\alpha=1}^N \sigma_\alpha(U_\alpha, e_n)$. Letting $\mathcal{S} = \bigcup_{|e-e_0|<\delta'} \Gamma(e)$, the x_ns lie in the compact set $\overline{\mathcal{S}}$, so they accumulate towards a certain $x^* \in \overline{\mathcal{S}}$. Going to a subsequence, again denoted by $x_n \to x^*$, one finds

$$\Phi(x^*) = \Phi\Big(\lim_{n\to\infty} x_n\Big) = \lim_{n\to\infty} \Phi(x_n) = \lim_{n\to\infty} e_n = e.$$

Therefore, x^* lies in $\bigcup_{\alpha=1}^N \sigma_\alpha(U_\alpha \times]e_0-\delta', e_0+\delta'[)$, while the x_n's do not, contradicting the fact that this last region is open in $\mathbb{R}^d$.

Indeed,

Theorem 5.3. *There exists a $\delta \in]0, \delta'[$ such that for any $e \in]e_0-\delta, e_0+\delta[$, $\Sigma(x,e)$ is a real-analytic diffeomorphism between $\Gamma(e_0)$ and $\Gamma(e)$ (endowed with their surface structures).*

PROOF. The previous paragraph shows that $\Sigma(x,e)$ is surjective for e close enough to e_0. By a similar argument suppose there does not exist a $\delta > 0$ such that $\Sigma(x,e)$ is injective for every $e \in]e_0-\delta, e_0+\delta[$. Then, there exist a sequence, $\{e_n\}$, converging to e_0, and points, x_n, on $\Gamma(e_n)$, such that

$$x_n = \Sigma(y_n, e_n) = \Sigma(z_n, e_n)$$

for distinct points y_n and z_n on $\Gamma(e_0)$. Since the x_n's accumulate towards a certain $x^* \in \Gamma(e_0)$, y_n and z_n are eventually in the same coordinates neighborhood, contradicting the fact that each σ_α is a diffeomorphism. □

We limit our considerations to the system $\{(U_\alpha \times]e_0-\delta, e_0+\delta[, \sigma_\alpha)\}_{\alpha=1}^N$, where δ is specified by the previous theorem. We have proved:

Theorem 5.4. *Let $\omega_0 \in \mathrm{S}^{d-1}$, $e_0 \in]a', b'[$, and $\varepsilon > 0$ be arbitrarily fixed. Under Assumptions* A *and* B*, there exists a finite family of cubes of diameters less than ε,*

$$\{U_\alpha \times]e_0-\delta, e_0+\delta[\times B\}_{\alpha=1}^N,$$

where $[e_0-\delta, e_0+\delta] \subset]a', b'[$ and $\omega_0 \in B \subset \mathbb{R}^{d-1}$, and functions,

$$\sigma_\alpha \colon U_\alpha \times]e_0-\delta, e_0+\delta[\to \bigcup_{|e-e_0|<\delta} \Gamma(e)$$

such that the following holds:

(1) *For every α, $\sigma_\alpha(u,e)$ is a real-analytic diffeomorphism from a neighborhood of $\overline{U_\alpha} \times [e_0-\delta, e_0+\delta]$ to its image.*

(2) *For all $e \in]e_0-\delta, e_0+\delta[$, $\Gamma(e) = \bigcup_{\alpha=1}^N \sigma_\alpha(U_\alpha, e)$.*[7]

(3) *Let us denote by $I_f^{(\alpha)}(r; e, \omega)$ the oscillatory integral of amplitude $f(u; e, \omega)$ with respect to the phase $\varphi_\alpha(u; e, \omega) = \omega \cdot \sigma_\alpha(u, e)$. If $f(u; e, \omega)$ is smooth in the neighborhood of $\mathbb{R}^{d-1} \times \mathbb{R} \times \overline{B}$ and vanishes on $(U_\alpha \times]e_0-\delta, e_0+\delta[)^{\mathrm{c}} \times \overline{B}$, then*

$$I_f^{(\alpha)}(r; e, \omega) = O(r^{-\kappa/2})$$

and

$$\text{p. v.} \int_{\mathbb{R}} \frac{1}{\eta - e} I_f(r; \eta, \omega)\, \mathrm{d}\eta = O(r^{-\kappa/2}),$$

where both estimates are uniform in $(e, \omega) \in [e_0-\delta, e_0+\delta] \times \overline{B}$.

[7] In other words, $\{(U_\alpha, \sigma_\alpha(\cdot, e))\}_{\alpha=1}^N$ is a system of real-analytic parametrizations of $\Gamma(e)$ for *any* $e \in]e_0-\delta, e_0+\delta[$.

We now construct a smooth, joint "partition of unity" subordinated to the system of neighborhoods $\{\sigma_\alpha(U_\alpha, e)\}_{\alpha=1}^N$.

Theorem 5.5. *Let* $0 < \tilde{\delta} < \delta$ *and* $\mathcal{S} = \bigcup_{|e-e_0|<\tilde{\delta}} \Gamma(e)$. *There exists a family of functions,* $\{\chi_\alpha(x)\}_{\alpha=1}^N$, *smooth in* $x \in \mathbb{R}^d$, *satisfying*:

- $0 \leqslant \chi_\alpha \leqslant 1$,
- $\operatorname{supp} \chi_\alpha \subset \sigma_\alpha(U_\alpha \times]e_0 - \delta, e_0 + \delta[)$,
- $\sum_{\alpha=1}^N \chi_\alpha(x) = 1$ *for all* $x \in \bar{\mathcal{S}}$.

PROOF. Notice that $\bar{\mathcal{S}} \subset \bigcup_{\alpha=1}^N \sigma_\alpha(U_\alpha \times]e_0 - \delta, e_0 + \delta[)$, so in particular

$$\{\sigma_1(U_1 \times]e_0 - \delta, e_0 + \delta[), \dots, \sigma_N(U_N \times]e_0 - \delta, e_0 + \delta[), \mathbb{R}^d \setminus \bar{\mathcal{S}}\}$$

is a finite open covering of $\mathbb{R}^d$. After discarding $\chi_{N+1}(x)$, any partition of unity, $\{\chi_\alpha(x)\}_{\alpha=1}^{N+1}$, subordinated to the previous covering satisfies the stated properties. □

Fourier transforms. We are ready to compute the (uniform!) decays of Fourier transforms, $\mathcal{F}(\Gamma(e), f)$, for suitable amplitudes and derive an analogous result for Cauchy principal values.

Theorem 5.6. *Let* $n = r\omega$ *be the polar form of* $n \in \mathbb{Z}^d$, *where* $n \neq 0$, *and let the amplitude,* $f(x)$, *be smooth in* $x \in \mathbb{R}^d$. *Consider any interval* $[a, b] \subset]a', b'[$. *Under Assumptions* A *and* B,

$$\mathcal{F}(\Gamma(e), f)(n) = O(r^{-\kappa/2})$$

uniformly in $(e, \omega) \in [a, b] \times \mathrm{S}^{d-1}$.

PROOF. Let $e_0 \in [a, b]$ and $\omega_0 \in \mathrm{S}^{d-1}$ be fixed. By Theorems 5.4 and 5.5, there exist a $\delta > 0$, a cube, $B \ni \omega_0$, a joint system of parametrizations,

$$\{(U_\alpha \times]e_0 - 2\delta, e_0 + 2\delta[, \sigma_\alpha)\}_{\alpha=1}^N,$$

and a joint partition of unity, $\{\chi_\alpha\}_{\alpha=1}^N$, summing at 1 on $\bigcup_{|e-e_0|<\delta} \Gamma(e)$, such that the following holds: Let

$$I^{(\alpha)}(r; e, \omega) = \int_{U_\alpha} \mathrm{e}^{ir\omega\cdot\sigma_\alpha(u,e)} f(\sigma_\alpha(u, e)) \chi_\alpha(\sigma_\alpha(u, e)) J_\alpha(u, e)\, du,$$

where $J_\alpha(u, e)\, du$ is the element of surface of $\sigma_\alpha(U_\alpha, e)$. Then,

$$\mathcal{F}(\Gamma(e), f)(r\omega) = \sum_{\alpha=1}^N I^{(\alpha)}(r; e, \omega)$$

for every $(e, \omega) \in [e_0 - \delta, e_0 + \delta] \times \overline{B}$. Since $\chi_\alpha \circ \sigma_\alpha$, and hence the integrand of $I^{(\alpha)}(r; e, \omega)$ vanish outside $U_\alpha \times]e_0 - 2\delta, e_0 + 2\delta[$, by Theorem 5.4, $I^{(\alpha)}(r; e, \omega) = O(r^{-\kappa/2})$ uniformly in $(e, \omega) \in [e_0 - \delta, e_0 + \delta] \times \overline{B}$. Thus,

$$\mathcal{F}(\Gamma(e), f)(r\omega) = O(r^{-\kappa/2})$$

uniformly in $[e_0 - \delta, e_0 + \delta] \times \overline{B}$. Since $e_0 \in [a, b]$ and $\omega_0 \in \mathrm{S}^{d-1}$ are arbitrary, the result follows from the compactness of $[a, b] \times \mathrm{S}^{d-1}$. □

Cauchy principal value. An analogous result for principal values may be derived when affecting the amplitude by a cutoff function.

We establish first that the element of surface on any local part of $\Gamma(e)$ varies smoothly in (u,e). Having in hand a common system of parametrizations, $\sigma_\alpha(u,e)$, for $u \in U_\alpha$ and $|e-e_0| < \delta$ (where $e_0 \in]a',b'[$ is arbitrarily fixed), let us compute this element of surface. To this end, one considers the determinants, $M_\alpha^{(l)}$, of the submatrices of format $(d-1)\times(d-1)$ obtained from $[\partial_{u^{(k)}}\sigma_\alpha^{(l)}(u,e)]_{k,l}$ by removing its lth column, where $\sigma_\alpha = (\sigma_\alpha^{(1)},\ldots,\sigma_\alpha^{(d)})$. Let

$$J_\alpha(u,e) = \big(M_\alpha^{(1)}(u,e), -M_\alpha^{(2)}(u,e), \ldots, (-1)^{d-1}M_\alpha^{(d)}(u,e)\big).$$

By definition the element of surface is $\|J_\alpha(u,e)\|\,du$.

Theorem 5.7. $\|J_\alpha(u,e)\| = \big\|\nabla_x\Phi\big(\sigma_\alpha(u,e)\big)\big\|\,|\det \mathrm{D}_{(u,e)}\sigma_\alpha(u,e)|$.

Proof. The chain rule applied to $\Phi(\sigma_\alpha(u,e)) = e$ gives

$$\nabla_x\Phi\big(\sigma_\alpha(u,e)\big)\mathrm{D}_{(u,e)}\sigma_\alpha(u,e) = \begin{bmatrix}0 & \cdots & 0 & 1\end{bmatrix},$$

which we abbreviate $\nabla\Phi\mathrm{D}\sigma_\alpha = e_d^{\mathrm{t}}$. Since $\sigma_\alpha(u,e)$ is a diffeomorphism, $\mathrm{D}\sigma_\alpha$ is invertible. Thus,

$$\nabla\Phi = e_d^{\mathrm{t}}(\mathrm{D}\sigma_\alpha)^{-1} = \frac{1}{\det \mathrm{D}\sigma_\alpha} e_d^{\mathrm{t}}(\operatorname{adj}\mathrm{D}\sigma_\alpha),$$

where adj stands for the classical adjoint. The result follows from

$$e_d^{\mathrm{t}}(\operatorname{adj}\mathrm{D}\sigma_\alpha) = (-1)^{d-1}J_\alpha(u,e). \qquad \square$$

Corollary. *$\|J_\alpha(u,e)\|$ is smooth in $(u,e) \in U_\alpha \times]e_0-\delta, e_0+\delta[$.*

For fixed $\omega_0 \in \mathrm{S}^{d-1}$ and $e_0 \in]a',b'[$, let $\{(U_\alpha \times]e_0-3\delta, e_0+3\delta[, \sigma_\alpha)\}_{\alpha=1}^N$, $B \ni \omega_0$, and $\{\chi_\alpha\}_{\alpha=1}^N$, summing at 1 on $\bigcup_{|e-e_0|<2\delta}\Gamma(e)$, be given by Theorems 5.4 and 5.5. Let us define a cutoff function, $0 \leqslant \chi_{e_0}(x) \leqslant 1$, smooth in $x \in \mathbb{R}^d$, such that[8]

$$\chi_{e_0}(x) = \begin{cases} 1 & \text{if } |\Phi(x)-e_0| < \delta, \\ 0 & \text{if } |\Phi(x)-e_0| > 2\delta. \end{cases}$$

Notice that $\chi_{e_0}(x)$ is analytic in the neighborhood of $\Gamma(e_0)$—which will become important later. Under Assumptions A and B,

Theorem 5.8. *For $f(x)$ smooth in $x \in \mathbb{R}^d$ and $r > 0$*

$$\mathrm{p.\,v.}\int_{\mathbb{R}} \frac{1}{\eta - e}\mathcal{F}(\Gamma(\eta), \chi_{e_0}f)(r\omega)\,d\eta = O(r^{-\kappa/2})$$

uniformly in $(e,\omega) \in [e_0-\delta, e_0+\delta] \times \mathrm{S}^{d-1}$.

Proof. The considered principal value is equal to

$$\mathrm{p.\,v.}\int_{\mathbb{R}} \frac{1}{\eta - e}\sum_{\alpha=1}^{N} I_{F_\alpha}^{(\alpha)}(r;\eta,\omega)\,d\eta,$$

[8] Since the amplitude we will use does not depend on ω, the following cutoff function does not depend on ω_0; in other circumstances if it does, the following argument still holds.

where $F_\alpha(u,\eta) = (\chi_{e_0} f \chi_\alpha)(\sigma_\alpha(u,\eta))\|J_\alpha(u,\eta)\|$. Notice that

$$\chi_{e_0}(\sigma_\alpha(u,\eta)) = \begin{cases} 1 & \text{if } |\eta - e_0| < \delta, \\ 0 & \text{if } |\eta - e_0| > 2\delta. \end{cases}$$

Hence, $F_\alpha(u,\eta)$ is smooth on $\mathbb{R}^{d-1} \times \mathbb{R}$ and vanishes when

$$(u,\eta) \notin U_\alpha \times]e_0 - 2\delta, e_0 + 2\delta[.$$

By Theorem 5.4 the principal value under consideration is thus $O(r^{-\kappa/2})$ uniformly in $(e,\omega) \in [e_0 - \delta, e_0 + \delta] \times \overline{B}$. Since $\omega_0 \in \mathrm{S}^{d-1}$ is arbitrarily fixed, the result follows from the compactness of S^{d-1}. □

Analyticity of Fourier transform. We close this section by showing that for any fixed $n \in \mathbb{Z}^d$ the following Fourier transform, $\mathcal{F}(\Gamma(e), f)(n)$, is analytic at e_0 when the amplitude, $f(x)$, is analytic in the neighborhood of $\Gamma(e_0) \subset \mathbb{R}^d$ (where $e_0 \in]a', b'[$ is arbitrarily fixed). To this end the diffeomorphism between $\Gamma(e_0)$ and $\Gamma(e)$, $\Sigma(x,e)$, defined in the relation (5.5) is helpful, because, given a system of *disjoint* open neighborhoods on $\Gamma(e_0)$ covering all $\Gamma(e_0)$ except a set of area zero, its lifting to $\Gamma(e)$ via $\Sigma(x,e)$ also covers the whole $\Gamma(e)$ except a set of area zero.

In detail, let $\{(U_\alpha \times]e_0 - \delta, e_0 + \delta[, \sigma_\alpha)\}_{\alpha=1}^N$ be given by Theorem 5.4.

Theorem 5.9. *There exists a finite, joint system of local parametrizations,*

$$\{(V_\beta \times]e_0 - \delta, e_0 + \delta[, \sigma_\beta)\}_{\beta=1}^M,$$

such that:

- *For all $\beta \in \{1, \ldots, M\}$ there exists an $\alpha_\beta \in \{1, \ldots, N\}$ such that $V_\beta \subseteq U_{\alpha_\beta}$ and $\sigma_\beta = \sigma_{\alpha_\beta} \restriction V_\beta$;*
- *For all $e \in]e_0 - \delta, e_0 + \delta[$ the coordinates neighborhoods $\{\sigma_\beta(V_\beta, e)\}_{\beta=1}^M$ are mutually disjoints;*
- *For all $e \in]e_0 - \delta, e_0 + \delta[$ the area of $\Gamma(e) \setminus \bigcup_{\beta=1}^M \sigma_\beta(V_\beta, e)$ is zero.*[9]

PROOF. One may construct a system of disjoint, open neighborhoods of full area on $\Gamma(e_0)$ by considering all nonempty cells of the form[10]

$$\sigma_{\alpha_1}(U_{\alpha_1}, e_0) \cap \cdots \cap \sigma_{\alpha_l}(U_{\alpha_l}, e_0) \cap \sigma_{\alpha_{l+1}}(\overline{U_{\alpha_{l+1}}}, e_0)^c \cap \cdots \cap \sigma_{\alpha_N}(\overline{U_{\alpha_N}}, e_0)^c,$$

where $\{\alpha_1, \ldots, \alpha_N\} = \{1, \cdots, N\}$. These cells are open, disjoint, and cover all $\Gamma(e_0)$ except the set $\sigma_1(\partial U_1, e_0) \cup \cdots \cup \sigma_N(\partial U_N, e_0)$, whose area is zero. The diffeomorphism $\Sigma(x,e)$ lifts this exceptional set onto an exceptional set in $\Gamma(e)$ (which is not a surprise). The peculiarity of $\Sigma(x,e)$ is that *all points on $\Gamma(e)$ which are not the image of an exceptional $x \in \Gamma(e_0)$ are covered.* The result follows. □

Theorem 5.10. *Suppose $f(x)$ is analytic in a neighborhood of $\Gamma(e_0) \subset \mathbb{R}^d$. Then, for any fixed $n \in \mathbb{Z}^d$ the Fourier transform $\mathcal{F}(\Gamma(e), f)(n)$ is analytic at $e_0 \in]a', b'[$.*

PROOF. By the previous theorem, for any $e \in]e_0 - \delta, e_0 + \delta[$

$$\mathcal{F}(\Gamma(e), f)(n) = \sum_{\beta=1}^M \int_{V_\beta} \mathrm{e}^{\mathrm{i}n \cdot \sigma_\beta(u,e)} f(\sigma_\beta(u,e)) J_\beta(u,e)\, \mathrm{d}u,$$

---

[9]In other words, the integral of 1 over this surface is zero.

[10]The complements are taken with respect to $\Gamma(e_0)$.

where $J_\beta(u,e)\,du$ is the element of surface of $\sigma_\beta(V_\beta, e)$. Since the integrand in this last expression analytic in the neighborhood of $\overline{V_\beta} \times [e_0 - \delta/2, e_0 + \delta/2]$ (that is, converges to its Taylor's series), the result follows. □

6. Green's functions

We turn our attention to

(6.1)
$$G(n,z) = \int_{\mathbb{T}^d} \frac{\mathrm{e}^{\mathrm{i}n\cdot x}}{\Phi(x) - z}\,\mathrm{d}x,$$

where $z \in \mathbb{C}_+$, $n \in \mathbb{Z}^d$, and $\Phi(x)$ is real-valued, analytic, and *periodic* on $\mathbb{T}^d$. Here, $\mathbb{T}^d$ denotes the *torus* of dimension d, that is, the set $[-\pi,\pi]^d$ endowed with the quotient topology induced by congruence modulo 2π. In particular,

$$\{x \in \mathbb{T}^d; \Phi(x) = e\} = \Gamma(e)$$

is a compact manifold. It may be covered by a finite system of coordinates neighborhoods admitting a subordinated partition of unity. Since each coordinate neighborhood is embedded in $\mathbb{R}^d$, it is clear that all results in Section 5 apply under appropriate hypotheses, namely: Let $a' < b'$ and $\mathcal{S}' = \bigcup_{e\in]a',b'[} \Gamma(e)$. We assume:

Assumption C. • $\Phi(x)$ is real-valued, analytic, and periodic on $\mathbb{T}^d$;
- $\nabla\Phi(x) \neq 0$ for all $x \in \mathcal{S}'$;
- For every $e \in]a',b'[$, $\Gamma(e)$ admits at least κ nonvanishing principal curvatures at any point, where $\kappa \geqslant 1$ is a fixed integer.

For $e \in \mathbb{R}$ let

$$G(n,e) = \lim_{\substack{z\to e\\ z\in\mathbb{C}_+}} G(n,z),$$

where of course the existence of such a limit has to be established. In the present section we compute the decay of $G(n,e)$ as $|n| \to \infty$. We are especially interested in the case where e is in the range of $\Phi(x)$; otherwise, since by assumption $\Phi(x)$ admits a holomorphic extension, one may slightly change the domain of integration in (6.1) and deduce that $G(n,e)$ decays exponentially.

The proof of the following elementary lemma, based on Cauchy's theorem, is omitted.

Lemma. *Consider a function, $f(\eta)$, continuous in $\eta \in [a,b]$ and analytic at $e \in]a,b[$. Then,*

$$\lim_{\substack{z\to e\\ z\in\mathbb{C}_+}} \int_a^b \frac{f(\eta)}{\eta - z}\,\mathrm{d}\eta = \pi\mathrm{i} f(e) + \mathrm{p.\,v.} \int_a^b \frac{f(\eta)}{\eta - e}\,\mathrm{d}\eta.$$

For fixed $\omega_0 \in \mathrm{S}^{d-1}$ and $e_0 \in]a',b'[$, let $\{(U_\alpha \times]e_0 - 3\delta, e_0 + 3\delta[, \sigma_\alpha)\}_{\alpha=1}^N$, $B \ni \omega_0$, $\{\chi_\alpha\}_{\alpha=1}^N$, summing to 1 on $\bigcup_{|e-e_0|<2\delta} \Gamma(e)$, and χ_{e_0} be given by Theorems 5.4, 5.5, and 5.8.

Theorem 6.1. *Suppose $|e - e_0| < \delta$ and $\omega \in B$. Then, $G(n,e)$ exists and is equal to*

$$\pi\mathrm{i}\mathcal{F}(\Gamma(e), P)(n) + \mathrm{p.\,v.} \int_{-\infty}^{\infty} \frac{1}{\eta - e}\mathcal{F}(\Gamma(\eta), \chi_{e_0} P)(n)\,\mathrm{d}\eta + \int_{\mathbb{T}^d} \mathrm{e}^{\mathrm{i}n\cdot x} \frac{1 - \chi_{e_0}(x)}{\Phi(x) - e}\,\mathrm{d}x,$$

where $P(x) = \frac{1}{\|\nabla_x \Phi(x)\|}$.

PROOF. Notice that

$$\int_{\mathbb{T}^d} e^{in\cdot x} \frac{\chi_{e_0}(x)}{\Phi(x)-z}\, dx = \sum_{\alpha=1}^{N} \int_{\mathbb{T}^d} e^{in\cdot x} \frac{\chi_{e_0}(x)\chi_\alpha(x)}{\Phi(x)-z}\, dx.$$

The change of variables $x = \sigma_\alpha(u,\eta)$ applied to the above gives

$$\sum_{\alpha=1}^{N} \int_{\mathbb{R}} \int_{\mathbb{R}^{d-1}} e^{in\cdot\sigma_\alpha(u,\eta)} \frac{\chi_{e_0}(\sigma_\alpha(u,\eta))\chi_\alpha(\sigma_\alpha(u,\eta))}{\eta - z} |\det \mathrm{D}_{(u,\eta)}\sigma_\alpha(u,\eta)|\, du\, d\eta,$$

which, by Theorem 5.7, is equal to

$$\sum_{\alpha=1}^{N} \int_{\mathbb{R}} \frac{1}{\eta - z} \int_{\Gamma(\eta)} e^{in\cdot x} \chi_{e_0}(x)\chi_\alpha(x) P(x)\, dS(x)\, d\eta,$$

where $dS(x)$ denotes the element of surface. Hence,

$$\int_{\mathbb{T}^d} e^{in\cdot x} \frac{\chi_{e_0}(x)}{\Phi(x)-z}\, dx = \int_{\mathbb{R}} \frac{1}{\eta - z} \mathcal{F}(\Gamma(\eta), \chi_{e_0} P)\, d\eta.$$

Notice that $\chi_{e_0} P$ is analytic in the neighborhood of $\Gamma(e)$ when $|e - e_0| < \delta$. Hence, by Theorem 5.10, $\mathcal{F}(\Gamma(\eta), \chi_{e_0} P)$ is analytic at $\eta = e$. The result follows from the lemma and the dominated convergence theorem. □

The desired decay follows:

Theorem 6.2. *Let $n = r\omega$ be the polar form of $n \in \mathbb{Z}^d$, where $n \neq 0$, and consider any interval $[a,b] \subset]a',b'[$. Under Assumption* C

$$G(n,e) = O(r^{-\kappa/2})$$

uniformly in $(e,\omega) \in [a,b] \times \mathrm{S}^{d-1}$.

PROOF. For ω_0, e_0, etc., as above, by Theorems 5.6 and 5.8

$$\pi i \mathcal{F}(\Gamma(e), P)(n) = O(r^{-\kappa/2})$$

and

$$\text{p. v.} \int_{-\infty}^{\infty} \frac{1}{\eta - e} \mathcal{F}(\Gamma(\eta), \chi_{e_0} P)(n)\, d\eta = O(r^{-\kappa/2})$$

uniformly on $[e_0 - \delta, e_0 + \delta] \times \overline{B}$. Moreover, by the scholium of Theorem 3.1

$$\int_{\mathbb{T}^d} e^{in\cdot x} \frac{1 - \chi_{e_0}(x)}{\Phi(x) - e}\, dx = O(r^{-\infty})$$

uniformly on $[e_0 - \delta, e_0 + \delta] \times \overline{B}$. Since (e_0, ω_0) is arbitrary in $[a,b] \times \mathrm{S}^{d-1}$, which is compact, the result follows. □

7. Application to generalized Laplacians

In this section we apply the previous results to Green's functions of concrete Laplacians on $\mathbb{Z}^d$. We focus on two specific examples: the *standard Laplacian* and the *Molchanov – Vainberg Laplacian*.

At a higher level of generality let us consider a simple graph without loop, whose set of vertices is denoted by X (where X is allowed to be infinite). For $m, n \in X$, $\mathrm{d}(m,n)$ denotes the *graph distance* between m and n, that is, the length of the minimal chain joining m and n in the graph (∞ if m and n lie in two different connected components). Of course, d is a metric distance with values in $\mathbb{N} \cup \{\infty\}$.

Notice also that (X, d) determines the graph completely, since $\{m, n\}$ is an edge if and only if $\mathrm{d}(m, n) = 1$.

We are interested in the Hilbert space $l^2(X)$, consisting of square summable sequences indexed by X. Its usual basis is $\{\delta_n\}_{n \in X}$, where δ denotes the Kronecker delta:

$$\delta_n(m) = \mathbf{1}_{\{n\}}(m) = \begin{cases} 1 & \text{if } m = n, \\ 0 & \text{if } m \neq n. \end{cases}$$

The *adjacency operator* on $l^2(X)$ with respect to (X, d), sometimes called *Laplacian*, is defined as

$$(\Delta\psi)(n) = \sum_{\mathrm{d}(m,n)=1} \psi(m),$$

where $\psi \in l^2(X)$. In particular, $(\Delta\delta_n)(m) = \mathbf{1}_{\mathcal{V}_n}(m)$, where

$$\mathcal{V}_n = \{m \in X; \mathrm{d}(m, n) = 1\}.$$

In the sequel we suppose that the degrees of the vertices of the considered graph are bounded, in other words, $\sup_n \#\mathcal{V}_n \leqslant B$ for a certain $B < \infty$.[11] Then, Δ *is a bounded, selfadjoint operator on* $l^2(X)$, as one may easily verify.

We are now interested in the case where $X = \mathbb{Z}^d$ and the graph distance, $\mathrm{d}(m, n)$, is translational invariant. In these circumstances we call Δ a *generalized Laplacian.* Then, $\mathrm{d}(m, n)$ is a function of $m - n$ only. Hence, letting $\mathcal{V} = \mathcal{V}_0$ (where $0 \in \mathbb{Z}^d$ denotes the origin), the considered graph is clearly determined by $(\mathbb{Z}^d, \mathcal{V})$. In particular,

$$\Delta\psi(n) = \sum_{v \in \mathcal{V}} \psi(n + v),$$

where $\#\mathcal{V} \leqslant B$ is still assumed to be finite.

Recall that the Fourier transform of $\psi \in l^2(\mathbb{Z}^d)$ is the following function, defined for $x \in \mathbb{T}^d$:

$$\widehat{\psi}(x) = (\mathcal{F}\psi)(x) = (2\pi)^{-d/2} \sum_{n \in \mathbb{Z}^d} \mathrm{e}^{\mathrm{i}n \cdot x} \psi(n).$$

The *symbol* of Δ, denoted by $\widehat{\Delta}$, is the lifting of Δ via the Fourier transform:

$$\widehat{\Delta} = \mathcal{F}\Delta\mathcal{F}^{-1}.$$

Theorem 7.1. *Let Δ be the generalized Laplacian associated with a given $\mathcal{V} \subset \mathbb{Z}^d$. Then, its symbol is the multiplication by*

$$\Phi(x) = \sum_{v \in \mathcal{V}} \mathrm{e}^{\mathrm{i}v \cdot x} = \sum_{v \in \mathcal{V}} \cos(v \cdot x),$$

where $x \in \mathbb{T}^d$.

PROOF. $\widehat{\Delta}$ maps $\widehat{\psi}(x)$ to the following function:

$$(2\pi)^{-d/2} \sum_{n \in \mathbb{Z}^d} \mathrm{e}^{\mathrm{i}n \cdot x} \sum_{v \in \mathcal{V}} \psi(n+v) = (2\pi)^{-d/2} \sum_{v \in \mathcal{V}} \sum_{n \in \mathbb{Z}^d} \mathrm{e}^{\mathrm{i}(n-v) \cdot x} \psi(n) = \sum_{v \in \mathcal{V}} \mathrm{e}^{-\mathrm{i}v \cdot x} \widehat{\psi}(x).$$

Notice that

$$v \in \mathcal{V} \iff \mathrm{d}(v, 0) = 1 \iff \mathrm{d}(0, -v) = 1 \iff -v \in \mathcal{V}.$$

---

[11] The cardinality of a given set, A, is denoted by $\#A$.

Hence, denoting by $\mathcal{V}^+$ the set of $v \in \mathcal{V}$ whose first nonzero coordinate is positive,

$$\sum_{v\in\mathcal{V}} \mathrm{e}^{-\mathrm{i}v\cdot x} = \sum_{v\in\mathcal{V}} \mathrm{e}^{\mathrm{i}v\cdot x} = \sum_{v\in\mathcal{V}^+} 2\cos(v\cdot x) = \sum_{v\in\mathcal{V}} \cos(v\cdot x),$$

which completes the proof. □

Since the range of $\Phi(x)$ is equal to the spectrum of Δ, one obtains:

Corollary. $\operatorname{spec}(\Delta) = [\min\Phi, \#\mathcal{V}]$.

Notice that $\Phi(x)$ is analytic at each $x \in \mathbb{T}^d$. Hence,

$$E = \{x \in \mathbb{T}^d; \nabla_x\Phi(x) = 0\}$$

has Lebesgue measure zero. In particular, $\operatorname{spec}(\Delta)$ is purely absolutely continuous.[12]

The *Green's function* of Δ is defined as

$$G(m,n;z) = \langle \delta_m \mid (\Delta - z)^{-1}\delta_n\rangle$$

for $m, n \in \mathbb{Z}^d$ and $z \in \mathbb{C}_+$. Since Δ is translational invariant, this last function depends on $m - n$ and z only. In fact, $G(m,n;z) = G(0, n-m; z)$ so we denote this latter simply by $G(n-m, z)$. Since the Fourier transform is unitary, for any $n \in \mathbb{Z}^d$ and $z \in \mathbb{C}_+$

$$G(n,z) = \left\langle \widehat{\delta_0}(x) \mid (\widehat{\Delta} - z)^{-1}\widehat{\delta_n}(x)\right\rangle_2 = (2\pi)^{-d}\int_{\mathbb{T}^d} \frac{\mathrm{e}^{\mathrm{i}n\cdot x}}{\Phi(x) - z}\,\mathrm{d}x.$$

By Theorem 6.1 the limit

$$G(n,e) = \lim_{\substack{z\to e\\ z\in\mathbb{C}_+}} G(n,z)$$

exists for any $e \notin E$. We will compute its decay when $|n| \to \infty$ in two particular examples: the *standard discrete Laplacian* and the *Molchanov – Vainberg Laplacian*, which are described in the rest of the section. Both are important in mathematical physics, since both are discretizations of the continuous Laplacian on $\mathbb{R}^d$ (up to an additive constant).

Standard Laplacian. The *standard Laplacian* is the adjacency operator of the usual grid on $\mathbb{Z}^d$; it is specified by the graph distance

$$\mathrm{d}(m,n) = |m-n|_1 = \sum_{j=1}^{d} |m^{(j)} - n^{(j)}|,$$

so the set of immediate neighbors of the origin is

$$\mathcal{V} = \{(\pm1, 0, \ldots, 0), (0, \pm1, \ldots, 0), \ldots, (0, 0, \ldots, \pm1)\}.$$

[12]This may be seen in the following way: $L^2(\mathbb{T}^d) = L^2(\mathbb{T}^d \setminus E)$. Moreover, by the Inverse Function Theorem $\Phi(x)$ is invertible on $\mathbb{T}^d \setminus E$. Let $\Psi(y)$ be its inverse, so the change of variables $y = \Phi(x)$ gives a unitary equivalence between $L^2(\mathbb{T}^d)$ with the Lebesgue measure and $L^2(\Phi(\mathbb{T}^d \setminus E))$ with the measure $\|\nabla_y\Psi(y)\|\,\mathrm{d}y$. Via this unitary equivalence the operator of multiplication by $\Phi(x)$ is lifted to the operator of multiplication by y, so by definition $\|\nabla_y\Psi(y)\|\,\mathrm{d}y \restriction \Phi(\mathbb{T}^d \setminus E)$ is a spectral measure for Δ — and is clearly absolutely continuous.

Hence, by Theorem 7.1 the symbol of the standard Laplacian is the operator of multiplication by

$$\Phi(x) = 2\sum_{j=1}^{d} \cos x^{(j)}.$$

Thus, the spectrum of Δ is purely absolutely continuous and equal to $[-2d, 2d]$. Notice that

$$\nabla_x \Phi(x) = (-2\sin x^{(1)}, \dots, -2\sin x^{(d)}),$$

which vanishes only if $\cos x^{(j)} = \pm 1$ for all $j = 1, \dots, d$, in which case $\Phi(x) \in \{-2d, -2d+4, \dots, 2d-4, 2d\}$. In particular, the level surfaces of $\Phi(x)$,

$$\Gamma(e) = \{x \in \mathbb{T}^d; \Phi(x) = e\},$$

are regular for all $e \notin \{-2d, -2d+4, \dots, 2d-4, 2d\}$.

Let us show that for such e's the level surfaces are exempt of planarity, except for $e = 0$:

Theorem 7.2. *Let* $E = \{-2d, -2d+4, \dots, 2d-4, 2d\} \cup \{0\}$ *and suppose* $e \in [-2d, 2d] \setminus E$. *Then,* $\Gamma(e)$ *admits at least one nonvanishing principal curvature at any point.*

PROOF. Let $x_0 \in \Gamma(e)$ be fixed. By choice of e, $\Gamma(e)$ is regular, so there exists a j such that $\sin x_0^{(j)} \neq 0$. After renaming the variables, the equation defining $\Gamma(e)$ in a neighborhood of x_0 thus becomes

$$2\cos u^{(1)} + \cdots + 2\cos u^{(d-1)} + 2\cos h = e, \tag{7.1}$$

where $\sin h \neq 0$ on a certain interval. In particular, the derivative with respect to h of the left-hand side in (7.1) is not zero. Therefore, writing $u = (u^{(1)}, \dots, u^{(d-1)})$, there exists an implicit function, $\mathrm{h}(u)$, such that

$$2\cos u^{(1)} + \cdots + 2\cos u^{(d-1)} + 2\cos \mathrm{h}(u) = e$$

for all u in a neighborhood of u_0, where $\big(u_0, \mathrm{h}(u_0)\big)$ is the permuted x_0. The number of nonvanishing principal curvatures at x_0 is then given by the rank of $\mathrm{D}_u^2\,\mathrm{h}(u_0)$. Indeed, for any $j = 1, \dots, d-1$, differentiating the previous equation with respect to $u^{(j)}$ gives

$$-\sin u^{(j)} - \sin h\, \partial_{u^{(j)}} h = 0, \tag{7.2}$$

where $h = \mathrm{h}(u)$. Consequently,

$$-\sin h\, \mathrm{D}_u^2\,\mathrm{h}(u) = \mathrm{diag}(\cos u^{(1)}, \dots, \cos u^{(d-1)}) + \cos h\,[\partial_{u^{(j)}}\,\mathrm{h}(u)\partial_{u^{(k)}}\,\mathrm{h}(u)]_{j,k=1}^{d-1}.$$

Since $\sin h \neq 0$, it suffices to show that the right side in the previous equation does not vanish in a neighborhood of u_0. By the equation (7.2) this matrix is equal to

$$\mathrm{diag}(\cos u^{(1)}, \dots, \cos u^{(d-1)}) + \frac{\cos h}{\sin^2 h}[\sin u^{(j)} \sin u^{(k)}]_{j,k=1}^{d-1}. \tag{7.3}$$

Suppose by contradiction there exists an u in the considered neighborhood such that the above vanishes. Then, for all $j = 1, \dots, d-1$

$$\cos u^{(j)} \sin^2 h + \cos h \sin^2 u^{(j)} = 0,$$

which is equivalent to

$$(1 - \cos h \cos u^{(j)})(\cos u^{(j)} + \cos h) = 0.$$

Moreover, since $\sin h \neq 0$, it follows that $1 - \cosh h \cos u^{(j)} \neq 0$, so

$$\cos u^{(j)} = -\cosh h \quad \text{for } j = 1, \ldots, d-1$$

at such an u. Hence, by the equation (7.1), $2(2-d)\cosh h = e$. If $d = 2$, then $e = 0$, contrary to our assumption. Thus, the previous situation would occur only when $d > 2$, in which case $\cosh h = -e/(2d-4)$ and $\cos u^{(j)} = e/(2d-4)$ for $j = 1, \ldots, d-1$. Then, $\cos^2 h = \cos^2 u^{(j)}$ for any j, so $\sin u^{(j)} \sin u^{(k)} = \pm \sin^2 h$. Consequently, the (j,k)-th element of the considered matrix, (7.3), is 0 when $j = k$, but $\pm e/(2d-4)$ otherwise. This last quantity differs from zero (since $e \notin E$), which provides a contradiction.

In conclusion, at any point in a neighborhood of x_0, $\Gamma(e)$ admits at least one nonvanishing principal curvature. Since x_0 is arbitrary, this completes the proof. □

Theorem 6.2 thus gives a polynomial decay for the Green's function, $G(n,e)$, associated with the standard Laplacian:

Corollary. *Let* $E = \{-2d, -2d+4, \ldots, 2d-4, 2d\} \cup \{0\}$ *and suppose* $e \in [-2d, 2d] \setminus E$. *Then,*

$$G(n,e) = \lim_{\substack{z \to e \\ z \in \mathbb{C}_+}} G(n,z) = O(|n|^{-1/2})$$

when $|n| \to \infty$, *uniformly in* e *on each compact and uniformly in* $\omega \in \mathrm{S}^{d-1}$, *where* $n = |n|\omega$ *is the polar form of* $n \neq 0$.

Molchanov – Vainberg Laplacian. In order to avoid convexity problems, Molchanov and Vainberg have suggested to change the discretization of the Laplacian. They have based their construction on the 2^d full-diagonal neighbors of elements in $\mathbb{Z}^d$, instead of their $2d$ immediate neighbors. The constant-energy surfaces of the resulting operator are strictly convex in any dimension, as shown below.

Explicitly, the *Molchanov – Vainberg Laplacian* (or *diagonal Laplacian*) is the adjacency operator of the translational invariant graph specified by the following set of points adjacent to the origin:

$$\mathcal{V} = \{(v^{(1)}, \ldots, v^{(d)}); v^{(j)} \in \{1, -1\} \text{ for } j = 1, \ldots, d\}.$$

By an elementary combinatorial argument, $n \in \mathbb{Z}^d$ is in the component of the origin if and only if the $n^{(j)}$'s are all even or all odd. Indeed, the considered graph consists of 2^{d-1} connected components with set of representatives

$$\{(0, n^{(2)}, \ldots, n^{(d)}); n^{(j)} \in \{0,1\} \text{ for } j = 2, \ldots, d\}.$$

The graph is also specified by the following metric:

$$\mathrm{d}(m,n) = \begin{cases} |m-n|_\infty & \text{if the components of } m-n \text{ have the same parity,} \\ \infty & \text{otherwise,} \end{cases}$$

where $|n|_\infty = \max_{j=1}^{d} |n^{(j)}|$.

Remarkably, the symbol of the Molchanov – Vainberg Laplacian factorizes:

Theorem 7.3. $\widehat{\Delta}$ *is the operator of multiplication by*

$$\Phi(x) = 2^d \cos x^{(1)} \cdots \cos x^{(d)},$$

where $x = (x^{(1)}, \ldots, x^{(d)}) \in \mathbb{T}^d$.

PROOF. Let us denote by $\{e_1, \dots, e_d\}$ the standard basis of $\mathbb{Z}^d$. By Theorem 7.1 the symbol of Δ is the multiplication by

$$\Phi(x) = \sum_{v\in\mathcal{V}} \mathrm{e}^{\mathrm{i}x\cdot v} = \sum_{v\in\mathcal{V}} \mathrm{e}^{\mathrm{i}x\cdot\sum_{j=1}^{d} v^{(j)}e_j} = \sum_{v\in\mathcal{V}}\prod_{j=1}^{d} \mathrm{e}^{\mathrm{i}x\cdot v^{(j)}e_j}$$
$$= \prod_{j=1}^{d}(\mathrm{e}^{\mathrm{i}x\cdot e_j} + \mathrm{e}^{-\mathrm{i}x\cdot e_j}) = 2^d\prod_{j=1}^{d}\cos x^{(j)},$$

as claimed. □

Consequently, the spectrum of Δ is purely absolutely continuous and equal to $[-2^d, 2^d]$. Moreover,

Lemma. *Suppose* $0 < |e| < 2^d$. *Then, for all* $x \in \Gamma(e)$, $\nabla_x\Phi(x) \neq 0$. *In particular,* $\Gamma(e)$ *defines a regular surface for such an* e.

PROOF. If $x \in \Gamma(e)$, then $\cos x^{(j)} \neq 0$ for $j = 1, \dots, d$, so $\partial_{x^{(j)}}\Phi(x) = -2^d e\tan x^{(j)}$. Thus, $\|\nabla_x\Phi(x)\|^2 = 4^d e^2\sum_{j=1}^{d}\tan^2 x^{(j)}$, which differs from 0, since $e \neq \pm 2^d$. □

Let us investigate the constant-energy surfaces associated with $\Phi(x)$. Firstly, let us consider the covering $\widetilde{\Gamma}(e) = \{x \in \mathbb{R}^d; \Phi(x) = e\}$.

If $e = 0$, $\widetilde{\Gamma}(e)$ consists of the hyperplanes of equation $x^{(j)} = (2k+1)\frac{\pi}{2}$ for $k \in \mathbb{Z}$ and $j = 1, \dots, d$. These hyperplanes divide $\mathbb{R}^d$ into open hypercubes, which we call *cells*. The cells admit a good bicoloration in the following sense: starting from a set of two colors, say, red and blue, it is possible to paint each cell in such a way that the $2d$ neighbors of any red cell are blue and vice versa. Let us accomplish this, the cell containing the origin being painted in red.

If $e = 2^d$, then $\widetilde{\Gamma}(e)$ is a discrete set consisting of the centers of the red cells. On the other hand, if $e = -2^d$, then $\widetilde{\Gamma}(e)$ consists of the centers of the blue cells.

When x varies continuously, $\Phi(x)$ changes sign each time one of the previous hyperplanes is crossed. It follows that the connected components of $\widetilde{\Gamma}(e)$ are enclosed in the red cells when $e > 0$, each red cell containing one component. Moreover, these components are all congruent. The situation is the same when $e < 0$, but replacing the red cells with the blue ones.

Finally, $\Gamma(e)$ is obtained from the previous surface by restricting $\widetilde{\Gamma}(e)$ to the torus, where $e \in\,]-2^d, 2^d[\setminus\{0\}$ is fixed. It follows that $\Gamma(e)$ consists of 2^{d-1} identical connected components.

As Molchanov and Vainberg conjectured,

Theorem 7.4. *For* $0 < |e| < 2^d$ *any component of* $\Gamma(e)$ *is strictly convex.*

PROOF. Suppose $0 < e < 2^d$, the other case being similar. Then, without loss of generality the considered component is

$$\left\{x \in \left]-\frac{\pi}{2}, \frac{\pi}{2}\right[^d; 2^d\prod_{j=1}^{d}\cos x^{(j)} = e\right\}.$$

Let $m = d-1$ and $h = x^{(d)}$. The equation defining the previous component becomes

$$2^d\cos x^{(1)}\cdots\cos x^{(m)}\cos h = e.$$

Since each factor in the above is positive,

$$d \ln 2 + \ln \cos x^{(1)} + \cdots + \ln \cos x^{(m)} + \ln \cosh h - \ln e = 0. \tag{7.4}$$

Since the considered component is symmetric with respect to the hyperplanes $x^{(j)} = 0$ and $x^{(j)} = x^{(l)}$, where $j, l \in \{1, \ldots, d\}$ are distinct, it suffices to show the result on the fundamental domain $h \leqslant x^{(1)} \leqslant \cdots \leqslant x^{(m)} \leqslant 0$. There, $h \neq 0$ since $e \neq 2^d$.

The derivative with respect to h of the left side in (7.4) is $-\tanh h$, which does not vanish. Consequently, an implicit function, $h = \mathrm{h}(x, e)$, satisfying (7.4) in a neighborhood of an arbitrarily fixed point in the fundamental domain exists, is analytic, and induces a local parametrization of the previous component.

It thus suffices to show that $\mathrm{D}_x^2\,\mathrm{h}(x, e)$ is positive definite. Differentiating (7.4) with respect to $x^{(j)}$ gives

$$-\tan x^{(j)} - \tanh h\, \partial_{x^{(j)}} h = 0,$$

where $h = \mathrm{h}(x, e)$. Differentiating this last equation with respect to $x^{(l)}$ gives

$$\begin{aligned} -\sec^2 x^{(j)} - \operatorname{sech}^2 h\, (\partial_{x^{(j)}} h)^2 - \tanh h\, \partial^2_{x^{(j)}} h &= 0 && \text{if } l = j, \\ -\operatorname{sech}^2 h\, \partial_{x^{(j)}} h \partial_{x^{(l)}} h - \tanh h\, \partial_{x^{(j)}} \partial_{x^{(l)}} h &= 0 && \text{otherwise.} \end{aligned}$$

Let $a_{jl} = -\tanh h\, \partial_{x^{(j)}} \partial_{x^{(l)}} h$. Since $-\tanh h > 0$, it suffices to show $[a_{jl}] > 0$. By the above,

$$[a_{jl}] = \operatorname{diag}(\sec^2 x^{(1)}, \ldots, \sec^2 x^{(m)}) + \operatorname{sech}^2 h\, [\partial_{x^{(j)}} h \partial_{x^{(l)}} h].$$

Clearly, the first term of the right-hand side in the previous equation is strictly positive. Moreover, the second term is nonnegative, since all its principal minors are zero except the first, which is a square. Hence, $[a_{jl}] > 0$, which completes the proof. □

Theorem 6.2 then gives an optimal decay for the Green's function, $G(n, e)$, associated with the Molchanov – Vainberg Laplacian. Explicitly:

Corollary. *Let $E = \{-2^d, 0, 2^d\}$ and suppose $e \in [-2^d, 2^d] \setminus E$. Then,*

$$G(n, e) = \lim_{\substack{z \to e \\ z \in \mathbb{C}_+}} G(n, z) = O(|n|^{-(d-1)/2})$$

when $|n| \to \infty$, uniformly in e on each compact and uniformly in $\omega \in \mathrm{S}^{d-1}$, where $n = |n|\omega$ is the polar form of $n \neq 0$.

Appendix A. Estimate using Bessel function

Let Δ be the *standard* discrete Laplacian in dimension $d \geqslant 4$. Its Green's function is denoted by

$$G(n - m, z) = \langle \delta_m \mid (\Delta - z)^{-1} \delta_n \rangle,$$

where $m, n \in \mathbb{Z}^d$, $z \in \mathbb{C}_+$, and δ is the Kronecker delta. By Kato's formula

$$G(n, z) = \mathrm{i} \int_{-\infty}^{0} \mathrm{e}^{-\mathrm{i}tz} \langle \delta_0 \mid \mathrm{e}^{\mathrm{i}t\Delta} \delta_n \rangle \, \mathrm{d}t. \tag{A.1}$$

Recall that the symbol of Δ is the multiplication by $\Phi(x) = 2\sum_{j=1}^{d}\cos x^{(j)}$, where $x = (x^{(1)}, \dots, x^{(d)}) \in \mathbb{T}^d$. Hence,

$$\langle \delta_0 \mid \mathrm{e}^{\mathrm{i}t\Delta}\delta_n\rangle = (2\pi)^{-d}\prod_{j=1}^{d}\int_{-\pi}^{\pi}\mathrm{e}^{\mathrm{i}n^{(j)}k}\mathrm{e}^{2\mathrm{i}t\cos k}\,\mathrm{d}k = \prod_{j=1}^{d}\mathrm{i}^{(n^{(j)})}J_{n^{(j)}}(2t)$$

at any $n = (n^{(1)}, \dots, n^{(d)})$ and $z \in \mathbb{C}_+$, where $J_m(t) = (1/2\pi)\int_{-\pi}^{\pi}\mathrm{e}^{\mathrm{i}mk}\mathrm{e}^{-\mathrm{i}t\sin k}\,\mathrm{d}k$ is the Bessel function.

It is well known that there exists a universal constant, C, such that

$$|J_m(t)| \leqslant \frac{C}{|m|^{1/3}} \quad \text{and} \quad |J_m(t)| \leqslant \frac{C}{|t|^{1/3}}$$

for any $m \in \mathbb{Z}$ and $t \in \mathbb{R}$ (see [4]). Consequently, since $d \geqslant 4$, for an arbitrarily fixed $\varepsilon > 0$

$$|\langle \delta_0 \mid \mathrm{e}^{\mathrm{i}t\Delta}\delta_n\rangle| \leqslant C^d|t|^{-1/3}|t|^{-1/3}|t|^{-1/3}(|t|^{-\varepsilon/3}|n^{(j_4)}|^{-(1-\varepsilon)/3})|n^{(j_5)}|^{-1/3}\cdots|n^{(j_d)}|^{-1/3},$$

where $(n^{(j_1)}, \dots, n^{(j_d)})$ is a permutation of $(n^{(1)}, \dots, n^{(d)})$. This last estimate and the equation (A.1) then give

$$|G(n, e + \mathrm{i}0)| \leqslant \mathrm{Const}\,|n^{(j_4)}|^{-(1-\varepsilon)/3}|n^{(j_5)}|^{-1/3}\cdots|n^{(j_d)}|^{-1/3}.$$

Since the permutation of $(n^{(1)}, \dots, n^{(d)})$ is arbitrary, it follows a fortiori that $G(n, e+\mathrm{i}0) = O(|n|^{-1/3})$ in dimension greater than 4 and $G(n, e+\mathrm{i}0) = O(|n|^{-1/3+\delta})$ for an arbitrary $\delta > 0$ in dimension 4.

References

1. L. Erdős and M. Salmhofer, *Decay of the Fourier transform of surfaces with vanishing curvature*, preprint.
2. L. Erdős, M. Salmhofer, and H.-T. Yau, *Quantum diffusion of the random Schrödinger evolution in the scaling limit*, preprint.
3. V. Jakšić and P. Poulin, *Scattering from sparse potential on graphs*, preprint.
4. L. Landau, *Monotonicity and bounds on Bessel functions*, Electron. J. Differ. Equ. Conf. **4** (2000), 147 – 154.
5. W. Littman, *Fourier transforms of surface-carried measures and differentiability of surface averages*, Bull. Amer. Math. Soc. **69** (1963), 766 – 770.
6. S. Molchanov and B. Vainberg, *Scattering on the system of the sparse bumps: multidimensional case*, Appl. Anal. **71** (1999), no. 1-4, 167 – 185.
7. P. Poulin, *The Molchanov – Vainberg Laplacian*, Proc. Amer. Math. Soc. **135** (2007), no. 1, 77 – 85.
8. W. Shaban and B. Vainberg, *Radiation conditions for the difference Schrödinger operators*, Appl. Anal. **80** (2001), no. 3-4, 525 – 556.
9. E. M. Stein, *Harmonic analysis: real-variable methods, orthogonality, and oscillatory integrals*, Princeton Math. Ser., vol. 43, Princeton Univ. Press, Princeton, NJ, 1993.
10. B. R. Vainberg, *Asymptotic methods in equations of mathematical physics*, Gordon and Breach Science Publisher, New York, 1989.

Department of Mathematics and Statistics, McGill University, 805, rue Sherbrooke Ouest, Montréal, QC H3A2K6, Canada

Current address: Department of Mathematical Sciences, NTNU — Norwegian University of Science and Technology, 7491 Trondheim, Norway

E-mail address: `poulin.philippe@hotmail.com`

Centre de Recherches Mathématiques
CRM Proceedings and Lecture Notes
Volume **42**, 2007

Orthogonal Polynomials with Exponentially Decaying Recursion Coefficients

Barry Simon

ABSTRACT. We review recent results on necessary and sufficient conditions for measures on $\mathbb{R}$ and $\partial\mathbb{D}$ to yield exponential decay of the recursion coefficients of the corresponding orthogonal polynomials. We include results on the relation of detailed asymptotics of the recursion coefficients to detailed analyticity of the measures. We present an analog of Carmona's formula for OPRL. A major role is played by the Szegő and Jost functions.

1. Introduction: Szegő and Jost functions

In broad strokes, spectral theory concerns the connection between the coefficients in differential or difference equations and the spectral measures associated to those equations. The process of going from coefficients to the measures is the direct problem, and the other direction is the inverse spectral problem. The gems of spectral theory are ones that set up one-one correspondences between classes of measures and coefficients with some properties. Examples are Verblunsky's form of Szegő's theorem [25] and the Killip – Simon theorem [12]. In this paper, our goal is to describe (mainly) recent results involving such gems for orthogonal polynomials whose recursion coefficients decay exponentially. These are technically simpler systems than the L^2 results just quoted but have more involved details.

The two classes we discuss are orthogonal polynomials on the real line (OPRL) and on the unit circle (OPUC). For the OPRL case, we have a probability measure, $d\rho$, on $\mathbb{R}$ of bounded but infinite support whose orthonormal polynomials, $p_n(x)$, obey

$$xp_n(x) = a_{n+1}p_{n+1}(x) + b_{n+1}p_n(x) + a_np_{n-1}(x) \tag{1.1}$$

with $b_n \in \mathbb{R}$ and $a_n \in (0,\infty)$ and called Jacobi parameters. $\{a_n, b_n\}_{n=1}^{\infty}$ is a description of $d\rho$ in that there is a one-one correspondence between bounded sets of such Jacobi parameters and such $d\rho$s. For background discussion of OPRL, see [4, 9, 18, 24].

2000 *Mathematics Subject Classification.* Primary 42C05, 42C25, 34L05; Secondary 33C45, 33C47, 39A10.

This work is supported in part by NSF grant DMS-0140592.
This is the final form of the paper.

For the OPUC case, $\mathbb{D} = \{z \in \mathbb{C} \mid |z| < 1\}$, and $d\mu$ is a probability measure on $\partial\mathbb{D}$ whose support is not a finite set. The orthonormal polynomials, $\varphi_n(z)$, obey the Szegő recursion relation:

$$z\varphi_n(z) = \rho_n \varphi_{n+1}(z) + \bar{\alpha}_n \varphi_n^*(z) \tag{1.2}$$

$$\varphi_n^*(z) = z^n \overline{\varphi_n(1/\bar{z})} \tag{1.3}$$

$$\rho_n = (1 - |\alpha_n|^2)^{1/2} \tag{1.4}$$

with $\alpha_n \in \mathbb{D}$ and called Verblunsky coefficients. $\{\alpha_n\}_{n=0}^\infty$ is a description of $d\mu$ in that there is a one-one correspondence between sequences of α_n obeying $|\alpha_n| < 1$ and such $d\mu$s. For background discussion of OPUC, see [10, 17–19, 24].

The measure theoretic side of the equivalences will be in terms of a derived object, rather than the measures themselves. For OPUC, the object is $D(z)$, the Szegő function [18, Section 2.4]. One says the Szegő condition holds if and only if

$$d\mu(\theta) = w(\theta)\,\frac{d\theta}{2\pi} + d\mu_s \tag{1.5}$$

where $d\mu_s$ is singular and

$$\int \log\bigl(w(\theta)\bigr)\,\frac{d\theta}{2\pi} > -\infty \tag{1.6}$$

(which is known to be equivalent to $\sum_{n=0}^\infty |\alpha_n|^2 < \infty$). In that case, $D(z)$ is defined by

$$D(z) = \exp\biggl(\int \frac{e^{i\theta} + z}{e^{i\theta} - z}\,\log\bigl(w(\theta)\bigr)\,\frac{d\theta}{4\pi}\biggr) \tag{1.7}$$

which obeys

$$\varphi_n^*(z) \to D(z)^{-1} \tag{1.8}$$

if $|z| < 1$.

$D(z)$ does not uniquely determine $d\mu$, but it does if $d\mu_s = 0$, as it will be in our cases of interest, since

$$w(\theta) = \lim_{r\uparrow 1} |D(re^{i\theta})|^2 \tag{1.9}$$

for a.e. θ.

For OPRL, the object is the Jost function. The situation is not as clean as the OPUC case in that there are not simple necessary and sufficient conditions for existence in terms of the measure. There are necessary and sufficient conditions in terms of the Jacobi parameters (see [5; 19, Section 13.9]) but not for the measure. However, there are sufficient conditions for existence that suffice for us here. Suppose

$$d\rho(x) = f(x)\,dx + d\rho_s(x) \tag{1.10}$$

where f is supported on $[-2, 2]$, and outside this set, the singular part, $d\rho_s$, has only pure points $\{E_j^\pm\}_{j=0}^{N_\pm}$ with

$$E_1^- < E_2^- < \cdots < -2 < 2 < \cdots < E_2^+ < E_1^+ \tag{1.11}$$

and suppose

$$\sum_{j,\pm} (|E_j^\pm| - 2)^{1/2} < \infty \tag{1.12}$$

and that

$$\int_{-2}^{2} \log(f(x))(4-x^2)^{-1/2}\,\mathrm{d}x > -\infty \tag{1.13}$$

Then (originally in Peherstorfer – Yuditskii [15]; see also Simon – Zlatoš [23] and [19, Theorem 13.8.9]) there is an analytic function u on $\mathbb{D}$ so that its zeros are precisely those points $z_j^{\pm}$ in $\mathbb{D}$ given by

$$z_j^{\pm} + (z_j^{\pm})^{-1} = E_j^{\pm} \tag{1.14}$$

and if B is the Blaschke product (convergent by (1.12))

$$B(z) = \prod_{j=\pm 1} \frac{z - z_j^{\pm}}{1 - \bar{z}_j^{\pm} z} \tag{1.15}$$

then $Bu^{-1} \in H^2$ and the boundary values of u obey

$$|u(\mathrm{e}^{\mathrm{i}\theta})|^2 \operatorname{Im} M(\mathrm{e}^{\mathrm{i}\theta}) = \sin\theta \tag{1.16}$$

where

$$M(z) = \int \frac{\mathrm{d}\rho(x)}{z + z^{-1} - x} \tag{1.17}$$

(so $\operatorname{Im} M(\mathrm{e}^{\mathrm{i}\theta})$ is related to $f(2\cos\theta)$).

These properties determine u uniquely. Unlike the OPUC case, u does not determine $\mathrm{d}\rho$ even if $\mathrm{d}\rho_{\mathrm{s}} \restriction [-2,2] = 0$ for u only determines f and the localization of the pure points of $\mathrm{d}\rho$ on $\mathbb{R} \setminus [-2,2]$. To recover $\mathrm{d}\rho$, we also need to know the weights of the pure points; equivalently, the residues of the poles of M at the $z_j^{\pm}$.

The theme of this review is that detailed results on exponential decay of recursion coefficients are equivalent to analyticity results on D^{-1} in the OPUC case and u in the OPRL case. That exponential decay implies analyticity has been in the physics literature for Schrödinger operators for over fifty years. The subtle aspect is the strict equivalence — an idea that appeared first in Nevai – Totik [14].

In Section 2, we discuss some aspects of finite range potentials, and in Section 3, following Nevai – Totik [14] and Damanik – Simon [6], the initial equivalence. In Section 4, following Simon [20, 21], we discuss detailed exponential asymptotics and meromorphic S and u.

I would like to thank J. Christiansen, L. Golinskii, P. Nevai, and V. Totik for useful comments.

Stas Molchanov is a leading figure in spectral theory. It is a pleasure to present this birthday bouquet to him.

2. Finite range

In this section, we present new results on approximation by finite range "potentials." We begin with an OPRL analog of Carmona's result [3] on boundary condition averaging for Schrödinger operators. We will also see that Bernstein – Szegő measures for OPUC can be viewed through the Carmona lens. Carmona's proof relies on computing derivatives of Prüfer variables — our proof here is spectral averaging making the relation to [22] transparent.

Let J be the semi-infinite Jacobi matrix

$$J = \begin{pmatrix} b_1 & a_1 & 0 & \dots \\ a_1 & b_2 & a_2 & \dots \\ 0 & a_2 & b_3 & \dots \\ \dots & \dots & \dots & \dots \\ \dots & \dots & \dots & \dots \end{pmatrix} \tag{2.1}$$

associated to an OPRL with measure $d\rho$; $J_{n;F}$, the matrix obtained from the top n rows and n left columns; and J_n^b, the matrix $J_{n;F}$ with b_n replaced by $b_n + b$. Here $b \in \mathbb{C}$. Notice that if $\operatorname{Im} b \leq 0$, then $\operatorname{spec}(J_n^b) \subset \mathbb{C} \setminus \mathbb{C}_+$, so

$$m_n^{(b)}(z) = \langle \delta_0, (J_n^{(b)} - z)^{-1} \delta_0 \rangle$$

is analytic for $b \in \overline{\mathbb{C}}_-$ and z fixed in $\mathbb{C}_+$.

If $d\rho$ is a determinate moment problem, then J is essentially selfadjoint on finite sequences [16], so

$$m_n^{(b)}(z) \to m(z) \equiv \int \frac{d\rho(x)}{x - z} \tag{2.2}$$

for any b. Thus, if $d\nu_n$ is defined by

$$\begin{aligned} \widetilde{m}_n(z) &\equiv \frac{1}{\pi} \int_{-\infty}^{\infty} m_n^{(b)}(z) \frac{db}{1 + b^2} \\ &= \int \frac{d\nu^{(n)}(x)}{x - z} \end{aligned} \tag{2.3}$$

then

$$d\nu^{(n)} \to d\rho \tag{2.4}$$

weakly. $d\nu^{(n)}$ is thus the average over b of the pure point spectral measures of $J_n^{(b)}$.

Theorem 2.1. *If $p_n(x)$ are the orthonormal OPRL, then*

$$d\nu^{(n)}(x) = \frac{dx}{\pi \big(a_n^2 p_n^2(x) + p_{n-1}^2(x) \big)} \tag{2.5}$$

In particular, the right-hand side of (2.5) converges weakly to $d\rho$. More is true, for Gaussian quadrature implies that if $m_n^{(b)}(z) = \int d\rho_n^{(b)}(x)(x - z)^{-1}$, then $\int x^\ell \, d\rho_n^{(b)}(x) = \int x^\ell \, d\rho(x)$ for $\ell \leq 2n - 2$, and thus,

$$\int x^\ell \, d\nu^{(n)}(x) = \int x^\ell \, d\rho, \quad \ell = 0, \dots, 2n - 2 \tag{2.6}$$

Of course, $d\nu^{(n)}$ does not have *all* moments finite; indeed, $\int |x|^\ell \, d\nu^{(n)} = \infty$ for $\ell \geq 2n - 1$.

Proof of Theorem 2.1. It is well known (see [18, Section 1.2]) that

$$\det(z - J_{n;F}) = P_n(z) \tag{2.7}$$

the monic OPRL, and if $J_{n;F}^{(1)}$ is the matrix obtained by removing the top row and leftmost column (i.e., 11 minor), then

$$\det(z - J_{n;F}^{(1)}) = Q_n(z) \tag{2.8}$$

the monic second kind polynomial of degree $n - 1$.

By expanding $\det(z - J_n^{(b)})$ in minors, we see

$$\begin{aligned}\det(z - J_n^{(b)}) &= P_n(z) - bP_{n-1}(z) \\ &= (a_1 \cdots a_{n-1})\big(a_n p_n(z) - b p_{n-1}(z)\big)\end{aligned} \tag{2.9}$$

and thus,

$$m_n^{(b)}(z) = -\frac{\big(a_n q_n(z) - b q_{n-1}(z)\big)}{\big(a_n p_n(z) - b p_{n-1}(z)\big)} \tag{2.10}$$

As noted above, if $\operatorname{Im} b \leq 0$, $m_n^{(b)}(z)$ has its poles in $\operatorname{Im} z \leq 0$ and thus, if $\operatorname{Im} z > 0$, $m_n^{(b)}(z)$ is analytic in $\operatorname{Im} b \leq 0$. Thus, we can close the contour in the lower half-plane and find for $\operatorname{Im} z > 0$,

$$\widetilde{m}_n(z) = m_n^{(b=-\mathrm{i})}(z) = -\frac{\big(a_n q_n(z) + \mathrm{i} q_{n-1}(b)\big)}{\big(a_n p_n(z) + \mathrm{i} p_{n-1}(z)\big)} \tag{2.11}$$

Thus, $\widetilde{m}_n$ is analytic on $\overline{\mathbb{C}}_+$, so

$$d\nu_n(x) = \pi^{-1} \operatorname{Im} \widetilde{m}_n(x)\, dx$$

Since p_n, p_{n-1}, q_n, q_{n-1} are real on $\mathbb{R}$,

$$\operatorname{Im} \widetilde{m}_n(x) = \frac{a_n\big(p_{n-1}(x)q_n(x) - p_n(x)q_{n-1}(x)\big)}{\big(a_n^2 p_n(x)^2 + p_{n-1}(x)^2\big)} \tag{2.12}$$

which is (2.7) by a standard Wronskian calculation (see [18, (1.2.51)]). □

By this same calculation, one can recover Carmona's formula for the Schrödinger operator case.

One can ask about the analog of this for OPUC. Given a nontrivial measure, $d\mu$, on $\partial\mathbb{D}$ and $\omega = e^{i\theta} \in \partial\mathbb{D}$, we define $d\mu_n^{(\omega)}$ to be the trivial measure with Verblunsky coefficients

$$\begin{aligned}\alpha_j &= \alpha_j(d\mu), \quad j = 0, \ldots, n-1 \\ \alpha_n &= \omega\end{aligned}$$

Then $d\mu_n^{(\omega)}$ is the measure with $n+1$ pure points at the zeros of the paraorthogonal polynomial (POPUC),

$$\Phi_{n+1}^{(\omega)}(z) = z\Phi_n(z) - \bar{\omega}\Phi_n^*(z) \tag{2.13}$$

Theorem 2.2. $d\mu_n \equiv \int d\theta/(2\pi) d\mu_n(e^{i\theta})$ *is the Bernstein–Szegő measure*

$$d\mu_n = \frac{d\theta}{2\pi|\varphi_n(e^{i\theta})|^2} \tag{2.14}$$

PROOF. If ψ_n are the second kind polynomials, Geronimus' formula for $F(z)$ (see [18, Theorem 3.2.4]) implies ($F(z; d\mu) = \int \frac{e^{i\theta}+z}{e^{i\theta}-z} d\mu(\theta)$)

$$F(z; d\mu_n^{(\omega)}) = \frac{\psi_n^*(z) - \omega z \psi_n(z)}{\varphi_n^*(z) - \omega z \varphi_n(z)} \tag{2.15}$$

Averaging ω over $\frac{d\theta}{2\pi}$ gives the value at $\omega = 0$ since this function is analytic in ω for z fixed in $\mathbb{D}$. It follows that

$$F(z; d\mu_n) = \frac{\psi_n^*(z)}{\varphi_n^*(z)} \tag{2.16}$$

and yields (2.14) by [18, (3.2.35)]. □

The Bernstein – Szegő approximation also has the property of being the measure associated to extending the αs up to n to be free beyond n (i.e., $\alpha_j = 0$ for $j \geq n$). One can ask about the analogous approximation for OPRL. We will get the function S_n used by Dombrowski – Nevai [8]:

Let J_ℓ be the Jacobi matrix with parameters

$$a_n(J_\ell) = \begin{cases} a_n(J) & n = 1, \dots, \ell - 1 \\ 1 & n \geq \ell \end{cases} \tag{2.17}$$

$$b_n(J_\ell) = \begin{cases} b_n(J) & n = 1, \dots, \ell \\ 0 & n \geq \ell \end{cases} \tag{2.18}$$

According to Theorem 13.6.1 (with a_ℓ replaced by 1), its Jost function is ($x = z + 1/z$)

$$g_\ell(z) = z^\ell \left(p_\ell\left(z + \frac{1}{z}\right) - z p_{\ell-1}\left(z + \frac{1}{z}\right) \right) \tag{2.19}$$

Define $S_\ell(x)$ by

$$S_\ell\left(z + \frac{1}{z}\right) \equiv g_\ell(z) g_\ell\left(\frac{1}{z}\right) \tag{2.20}$$

Then, by (2.19),

$$S_\ell(x) = p_\ell(x)^2 + p_{\ell-1}(x)^2 - x p_\ell(x) p_{\ell-1}(x) \tag{2.21}$$

Taking into account the different normalization (for us, "free" is $a_k = 1$; for them, $a_k = \frac{1}{2}$), this is the function $S_\ell(x)$ of Dombrowski – Nevai [8]. The approximating measure has a.c. part related to $\mathrm{d}x/|g_\ell(z)|^2$ on $[-2, 2]$ which is $\mathrm{d}x/S_\ell(x)$. The eigenvalues of J_ℓ are zeros of $S_\ell(x)$ but not all zeros since S_ℓ also vanishes if $g_\ell(1/z) = 0$, that is, at antibound state and resonance energies.

For most purposes, (2.7) is a more useful representation than the one associated to S_ℓ.

3. Necessary and sufficient conditions on exponential decay

The starting point of the recent results on exponential decay is the following result of Nevai – Totik for OPUC:

Theorem 3.1 ([14]). *Let* $\mathrm{d}\mu$ *be a nontrivial probability measure on* $\partial\mathbb{D}$ *and* $R > 1$. *Then the following are equivalent*:

(a)

$$\limsup_{n\to\infty} |\alpha_n(\mathrm{d}\mu)|^{1/n} \leq R^{-1} \tag{3.1}$$

(b) $\mathrm{d}\mu_s = 0$, *the Szegő condition* (1.6) *holds, and* $D(z)^{-1}$ *has an analytic continuation to* $\{z \mid |z| < R\}$.

Remark. Since $R^{-1} < 1$, (3.1) is an expression of exponential decay.

The proof is easy. If (3.1) holds, Szegő recursion first implies inductively that for $|z| = 1$,

$$|\Phi_{n+1}(e^{i\theta})| \leq (1 + |\alpha_n|)|\Phi_n(e^{i\theta})| \tag{3.2}$$

so

$$\begin{aligned}\sup_{n,|z|\leq 1} |\Phi_n^*(z)| &= \sup_{n,\theta}|\Phi_n^*(e^{i\theta})| \quad \text{(by the maximum principle)} \\ &\leq \prod_{j=0}^{\infty}(1 + |\alpha_j|) \equiv C < \infty\end{aligned} \tag{3.3}$$

and thus, for $|z| > 1$,

$$|\Phi_n(z)| \leq C|z|^n \tag{3.4}$$

Iterating

$$\Phi_{n+1}^*(z) = \Phi_n^*(z) - \alpha_n z \Phi_n(z) \tag{3.5}$$

we get

$$\Phi_n^*(z) = 1 - \sum_{j=0}^{n-1} \alpha_j z \Phi_j(z) \tag{3.6}$$

(3.1), (3.4), and (3.6) imply that for any $\varepsilon > 0$,

$$\sup_{n,|z|<R-\varepsilon} |\Phi_n^*(z)| < \infty$$

which implies that $\varphi_n^*(z)$ has a limit for $|z| < R$. This limit defines the analytic continuation of $D(z)^{-1}$.

For the other direction, one can use either of two similar-looking but distinct formulas relating D to α_n. One can use a formula of Geronimus [10] and Freud [9] as Nevai – Totik [14] do (it requires $d\mu_s = 0$)

$$\alpha_n = -\kappa_\infty \int \overline{\Phi_{n+1}(e^{i\theta})}\, D(e^{i\theta})^{-1} d\mu(\theta) \tag{3.7}$$

or the following formula of Simon [20] derived from iterated Szegő recursion:

$$\alpha_n = -\kappa_\infty^{-1}\kappa_n^2 \int \overline{\Phi_n(e^{i\theta})}\, [D(e^{i\theta})^{-1} - D(0)^{-1}] e^{-i\theta}\, d\mu(\theta) \tag{3.8}$$

In these formulas,

$$\kappa_n = \prod_{j=0}^{n-1}(1 - |\alpha_j|^2)^{-1/2} \qquad \kappa_\infty = \lim_{n\to\infty} \kappa_n$$

To get exponential decay of α_n from (3.7) or (3.8), one uses $\int \overline{\Phi_n(e^{i\theta})e^{-ij\theta}}\, d\mu(\theta) = 0$ for $j < n$ and the Taylor series for D^{-1} to see that α_n is bounded by the tail of the Taylor series of $D(z)^{-1}$ which, of course, decays exponentially if $D(z)^{-1}$ is analytic in $|z| < R$.

For OPRL, the analogs of Theorem 3.1 are due to Damanik – Simon [6]. The result is simpler if there are no bound states or resonances where

Definition. We say a measure $\mathrm{d}\rho$ on $\mathbb{R}$ has no bound states or resonances if

$$\mathrm{d}\rho(x) = f(x)\,\mathrm{d}x + \mathrm{d}\rho_{\mathrm{s}} \tag{3.9}$$

where

$$\mathrm{supp}(\mathrm{d}\rho) \subset [-2,2] \tag{3.10}$$

and

$$\int (4-x^2)^{-1} f(x)\,\mathrm{d}x < \infty \tag{3.11}$$

Theorem 3.2 ([6]). *Let $R > 1$. Suppose $\mathrm{d}\rho$ has no bound states or resonances. Then $u(z)$ has an analytic continuation to $\{z \mid |z| < R\}$ if and only if*

$$\limsup[|a_n(\mathrm{d}\rho) - 1| + |b_n(\mathrm{d}\rho)|]^{1/2n} \le R^{-1} \tag{3.12}$$

[6] has several proofs, but the simplest one is in [21]. When (3.11) holds, there is a measure $\mathrm{d}\mu$ on $\partial\mathbb{D}$ given by

$$\mathrm{d}\mu(\theta) = w(\theta)\,\frac{\mathrm{d}\theta}{2\pi} + \mathrm{d}\mu_{\mathrm{s}} \tag{3.13}$$

where

$$w\left(\arccos\left(\frac{x}{2}\right)\right) = c(4-x^2)^{-1/2} f(x) \tag{3.14}$$

for suitable c and $\mathrm{d}\mu_{\mathrm{s}}$. The Verblunsky coefficients α_n for $\mathrm{d}\mu$ and Jacobi parameters for $\mathrm{d}\rho$ are related by ([2; 11; 19, Section 13.2])

$$b_{n+1} = \alpha_{2n} - \alpha_{2n+2} - \alpha_{2n+1}(\alpha_{2n} + \alpha_{2n+2}) \tag{3.15}$$

$$a_{n+1}^2 - 1 = \alpha_{2n+1} - \alpha_{2n+3} - \alpha_{2n+2}^2(1-\alpha_{2n+3})(1+\alpha_{2n+1}) - \alpha_{2n+3}\alpha_{2n+1} \tag{3.16}$$

and the Jost function for $\mathrm{d}\rho$ and Szegő function for $\mathrm{d}\mu$ by

$$u(z) = (1-|\alpha_0|^2)(1-\alpha_1)D(z)^{-1} \tag{3.17}$$

From this, it is easy to derive Theorem 3.2 from Theorem 3.1.

To understand the situation when J has bound states, we note the analytic continuation of (1.16) says

$$u(z)u\left(\frac{1}{z}\right)\left[M(z) - M\left(\frac{1}{z}\right)\right] = z - z^{-1} \tag{3.18}$$

(this uses also u, M real on $\mathbb{R}$). Recall that if $z_0 \in \mathbb{D}$ is such that $z_0 + z_0^{-1}$ is an eigenvalue of J, then $u(z_0) = 0$. An argument shows that if $|z_0| > R^{-1}$ and $|a_n - 1| + |b_n| \le CR^{-2n}$, then $u(z_0^{-1}) \neq 0$ and $M(1/z)$ is regular at z_0. Thus, (3.18) implies a relation between $u'(z_0)$, $u(1/z_0)$, and the residue of the pole of $M(z)$ at z_0. This leads to

Definition. Suppose u is analytic in $\{z \mid |z| < R\}$ for some $R > 1$ and $z_0 \in \mathbb{D}$ with $u(z_0) = 0$ and $|z_0| > R^{-1}$. We say the weight of the point mass at $z_0 + z_0^{-1}$ is canonical if

$$\lim_{z \to z_0} (z - z_0)M(z_0) = (z_0 - z_0^{-1})\left[u'(z_0)u\left(\frac{1}{z_0}\right)\right]. \tag{3.19}$$

Theorem 3.3 ([6]). *Fix $R > 1$. Then* (3.13) *holds if and only if*

(i) *$u(z)$ has an analytic continuation to $\{z \mid |z| < R\}$.*

(ii) *The point mass at each* $z_0 \in \mathbb{D}$ *with* $|z_0| > R^{-1}$ *and* $u(z_0) = 0$ *is a canonical weight.*

If u is entire and has m zeros in $\mathbb{D}$, the set of measures with that u has dimension $m - 1$. A single point on this space has decay at rate faster than any exponential. Similarly, if u is a polynomial, $\{a_n - 1, b_n\}$ has finite support if and only if all weights are canonical.

4. Detailed asymptotics

Let S be defined by

$$S(z) = -\sum_{j=0}^{\infty} \alpha_{j-1} z^j \tag{4.1}$$

where $\alpha_{-1} = -1$. Of course, when D exists, both $D(z)^{-1}$ and $S(z)$ are analytic near $z = 0$. Theorem 3.1 can be rephrased.

Theorem 4.1. *The Taylor series of* $D(z)^{-1}$ *and* $S(z)$ *have the same radius of convergence.*

Barrios Rolanía et al. [1] extend this to show $S(z)$ is meromorphic in $\{z \mid |z| < R + \varepsilon\}$ with a single simple pole at $z = R$ if and only if $D(z)^{-1}$ is meromorphic in a similar region. This condition on S is, of course, equivalent to

$$\alpha_n = CR^{-n} + O(R^{-n(1+\delta)}) \tag{4.2}$$

which is how they phrased their result. To go further, it is useful to define

$$r(z) = \overline{D(1/\bar{z})}\, D(z)^{-1} \tag{4.3}$$

which is analytic in $\{z \mid 1 - \varepsilon < |z| < R\}$ if (3.1) holds. Simon [18] proved that $r(z) - S(z)$ is analytic in $\{z \mid 1 - \varepsilon < |z| < R^2\}$ when (3.1) holds, thereby generalizing [1]. The ultimate result of this genre was found independently by Deift – Östensson [7] and Martínez-Finkelshtein et al. [13]; an alternate proof was then found by Simon [20].

Theorem 4.2. *If* (3.1) *holds for some* $R > 1$, *then* $r(z) - S(z)$ *is analytic in* $\{z \mid 1 - \varepsilon < |z| < R^3\}$.

This is optimal in that there are examples [13, 20] where S (and r) have a simple pole at $z = R$ but $S - r$ has a pole at $z = R^3$.

Motivated by this, Simon [20] proved:

Theorem 4.3. $S(z)$ *is an entire meromorphic function if and only if* $D(z)^{-1}$ *is.*

One can even relate the poles. Given a set S in $\{z \mid |z| > 1\}$ which is discrete, one defines $\mathbb{G}(S)$ to be the set of all products $z_1 \dots z_{n+1} \bar{z}_{n+2} \dots \bar{z}_{2n+1}$ where $z_j \in S$. Then

Theorem 4.4 ([20]). *Let* $S(z)$ *be entire meromorphic and let* P *be the poles of* $D(z)^{-1}$ *and* T *the set of poles of* $S(z)$. *Then* $P \subset \mathbb{G}(T)$ *and* $T \subset \mathbb{G}(P)$.

Simon [21] studies analogs of the results for OPRL. In the Jacobi case, define

$$B(z) = 1 - \sum_{n=0}^{\infty} [b_{n+1} z^{2n+1} + (a_{n+2} - 1) z^{2n+2}] \tag{4.4}$$

The analog of Theorem 4.2 is

Theorem 4.5 ([21]). *Suppose $R > 1$ and*

$$\limsup_{n\to\infty} (|a_n^2 - 1| + |b_n|)^{1/2n} = R^{-1}$$

Then $(1-z^2)u(z) + z^2u(1/z)B(z)$ is analytic in $\{z \mid R^{-1} < |z| < R^2\}$.

As explained there, R^2 is optimal. The analog of Theorem 4.3 is

Theorem 4.6. *$B(z)$ is an entire meromorphic function if and only if $u(z)$ is.*

The connection between poles, that is, the analog of Theorem 4.4 is complicated but appears in [21].

Notes added in proof

(1) Theorem 2.1 is not new. It appeared already (at least if $a_n = 1$) as Proposition 2.1 in: D. Krutikov and C. Remling, *Schrödinger operators with sparse potentials: Asymptotics of the Fourier transform of the spectral measure*, Comm. Math. Phys. **223** (2001), no. 3, 509–532. Remling informs me that their proof using non-selfadjoint boundary conditions (essentially equivalent to our calculation) was motivated by earlier work of Atkinson and Clark.

(2) A. Zlatoš has pointed out that in the discussion of the formulas of Dombrowski–Nevai, I left out a factor of $\sqrt{(4-x^2)}$.

I would like to thank Christian Remling and Andrej Zlatoš for their remarks.

References

1. D. Barrios Rolanía, G. López Lagomasino, and E. B. Saff, *Asymptotics of orthogonal polynomials inside the unit circle and Szegő–Padé approximants*, J. Comput. Appl. Math. **133** (2001), no. 1-2, 171–181.
2. E. Berriochoa, A. Cachafeiro, and J. García-Amor, *Generalizations of the Szegő transformation interval-unit circle*, preprint.
3. R. Carmona, *One-dimensional Schrödinger operators with random or deterministic potentials: New spectral types*, J. Funct. Anal. **51** (1983), no. 2, 229–258.
4. T. S. Chihara, *An introduction to orthogonal polynomials*, Math. Appl., vol. 13, Gordon & Breach, New York, 1978.
5. D. Damanik and B. Simon, *Jost functions and Jost solutions for Jacobi matrices. I: A necessary and sufficient condition for Szegő asymptotics*, Invent. Math. **165**, no. 1, 1–50.
6. ———, *Jost functions and Jost solutions for Jacobi matrices. II: Decay and analyticity*, Int. Math. Res. Not. **2006** (2006), Article ID 19396.
7. P. Deift and J. Östensson, *A Riemann–Hilbert approach to some theorems on Toeplitz operators and orthogonal polynomials*, J. Approx. Theory **139** (2006), no. 1-2, 144–171.
8. J. Dombrowski and P. Nevai, *Orthogonal polynomials, measures and recurrence relations*, SIAM J. Math. Anal. **17** (1986), no. 3, 752–759.
9. G. Freud, *Orthogonal polynomials*, Pergamon Press, Oxford–New York, 1971.
10. Ya. L. Geronimus, *Orthogonal polynomials: Estimates, asymptotic formulas, and series of polynomials orthogonal on the unit circle and on an interval*, Consultants Bureau, New York, 1961.
11. R. Killip and I. Nenciu, *Matrix models for circular ensembles*, Int. Math. Res. Not. **2004** (2004), no. 50, 2665–2701.
12. R. Killip and B. Simon, *Sum rules for Jacobi matrices and their applications to spectral theory*, Ann. of Math. (2) **158** (2003), no. 1, 253–321.
13. A. Martínez-Finkelshtein, K. T.-R. McLaughlin, and E. B. Saff, *Szegő orthogonal polynomials with respect to an analytic weight: canonical representation and strong asymptotics*, Constr. Approx. **24** (2006), no. 3, 319–363.
14. P. Nevai and V. Totik, *Orthogonal polynomials and their zeros*, Acta Sci. Math. (Szeged) **53** (1989), no. 1-2, 99–104.

15. F. Peherstorfer and P. Yuditskii, *Asymptotics of orthonormal polynomials in the presence of a denumerable set of mass points*, Proc. Amer. Math. Soc. **129** (2001), no. 11, 3213–3220.
16. B. Simon, *The classical moment problem as a self-adjoint finite difference operator*, Adv. in Math. **137** (1998), no. 1, 82–203.
17. ______, *OPUC on one foot*, Bull. Amer. Math. Soc. (N.S.) **42** (2005), no. 4, 431–460.
18. ______, *Orthogonal polynomials on the unit circle.* Part 1: *Classical theory*, Amer. Math. Soc. Colloq. Publ., vol. 54, Part 1, Amer. Math. Soc., Providence, RI, 2005.
19. ______, *Orthogonal polynomials on the unit circle.* Part 2: *Spectral theory*, Amer. Math. Soc. Colloq. Publ., vol. 54, Part 2, Amer. Math. Soc., Providence, RI, 2005.
20. ______, *Meromorphic Szegő functions and asymptotic series for Verblunsky coefficients*, Acta Math. **195** (2005), 267–285.
21. ______, *Meromorphic Jost functions and asymptotic series for Jacobi parameters*, Funct. Anal. Appl., to appear.
22. B. Simon and T. Wolff, *Singular continuous spectrum under rank one perturbations and localization for random Hamiltonians*, Comm. Pure Appl. Math. **39** (1986), no. 1, 75–90.
23. B. Simon and A. Zlatoš, *Sum rules and the Szegő condition for orthogonal polynomials on the real line*, Comm. Math. Phys. **242** (2003), no. 3, 393–423.
24. G. Szegő, *Orthogonal polynomials*, 3rd ed., Amer. Math. Soc. Colloq. Publ., vol. 23, Amer. Math. Soc., Providence, RI, 1967.
25. S. Verblunsky, *On positive harmonic functions (second paper)*, Proc. London Math. Soc. (2) **40** (1936), 290–320.

Mathematics 253-37, California Institute of Technology, Pasadena, CA 91125, USA
E-mail address: bsimon@caltech.edu

Centre de Recherches Mathématiques
CRM Proceedings and Lecture Notes
Volume **42**, 2007

Poisson Statistics for Eigenvalues: From Random Schrödinger Operators to Random CMV Matrices

Mihai Stoiciu

ABSTRACT. We give a brief overview of the results and methods used in the study of the statistical distribution of eigenvalues for random Schrödinger operators and random CMV matrices.

1. Introduction

The study of the spectra of random Schrödinger operators and the distribution of their eigenvalues was initiated by the very important paper of Anderson [4]. Anderson showed that certain random lattices exhibit absence of diffusion, a phenomenon which is now called Anderson localization.

The first rigorous mathematical proof of the Anderson localization was obtained by Goldsheid et al. [10]. They studied a special case of the continuous one-dimensional random Schrödinger operator, where the potential was defined using Brownian motion on a Riemannian manifold.

A few years later, Fröhlich and Spencer analyzed a model of discrete Schrödinger operator called the Anderson tight-binding model. In [8], they obtained the first proof of the Anderson localization for multidimensional Schrödinger operators. Different proofs of the Anderson localization were discovered later; we will only mention the the proof of Aizenman and Molchanov [3], which is relevant for the topic considered here.

Recent developments in the theory of orthogonal polynomials on the unit circle have emphasized the importance of a new class of unitary matrices, the CMV matrices. These matrices can be seen as unitary analogues of the self-adjoint Jacobi matrices (see [17,18] for a detailed presentation of the theory of orthogonal polynomials on the unit circle and CMV matrices; see also the recent review article [21]). Many of the techniques used in the study of random Schrödinger operators can be adapted to random CMV matrices. Localization results for random orthogonal polynomials on the unit circle and random CMV matrices can be found in [18,24].

In addition to the Anderson localization, one can also study the statistical distribution of the eigenvalues. It turns out that in the regimes where the Anderson localization is expected, the local statistical distribution of the eigenvalues of

2000 *Mathematics Subject Classification.* Primary 15A52; Secondary 42C05, 81Q10.
This is the final form of the paper.

the random Schrödinger operators is Poisson. This means that there is no local correlation between these eigenvalues.

The first result on the local structure of the spectrum of the one-dimensional Schrödinger operators was obtained by Molchanov in [16]. This result was extended to the multidimensional Anderson tight-binding model by Minami [14], using [3].

A similar result holds for the eigenvalues of random CMV matrices; see [22, 23]. More precisely, for some classes of random CMV matrices, the local statistical distribution of the eigenvalues converges to the Poisson distribution. This is equivalent to the fact that the local distribution of the zeros of random paraorthogonal polynomials on the unit circle converges to the Poisson distribution. The result was extended from paraorthogonal polynomials to regular orthogonal polynomials by Davies and Simon in [7].

2. Random Schrödinger operators

The one-dimensional continuum random Schrödinger operator studied by Goldsheid et al. was defined by:

$$H(\omega) = -\frac{\mathrm{d}^2}{\mathrm{d}t^2} + q(t,\omega), \quad t \in \mathbb{R}, \omega \in \Omega \tag{2.1}$$

where $q(t,\omega)$ is a stationary random potential.

More precisely,

$$q(t,\omega) = F\big(x_t(\omega)\big) \tag{2.2}$$

where $x_t(\omega)$ is the Brownian motion on a compact manifold K and $F\colon K \to \mathbb{R}$ be a smooth nonflat Morse function.

Goldsheid et al. proved in [10] that the operator $H = H(\omega)$ defined by (2.1) has pure point spectrum for almost all $\omega \in \Omega$ (i.e., a complete system of eigenvalues in $L^2(\mathbb{R})$). This result was strengthened in [15] by Molchanov, who showed that with probability 1, each eigenfunction of the operator $H = H(\omega)$ defined by (2.1) decreases exponentially. Thus, (2.1) was the first model of one-dimensional random Schrödinger operators for which a mathematical proof of the Anderson localization was obtained.

In addition to the Anderson localization, one can also study the local structure of the spectrum. The main question is whether the energy levels are correlated or not. This issue was settled by Molchanov, who proved in [16] that there is no local correlation between the eigenvalues of the operator defined by (2.1).

In order to describe Molchanov's result, let $H_V = H_V(\omega)$ be the random one-dimensional Schrödinger operator on $L^2(-V,V)$ with Dirichlet boundary conditions. Denote by $N_V(I)$ the number of eigenvalues of the operator H_V situated in the interval I. Then, for any fixed $a_1 < b_1 \le a_2 < b_2 \le \cdots \le a_n < b_n$ and any nonnegative integers $k_1, k_2, \ldots, k_n$,

(2.3)

$$\lim_{V\to\infty} \mathbb{P}\Big(N_V\Big(E_0 + \frac{a_1}{2V}, E_0 + \frac{b_1}{2V}\Big) = k_1, \ldots, N_V\Big(E_0 + \frac{a_n}{2V}, E_0 + \frac{b_n}{2V}\Big) = k_n \Big)$$
$$= \mathrm{e}^{-(b_1-a_1)n(E_0)} \frac{((b_1-a_1)n(E_0))^{k_1}}{k_1!} \cdots \mathrm{e}^{-(b_n-a_n)n(E_0)} \frac{((b_n-a_n)n(E_0))^{k_n}}{k_n!}$$

This means that the local statistical distribution of the eigenvalues of the operator H_V (rescaled near E_0) converges, as $V \to \infty$, to the Poisson point process with intensity measure $n(E_0)\,\mathrm{d}x$, where dx denotes the Lebesgue measure and $n(E_0)$ is the density of states at E_0. Hence, Molchanov's result [16] shows that there is no local correlation between the energy levels of the random one-dimensional operator (2.1).

The next challenge for mathematical physicists was to see whether a similar result holds for multidimensional Schrödinger operators. A positive answer to this question was given by Minami in [14], who showed that near energies where Anderson localization is expected (more precisely, in regimes where the Anderson localization holds via [3]), there is no correlation between the eigenvalues of the Anderson tight-binding model.

In order to describe Minami's result, let $H = -\Delta + V_\omega$ be the standard Anderson tight-binding model acting in $l^2(\mathbb{Z}^d)$ and let $H^\Lambda = \chi_\Lambda H \chi_\Lambda$ be the truncated Schrödinger operator corresponding to the hypercube $\Lambda \subset \mathbb{Z}^d$. Also let $G^\Lambda(z) = (H^\Lambda - z)^{-1}$ be the resolvent associated to H^Λ. If $E_1(\Lambda) \le E_2(\Lambda) \le \cdots \le E_n(\Lambda)$ are the eigenvalues of H^Λ, then we can define the integrated density of states:

$$N(E) = \lim_{\Lambda \uparrow \mathbb{Z}^d} \frac{1}{|\Lambda|} \#\{j, E_j(\Lambda) \le E\} \tag{2.4}$$

and the density of states:

$$n(E) = \mathrm{d}N(E)/\mathrm{d}E \tag{2.5}$$

Note that if $n(E) > 0$, then the average spacing of the eigenvalues near the energy E is of order $|\Lambda|^{-1}$. Consider the rescaled spectrum (near E):

$$\xi_j(\Lambda, E) = |\Lambda|(E_j(\Lambda) - E) \tag{2.6}$$

and the corresponding point process

$$\xi(\Lambda, E) = \sum_j \delta_{\xi_j(\Lambda,E)} \tag{2.7}$$

Minami proved in [14] that, under suitable conditions:

$$\xi(\Lambda, E) \xrightarrow{\Lambda \uparrow \mathbb{Z}^d} (\text{Poisson point process of intensity } n(E)) \tag{2.8}$$

This means exactly that there is no correlation between the eigenvalues of H^Λ if Λ is large.

We should mention here that one of the main ingredients of the proof is the following technical lemma:

$$\mathbb{E}\left[\det \begin{pmatrix} \operatorname{Im} G^\Lambda(z; x, x) & \operatorname{Im} G^\Lambda(z; x, y) \\ \operatorname{Im} G^\Lambda(z; y, x) & \operatorname{Im} G^\Lambda(z; y, y) \end{pmatrix}\right] \le C \tag{2.9}$$

where the constant C depends only on the distribution of the potential $V = V_\omega$. This result is referred to as the "Minami trick." It is worth to mention that (2.9) encodes a spectacular and somewhat mysterious cancellation that has found applications to other questions in mathematical physics; see for example [13].

3. Random CMV Matrices

Recent work of Cantero et al. [5] emphasized the importance of a new class of unitary random matrices called now CMV matrices. These matrices are intimately connected with the orthogonal polynomials on the unit circle (see the monographs [17, 18] on the theory of orthogonal polynomials on the unit circle).

The CMV matrix is a five-diagonal matrix realization for the unitary operator $z \to zf(z)$ on $L^2(\mathbb{T};\mu)$, where μ is a nontrivial probability measure on the unit circle $\mathbb{T}$ (we call a measure nontrivial if it is not supported on finitely many points). For any such measure μ we can apply the Gram-Schmidt procedure to the set of polynomials $\{1, z, z^2, \dots\} \in L^2(\mathbb{T},\mu)$ and get the set of monic orthogonal polynomials $\{\Phi_0(z, \mathrm{d}\mu), \Phi_1(z, \mathrm{d}\mu), \Phi_2(z, \mathrm{d}\mu), \dots\} \in L^2(\mathbb{T};\mu)$.

These polynomials obey the recurrence relation

$$\Phi_{k+1}(z, \mathrm{d}\mu) = z\Phi_k(z, \mathrm{d}\mu) - \bar{\alpha}_k \Phi_k^*(z, \mathrm{d}\mu) \quad k \geq 0 \tag{3.1}$$

where, for $\Phi_k(z) = \sum_{j=0}^k b_j z^j$, the reversed polynomial $\Phi_k^*(z)$ is defined by $\Phi_k^*(z) = \sum_{j=0}^k \bar{b}_{k-j} z^j$. The recurrence coefficients $\{\alpha_n\}_{n\geq 0}$ are called Verblunsky coefficients; they are complex numbers of absolute value < 1.

If we apply the Gram – Schmidt algorithm to the sequence $\{1, z, z^{-1}, z^2, z^{-2}, \dots\}$ we get the set $\{\chi_0(z), \chi_1(z), \chi_2(z), \dots\}$, which is a basis of $L^2(\mathbb{T};\mu)$. The CMV matrix associated to the measure μ is the matrix representation of the operator $f(z) \to zf(z)$ on $L^2(\mathbb{T};\mu)$. It has the form:

$$\mathcal{C} = \begin{pmatrix} \bar{\alpha}_0 & \bar{\alpha}_1\rho_0 & \rho_1\rho_0 & 0 & 0 & \dots \\ \rho_0 & -\bar{\alpha}_1\alpha_0 & -\rho_1\alpha_0 & 0 & 0 & \dots \\ 0 & \bar{\alpha}_2\rho_1 & -\bar{\alpha}_2\alpha_1 & \bar{\alpha}_3\rho_2 & \rho_3\rho_2 & \dots \\ 0 & \rho_2\rho_1 & -\rho_2\alpha_1 & -\bar{\alpha}_3\alpha_2 & -\rho_3\alpha_2 & \dots \\ 0 & 0 & 0 & \bar{\alpha}_4\rho_3 & -\bar{\alpha}_4\alpha_3 & \dots \\ \dots & \dots & \dots & \dots & \dots & \dots \end{pmatrix} \tag{3.2}$$

where $\rho_k = \sqrt{1 - |\alpha_k|^2}$

Note that the Jacobi matrices obtained in a similar way for orthogonal polynomials on the real line are tridiagonal matrices. As in the case of orthogonal polynomials on the real line, an important connection between CMV matrices and monic orthogonal polynomials is

$$\Phi_n(z) = \det(zI - \mathcal{C}^{(n)}) \tag{3.3}$$

where $\mathcal{C}^{(n)}$ is the upper left $n \times n$ corner of $\mathcal{C}$.

If $|\alpha_{n-1}| = 1$, then the CMV matrix decouples between $(n-1)$ and n. The upper left corner is an $(n \times n)$ unitary matrix

$$\mathcal{C}^{(n)} = \mathcal{C}^{(n)}_{\{\alpha_0, \alpha_1, \dots, \alpha_{n-1}\}} \tag{3.4}$$

We will consider random CMV matrices and study the statistical distribution of their eigenvalues. We will randomize the matrix $\mathcal{C}^{(n)}_{\{\alpha_0, \alpha_1, \dots, \alpha_{n-1}\}}$ by taking independent identically distributed random variables $\alpha_0, \alpha_1, \dots, \alpha_{n-2}$. The last variable, α_{n-1} will be chosen to be uniformly distributed on the unit circle. As in the case of random Schrödinger operators, for various classes of $n \times n$ random CMV matrices, the local statistical distribution of the eigenvalues of these matrices will converge (as $n \to \infty$) to the Poisson distribution. This fact indicates that, as n gets large, there is no local correlation between these eigenvalues. The following theorem can be found in [23] (and will be extended to singular distributions in Theorem 3.2):

Theorem 3.1. *Consider the random CMV matrices* $\mathcal{C}^{(n)} = \mathcal{C}^{(n)}_{\{\alpha_0, \alpha_1, \dots, \alpha_{n-1}\}}$ *where* $\alpha_0, \alpha_1, \dots, \alpha_{n-2}$ *are i.i.d. random variables distributed uniformly in a disk of radius* $r < 1$, *and* α_{n-1} *is another random variable independent of the previous ones and uniformly distributed on the unit circle.*

Consider the space

$$\Omega_n = \{\alpha = (\alpha_0, \alpha_1, \dots, \alpha_{n-2}, \alpha_{n-1}) \in D(0,r) \times D(0,r) \times \cdots \times D(0,r) \times \mathbb{T}\}$$

with the probability measure $\mathbb{P}_n$ *obtained by taking the product of the uniform (Lebesgue) measures on each* $D(0,r)$ *and on* $\mathbb{T}$. *Fix a point* $\mathrm{e}^{\mathrm{i}\theta_0} \in \mathbb{T}$ *and let* $\zeta^{(n)}$ *be the point process defined by* $\zeta^{(n)} = \sum_{k=1}^{n} \delta_{z_k}$, *where* $\{z_1, z_2, \dots, z_n\}$ *are the eigenvalues of the matrix* $\mathcal{C}^{(n)}$ *(each eigenvalue* z_i *depends on* $\alpha_0, \alpha_1, \dots, \alpha_{n-1}$*).*

Then, on a fine scale (of order $1/n$*) near* $\mathrm{e}^{\mathrm{i}\theta_0}$, *the point process* $\zeta^{(n)}$ *converges to the Poisson point process with intensity measure* $n\mathrm{d}\theta/(2\pi)$ *(where* $\mathrm{d}\theta/(2\pi)$ *is the normalized Lebesgue measure). This means that for any fixed* $a_1 < b_1 \leq a_2 < b_2 \leq \cdots \leq a_m < b_m$ *and any nonnegative integers* $k_1, k_2, \dots, k_m$, *we have*

(3.5)
$$\mathbb{P}_n\Big(\zeta^{(n)}\big(\mathrm{e}^{\mathrm{i}(\theta_0+2\pi a_1/n)}, \mathrm{e}^{\mathrm{i}(\theta_0+2\pi b_1/n)}\big) = k_1, \dots, \zeta^{(n)}\big(\mathrm{e}^{\mathrm{i}(\theta_0+2\pi a_m/n)}, \mathrm{e}^{\mathrm{i}(\theta_0+2\pi b_m/n)}\big) = k_m\Big)$$
$$\longrightarrow \mathrm{e}^{-(b_1-a_1)} \frac{(b_1-a_1)^{k_1}}{k_1!} \cdots \mathrm{e}^{-(b_m-a_m)} \frac{(b_m-a_m)^{k_m}}{k_m!}$$

as $n \to \infty$.

The result is similar to the ones obtained for random Schrödinger operators by Molchanov and Minami (see Section 2). All three proofs use the same road map towards obtaining the Poisson distribution in the limit:

(1) uniform estimates on the resolvent;
(2) control of the eigenfunctions;
(3) decoupling of the random matrix into smaller identical blocks;
(4) proving that the smaller blocks cannot contribute two or more eigenvalues in given intervals;
(5) obtaining the Poisson distribution as a limit of Bernoulli distributions.

We will give a few more details about this road map, following [23]. For the random CMV matrices considered in Theorem 3.1, we can prove that there exist two constants C and D such that for any $z \in \mathbb{C}$, $|z| < 1$ and any $s < 1$,

$$\mathbb{E}\big(\big|F_{kl}(z, \mathcal{C}_\alpha^{(n)})\big|^s\big) \leq C\mathrm{e}^{-D|k-l|} \tag{3.6}$$

where

$$F_{kl}(z, \mathcal{C}_\alpha^{(n)}) = \left[\frac{\mathcal{C}_\alpha^{(n)} + z}{\mathcal{C}_\alpha^{(n)} - z}\right]_{kl} \tag{3.7}$$

is the entry in row k and column l of the matrix $[(\mathcal{C}_\alpha^{(n)} + z)/(\mathcal{C}_\alpha^{(n)} - z)]$. The idea of controlling fractional moments of the resolvent originates from [3].

The steps for getting the exponential decay of $\mathbb{E}\big(\big|F_{kl}(z, \mathcal{C}_\alpha^{(n)})\big|^s\big)$ (relation (3.6)) are:

(a) Uniform upper bound for all the fractional moments:

$$\mathbb{E}\big(\big|F_{kl}(z, \mathcal{C}_\alpha^{(n)})\big|^s\big) \leq \frac{2^{2-s}}{\cos \pi s/2} \tag{3.8}$$

(b) Uniform decay on rows: For every $\varepsilon > 0$ there exists a $k_\varepsilon > 0$ such that for any $k \geq k_\varepsilon$ and for any j, we have

$$\mathbb{E}\big(\big|F_{j,j+k}(z, \mathcal{C}_\alpha^{(n)})\big|^s\big) \leq \varepsilon \tag{3.9}$$

(c) Spectral averaging estimate (i.e., evaluating the conditional expectation using a finite rank perturbation)

$$\mathbb{E}\big(\big|F_{kk}(z,\mathcal{C}_\alpha^{(n)})\big|^s \mid \{\alpha_i\}_{i\neq k}\big) \leq \frac{4}{1-s}\,32^s \tag{3.10}$$

(d) Uniform decay implies uniform exponential decay

$$\mathbb{E}\big(\big|F_{kl}(z,\mathcal{C}_\alpha^{(n)})\big|^s\big) \leq C\mathrm{e}^{-D|k-l|} \tag{3.11}$$

for some positive constants $C, D > 0$. This follows from a method developed by Aizenman et al. which shows that fast power decay implies exponential decay.

The bounds (3.6) can be used to prove that the eigenfunctions of the random matrix $\mathcal{C}_\alpha^{(n)}$ are exponentially localized with probability 1. We should mention that in this step we use Simon's extension to CMV matrices (see [19]) of a result of Aizenman on dynamical localization for Schrödinger operators (see [1]). We should mention that this is one of the places where our assumption that the common distribution of the Verblunsky coefficients α_k is rotation invariant is critical.

The exponential localization allows us to decouple the CMV matrix $\mathcal{C}^{(n)} = \mathcal{C}_\alpha^{(n)}$. More precisely, we can consider another unitary matrix $\widetilde{\mathcal{C}}^{(n)} = \widetilde{\mathcal{C}}_\alpha^{(n)}$ which decouples into the sum of $[\ln n]$ smaller identical unitary matrices:

$$\widetilde{\mathcal{C}}^{(n)} = \widetilde{\mathcal{C}}_1^{(n)} \oplus \widetilde{\mathcal{C}}_2^{(n)} \oplus \cdots \oplus \widetilde{\mathcal{C}}_{[\ln n]}^{(n)} \tag{3.12}$$

such that the statistical distribution of the eigenvalues of the matrix $\mathcal{C}^{(n)}$ is the same as the statistical distribution of the eigenvalues of the matrix $\widetilde{\mathcal{C}}^{(n)}$.

For any p, $1 \leq p \leq [\ln n]$, we consider the point process $\zeta^{(n,p)} = \sum_{k=1}^{[\ln n]} \delta_{z_k^{(p)}}$, where $z_1^{(p)}, z_2^{(p)}, \ldots, z_{[n/\ln n]}^{(p)}$ are the eigenvalues of the matrix $\widetilde{\mathcal{C}}_p^{(n)}$. Therefore $\zeta^{(n)}$ can be approximated by $\sum_{k=1}^{[\ln n]} \zeta^{(n,p)}$. This means that the point process $\zeta^{(n)}$ is infinitely divisible as $n \to \infty$. References for the general theory of point processes are [6, 11].

The next step is to prove that each small matrix $\widetilde{\mathcal{C}}_p^{(n)}$ contributes at most one eigenvalue in the interval of size $1/n$. This follows from the fact that for any $\mathrm{e}^{\mathrm{i}\theta} \in \mathbb{T}$ and any interval I_n of size $1/n$ near $\mathrm{e}^{\mathrm{i}\theta}$ we have

$$\mathbb{P}_n(\zeta^{(n,p)}(I_n) \geq 2) = O([\ln n]^{-2}) \quad \text{as } n \to \infty \tag{3.13}$$

which implies

$$\sum_{p=1}^{[\ln n]} \mathbb{P}_n\left(\zeta^{(n,p)}(I_n) \geq 2\right) \to 0 \quad \text{as } n \to \infty. \tag{3.14}$$

Hence an interval of size $\frac{1}{n}$ contains k eigenvalues if and only if exactly k of the matrices $\widetilde{\mathcal{C}}_1^{(n)}, \widetilde{\mathcal{C}}_2^{(n)}, \ldots, \widetilde{\mathcal{C}}_{[\ln n]}^{(n)}$ contribute one eigenvalue in the interval I_n. This is precisely a Bernoulli distribution and the limit as $n \to \infty$ is a Poisson distribution. Hence we conclude that the limit $\zeta^{(n)}$ is a Poisson point process.

Theorem 3.1 shows that if α_k, $0 \leq k \leq (n-2)$ are uniformly distributed on a disk of radius $r < 1$, then the eigenvalues of the corresponding CMV matrix $\mathcal{C}^{(n)}$ are not locally correlated for large n. A natural question is whether this result can be extended to different types of distributions for the Verblunsky coefficients α_k. It turns out that the rotation invariance was critical in several places, but we can allow distributions that are singular with respect to the Lebesgue measure. Thus, if for $r \in (0,1)$ we denote by $C(0,r)$ the circle centered at the origin and of radius r, we get:

Theorem 3.2. *Consider the random CMV matrices* $\mathcal{C}^{(n)} = \mathcal{C}^{(n)}_{\{\alpha_0,\alpha_1,\ldots,\alpha_{n-1}\}}$ *where* $\alpha_0, \alpha_1, \ldots, \alpha_{n-2}$ *are i.i.d. random variables distributed uniformly in a circle of radius* $r < 1$*, and* α_{n-1} *is another random variable independent of the previous ones and uniformly distributed on the unit circle.*

Consider the space

$$\Omega_n = \{\alpha = (\alpha_0, \alpha_1, \ldots, \alpha_{n-2}, \alpha_{n-1}) \in C(0,r) \times C(0,r) \times \cdots \times C(0,r) \times \mathbb{T}\}$$

with the probability measure $\mathbb{P}_n$ *obtained by taking the product of the uniform (one-dimensional Lebesgue) measures on each* $C(0,r)$ *and on* $\mathbb{T}$*. Fix a point* $\mathrm{e}^{\mathrm{i}\theta_0} \in \mathbb{T}$ *and let* $\zeta^{(n)}$ *be the point process obtained from the eigenvalues of the truncated CMV matrix* $\mathcal{C}^{(n)}$*.*

Then, on a fine scale (of order $1/n$*) near* $\mathrm{e}^{\mathrm{i}\theta_0}$*, the point process* $\zeta^{(n)}$ *converges to the Poisson point process with intensity measure* $n\mathrm{d}\theta/(2\pi)$ *(here* $\mathrm{d}\theta/(2\pi)$ *is the normalized Lebesgue measure). This means that for any fixed* $a_1 < b_1 \leq a_2 < b_2 \leq \cdots \leq a_m < b_m$ *and any nonnegative integers* $k_1, k_2, \ldots, k_m$*, we have*

(3.15)

$$\mathbb{P}_n(\zeta^{(n)}(\mathrm{e}^{\mathrm{i}(\theta_0+2\pi a_1/n)}, \mathrm{e}^{\mathrm{i}(\theta_0+2\pi b_1/n)}) = k_1, \ldots, \zeta^{(n)}(\mathrm{e}^{\mathrm{i}(\theta_0+2\pi a_m/n)}, \mathrm{e}^{\mathrm{i}(\theta_0+2\pi b_m/n)}) = k_m)$$
$$\to \mathrm{e}^{-(b_1-a_1)}\frac{(b_1-a_1)^{k_1}}{k_1!} \cdots \mathrm{e}^{-(b_m-a_m)}\frac{(b_m-a_m)^{k_m}}{k_m!}$$

as $n \to \infty$.

PROOF. The proof is similar to the proof of the Theorem 3.1, with some modifications required by the fact that we use a different probability distribution for the random Verblunsky coefficients.

With the exception of (3.10), all the steps outlined before hold when we replace the uniform distribution on the disk or radius r with the uniform distribution on the circle of radius r.

Lemma 3.3 (see below) shows that the conditional moments of order s with $s \in (0, \frac{1}{2})$ are uniformly bounded. A standard application of Hölder's theorem implies that the result in Lemma 3.3 holds for all $s \in (0,1)$.

We should also mention that Aizenman's theorem for orthogonal polynomials on the unit circle (see [19]) holds for Verblunsky coefficients uniformly distributed on the circle of radius r. As in the proof of Theorem 3.1, we can now derive the exponential decay of the fractional moments of the matrix elements of the resolvent of $\mathcal{C}^{(n)}_\alpha$.

Following the same route as in the proof of Theorem 3.1 (the eigenfunctions are exponentially localized, the matrix $\mathcal{C}^{(n)}_\alpha$ can be decoupled, and the decoupled matrices contribute at most one eigenvalue in each interval of size $O(1/n)$) we can conclude that (3.15) holds, that is, the local statistical distribution of the eigenvalues of $\mathcal{C}^{(n)}_\alpha$ is Poisson. □

We will now give the analog of (3.10) for the case of Verblunsky coefficients uniformly distributed on $C(0,r)$.

Lemma 3.3. *For any* $s \in (0, \frac{1}{2})$, *any* k, $1 \le k \le n$, *and any choice of* $\alpha = \alpha_0, \alpha_1, \dots, \alpha_{n-1}$ *as in Theorem 3.2,*

$$\mathbb{E}\left(\left|F_{kk}(z, \mathcal{C}_\alpha^{(n)})\right|^s \mid \{\alpha_i\}_{i \neq k}\right) \le K(s,r) \tag{3.16}$$

where a possible value for the constant is $K(s,r) = 4 \cdot 64^s \cdot \pi^{1-2s}/((1-2s)r^s)$.

PROOF. Using properties of orthogonal polynomials on the unit circle (see [9, 12] and also [22, 23]), the dependence of $F_{kk}(z, \mathcal{C}_\alpha^{(n)})$ on α_k is of the form

$$F_{kk}(z, \mathcal{C}_\alpha^{(n)}) = \frac{2}{1 - w_1 \frac{\alpha_k + w_2}{1 + \bar{\alpha}_k w_2}} \tag{3.17}$$

where $w_1, w_2 \in \mathbb{C}$ are complex numbers of absolute value strictly less than 1 that do not depend on α_k.

Therefore, in order to prove (3.16), it is enough to find a uniform bound for

$$I = \int_{C(0,r)} \left| \frac{4}{1 + \bar{\alpha}_k w_2 - w_1(\alpha_k + w_2)} \right|^s d\mu_r(\alpha_k) \tag{3.18}$$

for any $w_1, w_2 \in \mathbb{D}$, (μ_r denotes one-dimensional Lebesgue measure $d\theta/(2\pi)$ on $C(0,r)$ and $\mathbb{D}$ is the set of all complex numbers of absolute value < 1).

Let $\alpha_k = x + iy$, with $x, y \in \mathbb{R}$. Then

$$1 + \bar{\alpha}_k w_2 - w_1(\alpha_k + w_2) = x(-w_1 + w_2) + y(-iw_1 - iw_2) + (1 - w_1 w_2) \tag{3.19}$$

Let's denote by $M(x, y, w_1, w_2)$ the right-hand-side of the previous equation. Then

$$\int_{C(0,r)} \left| \frac{4}{M(x, y, w_1, w_2)} \right|^s d\mu_r(\alpha_k) \le \min\{I_1, I_2\} \tag{3.20}$$

where

$$I_1 = \int_{C(0,r)} \left| \frac{4}{\operatorname{Re} M(x, y, w_1, w_2)} \right|^s d\mu_r(\alpha_k) \tag{3.21}$$

and

$$I_2 = \int_{C(0,r)} \left| \frac{4}{\operatorname{Im} M(x, y, w_1, w_2)} \right|^s d\mu_r(\alpha_k) \tag{3.22}$$

Let $\varepsilon > 0$ (we will choose ε later). If $|w_2 - w_1| < \varepsilon$ and $|w_2 + w_1| < \varepsilon$, then $|M(x, y, w_1, w_2)| \ge 1 - 2\varepsilon - \varepsilon^2$. We immediately get

$$\int_{C(0,r)} \left| \frac{4}{M(x, y, w_1, w_2)} \right|^s d\mu_r(\alpha_k) \le \left(\frac{4}{1 - 2\varepsilon - \varepsilon^2} \right)^s. \tag{3.23}$$

If at least one of the complex numbers $(w_2 - w_1)$ and $(w_2 + w_1)$ is greater or equal in absolute value than ε, then at least one of

$$M_1(w_1, w_2) = \sqrt{(\operatorname{Re}(w_2 - w_1))^2 + (\operatorname{Re}(-iw_1 - iw_2))^2} \tag{3.24}$$

and

$$M_2(w_1, w_2) = \sqrt{(\operatorname{Im}(w_2 - w_1))^2 + (\operatorname{Im}(-iw_1 - iw_2))^2} \tag{3.25}$$

is greater than $\varepsilon/2$. Without loss of generality, we can assume that $M_1(w_1, w_2) \geq \varepsilon/2$. Let $\varphi_1(w_1, w_2) = \mathrm{Re}(-w_1 + w_2)/M_1(w_1, w_2)$, $\varphi_2(w_1, w_2) = \mathrm{Re}(-\mathrm{i}w_1 - \mathrm{i}w_2)/M_1(w_1, w_2)$ and $\varphi_3(w_1, w_2) = \mathrm{Re}(1 - w_1 w_2)/M_1(w_1, w_2)$. Then

(3.26)
$$\begin{aligned} I_1 &= \int_{C(0,r)} \left| \frac{4}{x\,\mathrm{Re}(-w_1 + w_2) + y\,\mathrm{Re}(-\mathrm{i}w_1 - \mathrm{i}w_2) + \mathrm{Re}(1 - w_1 w_2)} \right|^s d\mu_r(\alpha_k) \\ &= \int_{C(0,r)} \left| \frac{4}{x\varphi_1(w_1, w_2) + y\varphi_2(w_1, w_2) + \varphi_3(w_1, w_2)} \right|^s \left| \frac{1}{M_1(w_1, w_2)} \right|^s d\mu_r(\alpha_k) \\ &= \int_0^{2\pi} \frac{1}{r^s} \left| \frac{4}{(\cos\theta)\varphi_1(w_1, w_2) + (\sin\theta)\varphi_2(w_1, w_2) + \varphi_3(w_1, w_2)/r} \right|^s \\ &\qquad \times \left| \frac{1}{M_1(w_1, w_2)} \right|^s \frac{d\theta}{2\pi} \end{aligned}$$

Let $\theta_0 = \theta_0(w_1, w_2)$ be an angle such that $\sin\theta_0 = \mathrm{Re}(-w_1 + w_2)/M_1(w_1, w_2)$ and $\cos\theta_0 = \mathrm{Re}(-\mathrm{i}w_1 - \mathrm{i}w_2)/M_1(w_1, w_2)$ and let $K_0 = K_0(w_1, w_2, r)$ be defined by $K_0 = \mathrm{Re}(1 - w_1 w_2)/\big(rM_1(w_1, w_2)\big)$.

Using (3.26), we get

$$I_1 \leq \frac{8^s}{\varepsilon^s r^s} \int_0^{2\pi} \left| \frac{1}{\sin(\theta + \theta_0) + K_0} \right|^s \frac{d\theta}{2\pi} = \frac{8^s}{\varepsilon^s r^s} \int_0^{2\pi} \left| \frac{1}{\sin\theta + K_0} \right|^s \frac{d\theta}{2\pi} \tag{3.27}$$

Without loss of generality, we can assume that in the previous relation we have $K_0 \in [-1, 1]$ and therefore $K_0 = \sin\kappa$, for a $\kappa \in [0, 2\pi)$. Therefore,

(3.28)
$$\begin{aligned} I_1 &\leq \frac{8^s}{\varepsilon^s r^s} \int_0^{2\pi} \left| \frac{1}{2\left(\sin\big((\theta + \kappa)/2\big)\cos\big((\theta - \kappa)/2\big)\right)} \right|^s \frac{d\theta}{2\pi} \\ &= \frac{8^s}{\varepsilon^s r^s} \int_0^{2\pi} \left| \frac{1}{2\left(\sin\big((\theta + \kappa)/2\big)\sin\big((\pi - \theta + \kappa)/2\big)\right)} \right|^s \frac{d\theta}{2\pi}. \end{aligned}$$

The function $\sin x$ vanishes as fast as x at each zero. We actually have

$$|\sin x| \geq \frac{2}{\pi}|x| \quad \text{for each } x \in \left[-\frac{\pi}{2}, \frac{\pi}{2}\right] \tag{3.29}$$

so using (3.28) and the fact that the function sin vanishes twice in an interval of length 2π, we get

$$I_1 \leq 4 \cdot \frac{4^s}{\varepsilon^s r^s} \int_{-\pi/2}^{\pi/2} \frac{1}{|\theta|^{2s}} \frac{d\theta}{2\pi} = \frac{2^{2s+3}}{\varepsilon^s r^s} \frac{(\pi/2)^{1-2s}}{1-2s} = 2^{4s+2} \frac{\pi^{1-2s}}{\varepsilon^s r^s} \frac{1}{1-2s}. \tag{3.30}$$

We can therefore conclude that for I defined in (3.18), we have

$$I \leq \max\left\{ \left(\frac{4}{1 - 2\varepsilon - \varepsilon^2} \right)^s, 2^{4s+2} \frac{\pi^{1-2s}}{\varepsilon^s r^s} \frac{1}{1-2s} \right\} \tag{3.31}$$

For $\varepsilon = \frac{1}{4}$ we get

$$I \leq 4 \cdot 64^s \cdot \frac{\pi^{1-2s}}{r^s(1-2s)} \tag{3.32}$$

which is exactly (3.16). □

Theorems 3.1 and 3.2 show that if the Verblunsky coefficients are independent and identically distributed, then the statistical distribution of the eigenvalues of the random CMV matrices obtained is Poisson (no correlation). Recent work of Simon [20] showed that if the Verblunsky coefficients are independent random variables and with exponentially decaying distributions, then the eigenvalues of the CMV matrices repel each other giving a "clock distribution" on the unit circle. It would be very interesting to discover which distributions of the random Verblunsky coefficients give a transition in the statistical distribution of the eigenvalues from noncorrelation (Poisson) to repulsion ("clock").

References

1. M. Aizenman, *Localization at weak disorder: Some elementary bounds*, Rev. Math. Phys. **6** (1994), no. 5A, 1163–1182.
2. M. Aizenman, J. H. Schenker, R. M. Friedrich, and D. Hundertmark, *Finite-volume fractional-moment criteria for Anderson localization*, Comm. Math. Phys. **224** (2001), no. 1, 219–253.
3. Aizenman M. and S. Molchanov, *Localization at large disorder and at extreme energiestextup: an elementary derivation*, Comm. Math. Phys. **157** (1993), no. 2, 245–278.
4. P. W. Anderson, *Absence of diffusion in certain random lattices*, Phys. Rev. (2) **109** (1958), 1492–1505.
5. M. J. Cantero, L. Moral, and L. Velázquez, *Five-diagonal matrices and zeros of orthogonal polynomials on the unit circle*, Linear Algebra Appl. **362** (2003), 29–56.
6. D. J. Daley and D. Vere-Jones, *An introduction to the theory of point processes*, Springer Ser. Statist., Springer, New York, 1988.
7. E. B. Davies and B. Simon, *Eigenvalue estimates for non-normal matrices and the zeros of random orthogonal polynomials on the unit circle*, J. Approx. Theory **141** (2006), no. 2, 189–213.
8. J. Fröhlich and T. Spencer, *Absence of diffusion in the Anderson tight binding model for large disorder or low energy*, Comm. Math. Phys. **88** (1983), no. 2, 151–184.
9. L. Ya. Geronimus, *Orthogonal polynomials: Estimates, asymptotic formulas, and series of polynomials orthogonal on the unit circle and on an interval*, Consultants Bureau, New York, 1961.
10. I. Ya. Goldsheid, S. A. Molchanov, and L. A. Pastur, *A random homogeneous Schrödinger operator has a pure point spectrum*, Funkcional. Anal. i Priložen. **11** (1977), no. 1, 1–10 (Russian); English transl., Functional Anal. Appl. **11** (1977), no. 1, 1–8.
11. O. Kallenberg, *Random measures*, Akademie Verlag, Berlin, 1976.
12. S. Khrushchev, *Schur's algorithm, orthogonal polynomials, and convergence of Wall's continued fractions in $L^2(\mathbb{T})$*, J. Approx. Theory **108** (2001), no. 2, 161–248.
13. A. Klein, O. Lenoble, and P. Müller, *On Mott's formula for the AC-conductivity in the Anderson model*, Ann. of Math. (2), to appear, available at `arXiv:math-ph/0508007`.
14. N. Minami, *Local fluctuation of the spectrum of a multidimensional Anderson tight binding model*, Comm. Math. Phys. **177** (1996), no. 3, 709–725.
15. S. A. Molchanov, *Structure of the eigenfunctions of one-dimensional unordered structures*, Izv. Akad. Nauk SSSR Ser. Mat. **42** (1978), no. 1, 70–103 (Russian); English transl., Math. USSR-Izv. **42** (1978), no. 1, 69–101.
16. ———, *The local structure of the spectrum of the one-dimensional Schrödinger operator*, Comm. Math. Phys. **78** (1980/81), no. 3, 429–446.
17. B. Simon, *Orthogonal polynomials on the unit circle*. Part 1: *Classical theory*, Amer. Math. Soc. Colloq. Publ., vol. 54, Part 1, Amer. Math. Soc., Providence, RI, 2005.
18. ———, *Orthogonal polynomials on the unit circle*. Part 2: *Spectral theory*, Amer. Math. Soc. Colloq. Publ., vol. 54, Part 2, Amer. Math. Soc., Providence, RI, 2005.
19. ———, *Aizenman's theorem for orthogonal polynomials on the unit circle*, Const. Approx. **23** (2006), no. 2, 229–240.
20. ———, *Fine structure of the zeros of orthogonal polynomials*. I: *A tale of two pictures*, Electron. Trans. Numer. Anal. **25** (2006), 328–368.
21. ———, *CMV matrices: Five years after*, Proceedings of the W. D. Evans 65th Birthday Conference, to appear.

22. M. Stoiciu, *Zeros of random orthogonal polynomials on the unit circle*, Ph.D. thesis, Caltech, 2005.
23. ______, *The statistical distribution of the zeros of random paraorthogonal polynomials on the unit circle*, J. Approx. Theory **139**, no. 1-2, 29 – 64.
24. A. V. Teplyaev, *The pure point spectrum of random orthogonal polynomials on the circle*, Dokl. Akad. Nauk SSSR **320** (1991), no. 1, 49 – 53 (Russian); English transl., Soviet Math. Dokl. **44** (1992), no. 2, 407–411.

Mathematics and Statistics, Williams College, Williamstown, 01267, USA
E-mail address: mstoiciu@williams.edu

Titles in This Series

42 **Donald A. Dawson, Vojkan Jakšić, and Boris Vainberg, Editors,** Probability and mathematical physics: a volume in honor of Stanislav Molchanov, 2007

41 **André Bandrauk, Michel C. Delfour, and Claude Le Bris, Editors,** High-dimensional partial differential equations in science and engineering, 2007

40 **V. Apostolov, A. Dancer, N. Hitchin, and M. Wang, Editors,** Perspectives in Riemannian geometry, 2006

39 **P. Winternitz, D. Gomez-Ullate, A. Iserles, D. Levi, P. J. Olver, R. Quispel, and P. Tempesta, Editors,** Group theory and numerical analysis, 2005

38 **Jacques Hurtubise and Eyal Markman, Editors,** Algebraic structures and moduli spaces, 2004

37 **P. Tempesta, P. Winternitz, J. Harnad, W. Miller, Jr., G. Pogosyan, and M. Rodriguez, Editors,** Superintegrability in classical and quantum systems, 2004

36 **Hershy Kisilevsky and Eyal Z. Goren, Editors,** Number theory, 2004

35 **H. E. A. Eddy Campbell and David L. Wehlau, Editors,** Invariant theory in all characteristics, 2004

34 **P. Winternitz, J. Harnad, C. S. Lam, and J. Patera, Editors,** Symmetry in physics, 2004

33 **André D. Bandrauk, Michel C. Delfour, and Claude Le Bris, Editors,** Quantum control: Mathematical and numerical challenges, 2003

32 **Vadim B. Kuznetsov, Editor,** The Kowalevski property, 2002

31 **John Harnad and Alexander Its, Editors,** Isomonodromic deformations and applications in physics, 2002

30 **John McKay and Abdellah Sebbar, Editors,** Proceedings on moonshine and related topics, 2001

29 **Alan Coley, Decio Levi, Robert Milson, Colin Rogers, and Pavel Winternitz, Editors,** Bäcklund and Darboux transformations. The geometry of solitons, 2001

28 **J. C. Taylor, Editor,** Topics in probability and Lie groups: Boundary theory, 2001

27 **I. M. Sigal and C. Sulem, Editors,** Nonlinear dynamics and renormalization group, 2001

26 **J. Harnad, G. Sabidussi, and P. Winternitz, Editors,** Integrable systems: From classical to quantum, 2000

25 **Decio Levi and Orlando Ragnisco, Editors,** SIDE III—Symmetries and integrability of difference equations, 2000

24 **B. Brent Gordon, James D. Lewis, Stefan Müller-Stach, Shuji Saito, and Noriko Yui, Editors,** The arithmetic and geometry of algebraic cycles, 2000

23 **Pierre Hansen and Odile Marcotte, Editors,** Graph colouring and applications, 1999

22 **Jan Felipe van Diejen and Luc Vinet, Editors,** Algebraic methods and q-special functions, 1999

21 **Michel Fortin, Editor,** Plates and shells, 1999

20 **Katie Coughlin, Editor,** Semi-analytic methods for the Navier-Stokes equations, 1999

19 **Rajiv Gupta and Kenneth S. Williams, Editors,** Number theory, 1999

18 **Serge Dubuc and Gilles Deslauriers, Editors,** Spline functions and the theory of wavelets, 1999

17 **Olga Kharlampovich, Editor,** Summer school in group theory in Banff, 1996, 1998

16 **Alain Vincent, Editor,** Numerical methods in fluid mechanics, 1998

15 **François Lalonde, Editor,** Geometry, topology, and dynamics, 1998

14 **John Harnad and Alex Kasman, Editors,** The bispectral problem, 1998

13 **Michel C. Delfour, Editor,** Boundaries, interfaces, and transitions, 1998

TITLES IN THIS SERIES

12 **Peter C. Greiner, Victor Ivrii, Luis A. Seco, and Catherine Sulem, Editors,** Partial differential equations and their applications, 1997

11 **Luc Vinet, Editor,** Advances in mathematical sciences: CRM's 25 years, 1997

10 **Donald E. Knuth,** Stable marriage and its relation to other combinatorial problems: An introduction to the mathematical analysis of algorithms, 1997

9 **D. Levi, L. Vinet, and P. Winternitz, Editors,** Symmetries and integrability of difference equations, 1996

8 **J. Feldman, R. Froese, and L. M. Rosen, Editors,** Mathematical quantum theory II: Schrödinger operators, 1995

7 **J. Feldman, R. Froese, and L. M. Rosen, Editors,** Mathematical quantum theory I: Field theory and many-body theory, 1994

6 **Guido Mislin, Editor,** The Hilton Symposium 1993: Topics in topology and group theory, 1994

5 **D. A. Dawson, Editor,** Measure-valued processes, stochastic partial differential equations, and interacting systems, 1994

4 **Hershy Kisilevsky and M. Ram Murty, Editors,** Elliptic curves and related topics, 1994

3 **Rémi Vaillancourt and Andrei L. Smirnov, Editors,** Asymptotic methods in mechanics, 1993

2 **Philip D. Loewen,** Optimal control via nonsmooth analysis, 1993

1 **M. Ram Murty, Editor,** Theta functions: From the classical to the modern, 1993